W0227981

ALLE ZEIT WACH
SJ
1842

L. Deecke J.C. Eccles V.B. Mountcastle (Eds.)

From Neuron to Action

An Appraisal of Fundamental and Clinical Research

With 283 Figures and 36 Tables

Springer-Verlag Berlin Heidelberg GmbH

Professor Dr. med. Lüder Deecke
Neurological University Clinic Vienna
Lazarettgasse 14, 1090 Vienna, Austria

Professor Sir John C. Eccles
Ca' a la Gra'
6646 Contra (Locarno) TI, Switzerland

Professor Dr. med. Vernon B. Mountcastle
Department of Neuroscience
The Johns Hopkins University School of Medicine
725 North Wolfe Street, Baltimore, MD 21205, USA

Library of Congress Cataloging-in-Publication Data
From neuron to action : an appraisal of fundamental and clinical research / L. Deecke, J.C. Eccles, V.B. Mountcastle (eds.). p. cm. Papers presented at the International Symposium „From Neuron to Action" held on 24-27 April 1988 in Vienna in honor of the 60th birthday of Prof. Hans Helmut Kornhuber. Includes bibliographical references. Includes index.
ISBN 978-3-662-02603-8 ISBN 978-3-662-02601-4 (eBook)
DOI 10.1007/978-3-662-02601-4
1. Efferent pathways--Congresses. 2. Motor neurons--Congresses. 3. Kornhuber, H. H.--Congresses. I. Deecke, Lüder. II. Eccles, John C. (John Carew), Sir, 1903- . III. Mountcastle, Vernon B. IV. Kornhuber, H. H. V. International Symposium „From Neuron to Action" (1988 : Vienna, Austria) [DNLM: 1. Kornhuber, H. H. 2. Equilibrium--physiology--congresses. 3. Motor Activity--physiology--congresses. 4. Nervous System--physiology--congresses. 5. Nervous System Diseases--physiopathology--congresses. 6. Perception-physiology-congresses. WL 102 F931 1988] QP369.F76 1990
599'.01852--dc20 DNLM/DLC

Originally published by Springer-Verlag Berlin Heidelberg New York in 1990
Softcover reprint of the hardcover 1st edition 1990

25/3134-543210 – Printed on acid-free paper

(photographed by J. C. Aschoff)

The chapters of this book are dedicated
to Professor Dr. Dr. h. c. H.H. Kornhuber

1 Kornhuber
2 Rüdel
3 Bechinger
4 Mountcastle
5 Eccles
6 Deecke
7 Janssen
8 Lux
9 Seitelberger
10 Aschoff
11 ?
12 Eckener
13 Brooks
14 Pompeiano
15 G. Huber
16 Lehmann
17 Caspers
18 Iggo
19 Gross
20 Schreiber
21 Buser
22 R.F. Schmidt
23 Schüttler
24 Horn
25 Lang
26 Widder
27 Bauer
28 Marbach
29 Szirtes
30 Dichgans
31 Westphal
32 Florey
33 Scheich
34 Heibel
35 Cheyne
36 Witschel
37 Robinson
38 Backhaus
39 Mauch
40 Gonsette
41 E. Horn
42 W.J. Schmidt
43 Potthoff
44 Hülser
45 Freund
46 Warecka
47 Mergner
48 J. Kornhuber
49 M. Kornhuber
50 ?
51 Zimmermann
52 Wallesch
53 Danielczyk

Preface

The papers compiled in this book were presented at the International Symposium "From Neuron to Action" held on 24 - 27 April 1988 in Vienna in honor of the 60th birthday of Prof. Hans Helmut Kornhuber of the Department of Neurology, University of Ulm, Federal Republic of Germany.

The program of the symposium, and so the contents of the book, reflect the breadth of Kornhuber's own interests. This diversity was evident early in his career, when, after 4½ years in Russian prisoner of war camps, he studied medicine at five different universities, which he selected not so much for reasons of medicine as for their renowned philosophers, including Karl Jaspers. After graduating under Kurt Schneider, Professor of Psychiatry at Heidelberg, and neurological training with Richard Jung at Freiburg, he wrote his *Habilitationsschrift* for Neurology and Neurophysiology in 1962 at Freiburg. He then went to Baltimore to work with V.B. Mountcastle on the somatosensory system at the Department of Physiology of the Johns Hopkins University.

Back in Freiburg, Kornhuber developed the "Kornhuber stimulator", which is a force- and displacement-controlled mechanostimulator used by, among others, John Eccles. In 1966 Kornhuber accepted the Chair of Neurology at the University of Ulm, where he became one of the founding professors of that young university. His large number of scientific publications reflect an extraordinarily broad spectrum of interests.

After early work on cyclothymic depression and the psychology and psychiatry of prisoners of war, Kornhuber published seminal papers on the oculomotor system, equilibrium, and vestibular physiology and pathology. His excellent handbook articles are still milestones in the field.

In 1964 Kornhuber and Deecke discovered the readiness potential, the German word *Bereitschaftspotential* being taken up as a specific term even within the English-speaking world. A year later, in 1965, Kornhuber, Frederickson, and Schwarz discovered the vestibular cortical representation area in the monkey. In 1967, Kornhuber and Fuchs investigated the projection of the external eye muscle receptors to the cerebellum. During his period at John Hopkins University with Mountcastle, he published papers on touch and position sense, as well as on information flow (channel capacity) in man. Kornhuber also thought deeply about the mind-body problem and the question of human freedom, a subject on which he opened new dimensions by making it accessible to scientific investigation. In addition, Kornhuber developed challenging theories on the physiology of human voluntary movement and speech.

Later, Kornhuber devoted his scientific activities to the treatment of multiple sclerosis, and identified methods that really do help patients. For example, he developed effective methods for the treatment of spasticity and neurogenic bladder dysfunction, and he found that in the treatment of the acute exacerbation of MS, cortisone plus low-dose cyclophosphamide leads to longer-lasting benefits than cortisone alone. Kornhuber also became interested in schizophrenia, the glutamate theory deriving from him, and he directed attention to the reconsideration of will – a human capacity forgotten by scientists – the physiological basis of which he described. Even a booklet on religion appeared from his pen.

In neurology Kornhuber placed special emphasis on the prevention of illness, and it was he who coined the term "preventive neurology". He was the first person in Germany who direct attention to the fact that ethanol is a common cause of arterial hypertension. He went into schools and taught the students to measure their parents' blood pressure, which proved an effective method for the early detection of hypertension. However, Kornhuber was also the first to show that alcohol-related obesity in males is not the effect of caloric alcohol action but of direct toxic mechanisms which act via fatty liver hyperinsulinism even at so-called normal levels of alcohol consumption. Furthermore, the identification of daily "normal" alcohol as the main cause of hypercholesterolemia and type II diabetes was primarily due to Kornhuber. He is a man who not only writes but also acts. It was this attribute that led him to enter the field of health politics, making effective suggestions as to how pathogenic factors can be avoided and advocating that those who indulge in alcohol and cigarette consumption should compensate for the consequences by paying a health levy on alcoholic beverages and cigarettes, the revenue from which should go to health insurance.

Recently, Kornhuber and his group showed in a rat model that the calcium overload blocker flunarizine is of significant benefit in cerebral hemorrhage. Since it has a beneficial effect in ischemic stroke as well, the initial treatment measure in stroke is the injection of flunarizine by the physician who first sees the patient. This treatment is also to be considered in the neonate with traumatic hemorrhage, which was sonographically shown by Kornhuber and his coworkers to be more frequent than previously thought.

There is much more that could be said about the many contributions Hans Kornhuber has made to science and medicine, which are only partially reflected by the many prizes he has received, e.g., from the German EEG Society (Berger Prize), the City of Ulm, and the Bárány Society (Hallpike-Nylen Prize), and by other accolades bestowed upon him, such as his honorary memberships, the degree of doctor honoris causa from the University of Brussels, the Bundesverdienstkreuz, and the Lazarus-von-Schwendi medal.

The painting that appeared on the cover of the symposium program is reproduced in this book, as it appropriately depicts a motif specific to Ulm and its connection to Vienna, the location of the Symposium. The boats that can be seen on the Danube canal are of special interest since they are *Ulmer Schachteln!*, literally "Ulm crates", which were used to float downstream from Ulm to Vienna where they then sold as firewood. Also depicted are

"Ulmer Schachteln"

Ferdinand's Bridge and St. Stephan's cathedral, of which appears the clock that lends the painting its title – *Bilderuhr* ("Picture Clock"). In the background the slopes of the Vienna Woods are visible, including Kahlenberg and Leopoldsberg with their castles.

Several of our colleagues helped in organizing the symposium, in particular Dr. Wilfried Lang of the Neurological University Clinic, Vienna. Furthermore, we thank Prof. Eberhard Horn and Prof. Jürgen Aschoff of the Department of Neurology, University of Ulm. The invaluable secretarial help of Frau Elisabeth Ribar-Maurer, Vienna Medical Academy, and Frau Ottilie Kirschenhofer, Neurological University Clinic, Vienna, is gratefully acknowledged, as is the invaluable contribution of Dipl.-Ing. Gerald Lindinger, Neurological University Clinic of Vienna, in ensuring the excellent computer printing of the manuscript. We would also like to thank Mrs. S. Benko and Dr. M. Wilson of Springer-Verlag for their valuable cooperation in producing this volume.

The substantial financial support provided by the following companies was of particular help in making the symposium possible: Janssen GmbH, Neuß; Merckle GmbH, Blaubeuren; Ratiopharm GmbH, Ulm; BASF Ges.m.b.H., Vienna; Bayer Austria Ges.m.b.H., Vienna; Bayer AG, Leverkusen; Boehringer Ingelheim; Byk Gulden Constance; Ciba-Geigy GmbH, Wehr; Ciba-Geigy Ges.m.b.H., Vienna; Desitin Arzneimittel GmbH, Hamburg; Dornier-System GmbH, Friedrichshafen; Gerot Pharmazeutika Ges.m.b.H., Vienna; Hoechst Austria AG, Vienna; Hoffmann–La Roche Ges.m.b.H., Vienna; Janssen Pharmaceutica, Vienna; Dr. Madaus GmbH & Co, Cologne; Madaus Ges.m.b.H., Vienna; E. Merck, Darmstadt; Reck Maschinenbau GmbH, Betzenweiler; Dr. Rentschler GmbH & Co, Laupheim; Sandoz AG, Nuremberg; Schering Ges.m.b.H., Vienna; Siemens AG, Erlangen; Siemens Austria, Vienna; Thiemann Arzneimittel GmbH, Waltrop.

Further support was given by: Asta Pharma AG, Frankfurt; Beecham Pharma Ges.m.b.H., Vienna; Bender & Co., Vienna; Fidia Pharmaforschung GmbH, Munich; Fresenius AG, Oberursel; Frosst Pharma GmbH, Munich; Glaxo Pharmazeutica Ges.m.b.H., Vienna; Paul Hartmann AG, Heidenheim; Hoffmann–La Roche AG, Grenzach,Wyhlen; ICI Österreich GmbH, Vienna; Klinge Pharma GmbH, Munich; Ludwig Merckle GmbH, Vienna; Merck, Sharp & Dohme GmbH, Munich; Nattermann & Cie GmbH, Cologne; G. Pohl-Boskamp GmbH & Co, Hohenlockstedt; Promonta GmbH, Pforzheim-Würm; Schwarz Pharma GmbH, Monheim; Wander Pharma GmbH, Nuremberg.

We are most grateful to Fidia, Munich, who met the total printing costs of the book.

Vienna, Autumn 1990

Lüder Deecke
John C. Eccles
Vernon B. Mountcastle

Contents

Part 2: Oculomotor and Equilibrium

Part 3: Sensory and Cognitive

Part 4: Evolution

Part 5: Synaptic and Elementary Processes

Part 6: Neurological Sciences I (Psychiatry)

Part 7: Neurological Sciences II (Neurology and Neurosurgery)

Index of Authors

Part 1

Motor

Adaptations and Learning of Arm Movements

V. B. Brooks
Department of Physiology, University of Western Ontario, London, Ont. N6A 5C1 Canada

Introduction

Practice of intended movements produces adaptations that optimize their execution. Relatively little is known about their neural mechanisms except that normal cerebellar function is needed for the optimal composition of the muscle synergies used in moving or holding the arm (cf. review Brooks and Thach 1981; Brooks 1984b). We have studied the relation of self-optimizing adaptations of programmed arm movements to changes of the underlying muscle activity. The animals were rewarded for *what* they did: for a behavior that consisted of a correct movement into target followed by a steady arm posture outlasting the movement by several seconds, but not for *how* they did it. Our previous findings (Brooks et al. 1983; Brooks and Watts 1988) were confirmed: monkeys learn first to make well-programmed ('continuous') movements instead of poorly programmed ('discontinuous') ones, after which they adapt continuous movements towards optimal execution. Continuous movements are recognized by their single-peaked velocity profiles (Brooks et al. 1973; Polit and Bizzi 1979). Adaptive changes are observed in two ways: 1) the shape of the single-peaked velocity profiles become steeper, slimmer and more symmetrical; and 2) the scale of the profiles enlarges through increases of their peak and average velocities. Scales change in such proportion that their ratio (a 'shape index', Hogan 1984) tends to be lowered. Shapes of velocity profiles reflect largely on movement programs while scaling reflects more on the incidental details of their execution. How muscle use is reorganized in adapted postures is described and discussed in relation to neural mechanisms.

Methods

Motor Learning. Two cynomolgus monkeys were studied during operant reward training to carry out a small item of 'behavior' that consisted of an intended elbow movement followed by an intended elbow posture. The movements turned an unloaded handle to place a display cursor into a step-tracking target in one uninterrupted trajectory starting within 1 s after the target step that served as the cue to move. The postures had to hold the cursor in the target for a comparatively long time: 3-3.5 s (Figs. 1A,B). Both parts of the behavior, postures as well as movements, contributed to gathering fruit juice rewards because reward-yielding trials followed one another more quickly if

Monkey positions handle and cursor in. step-tracking task.

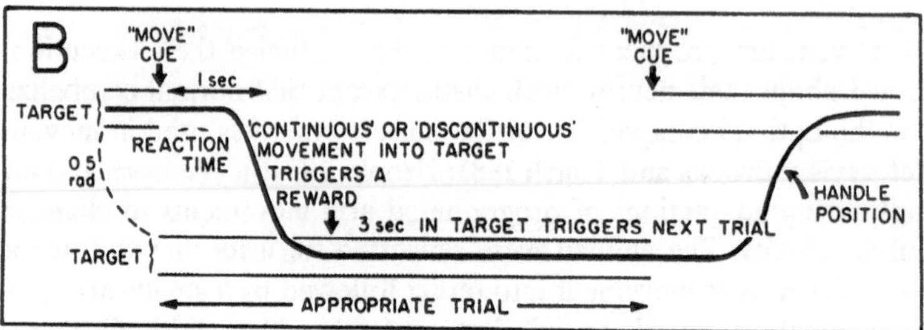

'Continuous' movements have only one velocity peak.
3 sec. hold in target must be repeated if interrupted.

Fig. 1. A Task set-up with monkey holding handle while looking at displays of handle position (cursor) and target position. Targets are indicated in the right drawing for the reader. **B** Task paradigm is explained by means of a schematic position trace representing a trial with appropriate behavior. (From Brooks and Watts 1988)

correct arm movements were followed by appropriately maintained arm postures. Therefore, use and adaptation of 'continuous' movements and of the underlying EMG changes reflect self-optimization of programs for movements rather than for behavior. For instance, although practice shortened movement durations by a fraction of a second, this could not increase the animals' rewards substantially because a much longer delay was imposed by the task requirement for steady arm postures of several seconds to trigger the next trial.

Data Base. Results obtained with two animals are presented (Brooks and Watts, in preparation). Properties of task-related behavior and of movement execution were classified for all movements made in several weeks' training (7300 movements for F36 and 23700 for F37), and reaction times as well as details of all programmed movements were measured as in previous studies (Brooks and Watts 1988; Brooks et al. 1983). Fig. 2 illustrates the usual manner of learning the task: 'continuous' movements came to be used almost exclusively in the context of 'appropriate behavior' ('appropriate trials': Figs. 1B, 2). The session means plotted in Fig. 2A against time confirm our previous results (Fig. 4 in Brooks and Watts 1988). Session means represent about 200 movements. A linearly rising relation is demonstrated in Fig. 2B between behavioral skill and motor

'Motor' skill in context tends to follow 'behavioral' skill.

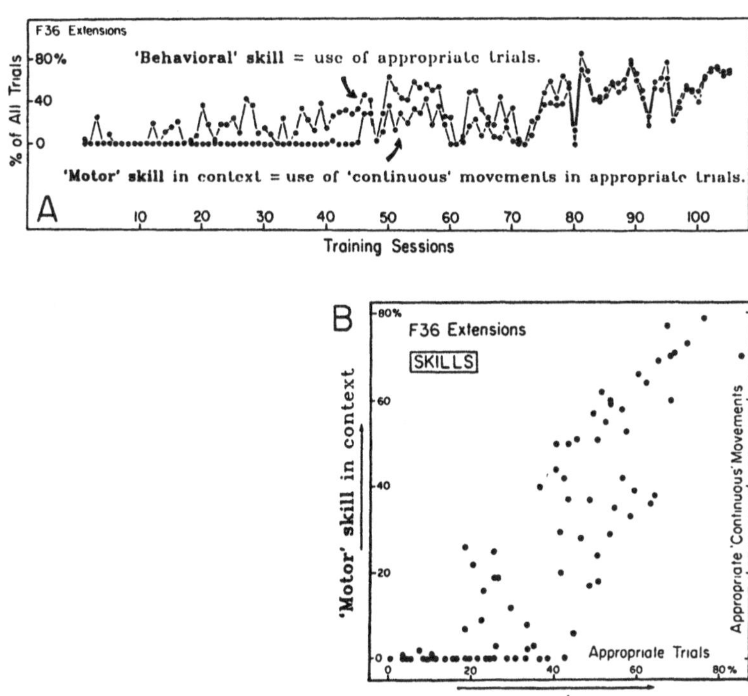

Fig. 2. Task performance skills consist of behavioral skill and motor skill used in context of task. **A** Progression of skill in successive training sessions (abscissa) plotted as mean incidence of skillful trials per session (ordinate). **B** Progression of motor skill in context (ordinate) plotted against behavioral skill (abscissa). (Monkey F36, continuous extensions made in appropriate trials.)

skill in context whose onset, however, may be delayed until behavioral insight occurs (cf. Brooks 1986b).

Kinematic Data and EMG. Movement onset was determined electronically as the point at which velocity reached 10 deg/s (0.17 rad/s). Velocity records were created by differentiation of the stored digitized position records. Multiunit EMG activity was recorded in biceps brachii and triceps brachii by means of fine wires pushed into the muscles through the skin for each recording session. EMG activity was differentially amplified, filtered (30 - 1000 Hz bandwidth), full-wave rectified, and digitized as 2 s bursts at 500 Hz. These data were smoothed by block-averaging 5 successive points. Off-line measurements of EMG were made electronically after visual determination of EMG onset in relation to movements as indicated in Fig. 3.

Results

During motor learning, movements changed in a consistent manner that is considered self-optimizing because the changes occurred without any tangible reward to the ani-

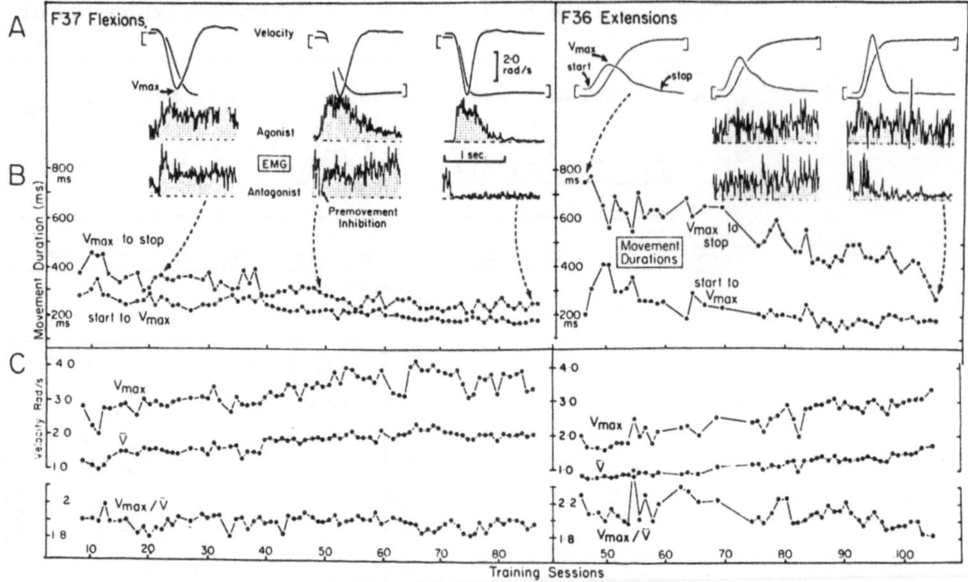

Fig. 3. During motor learning, cocontraction of opposing muscles gives way to finer control (A); as velocity profiles of single-peaked movements made in the context of appropiate task behavior become slimmer and more symmetrical (B); and peak- and average velocities (Vmax; V) both rise but Vmax/V falls to near 1.9-1.8 for both animals (C). **A** Three examples of velocity and position traces, with EMGs below, of representative single trials; brackets denote target positions; averages of 20 trials within session indicated by arrows and numbered on the abscissa). **B** Graphs of session means (abscissa) of movement durations (ms, ordinate) from start to peak velocity: Vmax and from Vmax to movement stop. **C** Graphs of session means of Vmax; average velocity; V; and Vmax/V. (From Brooks VB and Watts SL, in preparation).

mals. The top row in Fig. 3A shows progressive changes of shape and scale of programmed movements made in the task context. Although these movements accelerated without deviations to their peak velocities, their decelerations changed gradually from stepwise decrements to smooth descents. Loosely speaking, the profiles indicate that flights of the arm had firmly programmed take-offs but hesitant landings. With extended practice, however, the landings became as skilled as the take-offs.

Increased motor skill created movements with accelerations and decelerations of nearly equal durations. This change of shape towards symmetry was accompanied by changes of scale: velocities rose to higher peaks in briefer lengths of time. Thus, the animals learned programs that applied more force more quickly and also withdrew it just as quickly. The consequent shortening of durations from movement start to peak velocity and from peak velocity to movement stop is graphed in Fig. 3B: these periods of acceleration and deceleration converged towards equality near 0.2 s.

Skilled, smooth matching of move and hold programs was accomplished by changed use of the muscles that boosted take-off (agonist) and those that braked it (antagonist). The midsection of Fig. 3A illustrates how movements were implemented initially by step-changes of agonist and antagonist activity rather than by the triphasic burst pattern familiar from movements that are made as quickly as possible. As learning progressed, the elbow joint relaxed through lessened cocontraction of opposing muscles which reduced joint stiffness and increased net torque. Once the task was learned, in-target arm

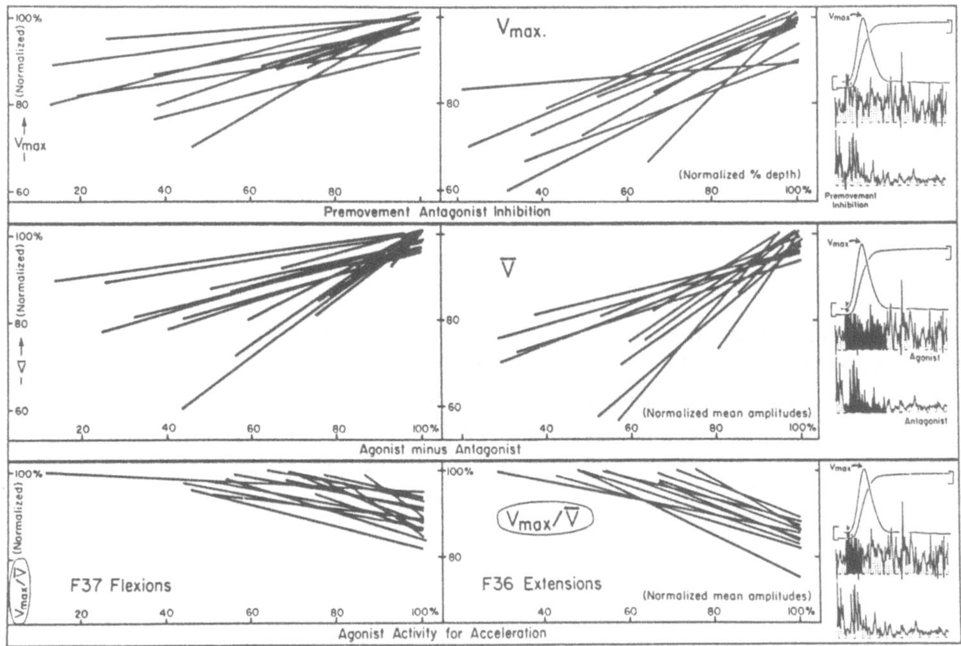

Fig. 4. Velocity profiles depend on intensity of premovement antagonist inhibition and of cocontraction of opposing muscles during the movement, factors that in turn determine agonist activity during acceleration (muscular activity periods indicated in insets on right and plotted along abscissae). Lines of best fit are plotted in the top row for Vmax against premovement antagonist inhibition; in the middle row for V against agonist-minus-antagonist activity; and in the bottom row for Vmax/V against agonist activity during acceleration. Each line represents best fir for 100-150 single-peaked movements made in behaviorally appropiate trials in one training session, calculated as means of 5 succesive movements per datum point; points not shown. (From Brooks VB and Watts SL, in preparation).

postures were no longer maintained by cocontraction of opposing muscles but they still required tonic agonist activity because the targets were lateral to the neutral elbow angle (Fig. 1). Periods of antagonist deactivation corresponded to periods of greater agonist activity.

Figure 3C shows that peak and average velocities both increased steadily, and in such proportion so as to lower their ratio: from near 2.2 to near 1.9. Changing movement shapes altered movement scaling: durations shortened for movements that retained constant amplitude because of the task requirement to land in the targets. Values for the most 'skilled' movements approximate the shape index criterion ratio of 1.8 that signals optimization of accelerative transients (minimal 'jerk'). That would fit the almost time-symmetrical shapes of these samples (Hogan 1984; see Fig. 3A).

EMG changes are related systematically to kinematic changes in Fig. 4. Each line represents the best fit for a number of points (up to 30, each representing 5 successive continous movements made in behaviorally appropriate trials in one session). The three correlations in Fig. 4 were all at least at the 1% level of significance, far better than other correlations examined. The upper two sets of graphs in Fig. 4 reflect two aspects of movement scaling: peak velocity (V_{max}) depended on premovement inhibition of the antagonist, i.e. on removing braking forces at take-off; and average velocity (V) was related to falling cocontraction of agonist-minus-antagonist throughout the movement, i.e. to average net torque. The lowermost set of graphs in Fig. 4 reflects an aspect of movement shaping: the index ratio of peak/average velocity (V_{max}/V) was correlated inversely to agonist activity that occurred up to peak velocity, i.e. strong acceleration tended to lower the shape index ratio.

Discussion

Our findings can be summarized simply: monkeys relaxed the elbow joint as they gained experience in its control during a step-tracking task, even though they were not rewarded for any particular mode of movement execution. The arm was relaxed by lessened cocontraction of opposing muscles during arm movements and postures. Experience is also manifest in the refined move program: velocity profiles tended towards time-symmetry with a ratio of peak/average velocity that suggests minimization of jerk for at least some movements. This refinement was executed by muscular activity controlled to time and scale application of more torque, an ability that was not present during early training when cocontraction ensured safety through stiff joints. How the brain and spinal cord engage muscular activity into skilled patterns (Fig. 3) is not obvious, but let us take stock.

The shape of the velocity profile reflects an aspect of the learned 'reference trajectory' of programmed movements with respect to time spent (Abend et al. 1982; Bizzi et al. 1982) or to distance traversed (Cooke 1980). Reference trajectories are ordered by the motor cortex as a sequence of force vectors (Georgopoulos et al. 1982) for the 'object of greatest attention for the brain': in this case the elbow joint (cf. Brooks 1984b). Trajectories are enacted by the spinal cord through learned sequences of joint torques and stiffness; that is, by shifting the participating muscles through learned sequences of length-tension relations (Polit and Bizzi 1979; Feldman 1980) suitable for contributing optimally to joint torques (Buchanan et al. 1986). These programs depend on the cerebellum, which composes skilled movements by selection, timing and scaling of the participating muscles. It does this by means of predictive, feedforward use of learned commands as well as by means of reafferent and reefferent signals (e.g. Meyer-Lohmann et al. 1975, 1977; Hore and Vilis 1984; cf. review Brooks and Thach 1981). This view was approached by Kornhuber (1974) with the suggestion that "...the cerebral cortex probably provides only a sequential program with specification of direction, extent, etc. and the cerebellum takes care of the timing, using reafferent signals from lower centers...".

Although cerebellar function is essential for generating optimal movements with their associated peak and average velocities, their shape index ratio is independent of cerebellar function as long as movement shape remains the same. Adaptive 'tuning' of agonist/antagonist cocontraction can scale movements without changing their shape, and cerebellar dysfunction can reverse these adaptations (Brooks 1984a). This point is recalled in Fig. 5, which illustrates lowering of peak and average velocities (for movements that remained 'continuous': not shown here but see Figs. 9-11 in Brooks 1984a) by disabling a cerebellar output or input. Cerebellar dysfunction first removes the adaptations of continuous movements and then decomposes the basic programs for conti-

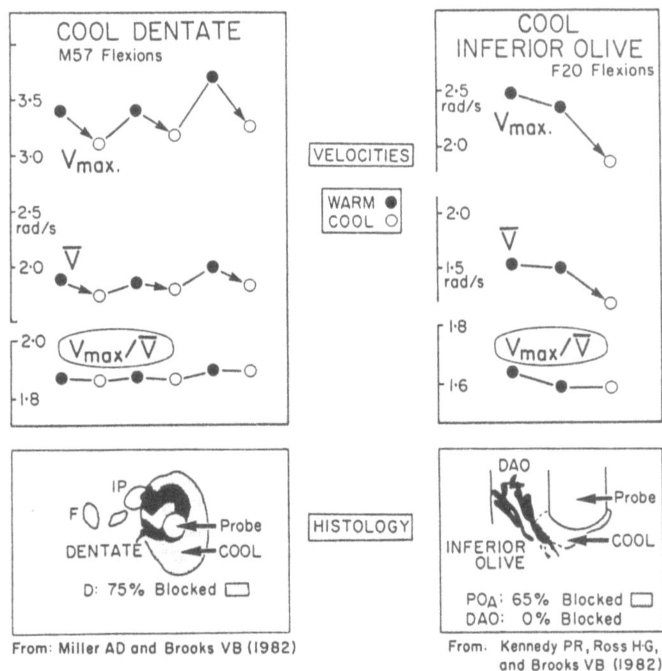

Fig. 5. Cerebellar dysfunction can depress two scale aspects of continuous movements (Vmax, V; upper and middle graphs) in such proportion so as not to affect their ratio (Vmax/V, lowermost graph). This shape index remained unchanged for those movements that remained continuous (not shown) during cerebellar dysfunction produced by local cooling (open symbols in graphs). Histology diagrams show % area that was blocked (stippled) for nuclei (black): D = dentate n.; POA = principal olivary n. anterior. (Data calculated from Miller and Brooks 1982; Kennedy et al. 1982; cf. Brooks 1984a)

nuous movements, which brings back the 'hesitant landings', i.e. stepwise decelerations, of early training (Brooks 1984a).

In this paper the distinction has been stressed between actions regarding *what* to do as distinct from those dealing with *how* to do it. What to do is covered by 'plans' for movements and postures that are carried out by 'programs' for how to do it. Practice establishes predictive routines for plans and programs. Planning is an activity of the 'highest' level of the motor hierarchy while the middle level, and particularly the cerebro-cerebellar circuit, deals with programming and with optimization of postures and movements. This view is summarized in Fig. 7, a derivation from Eccles' school put forward originally by Allen and Tsukahara (1974; cf. Brooks 1986a).

In Fig. 6 the reward systems of the brain are only alluded to by barely sketched-in connections of the limbic system and the hypothalamus. Yet these systems are of the essence for the interpretation of our paradigm, which distinguishes between rewarded and nonrewarded performance changes. Knowledge of results achieved by practice is essential for learning, which presumably involves the highest level. Kornhuber (1973) addressed this problem by assigning a 'selection' function to the limbic Papez circuit that connects with all levels, especially through the cingulate cortex. Prominent cingulate potentials have been observed during the critical transition from learning by trial and error to insightful learning in humans (Deecke et al. 1984) and in monkeys (Brooks

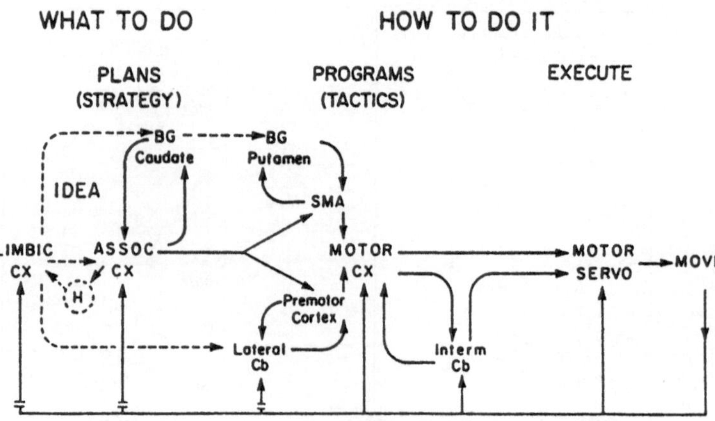

Fig. 6. A simplified scheme of information flow for voluntary movements. Thalamic and other nuclei are omitted and only some limbic connections shown (broken lines). BG, basal ganglia; Cb, cerebellum; CX, cortex; H, hypothalamus. (Modified from Brooks 1986a; derived from Allen and Tsukahara 1974)

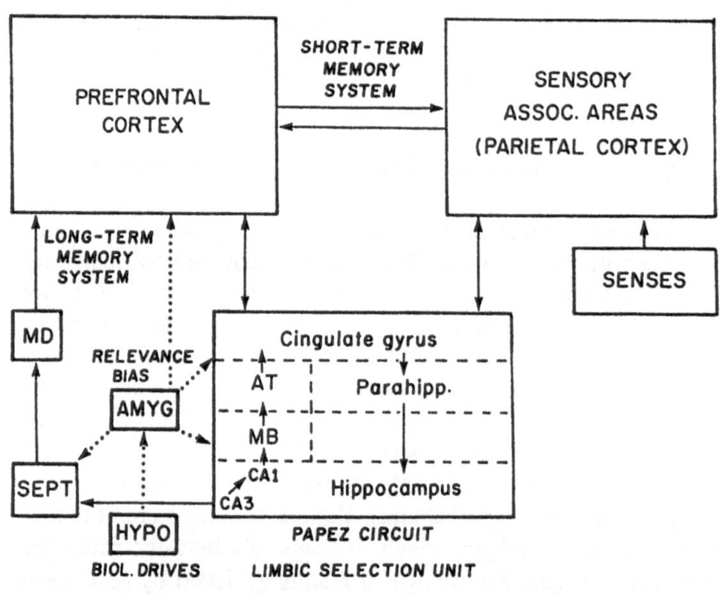

Fig. 7. A simplified scheme of information flow for selection of relevant events, memories, and motivations toward action in relation to the components of short-term and long-term memory systems. AMYG, amygdala; AT, anteroventral n. of thalamus; CA1, CA3, cell systems in hippocampus; HYPO, hypothalamus; MB, mamillary body; MD, mediodorsal thalamic nucleus; SEPT, septum. (Modified from Fig. 2 in Kornhuber 1973.)

1986b; Gemba et al. 1986). Fig. 7 presents a modification of Kornhuber's diagram (Fig. 2; 1973), retaining his insights but adding the circuits of the amygdala as likely mechanisms for detecting relevant events, memories and motivations as part of the limbic selection process (cf. Brooks 1986b).

Summary

Self-optimizing adaptations of programmed movements and postures were studied during motor learning. Monkeys were rewarded for knowing *what* to do (behavioral skill) in a step-tracking task in which an arm movement into a target had to be followed by a prolonged holding posture of the arm in target to trigger the next trial. The animals were not rewarded for any particular mode of movement execution or their adaptations (*how* to do it). Yet, motor skill, i.e. use of correct programmed movements, made in the context of the task, developed in proportion to the behavioral skill. Torque increased as the arm became less stiff during moving and holding through lessened cocontraction of opposing muscles. Peak velocities increased in proportion to joint relaxation measured as premovement inhibition of the antagonist; and average velocities increased in proportion to average net torque measured as agonist-minus-antagonist activity throughout the movement. A shape index (peak/average velocities) decreased because the two velocity scaling factors increased at slightly different rates. The index fell with rising acceleratory thrust measured as agonist activity up to peak velocity.

References

Abend W, Bizzi E, Morasso P (1982) Human arm trajectory formation. Brain 105: 331-348

Allen GI, Tsukahara N (1974) Cerebrocerebellar communication systems. Physiol Rev 54:957-1006

Bizzi E, Accornero N, Chapple W, Hogan N (1982) Arm trajectory formation in monkeys. Exp Brain Res 46: 139-143

Brooks VB (1984a) The cerebellum and adaptive tuning of movements. In: Creutzfeldt O, Schmidt RF, Willis WD, (eds) Sensory motor integration in the nervous system. Exp Brain Res, [Suppl] 9:170-183

Brooks VB (1984b) How are 'move' and 'hold' programs matched? In: Bloedel JL, Dichgans J, Precht W (eds) Cerebellar functions. Springer, Berlin Heidelberg New York, pp 1-23

Brooks VB (1986a) The neural basis of motor control. Oxford University Press, New York

Brooks VB (1986b) How does the limbic system assist motor learning? A limbic comparator hypothesis. Brain Behav Evol 29: 29-53

Brooks VB, Thach WT (1981) Cerebellar control of posture and movement. In: Brooks VB (ed) Motor control. American Physiological Society, Bethesda, pp 877-946 (Handbook of Physiology, sect 1, Vol 2)

Brooks VB, Watts SL (1988) Adaptive programming of arm movements. J Motor Behav 20:117-132

Brooks VB, Cooke JD, Thomas JS (1973) The continuity of movements. In Stein RB, Pearson KG, Smith RS, Redford JB (eds) Control of posture and locomotion. Plenum, New York, pp 257-272

Brooks VB, Kennedy PR, Ross HG (1983) Movement programming depends on understanding of behavioral requirements. Physiol Behav 31: 561-563

Buchanan TS, Almdale DP, Lewis JL, Rymer WZ (1986) Characteristics of synergic relations during isometric contractions of human elbow muscles. J Neurophysiol 56: 1225-1241

Cooke JD (1980) The organization of simple, skilled movements. In: Stelmach GE, Requin J (eds) Tutorials in motor behavior. Amsterdam: North Holland, Amsterdam, pp 199-212

Deecke L, Heise B, Kornhuber HH, Lang M, Lang W (1984) Brain potentials associated with voluntary manual tracking: Bereitschaftspotential, conditioned premotion positivity, directed attention potential, and relaxation potential. Anticipatory activity of the limbic and frontal cortex. Ann NY Acad Sci 425: 450-464

Feldman AG (1980) Superposition of motor programs. I. Rhythmic forearm movements in man. Neuroscience 5: 81-90
Gemba H, Sasaki K, Brooks VB (1986) 'Error' potentials in limbic cortex (anterior cingulate area 24) in monkeys. Neurosci Lett 70: 223-227
Georgopoulos AP, Kalaska JF, Caminiti R, Massey JT (1982) On the relations between the direction of two-dimensional arm movements and cell discharge in primate motor cortex. J Neurosci 2: 1527-1537
Hogan N (1984) An organizing principle for a class of voluntary movements. J Neurosci 4: 2745-2754
Hore J, Vilis T (1984) A cerebellar-dependent efference copy mechanism for generating appropriate muscle responses to limb perturbations. In: Bloedel J, Dichgans J, Precht W (eds) Cėrebellar functions. Springer, Berlin Heidelberg New York, pp 24-35
Kennedy PR, Ross HG, Brooks VB (1982) Participation of the principal olivary nucleus in neocerebellar motor control. Exp Brain Res 47: 95-104.
Kornhuber HH (1973) Neural control of input into long term memory: limbic system and amnestic syndrome in man. In: Zippel HP (ed) Memory and transfer of information. Plenum, New York, pp 1-22
Kornhuber HH (1974) Cerebral cortex, cerebellum, and basal ganglia: an introduction to their motor functions. In: Schmitt FO, Worden FG (eds) The neurosciences third study program. MIT Press, Cambridge, pp 267-280
Meyer-Lohmann J, Conrad B, Matsunami K, Brooks VB (1975) Effects of dentate cooling on precentral unit activity following torque pulse injections into elbow movements. Brain Res 94: 237-251
Meyer-Lohmann J, Hore J, Brooks VB (1977) Cerebellar participation in generation of prompt arm movements. J Neurophysiol 40: 1038-1050
Miller AD, Brooks VB (1982) Parallel pathways for movement initiation in monkeys. Exp Brain Res 45: 328-332.
Polit A, Bizzi E (1979) Characteristics of motor programs underlying arm movements in monkey. J Neurophysiol 42: 183-194

Emergent Issues in the Control of Multi-joint Movements

E. Bizzi and F.A. Mussa-Ivaldi

Department of Brain and Cognitive Sciences, Massachusetts Institute of Technology, Cambridge, Ma. 02139 U.S.A.

Coordinate Transformations and Arm Trajectory Planning

The necessary first step in any planning of an arm trajectory must be for the central nervous system (CNS) to represent the position of the target to be reached. This initial step involves transforming the retinal image of the target into head-centered and, ultimately, body-centered coordinates.

An insight into this process has been provided by Andersen et al. (1985, 1987), who recorded the activity of single neurons from area 7a of the posterior parietal cortex in monkeys. They found that the responsiveness of these cells' receptive field to retinotopic stimuli was influenced by the angle of gaze. Their experiment demonstrated that there was an interaction between the system that represented the eye's position and the system that represented the retinal position. This interaction tuned parietal neurons to represent the location of targets in head-centered coordinates.

It is important to emphasize that this coordinate transformation occurs in the posterior parietal cortex, an area that appears to be crucial in the planning of voluntary actions. Lesions of the parietal lobe in humans produce an array of symptoms, including the inability to localize visual targets, the loss of spatial memory and deficits in motor coordination.

In addition to the representation of the target in body-centered coordinates, the CNS must also represent the initial arm configuration in order to plan the arm trajectory.

Once the initial position and the final target are expressed in the same coordinate frame, then the CNS must solve the problem of representing the whole trajectory planning the path and the velocity of the hand in space. There is some evidence that this representation may be formed in the posterior parietal cortex and the medial regions of the frontal lobe.

Recent results obtained by Georgopoulos et al. (1980), based on recording single neurons from the parietal cortex of monkeys, indicated a correlation between neural activity and the direction of the arm's movement. These experiments consisted of recording activity from individual cortical cells as the monkey executed movements from a fixed location to a set of targets placed around the starting point. The results indicated that the curves relating the frequency of discharge of a cell to the direction of movement were characterized by a single maximum, corresponding to the "preferred direction" of the cell. In addition, the average direction of a cell ensemble (defined as the weighted sum of all the preferred directions with the weights given by the cell activities) was found to be in good agreement with the direction of hand movements.

It is tempting to speculate that the signals from the motor cortex may represent the physiological substrate of trajectory planning in spatial coordinates. However, the viability of an alternative hypothesis has recently been demonstrated (Mussa-Ivaldi 1988). According to the latter, it is also possible to interpret Georgopoulos's results as evidence that cortical cells encode muscle activations instead of the direction of hand movements.

Another difficult challenge for motor neurophysiologists is to understand the transformation from trajectory representation into the appropriate joint motion and joint torques. In the past, physiologists have not specifically addressed this question. The signals from "motor" areas were assumed to activate the segmental spinal cord apparatus and generate the desired movement. Very little attention was paid to the complex problem of deriving joint motions and joint torques and compensating for dynamic interactions.

In this chapter, we will discuss how the transformation from planning to joint motion and forces is handled by biological systems. We will also present a neurobiological view on torque generation based on exploiting muscle mechanical properties.

We believe that the issue of understanding the role of muscles' mechanical properties deserves special emphasis because it is likely that the features displayed by the neural controller have evolved as a result of the need not only to control, but also to take advantage of the mechanical properties of the musculoskeletal apparatus.

A case for this approach was made by Feldman (1974a,b), who investigated the spring-like properties of the human arm. Muscles do indeed behave like tunable springs in the sense that the force generated by them is a function of length and level of neural activation (Rack and Westbury 1974).

In addition, muscles are arranged about the joints in an agonist-antagonist configuration. If we attribute spring-like properties to muscles, then a limb's posture is maintained when the forces exerted by the agonist and antagonist muscle groups are equal and opposite. This fact implies that when an external force is applied, the limb is displaced by an amount that varies with both the external force and the stiffness of the muscle. When the external force is removed, the limb should return to the original position. Experimental studies of arm movements in monkeys have shown that a forearm posture is indeed an equilibrium point between opposing spring-like forces (Bizzi et al. 1976).

The observation that posture is obtained from the equilibrium between the length-tension properties of opposing muscles led to the idea that movements result from a shift of the equilibrium point. This hypothesis was first proposed by Feldman (1966). The studies by Bizzi et al. (1976), Kelso (1977) and Kelso and Holt (1980) provided the needed experimental evidence.

In particular, Bizzi et al. (1984) demonstrated that the transition from one arm posture to another is achieved by adjusting the relative intensity of neural signals directed to each of the opposing muscles. According to this result, a single-joint arm trajectory is obtained through neural signals which specify a series of equilibrium positions for the limb.

The experimental evidence supporting this important point derives from three sets of experiments, which will be briefly summarized here. The movements used in these experiments were single-joint elbow flexion and extension, which lasted approximately 700 ms for a 60° amplitude.

The first set of experiments was performed in both intact monkeys and in those deprived of a sensory feedback. The monkey's arm was briefly held in its initial position after a target that indicated final position had been presented. It was found that movements to the target after release of the arm were faster than control movements in the absence of a holding action. Figure 1 shows a plot of the accelerative transients against the durations of the holding period in the same animal before and after interruption of the nerves conveying sensory information. The time course of the increase in the amplitude of the accelerative transient was virtually identical in the two conditions.

It was found that the initial acceleration after release of the forearm increased gradually with the duration of the holding period, reaching a steady-state value no sooner than 400 ms after muscles' activation. These results demonstrated that the CNS had programmed a slow, gradual shift of the equilibrium position instead of a sudden, discontinuous transition to the final position.

The same conclusions are supported by a second set of experiments (Bizzi et al. 1984) based on forcing the forearm to a target position through an assisting torque pulse applied at the beginning of a visually triggered forearm movement. The goal of this experiment was to move the limb ahead of the equilibrium position with an externally imposed displacement in the direction of the target. It was found that the forearm, after being forced by the assisting pulse to the target position, returned to a point between the initial and the final position before moving to the endpoint. This outcome results from the fact that a restoring force is generated by the elastic muscle properties. Note that if muscles merely generated force or if the elastic properties were negligible, we would not have seen the same return motion of the limb. Since the same response to our torque pulse was also observed in monkeys deprived of sensory feedback (Bizzi et al. 1984), it was inferred that proprioceptive reflexes are not essential to the generation

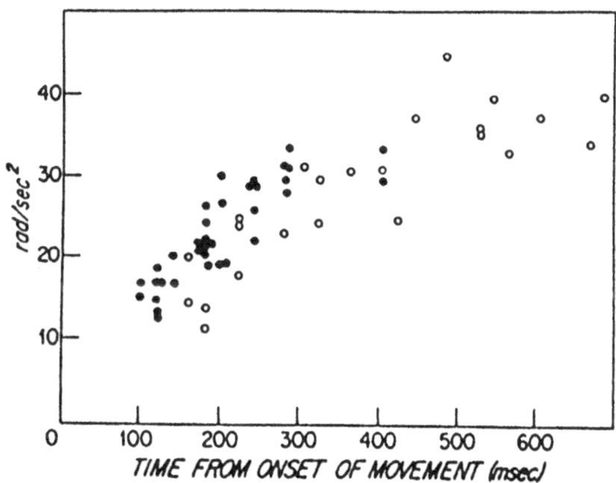

Fig.1. The forearm of intact and deafferented animals was held in its initial position while the animal attempted to move toward a target light. Then, the forearm was released at various times. This figure is a plot of acceleration (immediately following release) versus holding time. The abscissa shows time in milliseconds; the ordinate shows radians per second squared. Solid circles: intact animal; open circles: deafferented animal. (From Bizzi et al. 1984)

of restoring forces. Taken together, these results suggest that alpha motoneuronal activity specifies a series of equilibrium positions throughout the movement.

Finally, in a third set of experiments, the arm was not only driven to the target location, but also held there for a variable amount of time (1 to 3 s), after which the target light at the new position was activated. A cover prevented the animal from seeing its arm (Figure 2). After the reaction time to the presentation of the light, the monkey activated its muscles (flexors in the case of Figure 2) to reach the target position. At this point, usually shortly after the onset of muscle activity, the servo which held the arm was deactivated. The arm then returned to a point intermediate between the initial and the target positions before moving back to the target position. Note that, during the return movement that required extension, evident flexor activity was present. The amplitude of the return movement was a function of the duration of the holding action. If enough time elapsed between activation of the target light and deactivation of the servo, the arm remained in the target position upon release.

These observations provide further support for the view that motoneuronal activity specifies a series of equilibrium positions throughout the movement. If the muscles me-

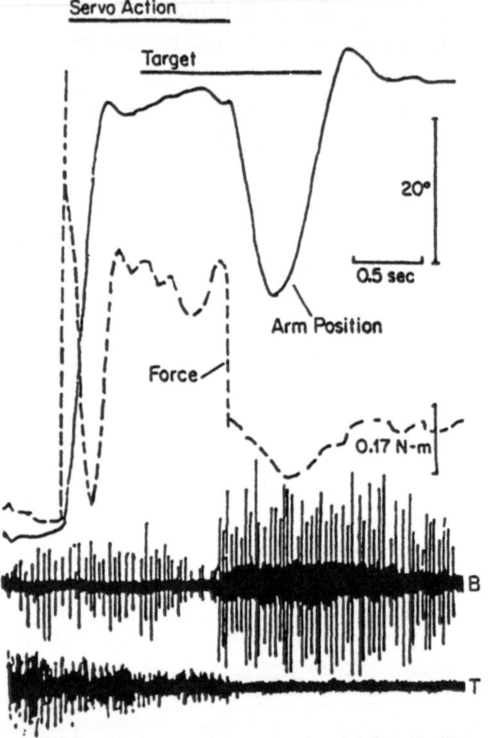

Fig.2. Forearm movements of deafferented monkeys with a holding action in the final position. While the target light remained off, the servo moved the arm to the target position. Then the target light was activated, and the servo was turned off. The arm returned to a position intermediate between the initial and target positions before moving back to the target position. Similar results were obtained in many trials in two monkeys. The upper bar indicates duration of servo action. The lower bar indicates onset of the target light. The broad trace shows arm position; the dashed trace shows torque. B: flexor (biceps); T: extensor (triceps). (From Bizzi et al. 1984)

rely generated force during the transient phase of a movement, we would not have seen the pronounced return motion of the limb during flexor muscle activity (Figure 2). This series of equilibrium positions has been termed the "virtual trajectory" (Hogan 1984).

The idea of a moving equilibrium point is a direct consequence of two known facts: one, that a limb is at static equilibrium when all the torques generated by opposing muscles cancel out, and two, that the neural input to each muscle has the effect of selecting a length-tension curve. It follows that at all times the neural activities directed to all the muscles acting on a limb can be "translated" into the corresponding equilibrium angle, which is given by the balance of the elastic torques.

During the execution of a movement, these equilibria or virtual positions act as "centers of attraction." At any time, the difference between actual and virtual position generates an elastic force directed towards the virtual position. The end course of the movement is determined by the interaction of this elastic force with limb inertia and viscosity. However, if the neural inputs were suddenly frozen, the limb would ultimately come to rest in the virtual position encoded by the current values of the inputs.

Traditionally, in motor control, behaviors like maintaining posture, generating movements, and generating forces have been considered separate endeavors requiring different control schemes. In contrast, we have shown here that these apparently diverse behaviors share a common scheme. We have also shown that this unification is contingent upon the mechanical properties of the musculoskeletal system.

Arm Multi-joint Posture and Movement

In the previous section, we have described the muscle's elastic behavior and presented a new view on the execution of single-joint movements. One of the main lessons to be derived from single-joint experiments is that movement is nothing more than a series of sequentially implemented postures. Movement derives from shifting the equilibrium point between agonist and antagonist length-tension curves. While it is easy to understand these experimental results in the context of single-joint motion, it is the multi-joint case that poses the most difficult challenge for this theory.

Multi-joint studies involve approaching qualitatively different aspects of motor control. For example, through single-joint studies, it was found that the CNS achieves a stable posture of the forearm by selecting appropriate length-tension curves of the elbow muscles so that, at the desired elbow angle, the torque generated by the flexors is equal and opposite to the torque generated by the extensors. As a small external perturbation displaces the limb $\delta\theta$ degrees from its equilibrium location, the elastic muscle properties generate a restoring torque δT. The ratio of this torque to the imposed displacement is a single number expressing the stiffness of the elbow. By contrast, in a multi-joint situation, if a displacement is externally imposed on the hand, rather than on a single articulation, the amount of stretch experienced by the muscles depends not only upon the amplitude of the perturbation, but also upon the direction of the forces. Then, a single number is no longer sufficient to describe the force/displacement relation.

To deal with this more complex situation, a new experimental approach to the study of posture and movement was developed (Mussa-Ivaldi et al. 1985). This approach was based on measuring the net spring-like behavior of the multi-joint arm by displacing the hand in several directions (see Figure 3). As the hand came to rest at the end of each displacement, the force, $F=(F_x,F_y)$, exerted by the subject on the handle was measured. Since the hand was stationary, this force had no viscous or inertial components and could only be due to muscle length-tension properties (including reflex components).

With a small displacement of the hand, $\delta r=(\delta x,\delta y)$, it is legitimate to assume a linear relation of the form:

$$F_x = K_{xx}\delta x + K_{xy}\delta y$$
$$F_y = K_{yx}\delta x + K_{yy}\delta y \qquad (1)$$

Then, by measuring forces and displacements in different directions, it is possible to estimate the K coefficients from a linear regression applied independently to both the

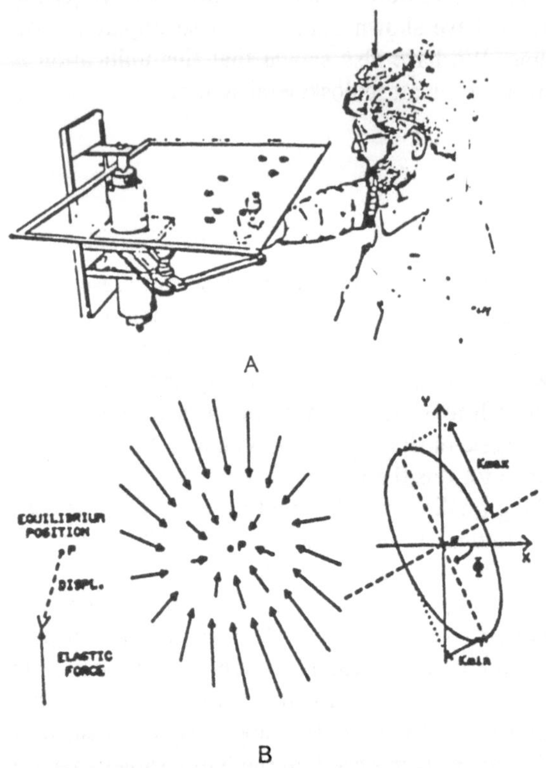

Fig.3. A. Experimental setup: sketch of the apparatus in a typical experimental situation. **B.** Stiffness representation. Left: when the hand is displaced from its equilibrium position, an elastic restoring force is observed which in general is not co-linear with the displacement vector. Center: several displacements of variable amplitude and direction are plotted together with the restoring forces computed from a measured hand stiffness. Right: the trajectory of the force vectors obtained by means of the previous procedure is an ellipse with the major and minor axes indicated, respectively, by Kmax and Kmin. The angle, Φ, between the major axis and the fixed x axis is the stiffness orientation. The shape is given by the ratio, Kmax/Kmin, and the size, or magnitude, is the area enclosed by the ellipse. (Modified figure from Mussa-Ivaldi et al. 1985)

above expressions. These coefficients can be represented by a single entity: a table, or matrix, expressing the multidimensional stiffness of the hand:

$$K = \begin{bmatrix} K_{xx} & K_{xy} \\ K_{yx} & K_{yy} \end{bmatrix}$$

With this notation, Equation 1 assumes a more compact form, $F = K\,\delta r$, which is analogous to the equation describing the behavior of a one-dimensional system such as a single joint. However, in the multi-joint situation, the stiffness matrix is a more complex entity than the single-joint stiffness and provides new insights into multi-joint posture.

The hand stiffness in the vicinity of equilibrium is represented by a matrix which was estimated by analyzing the force and displacement vectors. The hand stiffness was represented as an ellipse characterized by three parameters: magnitude (the total area derived from the determinant of the stiffness matrix); orientation (the direction of maximum stiffness); and shape (the ratio among maximum and minimum stiffness). This

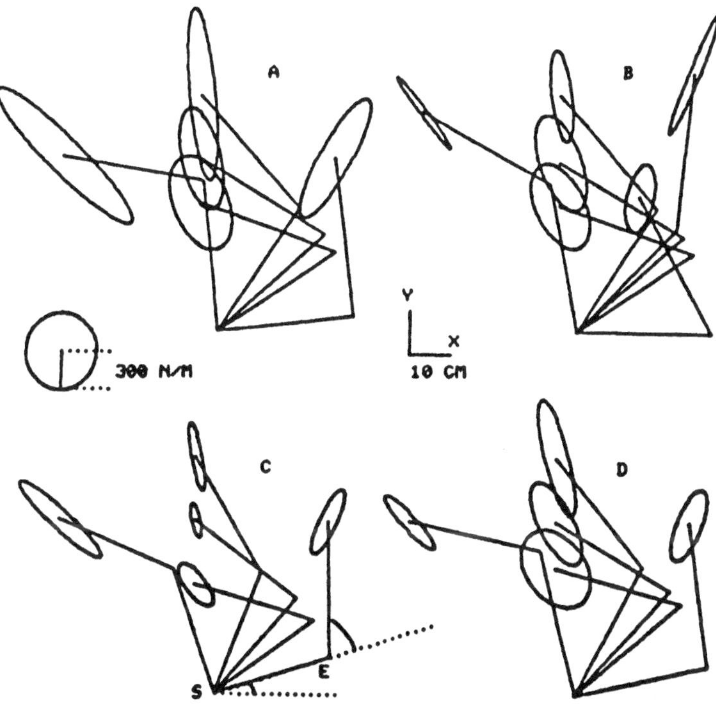

Fig.4. Stiffness ellipses obtained from four subjects during the postural task. Each ellipse has been derived by regression on about 60 force and displacement vectors. The upper arm and the forearm are indicated schematically by two line segments, and the ellipses are placed on the hand. The calibration for the stiffness is provided by the circle to the left which represents an isotropic hand stiffness of 300 N/m. (From Mussa-Ivaldi et al. 1985)

ellipse captures the main geometrical features of the elastic-force field associated with a given hand posture and provides an understanding of how the arm interacts with the environment.

The hand stiffness of four subjects maintaining the hand in a number of workspace locations is shown in Figure 4. The stiffness ellipses measured at given hand postures are also shown along with a schematic display of the corresponding arm configurations. A remarkable feature of these data is the similarity between the different subjects with respect to stiffness, shape and orientation. By contrast, the stiffness magnitude varies considerably. This graphical representation provides a "gestalt" and affords a qualitative understanding of the way in which the hand may interact with external forces that could change its posture.

Describing hand posture as an "oriented stiffness ellipse" helps us to determine which elements of motor behavior require accurate coordination of neural signals and which result from the arm's biomechanical design. To address this issue, Mussa-Ivaldi et al. (1987) measured the postural stiffness in four different conditions: with no load and with a 10 Newton force applied at each of the three following directions: 0°, 45° and 90°. These forces were applied in order to elicit different patterns of contraction in different sets of muscles and thus to test how variations in neural input affect the parameters of postural stiffness measured at a given location.

This procedure generated a surprising result (Figure 5): there were large changes in stiffness amplitude, but only small changes in orientation and shape. This result means that the parameters of orientation and shape are predominantly dependent upon musculoskeletal geometrical properties of the arm.

It was found that one way to affect the stiffness shape and orientation is to change the configuration of the arm while the hand remains in a given position. This finding suggests that an effective strategy for modifying all parameters of the postural stiffness may be to combine variations of neural input to the muscles with variations of configuration of the "extra" or redundant degrees of freedom of the limb. Changes in configuration also have a relevant effect on other components of motor impedance. Indeed, changing arm configuration is the only way the CNS can change the endpoint inertia of the limb (Hogan 1985). From this point of view, it can be seen that the configuration of the limb should be regarded as one of the "commanding inputs" available to the CNS for controlling posture. Redundancy of the musculoskeletal system is usually regarded as a problem to be overcome by the CNS in coordinating limb movements (Bernstein 1967); instead, the results reported here show that redundancy may also offer alternative ways to control postural dynamics.

To sum up, the experimental evidence indicates that the equilibrium position of the hand is established by the coordinated interaction of elastic forces generated by the arm muscles (Mussa-Ivaldi et al. 1985). According to the equilibrium trajectory hypothesis, which was tested first in the context of single-joint movements, the multi-joint arm trajectory is achieved by gradually shifting the arm equilibrium between the initial and final positions. In this control scheme, the hand tracks its equilibrium point, and torque is not an explicitly computed variable.

Evidence supporting this hypothesis in the context of multi-joint hand movements has been obtained by combining observations of hand movements with computer simulation studies. A model developed by Flash (1987) has successfully captured the kinematic features of measured planar arm trajectories.

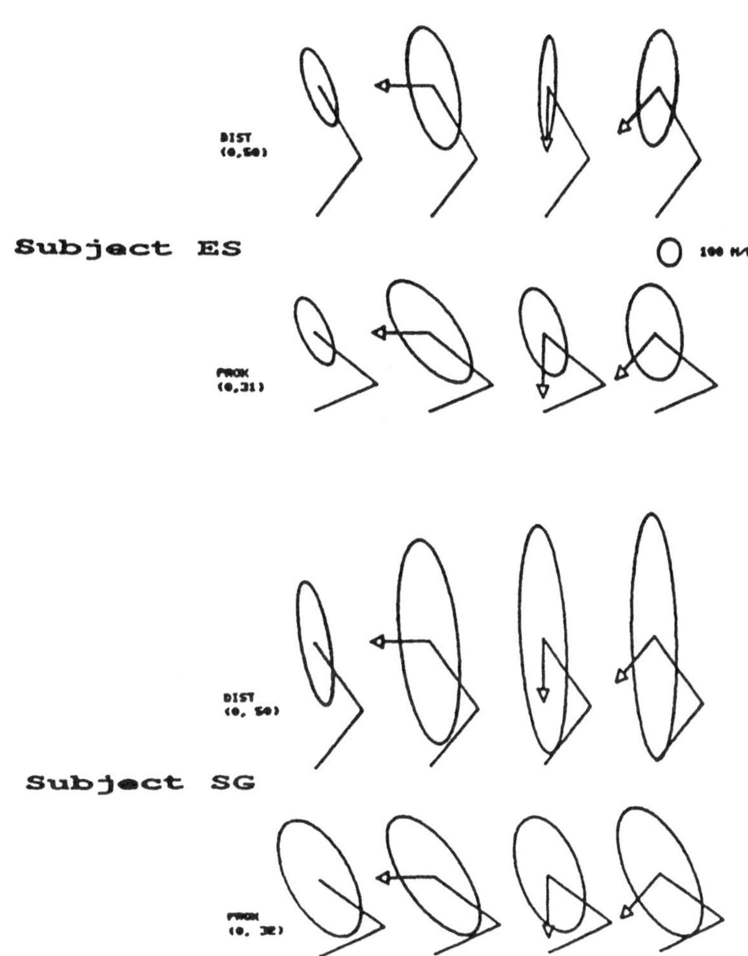

Fig.5. Postural stiffness without and with loads. The arrows indicate the directions of the 10 Newton load

A directly testable consequence of the equilibrium trajectory hypothesis would be the built-in stability of movements: if, during the execution of a hand trajectory, an unpredicted disturbance displaces the hand from the planned path, then according to the equilibrium trajectory hypothesis, the muscle's elastic properties and the proprioceptive reflexes generate a force attracting the hand towards the original path.

This prediction was experimentally confirmed by McKeon et al. (1984). They asked subjects to perform pointing movements between two targets while gripping the handle of a two-link manipulandum similar to that shown in Figure 3. A clutch mounted on the inner joint of the manipulandum was used to brake the inner link under computer control. As the clutch was activated at the onset of a movement, the hand trajectory was restricted to a circular path with a radius equal to the length of the outer link of the manipulandum. While the clutch was engaged, the handle force was found to be always strongly oriented so as to restore the hand to the unconstrained path and not to the endpoint of the path.

The success of the simulation in capturing the kinematic details of measured arm movements is important as a step towards providing us with a new intellectual frame for understanding trajectory formation in the multi-joint context. This work indicated a planning strategy whereby the motor controller may avoid complex computational problems such as the solution of inverse dynamics. According to the equilibrium trajectory hypothesis, the muscle's spring-like properties are responsible for generating the necessary joint torques, thus implicitly providing an approximated solution to the inverse dynamics problem. As the approximation becomes inadequate at higher speeds of acceleration, the stiffness can be increased and the equilibrium trajectory can be modified on the basis of the difference between the actual and the planned path. The task of the CNS is then to transform the planned trajectory into a different sequence of equilibrium positions and stiffnesses.

Summary

In this chapter we focus on the control of arm movements and, in particular, on the way in which the central nervous system (CNS) proceeds from the planning of trajectories to their execution. The investigations which are described here indicate that the brain can take advantage of the biological design of the arm in order to simplify some of the computational tasks involved in arm movements.

In this chapter we have considered the following points:

1. Coordinate transformations and arm trajectory planning
2. Arm multi-joint posture and movement

Acknowledgements. This research was supported by NIH Grants NS09343, AR26710 and EY02621 and Office of Naval Research Grant N00014/88/K/0372.

References

Andersen R.A., Essick G.K., Siegel R.M., (1985) Encoding spatial location by posterior parietal neurons. Science 230:456-458

Andersen R.A., Essick G.K., Siegel R.M. (1987) Neurons of area 7 activated by both visual stimuli and oculomotor behavior. Exp Brain Res 67:316-322

Bernstein M. (1967) The co-ordination and regulation of movements. Pergamon, Oxford

Bizzi E., Polit A., Morasso P. (1976) Mechanisms underlying achievement of final head position. J.Neurophysiol. 39:435-444

Bizzi E., Accornero N., Chapple W., Hogan N. (1984) Posture control and trajectory formation during arm movement. J.Neurosci.4:2738-2744

Feldman A.G. (1966) Functional tuning of the nervous system during control of movement or maintenance of a steady posture. III.Mechanographic analysis of the execution by man of the simplest motor tasks. Biophysics 11:766-775

Feldman A.G. (1974a) Change of muscle length due to shift of the equilibrium point of the muscle-load system. Biofizika 19:534-538

Feldman A.G. (1974b) Control of muscle length. Biofizika 19:749-751

Flash T. (1987) The control of hand equilibrium trajectories in multi-joint arm movements. Biol. Cybern. 57:257-274

Georgopoulos A.P., Kalaska J.F., Massey J.T. (1980) Cortical mechanisms of two-dimensional aiming arm movements. I.Aiming at different target locations. Neurosci. Abstr 6:156
Hogan N. (1984) An organizing principle for a class of voluntary movements. J. Neurosci. 4:2745-2754
Hogan N. (1985) The mechanics of multi-joint posture and movement control. Biol. Cybern. 52:315-331
Kelso J.A.S. (1977) Motor control mechanisms underlying human movement reproduction. J.Exp.Psychol. 3:529-543
Kelso J.A.S., Holt K.G. (1980) Exploring a vibratory system analysis of human movement production. J. Neurophysiol. 43:1183-1196
McKeon B., Hogan N., Bizzi E. (1984) Effect of temporary path constraint during planar arm movements. Neurosci. Abstr 10:337
Mussa-Ivaldi F.A. (1988) Do neurons in the motor cortex encode movement direction? An alternative hypothesis. Neurosci Lett 91: 106-111
Mussa-Ivaldi F.A., Hogan N., Bizzi E. (1985) Neural, mechanical and geometric factors subserving arm posture in humans. J. Neurosci. 5:2732-2743
Mussa-Ivaldi F.A., Hogan N., Bizzi E. (1987) The role of geometrical constraints in the control of multi-joint posture and movement. Neurosci. Abstr 13:347
Rack P.M.H., Westbury D.R. (1974) The short range stiffness of active mammalian muscle and its effect on mechanical properties. J. Physiol. (Lond.) 240:331-350

Looking Where the Action Is: Negative DC Shifts as Indicators of Cortical Activity

L. Deecke, W. Lang, F. Uhl and I. Podreka
Neurologische Universitätsklinik Wien, Lazarettgasse 14, 1090 Vienna, Austria

Introduction

Several means are now available for looking where the action is in the brain. One of these is the negative potential shift of cerebral cortex, which indicates activity is underway. Today we can also map the metabolic changes in the brain and the regional cerebral blood flow. The latter is possible using single photon emission computerized tomography (SPECT) and a new tracer substance, ^{99m}Tc-hexa-methyl-propylene-amine-oxime (HMPAO, Podreka et. al. 1987). This compound is trapped in metabolically active cells similar to 2-deoxyglucose. Finally, the magnetic fields in the brain can be recorded and mapped (magnetoencephalogram, MEG). Let us start with the latter.

Bereitschaftspotential and Bereitschaftsmagnetfeld

Figure 1 gives the typical shape of the Bereitschaftspotential (BP) or readiness potential in the EEG and a similar element in the MEG, which has been termed the Bereitschaftsmagnetfeld (BF) or readiness magnetic field. These comparative recordings have been carried out with a 1-channel MEG instrument in Vancouver; it is much easier now with our 14-channel MEG in Vienna. What does the MEG do? As we know from our physics lessons at school, every electrical field is accompanied by a magnetic field. So we see that the BP prior to voluntary movement has its MEG equivalent (BF), and the advantage is that the MEG has better localizing properties than the EEG.

In order to localize the current dipole in the brain (Fig. 2), one has to map the magnetic field lines on the head and try to define two maxima: one position where the field lines leaving the head have their maximum and another position where the field lines entering the head have their maximum. Once these maxima have been localized, we can say that an electrical dipole is midway between the two maxima 90° to their common line. The localizing ability of the MEG for dipoles is in the range of 1 mm, which can be considered excellent. In fact, the MEG gives the EEG, so to speak, the spatial dimension it does not otherwise have. Among the functional imaging techniques, it is the rCBF technique (in the ^{99m}Tc-HMPAO SPECT, see Fig. 4) which is very good in the space domain but has shortcomings in the time domain; the EEG is very good in the time domain - having a practically instantaneous time resolution - but is rather poor in the spatial domain. The MEG combines both attributes.

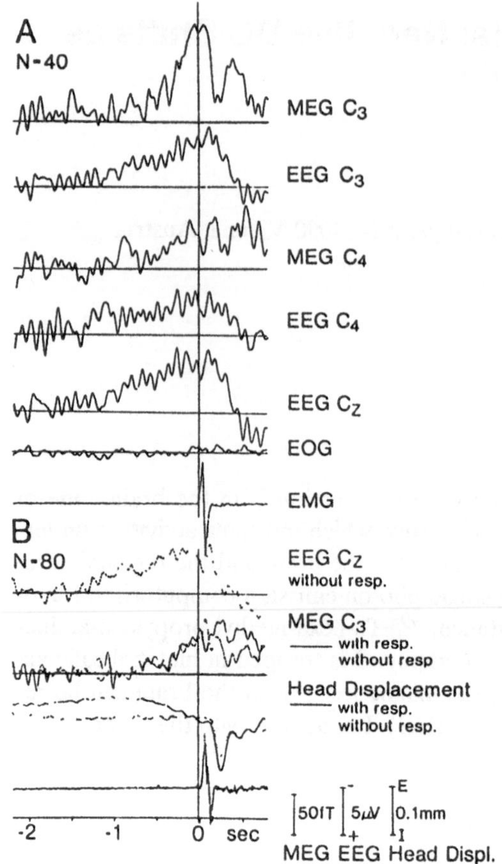

Fig. 1. Averaged magnetoencephalographic (MEG) and electroencephalographic (EEG) recordings accompanying finger movement. Subject L.D. **A** Typical experiment showing left precentral (C3) and right precentral (C4) MEG and corresponding EEG recordings, vertex EEG, electrooculogram (EOG) and electromyogram (EMG) for 40 self-paced right finger flexions. C3 MEG and C4 EEG were recorded simultaneously, as were C4 MEG and C3 EEG. Upward deflection in the MEG corresponds to magnetic field lines directed out of the head. Note the larger BFs at the left than at the right precentral positions. **B** Control experiments with normal respiration (solid curves) and without respiration (dotted curves), i.e. the subject held his breath for several seconds prior to, during and after finger movements. Vertex EEG recorded without respiration only. Head displacement measured with a remotely positioned strain gauge. Expiration (E) caused slight invisible anteversion (upward) of the head while inspiration (I) caused retroflexion. The finger movement caused slight phasic retroflexion of the head after the onset of movement. Note the normal BF and BP also in the absence of respiration. (From Deecke, et al. 1982)

Figure 2 shows the results when mapping the BF over the skull. Field lines leave the head (out, upward-going BF) in prerolandic regions, and fieldlines enter the head (in, downward-going BF) in retrorolandic regions. Thus, we know that in the rolandic region there is the electric dipole generator for this field. All these experiments were done by having the subject hold his breath, since breathing causes an artifact in the MEG, not only through the respiration wave (Grözinger et al. 1980) but simply by displacement of the head relative to the recording magnetometer: When we are sitting and inspire we make a slight retroversion and when expiring, a slight anteversion of the

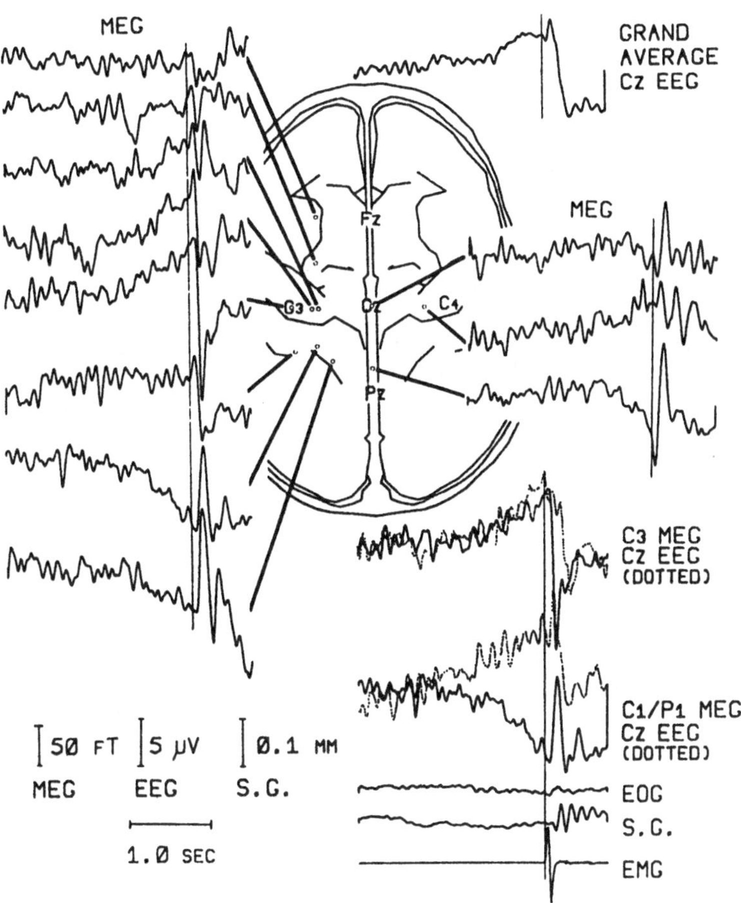

Fig. 2. Averaged MEG recordings over different locations on the scalp as indicated on the sketch showing upper cortical convolutions. Subject P.B. All averages are the result of successive blocks of 80 self-paced right finger flexions except for the grand average Cz EEG in the right upper corner. Vertical line indicates onset of electrical muscle activity (EMG). At the lower right is the MEG recorded over precentral (C3) and postcentral (C1/P1) locations with the simultaneous Cz EEG superimposed for comparison. Shown below are the EOG, head displacement (SG) and EMG for the postcentral average. Bandpass 0.003-15 Hz (EMG 5-70 Hz). All recordings were done with breath holding (from Deecke et al. 1982)

head. This albeit slight displacement is very critical in MEG recordings. So we had to control for breathing with the subject first holding his breath and then, after an irregular period of respiration arrest, performing the voluntary movement.

Figure 3 shows the BF preceding foot movements. With foot movements, the primary motor cortex (MI) generator for the BP is to be expected in the midline (on Penfield's motor homunculus; the foot region is, as we know, on the mesial cortical surface; Penfield and Rasmussen 1950). So with right-sided foot movements, as carried out here, one should expect an electrical dipole on the left mesial cortical surface, which points with its negative pole to the right (i.e. ipsilaterally, so to speak, in the wrong direction). As a matter of fact, our results are an indication that Penfield's homunculus really exists

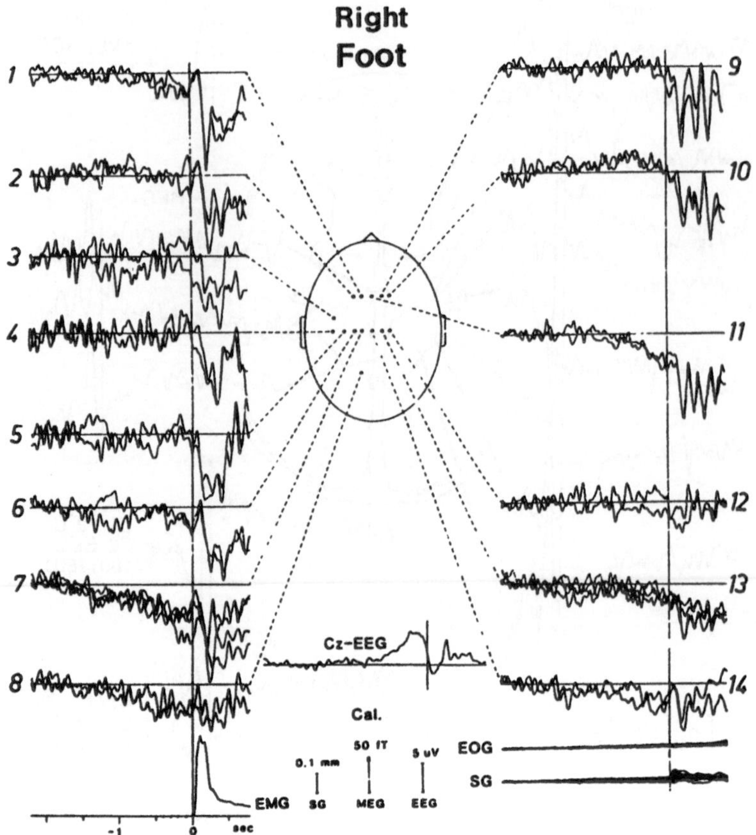

Fig. 3. Cortical magnetic fields accompanying voluntary plantar flexion of the right foot. Each trace represents the average of 40 artifact-free trials recorded at positions as follows (numbers in cm from Cz, in parentheses corresponding percentages of 10-20 system). 1:3.5 ant./2 left (10%/6%); 2:3.5 ant./3.5 left (10%/10%); 3:1.5 ant./7.5 left (4%/21%); 4:7.5 left (21%); 5:5.5 left (16%); 6:3.5 left (10%); 7:2 left (6%); 8:4.0 post./2 left (11%/6%); 9:3.5 ant./2.0 right (10%/6%); 10:3.5 ant./3.5 right (10%/10%); 11:3.5 ant. (10%); 12:3.5 right (10%); 13:2 right (6%); 14:Cz. SG = strain gauge; Cz EEG: grand average of the BP monitored during MEG recordings; Cal. = calibration (from Deecke et al. 1983)

physiologically - and not only with the unnatural electrical stimulation employed: The BF goes down over left retrorolandic regions, i.e. these field lines enter the head while it goes up over the right prerolandic region, i.e. these field lines leave the head. So the electric generator is midway between the two magnetic maxima. Our results confirm those of Hari et al. (1983), who also found field lines leaving the head over right prerolandic regions and field lines entering the head over left retrorolandic regions. Both experimental groups thus agree that the interconnecting line between the magnetic maxima is not straight, parallel to the midline, but oblique from left retrorolandic to right prerolandic. Hari et al. (1983) gave a plausible explanation: Part of area 4 for foot movements is at the mesial surface but the other part is in the anterior wall of the central sulcus and thus we are measuring the resultant oblique dipole.

From this we can speculate that the BP and, thus, the BF as well, must have at least two principal generators, since we know that the supplementary motor area (SMA) is

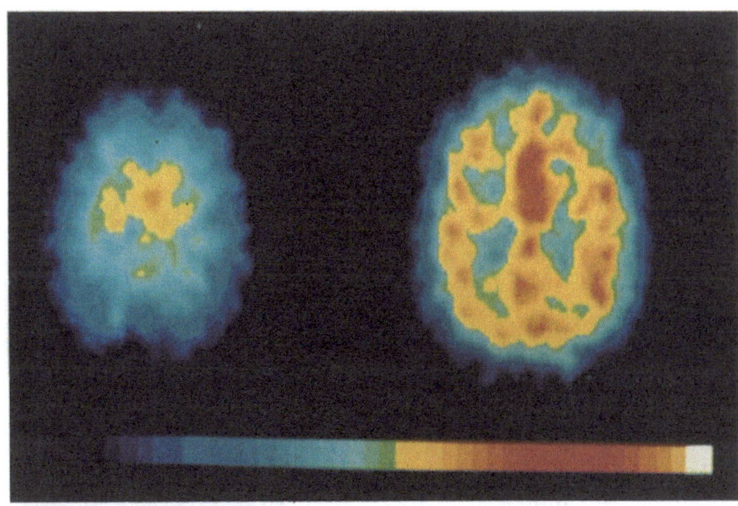

Fig. 4. ^{99m}Tc-HMPAO brain SPECT of subject J.B. 2 axial slices from high parietal sections. Voluntary tracking task of right hand with feedback. Relative tracer distribution displayed in colors ranging from blue (low) via yellow and red to white (high concentration, calibration scale on the right side). Note increased rCBF (activation) in both SMA regions with some more relative tracer accumulation in the left (contralateral) SMA (modified from Lang et al. 1988)

also active in voluntary movement. We know this quite well from both BP studies in parkinsonian patients (Deecke and Kornhuber 1978) and from rCBF studies in normal Ss (Lassen et al. 1978).

In Lassen et al.'s study rCBF was measured using the Xenon 133 method, in which SMA participation can only be concluded indirectly by "shining through" to the skull convexity, where the scintillation counters are positioned. In contrast, PET methods but also HMPAO SPECT subserve axial brain sections similar to CT, so that SMA activation can be localized within the brain. An example is given in Fig. 4, where right-sided hand movement in a voluntary tracking paradigm increased rCBF in both SMAs, the left showing more activation.

Figure 5 shows the BF with a more sophisticated finger tapping task which involved timing. The task consisted of sequential tapping with the fingers of the right hand onto the thumb. The subject performed the tapping very accurately, as can be seen in the rectified EMG, which shows the individual tapping components even in the average over 64 trials. The hypothesis was that with this precisely timed tapping task, the SMA should be activated so that we had to look for two generators. The SMA generator has to be expected on the mesial cortical surface of the left hemisphere, similar to the MI generator for foot movements as shown in the preceding illustrations. Anatomically, the SMA is juxtaposed to the foot region of the motor homunculus, extending just anterior to it. In fact, as can be seen in Fig. 5, there is an early generator defined by the early portion of the BF with its field lines leaving the head in the right prerolandic region (trace 7 in Fig. 5) and field lines entering the head in mid-retrorolandic regions (e.g. trace 10). Thus, the early BF reversal indicates activity of the SMA. In addition, over the hand area of the left precentral gyrus we see another, later, BF reversal with field lines entering and leaving around the rolandic area. In conclusion, there is evidence

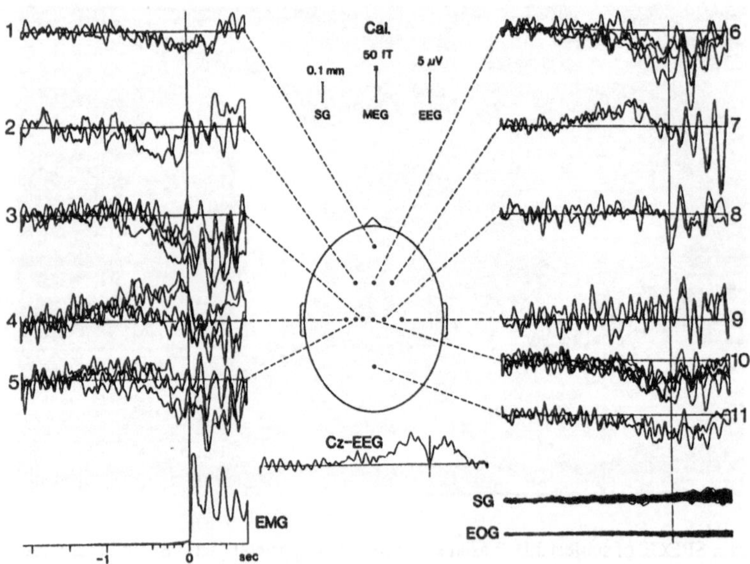

Fig. 5. Bereitschaftsmagnetfeld (BF) or readiness magnetic field preceding a complex right finger tapping task in a right-handed subject. Each trace represents an average of 40 artifact-free trials. Recording positions (in percentages of 10-20 system, A = anterior, P = posterior, R = right, L = left) from Cz (vertex) as follows: (1)20%A = Fz; (3)6%L; (4)16%L = roughly C3; (5)10%A = FCz; (7)10%A, 10%R; (8)6%R; (9)16%R = roughly C4; (10)Cz; (11)11%P. SG = strain gauge recordings of head displacement. Bandwidth 0.03 - 15 Hz for all, except EMG 5 - 70 Hz. Upward deflections correspond to fields leaving the head (MEG) or negative potentials (EEG)

that the early portion of the BF stems from the SMA area, while the second portion is derived from area 4. Recently, Cheyne (1988) has studied the BF preceding finger movements using many experiments with extensive mapping. He did not find just a contralateral BF reversal but also a distinct second reversal over the ipsilateral hand area of the motor cortex, thus confirming the bilaterality of the BP as already described by Kornhuber and Deecke in their original publications of 1964 and 1965.

In further experiments using the MEG, we also investigated the BF preceding speech (Fig. 6) and had great help with our method of breath holding, which we had already employed in the finger and foot movements. After an irregular interval of holding the breath, the subjects uttered words beginning with the letter P, such as Paul, Peter, pole, etc. This was done since we knew from experiments by Grözinger et al. (1980) that words beginning with a P and a following vowel offer excellent triggering conditions, when using the rectified EMG of the orbicularis oris muscle as the trigger. In the lower right corner of Fig. 6, the rectified orbicularis EMG is shown in the top trace, the (rectified) orbicularis oris mechanogram is shown in the middle trace, and the (rectified) microphone signal of the subject's phonation is seen in the bottom trace. As can be seen, phonation comes after articulation (as Grözinger et al. (1980) have already shown previously, pointing out that the trigger on phonation will be delayed). When studying speech potentials, one has to make sure to trigger on the first articulation and not on phonation, which may begin as late as 1/3 s after the former. The results of Fig. 6 show for a right-handed subject that it is only over the left hemisphere that the BF re-

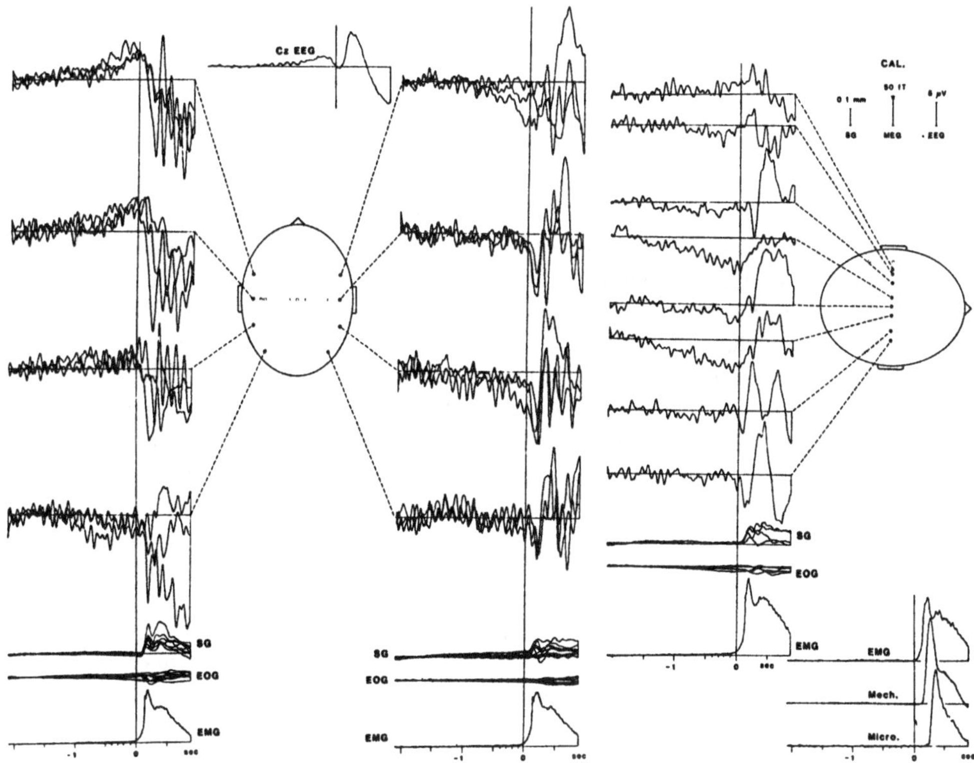

Fig. 6. Results from vocalization of different words beginning with the letter p and followed by a vowel. Subject LD. Superimposed averages of 80 trials. Upward Bereitschaftsmagnetfeld (BF) indicates field lines leaving the head, downward BF those entering the head. Inset at the bottom right indicates onset of speech utterance at s 0, coinciding with the beginning of rectified EMG of orbicularis oris muscle. Mech. = mechanogram of this muscle, and Micro. = rectified microphone signal (phonation). Note that phonation comes 1/3 s after articulation. Further note that a BF reversal is seen only over the left hemisphere, indicating activity of the speech areas Broca and Wernicke. (From Weinberg et al. 1983)

verses, with field lines leaving the head over anterior left hemispheric regions and entering the head over posterior left hemispheric areas, so that the active electrical generator is to be assumed between the two, being indicative of activation of the speech centers but not distinguishing between Broca's and Wernicke's areas.

Bereitschaftspotential Accompanying Verbal and Spatial Tasks

After returning from Vancouver to Ulm to work together with Marion Engel and H.H. Kornhuber we wished to record the BP preceding speech in the EEG without the superposition of the respiration wave by means of our breath holding method. In a rather large study of 36 right-handed subjects, we investigated the BP preceding the uttering of P-words. In pilot experiments (cf. Fig. 1 in Deecke et al. 1986), we proved that the breath holding procedure per se did not cause negative DC shifts. This was not trivial in view of the work of Caspers and Speckmann (1969). However, in our epoch of 2 s before through 4 s after the start of holding the breath there was an activity neither before nor after the beginning of breath holding.

Figure 7 proves that we were able to record the BP prior to speech without "contamination" by the respiration wave (cf. Grözinger et al. 1980). As you can see, we had to terminate the graphs at s 0, which is the start of the rectified orbicularis EMG, because after this a huge artifact from the tongue appears. This "glossokinetic" potential had already been shown by Grözinger et al. (1980). The tongue movement artifact can be seen even in the EOG (top trace of Fig. 7), where it causes a large bump after the onset of speech, so that one can make no slow potential measurements after speech onset. However, before speech onset brain potentials can be well recorded: We see a large BP preceding speech, particularly over the midline in recordings from Cz and FCz, which is the vertex, and 10% anterior to it, which indicates a strong participation of the SMA. When comparing recordings from the left and right hemispheres, we noticed that the BP is bilaterally symmetrical in its early component. Only in the later foreperiod did it become lateralized towards the left (speech-dominant) hemisphere. This is important since speech is perhaps the most lateralized function of man, being processed only in the left hemisphere of right-handed subjects, yet still the voluntary initiation of speech starts bilaterally, i.e. in both hemispheres. Only later, when channeled into the "final pathway," does the negativity shift to the left hemisphere.

The investigation of verbal and spatial tasks using the BP paradigm was continued in an experiment by Schreiber et al. (1983). Schreiber chose his doctoral thesis according to the meaning of his name (in German "Schreiber" means "writer"), and he investigated

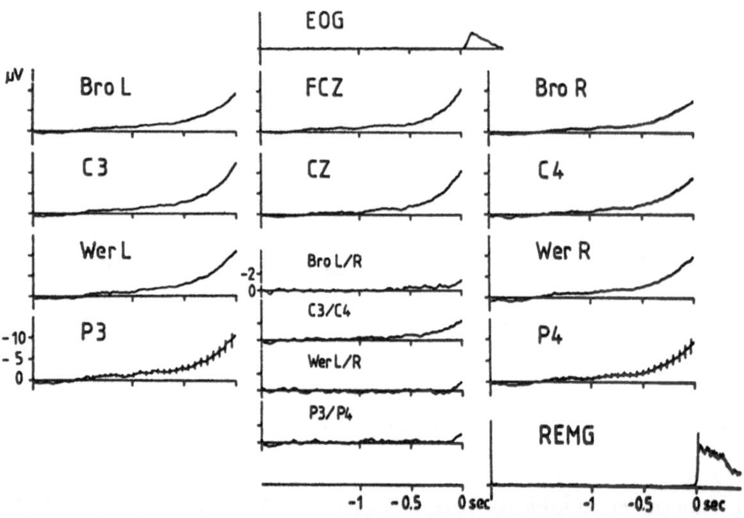

Fig. 7. Grand averages of the BP prior to speech after holding the breath. Each graph represents the mean of 4608 artifact-free utterances of words beginning with the letter p (36 subjects x 128 trials). Calibration in μV on the ordinate of P3 (left) for the monopolar recordings and on the ordinate of Bro L/R for the bipolar recordings (reconstructed by computer) in the lower center section. The vertical bars in P3 and P4 give the double standard error (2 SE). The EOG shows no eye movement during the entire foreperiod; the upward deflection after the onset of speaking is not eye movement either but the glossokinetic artifact. Abrupt onset of speaking (trained) seen in the rectified EMG (REMG) of the orbicularis oris muscle used for trigger. (From Deecke et al. 1986)

the BP preceding writing. Since this is a typical left hemisphere task, he contrasted it with drawing (of a pentagram), which is a typical task of the right hemisphere, and for control, he chose meaningless scribbling.

Figure 8 gives the results for writing. As may be noticed, for this task, which definitely is more complicated than simple finger movements, a large early BP - starting ca. 2½ s before the onset of movement - is recorded over the SMA. Furthermore, the BP extends to the frontal lobes, where it is generally absent or even positive for simple finger movements. On a closer inspection in the period preceding writing there is a larger BP over the left frontal cortex than there is over the right. Since the writing was carried out with the right hand, we would expect a left hemisphere preponderance of the BP. But this effect was controlled by doing the same experimental task except that the signature was replaced by nonsense scribbling (cf. later, Fig. 10).

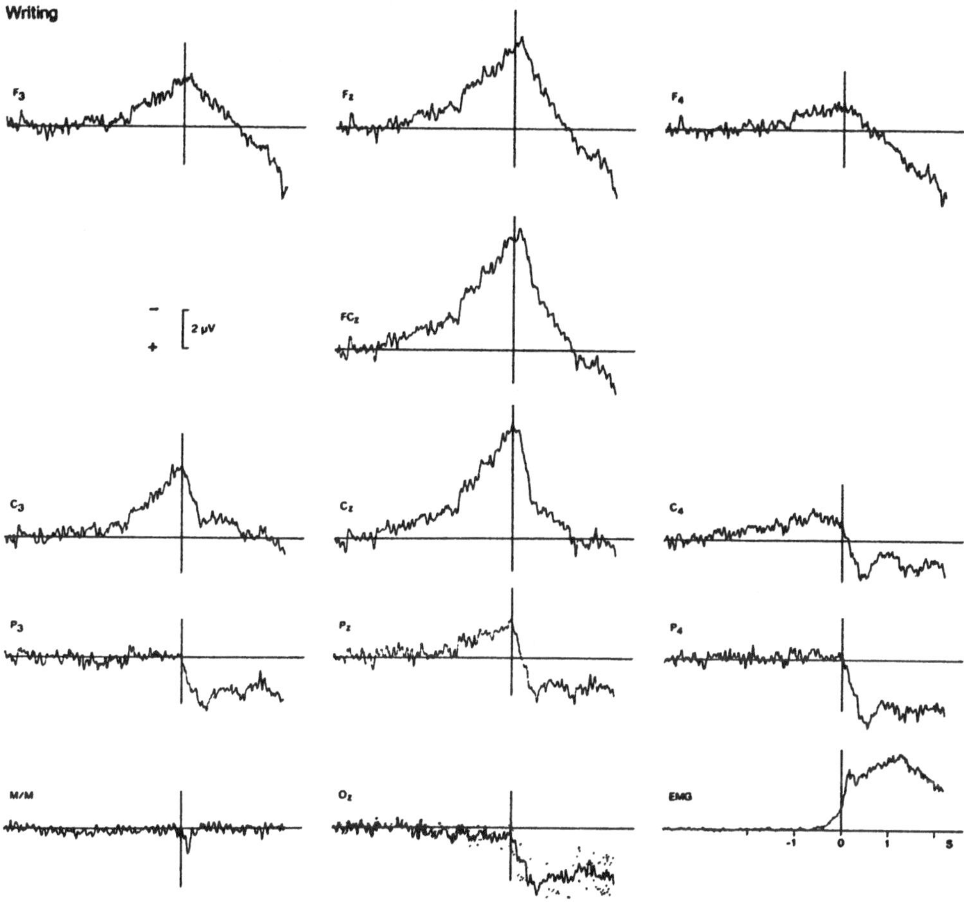

Fig. 8. Grand averages (N = 20 subjects) of cerebral potentials related to writing one's own signature repeatedly in a self-initiated manner. Negative up. Leads arranged according to their topography on the scalp as viewed from above. M/M, bipolar recording derived from left versus right mastoid, in order to check for artifacts. EMG, rectified electromyogram from right flexor carpi radialis muscle. The computer calculated the average curves together with ±1 SE as shown in the recording of Oz by way of example. Time base 6 s, unsmoothed data. Note larger BP in F3 than in F4. (From Schreiber et al. 1983)

With drawing (Fig. 9), there is more negativity over the right frontal area than there is over the left, and the difference is statistically significant. On closer inspection, we see that the negativity persists past the onset of the drawing, indicating right frontal lobe activity during the performance of completing the pentagram (i.e. to end it at the beginning in order to "close" it, which was the requirement). This right frontal preponderance of the BP accompanying the spatial task of drawing is remarkable, since it had to compete with the left central BP preponderance due to the performing right hand.

As mentioned before, Fig. 10 gives the results of scribbling for control: There is no difference between the left and right frontal lobes, but a similar SMA participation as in the other tasks and the preponderance of the performing hand contralaterally.

Finally, we tried to investigate - together with Frank Uhl, the Lang brothers and H.H. Kornhuber - the cerebral potentials accompanying writing as opposed to drawing on dictation. This more complicated cerebral dominance paradigm is depicted in the insets of the last four illustrations (Figs. 11-14), where the scheme is adopted from Ro-

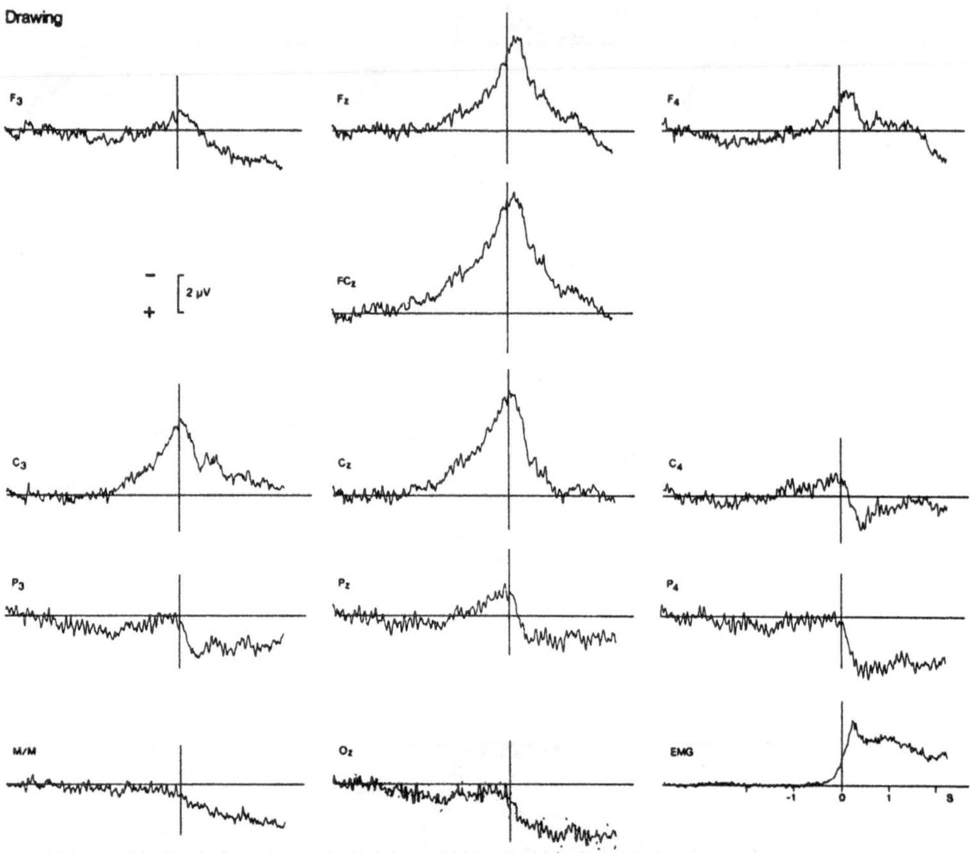

Fig. 9. Grand averages (N = 20 subjects) of cerebral potentials related to drawing a pentagram repeatedly in a self-initiated fashion. Conventions similar to those in figure 8. Note larger BP in F4 as compared to F3 outlasting the onset of drawing (right frontal performance negativity during drawing, due to difficulty of task in ending at the beginning of the pentagram ("closing" the figure) without visual control (eyes were fixed on a fixation point). (From Schreiber et al. 1983)

ger Sperry's work. However, our subjects were normal and had an intact corpus callosum in contrast to Sperry's subjects. In our studies the Ss fixed their gaze on a fixation point (Fix) and had a stylus in their right hand. When they lowered the stylus to the plate in a voluntary self-initiated manner, a slide containing a word appeared. The slides were negatives, i.e. the words appeared in white on a black background in a quasi-tachistoscopic way, short enough to prevent eye movements. Furthermore monohemispherical application of the verbal material was ascertained by controlling subject's fixation. In the first condition, VRR, the subject received the verbal material (word) in the right visual field and responded with his right hand, i.e. he had to write down the word with the stylus on the plate. So the task is writing on dictation, not oral dictation but visual dictation ("copying" so to speak). The VRR task is the easiest for right-handers, which all our 22 subjects were, since a right-hander needs only the left hemisphere to do the VRR task: The word is projected into the right visual field, from where it enters the left hemisphere, which is capable of reading it and then transfers the infor-

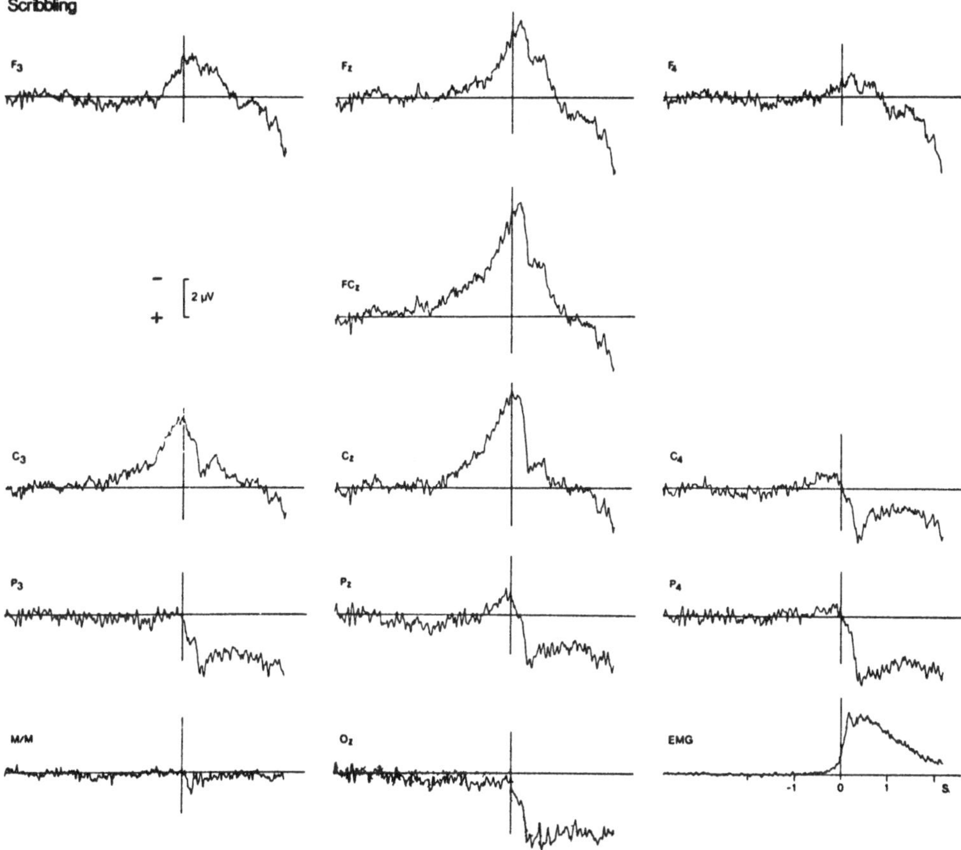

Fig. 10. Grand averages (N = 20 subjects) of cerebral potentials related to meaningless scribbling repeatedly in a self-initiated fashion. Conventions similar to those in figure 8. Note CPN of the BP (contralateral preponderance of negativity) in C3 as compared to C4 associated with use of the right hand, even slightly electrotonically conducted to F3 in its later component. However, the left frontal preponderance of the BP for writing was significantly larger than the one for scribbling ($2p < 0.05$). (From Schreiber et al. 1983)

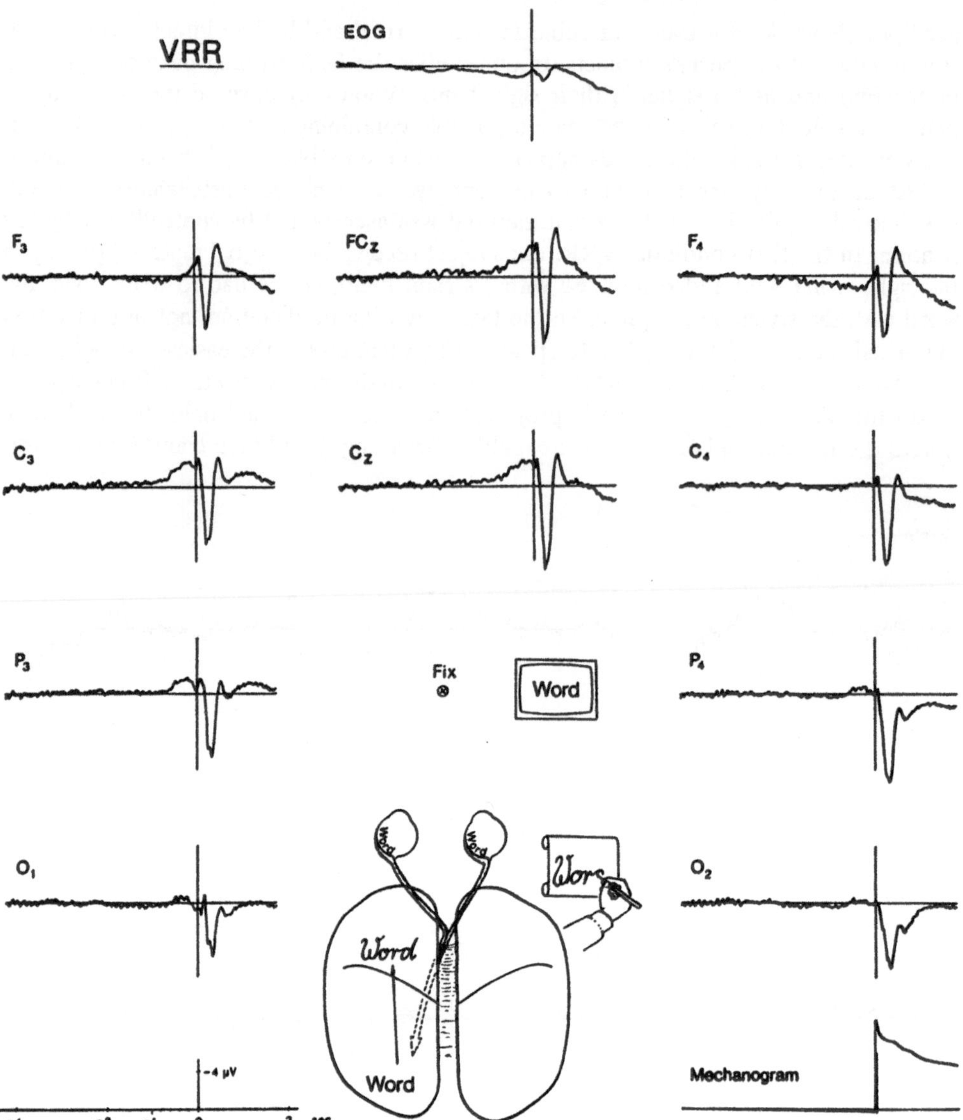

Fig. 11. Topographic arrangement of grand averages across 15 right-handed subjects. As sketched in the lower part of the figure, the verbal material was presented within the right hemifield of vision and the writing performed by the right hand (VRR). Note that BP only appears over the left hemisphere, since this task can - by the right-hander - be performed monohemispherically. (From Deecke et al. 1987)

mation to the left motor cortex, which writes it out via the right hand. This investigation tries to study, so to speak, the "Wernicke-Geschwind-model" of language processing by the slow potential method (Wernicke 1874; Geschwind and Galaburda 1984).

It is very interesting to note that with the VRR task we see a BP only over the left hemisphere - and over the midline, of course, which indicates SMA activity - but not over the right hemisphere because the right-hander only needs the left hemisphere in

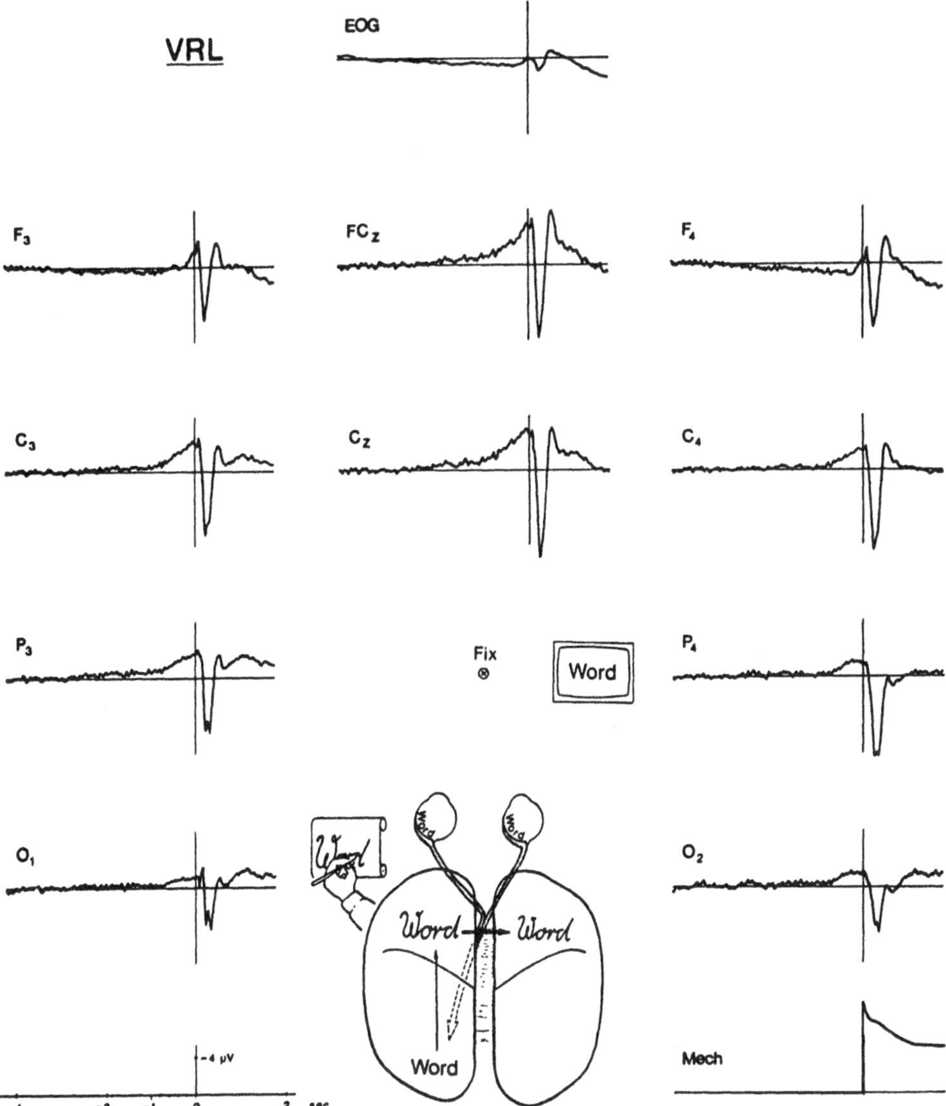

Fig. 12. Topographic arrangement of grand averages across 15 right-handed subjects. As sketched in the lower part of the figure, the verbal material was presented within the right hemifield of vision as in Fig. 11, but now the writing was performed by the left hand (VRL). Note that BP now also appears over the right hemisphere. (From the experiments of Deecke et al. 1987)

this task. The absence of the BP over the right hemisphere is very instructive. Since subjects performed the different tasks in blocks of 128 trials they were in a fixed set while performing the task and knew what kind of stimuli were coming. They sent their right hemisphere on vacation, so to speak, because it was not needed.

Figure 12 shows the situation VRL, which involves anterior corpus callosum crossing: The verbal material is projected in the right visual field, thus entering the left he-

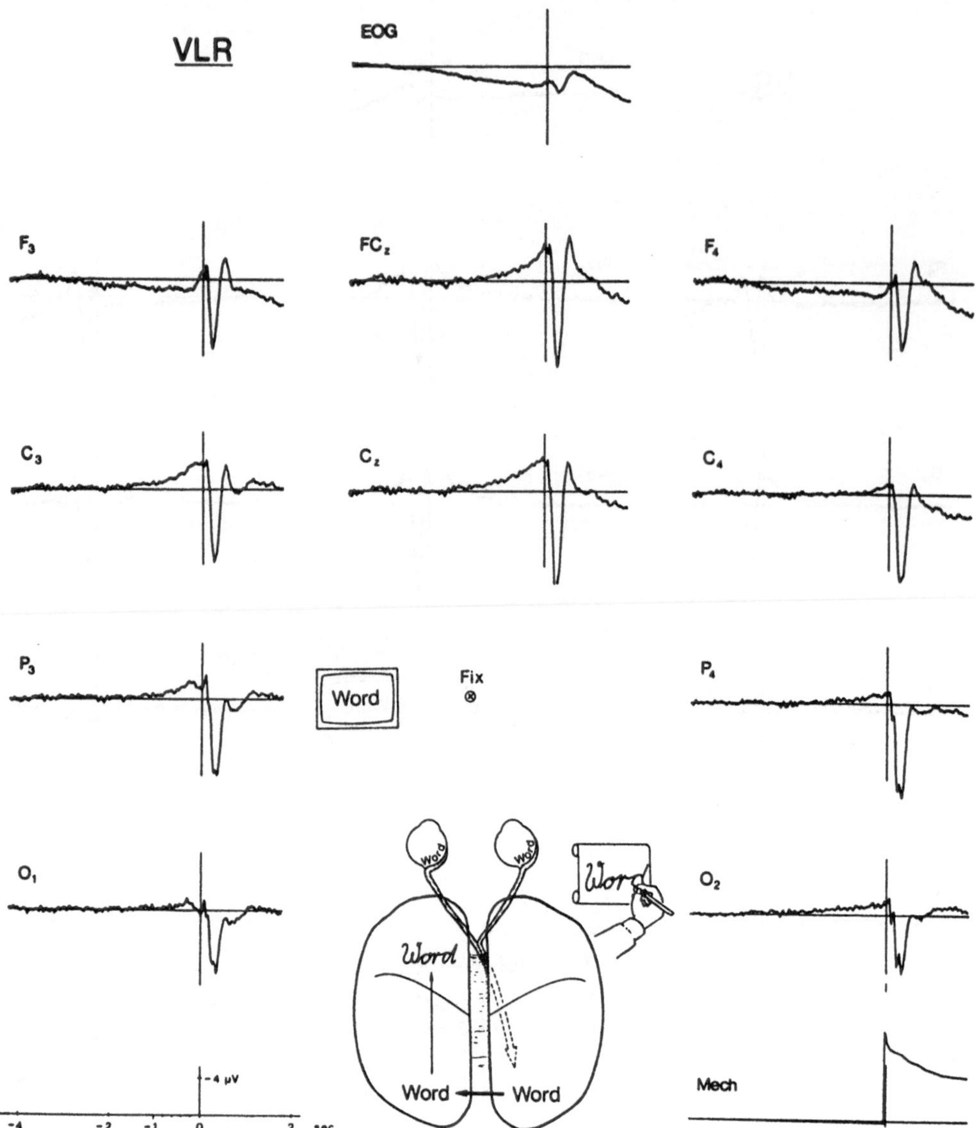

Fig. 13. Topographic arrangement of grand averages across 15 right-handed subjects. As sketched in the lower part of the figure, the verbal material was presented within the left hemifield of vision and the writing performed by the right hand (VLR). Note that an early BP now appears over the right occipital cortex in anticipation of the unusual verbal material that cannot be read by the right (illiterate) hemisphere. (From the experiments of Deecke et al. 1987)

misphere as previously, but it now has to be responded to with the left hand. In contrast to Fig. 11 (VRR) with the extreme right-hander's favorite condition, we do see some activity in the right hemisphere, which now has the output, and very close to movement onset there is also a BP in the right hemisphere.

Figure 13 gives the results of another condition, which is VLR, i.e. now the verbal material is projected into the left visual field and responded to by the right hand. This

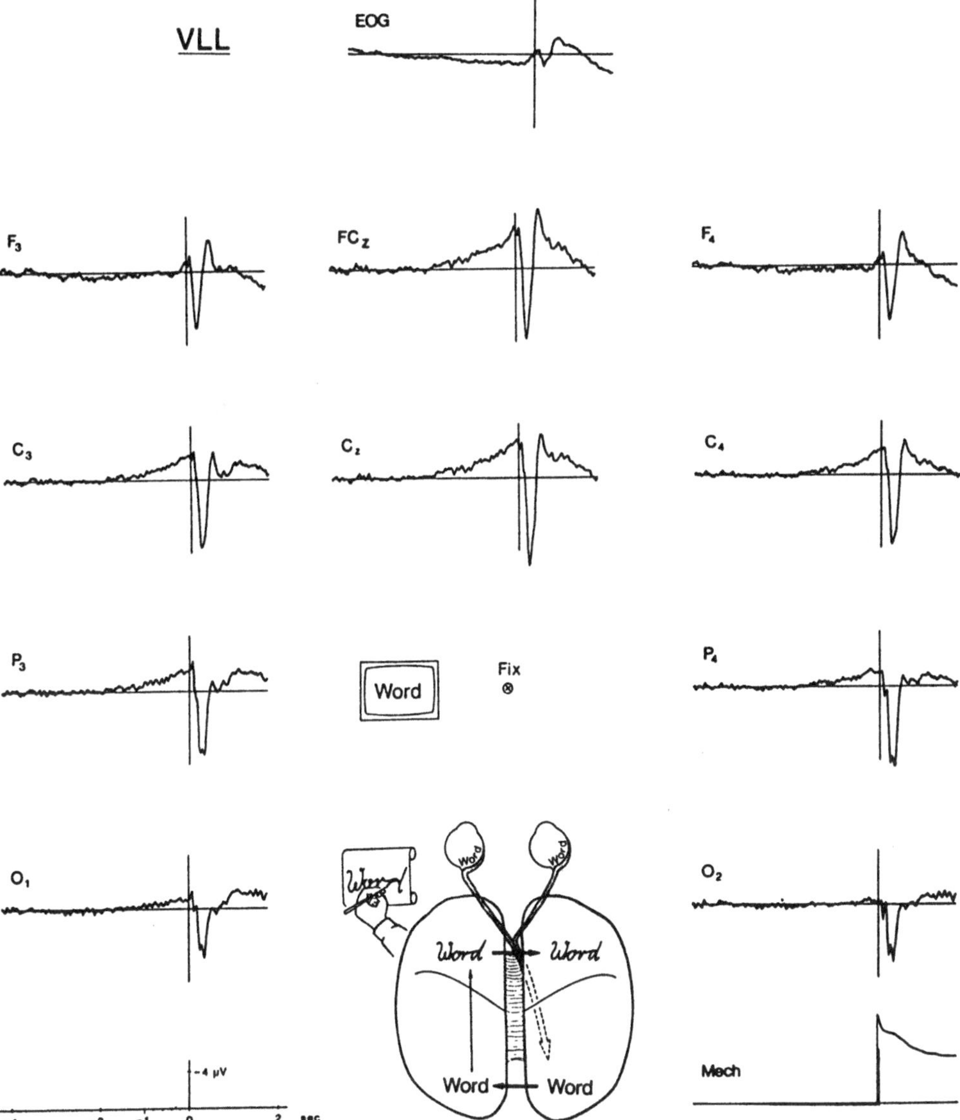

Fig. 14. Topographic arrangement of grand averages across 15 right-handed subjects. As sketched in the lower part of the figure, the verbal material was presented within the left hemifield of vision and the writing performed by the left hand (VLL). Note that this paradigm involved two transcallosal crossings, a posterior and an anterior one, generating a BP over both hemispheres and the midline. (From the experiments of Deecke et al. 1987)

condition also involves crossing the corpus callosum (namely a posterior callosal transfer). Keeping this in mind, we see a very interesting phenomenon: The word is perceived first by the right visual cortex, which unfortunately cannot read. So it has to expedite the material immediately to the left visual association cortex, which is capable of reading. However, if we look again to the right - illiterate - occipital cortex, we see a very early BP exclusively over this area. This is a sign of early anticipation and preparation of the right occipital cortex for the difficult task of handling verbal material, for which it is not skilled. And the right occipital cortex obviously has to try hard to handle the material properly in order to shift it to the language-dominant hemisphere, where it can be read by the angular gyrus and is then transmitted prerolandically to the motor cortex, which puts it into action.

Figure 14 finally gives the situation VLL, where we have two corpus callosum crossings, one posterior and one anterior, and this is when the right-hander receives the verbal material in his left visual field (entering the right occipital cortex) and responds with his left hand. Since the right occipital cortex cannot read, the material has to be shifted over to the dominant left hemisphere, from where it is transmitted prerolandically to the motor area and then has to cross the corpus callosum again to the right hemisphere, which controls the left hand. And here again we see that both hemispheres are active, showing a BP. Four similar tasks were carried out in the same experiments with spatial material (drawings of 3-dimensional geometrical figures) SRR, SRL, SLR, and SLL, which will be omitted here for reasons of space

In summary, we think that it is worthwhile to investigate negative shifts of the cortical steady potential in a topographic manner to design experimental settings which ask certain questions which can then be answered by a variety of techniques that we are now applying in Vienna in the same population of subjects, e.g. SPECT, MEG, and slow potentials together, since they complement one another.

References

Caspers H, Speckmann EJ (1969) DC Potential shift in paroxysmal states. In: Jasper HH et al. (eds): Basic mechanisms of the epilepsies. Little, Brown, Boston, pp 375-388

Cheyne DO (1988) Magnetic and electric field measurements of brain activity preceding voluntary movements: Implications for supplementary motor area function. PhD thesis, Psychology Department, Simon Fraser University, Vancouver

Deecke L, Kornhuber HH (1978) An electrical sign of participation of the mesial "supplementary" motor cortex in human voluntary finger movements. Brain Res 159: 473-476

Deecke L, Weinberg H, Brickett P (1982) Magnetic fields of the human brain accompanying voluntary movement. Bereitschaftsmagnetfeld. Exp Brain Res 48: 144-148

Deecke L, Boschert J, Weinberg H, Brickett P (1985) Magnetic fields of the human brain (Bereitschaftsmagnetfeld) preceding voluntary foot and toe movements. Exp Brain Res 52: 81-86

Deecke L, Boschert J, Brickett P, Weinberg H (1985) Magnetoencephalographic evidence for possible supplementary motor area participation in human voluntary movement. In: Weinberg H, Stroink G, Katila T (eds): Biomagnetism: applications and theory. Pergamon, New York pp 369-372

Deecke L, Engel M, Lang W, Kornhuber HH (1986) Bereitschaftspotential preceding speech after holding breath. Exp Brain Res 65: 219-223

Deecke L, Uhl F, Spieth F, Lang W, Lang M (1987) Cerebral potentials preceding and accompanying verbal and spatial tasks. In: Johnson R Jr, Rohrbaugh JW, Parasuraman R (eds) Current trends in event-related potential research. Electroenceph. Clin. Neurophysiol [Suppl] 40: 17-23

Geschwind N, Galaburda A (eds) (1984) Cerebral dominance: the biological foundations. Harvard University Press, Cambridge

Grözinger B, Kornhuber HH, Kriebel J, Szirtes J, Westphal (1980) The Bereitschaftspotential preceding the act of speaking. Also an analysis of artifacts. In: Kornhuber HH, Deecke L (eds) Motivation, motor and sensory processes of the brain. Electrical potentials, behaviour and clinical use. Prog. Brain Res 54: 798-80

Hari R, Antervo A, Katila T, Poutanen T, Seppanen M, Tuomisto T, Varpula T(1983) Cerebral magnetic fields associated with voluntary limb movements. Il Nuovo Cimento 2 D: 484-494

Kornhuber HH, Deecke L (1964) Hirnpotentialänderungen beim Menschen vor und nach Willkürbewegungen, dargestellt mit Magnetbandspeicherung und Rückwärtsanalyse. Pflügers Arch. 281: 52

Kornhuber HH, Deecke L (1965): Hirnpotentialänderungen bei Willkürbewegungen und passiven Bewegungen des Menschen: Bereitschaftspotential und reafferente Potentiale. Pflügers Arch. 284: 1-17

Lang W, Lang M, Podreka I, Steiner M, Uhl F, Suess E, Müller C, Deecke L (1988) DC-potential shifts and regional cerebral blood flow reveal frontal cortex involvement in human visuomotor learning. Exp. Brain Res 71: 353-364

Lassen NA, Ingvar DH, Skinhoj E (1978) Brain function and blood flow. Sci.Am. 239: 62-71

Penfield W, Rasmussen T (1950) The cerebral cortex of man. Macmillan, New York (1950).

Podreka I, Suess E, Goldenberg G, Steiner M, Brücke T, Müller C, Lang W, Neirinckx RD, Deecke L (1987) Initial experience with technetium-99m-HMPAO brain SPECT. J. Nucl. Med. 28: 1657-1666

Schreiber H, Lang M, Lang W, Kornhuber A, Heise B, Keidel M, Deecke L, Kornhuber HH (1983): Frontal hemispheric differences of the Bereitschaftspotential associated with writing and drawing. Hum Neurobiol. 2: 197-202

Uhl F, Lang W, Lang M, Kornhuber A, Deecke L (1988) Cortical slow potentials in verbal and spatial tasks - The effect of material, visual hemifield and performing hand. Neuropsychologia 26: 769-775

Weinberg H, Brickett P, Deecke L, Boschert J (1983) Slow magnetic fields of the brain preceding movements and speech. Il Nuovo Cimento 2: 495-504

Wernicke C (1874) Der aphasische Symptomencomplex. Max Cohn and Weigert, Breslau

Negative DC Shifts of the Supplementary and Motor Area Preceding and Accompanying Simultaneous and Sequential Finger Movements

W. Lang, C. Koska, F. Uhl, G. Lindinger, L. Deecke
Neurologische Universitätsklinik Wien, Lazarettgasse 14, 1090 Wien

Introduction

Volitional movements are preceded by the Bereitschaftspotential (BP: Kornhuber and Deecke 1965) and accompanied by sustained negative DC shifts (performance-related, negative DC shifts; Lang et al. 1988).

The contribution of the primary motor cortex to the BP is well established (Boschert and Deecke 1986). The mesial, fronto-central cortex including the supplementary motor area (SMA) is believed to be a major source for movement-related DC shifts in recordings of the central midline (Cz; Deecke and Kornhuber 1978). This hypothesis is based on two observations: (1) In parkinsonian patients amplitudes of BP in Cz vary independently from amplitudes in recordings of the lateral central cortex (C3, C4). (2) BP has its maximum in Cz in different movements such as speech and movements of eyes or fingers.

In a first experiment of the present study simple and complex motor sequences were performed by either the right or the left hand. Two hypotheses were tested: (1) The SMA contributes to movement-related negative DC shifts in Cz. BP and/or performance-related DC shifts in Cz vary with task difficulty but not with the side of the performing hand. (2) DC shifts in C3 and C4 vary with the side of the performing hand but not with task complexity. In the second experiment, various bimanually performed motor sequences and the accompanying DC shifts were compared.

Methods

Eighteen right-handed subjects (Ss) participated in experiment I; 20 Ss in experiment II. Four different tasks were employed in experiment I: (1) SI-RH: Ss had to flex and extend the forefinger of their right hand (RH) repeatedly as fast as possible (simple movements, SI). (2) SI-LH: Same as in (1) but performed by the left hand. (3) CO-RH: The complex (CO) task consisted of a sequence of four different movements: Ss again started the performance by flexing the forefinger; subsequently they had to extend the hand at the wrist, then to extend the forefinger and finally to flex the hand at the wrist. The complex task had to be performed either by the right hand (RH) or (4) by the left hand (CO-LH).

Another four tasks were investigated in experiment II: In all tasks, Ss held their index fingers in an intermediate position during the resting period and started to move the fingers in order to repeatedly reach three positions, a flexed, an intermediate and an extended one. In SI-S, movements were performed simultaneously (SI) in the same (S) direction. In SE-S, the right index finger started, the left finger followed with a delay of one movement (SE; sequential). In SI-D, the sequence was initiated by simultaneously flexing the right and extending the left finger. Thus, the two index fingers moved simultaneously but in different (D) directions. In SE-D, movements were performed sequentially and in different directions.

The experimental procedure was identical in the two experiments. Ss started the motor tasks at their own volition and moved for at least 6 s. The four tasks were performed in a random sequence; movements were measured by goniometers. EEG was recorded from F3, Fz, F4, C3* (1 cm anterior to C3 to overlie the hand area of the primary motor cortex), Cz* (1 cm anterior to Cz), C4* (1 cm anterior to C4), C1* (10% left of Cz*), C2* (10% right of Cz*), P3, Pz, and P4 with linked ears as reference. Frequency band of amplification ranged from DC to 70 Hz (upper cut-off frequency). In addition, EOG and EMG of left and right M. flexor indicis were recorded. In each condition at least 48 artifact-free trials were averaged. Analysis period started 5 s prior to movement onset and lasted 11 s; baseline was calculated from the first 1.5 s.

Two parameters were taken to describe negative DC shifts: N-BP (mean negativity over the last 250 ms prior to the voluntary initiation of the trial) to describe the Bereitschaftspotential and N-P (mean negativity measured between $t=2$ s (2 s after movement onset) and $t=4$ s) to describe the performance-related negativity. In experiment I, effects of within factors "performing hand (right or left)" and "task complexity (simple or complex)" were tested by two-way ANOVAs. Correspondingly, effects of within factors "temporal organization (sequential or simultaneous)" and "spatial organization (same or different directions)" were tested in experiment II by ANOVAs.

Results

Main results of experiment I are demonstrated in Fig. 1: A negative DC shift (Bereitschaftspotential) precedes movement onset ($t=0$). A sustained negative DC shift (performance-related negativity) accompanies task performance. As demonstrated in part A of the figure, DC shifts differ by the "performing hand". When the right hand (RH) performs the task, BP and N-P are larger in C3*, which is contralateral to the performing hand, than in C4*. The opposite is true in movements of the left hand (LH). Now, negative DC shifts are larger in C4* as compared to C3*. This lateralization effect was highly significant for N-BP and N-P ($F=$ 81.7 and 68.5 resp.; $p<0.00001$). The factor "task complexity" had no effect on N-BP and N-P in C3* and C4* ($F < 0.5$). As shown in part B of Fig. 1, "task complexity" significantly affected performance-related DC shifts in Cz*. N-BP and N-P were larger in the complex task (CO) than in the simple task (SI). This difference was highly significant for N-P ($F=18.2$, $p \leq 0.0001$). DC shifts in Cz* did not vary with the side performing the movement. Recordings in C1* and C2* varied significantly by task complexity but also by the side of the performing hand.

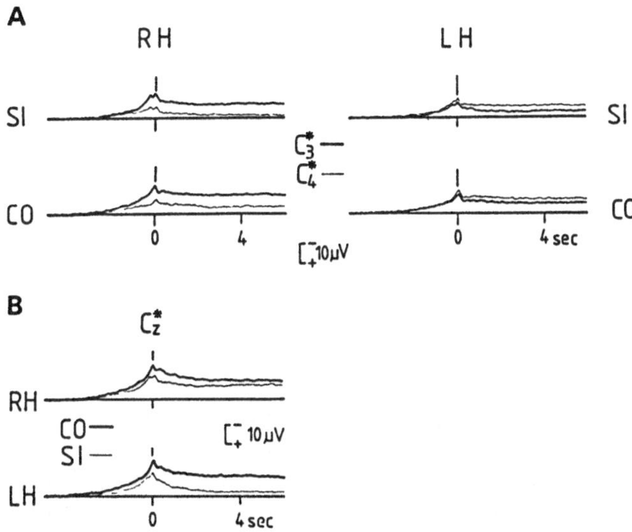

Fig. 1. A Grand averages across all subjects as obtained in C3* (thick line) and C4* (thin line). Movement onset at t=0. In both tasks [simple (SI) or complex (CO) task] DC shifts in in C3* and C4* depend on the side of the performing hand (RH: right hand; LH: left hand). **B** In Cz*, negative DC shifts are larger in the complex task (CO, thick line) as compared to the simple task (SI, thin line)

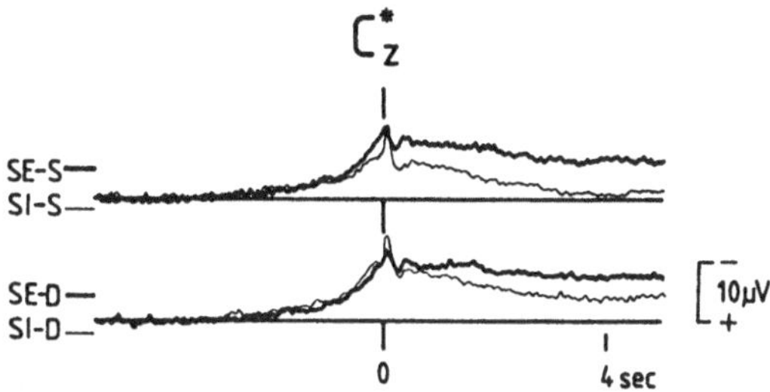

Fig. 2. Grand averages obtained in Cz*. Movement onset at t=0. Sequential tasks are compared to corresponding simultaneous conditions, SE-S to SI-S (upper line), SE-D to SI-D (lower line)

Main results of experiment II are shown in Fig. 2: When subjects moved their index fingers in a sequential manner (SE-S: sequential, same direction; SE-D: sequential, different directions) there was a large and sustained negative DC shift in recordings over the mesial frontocentral cortex (C1*, C2* and Cz*). In contrast, in simultaneous tasks (SI-S and SI-D), performance-related negativity (N-P) declined during performance. The difference between SE and SI tasks was restricted to recordings of the mesial frontocentral cortex and had its maximum in Cz* ($F = 18.9$; $p < 0.0001$). Performance-related DC shifts did not vary by the factor "spatial organization", i.e. there was no difference if Ss moved their index fingers (either simultaneously or sequentially) in the same or in different directions.

Discussion

The most important finding in experiment I is that variations of task complexity and movement execution ("performing hand") have dissociative effects on movement-related potentials in central leads: Recordings over lateral parts of the central hemispheres (C3* and C4*) are selectively sensitive to the side of the performing hand, whereas the recording over the central midline, Cz*, varies only and to a large degree by "task complexity". These data can be taken as evidence that two sources having different functions and locations contribute to movement-related potentials as picked up in C3*/C4* and in Cz*. The finding that complex sequences have larger DC shifts in recordings of the mesial frontocentral cortex than simple ones corresponds to data of Orgogozo and Larsen (1979), who measured greater regional cerebral blood flow in the SMA during complex motor sequences than during repetitive finger movements. The main result of experiment II is the highly significant effect of the temporal factor on performance-related negative cortical DC shifts. The influence was not a general phenomenon but was clearly restricted to those electrodes that collect activity from the central midline, including SMA, C1*, Cz* and C2*. In these recordings amplitudes were considerably larger in sequential than in simultaneous motor tasks. The fact that the spatial factor (direction of movement) had no effect on performance-related brain potentials supports the concept of Kornhuber and Deecke (Kornhuber 1984; Deecke et al. 1985) that the SMA may not be generally involved in linkage and execution of motor subroutines but in its temporal organization.

Summary

Cortical DC shifts were measured by scalp electrodes in various motor sequences. In these tasks, simple movements such as flexions or extensions of index fingers or hands had to be linked in different manners. The main findings are: Movement-related DC shifts in recordings over the mesial fronto central cortex (Cz) can functionally be separated from recordings over lateral parts of the primary motor cortex (C3 and C4). DC shifts in C3/4 vary with the side performing the task (either right or left hand) whereas DC shifts in Cz vary with structure and complexity of motor sequences.

References

Boschert J, Deecke L (1986) Cerebral potentials preceding voluntary toe, knee and hip movements and their vectors in human precentral gyrus. Brain Res 376: 175-179

Deecke L, Kornhuber HH (1978) An electrical sign of participation of the mesial "supplementary" motor cortex in human voluntary finger movements. Brain Res 159: 473-476

Deecke L, Kornhuber HH, Lang W, Lang M, Schreiber H (1985) Timing function of the frontal cortex in sequential motor and learning tasks. Hum Neurobiol 4: 143-154

Kornhuber HH (1984) Attention, readiness for action and the stages of voluntary decision - some electrophysiological correlates in man. In: Creutzfeldt O, Schmidt RF, Willis WD (eds.) Sensory-motor integration in the nervous system. Exp Brain Res [Suppl] 9: 420-429

Kornhuber HH, Deecke L (1965) Hirnpotentialänderungen bei Willkürbewegungen und passiven Bewegungen des Menschen: Bereitschaftspotential und reafferente Potentiale. Pflügers Arch 284: 1-17

Lang W, Lang M, Uhl F, Koska C, Kornhuber A, Deecke L (1988) Negative cortical DC shifts preceding and accompanying simultaneous and sequential finger movements. Exp Brain Res 71: 579-587

Orgogozo JM, Larsen B (1979) Activation of the supplementary motor area during movement in man suggests it works as a supramotor area. Science 206: 847-850

DC Potential Shifts and Regional Cerebral Blood Flow Reveal Frontal Cortex Involvement in Human Visuomotor Learning

M. Lang*, W. Lang, I. Podreka, L. Deecke
*Neurologische Universitätsklinik Ulm,Oberer Eselsberg 45, 7900 Ulm
Neurologische Universitätsklinik Wien, Lazarettgasse 14, 1090 Wien

Introduction

The human cortex integrates perceptions and motivations and is involved in the development of goal-directed behavior. Patterns of task-related cortical activation can be measured by changes of neuromagnetic and electrical fields or of regional blood flow and metabolism. The aim of the present experiment was to identify brain areas activated during visuomotor learning. Two different physiological parameters were assessed, (1) cortical DC shifts preceding (Bereitschaftspotential, BP) and accompanying motor performance and (2) regional cerebral blood flow (rCBF).

Methods

In continuation of previous experiments (Lang et al. 1983, 1986; Kornhuber 1984), visuomotor learning was investigated in a joint study by measuring both rCBF and performance-related DC shifts (Lang et al. 1988). rCBF was measured using single photon emission computer tomography (SPECT) and a new technetium-labeled compound (hexamethyl-propylene-amine-oxime, HMPAO), which crosses the blood-brain barrier and becomes trapped in brain cells within about 2 min. Tracer concentration correlates well with rCBF (Podreka et al. 1987). In order to study visuomotor learning, 16 subjects (Ss) were exposed to a conflicting situation: a visual target moved on a TV screen and had to be tracked by moving the right hand in a mirrored fashion, i.e. movements of the target to the right side required hand moving to the left and vice versa, whereas up and down was not inverted. This visuomotor learning task was compared to normal, non-inverted tracking for control. To provide feedback information, tracking performance was coupled back to the screen. In the rCBF study, Ss were examined twice under both conditions. Ss continuously performed either the mirrored or the simple tracking over a 10 min period. 3 min after the beginning of the performance ^{99m}Tc-HMPAO was applied at dosages of 0.2 mCi/kg body weight. 3.125 mm axial slices were reconstructed after filtering and attenuation. Five adjacent, transversal cross sections were used for evaluation. Count rates of regions of interest were divided by the mean count rate across all regions (RI: relative regional index). In the DC potential study Ss fixed their gaze on a point positioned in the center of the upper frame of the TV screen. They voluntarily initiated movements of a target circle on the screen by pressing a stylus onto a photo detector plate. The visual target moved at constant speed in three successive random directions with each trajectory lasting 1.5 s. Movement-related DC potentials were recorded from scalp electrodes with linked ears serving as reference. 96 artifact-free trials were averaged in each condition (mirrored tracking and simple tracking). The 4.5 s lasting tracking performance was associated with a sustained negative DC shift. The mean amplitude (N) of this interval was calculated and compared between learning task and control task (dN; difference of performance-related negativity).

Results

Figure 1 displays performance-related DC shifts. Recordings obtained in the visuomotor learning task are shown in upper lines, results of the simple tracking task in lower lines. Performance-related DC shifts were larger in the learning task as compared to the control situation. Significant differences (dN) were found in Cz and FCz ($p<0.0001$, paired t-tests) but also in F3, Fz and F4 ($p<0.01$). There were almost no differences between the two tasks in parietal (P3, P4) and occipital (Oz) recordings. Between subjects, dN was significantly correlated to the individual success of visuomotor learning (reduction of error; Lang et al. 1988) in F3, Fz, F4, FCz, Cz and C4.

Relative tracer concentrations (RIs) were larger in the visuomotor learning tasks as compared to the tracking control in regions related to the left and right middle frontal gyrus ($p \leq 0.002$ and $p<0.0001$ resp., paired t-tests), the frontomedial cortex ($p \leq 0.003$), basal ganglia ($p \leq 0.050$), and cerebellum ($p \leq 0.003$). Between subjects, differences of

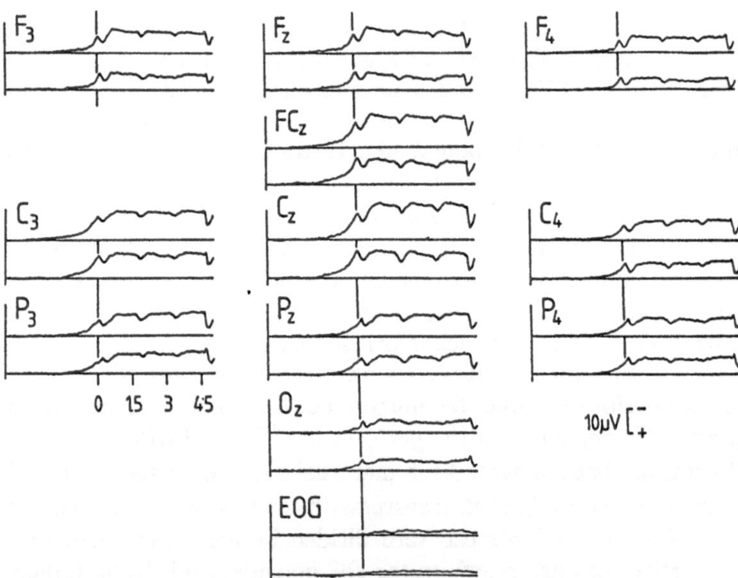

Fig. 1. Grand averages across all 16 subjects. Results of the learning task are shown in upper lines, results of the simple tracking control in lower lines. Voluntary movement onset at $t=0$

count rates (dRIs) in the right middle frontal gyrus were correlated with dRIs in the basal ganglia (r:0.583, $p < 0.01$).

Discussion

The present experiment gives evidence for frontal lobe involvement in visuomotor learning. In the learning task, Ss were forced to interpose an additional cognitive operation (left-right inversion) between sensory and motor processes. Furthermore, effort was needed in order to reduce tracking error and to withstand against falling back into direct pursuit. The ability to develop new patterns of sensory-motor interaction and to withstand interference is obviously a function of the frontal lobe. Basal ganglia and cerebellum are also involved in this learning process, as demonstrated by the rCBF study.

Summary

In the present study, measurements of the regional cerebral blood flow and of cortical DC shifts (using scalp electrodes) give evidence that the frontal lobe is activated during visuomotor learning.

References

Kornhuber HH (1984) Attention, readiness for action and the stages of voluntary decision - some electrophysiological correlates in man. In: Creutzfeldt O, Schmidt RF, Willis WD (eds) Sensory-motor integration in the nervous system. Exp Brain Res [Suppl] 9: 420-429

Lang W, Lang M, Kornhuber A, Deecke L, Kornhuber HH (1983) Human cerebral potentials and visuomotor learning. Pflügers Arch 399:342-344

Lang W, Lang M, Kornhuber A, Kornhuber HH (1986) Electrophysiological evidence for right frontal lobe dominance in spatial visuomotor learning. Arch Ital Biol 124:1-13

Lang W, Lang M, Podreka I, Steiner M, Uhl F, Suess E, Müller Ch, Deecke L (1988) DC-potential shifts and regional cerebral blood flow reveal frontal cortex involvement in human visuomotor learning. Exp Brain Res 71: 353-364

Podreka I, Suess E, Goldenberg G, Steiner M, Brücke T, Müller C, Lang W, Neininckx RD, Deecke L (1987) Initial experience with technetium-99m-HMPAO brain SPECT. J Nucl Med 28:1657-1666

Event-Related Slow Potentials Recorded from Cortex and Depth of the Human Brain

J. A. Ganglberger
Gallitzinstraβe 108/3/2, 1160 Wien, Österreich

The discovery of the contingent negative variation (CNV) in 1963 by Walter et al. (1964) and of the Bereitschaftspotential (BP) in 1964 by Kornhuber and Deecke (1964, 1965) led to vast progress in EEG research, increasing the interest in neuropsychological research. During stereotactic interventions under local anaesthesia it was possible to record slow potential changes from cortex and various deep structures. The on-line recording of BPs was done by instantaneous opisthochronic analysis (Walter 1967). For cortical recording the method of Housepian and Pool (1962) was adopted. The flexible Teflon-isolated Ag/AgCl electrodes were produced and chlorided by myself. They were flexible enough to be introduced through the small high-frontal stereotactic burr hole, extreme care being taken to avoid damage to small subdural vessels. Usually four subdural electrodes were introduced, two in an anterior and two in a posterior direction, the exact location always being more or less a function of chance, since the slightest resistance was an absolute indication to stop, which could sometimes be overcome by retracting the electrode and giving it another direction for advancement. The depth electrode (a coaxial probe with a pole distance of 5 mm), using the Freiburg stereotactic equipment, reaches the calculated target within a maximal error of ±0.5 mm. The recording electrodes were initially referred to the stereotactic frame (being fixed to the skull on at least four points), and later to linked mastoids. In addition to the subdural electrodes, subgaleal needle electrodes were used on the vertex and over both hemispheres.

The patients had to give their special consent for these studies. The majority were suffering from Parkinson's disease, some from temporal lobe epilepsy, fewer from different hyperkinetic disorders; some had intractable pain. By 1979 we had already studied CNV and BP in more than 300 cases. Evoked potentials (EPs) of different sense modalities and cortical responses to depth stimulation were also analyzed.

Before giving results concerning the main subject it should be mentioned that it was possible to demonstrate the primary motor response in man within an onset of deflection latency of 2-4 ms (Ganglberger and Haider 1969), as predicted by Hassler (pers. comm.). The onset of deflection latency of auditory EPs (AEPs) within the medial geniculate body could be determined at 7 ms. It could be shown that the electrophysiological phenomenon of extinction (Dusser de Barenne and McCulloch 1937) was not only a local one, but also a distant one (Ganglberger 1970). Upon stimulation of the amygdala, long-latency cortical responses were found over the prefrontal cortex (Ganglberger et al. 1971), the projection fields of the nc. medialis (M of Hassler 1959), which suggested a polysynaptic detour from the amygdala via the nc. medialis. Since

Klingler and Gloor (1960) demonstrated a pathway from the amygdala via the inferior thalamic peduncle to the nc. medialis, this led to the introduction of two new target points for treatment of temporal lobe epilepsy (Ganglberger 1976, 1984). With the lateral drive-out electrode it could be shown that motor relay nuclei respond considerably earlier than the somatosensory relay nuclei (Ganglberger and Haider 1969), although the latter are closer to the medial geniculate body. This may be explained by the importance of the auditory system as a warning system in wildlife.

When studying CNV and BP it was especially favourable when cortical electrodes could be placed over motor and premotor regions with the depth electrode in the motor thalamus (V.o.a. or V.o.p of Hassler). In such cases some insight into the thalamo-cortical relationship of CNV and BP could be gained.

When during a simple CNV paradigm (using paired clicks with an interstimulus interval of 1.5 s at irregular intervals) no adaptation to the second (imperative) stimulus occurred in the thalamus, then the CNV also failed to appear on the cortex. When paired stimuli outside of a CNV paradigm were presented at irregular intervals, adaptation to the second stimulus was missing.

There were a lot of individual differences in the thalamic recordings during a simple CNV paradigm. Quite often it was difficult to interpret some sort of CNV into the thalamic traces (Fig. 1). With implanted multiple electrodes McCallum et al. (1976) found CNVs around the thalamus to be more prominent; rostrally and dorsally they were positive, but in subthalamic and midbrain electrode sites prominent negative CNVs were seen. In the brainstem BPs seemed to be more prominent than CNVs.

Since the BP precedes motor action it was to be expected that it should be found in the motor thalamus. Sometimes one could even detect a premotion positivity (Fig. 2), as described in detail by Deecke et al. (1969) on the scalp.

In quite a number of cases the BP over motor and premotor fields was not as prominent as expected. There is not much doubt that this is due to the fact that a large majority of our cases were suffering from Parkinson's disease. Deecke et al. (1973, 1976) have shown that the BP is reduced in akinetic Parkinsonian patients and what is most convincing is that in hemi-parkinsonism the BP is reduced only over the afflicted hemisphere. (The patients selected for operation had mostly rigidity or tremor and preferably nearly no akinesia.)

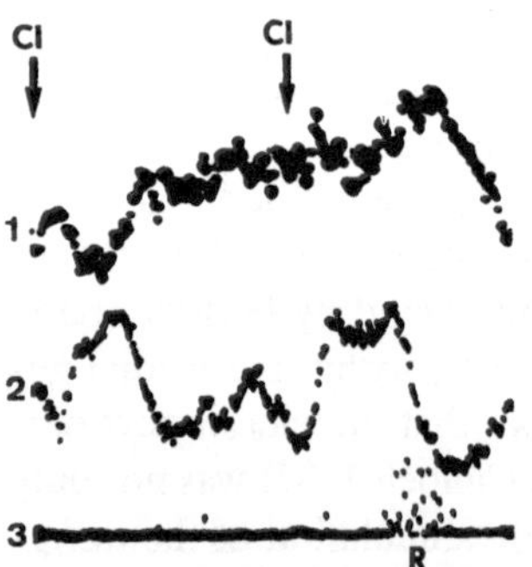

Fig. 1. Simple CNV paradigm, reaction to imperative stimulus thumb press of microswitch. ISI of clicks 1.5 s, 30 runs. Channel 1: motor cortex, channel 2: motor thalamus (V.o.p), channel 3: marking reaction. There is an unquestionable CNV on the cortex, but the thalamic slow potential changes are open to discussion

The most prominent BPs were seen in the nc. medialis with an early onset. In the nc. centralis (Ce of Hassler) the onset was much later than in the nc. medialis. In cases of intractable pain it was possible to record slow potential changes from both of these nuclei. In the motor thalamus, zona incerta, Forel's field H and the fornix the BP was sometimes very prominent, sometimes less prominent. Over the frontopolar region nearer the base the polarity was positive; the more dorsal part of area 10 was negative, like the larger part of the frontal cortex (Fig. 3). In a few cases in which electrodes were

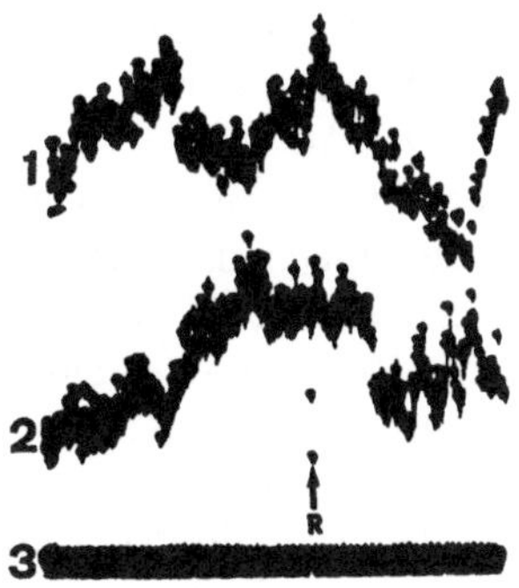

Fig. 2. On-line analysis of a Bereitschaftspotential. Total analysis time 4 s. Channel 1: motor cortex, channel 2: motor thalamus (V.o.a/p), channel 3: marking voluntary movement. One may even see a premotion positivity in the thalamus as well as on the cortex

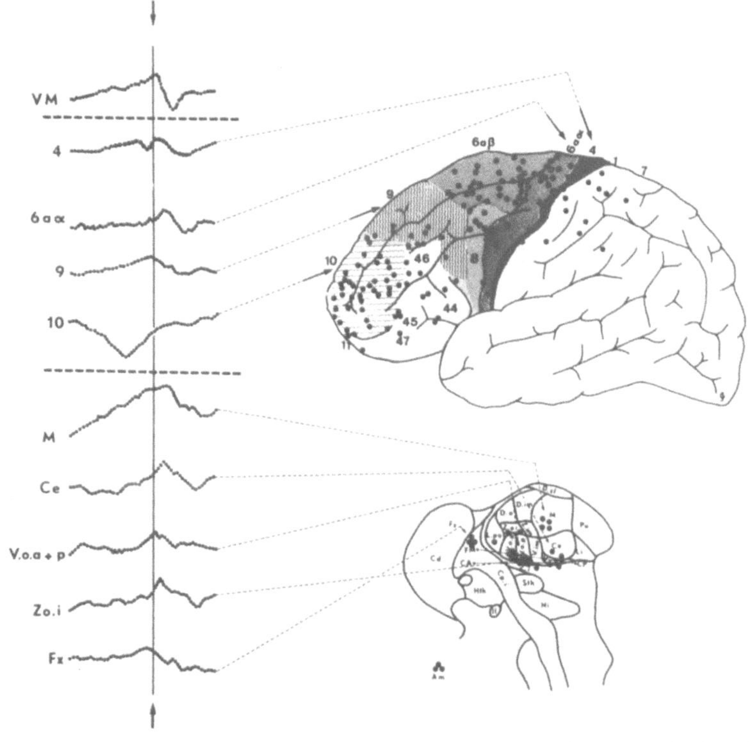

Fig. 3. Schematic synopsis of Bereitschaftspotentials from various cortical and subcortical electrode sites of 100 subjects. Overlap of electrode position not shown

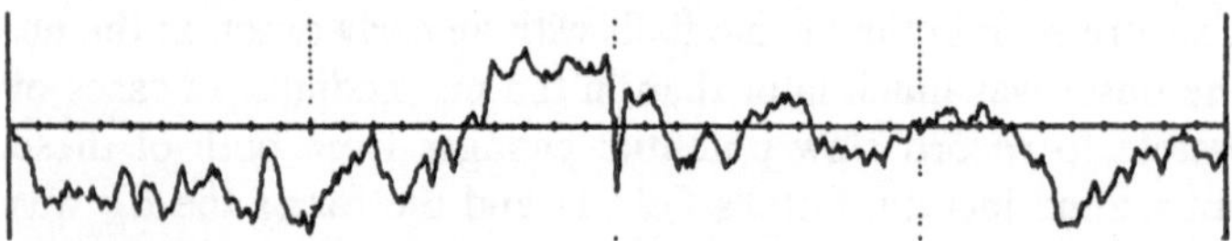

Fig. 4. Grand mean of BPs recorded from fornix in 7 patients. Voluntary movement thumb press of microswitch; analysis time 2 s prior to and 2 s after movement

just posterior to the rolandic sulcus the polarity was inversed again, but not over the parietal areas 5 and 7.

Deecke et al. (1971, 1973) suggested an internal feedback loop for the generation of the BP with initiation of movement in parietal regions, leading via cerebellum to the motor thalamus motor cortex. In a dozen patients who gave special consent, epidural electrodes could be placed over area 7 on both hemispheres. Very early onset, prominent BPs without premotion positivity could be recorded. CNV and BP could also be recorded from the fornix (Fig. 4).

In the meantime a new multiple semi-micro-DC probe for intracerebral recording has been developed at my suggestion at the Technical University of Vienna. With this new electrode it was possible to record a Bereitschaftspotential in the V.o.a/p within the limited space of only 3 mm in bipolar mode (Urban et al. in press).

Summary

A short survey is given, covering 20 years of experience in short-time recording of slow potential phenomena from cortex and depth of the human brain during stereotactic interventions under local anaesthesia. Differences of contingent negative variation and Bereitschaftspotential already known from scalp recordings could be confirmed, e.g. the frequently found positivity of the Bereitschaftspotential over fronto-polar regions. Some new facts could be added to the vast knowledge accumulated by the various groups involved in slow potential research. The data were collected at my former Department of Functional Neurosurgery and Clinical Neurophysiology, University of Vienna.

References

Deecke L, Scheid P, Kornhuber HH (1969) Distribution of readiness potential, pre-motion positivity, and motor potential of the human cerebral cortex preceding voluntary finger movements. Exp Brain Res 7: 158-168.

Deecke L, Becker W, Grözinger B, Scheid P, Kornhuber HH (1973) Human brain potentials preceding voluntary limb movements. Electroencephalogr Clin Neurophysiol [Suppl] 33: 85-94

Deecke L., Kornhuber HH, Schmitt G (1976) Bereitschaftspotential in parkinsonian patients. In: McCallum WC, Knott JR (eds) The responsive brain., Wright, Bristol, pp 169-171

Dusser de Barenne JG, McCulloch WS (1937) Local stimulatory inactivation within the cerebral cortex, the factor for extinction. Am J Physiol 118: 510-524

Ganglberger JA (1970) Stereotaktische Operationen und neuere Hirnforschung, Hollinek, Wien

Ganglberger JA (1976) New possibilities of stereotactic treatment of temporal lobe epilepsy (TLE). Acta Neurochir [Suppl] 23: 211-214

Ganglberger JA (1984) Additional new approach in treatment of temporal epilepsy. Acta Neurochir [Suppl] 33: 149-154
Ganglberger JA, Haider M (1969) Computer analysis of cortical responses to thalamic stimulation and of thalamo-cortical relationship of contingent negative variation in man. In: Gillingham FJ, Donaldson IML (eds) 3rd Symposium on Parkinson's disease. Livingstone, Edinburgh, pp 138-141
Ganglberger JA, Groll-Knapp E, Haider M (1971) Computer analysis of electrophysiological phenomena during stereotactic fornico- and amygdalotomy. In: Umbach W (ed) Special topics in stereotaxis. Hippokrates, Stuttgart, pp 149-155
Hassler R (1959) Anatomy of the thalamus. In: Schaltenbrand G, Bailey P (eds) Introduction to stereotaxis with an atlas of the human brain, vol 1. Thieme, Stuttgart, pp 230-288
Housepian EM, Pool JL (1962) Application of stereotaxic methods to histochemical, electronmicroscopic and electrophysiological studies of human subcortical structures. Confin Neurol 22: 171-177
Klingler J, Gloor P (1960) The connections of the amygdala and of the anterior temporal cortex in the human brain. J Comp Neurol 115, 3: 333-369
Kornhuber HH, Deecke L (1964) Hirnpotentialänderungen beim Menschen vor und nach Willkürbewegungen, dargestellt mit Magnetbandspeicherung und Rückwärtsanalyse. Pflügers Arch 281: R2
Kornhuber HH, Deecke L (1965) Hirnpotentialänderungen bei Willkürbewegungen und passiven Bewegungen des Menschen: Bereitschaftspotential und reaffente Potentiale. Pflügers Arch 281: 1-17
McCallum WC, Papakostopoulos D, Griffith HB (1976) Distribution of CNV and other slow potential changes in human brainstem structures. In: McCallum WC, Knott JR (eds) The responsive brain. Wright, Bristol, pp 205-210
Urban G, Ganglberger JA, Olcaytug F, Kohl F, Schallauer R, Trimmel M, Schmid H, Prohaska O, (in press) Development of a multiple thin film semi-micro-DC probe for intracerebral recordings. IEEE
Walter WG (1967) Slow potential changes in the human brain associated with expectancy, decision and intention. Electroencephalogr Clin Neurophysiol [Suppl] 33:123-130
Walter WG, Cooper R, Aldridge VJ, McCallum WC (1964) The contingent negative variation: an electrocortical sign of sensorimotor association and expectancy in the human brain. Nature 203: 380-384

Cortical DC-Shifts Related to Sustained Sensory Stimulation and Motor Activity

M. Haider, E. Groll-Knapp, M. Trimmel
Institute of Environmental Hygiene, Univ. of Vienna, Kinderspitalg. 15, 1095 Vienna, Austria

Introduction

On the question of event-related slow (DC) potentials in the brain, Haider et al. (1981) published a review paper some years ago. Since that time our group has tried to deal with brain DC shift related to longer term information processing and motor activity. This is of course only possible with the use of special DC electrodes and special DC amplification systems.

Event-related slow potentials like the contingent negative potentials described by Walter (1964) or the Bereitschaftspotential described by Kornhuber and Deecke (1964) normally occur within the range of a few seconds. One of the aims of our research during recent years was to find out which kinds and types of DC shift in the brain may be related to longer lasting (20 to 30 s) sensory stimulation and motor activity. Some results of that research will be described in this paper.

Information-Related Cortical DC Shifts in the Rat

In one of the experiments (Haider et al. 1982; Haselberger 1982a,b) information- and intensity-related cortical DC shifts in the rat have been found. 12 free movable rats have been implanted with special agar-embedded Ag/AgCl sintered electrodes (midline cortex versus nosebone). DC shifts were registered with the help of a special DC amplification system which allowed a resolution of 1 μV in the range of 1 V. As biologically significant stimuli series of cries of rats during a fight ("battle cries") with a mean frequency of 24 kHz and a sound intensity of 90 dB were applied over a duration of 27 seconds. DC shifts were summed up over 10 trials. 3 control conditions were used: in one of these conditions the rats were exposed over 27 s to a series of frequency-transformed battle cries (transformed from a mean frequency of 24 kHz to a mean frequency of 8 kHz). Here the assumption was made that the transformation of sound stimulation will also change its information content, since it has been found that a rat cry when snapped by a snake has a frequency of 8 kHz. In the 2 other control conditions the rats were exposed for 27 s to energy-equivalent artificial tonestimuli of the 2 frequencies (24 kHz and 8 kHz). Some results are shown in Fig. 1.

During the 27 s exposure to a series of repeated battle cries a negative DC shift (negative is down in Fig. 1) over the whole exposure time with a clear on - and - off ef-

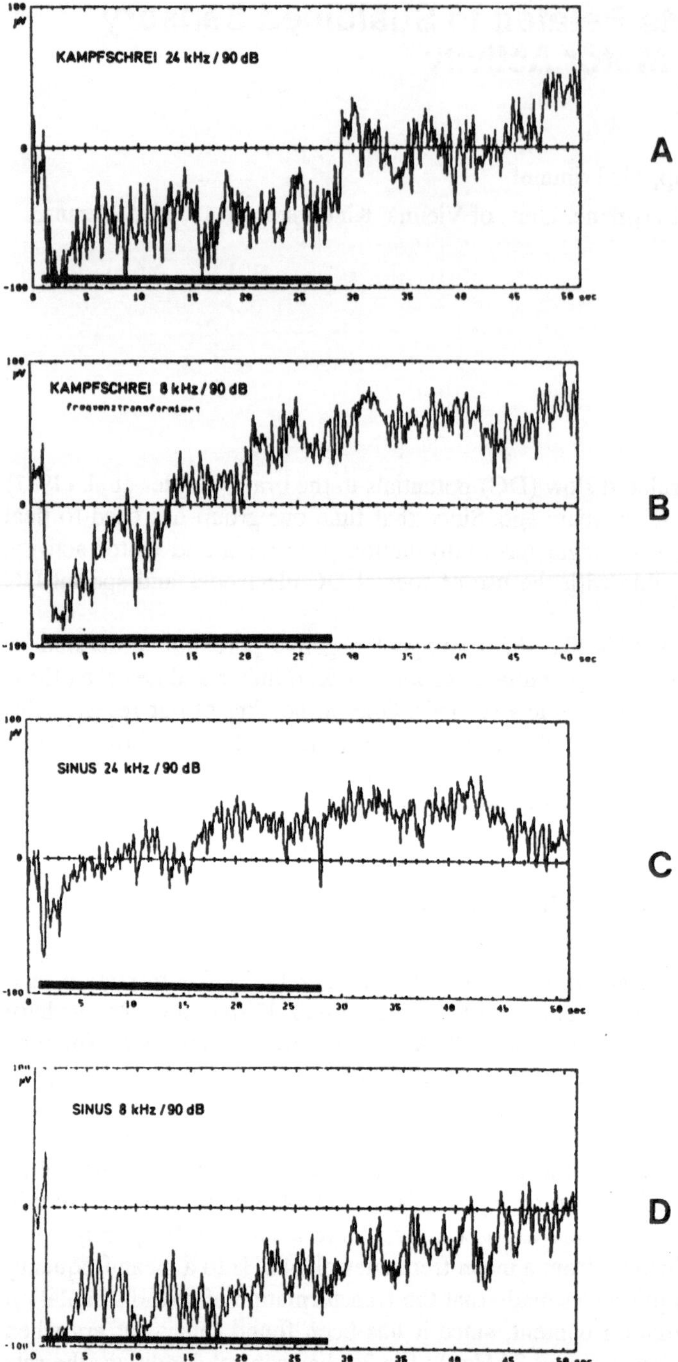

Fig. 1. Cortical DC shifts in the rat during biological significant stimuli (**A**) and control conditions (**B-D**). **A.** Biological significant stimuli ('battle cries' of 24 kHz mean frequency and 90 dB over 27 s). **B.** Frequency transformed battle cries (transformed from a mean frequency of 24 kHz to 8 kHz). C. Energy-equivalent artificial tonestimuli of 24 kHz. **D.** Energy-equivalent artificial tonestimuli of 8 kHz. Negativity is down in all parts of Fig. 1

fect could be observed (see part A in Fig. 1). The exposure to the 27 s series of frequency-transformed battle cries showed a very different picture: a clear negative on-effect was followed by an overshooting positivity, already occurring during the exposure time (see part B in Fig. 1). This overshooting positivity was also seen when rats were exposed to single "death cries" (the cry when a rat was snapped by a snake). In those instances a large negative on-effect was followed by an overshooting, longlasting positivity.

The control situations with energy-equivalent artifical tone series showed an on-effect with a quick (24 kHz tone series, part C in Fig. 1) or slower (8 kHz tone series, part D in Fig. 1) return to baselines.

The situations with the frequency-transformed signals obviously represent very different situations to the animals. The overshooting longlasting positivity may be the reaction to the biologically meaningful signals. Since it has the same frequency as a "death cry" when snapped by a snake this situation might be signalling something dangerous. The overshooting positivity may, then, be related to frightening and/or alienating biological signals.

DC Shifts in the Human Brain, Related to Cued Motor Activity

From the many stereoelectroencephalographic studies on event-related slow potentials in the human brain, which we performed together with Ganglberger, we will mention only one here, in which we used the paradigm of cued motor activity (Knapp et al. 1980). It was a study during stereotactic surgey in a patient with temporal lobe epilepsy, with a cortical electrode over area 4 and another electrode in the fornix. Some results are demonstrated in Fig. 2.

A cued goal-directed movement was to be started after the first of 2 cuing signals (S1) and to be ended by a button press after a second signal (S2). The contingent movement potentials have been averaged either from the button press as a trigger (left side) or from the cuing signal S1 as a trigger. In the cortex the contingent goal-directed movement potentials may be seen with clear on- and off-effects and a longlasting negativity when triggered from the reaction as well as from the cuing signal. In the fornix the pictures are different: a clear movement-related potential is seen when triggered from the reaction and a more complex potential when triggered from the signal. Generally we found that the more cognitive elements were included in movement performance, the more prominent and distinct were the movement-related potentials during initiation and execution of movements.

In an experimental study (Trimmel et al. in press) with 30 subjects the DC potentials of the brain were recorded from C3 and C4 vs. linked mastoids during lifting, holding for 20 s and lowering different weights with the right arm. The time intervals were cued by short imperative tones. Electromyographic activity was recorded simultaneously from the right biceps branchii and integrated. Each record of averaged brain DC potentials and integrated EMG contained 2 s baseline, 20 s muscle contraction and a period of 33 s after contraction. In Fig. 3 some results are demonstrated. Grand averages across subjects and 2 weight conditions are shown (30 trial averages).

Following the acoustic evoked potential to S1 there is a prominent negative shift (peaking about 1,5 s after S1) accompanying the period of lifting the weight. During

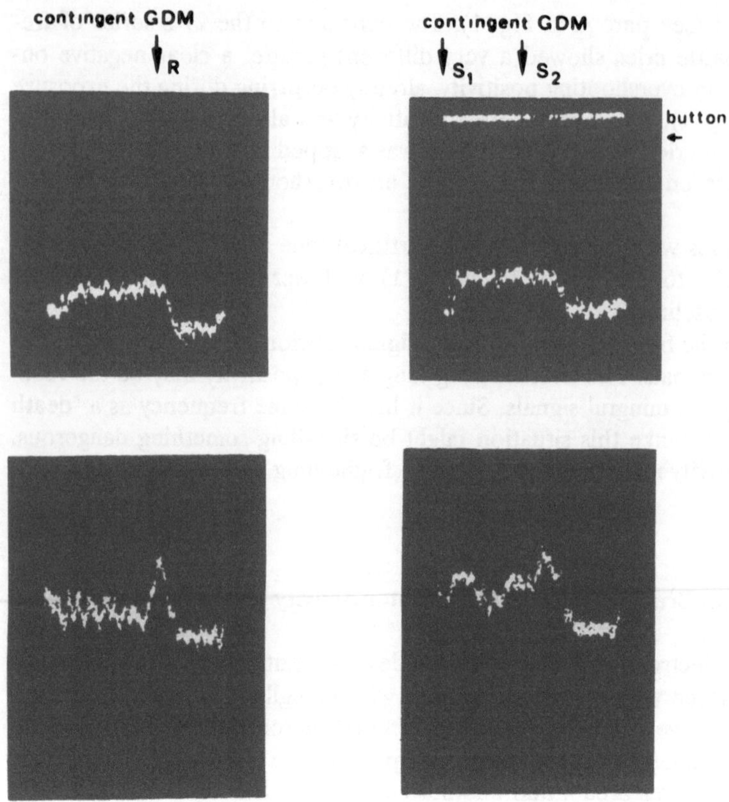

Fig. 2. Cortical (area 4, upper row) and fornix (lower row) potentials related to cued motor activity (negativity is upwards, analysis time 4 s). Contingent goal-directed potentials (contingent GDM) are triggered from the reaction (left side) as well as from the cuing signal (right side)

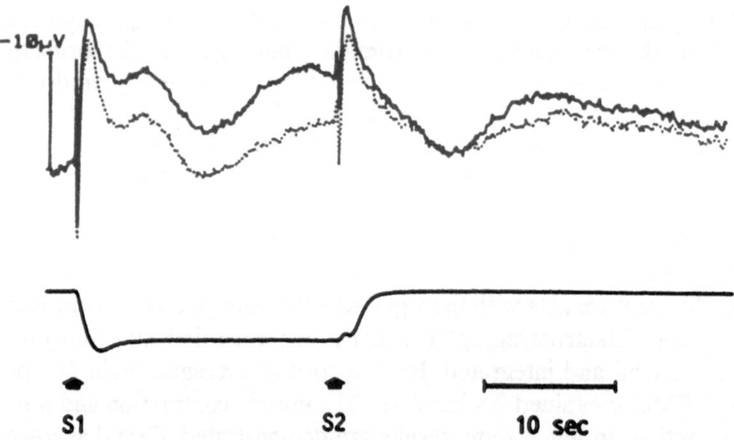

Fig. 3. Upper part: DC potentials from C3 (solid line) and C4 (dotted line) during (from S1 to S2) and after sustained isotonic contraction. Grand averages across Ss and 2 weight conditions (30 trial averages). Lower part: Integrated EMG from the right biceps brachii muscle averaged across Ss, 2 weight conditions and 30 trials (negative up). S1 indicates the signal to lift the weight, S2 indicates the signal to lower the weight

sustained contraction the negativity is not kept straight over the whole 20 s but shows some shifting towards positivity in between (V-shaped form of potential). After the acoustic evoked potential to S2 there is a negative shift during the lowering of the weight, followed by a longer lasting (10 s) positive after-effect. The movement-related potential shifts were more negative at the contralateral locations. Comparable findings have been reported by Deecke et al. (1969), Papakostopoulos (1978) as well as Grünewald et al. (1979). We found no differences between different weight conditions (therefore Fig. 3 shows averages over weight conditions). This could means that the amount of force may be reflected in the Bereitschaftspotential (Hink et al. 1983), but not so much during execution of weight lifting.

Discussion

Concerning the study with sustained sensory stimulation it is clear that continuous biologically meaningful stimulation may maintain long-term negativity over large areas of the cortex. What seems especially interesting is the fact of that with frightening and alienating signals one might get overshooting positivity. One might speculate that this overshooting positivity may represent some kind of "protective inhibition". More research should be directed towards these problem areas before more definite conclusions can be drawn. Concerning the study with cued, sustained motor activity it seems that the DC shifts during initiation, execution and termination of movements follow a certain pattern: Negative potentials accompany the initiation and termination of weight lifting, but some shifts towards positivity occur during sustained contraction and after the termination. In this respect it might be mentioned that the brain macropotentials studied here may be similarly discussed as in the work of Papakostopoulos and Jones (1980), who concluded "... that the brain macropotentials studied are not indices of the physical or mechanical parameters of the movement nor they are indices of general unspecific functions. What they probably indicate are ideokinetic functions related to the particular decision upon various aspects of the organization of the physical parameters of the movement." (p. 201). One might speculate here, that during cued sustained motor activity the earlier negative parts contain elements of orientation (orienting potentials) and the later negative parts contain elements of expectation (expectation potentials).

It is clear from all our results that cortical DC shifts depend on many variables, amongst them the state of the brain as well as psychological states and processes. One aspect of the involved relations seems to be that the greater the biological significance of the signal and the higher the number of cognitive elements and particular decisions involved in the organization of movements, the more prominent and distinct are the DC shifts.

Summary

Some studies on sustained sensory stimulation and sustained cued motor activity are reported. Cortical DC shifts in rats, related to 27 s exposure to biologically meaningful and neutral signals showed different patterns: longlasting negativity for the meaningful signals, only short on-effects for the neutral control stimuli and some overshooting positivity for frightening and alienated signals. During sustained (20 s) weight lifting we observed negative DC shifts during initiation and termination of movements but shifts towards positivity during sustained contraction and after its termination. In conclusion it seems that the greater the biological significance of the signals and the more cognitive elements and particular decisions are involved, the more prominent and distinct the cortical DC shifts become.

References

Deecke, L., Scheid, P., Kornhuber, H.H. (1969) Distribution of readiness potential, pre-motion positivity and motor potentials of the human cerebral cortex preceding voluntary finger movements. Exp. Brain Res 7:158-169.

Grünewald, G., Grünewald-Zuberbier, E., Hömberg, V., Netz, J. (1979) Cerebral potentials during smooth goal-directed hand movements in right-handed and left-handed subjects. Pflügers Arch. 381:39-46

Haider, M., Groll-Knapp, E., Ganglberger, J. (1981) Event-related slow (DC) potentials in the human brain. Rev. Physiol. Biochem. Pharmacol. 88:125-197

Haider, M., Haselberger, A., Knapp, E., Trimmel, M. (1982a) Information-related cortical DC shifts in the rat. XXIV. Alpines EEG-Meeting, 1-3 February 1982, St. Moritz (Abstract)

Haider, M., Groll-Knapp, E., Koller, M., Trimmel, M. (1982b) Neuere psycho- und elektrophysiologische Untersuchungen über Auswirkungen von Lärm auf den Menschen. In: Österreichischer Arbeitsring für Lärmbekämfpung (ed), XII. AICB-Kongreβ 1982: Erfolge und Prognosen der Lärmbekämpfung. Bohmann, Wien, pp 29-32

Haselberger, A. (1982) Aktivierungsbedingte DC shifts vom Rattencortex und Verhaltensaspekte bei Lärm und Signalreizen. Dissertation, Wien

Hink, R.F., Deecke, L., Kornhuber, H.H. (1983) Force uncertainty of voluntary movements and human movement-related potentials. Biol. Psychol. 16:197-210.

Knapp, E., Schmid, H., Ganglberger, J., Haider, M. (1980) Cortical and subcortical potentials during goal-directed and serial goal-directed movements in humans. Prog. Brain Res. 54:66-69

Kornhuber, H.H., Deecke, L. (1964) Hirnpotentialänderungen beim Menschen vor und nach Willkürbewegungen, dargestellt mit Magnetbandspeicherung und Rückwärtsanalyse. Pflügers Arch. 281:52

Papakostopoulos, D. (1978) The present state of brain macropotentials in motor control research: A summary of issues. In: Otto D.A. (ed) Multidisciplinary Perspectives in event-related brain potential research. U.S. Environmental Protection Agency, Washington DC pp 77-81.

Papakostopoulos, D., Jones, J.G. (1980) The impact of different levels of muscular force on the contingent negative variation (CNV). Prog. Brain Res 54:195-202

Trimmel, M., Streicher, F., Groll-Knapp, E., Haider, M. (1989) Brain DC potentials during and after cued, sustained motor activity (muscle tension). J Psychophysiology 3: 349-359

Walter, W.G. (1964) Slow potential waves in the human brain associated with expectancy attention and decision. Arch Psychiatry 206:309-322

Coordination Between Posture and Movement in Parkinsonism and SMA Lesion

F. Viallet, J. Massion*, R. Massarino*, R. Khalil
Service de Neurologie, CHU La Timone, 13005 Marseille
*Laboratoire de Neurosciences Fonctionnelles, CNRS, 31 chemin Joseph Aiguier, 13402 Marseille, Cedex 9

Introduction

One of the most interesting neuroanatomical findings in the last ten years is the link between the basal ganglia and the supplementary motor area (SMA). Schell and Strick (1984) have described the SMA as the projection area from parts of the ventrolateral thalamic complex such as the VLO or VA which do not receive projections from the cerebellar nuclei but from the pallidum. This pallido-thalamocortical pathway is part of the basal ganglia motor loop described by Alexander et al. (1986) which is involved in motor control.

The link between the basal ganglia and the SMA suggests that the motor defects described in parkinsonians should be largely the same as those caused by SMA lesion. In fact, cortical lesion in the SMA region results temporarily in a motor neglect syndrome, with a loss of spontaneous movements and difficulty in initiating voluntary movements (LaPlane et al. 1977). A comparable motor neglect syndrome has also been observed in monkeys after unilateral lesion of the substantia nigra, pars compacta (SNc; Viallet et al. 1981) and the question arises as to the link between the motor neglect and akinesia. The SMA is involved in the planning of sequential actions such as apposing the thumb to other fingers in a given order, but not in simple movements (Roland et al. 1980; Deecke and Kornhuber 1978; Lang et al. 1989). It is also involved in bimanual tasks (LaPlane et al. 1977; Brinkman 1984; Goldberg 1986). Parkinsonian patients are also more severely impaired when performing complex movements (Benecke et al. 1986), and it was suggested by Marsden (1984) that one of the main defects observed in these patients has to do with initiating learned motor plans and adapting them to new environmental conditions.

One of the coordinated actions that are impaired in parkinsonian patients and remain so even after L-DOPA treatment is coordination between posture and movement in a bimanual load lifting task; the ability to maintain the forearm posture notwithstanding the postural perturbation introduced by the lifting of the load by the other arm that normally occurs in normal subjects is impaired in parkinsonian patients (Viallet et al.1987). The aim of the present series is to explore the same task in patients with SMA lesions in order to determine whether defective coordination of this kind is a common indication of basal ganglia and SMA dysfunctions.

Methods

The task used here was a bimanual load lifting task. A sitting subject was instructed to maintain one forearm (postural forearm) horizontal. The other arm, the voluntary arm, was used to lift the load. The following parameters were recorded: the elbow position, the weight exerted by the load on the forearm, the integrated EMG from the brachioradialis of the postural forearm and from the biceps of the voluntary arm. Five patients with unilateral SMA area lesion were examined (Fig. 1). Each arm was used in turn as the postural forearm and the voluntary arm. This made it possible to compare the subject's performances between the arm on the side of the lesion and the contralateral arm.

Results

In normal subjects, when unloading is triggered by the experimenter, a forearm upward rotation ensues (Hugon et al. 1982), due to the spring properties of the unloaded forearm flexors, whereas when unloading is initiated by the subject's own voluntary movement, the forearm remains in roughly the same position despite the postural perturbation. This postural maintenance results from an inhibition of the forearm flexors prior to the onset of unloading, which is time locked with the activation of the biceps of the voluntary arm.

Concerning the central organization of the coordination between posture and movement, we have formed the hypothesis, on the basis of animal experiments, that the phasic postural command might result from collaterals from the movement control pathway. We have also suggested that the functioning and gain adjustment of the collateral pathway might be based on a gain and gate control (Paulignan et al. 1989; Fig. 2).

The results obtained with the patients with a unilateral lesion including the SMA are given in Fig 3. With these patients, the postural stabilization was preserved when the lesion was ipsilateral to the postural forearm, and a clearcut anticipatory deactivation of the forearm flexors was observed as in normal subjects. When the lesion was contralate-

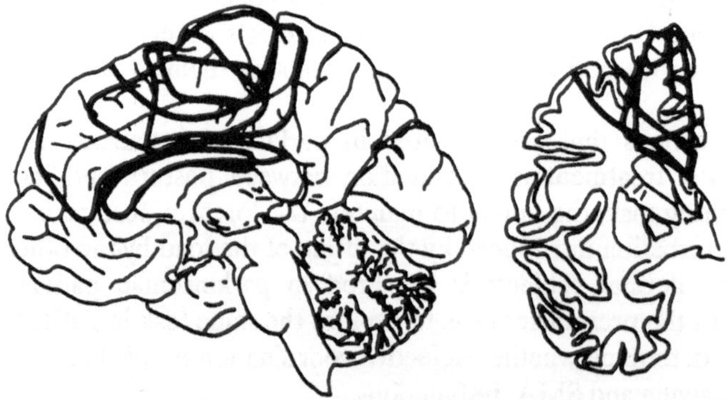

Fig. 1. Superimposition of the areas with lesions in the five patients with lesion of the SMA region. Left: extent of the lesion on the medial interhemispheric part; Right: extent of the lesion in frontal cross section of the damaged area

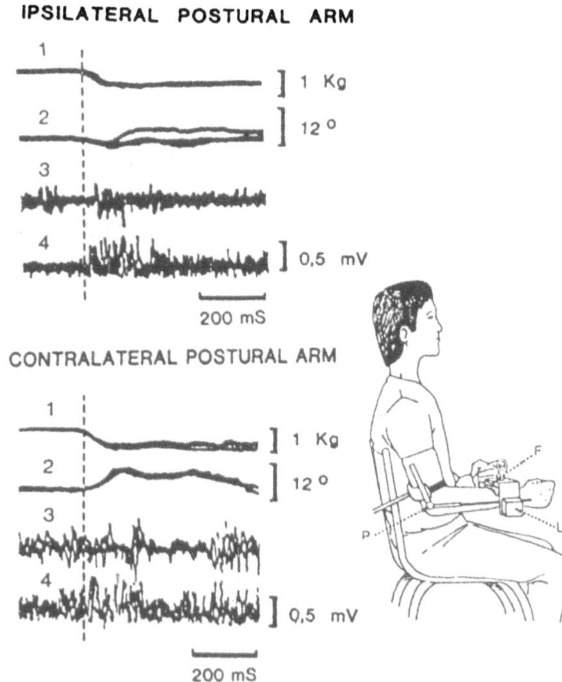

Fig. 2. Effect of the SMA region lesion on one side on the forearm posture during the load lifting task. 1: Force changes recorded by a force platform attached to the wrist (F); 2: Elbow angle recorded by a potentiometer (P); 3: Integrated EMG of the forearms flexors (brachioradialis); 4: Integrated EMG from the biceps of moving arm

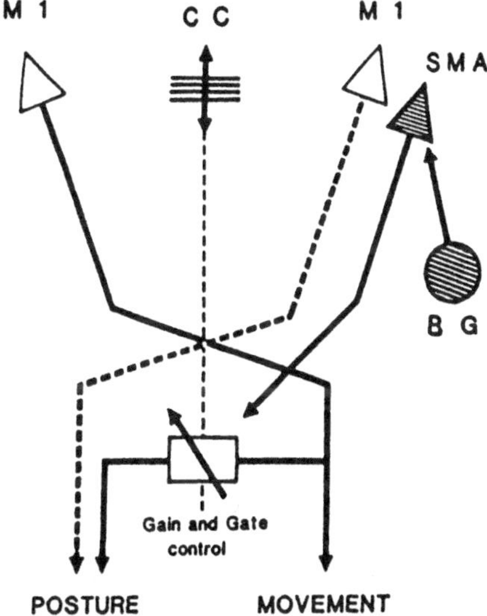

Fig. 3. Scheme of the central organization of the coordination in this load lifting task. M1: motor cortex; CC: corpus callosum; SMA: supplementary motor area; BG: basal ganglia. The left M1 controls both the contralateral movement and, through collaterals, the phasic postural changes in the ipsilateral postural forearm. The SMA and BG contralateral to the postural arm control the gain of the internal collateral pathway

ral to the postural forearm, then the maintenance of the forearm posture during the load lifting task was impaired: a forearm rotation of the postural arm occurred, and the anticipatory deactivation of the postural forearm flexors was not clearly apparent. This indicates first that the SMA region plays a role in the postural maintenance during the load lifting task, and secondly that it is not the SMA area contralateral to the perturbing voluntary movement but that contralateral to the postural forearm, whose position has to be stabilized, which is involved in the postural stabilization.

Comparable results were obtained with two spastic patients. In these patients a deficit in the postural maintenance was observed when the postural forearm was contralateral to the lesion; the forearm rotation was comparable to the situation in which the load was lifted by the experimenter, and no anticipatory inhibition of the forearm flexors was observed. This indicates that not only the SMA region but probably also other premotor and motor areas are involved in the maintenance of forearm posture during load lifting.

Discussion

The coordination between posture and movement is necessary in order to minimize the postural disturbance provoked by a voluntary movement (see Belenkiy et al. 1967; Massion 1984). This is achieved by means of a feedforward postural control, associated with the voluntary movement which corrects in advance the forthcoming effects of the postural disturbance due to the movement. The anticipatory postural adjustments observed in the present bimanual load lifting task had the same general characteristics as those associated with voluntary movements in standing subjects.

The finding that the SMA region is involved in the coordination between posture and movement is not unexpected. Wiesendanger (1986) suggested earlier on that SMA might play a role in this coordination. Gurfinkel and Elner (1988) have shown that standing patients who showed defective anticipatory postural adjustments when performing arm movements had lesions which were predominantly located in the SMA region. In addition, Brinkman (1984) has noted that coordination between movement and posture during a bimanual task was impaired in monkeys after unilateral SMA lesion. The impaired coordination observed between posture and movement in our bimanual load lifting task in subjects with SMA lesion is in fact in agreement with the previous results. This deficit exists not only after lesion of the SMA region, which is in fact the cortical projection area of the basal ganglia motor loop (Alexander et al. 1986), since it can also be observed in parkinsonian patients (Viallet et al. 1987). It is worth noting that in a hemiparkinsonian patient, the postural maintenance deficit was observed only when the postural forearm was on the disabled side (unpublished observation). These results suggest that, as in the case of SMA lesion, the basal ganglia contralateral to the postural forearm are responsible for the postural stabilization in this task.

It is quite surprising that the anticipatory postural adjustment of the postural forearm associated with the voluntary movement of the other arm was impaired only when the arm contralateral to the lesion was the postural forearm, since the SMA contralateral to the voluntary arm performing the load lifting might be expected to take over both the voluntary movement and the associated postural adjustment.The following is a possible explanation for this finding based on the hypothesis that a hierarchical organization is involved in this motor act. The position of the postural forearm may serve as a reference to calculate the movement parameters for the other arm. The stabilization of posture is a priority task, since it is a prerequisite for an accurate movement. This stabilization is achieved through the phasic postural adjustments associated with the voluntary movement. The role of the SMA region contralateral to the postural forearm during movement preparation might be to select the appropriate learned postural circuits depending on the movement to be performed. During the execution of the movement, the motor cortex contralateral to the moving arm might then control both the movement performance and, via control pathway collaterals, the postural adjustment of the postural forearm (Fig. 3). This would result in postural maintenance during the postural disturbing load lifting movement.

Summary

The SMA region receives projections from the basal ganglia motor loop. Similar deficits are to be observed in parkinsonians and in patients with SMA lesions. The role played by the SMA region in postural stabilization was explored during performance of a load lifting task. During the experiments, one forearm was maintained horizontal and loaded (postural forearm) while the other arm lifted the load. In healthy subjects, the forearm position remained quite stable despite the perturbations resulting from the voluntary load lifting movement, because a phasic anticipatory inhibition of the postural forearm flexors takes place at the time of the movement onset. Postural maintenance is impaired after lesion of the SMA region contralateral to the postural forearm whereas it is unimpaired when the postural forearm is ipsilateral to the lesioned SMA. This suggests that the SMA region plays a role, together with the basal ganglia, in presetting the phasic postural circuits which are responsible for the feedforward forearm postural control during voluntary load lifting. The SMA critical for this coordination is the SMA contralateral to the postural forearm to be stabilized when the load is lifted by the other arm.

References

Alexander GE, Delong MR, Strick PL (1986) Parallel organization of functionally segregated circuits linking basal ganglia and cortex. Annu Rev Neurosci 9: 357-381

Belenkiy VE, Gurfinkel VS, PALTSEV EI (1967) On elements of control voluntary movements (in Russian). Biofizika 12: 135-141

Benecke R, Rotwhell JC, Dick JPR, Day BL, Marsden CD (1986) Performance of simultaneous movements in patients with Parkinson's disease. Brain 109: 739-758

Brinkman C (1984) Supplementary motor area of the monkey's cerebral cortex: short- and long-term deficits after unilateral ablation, and the effects of subsequent callosal section. J Neurosci 4: 918-929

Deecke L, Kornhuber HH (1978) An electrical sign of participation of the mesial "supplementary" motor cortex in human voluntary finger movements. Brain Res 159:473-476

Goldberg G (1986) Supplementary motor area structure and function: review and hypotheses. Behav Brain Sci 8: 567-616

Gurfinkel VS, Elner AM (1988) Participation of secondary motor area of the frontal lobe in organization of postural components of voluntary movements in man. Neurophysiologia 20: 7-15

Hugon M, Massion J, Wiesendanger M (1982) Anticipatory postural changes induced by active unloading and comparison with passive unloading in man. Pflugers Arch 393: 292-296

Lang W, Zilch O, Koska C, Lindinger G, Deecke L (1989) Negative cortical DC shifts preceding and accompanying simple and complex sequential movements . Exp Brain Res 74:99-104

LaPlane D, Orgogozo JM, Meinenger V, Degos JD (1977) Clinical consequences of corticectomies involving the supplementary motor area in man. J Neurol Sci 34: 301-314

Marsden CD (1984) Which motor disorder in Parkinson's disease indicates the true motor function of the basal ganglia. In: Function of the basal ganglia. Wiley, Chichester, pp 225-241 (Ciba symposium, vol 106)

Massion J (1984) Postural changes accompanying voluntary movements. Normal and pathological aspects. Hum Neurobiol 2: 261-267

Paulignan Y, Dufosse M, Hugon M, Massion J (1989) Acquisition of a coordination between posture and movement in a bimanual task. Exp Brain Res (in press)

Roland PE, Larsen B, Lassen NA, Skinhou E (1980) Supplementary motor area and other cortical areas in organization of voluntary movements in man. J Neurophysiol 43: 118-136

Schell GR, Strick P (1984) The origin of thalamic inputs to the arcuate premotor and supplementary motor areas. J Neurosci 4: 539-560

Viallet F, Trouche E, Beaubaton D, Nieoullon A, Legallet E (1981) Bradykinesia following unilateral lesions restricted to the substantia nigra in the baboon. Neurosci Lett 24: 97-102

Viallet F, Massion J, Massarino R, Khalil R (1987) Performance of a bimanual load-lifting task by Parkinsonian patients. J Neurol Neurosurg Psychiatry 50: 1274-1283

Wiesendanger M (1986) Recent developments in studies of the supplementary motor area of primates. Rev Physiol Biochem Pharmacol 103: 1-59

Feedback Mechanisms Controlling Skeletal Muscle Tone

A. Struppler
Department of Neurology and Clinical Neurophysiology, Technical University, Möhlstraße 28, 8000 München 80, FRG

We are interested here in the systems controlling both tonic innervation and stiffness of arm and hand muscles. Stereotactic interventions for treatment of pathological hyperkinesias provide the unique chance to investigate sensory motor systems in alert man. In addition, such physiological investigation may serve to improve functional neurosurgery. There are two interesting phenomena following therapeutic lesions:

1. Decreased muscle stiffness, i.e. the resistance of the relaxed muscle to passive stretch is diminished.
2. Failure in load compensation, e.g. holding the arm against gravity, controlled only by sensory feedback without visual control or during contraction against external load.

Additionally, the sense of muscle force (or effort) seems to be disturbed. Figure 1 demonstrates a set of our target points. The therapeutic lesions (target points) are anatomically correlated to subcortical structures by computer-assisted techniques (Lipinski et al. 1987; Lipinski and Struppler 1989). The functions of the neuronal systems concerning muscle tone can be tested by:

1. Intracerebral stimulation
2. Post-operative examination of motor performance.

Repetitive stimulation at the target point below the threshold for direct activation of the descending motor pathways augments the tonic EMG activity associated with resting tremor and rigidity (Fig. 2). Such an increase of the ongoing EMG activity could be evoked via the alpha and/or gammasystem.

Microrecordings from Ia afferents (MNG) originating in a hand flexor muscle showed that the sensitivity of muscle spindles to stretch can be augmented by analogous repetitive stimulation at the target point (Fig. 3):

These findings provide evidence that at the target point fibers are represented as facilitating muscle tone, at least where the gamma system is involved.

What are the consequences for motor control, particularly in muscle tone, following lesions in the target area? We investigated how immediate load perturbations applied to the contracting flexor muscles can be compensated. We used torque motors, as shown in Fig. 4.

First we investigated the flexor muscles of the forearm (Figs. 5, 6) because they contribute to posture, particularly in the upright position. Moreover, these muscles are used predominantly for development of force and for compensation of external perturbation during goal-directed movements.

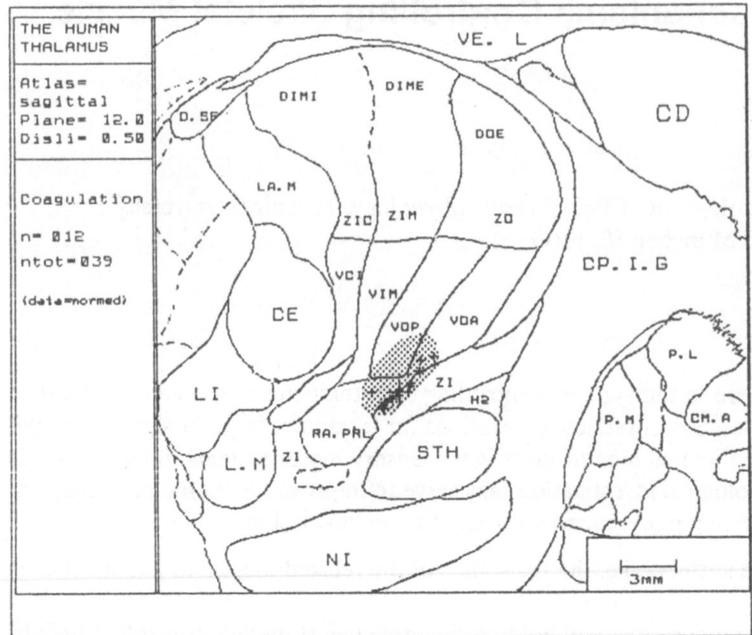

Fig. 1. Set of target points in reference to Schaltenbrand and Wahren's stereotaxic atlas (atlas plane saggittal 12.0, n = 12 of total 39 lesions correlated with this atlas plane). The correlation was done by computer-assisted techniques

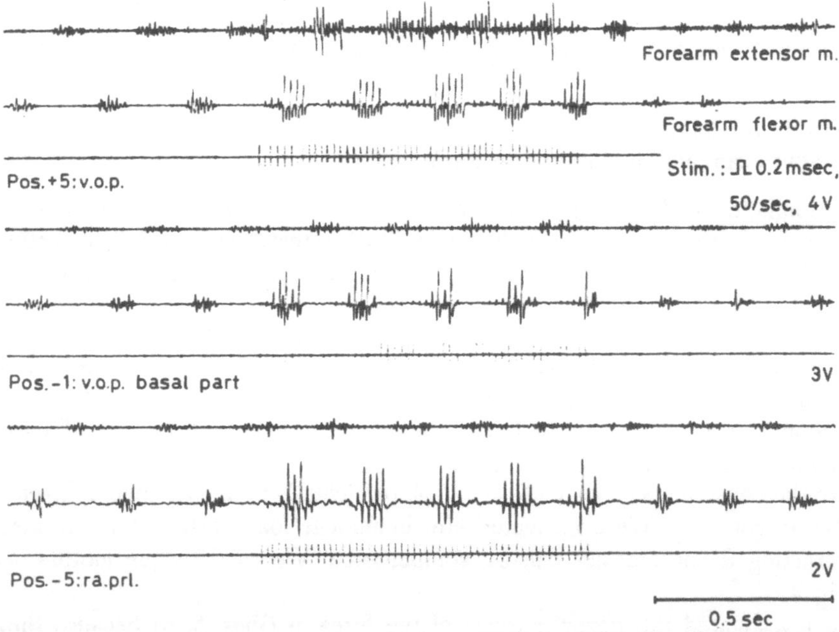

Fig. 2. Recording of resting tremor activity in antagonistic muscle groups during and after stimulation in the thalamus and subthalamus. Note that the spontaneous tremor activity can be facilitated by subthreshold stimuli within nucleus ventralis oralis posterior (V.O.P.) and radiatio prelemniscalis (Ra. prl.), (From Struppler et al. 1974)

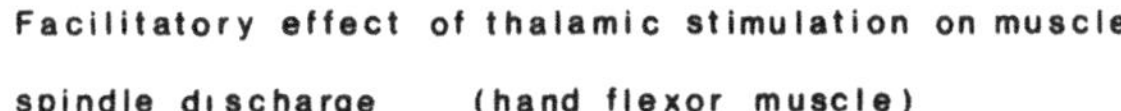

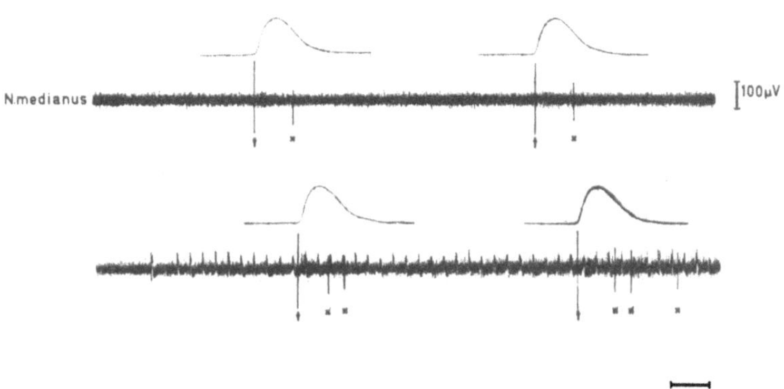

Fig. 3. Effect of thalamic stimulation on muscle spindle discharges elicited by electrically induced twitch contraction. Upper trace: electrically evoked twitch contraction activity of one muscle spindle (X) from a flexor muscle of the hand, firing during relaxation. Lower trace: the activity is increased during subthreshold repetitive stimulation at the target point. Arrows indicate electrical stimulus artifact. (From Struppler et al. 1978)

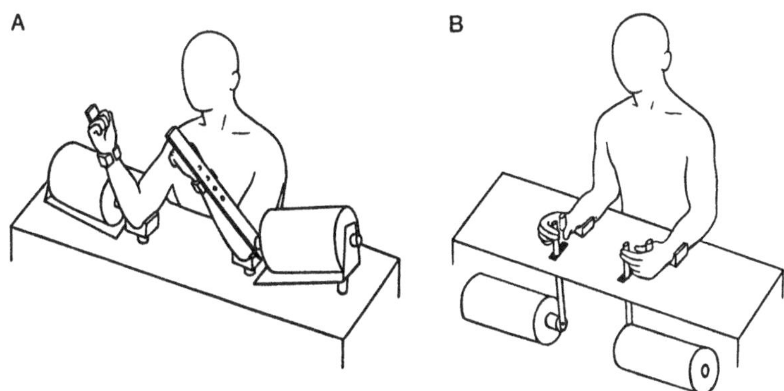

Fig. 4. Torque motor device used for quantitative assessment of muscle tone in the contracting forearm flexors (left) and in the finger flexors (right). (From Struppler et al. 1986)

Additionally we investigated the finger flexor muscles because: (a) one observes the same hypotonia following stereotactic lesions; (b) one can record muscle spindle activity in the median or ulnar nerve; (c) all motor performances with the tip of the fingers are highly controlled; (d) one can block the skin and joint afferents originating in the fingers; and (e) the inertias of the fingers are much lower as compared to the forearm.

The time course of the displacement induced by the perturbation can be analyzed with respect to reflex and prereflex components involved in rapid load compensation. The biomechanical properties of the system (inertia, elasticity, viscosity) must also be considered.

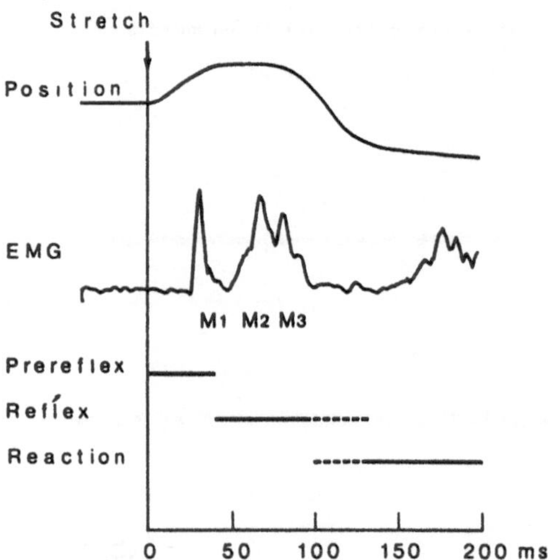

Fig. 5. Components in EMG ruling the mechanical behavior of a limb following a perturbation during load compensation

All experiments were performed under the instruction "Do not intervene" in order to keep the central drive constant; under these conditions there is no enhanced reflex excitability due to a "preset" and no coactivation of antagonistic muscle.

We measured the following: the prereflex stiffness, depending on the mechanical properties of the contracting and the noncontracting muscle fibers, the action of the stretch reflex, and the conditioning effect of passive finger extension on these parameters.

In parkinsonian patients the stretch-induced late component (M2) is increased as shown by Tatton and Lee (1975). It is normalized following stereotactic lesions associated with a larger displacement of the elbow flexors at a pronounced overshoot (Stewart-Holmes).

This overshoot could be caused by a decreased antagonistic triceps stretch reflex and/or by a modified pattern of innervation in the forearm flexor muscle during isometric contraction. During differential block, torque-induced elongation of the flexor muscle was increased, and the curve of the displacement revealed an overshoot, analogous to hypotonia following the stereotactic lesions. Therefore one can assume that this effect was caused by a change in the control of the flexor muscles and not by hypotonia of the antagonistic triceps muscle alone.

We could observe that angular displacement following external perturbations appeared earlier on the operated hypotonic side even before onset of any reflex activity. Therefore the question arises whether this kind of hypotonia is primarily due to reduced nonreflex muscle stiffness.

In principle, two mechanisms can be considered:

1. The original population of motor units generating the control motor performance is replaced by a different one with different intrinsic mechanical properties.

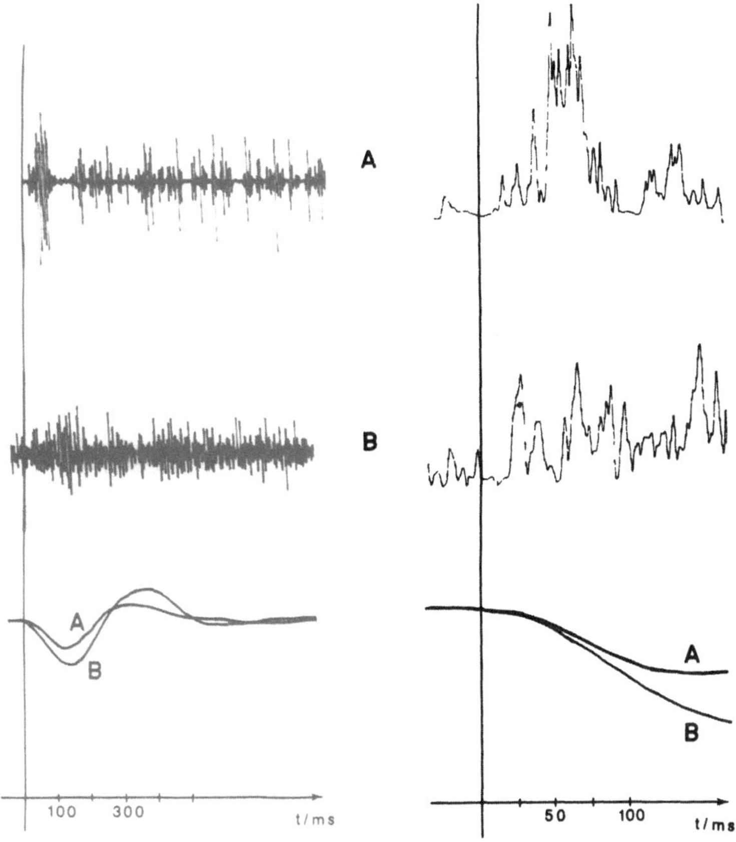

Fig. 6. Electromyographic (upper and intermediate traces) and mechanographic (lower traces) recordings of stretch-induced responses before (A) and after (B) operation. The perturbation begins at 0 ms. EMG recordings from the brachial muscle, angular displacement of the forearm measured at the elbow joint. Note the larger mechanical displacement following the operation. Left: single sweep, right: average of 8 sweeps. EMG activity rectified. (From Lehmann-Horn et al. 1982)

2. The component of muscle stiffness that is due to the intrafusal muscle tone is reduced because of a decreased static gamma activity.

Holding continuously against external load probably needs a special force and length feedback originating in the skeletal muscles. This feedback system can be blocked by lesions at various levels in the somatosensory system resulting in hypotonia. In hypotonia of peripheral origin, i.e. the polyneuropathy syndrome, there are usually deficiencies in various somatosensory modalities, such as sense of touch and vibration, and the tendon reflexes are abolished. In hypotonia of subthalamic origin, however, the so-called epicritic sensibility remains unchanged while the sensation of muscle force seems to be diminished. This may be due to a disconnection of muscle afferents, which are particularly concentrated in this area, according to neuroanatomical and physiological findings (Asanuma et al. 1983; Jones. 1983) (Fig.7).

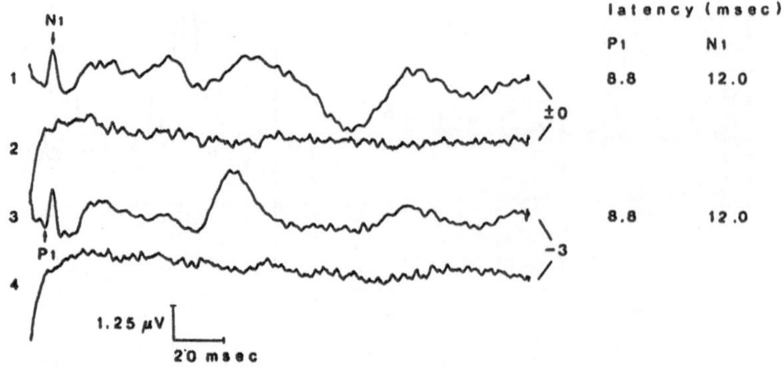

Fig. 7. In contrast to stimulation of the mixed nerve, stimulation of skin afferents evokes no significant potential within the target region. +0 and -3 symbolize the depth according to the stereotacic system. (Birk et al. 1986)

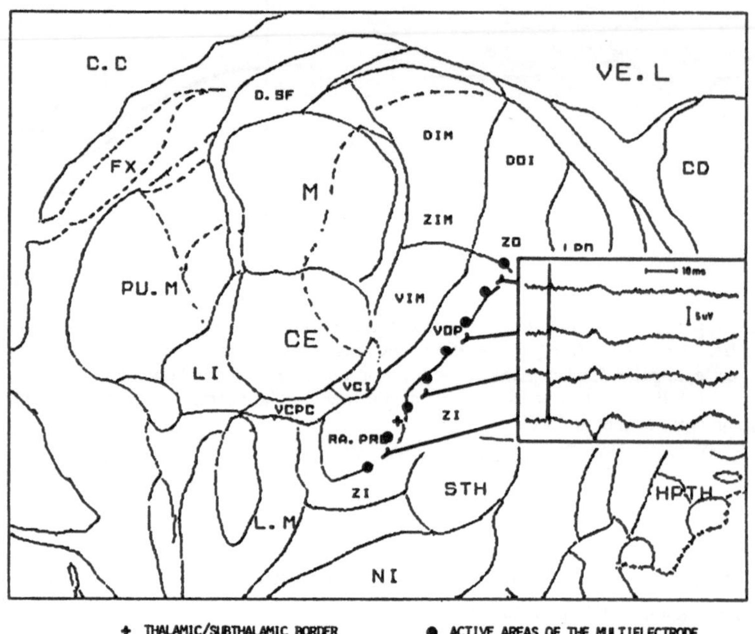

Fig. 8. The short-onset latency of about 12 ms and the phase reversal from subthalamic to thalamic nuclear border region indicates the generator located here and a direct spinal input

This is in keeping with the fact that evoked potentials of short-onset latency could be recorded within the target area following electrical stimulation of contralateral mixed peripheral nerves. In contrast, no significant evoked potentials were recorded in the same area following electrical stimulation of skin afferents (Birk et al. 1986).

By means of multielectrode recordings in the area of the calculated target point we estimate the short-onset latency at about 12 ms, and the phase reversal from subthalamic to thalamic nuclear border region indicates the generator located here and a direct spinal input (Fig. 8). The afferents presumably originate from skeletal muscle spindles.

The tentative scheme in Fig. 9 demonstrates somatosensory feedback loops responsible for regulation of muscle tone and estimation of force. Stereotactic disruption of proprioceptive afferents on the thalamic-subthalamic level may cause a reduced fusimotor drive and a reduced muscle spindle discharge in turn. Therefore, compensation of muscle stretch, dependent on tonic and phasic spinal reflexes, becomes insufficient and results in a decreased muscle stiffness (hypotonia). On the other hand, a decreased proprioceptive inflow leads to an underestimation of the voluntarly developed force under isometric conditions (comparison of the nominal value of force with the afferent signal in the associative cortex). Attempting to reproduce a preset value of force under isometric conditions, the subject tends to develop too much force.

As is known muscle tone is dependent on the intrinsic mechanical properties of the muscle fibers, on the one hand, and all various innervatory influences, such as spinal

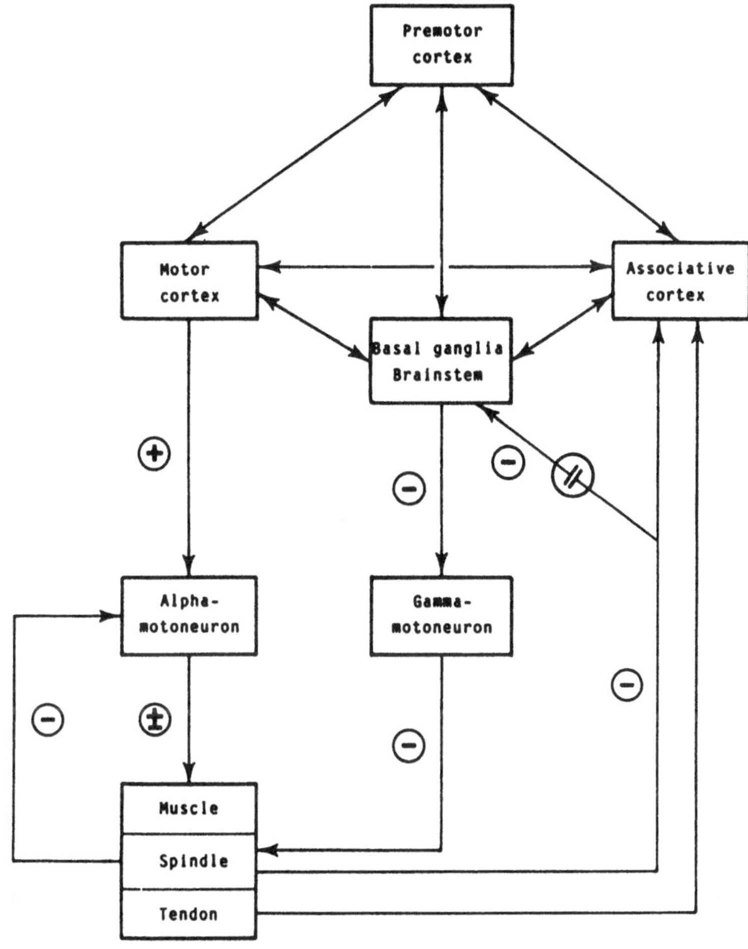

Fig. 9. Tentative scheme demonstrating somatosensory feedback loops responsible for regulation of muscle tone and estimation of force

and supraspinal reflexes and central commands, on the other (Fig. 10). Furthermore, mechanical properties of extra and intrafusal muscle fibers at innervation are mutually dependent. For example, stiffness of the intrafusal fibers can modulate the threshold of the muscle spindle with subsequent effects on the stretch reflex.

To investigate the role of the viscoelastic properties on the muscle spindle sensitivity we applied isometric contraction or passive extension conditioning. Muscle stiffness was always markly reduced; the same test pulse caused a more pronounced displacement (Fig. 11).

The effect cannot be attributed exclusively to changes in stretch reflex activity because the two curves start to diverge already before reflex onset. This means that the nonreflex stiffness is modified. Investigating muscle spindle activity, we used the identification scheme shown in Fig. 12.

At the same time, passive extension reduced the dynamic sensitivity of muscle spindle primary endings, which can be demonstrated by microneurography. As can be seen in Fig. 13 the firing rate of the Ia afferents to an abrupt load perturbation was lo-

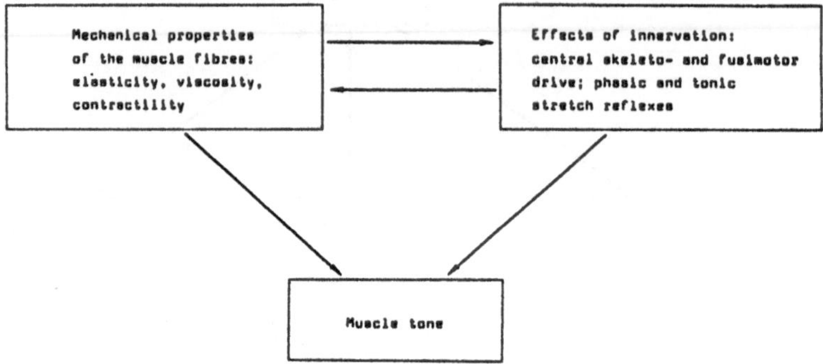

Fig. 10. Muscle tone depends on the inherent mechanical properties of the muscle fibers as well as on the effects on innervation

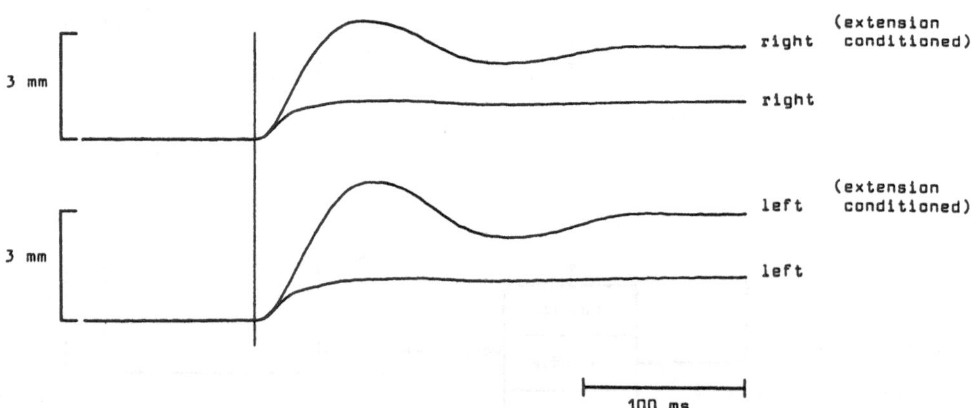

Fig. 11. Displacement of the fingertips following a torque pulse (4N). Upper traces in each pair of tied records show the response after conditioning by passive extension of the fingers. Each trace represents the average of 20 trials

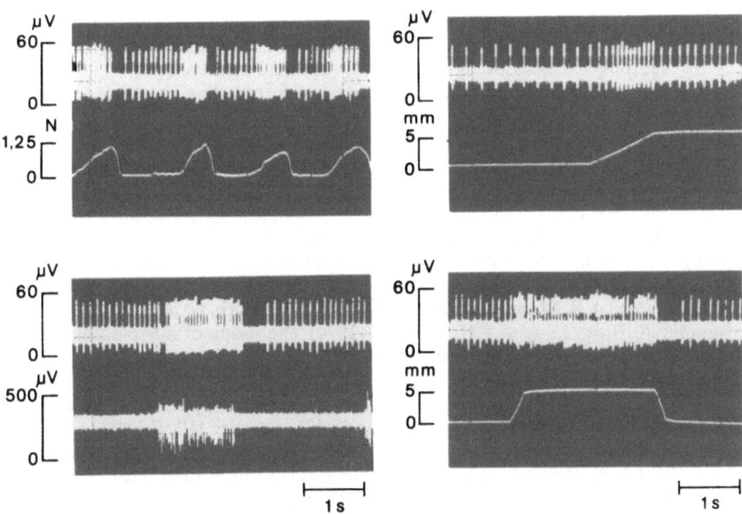

Fig. 12. Identification process of Ia afferent units: local pressure on the muscle belly (upper left), moderate voluntary isometric contraction (lower left) and ramp-and-hold stretches of the muscle in the relaxed state (upper and lower right)

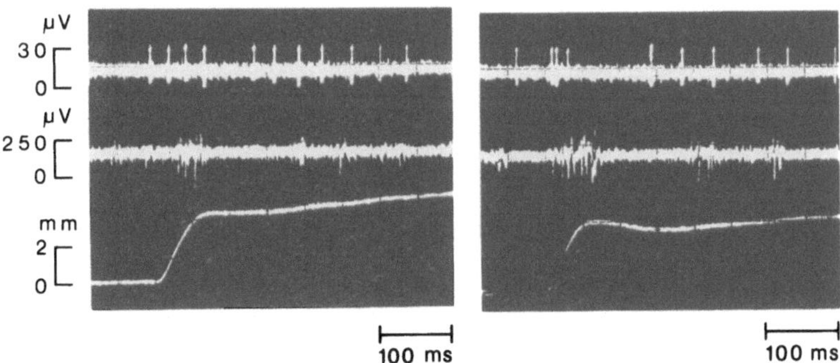

Fig. 13. Two single trails showing the effects of stretch (left) and contraction conditioning (right) on the Ia discharge (upper traces), EMG (intermediate traces), and displacement (lower traces). Standing load = 1.3 N, additional load = 1.56 N

wer following extension (left) compared to isometric contraction conditioning (right). The M1 component of the compound stretch reflex appeared slightly reduced following extension conditioning.

To gain more insight in the components of the feedback mechanisms we need further investigations concerning the static sensitivity of muscle spindles and recordings of the type of innervation (pattern of activated motor units).

References

Asanuma H, Thach W, Jones E (1983) Distribution of cerebellar terminations and their relations to other afferent terminations in the ventral lateral thalamic region of the monkey. Brain Res. 5: 237-265.

Birk P, Riescher H, Struppler A, Keidel M (1986) SEP and muscle responses related to thalamic and subthalamic structures in man. In: Struppler A, Weindl A (eds) Sensory-motor integration: implications for neurological diseases. Springer, Berlin Heidelberg New York

Jones E (1983) The nature of tzhe afferent pathways conveying short latency inputs to primate motor cortex. In: Desmedt J (ed) Raven, New York, pp 263-285

Lehmann-Horn F, Struppler A, Klein W, Lücking C, Burgmayer B, Deuschl G (1982) Veränderungen der motorischen Kontrolle bei Parkinson-Patienten. In: Struppler A (ed) Elektrophysiologische Diagnostik in der Neurologie. Thieme, Stuttgart, pp 236-237

Lipinski HG, Struppler A (1989) New trends in computer graphics and computer visions to assist functional neurosurgery. Stereotac Funct Neurosurg 52:234-241

Lipinski HG, Birk P, Struppler A (1987) Computerized stereotaxic neurosurgery. In: Lemke HU, Rhodes ML, Jaffee CC, Felix R (eds) Computer assisted radiology. Proceedings of the International Symposium C.A.R. Springer, Berlin Heidelberg New York, pp 348-352

Struppler A, Burg, Lücking CH, Velho F (1974) The mode of innervation following thalamotomy and subthalamotomy. Confin Neurol 36 4-6:347-354

Struppler A, Gerilovski L, Velho F, Erbel F, Altmann H (1978) Mode of innvervation following stereoencephalotomy. Contemp Clin Neurophysiology (EEG Suppl 34): 493-500

Struppler A, Riescher H, Lorenzen HW, Chen XZ, Groeter HP, Schaller I (1986) Torque motor for investigation of functional stretch reflex. ICEM 1986, Proceedings III. pp 949-943

Tatton W, Lee R (1975) Motor responses to sudden upper limb displacements of the normal humans and parkinsonian patients. Clin Res Rev 22:755a

Significance of Carbonic Anhydrase in the Function of Skeletal Muscle

P. Scheid, W. Siffert, K. Mückenhoff, B. Pelster, N. Clemens
Institut für Physiologie, Ruhr-Universität Bochum, 4630 Bochum 1, FRG

Carbonic Anhydrase and Isometric Force in Frog Muscle

To look into the function of CA, we have started our work with the excised frog gastrocnemius muscle with its nerve supply intact (Scheid and Siffert 1985). The muscle was suspended in air in an apparatus that allowed measurement of single-twitch isometric force. After a control, the muscle of one body side was incubated in frog Ringer's solution containing one of the CA inhibitors, acetazolamide, ethoxzolamide, or methazolamide; the muscle of the other body side was incubated, as control, in Ringer's solution without inhibitor.

Figure 1 shows the results obtained with various levels of ethoxzolamide in indirectly or directly stimulated muscle. There appears to be a dose-dependent reduction in iso-

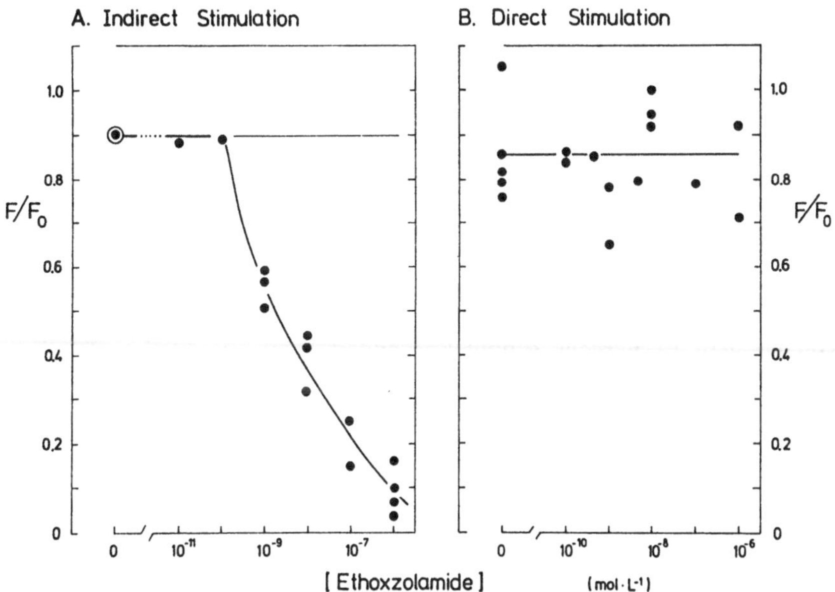

Fig. 1. Effect of carbonic anhydrase inhibition by ethoxzolamide on isometric force developed in frog gastrocnemius muscle after indirect (A) or direct (B) stimulation. Symbols represent individual measurements; open circle in A represents average of control tissues. See text for details

metric force with indirect stimulation, i.e., via supramaximal nerve stimulation, whereas no such reduction was observed with direct (= muscle) stimulation. Similar results were obtained with the other inhibitors, although the absolute concentrations were different, corresponding well with differences in the lipid solubility of these agents.

These data led us to conclude that CA inhibitors interfere with neuromuscular transmission (Scheid and Siffert 1985), particularly since no effects were observed in the electric properties of the motor nerve.

Carbonic Anhydrase and Isometric Force in Rat Skeletal Muscle

Before considering the site of action within the end-plate, we looked at the significance in mammals. These studies were executed at physiologic temperatures, despite the technical problems deriving from the higher O_2 consumption rate and from potential diffusional O_2 delivery problems in the excised muscles (Clemens et al. 1986).

We used the rat diaphragm, extensor digitorum longus (EDL), and soleus muscles and applied ethoxzolamide as CA inhibitor. Again, single-twitch isometric force was measured in response to indirect and direct stimulation.

In none of the three muscle types did CA blockade exert an effect in directly stimulated preparations (same result as in frog muscle). For indirect stimulation, the muscles differed, however, qualitatively. CA blockade in EDL reduced isometric force, while it was without effect in soleus muscle and had an intermediate effect in diaphragm. These differences correlate with the fiber type: fast-twitch (white) in EDL, slow-twitch (red) in soleus, and mixed in diaphragm.

We concluded from this series that CA exerts similar effects on neuromuscular transmission in rat as in frog muscle, but that there are distinct differences between fiber types.

Carbonic Anhydrase and End-Plate Function

Experiments are currently in progress to elucidate the function of CA at the neuromuscular end-plate, and preliminary results are reported here.

We first used the frog cutaneous pectoris muscle, which is a thin muscular sheet well suited for intracellular recordings of postsynaptic muscle potentials. The muscle was carefully dissected with its nervous supply, suspended in an appropriate Ringer's solution for insertion of glass micropipettes (3 M KCl, 10 MΩ) into the fiber near its end-plate region. MEPP and EPP were recorded in response to nerve stimulation by short rectangular pulses. Muscle twitch was prevented by adding either curare (5×10^{-7} M) to the bathing solution or by replacing Ca^{2+} by Mg^{2+} in it.

Figure 2 shows recordings of MEPPs in control and after inhibiting CA by acetazolamide. It is evident that the amplitude of the MEPPs is not altered by blockade of CA, as can also be seen from the amplitude histogram in the same preparation. The frequency of the spontaneous MEPPs remained unaffected by acetazolamide.

We conclude from the results that the rate of quantal transmitter release is not affected by CA inhibition, and that in particular the amount of transmitter molecules per

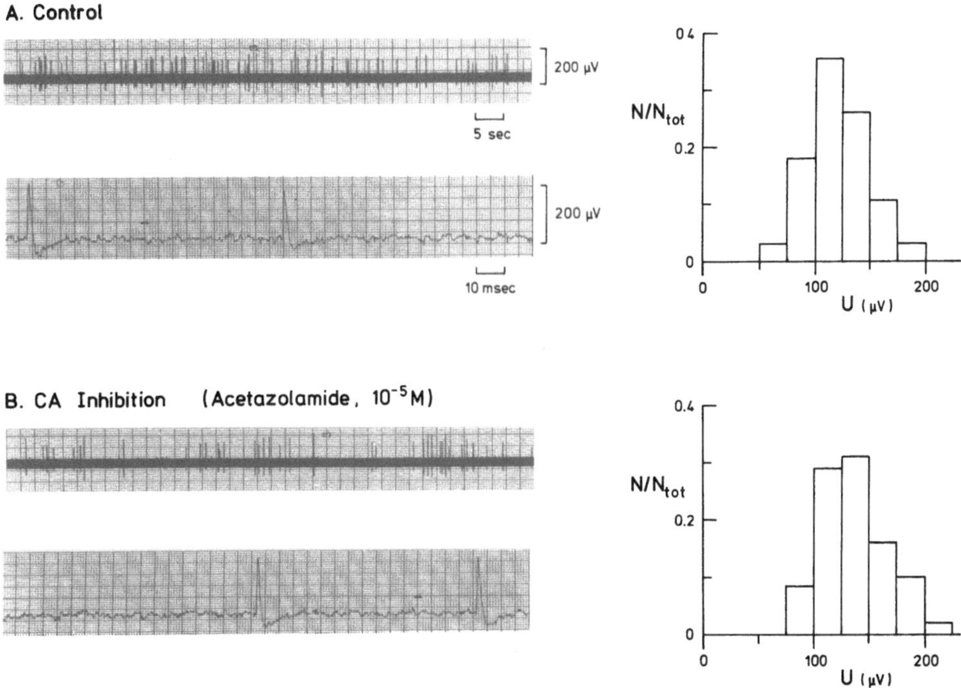

Fig. 2. Effect of carbonic anhydrase inhibition by acetazolamide (10^{-5} M) on miniature end-plate potentials (MEPP) in a typical preparation of frog cutaneous pectoris muscle. Sample records, at low and higher speed. Histograms based on 300 MEPPs in the given preparation

quantum is unchanged when CA is blocked; further, the subsynaptic action of the transmitter is not changed when CA is inhibited. This leads us to postulate that the action of CA on transmitter release is presynaptic.

These results are further corroborated by EPP in curarized muscle end-plates (Fig. 3). In most preparations, acetazolamide led to a reduction in the EPP amplitude, suggestive of a reduction in the number of transmitter quanta released upon nerve stimulation.

Experiments are now in progress in rat diaphragm as well, applying the same experimental techniques. The results appear to conform to those in frog muscle, suggesting that the mechanisms are of a more general type.

Physiologic Significance

Our experiments suggest an action of CA in the coupling between the presynaptic potential and transmitter release in the neuromuscluar endplate, and that differences may be present between nerves supplying different fiber types. Indirect support for our data is derived from histochemical studies, which have indeed shown CA to occur in motor nerves of the rat (Riley et al. 1982, 1984; Dermietzel et al. 1985).

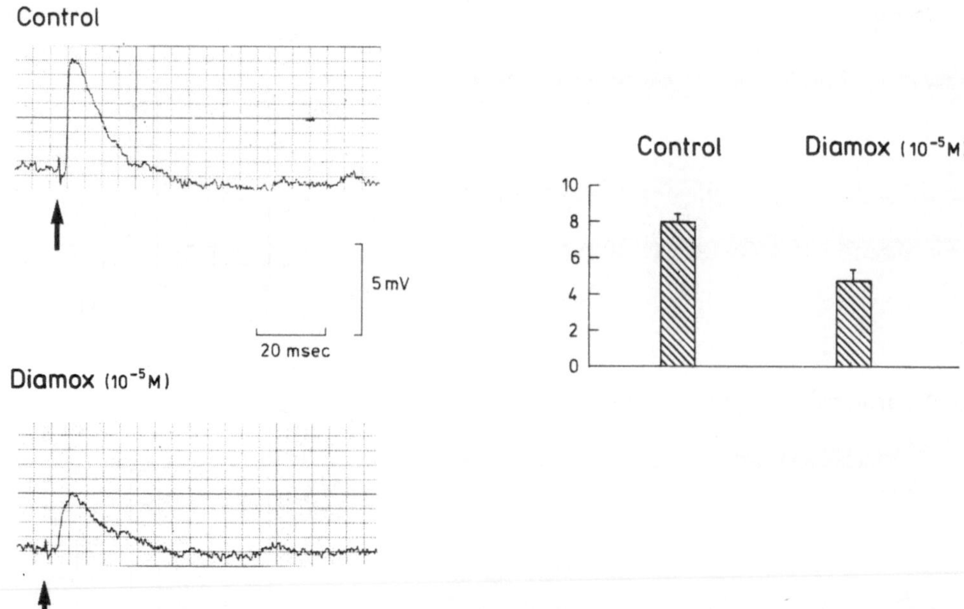

Fig. 3. Effects of carbonic anhydrase inhibition by acetazolamide (10^{-5} M) on evoked end-plate potentials (EPP) in a typical preparation of frog cutaneous pectoris muscle. The columns represent mean values (± SE) of 100 EPPs

Our data do not, however, allow determination of the mechanisms of this action of CA in neuromuscular transmission. It is generally believed that CA plays a role in increasing the kinetics of the $H^+/CO_2/HCO_3$ equilibration. We have recently shown that intracellular alkalinization is an important factor for Ca^{2+} release in stimulated human platelets (Siffert et al. 1987). A similar effect of fast pH changes upon membrane depolarization may be important in the presynaptic ending as well. Experiments are in progress in our laboratory that involve fast-responding pH- and Ca^{2+}-sensitive microelectrodes for intracellular recording in large presynaptic endings.

Inhibitors of CA, particularly acetazolamide, have been used in patients, e.g., as a diuretic drug, but ill effects on muscular performance have, to our knowledge, not been reported. That the effects are mild may be due to the low level of acetazolamide reached with therapeutic doses at the neuromuscular end-plate, and to the low lipid solubility of this drug. In patients suffering hypokalemic periodic paralysis or myasthenia gravis, treatment with acetazolamide has been reported to dimish muscle performance (Dalakas and Engel 1983; Carmignani et al. 1984).

Summary

The significance of carbonic anhydrase (CA) for skeletal muscle function has been studied by blocking CA with specific inhibitors. In frog gastrocnemius, isometric force of contraction was reduced by CA blockade when the muscle was stimulated indirectly (by its motor nerve) but was unaffected by direct stimulation. Similar results were obtained in various rat muscles. In frog cutaneous pectoris muscle, intracellularly recorded evoked end-plate potentials (EPP) were diminished in amplitude after CA blockade, whereas miniature end-plate potentials (MEPP) were unaffected both in amplitude and frequency. The data suggest that CA is involved in neuromuscular transmission, probably in the coupling between presynaptic depolarization and transmitter release.

CA, the enzyme that catalyzes the CO_2 hydration/dehydration reaction, has been observed in a number of tissues. Isoenzymes have been demonstrated to occur in skeletal muscle (cf. Scheid and Siffert 1985). The low-activity, sulfonamide-resistant CA III appears to exist in slow-twitch oxidative (red) fibers only, while the high-activity, sulfonamide-sensitive CA II has been shown to occur in fast-twitch (white) muscle fibers as well. Whereas the biochemical properties of the enzymes have been well described, there exists very little information about the functional significance of the enzymes.

References

Carmignani M, Scopetta C, Ranelletti FO, Tonali P (1984) Adverse interaction between acetazolamide and anticholinesterase drugs at the normal and myasthenic neuromuscular junction level. Int J Clin Pharmacol Ther Toxicol 22: 140-144

Clemens N, Siffert W, Scheid P (1986) Effects of carbonic anhydrase inhibitors on isometric force of rat skeletal muscle. Pflügers Arch 406: R 29

Dalakas MC, Engel W (1983) Treatment of 'permanent' weakness in familial hypokalemic periodic paralysis. Muscle Nerve 6: 182-186

Dermietzel R, Leibstein A, Siffert W, Zamboglou N, Gros, G (1985) A fast screening method for histochemical localization of carbonic anhydrase. J Histochem Cytochem 33: 93-98

Riley DA, Ellis S, Bain J (1982) Carbonic anhydrase activity in skeletal muscle fiber types, axons, spindles, and capillaries of rat soleus and extensor digitorum longus muscles. J Histochem Cytochem 30: 1275-1288

Riley DA, Ellis S, Bain (1984) Ultrastructural cytochemical localization of carbonic anhydrase activity in rat peripheral sensory and motor nerves, dorsal root ganglia, and dorsal column nuclei. Neuroscience 13: 189-206

Scheid P, Siffert W (1985) Effects of inhibiting carbonic anhydrase on isometric contraction of frog skeletal muscle. J Physiol (Lond) 361: 91-101

Siffert W, Scheid P, Akkerman JWN (1987) Role of cytoplasmic pH for stimulus-induced Ca2+ mobilization in human platelets. In: Heilmeyer LMG (ed) Signal Transduction and Protein Phosphorylation. Plenum, New York, pp 317-321

Part 2

Oculomotor and Equilibrium

Holding the Eye Still After a Saccade

D.A. Robinson, Z. Kapoula* and H.P. Goldstein+
Department of Ophthalmology, The Johns Hopkins University, School of Medicine, Baltimore, Maryland, USA
* Laboratoire de Psychologie Experiméntale, C.N.R.S. Université René Descartes, 28 rue Serpente, 75006 Paris, France
+ Foerderer Center, Wills Eye Hospital, Philadelphia, Pennsylvania, USA

Introduction

After a saccadic eye movement, it is desirable for clear vision that the eyes come quickly to rest and remain still. This does not happen automatically; the high velocity of a saccade is created by an intense burst of motoneuron activity, and when it ends, not all of the viscoelastic elements in the orbit have reached a steady state.The orbital contents must still progress to their new states of equilibrium. If something is not done during this postsaccadic period, eye position will drift as these internal forces readjust. To prevent this, the innervation sent to the muscles must take corrective action.

Various models of eye mechanics have been proposed, each demanding its own time course of innervation to achieve postsaccadic stability. Experimental recordings from oculomotor motoneurons in behaving primates capable of addressing this issue have been scant. This note reviews three proposed models, the postsaccadic waveforms that each demands, and the extent to which experimental recordings support these hypotheses.

The Models

Model I

Early studies of the discharge-rate modulation of oculomotor motoneurons in monkeys (e.g. Fuchs and Luschei 1970) found the innervation pattern to consist basically of a pulse and step. This was compatible with a single viscoelasticity described by a first-order, linear differential equation as shown in Fig. 1A. The pulse drives the eye quickly to the new position, overcoming viscosity; the step holds it there against the elastic restoring forces of the orbit. The detailed nature of the pulse is not considered here; all agree that the change in firing rate (ΔR) falls during the pulse, but in our view, the rapid displacement of the eye largely reflects the area under the pulse rather than its exact waveshape.We concentrate on events just after the pulse.

This model is, of course, too simple, although it does explain many of the postsaccadic drifts that one sees on clinical examination. If, for example, the amplitude of the pulse is too large for the step, the eye initially goes too far and then drifts back to the steady-state position determined by the step with the time constant (T_e) of about 0.25 s. If the pulse is too small, the eye falls short and drifts onward (e.g. Zee and Robinson 1979). One sees such movements in, for example, internuclear ophthalmoplegia.

Because this model is only first-order, there is only one adjustable parameter: the ratio of the pulse to the step. This ratio is, fortunately, under parametric control (Optican and Robinson 1980) so that pulse-step mismatches due to lesions can be corrected, but the prediction for postsaccadic drift is that if the pulse and step are appropriate, a steady postsaccadic discharge rate is needed to maintain eye position. This is not correct.

Model II

Even before neural recordings were available, one could anticipate that a first-order model would be inadequate (Robinson 1964), and Keller (1973) showed that at least a second-order term at least should be added. This is the term a in Fig. 1B. The first thing one thinks of as its physical basis is the moment of inertia of the eyeball (J in Fig. 1B, model IIa). This nineteenth century idea led to the hypothesis of active braking: once a mass has been accelerated, an opposing force is required to stop it. This, however, turned out not to be significant. The moment of inertia of the eyeball is negligibly small (Robinson 1964) and model IIa in Fig. 1B is not correct.

A more realistic arrangement with the same equation is model IIb. The pattern of innervation needed for this model is also a large pulse in one direction followed by a small pulse in the other (active braking).Thus, the braking pulse is not required because of inertia, but simply because the system is second-order; in this case, because it has two elasticities and viscosities. The upper pair, k_1 and r_1 , may be thought of as representing the series-elastic element and force-velocity relationship of muscle.

Stark and colleagues (Cook and Stark 1967; Clark and Stark 1974; Bahill et al. 1980) also developed a model which, when simplified and linearized, resembles model IIb. It too, of course, requires an active braking pulse, and these authors emphasized that this signal was needed for time optimality. This means that not only should the saccade be fast but there should also be no postsaccadic drift. This sort of model has two time constants, T_{e1} and T_{e2} (typical values are, roughly, 200 and 20 ms respectively). To oversimplify, proper adjustment of the main pulse and step prevents postsaccadic drift with a time constant near T_{e1} while adjustment of the small backward pulse prevents drifts with a time constant near T_{e2}

Model III

When one displaces a human or monkey eye and releases it, the returning movement is obviously composed of several time constants, some large, and depending on whether the subject is awake or anesthetized, and wether the horizontal recti are attached or not. This means that the passive tissues (k_2, r_2) in model IIb should be replaced by at least two viscoelasticities. Model III (Fig. 1C) emphasizes this aspect. This too is oversimplified because the known properties of muscle (series-elastic element, force-velocity relationship, length-tension relationship) are not present in this simple format. Ne-

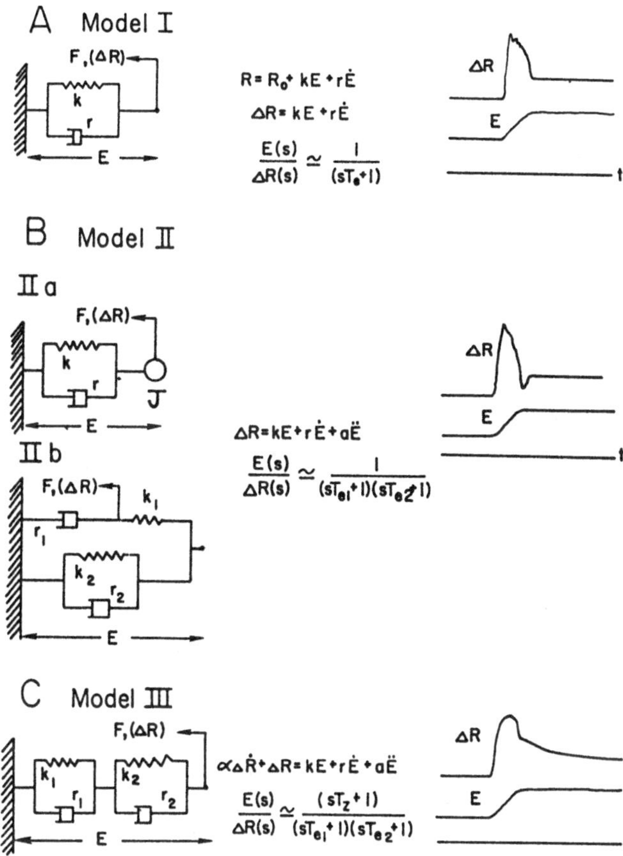

Fig. 1. Three models of eye mechanics. The modulation (ΔR) of the discharge rate (R) of motoneurons around the background rate (Ro) is equivalent to the driving force (F). In the center is the differential equation relating R to eye position E and the same relationship expressed as a transfer function in Laplace notation. To the left is a mechanical model composed of elasticities (k's) and viscosities (r's) described by the equation. To the right is a sketch of a saccade (trace E) and the discharge modulation (R) needed by this model to make that saccade. See text for details

vertheless, it brings to light a new aspect of saccadic innervation, proposed long ago (Robinson 1964) but subsequently neglected. The wave form, R(t), to achieve no post-saccadic drift, is a pulse-slide-step; at the end of the pulse, R is still considerably larger than is required in the steady state (the step), and it decreases exponentially to that value (the slide) with the time constant T_z (Fig. 2B; Goldstein 1983).

The rationale for the slide is obvious. The pulse suddenly stretches the two viscoelasticities. At the end of the saccade, the springs are not stretched by the ratios required for steady state. As they readjust, the net force of these two elements decreases, and the eye would drift onward. To offset this, the active-state tension created by R must decrease (the slide) so that the eye does not drift.

This new feature shows up mathematically in the zero $(sT_z + 1)$ of the transfer function and, in the differential equation, in the term on the left; this produces very important differences between model III and model IIb. A major difference is that active braking is not required. Thus, while a small backward pulse is required for time optimality in model IIb, it is not needed in model III even though both systems are second-order. Both models are oversimplified. Simulations that combine features of both still require a slide but not active braking (Robinson 1964). Nevertheless, only neural recordings can settle the issue.

The Evidence

The Slide

The evidence for the slide is strong and unequivocal. Goldstein (1983) found its amplitude - the change in R from the end of the burst to the steady state - to be typically 40 spikes/s. This is not small. The time constant (T_z) was 92 ms. After a saccade during which a motoneuron was inhibited, R rose to its steady state, similarly reflecting reciprocal innervation. Examples are shown in Fig. 2.

In the control of postsaccadic drift it is important to consider adaptive plasticity. If some component of the innervation is proposed as useful in controlling such drift, one should be able to show plasticity of that component. Optican and Miles (1985) induced postsaccadic drift in monkeys by drifting the visual scene as though the eye were drifting. They found that they could change the time constant of the induced postsaccadic drift, and that this could be done only if they used a model with a pulse-slide-step. This indicates that both the amplitude and time constant of the slide are under adaptive control; the slide is not an epi-phenomenon such as post tetanic potentiation but a component deliberately manipulated by the central nervous system, probably the flocculus, to eliminate postsaccadic drift. Similar results have been seen in humans (Kapoula et al. 1989).

In summary, the slide is established as an important part of the pattern of saccadic innervation to control postsaccadic drift.

Active Braking

The evidence here is marginal and confusing. van Gisbergen et al. (1981) and Goldstein (1983), recording from motoneurons in monkeys, and Sindermann et al. (1978), observing the human electromyogram, found a small "blip" or pulse of excitation in antagonist motoneurons just as they came out of inhibition at the end of the main pulse. Goldstein detected an even smaller dip in the agonist rate just after the main pulse, not seen by the others. An example is shown in Fig. 2. They could be simply posttetanic fatigue and postinhibitory rebound. The main question is whether these blips, constituting 0.1 - 0.2 spikes, are large enough to constitute a braking pulse.

Dynamic Overshoot

Additional evidence concerns a phenomenon called dynamic overshoot. This term describes a tiny (0.2° - 0.5°) saccade immediately following a main saccade, but in the op-

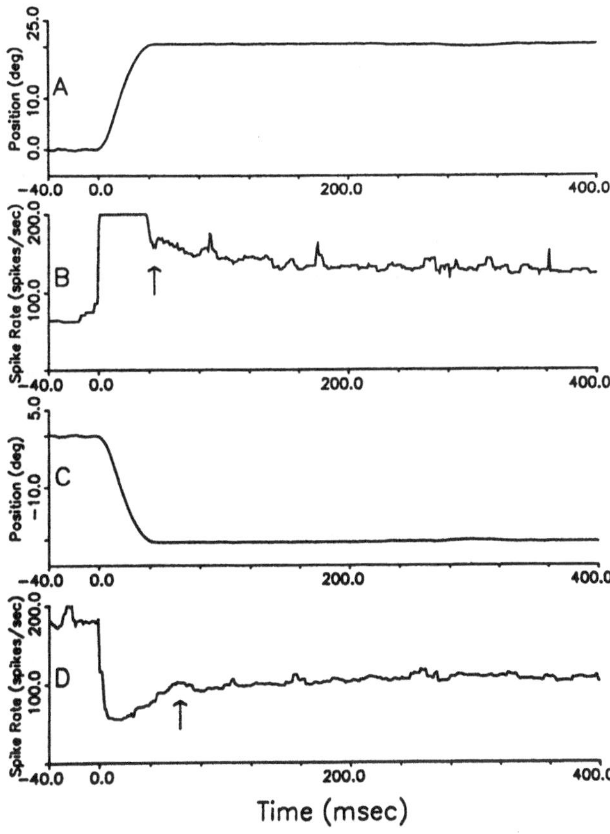

Fig. 2. Examples, at arrows, of a tiny but consistent dip in discharge rate after the main burst (mostly cut off due to scaling) in B accompanying a 20° saccade in A and a bump following inhibition in D after a 20° saccade in the other direction in C. The neuron was in the abducens nucleus of a trained monkey. Records were averaged over 5-8 trials. Time shift between start of saccade and burst is removed. In D, intrasaccadic rate was zero; rate shown is reciprocal of saccade duration

posite direction (Bahill et al. 1975; Kapoula et al. 1986). Its significance here is that it is probably created by a small reversed pulse just at the end of the saccade. Stark, Bahill, and colleagues felt that its discovery supported their hypothesis of active braking. Of course, dynamic overshoot by itself must be a mistake; after all, the point of the braking pulse is to bring the eye to rest, not create yet another saccade that delays stable vision by an additional 20 ms (the duration of dynamic overshoot). Nevertheless, if the braking pulse is small, noisy fluctuations could cause such errors. Is dynamic overshoot therefore evidence for active braking?

In the studies of Sindermann et al. (1978) and Goldstein (1983), eye movements were either not measured or not examined with a sensitivity required to detect dynamic overshoot so its presence could not be correlated with the intensity of a braking pulse. Two monkeys were studied by van Gisbergen et al. (1981). One of them had no dynamic overshoot, the other did but only for saccades smaller than 3°. Figure 3 shows examples of this dynamic overshoot and the corresponding discharges of an abducens neuron. In this case, dynamic overshoot was clearly created by a burst of neural discharges. Yet one hesitates to build an hypothesis on two neurons in one monkey. Some of the dynamic overshoots in this monkey were quite large (almost 1°, Fig. 3A). Is this pathological? Clearly it needs to be established that the monkey is a suitable animal model; we think, from casual observations, that monkeys do have dynamic overshoot, but that as with humans (Kapoula et al. 1986), it is quite idiosynchratic. If so, can dynamic overshoot be correlated with an excessive braking pulse, and can a braking pulse be observed in the absence of dynamic overshoot? Figure 3D is an example. The first interspike interval after the saccade is clearly smaller than the rest. This could create the blip in Fig. 2D, but whether it is qualitatively merely a smaller version of the clear pulses in Fig. 3A-C is difficult to say.

If a braking pulse is deliberately used to lessen postsaccadic drift, one would presume it to be under adaptive, plastic control. The only evidence bearing on this is previously unpublished data gathered in conjunction with a study of Kapoula et al. (1987). These authors patched one eye for 3 days in five subjects to see wether the onward postsaccadic drift normally found in the adducting eye of humans (Kapoula et al. 1986) decreases in the unpatched eye. It did, while no significant changes occurred in postsaccadic drift for abducting saccades or for either direction in the patched eye. The incidence of dynamic overshoot was examined before and after patching. The results are shown in Table 1.

Table 1B reaffirms that before patching, onward postsaccadic drift is larger in the adducting than the abducting eye (Kapoula et al. 1986); 1.67°/s vs. 0.8°/s in this case. After patching, this postadduction drift decreased from 1.67°/s to 0.44°/s. Table 1A reconfirms that dynamic overshoot is largely monocular and idiosynchratic. In this group, for the abducting eye, it ranged in frequency from 17 to 71%. Amplitudes were, again, about 0.2° as found earlier (ibid). Frequency was always higher, usually much higher, on abduction than on adduction. Subject 5 is an exception; for the other four subjects, on average, 82% of their dynamic overshoots were monocular, occurring only in the abducting eye. This fact makes it theoretically difficult to attribute dynamic overshoot to a mechanism that is supposed to control postsaccadic drift.

The main point of Table 1A is to show that after patching there was no consistent trend in the incidence of dynamic overshoot. For abducting saccades, the frequency went up in some subjects, down in others, the group mean changing only from 46% to 41%. Similar changes occurred for adduction. If, for example, one wished to decrease the onward postsaccadic drift after adduction, one would increase the amplitude of the braking pulse, which would raise the frequency of dynamic overshoot in both eyes. This

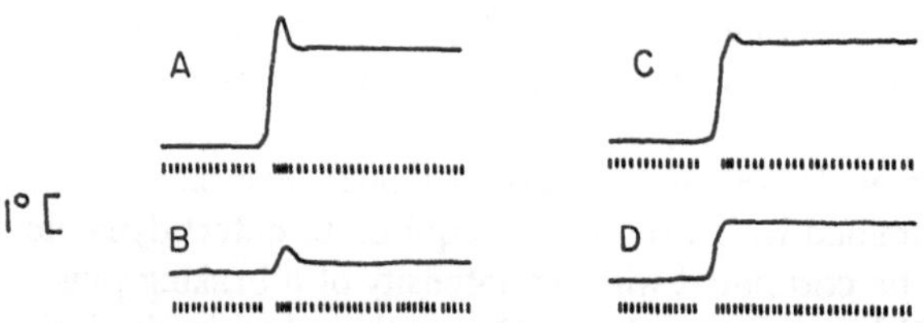

Fig. 3. Eye position and action potentials of an abducens neuron in one rhesus monkey that had dynamic overshoot, often severe, for saccades less than about 3°. Examples with large, medium, and no dynamic overshoot are shown. In each case the dynamic overshoot is driven by neural activity. Traced by hand by J.A.M. van Gisbergen about 1977

Table 1. The influence of patching one eye for 3 days on the amplitude of postsaccadic drift and the frequency of dynamic overshoot. Data are for the nonpatched eye before (Pre-) and after (Post-) the three days. Each datum is the mean (and standard deviation where relevant) of about 60 saccades. An asterisk indicates a statistically significant change after patching. Postsaccadic drift is the average velocity of the eye in the 100 ms immediately after the saccade. Abd., abducting saccades; Add., adducting saccades.The data were averaged over centrifugal and centripetal saccades of sizes 3°, 5°, 10°, and 20°

Subject	A Frequency of Dynamic Overshoot, %				B Post-saccadic Drift, deg/sec			
	Pre-		Post-		Pre-		Post-	
	Abd.	Add.	Abd.	Add.	Abd.	Add.	Abd.	Add.
1	71	16	76	43	0.59 ± 0.96	1.15 ± 2.02	0.18 ± 1.14*	0.07 ± 1.07*
2	48	4	20	0	2.61 ± 5.89	4.01 ± 6.47	0.89 ± 0.7*	0.54 ± 0.69*
3	26	6	10	0	0.16 ± 0.58	1.4 ± 1.42	0.18 ± 0.59	0.35 ± 0.69*
4	17	6	9	3	0.64 ± 1.19	1.56 ± 1.83	0.6 ± 1.31	1.11 ± 1.09
5	66	49	84	57	0.01 ± 0.57	0.25 ± 0.61	0.07 ± 0.76	0.11 ± 0.63
Means	46	16	41	21	0.80 ± 0.94	1.67 ± 1.25	0.38 ± 0.31	0.44 ± 0.38

did not occur. After 3 days, the postsaccadic drift velocity after adducting saccades was about equal to that after abducting saccades. If the lower postsaccadic drift in the abducting eye before patching were due to a large braking pulse, then the same should be true of the adducting eye after patching, and the incidence of dynamic overshoot should be equal for saccades in the two directions. This clearly is not the case. When Kapoula et al. (1989) induced postsaccadic drift in humans by drifting a visual scene as though the eye were drifting, they also found no consistent change in the frequency of dynamic overshoot. Consequently, we feel that these results indicate no obvious relationship between dynamic overshoot, postsaccadic drift, and plasticity.

Another explanation, an extension of one by Bahill et al. (1975), is to suppose that the medial rectus muscle (or motoneurons) exhibited a common phenomenon called postinhibitory rebound (a small overshoot of activity after release from inhibition), and that for some reason the lateral rectus system did not, or did so less. The result would be: a "braking pulse" for abducting saccades, no "braking pulse" for adducting saccades, monocular dynamic overshoot, and no relationship of dynamic overshoot with plastic modifications of postsaccadic drift. In a similar vein, Goldstein (1987) believes that some dynamic overshoots are not saccades at all but reflect the effect of pulse-step mismatches on mechanical elements with small time constants.

These problems and speculations indicate that while the slide in the pulse-slide-step of innervation plays an important role in regulating postsaccadic drift, many questions surround a similar conclusion for a putative braking pulse.

Summary

To bring the eye quickly to rest after a saccade, special components are required in the innervation pattern. Two suggested components have been active braking by a late reversed pulse and a slide or smooth transition between the main saccadic pulse and the subsequent step. Electrophysiological and behavioral studies show the slide to be a major element that is used by adaptive mechanisms to control postsaccadic eye drift. Evidence for a braking pulse is weak. Its relation to dynamic overshoot is uncertain. The latter appears not to be used by adaptive mechanisms. The role of a putative braking pulse remains questionable.

Acknowledgements. This work was supported in part by grant EY00598 from the Eye Institute of the National Institutes of Health, Bethesda MD, USA (D.A.R.), by the Philippe Foundation, Paris France (Z.K.), and by the Foerderer Eye Movement Center for Children, Philadelphia PA, USA (H.P.G).

References

Bahill AT, Clark MR, Stark L (1975) Dynamic overshoot in saccadic eye movements is caused by neurological control signal reversals. Exp Neurol 48: 107-122

Bahill AT, Latimer JR, Troost BT (1980) Linear homeomorphic model for human movement. IEEE Trans Biomed Eng 27: 631-639

Clark MR, Stark L (1974) Control of human eye movements I. Modelling of extraocular muscles II A model for the extraocular plant mechanism III Dynamic characteristics of the eye tracking mechanism. Math Biosci 20: 191-265

Cook G, Stark L (1967) Derivation of a model for the human eye-positioning mechanism. Bull Math Biophys 29: 153-174

Fuchs AF, Luschei ES (1970) Firing patterns of abducens neurons of alert monkeys in relationship to horizontal eye movement. J Neurophysiol 33: 382-392

Goldstein HP (1983) The neural encoding of saccades in the rhesus monkey. Thesis, The Johns Hopkins University, Baltimore

Goldstein HP (1987) Modelling postsaccadic drift: dynamic overshoot may be passive. In Foster KR (ed) Proceedings of the 13th annual NE Bioengineering Conference, vol 1. IEEE Inc, pp 245-248

Kapoula ZA, Robinson DA, Hain TC (1986) Motion of the eye immediately after a saccade. Exp Brain Res 61: 386-394

Kapoula Z, Hain TC, Zee DS, Robinson DA (1987) Adaptive changes in postsaccadic drift induced by patching one eye. Vision Res 27: 1299-1307

Kapoula Z, Optican LM, Robinson DA (1989) Visually induced plasticity of postsaccadic ocular drift in normal humans. J. Neurophysiol 61: 879-891

Keller EL (1973) Accommodative vergence in the alert monkey. Vision Res 13: 1565-1575

Optican LM, Miles FA (1985) Visually induced adaptive changes in primate saccadic oculomotor control signals. J Neurophysiol 54: 940-958

Optican LM, Robinson DA (1980) Cerebellar-dependent adaptive control of the primate saccadic system. J Neurophysiol 44: 1058-1076

Robinson DA (1964) The mechanics of human saccadic eye movement. J Physiol (Lond) 174: 245-264

Sindermann F, Geiselmann B, Fischler M (1978) Single motor unit activity in extraocular muscles in man during fixation and saccades. Electroencephalogr Clin Neurophysiol 45: 64-73

Do the Pretectum and Accessory Optic System Play Different Roles in Optokinetic Nystagmus?

M.J. Mustari and A. F. Fuchs
Department of Physiology and Biophysics and Regional Primate Research Center, University of Washington, Seattle, WA 98195

Introduction

Recent studies in birds and lower mammals have implicated the accessory optic system (AOS) and NOT in the generation of slow eye movements, especially OKN (reviewed by Simpson 1984). Single units in these structures respond to both the direction and velocity of full-field visual stimuli, and lesions of these structures compromise OKN. Both the AOS, which consists of lateral, medial, and dorsal terminal nuclei (LTN, MTN, DTN), and the NOT receive direct retinal input from the contralateral eye. Single units in the LTN and MTN prefer vertical visual motion (Simpson et al. 1979; Grasse and Cynader 1982, 1984; Mustari et al. 1986) whereas those in the DTN and NOT prefer horizontal visual motion (Grasse and Cynader 1984; Collewijn 1975; Hoffmann and Schoppmann 1981; Hoffmann and Distler 1986; Mustari and Fuchs 1988). One pathway by which the AOS and NOT could influence eye movements is via a projection to that part of the inferior olive (IO), the dorsal cap of Kooy, that supplies climbing fibers to the flocculus (Maekawa and Simpson 1973). The flocculus is known to have a role in smooth-pursuit (Lisberger and Fuchs 1978) and optokinetic (Zee et al. 1981; Waespe et al. 1985) eye movements.

Although the role in the NOT of lower mammals has been studied extensively by recording, lesion, and anatomical techniques, the discharge properties and connectivity of units in the primate NOT and AOS have received little attention. To remedy this situation we report preliminary results that document differences in the discharge patterns of LTN (Mustari et al. 1986) and NOT (Mustari and Fuchs 1988) units in the behaving rhesus macaque (*Macaca mulatta*) and demonstrate a rather direct pathway from the retina to both the LTN and NOT and then to the dorsal cap of Kooy.

Methods

A detailed description of our behavioral training paradigms and recording techniques can be found elsewhere (Mustari et al. 1988). All monkeys were trained to fixate a stationary target spot on a tangent screen while visual test stimuli were moved and to track a jumping or smoothly moving target spot over a large-field patterned or dark background. The target spot and visual test stimuli were moved in various directions with either sinusoidal or triangular waveforms and velocities ranging from 0 to 200°/s. The direction selectivity of units was assessed by moving the background horizontally, vertically, or obliquely (i.e., in directions ±45° from horizontal) while the monkey fixated a stationary target spot. The preferred ("on") direction for each unit was calculated by obtaining 10-cycle averages of firing rate in each of the eight directions and then fitting a curve through the eight data points by means of an FFT algorithm (Wallman and Velez 1985). This procedure calculated a best direction for the unit which often was one not actually tested (see Fig. 3). Sensitivity to stimulus velocities was tested by moving a large-field pattern in the unit's preferred direction while the monkey fixated the stationary target spot. The mean firing rate minus the resting rate for the "on" half-cycle (averaged from at least 10 cycles) was then plotted against stimulus velocity.

The retinorecipient zones of the pretectum and AOS were revealed by placing intraocular injections of tritiated proline and leucine in one eye. Two animals were allowed to survive for 1 and 5 days postinjection so that mainly terminal fields and fibers of passage, respectively, could be visualized; the brains were processed with standard autoradiographic methods (Cowan et al. 1972). Two other animals received unilateral injections of horseradish peroxidase (HRP; sigma-type VI) in the vicinity of the dorsal cap of Kooy of the IO. After 24 h the animals were sacrificed by a lethal dose of barbiturate and processed for HRP histochemistry (Mesulum 1978).

Results and Discussion

Figure 1 shows that the monkey, like lower mammals, has a pathway from the retina to the pretectum to the IO and has a retinal-LTN-IO pathway as well. The pattern of retinal termination (contralateral to the injected eye) as revealed by orthograde transport of tritiated amino acids is illustrated in dark-field autoradiographs (Fig. 1A, C). The corresponding light-field sections show that the small patch of terminal labeling on the lateral brainstem lay in the LTN (Fig. 1A). The LTN was prominent at the level of the oculomotor nucleus (III) and no longer present at the level of the pretectal olivary nucleus (Fig. 1C). Label was not found at the putative site of the MTN (but see Cooper and Magnin 1986; Weber and Giolli 1986). At the level of the pretectum (Fig. 1C), retinal input extended across the entire mediolateral extent of the brainstem. The heaviest retinal termination was found in the pretectal olivary nucleus (NO), a structure thought to be involved in the pupillary light reflex (Fig. 1C). Substantial labeling was also found in the NOT, whose approximate border is indicated by the dashed outline in Fig. 1C. Although a lateral area of high labeling intensity was present at the appropriate location for the DTN, cytoarchitectonic features did not allow us to delineate the DTN from the lateral edge of the NOT.

The distribution of retrogradely labeled cells following HRP injections in the vicinity of the dorsal cap of the IO is shown in Fig. 1B and 1D. Labeled cells were found over the entire rostrocaudal extent of the LTN and NOT, although many cells in both structures remained unlabeled. The locations of labeled cells in both the LTN and NOT overlapped with the zones of retinal termination, supporting the suggestion that these structures are, at least in part, the sources of visual input to the IO and ultimately of the visual climbing fiber input to the flocculus. Retrogadely labeled cells were also found in regions outside the LTN and NOT, but usually not within regions of direct retinal input.

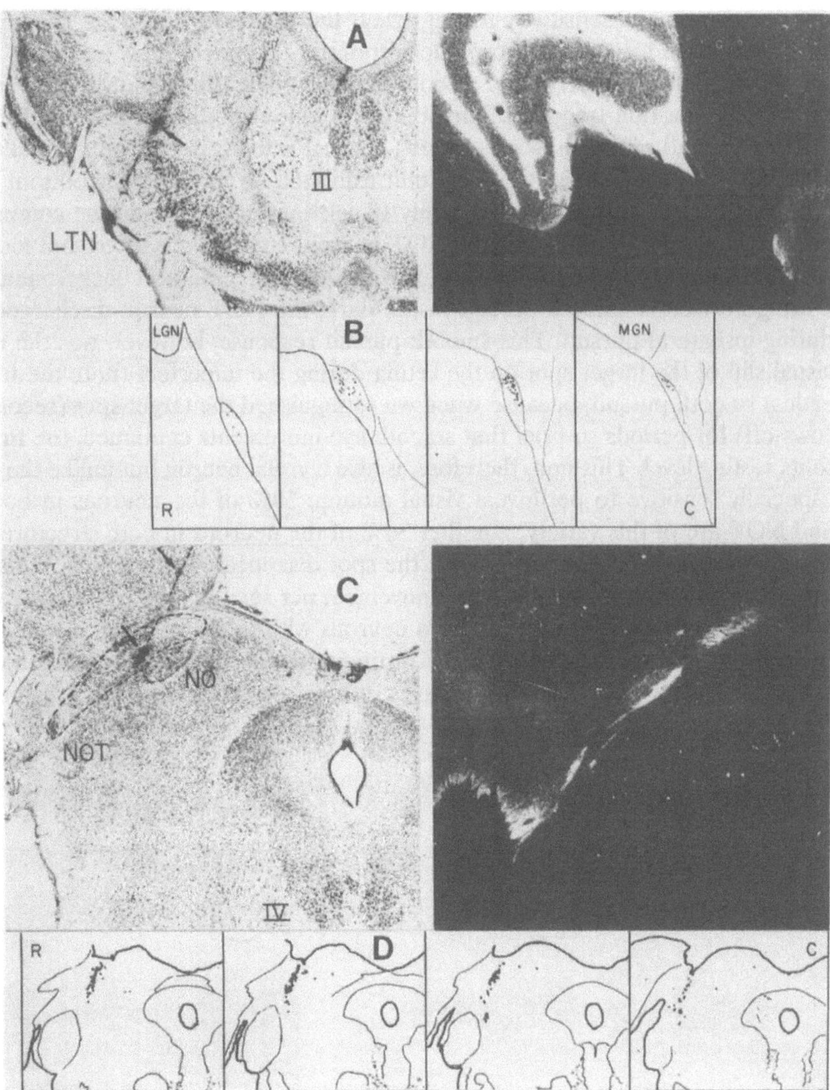

Fig. 1. Anatomical demonstration of a retinal-NOT/LTN-IO pathway in the monkey. The pattern of retinal termination following unilateral intraocular tritiated amino acid injections is shown in dark field for the LTN and pretectum (A,C; right). The locations of retrogradely labeled cells after HRP injections in the vicinity of the dorsal cap of Kooy in the IO are shown in several sections through the LTN (B) and pretectum (D); each dot represents a labeled cell. Examples of two recording tracks through the LTN and NOT are shown on the left side of A and C, respectively, with the deeper of two marking lesions (arrows) placed at the sites of typical units. NOT, nucleus of the optic tract; NO, pretectal olivary nucleus; LGN, lateral geniculate nucleus; MGN, medial geniculate nucleus; III, oculomotor complex; IV, trochlear nucleus

To determine how the LTN and NOT might participate in the generation of OKN, we recorded the activity of their neurons (a) during visual sensitivity testing when the monkeys fixated a stationary target spot while a large-field visual background was moved in a variety of directions and speeds to simulate optokinetic stimulation and (b) during eye-movement sensitivity testing when the monkeys tracked a spot moving on a dark background. Figure 2 shows the response of a typical LTN and NOT neuron during background movement and smooth pursuit. While the monkey fixated (note the flat vertical eye-position trace, top left), the LTN unit exhibited an increased firing rate as the background moved upward. During vertical smooth pursuit of a small target in an otherwise darkened room, the LTN unit exhibited no modulation (bottom left). These data suggest that this neuron is purely visual; purely visual neurons comprise 30% of both LTN and NOT units. For the NOT neuron, large-field background movement during fixation caused a sporadic but clear response for ipsilateral background movement. During horizontal smooth pursuit in the dark, the NOT neuron discharged vigorously during ipsilateral pursuit. This smooth-pursuit response, however, was the result of the visual slip of the target spot on the retina during the imperfect (note the frequent saccades) smooth pursuit, because when we extinguished the target spot (second cycle, arrows-off) for periods so brief that smooth eye movements continued, the firing rate fell to its resting level. This unit, therefore, is also a visual neuron but unlike the LTN unit is especially sensitive to perifoveal visual motion; 30% of the neurons in both the LTN and NOT are of this variety. Another 30% of the neurons in both structures continued their smooth-pursuit discharge when the spot disappeared, suggesting that they have a discharge component related to eye movement per se.

Although the LTN and NOT have neurons with the same types of visual responses, neurons in the two structures can be distinguished by both their direction and velocity sensitivities. The directional preferences of LTN and NOT neurons are summarized in

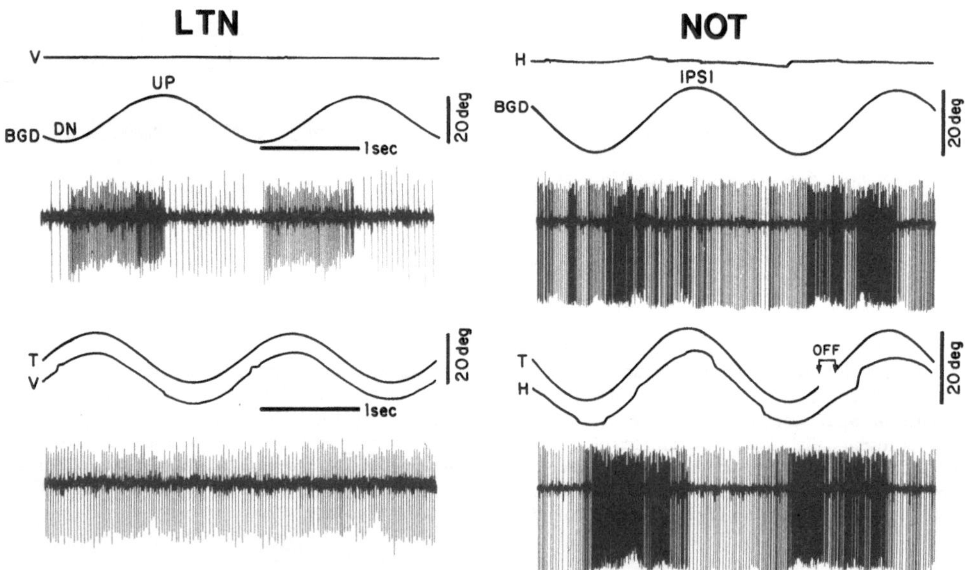

Fig. 2. Discharge of a representative LTN (left) and NOT (right) unit during large-field visual stimulation (top) and during smooth pursuit (bottom). Target (T) and background (BGD) movement is vertical for the LTN and horizontal for the NOT unit. At the lower right, the target spot is turned off briefly (arrows) during smooth pursuit. Vertical eye, horizontal eye, background and target positions signals are indicated by V, H, BGD and T

Fig. 3. The 10-cycle average discharge patterns for a typical LTN unit are shown by the solid histograms in Fig. 3A. The discharge was elevated above the resting rate for movement in the "on" (UP) direction and was driven below the resting level for "off" (down, DN) direction movement. The calculated preferred direction is indicated by the arrow, but the tuning for this and other LTN units was rather broad. All our LTN units preferred mainly upward visual motion (Fig. 3B); in contrast, NOT units preferred mainly ipsilaterally directed horizontal visual motion, e.g. units in the right NOT preferred rightward visual motion (Fig. 3C).

The velocity sensitivity of LTN and NOT neurons was tested by moving a large-field visual background in their preferred directions (Fig. 4). LTN neurons typically had low thresholds (i.e., <2°/s) and most reached their maximum discharge rates for speeds less than 13°/s. For example, the single LTN unit illustrated in Fig. 4 already was deeply modulated at 2°/s and reached its maximum firing rate at 4°/s, after which the response

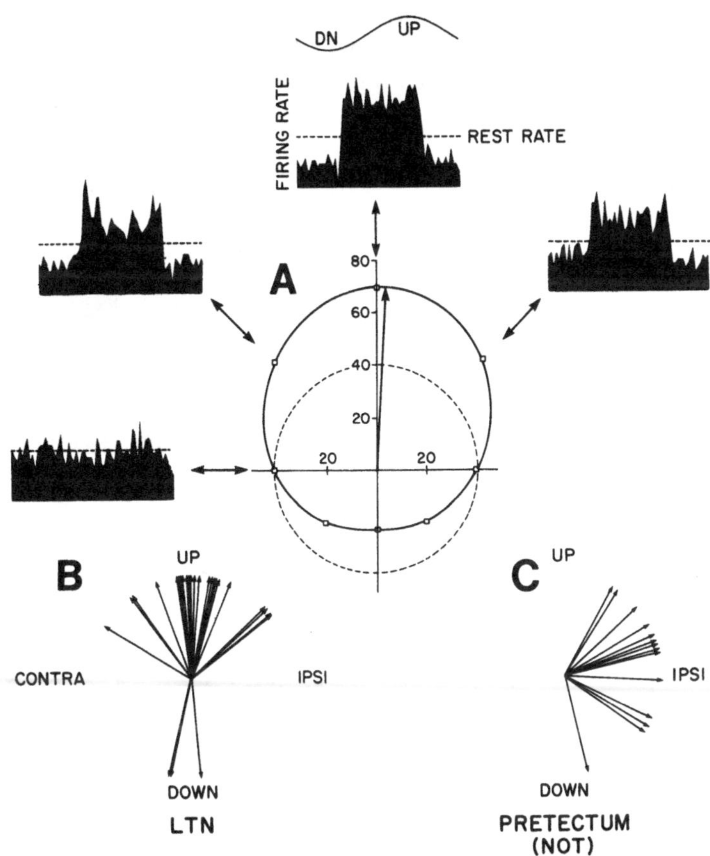

Fig. 3. Directional tuning during large-field visual stimulation for units in the LTN and NOT. A, an example of an LTN directional tuning curve. The average firing rate data (solid histograms) at each tested direction (open squares on plot) are fit with a smooth solid curve; the dashed circle represents the resting rate. The numbers on the axes are in spikes/s. The calculated preferred direction for the unit is shown by the arrow. The calculated preferred directions obtained from similar tuning curves are shown by arrows for the entire population of LTN neurons in B and for NOT neurons in C

modulation declined with increasing stimulus velocity; these data (solid triangles) indicate that such cells are tuned for low velocities. About half of the LTN units exhibited similar velocity tuning. The other LTN units (open symbol plots) had similarly low velocity thresholds and also reached their peak firing for low velocities, but then continued to fire more or less at their maximum rates over the entire range of velocities tested (0°-80°/s). The relation of firing rate and background velocity for representative NOT units is shown in the bottom portion of Fig. 4. In contrast to LTN neurons, NOT units showed a monotonic increase in firing rate over a wide range of stimulus velocities. For example, the NOT unit illustrated (histograms) showed a steady increase in firing rate up to 80°/s; at still higher velocities (solid squares at left), the response saturated. Therefore, NOT units have low velocity thresholds similar to those of LTN units, but most are capable of reliably encoding stimulus velocity over a much greater range.

These data suggest that NOT neurons convey information about both the direction and velocity of large-field background movement and thus seem ideally suited to be part of the afferent limb of the horizontal optokinetic response. In contrast, LTN neurons seem capable of encoding stimulus velocity over a much lower range of velocities. Since vertical OKN, like horizontal OKN, operates to velocities of at least 60°/s (Matsuo and Cohen 1984), LTN units seem unsuitable to support the vertical optokine-

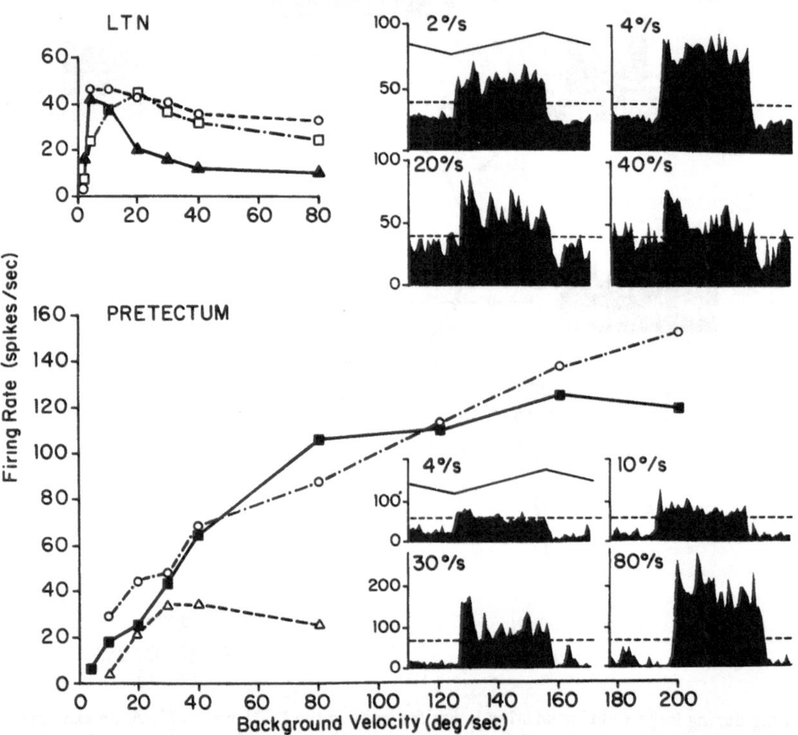

Fig. 4. Firing rate versus visual background velocity for representative LTN (top) and NOT (bottom) units. Solid histograms (right) show examples of averaged firing rates for the indicated velocities. The firing rate above resting level for the two illustrated units is plotted as solid data points in graphs to the left together with four other representative units plotted with open symbols

tic response. At least if the LTN does participate in vertical OKN, it must play a significantly different role than does the NOT in generating horizontal OKN.

Further data supporting a role for the NOT in the generation of OKN can be obtained by electrical stimulation (see also Schiff et al. 1988). Figure 5 shows that application of weak (40 μA; 300 Hz) electrical stimulation for 1 min at the site of NOT units (i.e., the deeper marking lesion in Fig. 1C) caused a gradual buildup of nystagmus, which gradually declined after the stimulus was turned off. The velocity record indicates that the nystagmus built up gradually over 10 s and declined over 25 s, a time course that closely resembles the OKN and optokinetic after-nystagmus patterns obtained if a monkey is subjected to a striped rotating cylinder (Cohen et al. 1977). However, unlike the OKN generated by cylinder rotation, the nystagmus generated by electrical stimulation does not exhibit the initial rapid rise in smooth eye velocity that is often attributed to smooth pursuit. Therefore, the nystagmus elicited by NOT stimulation seems to be a pure optokinetic response. Curiously, electrical stimulation applied at the site of typical LTN neurons, at even higher currents, produced no eye movements. Such negative results cannot be viewed as conclusive, however, until additional sites and perhaps bilateral stimulations, as suggested by Dr. Kornhuber at this symposium, have been tried.

Summary

Recordings of single-unit activity in the pretectal nucleus of the optic tract (NOT) and the lateral terminal nucleus (LTN) of the accessory optic system in trained monkeys have revealed the visual and/or eye-movement sensitivity of neurons in these structures.

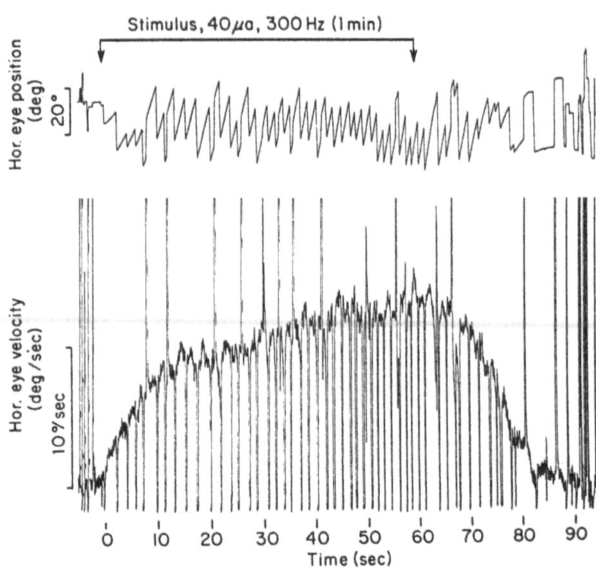

Fig. 5. Example of the nystagmus elicited by electrical stimulation of the NOT (between arrows, top) delivered at the site of the deeper arrow shown in Fig. 1C. Eye position is shown in the top panel and eye velocity in the bottom panel.

The NOT and LTN had similar types of units that discharged either to large-field background movement, during smooth pursuit in the dark, or in both behavioral situations. However, LTN neurons preferred visual stimulus motion and/or eye movements in the vertical direction whereas NOT neurons preferred horizontal movements. Furthermore, LTN neurons responded only in relation to very low velocities whereas NOT neurons exhibited a monotonic increase in firing rate with stimulus velocity over the entire range of horizontal optokinetic nystagmus (OKN). In addition, unilateral electrical stimulation at the site of NOT neurons elicited horizontal OKN whereas similar stimulation in the LTN elicited no eye movements. These data suggest that if the LTN is involved in the generation of vertical nystagmus at all, it plays a different role than the NOT does for horizontal OKN.

Acknowledgements: This study was supported by NIH grants EY00745, EY0609, and RR00166.

References

Cohen B, Matsuo V, Raphan T (1977) Quantitative analysis of the velocity characteristics of optokinetic nystagmus and optokinetic after-nystagmus. J Physiol (Lond) 270: 321-344

Collewijn H (1975) Direction-selective units in the rabbit's nucleus of the optic tract. Brain Res 100: 489-508

Cooper HM, Magnin M (1986) A common mammalian plan of accessory optic system organization revealed in all primates. Nature 324: 457-459

Cowan WM, Gottlieb DI, Hendrickson AE, Price JL, Woolsey TA (1972) The autoradiographic demonstration of axonal connections in the CNS. Brain Res 37: 21-51

Grasse KL, Cynader MS (1982) Electrophysiology of the medial terminal nucleus of the accessory optic system in the cat. J Neurophysiol 48: 490-504

Grasse KL, Cynader MS (1984) Electrophysiology of the lateral and dorsal terminal nuclei of the cat accessory optic system. J Neurophysiol 51: 276-293

Hoffmann KP, Schoppmann A (1981) A quantitative analysis of the directionspecific response of neurons in the cat's nucleus of the optic tract. Exp Brain Res 51: 236-246

Hoffmann KP, Distler C (1986) The role of direction selective cells in the nucleus of the optic tract of cat and monkey during optokinetic nystagmus. Adv Biosci 57: 261-266

Lisberger SG, Fuchs AF (1978) Role of the flocculus in rapid behavioral modification of the vestibuloocular reflex. I. Purkinje cell activity during visually guided horizontal smooth-pursuit eye movements and passive head rotation. J Neurophysiol 41: 733-763

Maekawa K, Simpson JI (1973) Climbing fiber responses evoked in the vestibulocerebellum of the rabbit from the visual system. J Neurophysiol 36: 649-666

Matsuo V, Cohen B (1984) Vertical optokinetic nystagmus and vestibular nystagmus in the monkey: up-down asymmetry and effects of gravity. Exp Brain Res 53: 197-216

Mesulum MM (1978) TMβ for HRP neurochemistry: A non-carcinogenic blue reaction product with superior sensitivity for visualizing neural afferents and efferents. J Histochem Cytochem 26: 106-117

Mustari MJ, Fuchs AF (1988) The response properties of units in the pretectal nucleus of the optic tract in the behaving monkey. Soc Neurosci 14: 955

Mustari MJ, Fuchs AF, Wallman J, Langer TP, Kaneko CRS (1986)Visual and oculomotor response properties of single units in the lateral terminal nucleus (LTN) of the behaving rhesus monkey. Soc Neurosci 12: 459

Mustari MJ, Fuchs AF, Wallmann J (1988) Smooth-pursuit-related units in the dorsolateral pons of the rhesus macaque: J Neurophysiol 60: 664-686

Schiff D, Cohen B, Raphan T (1988) Stimulation of the nucleus of the optic tract (NOT) induces nystagmus in the monkey. Exp Brain Res 70: 1-14

Simpson JI (1984) The accessory optic system. Ann Rev Neurosci 7: 13-41

Simpson JI, Soodak RE, Hess R (1979) The accessory optic system and its relationship to the vestibulo-cerebellum. In: Progr Brain Res 50: 715-724

Waespe W, Rudinger D, Wolfensberger M (1985) Purkinje cell activity in the flocculus of vestibular neurectomized and normal monkeys during optokinetic nystagmus (OKN) and smooth pursuit eye movements. Exp Brain Res 60: 243-262

Wallman J, Velez J (1985) Directional asymmetries of optokinetic nystagmus: developmental changes and relation to the accessory optic system and the vestibular system. J Neurosci 5: 317-329

Weber JT, Giolli RA (1986) The medial terminal nucleus of the monkey: evidence for a complete accessory optic system. Brain Res 365: 164-168

Zee DS, Yamazaki A, Butler PH, Gucer G (1981) Effects of ablation of flocculus and paraflocculus on eye movements in primate. J Neurophysiol 46: 878-899

Excitatory and Inhibitory Mechanisms Involved in the Dynamic Control of Posture During the Vestibulospinal Reflexes

O.Pompeiano

Dipartimento di Fisiologia e Biochimica, Università di Pisa, Via S.Zeno 31, 56100 Pisa, Italy

Introduction

In decerebrate cats, the vestibulospinal (VS) reflexes elicited by slow rotation about the longitudinal axis of the animal, leading to sinusoidal stimulation of macular, utricular receptors, are characterized by contraction of limb extensors during ipsilateral (side-down) tilt of the animal and relaxation during contralateral (side-up) tilt (Schor and Miller 1981; Manzoni et al. 1983a). These postural changes were originally attributed to the activity of the three-neuronal VS reflex arc, characterized by primary vestibular afferents, second-order VS neurons originating from the lateral vestibular nucleus (LVN), which exert a direct excitatory influence on ipsilateral limb extensor motoneurons (Lund and Pompeiano 1968) and their spinal motoneurons. In fact, most of the LVN neurons (Boyle and Pompeiano 1980; Schor and Miller 1982), including those projecting to the lumbosacral segments of the spinal cord (Marchand et al. 1987), responded to the positional signal during slow rotation of the animal with a predominant response pattern characterized by an increased discharge during side-down tilt and a decreased discharge during side-up tilt (α-response). Surprisingly, in spite of the good decerebrate rigidity, the gain of the VS reflexes was very low in forelimb extensors (Manzoni et al. 1983a) and almost negligible or absent in hindlimb (Boyle and Pompeiano 1984; Manzoni et al. 1984; Pompeiano et al. 1985b). In these instances the activity of the extensor motoneurons produced by the excitatory VS volleys could, at least in part, be limited by the simultaneous discharge of Renshaw (R) cells driven by the recurrent collaterals of the corresponding motoneurons.

Since these inhibitory interneurons can be influenced not only by the corresponding motoneurons via the recurrent collaterals but also by supraspinal sources acting directly on them (Hultborn et al. 1979; cf. Pompeiano 1984), we have postulated that in addition to the lateral VS tract, which acts directly on ipsilateral limb extensor motoneurons, there are other descending pathways which may modify the response gain of these motoneurons to the excitatory VS volleys by acting on R cells anatomically coupled with them.

The pathways which can be involved in the gain regulation of the VS reflexes are the coeruleospinal (CS) tract, which originates from norepinephrine (NE)-containing locus coeruleus (LC)-complex neurons, and the reticulospinal (RS) tract, originating from the inhibitory area of the medullary reticular formation (mRF).

Stimulation experiments have in fact shown that the CS pathway exerts a facilitatory influence on ipsilateral limb extensor (and flexor) motoneurons (Fung and Barnes

1987), an effect which has been attributed, at least in part, to inhibition of the R cells anatomically linked with them (Fung et al. 1987a). This is in agreement with early findings showing that R cells activity is suppressed by iontophoretically applied NE (Biscoe and Curtis 1966; Engberg and Ryall 1966; Weight and Salmoiraghi 1966), the putative transmitter of CS synapses. On the other hand, the medullary RS pathway exerts an inhibitory influence on ipsilateral limb extensor (and flexor) motoneurons (Jankowska et al. 1968) by exciting, at least in part, the corresponding R cells (see Pompeiano 1984 for references).

Responses of Pontine Coeruleospinal and Medullary Reticulospinal Neurons to Labyrinth Stimulation

In precollicular decerebrate cats we recorded the activity of single units located either in the LC-complex, which includes both the LC and the subcoeruleus (SC) (Jones and Friedman 1983), or in the medial inhibitory aspect of the mRF (Jankowska et al. 1968).

1. Responses of Locus Coeruleus-Complex Neurons.

Pompeiano et al. (1988) have recorded the activity of 141 neurons from the dorsolateral pontine tegmentum. This is the region where CS neurons are located, as shown by the retrograde cell marker horseradish peroxidase (Holstege and Kuypers 1987). Moreover, a large number of these spinally projecting neurons are noradrenergic in nature, as shown by immunocytochemical staining (Westlund and Coulter 1980; Westlund et al. 1983, 1984). In agreement with the results of previous studies (Foote et al. 1983), a large number of recorded units had the physiological characteristics attributed to the NE-containing LC-complex neurons, namely: (a) a prolonged extracellular spike wave duration ($\geq$1.5 ms); (b) a very slow and regular resting discharge; and (c) a response to a pinch stimulus applied to the ipsilateral paws characterized by a transient excitation followed by a prolonged inhibition. This delayed component of the response has been partially attributed to the fact that the LC neurons are not only noradrenergic but also NE-sensitive, due to the existence of self-inhibitory synapses which act on α_2-adrenoceptors by utilizing mechanisms of recurrent and/or lateral inhibition (Ennis and Aston-Jones 1986a). In addition, 16 out of the recorded LC-complex neurons were activated antidromically from the ipsilateral spinal cord at T12-L1, thus projecting to the lumbosacral segments of the spinal cord.

Among the 141 neurons located in the LC-complex, 80 (56.3%) responded to roll tilt of the animal at 0.15 Hz, $\pm$10°. In these instances the average gain of the first harmonic of responses corresponded to 0.19$\pm$0.26, S.D., imp./s/deg. Most of these units responded to the extreme animal displacements, thus being attributed to stimulation of macular, utricular receptors (Fig. 1B). In particular, 45 units (56.2%) were excited during side-up tilt (β-responses), while 20 units (25.0%) by side-down tilt of the animal (α-responses). Both populations of neurons fired with an average phase lead of the first harmonic responses of +17.9$\pm$28.9, S.D., deg with respect to the extreme animal displacements; moreover, the former units showed more than a twofold larger gain

(0.22±0.24, S.D., imp./s/deg) with respect to the latter units (0.09±0.11, S.D., imp./s/deg). The remaining 15 units (18.7%) showed intermediate phase angle of the responses. Similar response properties were also found for 11 out of the 16 antidromically identified CS neurons which responded to animal tilt.

2. Responses of Medullary Reticular Formation Neurons.

The activity of 168 neurons was recorded from the inhibitory aspect of the mRF, particularly from the nucleus reticularis magnocellularis, gigantocellularis, and ventralis (Manzoni et al. 1983c). Among these neurons, 93 were RS units activated antidromically from the ipsilateral spinal cord at T12-L1, thus projecting to the lumbosacral segments of the spinal cord. From the 168 recorded mRF units, 113 (67.3%) were affected by roll tilt of the animal at 0.026 Hz, ±10°. The gain of the first harmonic responses corresponded on the average to 0.32±0.36,S.D., imp./sec/deg. As shown for the LC-complex units, most of these reticular units responded to the extreme animal displacements, thus being attributed to stimulation of macular, utricular receptors (Fig. 1C). In particular, 71 units (62.8%) increased their discharge rate during side-up tilt (β-responses), while 24 units (21.3%) were excited by side-down tilt of the animal (α-responses). Both populations of neurons fired with an average phase lead of the first harmonic responses of +25.3±28.3, S.D., deg with respect to the extreme side-up or side-down animal displacement; moreover, the former units showed about a twofold larger gain (0.33±0.34, S.D., imp./s/deg) with respect to the latter units (0.18±0.16, S.D., imp./s/deg). The remaining 18 units (15.9%) showed intermediate phase angle of the responses. There were no great differences between the antidromic and the nonantidromic mRF neurons with respect to either the proportions of responsive units or the characteristics of unit responses to standard parameters of labyrinth stimulation.

3. Comparison with the Responses of Vestibulospinal Neurons.

The observation that most of the LC-complex neurons as well as of the mRF neurons, including those projecting to the lumbosacral segments of the spinal cord, were excited during side-up tilt (β-responses) differs from that obtained from unidentified LVN neurons (Boyle and Pompeiano 1980; Schor and Miller 1982) as well as from the identified LVN neurons projecting to the same segments of the spinal cord (Marchand et al. 1987). Among the whole population of 129 lateral VS units, 76 (58.9%) were affected by roll tilt of the animal at 0.026 Hz, ±10°. The gain of the first harmonic responses corresponded on the average to 0.47+0.44, S.D., imp./s/deg. Most of the units responded to the extreme animal displacements (Fig.1A). However, 51 units (67.1%) were excited during side-down tilt of the animal (α-responses) while 15 units (19.7%) by side-up tilt (β-responses). Both populations of units fired with an average phase lead of the first harmonic responses of +21.0±27.2, S.D., deg with respect to the extreme animal displacements; moreover, the former units showed, on the average, a larger gain (0.57±0.49, S.D., imp./s/deg) than the latter (0.26±0.19, S.D., imp./s/deg). The remaining 10 units (13.2%) showed intermediate phase angle of the responses.

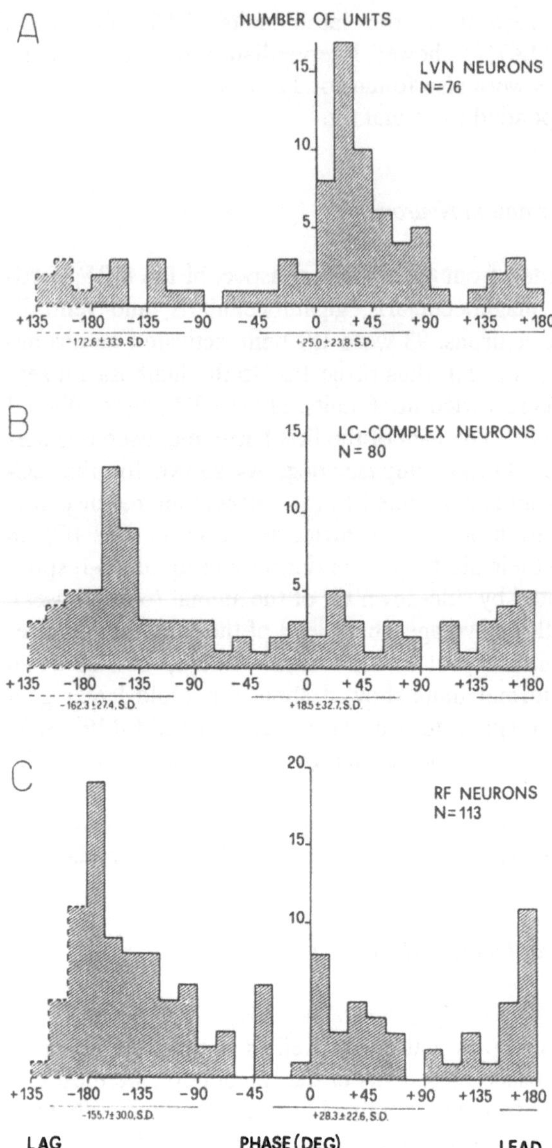

Fig. 1. Distribution of the phase angle of the first harmonic responses of different populations of neurons tested during roll tilt of the animal at 0.026 Hz (A,C) and 0.15 Hz (B), ±10°. All the experiments were performed in precollicular decerebrate cats. The upper histogram (A) illustrates the distribution of the phase angle of responses to animal tilt of 76 LVN neurons, antidromically activated by stimulation of the spinal cord at T12-L1, thus projecting to the lumbosacral segments of the spinal cord; the middle histogram (B) illustrates the response of 80 LC-complex neurons, 11 of which projecting to the lumbosacral segments of the spinal cord, while the lower histogram (C) illustrates the responses of 113 medullary RF neurons, 64 of which projecting to the lumbosacral segments of the spinal cord. Positive numbers in the abscissas indicate in degrees the phase lead, whereas negative numbers indicate the phase lag of responses with respect to the extreme side-down position of the animal, as indicated by 0°. Responses of the neurons to tilt, underlined by horizontal bars, were used to evaluate the average phase angle of units excited during or near the side-down (0°) or side-up displacement of the animal (180°). Most of the LVN neurons (51/76) were excited during side-down tilt of the animal, while most of the LC-complex (45/80) and the medullary RF (71/113) neurons were excited during side-up tilt. (A, From Marchand et al. 1987; B, Pompeiano et al. 1988; C, Manzoni et al. 1983c)

The responses of the LC-complex neurons, as well as of the mRF neurons can be attributed to direct projections from the vestibular nuclei to these pontine (Fung et al. 1987b; cf. Aston-Jones et al. 1986) and medullary structures (Ladpli and Brodal 1968). An alternative possibility, however, is that the macular input of one side is transmitted not only to the ipsilateral LVN but also, via a crossed pathway, to the ventral aspect of the contralateral mRF (Pompeiano 1979). In this region there are, in fact, not only inhibitory RS neurons (see Manzoni et al. 1983c for references) but also neurons which project to the LC-complex (Aston-Jones et al. 1986), where they exert a prominent excitatory influence on NE-containing neurons (Ennis and Aston-Jones 1986b, 1987). The labyrinthine volleys originating from macular receptors of one side during side-down tilt would then be responsible not only for the α-responses of ipsilateral VS neurons but also for the β-responses of CS as well as of medullary inhibitory RS neurons originating from the corresponding structures of the contralateral side.

Reciprocal Discharges of Locus Coeruleus-Complex Neurons and Medullary Reticular Formation Neurons and Their Influences on the Gain of Vestibulospinal Reflexes

The observation that both the CS neurons and the presumably inhibitory RS neurons display the same response pattern to tilt, in spite of their opposite influences on ipsilateral limb extensor motoneurons and the related R cells (see "Introduction"), can be understood only if we consider that from time to time the activity of the CS neurons predominates over that of the medullary RS neurons or vice versa. This conclusion is supported by the fact that, in addition to the direct CS projection, the noradrenergic and NE-sensitive LC neurons send afferents to the dorsal pontine reticular formation (pRF; Scheibel and Scheibel 1973; Sakai et al. 1977; Jones and Yang 1985), as proved also by the presence of NE-containing varicosities in this region (Jones and Friedman 1983), where they exert a tonic inhibitory influence (Sakai et al. 1981; Hobson and Steriade 1986). On the other hand, the dorsal pRF, which contains presumably cholinergic and self-excitatory cholinoceptive neurons (Pompeiano 1980; Vivaldi et al. 1980), as shown also by histochemical (Kimura and Maeda 1982) and electrophysiological studies (Shiromani and McGinty 1986), exerts via the lateral tegmentoreticular tract a tonic excitatory influence on the medullary inhibitory RS neurons (Sakai et al. 1979; Jones and Yang 1985; see also Sakai et al. 1981; Hobson and Steriade 1986). This connection between noradrenergic and presumably cholinergic pontine neurons is illustrated schematically in Fig. 2.

There is now evidence that in intact animals during waking as well as in precollicular decerebrate animals, the postural activity is present in so far as the LC-complex neurons show a regular discharge. This discharge acts not only directly through the CS projection on the spinal cord but also indirectly by inhibiting both the dorsal pRF neurons and the related medullary inhibitory RS neurons, as shown by their low resting discharge both in intact animals during waking (Hobson et al. 1974; Kanamori et al. 1980), and in decerebrate animals (Hoshino and Pompeiano 1976; Manzoni et al. 1983c). However, as soon as the discharge of the LC neurons decreases, as shown during desynchronized sleep in intact animals (Sakai et al. 1981; Hobson and Steriade 1986) or after systemic injection of an anticholinesterase in decerebrate cats (Pompeiano and Hoshino 1976; cf. Pompeiano 1980), the activity of the dorsal pRF

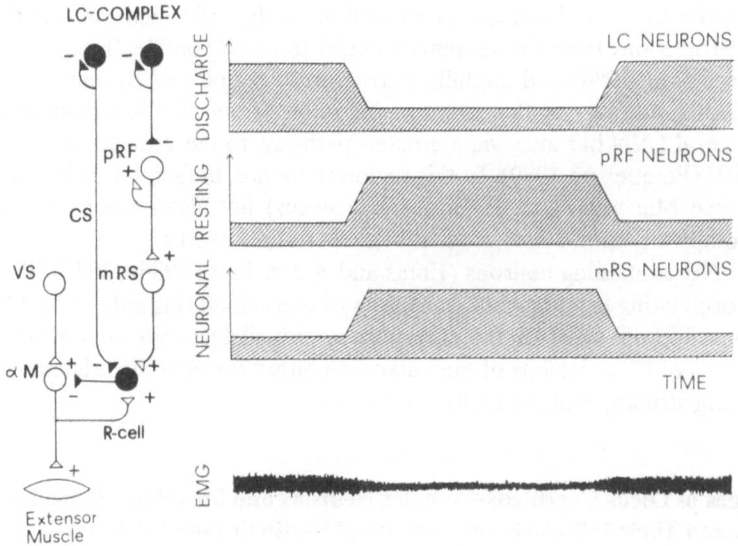

Fig. 2. Scheme illustrating the reciprocal changes in firing rate of the LC-complex neurons and the dorsal pRF neurons leading to changes in tonic activity of ipsilateral limb extensor muscles. Left side: inhibitory or excitatory neurons are indicated by filled or empty symbols, respectively. The NE-containing LC-complex neurons, which are self-inhibitory, also inhibit, through the CS projection, the R cells coupled with ipsilateral limb extensor motoneurons (αM) as well as the presumably cholinergic pRF neurons. On the other hand, the latter neurons, which are self-excitatory, also excite the medullary RS neurons and the related R cells. The VS neurons do not activate these R cells directly, but only through ipsilateral limb extensor motoneurons (αM) and their recurrent collaterals. Right side: a reduced discharge of LC neurons produces both disinhibition and facilitation of the R cells by utilizing the CS and the mRS projections, respectively; on the other hand, the increased discharge of the R cells reduces the activity of ipsilateral limb extensor motoneurons

(Hobson et al. 1974; Hoshino and Pompeiano 1976) and the related inhibitory RS neurons (Kanamori et al. 1980; Srivastava et al. 1982) increases, thus leading to a decrease or suppression in postural activity (Pompeiano 1967, 1980; Fig. 2). We postulate that in the first instance the firing rate of R cells linked with ipsilateral limb extensor motoneurons would decrease, due to both a sustained discharge of CS neurons which are inhibitory on them and a reduced discharge of the medullary RS neurons which are excitatory on them; on the other hand, just the opposite result would occur in the second instance.

In addition to these changes in posture, the neuronal system described above may also play an important role in the control of the VS reflexes. In particular, if the activity of the CS neurons predominates over that of the dorsal pRF neurons and the related RS neurons, the motoneurons innervating the ipsilateral limb extensors would be excited by an increased discharge of VS neurons during side-down tilt, but this effect would be, at least in part, attenuated by the reduced discharge of CS neurons leading to disinhibition of the related R cells. In this instance the R cells would be more prominently coupled with the ipsilateral limb extensor motoneurons, a finding which should reduce the response gain of the corresponding limb extensors to labyrinth stimulation. However, if the activity of the dorsal pRF neurons and the related RS neurons is more prominent than that of the CS neurons, for the same direction of animal orientation the response of the ipsilateral limb extensor motoneurons to the excitatory VS volleys would be enhanced by the reduced discharge of the medullary RS neurons, leading to disfacilitation of the related R cells. In this instance the R cells would be decoupled from the ipsilateral limb extensor motoneurons, a finding which should increase the response gain of the corresponding limb extensors to labyrinth stimulation.

The following experiments were performed to find out whether local injection in dorsal pontine structures of substances, which lead either to inactivation of the noradrenergic and NE-sensitive LC neurons or to activation of presumably cholinergic and cholinosensitive pRF neurons could not only decrease the postural activity but also increase the response gain of limb extensors to sinusoidal stimulation of labyrinth receptors.

1. Inactivation of NE-Sensitive Neurons Located in the Locus Coeruleus-Complex

In precollicular decerebrate cats, we (Pompeiano et al. 1987) studied the changes in posture as well as in VS reflexes produced by local injection into the LC-complex of the α_2-adrenergic agonist clonidine, which inhibits the activity of the corresponding noradrenergic and NE-sensitive neurons (Svensson et al. 1975). In particular, injection into the LC-complex of one side of 0.25 μl of clonidine solution at the concentration of 0.012-0.15 μg/μl of saline (marked with 5% pontamine) decreased the postural activity in the ipsilateral limbs but greatly increased the gain of the averaged multiunit EMG responses of the triceps brachii to animal tilt at 0.15 Hz, ± 10°; however, only slight changes in the phase angle of the responses were observed. The increased gain of the VS reflexes affected not only the ipsilateral but also the contralateral triceps brachii. The effects described above were first observed 10-15 min after injection of clonidine, reached the highest value in 30-60 min, and persisted for more than 2 h after the injection.

The postural and reflex changes produced by clonidine depended on the dose of injected substance. Moreover, histological controls demonstrated that the effective area included the LC and the SC as well as the dorsal pRF located immediately ventral to the LC-complex. This finding can easily be explained, since in the cat noradrenergic and NE-sensitive neurons are not precisely concentrated in the LC and the SC. Moreover, clonidine may act on α_2-receptors located not only on the cell body but also on terminals of NE-containing neurons.

Inactivation of the LC-complex neurons following clonidine injection not only suppressed the discharge of the CS neurons, but also released from inhibition the activity of the cholinoceptive pRF neurons and the related medullary inhibitory RS neurons. In fact, the postural and reflex changes following clonidine injection into the dorsal aspect of the pRF were greatly reduced or suppressed by local administration into the same pontine region of 0.25 μl of the cholinergic blocker atropine sulfate at the concentration of 6 μg/μl. This last effect was strictly ipsilateral.

2. Activation of Cholinosensitive Neurons Located in the Dorsal Pontine Reticular Formation.

In decerebrate cats, Barnes et al. (1987) have shown that local injection into the dorsal aspect of the pRF of 0.1-0.2 μl of a cholinergic agonist at the concentration of 0.01-0.2 μg/μl, while decreasing the decerebrate rigidity in the ipsilateral limbs, greatly increased the gain of the multiunit EMG responses of the ipsilateral triceps brachii to animal tilt at 0.15 Hz, ± 10°. However, only slight changes in the phase angle of the responses were observed. Similar results were obtained by injecting either carbachol, which is a mixed muscarinic and nicotinic agonist, or bethanechol, which is a pure muscarinic agonist. In both instances the changes in posture as well as in response gain appeared a few minutes after injection of the cholinergic agonist and persisted after a partial decline up to 3 h following the injection. Local administration of 0.25 μl of a solution of the muscarinic blocker atropine sulfate at the concentration of 6 μg/μl, not only produced a recovery in the postural activity of the ipsilateral limbs but also suppressed the increased gain of the responses produced by the cholinergic substances, returning them to the control values.

The effects described above were dose dependent. Moreover, the structure which was critically responsible for the postural and reflex changes described above was located in the dorsal pontine tegmentum immediately ventral to the LC. This area corresponded to the peri-LC_{α} and the surrounding pRF, from which a tegmentoreticular tract ending in the medullary inhibitory area originates (Sakai et al. 1979; Jones and Yang 1985).

It is of interest that effects similar to those elicited by local injection of cholinergic agonists into the dorsal pRF were also obtained after systemic injection in decerebrate cats of the anticholinesterase eserine sulfate at a dose (0.05-0.10 mg/kg, i.v.) lower than that which abolished the decerebrate rigidity (Pompeiano et al. 1983). In this instance, in which there was only a slight decrease in decerebrate rigidity, either a distinct modulation of the multiunit activity of extensor muscles occurred during animal tilt if the EMG response was absent prior to the injection, as for the triceps surae (Boyle and Pompeiano 1984; Manzoni et al. 1984; Pompeiano et al. 1985b), or else the gain of the EMG response to animal tilt greatly increased if the labyrinth input was only weakly effective, as for the triceps brachii (Pompeiano et al. 1983); however, no significant changes in the phase angles of the responses were observed (Fig. 3). The increase in response gain induced by the anticholinesterase started 5-10 min after injection, reached the peak in about 1 h, and then slowly declined. Interestingly, somatosensory or acoustic stimulations produced both a transient recovery of postural activity in the four limbs and a decrease or suppression of the EMG modulation of limb extensors during animal tilt (Pompeiano et al. 1983, 1985b), which contrast with the great stability of the effects induced by local injection of cholinergic agonists into the dorsal pRF.

Electrolytic lesions limited to the dorsal aspect of the pRF not only decreased the response gain of the triceps brachii to labyrinth stimulation in normal decerebrate cats, but also prevented this gain from increasing after eserine injection (D'Ascanio et al. 1985). These findings indicated that the cholinoceptive neurons activated by the anticholinesterase were located in the dorsal pRF, thus in agreement with the results obtained by local injection of cholinergic agonists into this pontine region (Barnes et al. 1987). Indeed, injection into the dorsal pRF of minute doses of eserine sulfate produ-

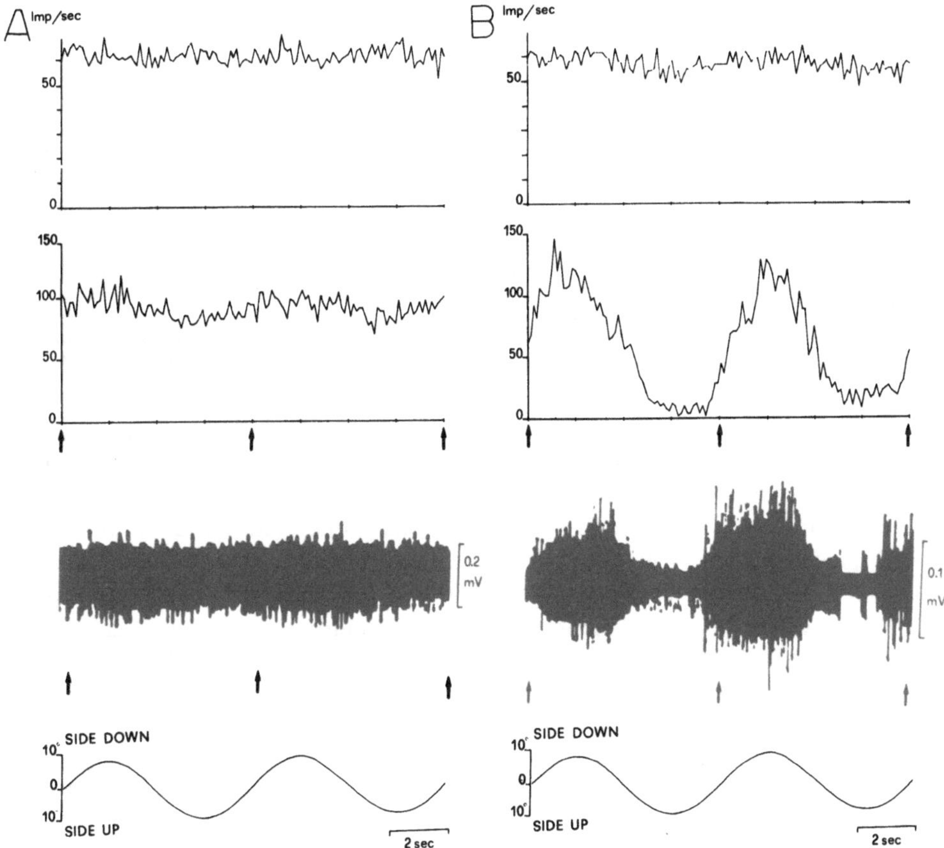

Fig. 3. Increase in response gain of limb extensors to animal tilt after injection of an anticholinesterase. Precollicular decerebrate cats. Upper and middle records: sequential pulse density histograms (SPDHs) showing the averaged multiunit EMG responses of the left triceps surae (upper records) and the left triceps brachii (middle records) to roll tilt of the animal at 0.15 Hz, ±10° (average of 7-12 sweeps, using 128 bins with 0.1 s bin width). Lower records: individual EMG traces contributing to the averaged responses illustrated in the middle records. Animal displacement is indicated below the EMG traces. The responses were recorded before (A) and after (B) individual or repetitive i.v. injection of 0.05-0.10 mg/kg eserine sulfate. Before injection of the anticholineserase no modulation (A, upper record) or a slight modulation (A,middle record) of the EMG activity was observed. Injection of the anticholinesterase, which produced only a slight reduction in the spontaneous EMG activity of the limb extensors, brought to the light (B, upper record) or greatly increased (B, middle record) the amplitude of the EMG modulation to animal tilt. In particular, the mean gain of the first harmonic response of the triceps surae increased from 0 to 0.41 imp./s/deg, while that of the triceps brachii increased from 0.79 to 5.41 imp./s/deg. (From Pompeiano et al. 1983, 1985b)

ced results similar to those induced by carbachol or bethanechol (unpublished observations).

It is of interest that the increase in gain of the VS reflexes occurred only when inactivation of the NE-sensitive LC-complex neurons or activation of the cholinoceptive pRF neurons produced only a decrease in postural activity; when the effects were so prominent to abolish the postural activity, the multiunit EMG responses of the limb extensor to animal tilt were suppressed (Manzoni et al. 1983b; D'Ascanio et al. 1988).

Responese of Renshaw Cells to Labyrinth Stimulation

The demonstration that in precollicular decerebrate cats the activity of the CS neurons predominates over that of the medullary inhibitory RS neurons, while just the opposite result occurs after tonic activation of the cholinergic pontine reticular system, has led us to investigate the role that R cells anatomically coupled with gastrocnemius-soleus (GS) motoneurons exert in the regulation of the response gain of the corresponding hindlimb extensors to sinusoidal labyrinth stimulation in these two different experimental conditions (Pompeiano et al. 1985a,b). All the experiments were performed in precollicular decerebrate animals submitted to bilateral neck deafferentation, immobilized with pancuronium bromide and artificially ventilated. Moreover, both hindlimbs were denervated except for the nerves supplying the left GS muscle, which was de-efferented by section of the ventral roots L6 - S2.

In a first group of decerebrate cats (Pompeiano et al. 1985a), the electrical activity of 47 R cells was recorded from L7 and upper S1 spinal cord segments, both in the animal at rest as well as during rotation of the head on the coronal plane (at 0.026-0.15 Hz, ± 10°), while the body remained fixed horizontally; due to neck deafferentation, this head displacement led to a selective stimulation of labyrinth receptors. Averaged responses (AR) of single units were recorded during successive cycles of rotation. Among these R cells which responded monosynaptically to antidromic ventral root stimulation, 22 units were disynaptically excited by the orthodromic group I volleys induced by single shock stimulation of the GS nerve, thus indicating that they were under the direct control of the recurrent collaterals of GS motoneurons monosynaptically driven by the orthodromic group Ia volleys (GS R cells). A large proportion of spontaneously firing R cells (i.e., 31/38 units, which included 16/20 GS R cells) responded to head rotation at the parameters indicated above. Moreover, most of the ARs were characterized by an increased discharge during side-down head rotation and a decreased discharge during side-up rotation (α-responses). This was true for 27 out of 33 groups of AR (i.e., 77.1%) recorded from 17 out of 24 R cells whose phase angle was carefully evaluated; similar result was also obtained for 22 out of 25 groups of AR (i.e., 88.0%) recorded from 13 of 15 GS R cells.

After these experiments had been performed, Pompeiano et al. (1985b) recorded the activity of 9 R cells, 7 of which anatomically coupled with the GS motoneurons, before and after systemic injection of an anticholinesterase. In the absence of drug injection 1 unit was silent, while the remaining 8 units fired at low rate at rest. The mean base frequency evaluated for the 9 R cells corresponded to 9.9 ± 9.6, S.D., imp./s (18 groups of AR). Four of the 9 R cells (2 of which coupled with the GS motoneurons) responded to head rotation at the parameters indicated above; the remaining 5 R cells coupled with the GS motoneurons did not respond to the stimulus. The first harmonic response of the R cells affected by head rotation showed on the average a gain of 0.074 ± 0.039, S.D., imp./s/deg and a phase lag of -10.4 ± 31.1, S.D., deg with respect to the extreme side-down head position (9 groups of AR; Fig. 4A). This last value closely corresponded to that recorded in previous experiments from a larger population of GS R cells (Pompeiano et al. 1985a).

After systemic injection of the anticholinesterase eserine sulfate (0.05-0.10 mg/kg, i.v.) the R cell which was silent in the normal decerebrate animal started to fire, whereas those which fired at low rate prior to the injection increased their discharge rate

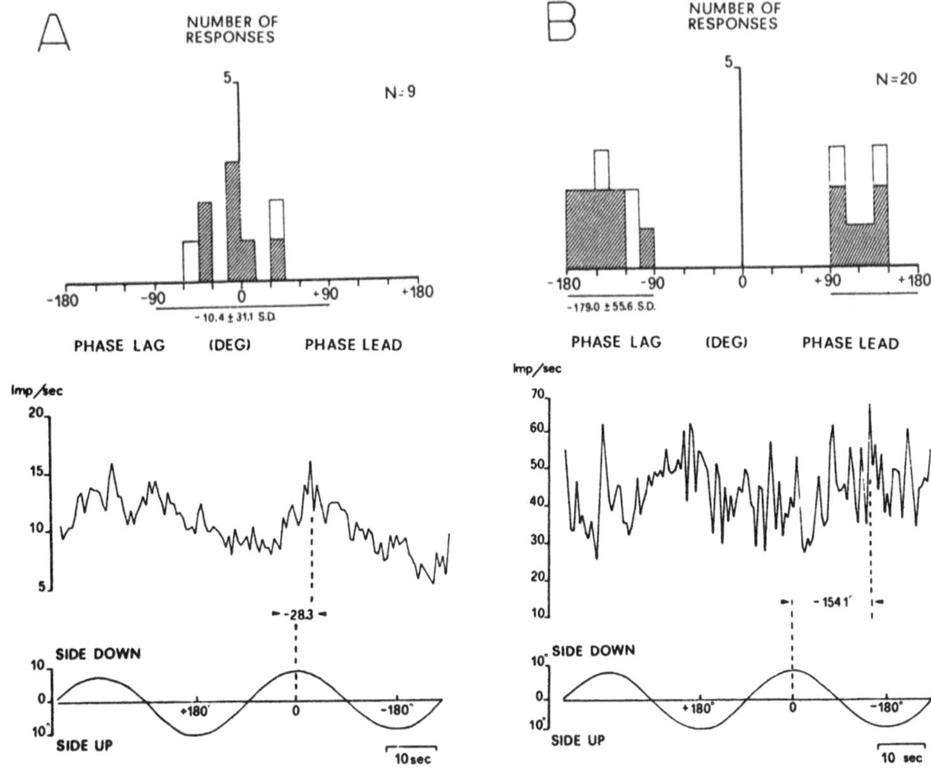

Fig. 4. Distribution of the phase angle of the first harmonic responses of R cells to head rotation recorded before (A) and after (B) injection of an anticholineserase. Precollicular decerebrate cats with bilateral neck deafferentation. The activity of R cells was recorded from L7 and upper S1 spinal cord segments. Upper histogram in A refers to 9 groups of averaged responses (9 AR) recorded from 4 R cells tested at the frequencies of head rotation of 0.026-0.15 Hz, ±10°. In particular, 7 AR were recorded from 2 R cells coupled with the GS motoneurons (hatched columns) and 2 AR were elicited from 2 unidentified R cells (white columns); 5 other R cells coupled with the GS motoneurons were unresponsive to head rotation. Upper histogram in B refers to 20 AR recorded from 9 R cells tested at the same parameters of head rotation indicated above, at least 15-20 min after i.v. injection of 0.05-0.10 mg/kg eserine sulfate; in particular, 15 AR were recorded from 7 R cells coupled with the GS motoneurons (hatched columns) and 5 AR were elicited from 2 unidentified R cells (white columns). Positive numbers on the abscissa indicate the phase lead in degrees, whereas negative numbers indicate the phase lag of responses with respect to the extreme side-down displacement of the head. Responses of R cells underlined by horizontal bars were used to evaluate the average phase angle of the units excited during side-down (0°) or side-up head displacement (180°), as indicated below each histogram. All the unit responses recorded in normal decerebrate cats showed an α-response (phase angle from ±90° to 0°). The same units, as well as those which were unresponsive to head rotation in normal decerebrate cats, displayed a β-response (phase angle from ±90° to 180°) after injection of the anticholinesterase. The lower records are SPDHs showing the responses of single R cells disynaptically excited by single shock stimulation of the GS nerve during head rotation at the frequency of 0.026 Hz, ±10° (average of 8 sweeps using 128 bins with 0.6 s bin width). In A the unit activity was recorded in a normal decerebrate cat and showed an average base frequency of 11.0 imp./s, a gain of 0.24 imp./s/deg and a phase lag of -28.3° with respect to the extreme side-down displacement of the head. In B the unit activity was taken 28 min after an i.v. injection of 0.10 mg/kg of eserine sulfate; in this instance the unit showed an average base frequency of 41.4 imp./s, a gain of 0.39 imp./s/deg, and a phase lag of -154.1° with respect to the extreme side-down displacement of the head. (From Pompeiano et al. 1985 a,b)

later on. Therefore, the mean base frequency evaluated for the 9 R cells tested to head rotation increased to 13.0 ± 12.7, S.D., imp./s (for 20 groups of AR). All R cells tested, including those which did not respond to labyrinth stimulation, now showed a clearcut modulation of their firing rate to the parameters of head rotation reported above. In this instance the gain of the first harmonic responses corresponded on the average to 0.096 ± 0.087, S.D., imp./s/deg. The most striking finding, however, was that all units showed β-responses, regardless whether the R cells were unresponsive or displayed an α-response prior to the injection; in particular, the phase angle of the responses corresponded on the average to a lag of -179.0 ± 55.6°, S.D., deg with respect to the extreme side-down head position (20 groups of AR; Fig. 4B). Similar results were also obtained if we considered only the AR recorded from R cells anatomically coupled with GS motoneurons.

The increase in mean firing rate and the appearance of the β-pattern of response were first detected 5 min after injection and persisted almost unmodified throughout the recording period (up to 35 min). It is of interest, however, that a change in the animal state following somatosensory or acoustic stimulations decreased the background discharge of the units and either suppressed the unit responses to head rotation or produced the reappearance of an α-response. These findings can be related to the recovery of postural activity in the four limbs as well as to the reduced gain of the VS reflexes which occurred after these stimuli in spite of the injection of anticholinesterase (Pompeiano et al. 1983).

Control experiments indicated that injection of the same doses of anticholinesterase, performed after transection of the spinal cord at T12, did not modify the activiy of GS R cells, which were silent or fired irregularly at a very low rate in the animal at rest. Moreover, these units did not respond to head rotation at the parameters reported above.

It is of interest that in decerebrate cats, in which the discharge of LC neurons predominated over that of the dorsal pRF neurons and the related medullary inhibitory RS neurons, the R cells linked with the GS motoneurons either fired at a low rate or were silent at rest (Pompeiano et al. 1985a). This finding can be attributed to the relatively high level of resting discharge of the CS neurons leading to inhibition of R cells activity (Fung et al. 1987a), as well as to the low level of discharge of medullary RS neurons leading to disfacilitation of these interneurons. The same R cells either did not respond to standard parameters of head rotation or showed only a small amplitude α-response (Pompeiano et al. 1985a). This response pattern can be attributed to the increased discharge during side-down rotation of VS neurons, which exert an excitatory influence on GS motoneurons and through their recurrent collaterals, on the related R cells (Marchand et al. 1987). Moreover, the simultaneous decrease in firing rate of the CS neurons for the same direction of rotation (Pompeiano et al. 1988) would reduce the inhibitory influence that these neurons exert on R cells, thus enhancing the functional coupling of these inhibitory interneurons with their own extensor motoneurons. This finding explains why the response gain of limb extensors was quite small or absent in these preparations (Manzoni et al. 1983a, 1984; Boyle and Pompeiano 1984; Pompeiano et al. 1985b).

After systemic administration of the anticholinesterase, in which a reduced discharge of LC neurons was associated with tonic activation of pRF neurons and the related medullary RS neurons in the animal at rest, the firing rate of the R cells linked with the GS motoneurons increased (Pompeiano et al. 1985b). This finding depends not only upon disinhibition of the R cells following inactivation of LC neurons but also on facilitation of the same R cells following activation of the pRF neurons and the related medullary RS neurons. The same R cells also responded to standard parameters of head rotation, but now all the tested units decreased their discharge during side-down head displacement (β-response), thus displaying a response pattern opposite to that obtained prior to the injection (Pompeiano et al. 1985b). The increased discharge of the RS neurons and the related R cells following injection of the anticholinesterase would keep under its inhibitory control the limb extensor motoneurons in the animal at rest, which accounts for the reduced postural activity that occurs in this experimental condition. However, the decrease in firing rate of the RS neurons during side-down head rotation (Manzoni et al. 1983c) would reduce the excitatory influence that these neurons exert on R cells (see Pompeiano 1984 for references). The resulting disinhibition of limb extensor motoneurons which occurs for the same direction of head displacement may thus increase the response gain of the corresponding extensor muscles to the same parameters of rotation.

Conclusions

The VS reflexes elicited by slow roll tilt of the animal, leading to stimulation of macular, utricular receptors, are characterized by contraction of ipsilateral limb extensors during side-down roll tilt of the animal and relaxation during side-up tilt. These postural changes were originally attributed to the activity of lateral VS neurons, which exert a direct excitatory influence on ipsilateral limb extensor motoneurons; in fact, most of the VS neurons, including those projecting to the lumbosacral segments of the spinal cord, showed an increased discharge during side-down tilt and a decreased discharge during side-up tilt (α-responses). However, in addition to the lateral VS neurons, there are other descending pathways which contribute to the VS reflexes by acting on R cells anatomically linked with the limb extensor motoneurons. These pathways are the CS tract originating from the NE-containing neurons located in the LC-complex, which exerts an inhibitory influence on R cells and the RS tract originating from the inhibitory area of the mRF, which exerts an excitatory influence on them. Experiments of unit recording have, in fact, proved that a large proportion of LC-complex neurons as well as of mRF neurons, including those projecting to the lumbosacral segments of the spinal cord, responded to slow roll tilt of the animal. However, most of these units showed an increased discharge during side-up tilt and a decreased discharge during side-down tilt (β-responses). The fact that CS and medullary RS neurons displayed the same response pattern to tilt, in spite of their opposite influence on ipsilateral limb extensor motoneurons and the related R cells, can be understood only if we consider that from time to time the activity of the CS neurons predominates over that of the medullary RS neurons or vice versa. This hypothesis is supported by the fact that in addition to a direct projection to the spinal cord, the NE-containing LC-complex neurons, which are self-inhibitory due to the existence of synaptic contacts acting on α_2-adrenoceptors, also project inhibitory afferents to the dorsal pRF; on the other hand the presumably cholinergic and cholinoceptive neurons located in this pontine reticular region send afferents to the medullary inhibitory RS system.

In precollicular decerebrate cats, in which the activity of the LC-complex neurons and thus of the CS neurons predominates over that of the dorsal pRF and the related medullary inhibitory RS neurons, the gain of the VS reflexes is very low in forelimb extensors and almost negligible or absent in hindlimbs. However, an increased discharge of the pRF neurons and the related medullary inhibitory RS neurons elicited either by local injection into the LC-complex of the α_2-adrenergic agonist clonidine, leading to functional inactivation of the noradrenergic LC-complex neurons, or by local injection of cholinergic agonists (e.g., carbachol or bethanechol) into the dorsal pRF, decreased the postural activity in the ipsilateral limbs but greatly enhanced the amplitude of the EMG modulation and thus the response gain of ipsilateral limb extensors to animal tilt. Similar results were also obtained either after systemic injection of the anticholinesterase eserine sulfate, an effect which was suppressed by electrolytic lesion of the dorsal pRF, or after local injection of eserine sulfate into the dorsal pRF, where presumably cholinergic and cholinoceptive neurons are located.

The effects described above are, at least in part, mediated through R cells acting on limb extensor motoneurons. In particular, in precollicular decerebrate cats the activity of R cells anatomically coupled with hindlimb extensor motoneurons was recorded both in the animal at rest as well as during selective stimulation of labyrinth receptors elicited by sinusoidal head rotation after neck deafferentation. In these preparations the R cells linked with the GS motoneurons showed a low level of activity, due to tonic inhibition induced by the discharge of the CS neurons. Moreover, the same R cells either did not respond to head rotation or showed only a small-amplitude α_2-response. This response pattern was attributed in part to an increased discharge during side-down rotation of VS neurons, which excite GS motoneurons and through their recurrent collaterals the related R cells, in part to a simultaneous reduced discharge of CS neurons leading to disinhibition of these R cells. This would enhance the functional coupling of these inhibitory motoneurons with their related interneurons, thus limiting the response gain of limb extensors to labyrinth stimulation.

After systemic injection of the anticholinesterase leading to tonic activation of cholinoceptive pRF neurons and the related medullary RS neurons, the firing rate of the R cells linked with the GS motoneurons increased, thus reducing the postural activity in the animal at rest. The same R cells also responded to head rotation, but now all the tested units showed a response pattern opposite to that obtained prior to the injection (β-response). In these instances, the reduced discharge of the RS neurons during side-down head rotation would reduce the excitatory influence that these neurons exert on R cells. The firing rate of these inhibitory interneurons would then decrease just at the time in which the GS motoneurons are driven by the VS volleys. The resulting disinhibition of limb extensor motoneurons which occurs for this direction of head displacement may thus increase the response gain of the corresponding muscles to the same parameters of head rotation.

The recurrent inhibitory circuit in the spinal cord could then be used by the CS and the RS pathways as a variable gain regulator at the motoneuronal level during the VS reflexes. Since the NE-containing LC-complex neurons as well as the related cholinergic pontine reticular neurons and the medullary RS neurons undergo spontaneous fluctuations in their firing rate (Hobson and Steriade 1986) leading to changes in postural activity (Pompeiano 1967) during the sleep-waking cycle, they may intervene in order to adapt to the animal state the response gain of limb extensors to labyrinth stimulation.

Summary

The vestibulospinal (VS) reflexes elicited by slow roll tilt of the animal leading to sinusoidal stimulation of macular, utricular receptors, are characterized by contraction of limb extensors during side-down tilt of the animal and relaxation during side-up tilt. These postural responses were originally attributed to the activity of lateral VS neurons which exert a direct excitatory influence on ipsilateral limb extensor motoneurons, since their predominant response pattern was characterized by an increased discharge during side-down tilt and a decreased discharge during side-up tilt (α-responses). There are, however, other descending pathways which may modify the responses of these motoneurons to VS volleys by acting through the related Renshaw (R) cells, i.e., the coeruleospinal (CS) tract originating from noradrenergic neurons of the locus coeruleus (LC)-complex, which is inhibitory on these R cells, and the reticulospinal (RS) tract originating from the medullary inhibitory area, which is excitatory on them. Both CS and RS neurons projecting to the lumbosacral cord responded to slow roll tilt of the animal; however, the majority of these two populations of neurons showed a response pattern opposite to that of the lateral VS neurons projecting to the same segments of the spinal cord, since they were excited during side-up tilt and depressed during side-down tilt (β-responses). Experiments indicated that when the activity of the CS neurons predominated over that of the medullary RS neurons, the R cells fired in phase with ipsilateral limb extensor motoneurons during side-down tilt, thus reducing the response gain of the corresponding limb extensors to labyrinth stimulation. However, when the activity of the medullary RS neurons predominated over that of the CS neurons, the same R cells fired out of phase with respect to ipsilateral limb extensor motoneurons, thus enhancing the response gain of the corresponding limb extensors to labyrinth stimulation. In conclusion it appears that, by modifying the functional coupling of R cells with the ipsilateral limb extensor motoneurons, the neuronal system involving the CS and the RS pathways operates as a variable gain regulator at the motoneuronal level during the VS reflexes.

Acknowledgements. This study was supported by the National Institute of Neurological and Communicative Disorders and Stroke Research grant NS 07685-21 and by a grant from the Ministero della Pubblica Istruzione, Rome, Italy.

References

ASTON-JONES G, ENNIS M, PIERIBONE VA, NICKELL WT, SHIPLEY MT (1986) The brain nucleus locus coeruleus: restricted afferent control of a broad efferent network. Science 234: 734-737

BARNES CD, D'ASCANIO P, POMPEIANO O, STAMPACCHIA G (1987) Effects of microinjection of cholinergic agonists into the pontine reticular formation on the gain of vestibulospinal reflexes in decerebrate cats. Arch Ital Biol 125: 71-105

BISCOE TJ, CURTIS DR (1966) Noradrenaline and inhibition of Renshaw cells. Science,151: 1230-1231

BOYLE R, POMPEIANO O (1980) Reciprocal responses to sinusoidal tilt of neurons in Deiters' nucleus their dynamic characteristics. Arch Ital Biol 118: 1-32

BOYLE R, POMPEIANO O (1984) Discharge activity of spindle afferents from the gastrocnemius-soleus muscle during head rotation in the decerebrate cat. Pflügers Arch 400: 140-150

D'ASCANIO P, BETTINI E, POMPEIANO O (1985) Tonic facilitatory influences of dorsal pontine reticular structures on the response gain of limb extensors to sinusoidal labyrinth and neck stimulations Arch Ital Biol 123: 101-132

D'ASCANIO P, POMPEIANO O, STAMPACCHIA G, TONONI G (1988) Inhibition of vestibulospinal reflexes following cholinergic activation of the dorsal pontine reticular formation. Arch.Ital Biol 126: 291-316

ENGBERG I, RYALL RW, (1966) The inhibitory action of noradrenaline and other monoamines on spinal neurons. J Physiol (Lond) 185: 298-322

ENNIS M, ASTON-JONES G (1986a) Evidence for self- and neighbor-mediated postactivation inhibition of locus coeruleus neurons. Brain Res 374: 299-305

ENNIS M, ASTON-JONES GA (1986b) A potent excitatory input to nucleus locus coeruleus from the ventrolateral medulla. Neurosci Lett 71: 299-305

ENNIS M, ASTON-JONES G (1987) Two physiologically distinct populations of neurons in the ventrolateral medulla innervate the locus coeruleus. Brain Res 425: 275-282

FOOTE SL, BLOOM FE, ASTON-JONES G (1983) Nucleus locus coeruleus: new evidence of anatomical and physiological specificity. Physiol Rev 63: 844-914

FUNG SJ, BARNES CD (1987) Membrane excitability changes in hindlimb motoneurons induced by stimulation of the locus coeruleus in cats. Brain Res 402: 230-242

FUNG SJ, POMPEIANO O, BARNES CD (1987a) Suppression of the recurrent inhibitory pathway in lumbar cord segments during locus coeruleus stimulation in cats. Brain Res 402: 351-354

FUNG SJ, REDDY VK, BOWKER RM, BARNES CD (1987b) Differential labeling of the vestibular complex following unilateral injections of horseradish peroxidase into the cat and rat locus coeruleus Brain Res 401: 347-352

HOBSON JA, McCARLEY RW, PIVIK RT, FREEDMAN R (1974) Selective firing by cat pontine brain stem neurons in desynchronized sleep J Neurophysiol 37: 497-511

HOBSON JA, STERIADE M (1986) Neuronal basis of behavioral state control In: Bloom FE(ed) Intrinsic regulatory system of the brain. American Physiological Society, Bethesda, pp 701-823 (Handbook of physiology, sect 1,vol 4)

HOLSTEGE JC, KUYPERS HGJM (1987) Brainstem projections to spinal motoneurons: an update. Neuroscience 23: 809-821

HOSHINO K, POMPEIANO O (1976) Selective discharge of pontine neurons during the postural atonia produced by an anticholinesterase in the decerebrate cat. Arch Ital Biol 114: 244-277

HULTBORN H, LINDSTRÖM S, WIGSTRÖM H (1979) On the function of recurrent inhibition in the spinal cord. Exp Brain Res 37: 399-403

JANKOWSKA E, LUND S, LUNDBERG A, POMPEIANO O (1968) Inhibitory effects evoked through ventral reticulospinal pathways. Arch Ital Biol 106: 124-140

JONES BE, FRIEDMAN L (1983) Atlas of catecholamine perikarya varicosities and pathways in the brainstem of the cat. J Comp Neurol 215: 382-396

JONES BE, YANG T-Z (1985) The efferent projections from the reticular formation and the locus coeruleus studied by anterograde and retrograde axonal transport in the rat. J Comp Neurol 222: 56-92

KANAMORI N, SAKAI K, JOUVET M (1980) Neuronal activity specific to paradoxical sleep in the ventromedial medullary reticular formation of unrestrained cats. Brain Res 189: 251-255

KIMURA H, MAEDA T (1982) Aminergic and cholinergic systems in the dorsolateral pontine tegmentum. Brain Res Bull 9: 493-499

LADPLI R, BRODAL A (1968) Experimental studies of commissural and reticular formation projections from the vestibular nuclei of the cat. Brain Res 8: 65-96

LUND S, POMPEIANO O (1968) Monosynaptic excitation of alpha-motoneurons from supraspinal structures in the cat. Acta Physiol Scand 73: 1-21

MANZONI D, POMPEIANO O, SRIVASTAVA UC, STAMPACCHIA G (1983a) Responses of forelimb extensors to sinusoidal stimulation of macular labyrinth and neck receptors. Arch Ital Biol 121: 205-214

MANZONI D, POMPEIANO O, SRIVASTAVA UC, STAMPACCHIA G (1983b) Inhibition of vestibular and neck reflexes in forelimb extensor muscles during the episodes of postural atonia induced by an anticholinesterase in decerebrate cat. Arch Ital Biol 121: 267-283

MANZONI D, POMPEIANO O, SRIVASTAVA UC, STAMPACCHIA G (1984) Gain regulation of vestibular reflexes in fore- and hindlimb muscles evoked by roll tilt. Boll Soc Ital Biol Sper 60 (suppl 3): 9-10

MANZONI D, POMPEIANO O, STAMPACCHIA G, SRIVASTAVA UC (1983c) Responses of medullary reticulospinal neurons to sinusoidal stimulation of labyrinth receptors in decerebrate cat. J Neurophysiol 50: 1059-1079

MARCHAND AR, MANZONI D, POMPEIANO O, STAMPACCHIA G (1987) Effects of stimulation of vestibular and neck receptors on Deiters neurons projecting to the lumbosacral cord. Pflügers Arch 409: 13-23

POMPEIANO O (1967) The neurophysiological mechanisms of the postural and motor events during desynchronized sleep. Res Publ Assoc Nerv Ment Dis 45: 351-423

POMPEIANO O (1979) Neck and macular labyrinthine influences on the cervical spino-reticulocerebellar pathway. In Granit R, Pompeiano O (eds) Reflex control of posture and movement. Elsevier/North-Holland, Amsterdam, pp 501-514 (Progress in brain research vol 50)
POMPEIANO O (1980) Cholinergic activation of reticular and vestibular mechanisms controlling posture and eye movements. In Hobson J A, Brazier M A B (eds) The reticular formation revisited. Raven, New York, pp 473-512 (IBRO monograph series, vol 6)
POMPEIANO O (1984) Recurrent inhibition. In: Davidoff RA (ed) Handbook of the spinal cord, vols 2 and 3. Marcel Decker, New York, pp 461-557
POMPEIANO O, D'ASCANIO P, HORN E, STAMPACCHIA G (1987) Effects of local injection of the α2-adrenergic agonist clonidine in the locus coeruleus complex on the gain of vestibulospinal and cervicospinal reflexes in decerebrate cats. Arch Ital Biol 125: 225-269
POMPEIANO O, HOSHINO K (1976) Tonic inhibition of dorsal pontine neurons during the postural atonia produced by an anticholinesterase in the decerebrate cat. Arch Ital Biol 114: 310-340
POMPEIANO O, MANZONI D, BARNES CD, STAMPACCHIA G, D'ASCANIO P (1988) Labyrinthine influences on locus coeruleus neurons. Acta Otolaryngol (Stockh) 105: 576-581
POMPEIANO O, MANZONI D, SRIVASTAVA UC, STAMPACCHIA G (1983) Cholinergic mechanisms controlling the response gain of forelimb extensor muscles to sinusoidal stimulation of macular labyrinth and neck receptors. Arch Ital Biol 121: 285-303
POMPEIANO O, WAND P, SRIVASTAVA UC (1985a) Responses of Renshaw cells coupled with hindlimb extensor motoneurons to sinusoidal stimulation of labyrinth receptors in the decerebrate cat. Pflügers Arch 403: 245-257
POMPEIANO O, WAND P, SRIVASTAVA UC (1985b) Influences of Renshaw cells on the gain of hindlimb extensor muscles to sinusoidal labyrinth stimulation. Pflügers Arch 404: 107-118
SAKAI K, SASTRE JP, KANAMORI N, JOUVET M (1981) State specific neurons in the ponto-medullary reticular formation with special reference to the postural atonia during paradoxical sleep in the cat. In Pompeiano O, Ajmone-Marsan C (eds) Brain mechanisms of perceptual awareness purposeful behavior. Raven, New York, pp 405-429 (IBRO monograph series, vol 8)
SAKAI K, SASTRE JP, SALVERT D, TOURET M, TOHYAMA M, JOUVET M (1979) Tegmento-reticular projections with special reference to the muscular atonia during paradoxical sleep in the cat: an HRP study. Brain Res 176: 233-254
SAKAI K, TOURET M, SALVERT D, LEGER L, JOUVET M (1977) Afferent projections to the cat locus coeruleus as visualized by the horseradish peroxidase technique. Brain Res 119: 21-41
SCHEIBEL ME, SCHEIBEL AB (1973) Discussion. In: Brain information conference report No 32.Brain Inormation Service/Brain Research Institute, UCLA, Los Angeles, pp 12-17
SCHOR RH, MILLER AD (1981) Vestibular reflexes in neck and forelimb muscles evoked by roll tilt. J Neurophysiol 46: 167-178
SCHOR RH, MILLER AD (1982) Relationship of cat vestibular neurons to otholith-spinal reflexes. Exp Brain Res 47: 137-144
SHIROMANI P, McGINTY DJ (1986) Pontine neuronal response to local cholinergic infusion: relation to REM sleep. Brain Res 386: 20-31
SRIVASTAVA UC, MANZONI D, POMPEIANO O, STAMPACCHIA G (1982) State-dependent properties of medullary reticular neurons involved during the labyrinth and neck reflexes. Neurosci Lett 10: S461
SVENSSON TH, BUNNEY BS, AGHAJANIAN GK (1975) Inhibition of both noradrenergic and serotonergic neurons in brain by the α-adrenergic agonist clonidine. Brain Res 92: 291-306
VIVALDI E, McCARLEY RW, HOBSON JA (1980) Evocation of desynchronized sleep signs by chemical microstimulation of the pontine brainstem. In: Hobson JA, Brazier MAB (eds) The reticular formation revisited. Raven, New York pp 513-529 (IBRO monograph series, vol 6)
WEIGHT FF, SALMOIRAGHI GC (1966) Adrenergic responses of Renshaw cells. J Pharmacol Exp Ther 154: 391-397
WESTLUND KN, BOWKER RM, ZIEGLER MG, COULTER JD (1983) Noradrenergic projections to the spinal cord of the rat. Brain Res 263: 15-31
WESTLUND KN, BOWKER RM, ZIEGLER MG, COULTER JD (1984) Origins and terminations of descending noradrenergic projections to the spinal cord of monkey. Brain Res 292: 1-16
WESTLUND KN, COULTER JD (1980) Descending projections of the locus coeruleus and subcoeruleus/medial parabrachial nuclei in monkey: axonal transport studies and dopamine-β-hydroxylase immunocytochemistry. Brain Res Rev 2: 235-264

Motion Perception with Moving Eyes

T. Brandt, M. Dieterich, T. Probst
Neurologische Universitätsklinik, Klinikum Groβhadern, Marchioninistraβe 15, 8000 München, FRG

Eye Movements Impair Object Motion Perception

We have shown that the thresholds for egocentric perception of object motion are significantly raised during concurrent head oscillations of ± 20° about the vertical Z-axis and fixation of the target (Degner and Brandt 1981; Brandt 1982). Subjects were exposed to the target which randomly moved either to the right or to the left at a constant angular velocity of 24' arc/s with a stepwise increase in exposure times from 0.25 to 10 s (20 repetitions of each stimulus condition). Conservative determination of threshold was based on 18 out of 20 possible correct perceptions of movement as well as direction. Sinusoidal active head oscillations raised the detection threshold for object motion by a factor of 2.9 at 1 Hz and 6.4 at 2 Hz oscillations (Fig. 2) despite intended stabilization of the target on the retina. This effect increases disproportionately with increasing eccentricity of the image of the moving stimulus on the retina (Fig. 3). Independently Wertheim (1981) was able to demonstrate that during smooth pursuit of a target (head

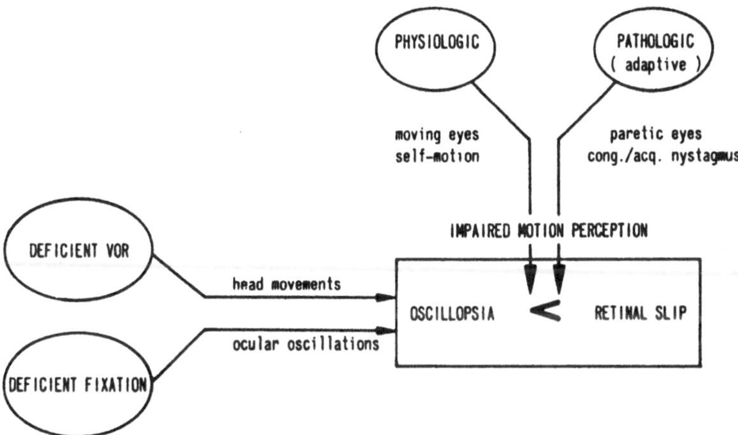

Fig. 1. Two mechanisms - a physiological and a pathological - involving motion perception. Oscillopsia is caused either by inappropriate compensatory eye movements (VOR) during head motion or by ocular oscillations which override fixation. Oscillopsia amplitudes are always smaller than net retinal slip because of partial suppression of motion perception under these conditions. Suppression of motion perception is due to the summation of a physiological and a pathological ("adaptive") elevation of thresholds to detect retinal image shifts of the viewed visual scene

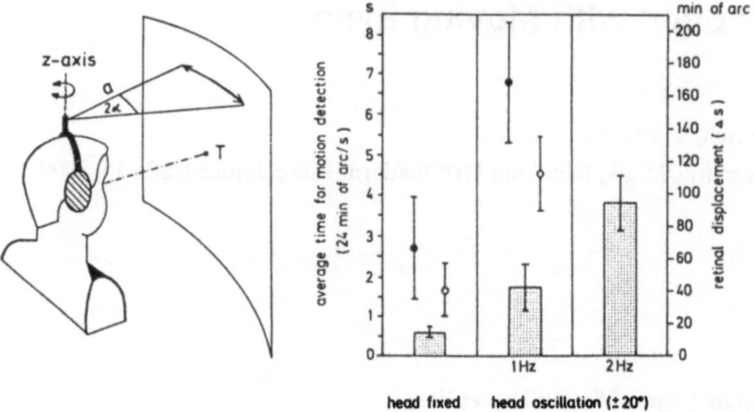

Fig. 2. Object motion perception with horizontal head oscillations (VOR). Thresholds for detection of object motion (means and standard deviations) in 12 normal subjects (columns) as compared to 8 patients suffering from acquired peripheral ocular motor palsies (●: paretic eye; O: unaffected eye). During the measurements the moving target (24' arc/s; left or right) was fixated by the subject, and the head was either fixed by a bite-board or voluntarily oscillated about the vertical z-axis at 1 or 2 Hz with an amplitude of ± 20° (motion perception during vestibulo-ocular reflex). Physiologically, thresholds for object motion detection increase significantly in normal subjects with increasing frequency of head oscillation (columns). With peripheral ocular-motor palsies a further pathological elevation of thresholds can be obtained for both the head-fixed condition and head oscillation in both eyes and is obviously more pronounced in the affected eye

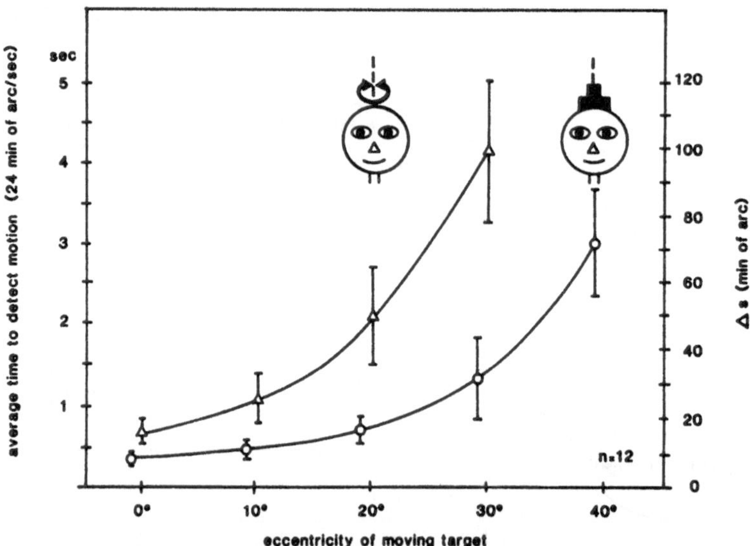

Fig. 3. Object motion perception as a function of retinal eccentricity and horizontal head oscillation. Thresholds to detect motion (in seconds and minutes of arc) as a function of the eccentricity of the target on the retina, of a moving target (24' of arc/s) projected onto a screen with either head fixed (0) or with ± 20°, 1 Hz head oscillations in yaw (Δ). Thresholds increase with increasing retinal eccentricity. Head oscillations raise thresholds disproportionally over the whole range

stationary) the threshold to detect motion of a visual background increases proportionally to ocular velocity irrespective of whether the stimulus and the eyes move in the same or opposite directions.

The "new" phenomenon of visual motion perception suppression during eye movements may reflect a basic sensory-motor mechanism because it has a somatosensory analogue. Elevated thresholds are reported for the perception of electrical stimuli applied to a fingertip as well as partially suppressed somatosensory evoked potentials with simultaneous movement of the stimulated finger in man (Coquery 1978; Rushton et al. 1981).

Self Motion Perception Impairs Object Motion Perception

That real motion of the eyes or the head is not the essential stimulus to suppress object motion perception was demonstrated by slow trunk oscillations (cervical stimulation) relative to the head fixed by a biteboard (Probst et al. 1986) and by circularvection studies. In these experiments (Figs. 4, 5) mean response times (n = 20) to the detection of object motion of a projected object (24' arc/s) were determined instead of threshold measurements. The measured times to indicate subjective motion onset of a moving target are the sum of the reaction and the perception times. For cervical stimulation the subjects sat on a rotary chair with the head firmly fixed in a helmet. The helmet itself was rigidly attached to the ceiling. The subject's trunk and legs were strapped onto the chair and sinusoidally oscillated at frequencies of 0.5 and 1 Hz with an amplitude of ±

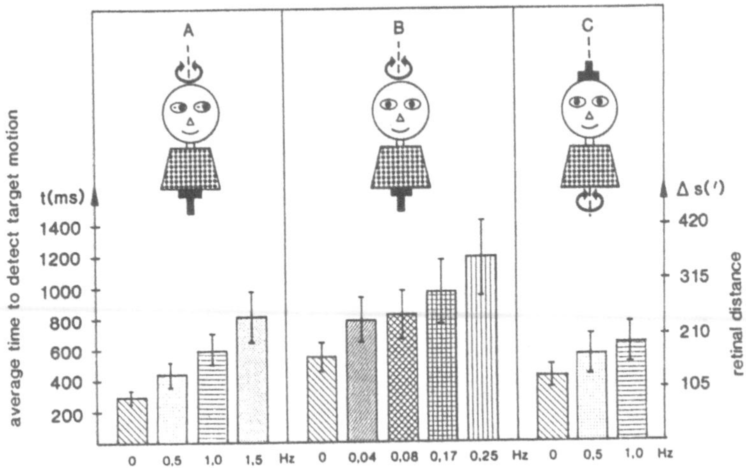

Fig. 4. Object motion perception with head or trunk oscillations. Means and standard deviations of the response times (milliseconds) or retinal distance (Δs in minutes of arc) necessary to detect target motion (speed: 5°/s) during different modes of simultaneous body motion. The target was fixated during horizontal head oscillation with VOR (A) or with fixation suppression of the VOR (B) by use of an optokinetic helmet (laboratory model as used by astronauts in the Spacelab-1 mission) or with the head fixed by the helmet and pure cervical stimulation provided by trunk oscillations (C). Response time to detect object motion increased with increasing frequency of either head or trunk oscillations

25°. The average time to detect motion under "static conditions" (no body motion: 436 ms) significantly increased through body oscillations by a factor of 1.34 at 0.5 Hz and 1.48 at 1.0 Hz (Fig. 4). Thus, pure trunk oscillation with the head and the eyes fixed in space (cervical stimulation) significantly impairs detection of motion of a fixated target.

Even stronger effects were seen with objectively stationary subjects for whom apparent self-motion was visually induced by full-field optokinetic stimulation with the head fixed, the subjects experiencing horizontal optokinetically induced circular vection (Fig. 5). The thresholds for perceiving horizontal target motion were raised by a factor of 5.5; here the target motion was perpendicular to the direction of the vertical pattern motion which induced pitch vection. However, with concurrent horizontal target motion during horizontal pattern motion, the thresholds were raised by a factor of 17.8. The physiological inhibitory interaction between object motion and self-motion perception may reflect lack of specifity (or a side effect) of a space constancy mechanism (efference copy?) which provides us with a stable picture of the world during locomotion. It has practical implications when one is riding a vehicle with the twofold perceptual task of controlling self-motion (by linear vection) and perceiving object motion simultaneously. Authorities road traffic accidents should consider an additional perceptual time of at least 300 ms for detecting critical changes in intercar distances beyond the usual reaction time as expected from laboratory data with the subject's head fixed by a biteboard (Probst et al. 1984; 1987).

This hypothesis was proven with a field study (vehicle guidance under natural conditions) and a corresponding simulation in the laboratory, in which a stationary sur-

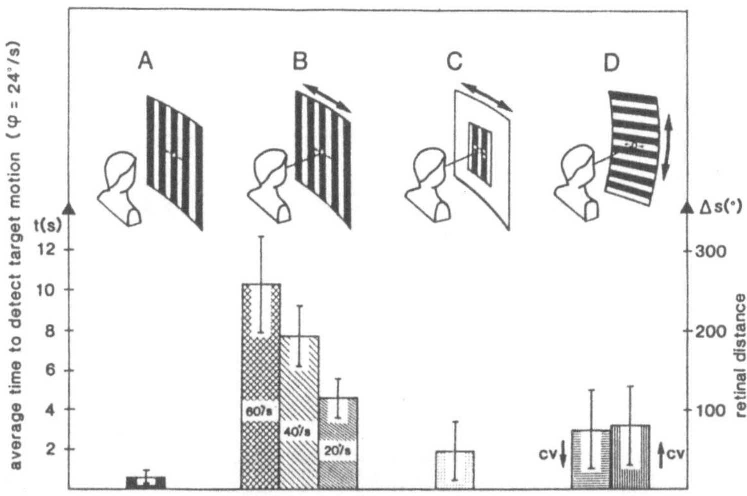

Fig. 5. Object motion perception as affected by simultaneous pattern motion. Thresholds (means and standard deviations in milliseconds) or retinal distance (Δs in minutes of arc) to detect motion of the 1° target moving horizontally with 24' arc/s during concurrent visually induced self-motion (circular vection, CV) about the vertical z-axis ("yaw vection", B) or the horizontal y-axis ("pitch vection", D) and during small field background motion, which does not induce the sensation of self-motion (C). Thresholds were raised significantly with CV (increasing with increasing CV velocity, B) as compared to the small field condition without CV (C) and the stationary condition (A). CV perpendicular to object motion also impaired object motion perception (D)

rounding eliminated the perception of self-motion (Fig. 6). As demonstrated by comparing the results of the vehicle experiments with those of the corresponding laboratory simulation, object motion perception is markedly impaired during a car ride.

Binocular Impairment of Motion Perception Caused by Monocular External Eye Muscle Paresis

Motion perception was investigated separately for the affected and unaffected eye in patients suffering from abducens, oculomotor, or trochlear palsy. Thresholds for the detection of object motion with the head fixed were significantly raised up to a factor of 5 for the paretic eye and a factor of 3.3 for the normal eye (Fig. 2). With sinusoidal head oscillations at 1 Hz (which physiologically elevates thresholds) the ratio between thresholds in patients and normals is about 4 for the paretic eye and 2.7 for the unaffected eye (Fig. 2). This clearly suggests that in patients with acquired peripheral ocular motor palsies, the physiological and the adaptive (pathological) impairments of motion perception are additive when the moving target is fixated during voluntary head motion (Dieterich and Brandt 1987; Brandt and Dieterich 1986).

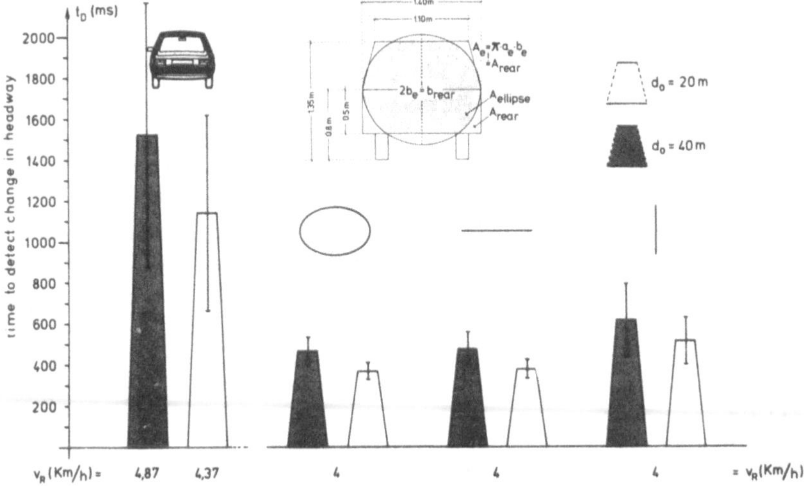

Fig. 6. Object motion perception under real road and simulated conditions. Mean response times were determined for the perception of changes in headway at distances of 20 and 40 m under real and simulated conditions without concurrent self-motion. An approximation of the perceptual effective area of the rear of the leading car was simulated by an ellipse of equivalent retinal size that was electronically generated by two phase-displaced sign waves and an oscilloscope operating in the X, Y mode. Headway changes were simulated by adjusting the retinal ellipse area with a triangle-wave generator. The times to detect changes in headway were significantly higher for the actual road condition (linear vection). Under static conditions in the laboratory there was no difference between the detection of a gradual change in area of the ellipse and a horizontal bar with the same but one-dimensional movement. Detection times, however, significantly increase ($p < 0.05$) if object movement occurred in corresponding but vertical dimensions only (vertical bar)

Summary

There is a physiological inhibitory interaction between concurrent self-motion and object-motion perception which has been demonstrated for optokinetic as well as vestibular or somatosensory (cervical) stimulation. Furthermore, there is a physiological mechanism of impaired motion perception with moving eyes which contributes to the suppression of oscillopsia in ocular motor disorders but does not completely account for it. An additional adaptive binocular impairment of motion perception must be involved which is separate from the physiological phenomenon and initiated by the particular ocular motor disorder (Fig. 1). In all instances of ocular motor disorders investigated so far (either supranuclear or infranuclear; either congenital or acquired) the amplitude of perceived motion of the visual scene was considerably smaller than the calculated net retinal slip. This partial suppression of distressing oscillopsia was inevitably linked to impaired motion perception in general.

Evidence is presented that the perception of object motion is impaired either by concurrent self-motion perception, by eye movements or by ocular motor disorders. Under natural, environmental conditions one moves freely with the twofold perceptual task of controlling self-motion and perceiving object motion simultaneously. Experimental studies of motion perception, however, have a laboratory tradition in which thresholds for the detection of object motion are determined with the head stationary.

It was our incidental observation that one has considerable difficulties to see the treetops moving in the wind while driving a vehicle. This led us to perform experiments on thresholds for the detection of single object motion under various stimulus conditions which simultaneously cause eye movements or induce the sensation of self-motion (Probst et al. 1986). The inhibitory interaction between self- and object motion perception has practical implications when one is riding a vehicle because it impairs detection of critical changes in headway (Probst et al. 1984; Probst et al. 1987).

Independently, we were able to demonstrate impaired motion perception in patients with congenital and acquired ocular motor disorders. Thresholds to detect object motion are increased in these patients, and consequently detection of oscillopsia due to involuntary retinal slip is reduced. In fact, angular displacement of perceived motion of the visual scene due to the nystagmus or the defective vestibulo-ocular reflex does not quantitatively match the net retinal slip (Brandt 1982; Wist et al. 1983; Büchele et al. 1983). Amplitude of perceived motion (oscillopsia) is always smaller than would be expected from the amplitude of the nystagmus. It can be demonstrated that the dissociation between the two can be explained by the combination of two separate mechanisms which involve motion perception (Fig. 1):

- A physiological elevation of thresholds to detect object motion with moving eyes
- A pathological elevation of thresholds to detect object motion with either infranuclear ocular motor palsy or supranuclear ocular oscillations

In a teleological sense, this "adaptive suppression" of the detection of retinal image motion is beneficial to the organism to the extent that it alleviates the distressing oscillopsia with the disadvantageous side effect of impaired motion perception in general.

References

Brandt T (1982) The relationship between retinal image slip, oscillopsia and postural imbalance. In: Lennerstrand G, Zee DS, Keller EL (eds) Functional bases of ocular motility disorders. Pergamon, Oxford, pp 379-385

Brandt T, Dieterich M (1986) Peripheral ocular motor palsy impairs motion perception. In: Keller EL, Zee DS (eds.) Adaptive processes in visual and oculomotor systems. Pergamon, Oxford, pp 457-463

Büchele W, Brandt T, Degner D (1983) Ataxia and oscillopsia in downbeat-nystagmus vertigo syndrome. Adv. Otorhinolaryngol 30: 291-297

Coquery JM (1978) Role of active movement in control of afferent input form skin in cat and man. In: Gordon G (ed) Active touch - the mechanism of recognition of objects by manipulation. Elmsford, New York, pp 161-169

Degner D, Brandt T (1981) Interaction between self- and object motion perception. Pflügers Arch. [Suppl] 389, R 30: 118

Dieterich M, Brandt T (1987) Impaired motion perception in congenital nystagmus and acquired ocular motor palsy. Clin. Vision Sci. 4: 337-345

Probst T, Krafczyk S, Brandt T, Wist ER (1984) Interaction between perceived self-motion and object motion impairs vehicle guidance. Science 2255: 536-538

Probst T, Brandt T, Degner D (1986) Object motion detection affected by concurrent self-motion perception: psychophysics of a new phenomenon. Behav. Brain Res. 22: 1-11

Probst T, Krafczyk S, Brandt T (1987) Object motion detection affected by concurrent self-motion perception: applied aspects for vehicle guidance. Ophthalmic Physiol. Opt. 7: 309-314

Rushton DN, Rothwell JC, Craggs MD (1981) Dating of somatosensory evoked potentials during different kinds of movement in man. Brain 104: 465-491

Wertheim AH (1981) On the relativity of perceived motion. Acta Psychol (Amst) 48: 97-110

Wist ER, Brandt T, Krafczyk S (1983) Oscillopsia and retinal slip: evidence supporting a clinical test. Brain 106: 153-168

Does the System for Smooth-Pursuit Eye Movements Rely on a Neuronal Representation of Target Motion in Space?

P. Thier, W. Koehler, U. W. Buettner, J. Dichgans

Department of Neurology, University of Tübingen, Hoppe-Seyler-Str.3, 7400 Tübingen, FRG

Recent thinking on the neuronal implementation of smooth-pursuit eye movements has been led largely by the considerable progress made in the analysis of cortical mechanisms of motion processing. One of the most influential concepts in recent cortical neurobiology has been the idea, originally formulated by Ungerleider and Mishkin (1982), that the primate visual cortex contains distinct parallel streams of processing that diverged as early as in the retina. One of the multiple pathways seems to be specialized in the processing of visual object motion and finally feeds the posterior parietal cortex (Deyoe and Van Essen 1988; Maunsell and Newsome 1987). The middle temporal area (MT), a small visual area on the posterior bank of the superior temporal sulcus (STS), is an early element of this motion pathway, which has been studied most intensively, starting with its original description by Dubner and Zeki in 1971 as a V1 recipient area containing a large number of direction-selective cells. The notion that MT is an integral part of the afferent limb of the neuronal machinery for smooth-pursuit eye movements is based mainly on the effect of small chemical lesions in MT (Newsome et al. 1985) which produce impaired smooth-pursuit eye movements to targets moving in any direction within the defective portion of the visual field. The significance of cortical motion processing in MT for smooth-pursuit eye movements was further supported by Lisberger and his colleagues (Lisberger and Westbrook 1985; Tychsen and Lisberger 1986), who were able to demonstrate convincingly that in humans and in subhuman primates the basic phenomenology of smooth-pursuit initiation in the monkey can be attributed to properties of MTcells. MT and other elements of the cortical motion pathway are known to project heavily onto parts of the pontine nuclei (Fries 1981; Brodal 1978; Glickstein et al. 1980, 1985). Hence the demonstration of motion sensitive visual cells in the DLPN of the monkey by Suzuki and Keller (1984) finally suggested that information on visual motion originating in MT and neighbouring areas of the STS was handed over to the cerebellum via the pontine nuclei (Fig. 1A), a circuitry for smooth pursuit, which we refer to as the sensory model.

Although there is no reason to dispute the significance of cortical processing of visual motion for smooth-pursuit, it cannot be overlooked that the sensory model is not able to account for many of the basic properties of smooth-pursuit eye movements and moreover ignores the contribution of cortical areas certainly not involved in visual motion processing.

The sensory model implies that the stimulus for smooth-pursuit eye movements is a retinal image. However, a number of experiments have demonstrated convincingly that retinal stimulation may not be necessary for smooth-pursuit eye movements to occur.

As early as 1916, for instance, Gertz showed that subjects are able to track sound sources and also tactile stimuli smoothly with their eyes. These observations were confirmed and extended by later work and led investigators such as Young (1977) and Steinbach (1976) to propose that we pursue an amodal percept of a target rather than an image on the retina. Although the proponents of the sensory model of smooth-pursuit eye movements seem to be aware of these ideas (see the recent review by Lisberger et al. 1987), their sensory model does not try to offer an explanation for the ability to track

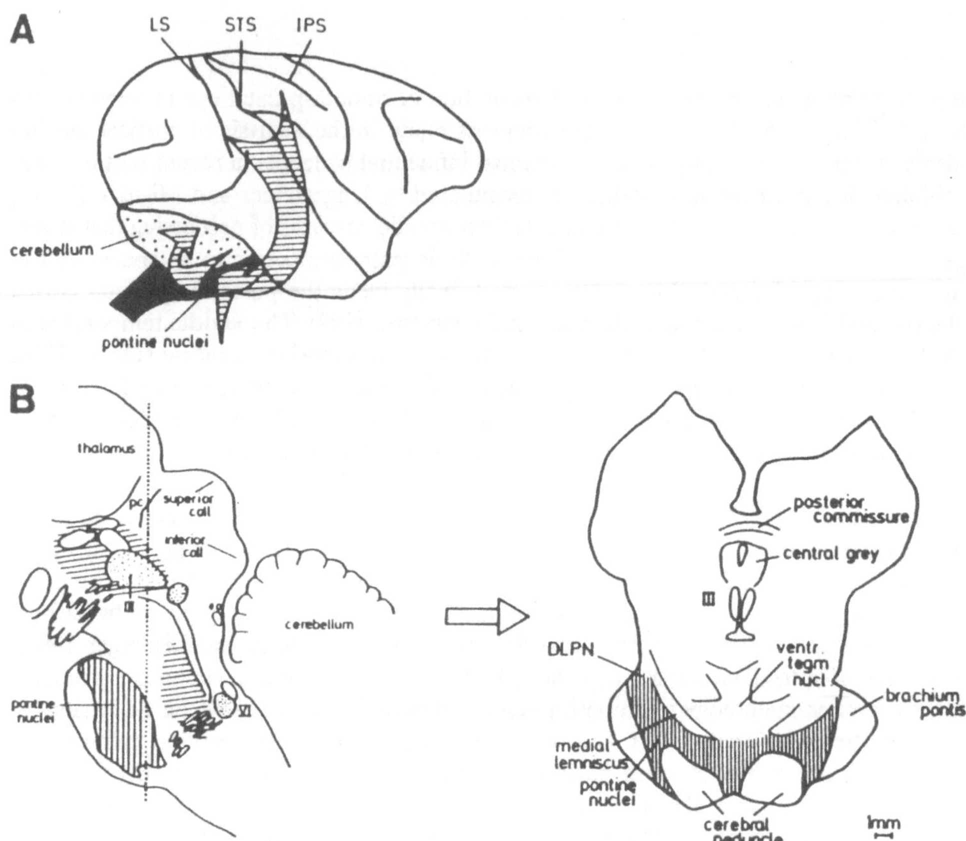

Fig. 1. A Sketch of the macaque brain emphasizing those structures currently supposed to contribute to the generation of smooth-pursuit eye movements elicited by moving visual targets. Image motion on the retina is analyzed in visual areas of the superior temporal sulcus like area MT. Information on visual motion is then handed over to the cerebellum (stippled) via the pontine nuclei, localized in the middle brainstem (black). These structures make up the elements of what is called the 'sensory model' of smooth-pursuit eye movements in the text. The way in which smooth-pursuit related signals from the cerebellum finally reach the oculomotor nuclei remains an unsolved puzzle. STS: superior temporal sulcus, LS: lunate sulcus, IPS: intraparietal sulcus. **B** Schematic parasagittal (left) and frontal section (right) through the brainstem showing the localization of the pontine nuclei, a huge mass of cells, surrounding the fibers of the cerebral peduncle along their descending way through the pons. The dorsolateral part of the pontine nuclei (DLPN) integrates afferents from several sources, among them the parieto-occipital cortex, the frontal eye fields and the superior colliculus (Brodal 1978; Fries 1981; Harting 1977; Stanton et al. 1988). Electrophysiological recordings were made from the DLPN and its neighbourhood in three awake rhesus monkeys, participating in a battery of visual, vestibular, and oculomotor tasks. For details of the methodology refer to Thier et al. (1988)

nonvisual targets smoothly with our eyes. Another essential problem that the sensory model faces is the existence of what has been called a cortical motor system for smooth-pursuit eye movements (Tusa and Ungerleider 1988), consisting of parts of the posterior parietal cortex, for instance, the so called middle superior temporal area (MST) on the anterior bank of the STS and the frontal eye fields (FEF). These areas contain neurons that discharge in relation to smooth-pursuit eye movements, and unilateral lesions of these areas result in a directional deficit of smooth-pursuit eye movements whether the target is localized in the ipsilateral or the contralateral visual field (Bruce and Goldberg 1985; Keating et al. 1985; Lynch 1987; Dürsteler et al. 1987). Like MT also the cortical motor system for smooth-pursuit eye movements also projects to the DLPN and its neighbourhood (Stanton et al. 1988; Tusa and Ungerleider 1988). Therefore it was not too surprising to find a certain percentage of cells in the DLPN discharging in relation to some motor aspect of smooth-pursuit (Mustari et al. 1986; Thier et al. 1988). Obviously, at least a part of the sensory-motor transformation must have already been achieved on a cortical level. However, unlike the role of the cortical sensory system, the role of the cortical motor system and correspondingly the properties of the sensory-motor transformation remain ill-defined .

Stimulated by observations that we made during an electrophysiological exploration of the DLPN and its neighbourhood, we propose here a specific solution of the problem of the cortical sensory-motor transformation for smooth-pursuit eye movements. We suggest as the nature of the cortical sensory-motor transformation the evaluation of a neuronal representation of target motion in space. This hypothesis is at present based on a limited sample of pontine cells and certainly needs a broader basis both with respect to the number of cells tested and also with respect to the paradigms applied to become fully conclusive. We present it at this early stage of investigation in order to stimulate thought about the limitations of smooth-pursuit models currently favoured.

The pontine nuclei surround the fibers of the cerebral peduncle along their descending path through the pons (Fig. 1B). In recent years we have explored electrophysiologically the dorsolateral part and its neighbourhood of the pontine nuclei in awake monkeys, trained to collaborate in a battery of vestibular and visuo-oculomotor tasks. Details of the methodology and an overall view of the cell types encountered in this part of the brainstem have been presented recently (Thier et al. 1988). In this report we concentrate on the properties of a minority of cells that we tentatively called visual-tracking (VT) neurons, being struck by their resemblance to the so called VT neurons found in parts of the parieto-occipital cortex (Mountcastle et al. 1975; Sakata et al. 1983; Wurtz and Newsome 1985). The linking feature of VT-neurons in both parieto-occipital cortex and the pontine nuclei is that they discharge whenever the eyes track the target in a specific preferred direction. This property is readily assessed by plotting mean discharge rate as a function of tracking direction, as exemplified in Fig. 2 for a pontine VT neuron which preferred tracking to the lower left. Since some of these VT neurons may also be driven by visual stimuli during stationary fixation (Mustari et al. 1986; Thier et al. 1988), it is first of all reasonable to suppose that these cells respond to retinal slip of the images of the visual environment or of the target . Whereas a contribution of the visual environment can easily be ruled out by darkening the room, retinal slip of the target image is much harder to control for since these cells might be sensitive to very small amounts of retinal slip beyond the limits of resolution of the equipment used to

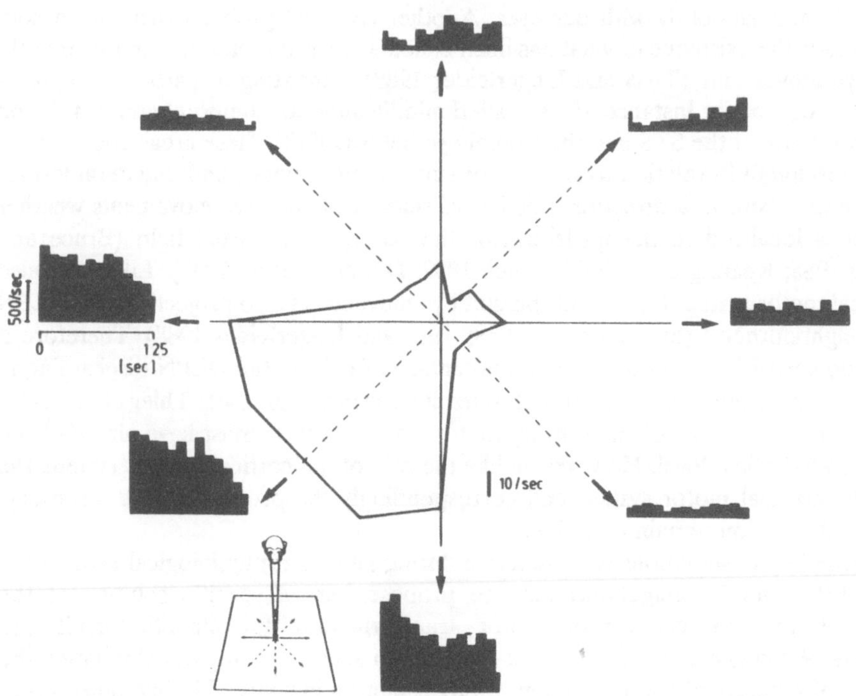

Fig. 2. Polar plot of mean discharge rate as a function of tracking direction ('direction tuning curve') for a typical pontine VT neuron. The monkey tracked a small laser target rear-projected on a tangent screen moving to and fro between two positions with constant velocity according to a triangular waveform (amplitude 10°; period 2.5 s). The orientation of the axis of movement was varied between 0° and 180° in steps of 22.5° (0° horizontal movement; 90° vertical movement). The direction tuning curve is supplemented by poststimulus time histograms showing the averaged neuronal responses for the indivual directions of tracking (bin width 100 ms). This VT neuron preferred tracking to the lower left

record eye position. A closer inspection of discharge profiles of pontine VT neurons, however, reveals that these cells are most probably not interested in retinal slip. Figure 3A shows a VT neuron which discharged tonically during downward tracking. Fortunately, however, as we are inclined to look at it now, tracking performance was rather bad. Smooth-pursuit eye movements were again and again interrupted by saccadic excursions introducing retinal slip of the target image with varying directions and amplitudes. The fact that the cell nevertheless continued firing tonically irrespective of what the eyes were actually doing shows two different things. Firstly, this VT neuron did not carry an efferent signal related to eye movements, and secondly this cell did not encode the movement of any image on the retina, neither that of the target nor that of stationary objects in the visual environment, possibly failed to have been noticed by the investigator.

One might argue that retinal slip velocities introduced by the saccadic excursions were simply too high to influence the discharge rate of a supposedly direction-selective visual cell. This objection can be weakened by looking at another VT neuron from the pontine nuclei, shown in Fig. 3B. This cell preferred smooth-pursuit eye movements to the left. While the eyes tracked the laser target moving on a circle, we temporarily prevented the eyes from keeping up with the target by opening the feedback loop electro-

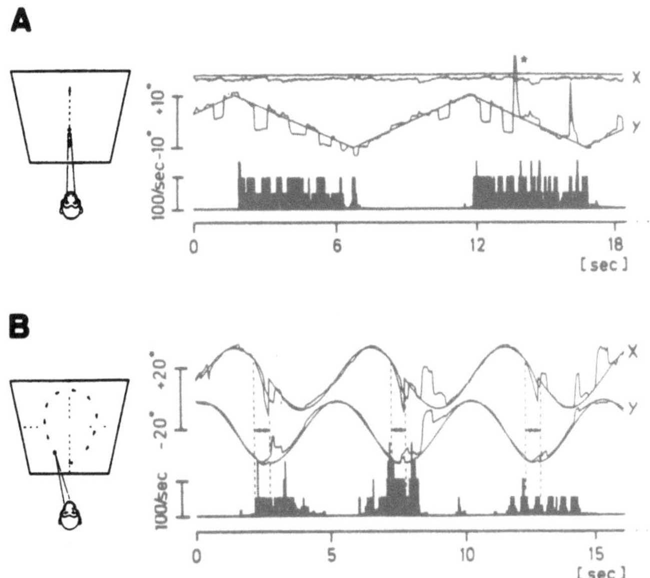

Fig. 3. A Vertical tracking eye movements of a monkey, elicited by a laser target moving up and down and back again according to a triangle waveform (amplitude 10°; period 10 s). Horizontal and vertical components of laser and eye position are superimposed. Eye position in this monkey was recorded with the electro-oculogram technique (compare Thier et al. 1988, for details). Asterisk marks distortion of the vertical eye position trace due to an eye lid blink. The pontine VT neuron, whose momentary discharge rate is shown, preferred downward tracking. Note that firing is not influenced by the frequent saccadic excursions introducing considerable retinal slip of varying amplitude and direction. **B** Smooth-pursuit eye movements elicited by a laser target moving in a clockwise direction on a circle of 20° radius with constant angular velocity (72°/s). Eye position in this monkey was recorded with the search coil technique (Robinson 1963). The pontine VT neuron, whose momentary discharge rate is plotted below the eye and target position traces, preferred leftward tracking. Whenever eyes and target were directed leftwards on the circle, as indicated by the pair of dashed lines, the eyes were temporarily prevented from keeping up with the target by opening the feedback loop electronically, i.e. by moving the target according to the sum of eye position (E) and target position (T), the latter as determined by the circle equation. During open loop the following relation held: target position $T^*(t_n)$ at time t_n was: $T^*(t_n) = E(t_{n-1}) - E(t_0) + T(t_n)$, with $T(t_n)$, target position as determined by the circle equation for time tn, $E(t_0)$, eye position in the last bin prior to opening of the feedback loop. The smoothly increasing extra retinal slip during open loop did not affect the cell's firing

nically. Opening of the feedback loop was timed to occur when target and eyes were directed leftward while moving on the circle, i.e. in the preferred direction of a putative direction-selective visual neuron. However, the additional retinal slip introduced in this way was without any consistent effect on the firing of the VT neuron under study, further supporting our thesis that at least some pontine VT neurons do not encode image motion on the retina. Whereas in the first case discharge rate was best related to the actual target trajectory, discharge rate in the second case did not reflect the temporary acceleration of the target during opening of the feedback loop. If we assume that short interruptions of maintained pursuit on a circle due to temporarily opening the feedback-loop are not sufficient to change the monkeys percept of a target moving with constant angular velocity, a comprehensive interpretation of the responses of the two VT neurons presented in Fig. 3 arises: These cells reflected the monkey's percept of a target moving in extrapersonal space rather than the movement of a target defined in physical terms!

Let us look at two further observations which support the hypothesis of a representation of a perceptual target motion by these cells.

If the small laser target moving on a screen is periodically blinked out while moving with constant velocity, a human observer has the impression of a continuously moving target running through a series of short tunnel segments (target off) separated by segments where the target is running above the ground (target on), so to speak. Monkeys, when confronted with this paradigm most probably also experience the percept of a continuously moving although not constantly visible target. We may infer this from the fact that they are able to track the physically discontinuous target smoothly for up to 1-s periods of blinking out. Figure 4 shows that tracking eye movements elicited by a blinking target moving on a circle in front of the monkey (Fig. 4A) are not worse than tracking eye movements elicited by a continuously visible target (Fig. 4B). And finally, what is most important for our hypothesis is that the responses of the few pontine VT neurons studied with this design also did not discriminate between the two situations. This is shown for a representative pontine VT neuron in Fig. 4. Again discharge rate seemed to be related to the percept of the continuously moving target rather than to image motion on the retina, which was discontinuous in the case of the blinked-out target.

In the standard laboratory situation subjects track visual stimuli with their eyes while trunk and head are stationary (eye tracking). During visual suppression of the vestibulo-ocular reflex (VOR) the opposite holds. The subject fixates a visual target, rotating about a common axis with the body and the head, thus suppressing the VOR which

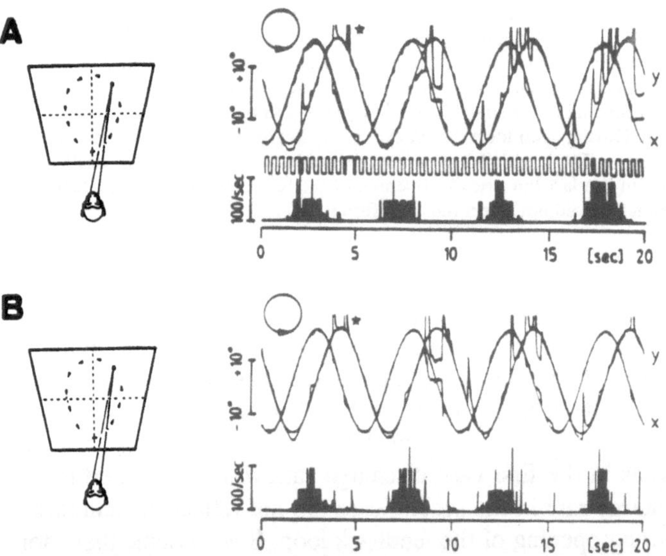

Fig. 4. Responses of a pontine VT neuron during visual tracking of a laser target moved counterclockwise over a homogeneous tangent screen (radius 20°; angular velocity 72°/s). Horizontal and vertical components of laser and eye position are superimposed. Eye position in this monkey was recorded with the electro-oculogram technique. Asterisks mark distortion of the vertical eye position trace due to eye lid blinks. **A** With periodic (410 msec on, 410 msec off) blinking of the continously moving target. **B** Response to tracking of the same target without blinking out the target (From Thier et al. 1988).

would otherwise be elicited to stabilize the image of the visual environment on the retina. Hence during visual suppression of the VOR the target is tracked with the moving head (head tracking) while the eyes are more or less stationary within the orbit. A VT neuron encoding target movement in extrapersonal space should be excited equally well, independent of the target being tracked by eye or head movement, as long as the target is moving in the preferred direction of that cell. The few VT neurons which could be studied under conditions of eye versus head tracking actually confirmed this expectation, as demonstrated in Fig. 5. During eye tracking the pontine VT neuron shown discharged whenever the target was directed leftward (Fig. 5A,B). As predicted by our hypothesis, during visual suppression of the horizontal VOR the cell fired whenever the target, which was now oscillated sinusoidally in synchrony with the head (and the trunk) about the animal's yaw axis, was directed leftward (Fig. 5C). Finally, suppression of the

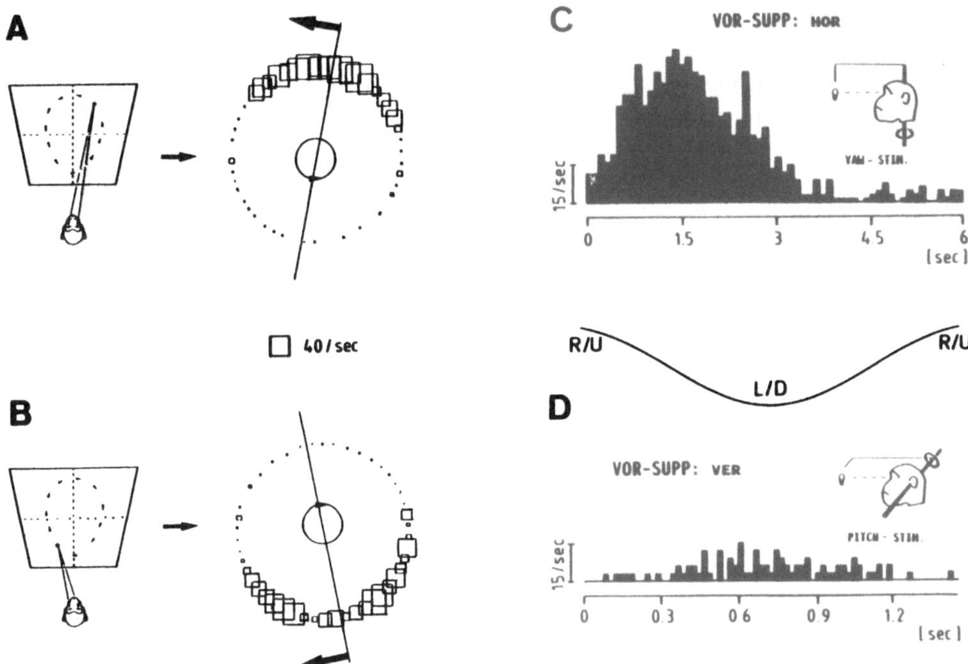

Fig. 5. Responses of a pontine VT neuron during 'eye-' (**A,B**) and 'head tracking' (**C,D**) compared. A,B: Smooth-pursuit eye movements were evoked by a laser target which moved with constant angular velocity (72 °/s, radius 20°) in a clockwise (**A**) or counterclockwise direction (**B**). Responses are represented by two-dimensional histograms plotting average discharge rate (bin width 100 ms) as a function of target position on the circle. Rectangle length and height are proportional to mean discharge rate. Rectangles are centered at the mean target position of that bin. Independent of the target moving in a clockwise or counterclockwise direction on the circle the discharge was largest whenever the vector of momentary target velocity pointed leftward. **C** Suppression of the horizontal VOR by fixation of a small red light oscillating in synchrony with the head about the yaw axis (amplitude 30°; period 6s). Firing rate is represented in form of a poststimulus time histogram (bin width 100 ms). Again the VT neuron discharged during leftward movement of the target. **D** Suppression of the vertical VOR. Oscillation of the small red light in synchrony with the head about the pitch axis (amplitude 7.5°; period 1.46 s). No or at best vague modulation of discharge rate. R/U: head maximal to the right (suppression of the horizontal VOR) or up (suppression of the vertical VOR); L/D: head maximal to the left or down, respectively

vertical VOR (Fig. 5D) and (not shown) oscillating the animal about either axis without visual target and in darkness (VOR in darkness) were ineffective in driving this cell.

In conclusion, various observations on pontine VT neurons suggest that at least some of them might encode the motion of a perceptual target in extrapersonal space. This hypothesis first of all offers a parsimonious conceptual basis which allows to predict the responses of pontine VT neurons correctly in all the paradigms studied by us so far. This is a major advantage when compared with alternative explanations that one might consider, which all suffer from the fact that they are usually useful only to predict the responses of a VT neuron in some of the paradigms applied. For instance, a continuous response to a periodically blinked out moving laser target as reported above would not necessarily require a neuronal representation of the motion of a perceptual target. One might perhaps also expect a continuous neuronal response to the blinked out target if the VT neuron were fed by an elementary visual movement detector of the Reichardt type (Reichardt and Varjú 1959) supplied with sufficiently different time constants between neighboring receptor inputs to account for the observed ability to tolerate blinking out of the target up to periods of 1 s without significantly impairing smooth-pursuit eye movements. Also an elementary movement detector would require a continuous displacement of the discontinuous target image into a consistent direction on the retina in order to yield a tonic output. During steady state tracking of a target moving in a predictable way, consistent retinal slip, however, may not be available and if present does not seem to be used by pontine VT neurons, which as we demonstrated are able to ignore even considerable amounts of retinal image displacement.

To calculate information on the movement of a perceptual target in space might at first sight look like an almost insuperable and moreover unnecessarily complex problem for the central nervous system. However, recent research on the saccadic system has demonstrated (Sparks and Mays 1983; Guthrie et al. 1983) that information on target position in a spatial rather than a retinal frame of reference is in fact attainable to the brain. To supply subsequent motor centers with a unified representation of target motion in extrapersonal space, i.e. independent of a specific sensory coordinate system, might in the end prove to be a very economical solution favored by evolution to support not only smooth-pursuit eye movements but goal-directed behavior in general.

Summary

The responses of a minority of neurons recorded from the dorsolateral pontine nucleus (DLPN) of monkeys trained to participate in a variety of visual and oculomotor tests are most parsimoniously described if one assumes that these neurons represent the motion of a perceptual target in extrapersonal space. Considering recent experimental evidence for a contribution of parts of the cerebral cortex to the generation of smooth-pursuit eye movements, which cannot be attributed to motion processing in a retinal frame of reference, it is furthermore suggested that this representation of target motion in space is already established on a cortical level. Hence with the DLPN serving as an anatomical interface the cerebral cortex seems to feed those parts of the cerebellum known to be essential for smooth-pursuit eye movements with a unified representation of target motion independent of specific sensory frames of reference.

References

Brodal P (1978) The corticopontine projection in the rhesus monkey. Origin and principles of organization. Brain 101:251-283

Bruce CJ, Goldberg ME (1985) Primate frontal eye fields. I. Single neurons discharging before saccades. J Neurophysiol 53:603-635

Dubner R, Zeki SM (1971) Response properties and receptive fields of cells in an anatomically defined region of the superior temporal sulcus in the monkey. Brain Res 35:528-532

DeYoe EA, Van Essen D (1988) Concurrent processing streams in monkey visual cortex. TINS 11:219-226

Dürsteler MR, Wurtz RH, Newsome WT (1987) Directional pursuit deficits following lesions of the foveal representation within the superior temporal sulcus of the macaque monkey. J Neurophysiol 57:1262-1287

Fries W (1981) The projection from striate and prestriate visual cortex onto the pontine nuclei in the macaque monkey. Soc Neurosci Abstr 7:762

Gertz H (1916) Über die gleitende (langsame) Augenbewegung. Z Psychol Physiol Sinnesorg Abt 2: Z Sinnesphysiol 49:29-58

Glickstein M, Cohen JL, Dixon B, Gibson A, Hollins M, Labossiere E, Robinson F (1980) Corticopontine visual projections in macaque monkeys. J Comp Neurol 190:209-229

Glickstein M, May JG, Mercier BE (1985) Corticopontine projection in the macaque: the distribution of labelled cortical cells after large injections of horseradish peroxidase in the pontine nuclei. J Comp Neurol 235:343-359

Guthrie BL, Porter JD, Sparks DL (1983) Corollary discharge provides accurate eye position information to the oculomotor system. Science 221:1193-1195

Harting JK (1977) Descending pathways from the superior colliculus: an autoradiographic analysis in the rhesus monkey (Macaca mulatta). J Comp Neurol 173:583-612

Keating EG, Gooley SG, Kenney DV (1985) Impared tracking and loss of predictive eye movements after removal of the frontal eye fields. Soc Neurosci Abstr 11:472

Lisberger SG, Westbrook LE (1985) Properties of visual inputs that initiate horizontal smooth-pursuit eye movements in monkeys. J Neurosci 5:1662-1673

Lisberger SG, Morris EJ, Tychsen L (1987) Visual motion processing and sensory-motor integration for smooth-pursuit eye movements. Ann Rev Neurosci 10:97-129

Lynch JC (1987) Frontal eye field lesions in monkeys disrupt visual pursuit. Exp Brain Res 68:437-441

Maunsell JHR, Newsome WT (1987) Visual processing in monkey extrastriate cortex. Ann Rev Neurosci 10:363-402

Mountcastle VB, Lynch JC, Georgopoulos A, Sakata H, Acuna C (1975) Posterior parietal association cortex of the monkey: command function for operations within extrapersonal space. J Neurophysiol 38:871-908

Mustari MJ, Fuchs AF, Wallman J (1986) The physiological response properties of single pontine units related to smooth pursuit in the trained monkey. In: Keller EL, Zee DS (eds) Adaptive processes in visual and oculomotor systems. Pergamon, Oxford, pp 253-260

Newsome WT, Wurtz RH, Dürsteler MR, Mikami A (1985) Deficits in visual motion processing following ibotenic acid lesions of the middle temporal visual area of the macaque monkey. J Neurosci 5:825-840

Reichardt W, Varjú D (1959) Übertragungseigenschaften im Auswertesystem für das Bewegungssehen (Folgerungen aus Experimenten an dem Rüsselkäfer Clorophanus viridis) Z Naturforschg 146: 674-689

Robinson DA (1963) A method of measuring eye movement using a scleral search coil in a magnetic field. IEEE Trans Biomed Eng 10:137-145

Sakata H, Shibutani H, Kawano K (1983) Functional properties of visual tracking neurons in posterior parietal association cortex of the monkey. J Neurophysiol 49:1364-1380

Sparks DL, Mays LE (1983) Spatial localization of saccade targets. I. Compensation for stimulation-induced perturbations in eye position. J Neurophysiol 49:45-63

Stanton GB, Goldberg ME, Bruce CJ (1988) Frontal eye field efferents in the macaque monkey: topography of terminal fields in midbrain and pons. J Comp Neurol 271:493-506

Steinbach MJ (1976) Pursuing the perceptual rather than the retinal stimulus. Vision Res 16:1371-1376

Suzuki DA, Keller EL (1984) Visual signals in the dorsolateral pontine nucleus of the alert monkey: their relationship to smooth-pursuit eye movements. Exp Brain Res 53:473-478

Thier P, Koehler W, Buettner UW (1988) Neuronal activity in the dorsolateral pontine nucleus of the alert monkey modulated by visual stimuli and eye movements. Exp Brain Res 70:496-512

Tusa RJ, Ungerleider LG (1988) Fiber pathways of cortical areas mediating smooth pursuit eye movements in monkeys. Ann Neurol 23:174-183

Tychsen L, Lisberger SG (1986) Visual motion processing for the initiation of smooth-pursuit eye movements in humans. J Neurophysiol 56:953-968

Ungerleider LG Mishkin M (1982) Two cortical visual systems. In: Ingle DJ, Goodale MA, Mansfield RJW (eds) The analysis of visual behavior. MIT Press, Cambridge, MA, pp 549-586

Wurtz RH, Newsome WT (1985) Divergent signals encoded by neurons in extrastriate areas MT and MST during smooth pursuit eye movements. Soc Neurosci Abstr 11:1246

Young LR (1977) Pursuit eye movement-what is being pursued? Control of gaze by brain stem neurons. Dev Neurosci 1:29-36

Optokinetic and Smooth-Pursuit Response After Adaptive Modification of the Vestibulo-Ocular Reflex

G. D. Paige, J. M. Fredrickson
Department of Otolaryngology, Washington University School of Medicine, St. Louis, MO 63110

Introduction

The vestibulo-ocular reflex (VOR) serves to maintain retinal image stability during rotation by utilizing semicircular canal input to generate conjugate eye movements that are equal but opposite to head movements. The VOR is related to the optokinetic reflex (OKR) which maintains retinal image stability by generating eye movement which follows large field visual motion. Another visual system, that of smooth pursuit (SP), allows accurate tracking of small central retinal targets. The VOR, OKR, and SP systems work together during natural motion to collectively assure stability of the desired visual target image. When VOR performance fails, and image stability is lost during head movements, visual-vestibular interaction is employed to adaptively modify and improve the VOR.

Adaptive modification of the primate and human VOR has been demonstrated following visual-vestibular mismatch induced by optical manipulations, such as 2x lenses (Istl-Lenz et al. 1985; Miles and Eighmy 1980; Lisberger et al. 1981), or vestibular lesions (Paige 1983b). Further, VOR changes are accompanied by changes in OKR responses (Lisberger et al. 1981; Paige 1983b) which presumably reflect shared circuitry between the VOR and OKR (Waespe and Henn 1977). This interaction is thought to influence primarily low-frequency (< 0.1 Hz) responses and not higher frequency or SP responses. Direct confirmation is limited, however, particularly in humans. Further, comparison of the OKR and SP frequency response before and after adaptive VOR modification is lacking.

This study evaluates the response characteristics of the human OKR and SP systems above 0.1 Hz before and after adaptive VOR modification induced by 2x lenses. The data allow a direct comparison between OKR and SP dynamics as well as potential changes in either or both system which might parallel VOR modification.

Methods

Eye movements were recorded (search coil technique) from 12 normal human volunteers (aged 22-88) before and after an 8 hour period of natural activity while wearing 2x binoculars (Nikon). The OKR was measured during oscillation of an alternating black and white striped pattern projected by a servocontrolled planetarium onto an 8' diameter cylindrical screen which surrounded the subject. SP was measured during tracking of an oscillating small red laser target reflected by an X-Y galvo system onto the same screen. The subject sat on a servocontrolled rate table with the head firmly restrained, which also allowed evaluation of the VOR in darkness.

The VOR, OKR, and SP systems were studied during sinusoidal oscillations in the range 0.25-4.0 Hz (limited to 1 Hz for OKR), at peak velocities of between 25 °/s and 100 °/s (see Table 1).

For each stimulus presentation, eye position and stimulus velocity signals were recorded digitally (200 Hz sample rate). Eye signals were differentiated and smoothed, and "fast" eye movements were removed. The resulting eye velocity response and the stimulus velocity signals were averaged over selected cycles, requiring at least 3 usable cycles for a trial to be accepted. The averaged eye and stimulus velocity signals were then fit to a sinusoid, and gain (eye/stimulus peak velocities) and phase (eye-stimulus phase angles) were calculated.

Results

Vestibulo-ocular Reflex. Although the focus of this report is on OKR and SP responses, the motivation for study requires acquisition of adaptive modification of the VOR. For purposes here, the VOR showed a near uniform gain increase averaging 37% in the studied bandwidth (0.25-4.0 Hz) after the 8 hour period of 2x lens adaption.

Optokinetic Reflex and Smooth Pursuit. Gain and phase of OKR and SP responses for all stimuli presented, both before and after 2x lens adaptation, are listed in Table 1. For each frequency and amplitude, and for both OKR and SP, no significant change in gain or phase was observed after wearing 2x lenses for 8 hours.

Potential differences between OKR and SP were also assessed, regardless of 2x lens effects. Differences were indeed observed, particularly at 1.0 Hz. OKR gains tended to be larger than SP gains while phase lags of OKR were less than those of SP. The effects

Table 1. OKR and SP gain and phase before and after 2x lens adaption

STIM.	AMP. o/s	FREQ. Hz	N	GAIN Before	SD	GAIN After	SD	PHASE Before	SD	PHASE After	SD
OKR	25	0.25	5	0.89	0.15	0.84	0.11	5.6	1.5	4.5	2.0
OKR	50	0.25	12	0.80	0.13	0.85	0.13	1.4	3.4	2.4	4.3
OKR	100	0.25	6	0.59	0.13	0.55	0.10	6.7	8.1	5.5	7.6
OKR	25	1.00	8	0.80	0.20	0.79	0.13	-0.5	6.3	0.9	4.8
OKR	50	1.00	12	0.64	0.17	0.70	0.20	-5.5	9.0	-3.3	8.9
OKR	100	1.00	9	0.60	0.16	0.60	0.20	-2.1	5.4	-0.2	6.1
SP	25	0.25	3	0.97	0.01	0.90	0.08	0.4	0.8	0,8	1.6
SP	50	0.25	12	0.85	0.11	0.84	0.10	3.0	2.5	2.4	2.1
SP	25	1.00	9	0.68	0.11	0.68	0.10	-8.1	5.2	-7.8	5.7
SP	50	1.00	11	0.60	0.15	0.63	0.17	-9.7	4.8	-8.6	5.8
SP	100	1.00	8	0.45	0.14	0.45	0.16	-12.5	4.9	-11.5	6.9
SP	25	2.50	7	0.21	0.13	0.26	0.17	-47.6	9.2	-43.2	11.3
SP	10	4.00	5	0.11	0.04	0.09	0.03	-81.1	35.3	-65.9	28.3

held at all stimulus amplitudes and were greatest at 100 °/s. Differences were highly significant for phase ($p < 0.005$, t-test on paired observations), but more variable for gain.

Discussion

After wearing 2x binoculars for 8 hours, normal human subjects experienced an average increase in VOR gain of 37% in the frequency bandwidth 0.25-4.0 Hz. This is in keeping with other studies utilizing similar methods in humans (Istl-Lenz et al. 1985; Collewijn et al. 1983). The robust change in VOR response characteristics was not accompanied by changes in either the OKR or SP in the same bandwidth, even when the systems were stressed by increasing stimulus velocity to 100 °/s; a technique useful for enhancing potential differences, as employed in showing low-frequency OKR effects (Lisberger et al. 1981; Paige 1983b).

The OKR in primates includes two components; a fast component thought responsible for responses to high frequency (>0.1 Hz) and rapid OKR stimuli and a slow component which is thought to reflect shared circuitry with the VOR (Cohen et al. 1977; Lisberger et al. 1981; Paige 1983a). The OKR changes induced during adaptive modification of the VOR seem limited to the OKR slow component, as implied by studies in monkeys (Lisberger et al. 1981; Paige 1983b). These studies failed to show changes in OKR responses to rapid- or high-frequency stimuli. The current study extends observations to humans and in a more detailed fashion in the frequency domain. Further, the SP system was shown to be clearly resistent to change despite VOR modification. While it is known that SP is unaltered by prolonged adaptation to reversing prisms (Melvill Jones and Gonshor 1982), this optically induced phenomenon is different from magnification, even in its effects on the VOR.

A seemingly unexpected finding is that the OKR performs better than SP as both systems are stressed by higher frequencies and/or stimulus velocities. Thus, while the two systems respond nearly indistinguishably at 0.25 Hz, differences between them are noticeable at 1.0 Hz. Compared to OKR responses, SP gain is lower while phase lag is greater for all stimulus velocities. The differences are largest for the greatest stimulus velocity. A similar phenomenon was described by Van den Berg and Collewijn (1986), who found that OKR gain was greater than SP of a small target during constant velocity stimuli. Presumably, the performance enhancement of OKR over SP reflects the considerably greater retinal stimulation provided by the full-field striped OKR stimulus as compared to the small and discrete SP target.

Summary

Optokinetic reflex (OKR) and smooth-pursuit (SP) responses were studied during modest- to high-frequency oscillatory stimuli before and after modification of the vestibulo-ocular reflex (VOR) induced by wearing 2x lenses. Although robust changes occurred in the VOR, no alteration in OKR or SP response characteristics were observed. However, differences between OKR and SP dynamics were identified, in that the OKR performed better than SP when both systems were stressed by high-stimulus frequencies and amplitudes.

References

Cohen B, Matsuo V, Raphan T (1977) Quantitative analysis of the velocity characteristics of optokinetic nystagmus and optokinetic after-nystagmus. J. Physiol. 270: 321-344.

Collwwijn H, Martins A J, Steinman, R M (1983) Compensa-tory eye movements during active and passive head movements: Fast adaptation to changes in visual magnification. J. Physiol. 340: 259-286.

Istl-Lenz Y, Hyden Dr., Schwarz D W F (1985) Response of the human vestibulo-ocular reflex following long-term 2x magnified visual input. Exp. Brain Res. 57: 448-455.

Lisberger S G, Miles F A, Optican L M, Eighmy B B (1981) Optokinetic response in the monkey: underlying mechanisms and their sensitivity to long-term adaptive changes in vestibuloocular reflex. J. Neurophysiol. 45: 869-890.

Melvill Jones G, Gonshor A (1982) Oculomotor response to rapid head oscillation (0.5-5.0 Hz) after prolonged adaptation to vision-reversal. Exp. Brain Res. 45: 45-58.

Miles F A, Eighmy B B (1980) Long-term adaptive changes in primate vestibuloocular reflex. I. Behavioral observations. J. Neurophysiol. 43: 1406-1425.

Paige G D (1983a) Vestibulo-ocular reflex and its interactions with visual following mechanisms in the squirrel monkey. I. Response characteristics in normal animals. J. Neurophysiol. 49: 134-151.

Paige G D (1983b) Vestibulo-ocular reflex and its interactions with visual following mechanisms in the squirrel monkey. II. Response characteristics and plasticity following unilateral inactivation of horizontal canal. J. Neurophysiol. 49: 152-168.

Van Den Berg AV, Collewun H (1986) Human smooth pursuit: effects of stimulus extent and of spatial and temporal constraints of the pursuit trajectory. Vis. Res. 26: 1209-1222.

Waespe W, Henn V (1977) Neuronal activity in the vestibular nuclei of the alerk monkey during vestibular and optokinetic stimulation. Exp. Brain Res. 27: 532-538.

The Detection of Motion by the Vestibular System

V. Henn
Department of Neurology, University Hospital, Zürich, Switzerland

In Vienna, in 1873, Ernst Mach and Josef Breuer, independently of each other, gave the first valid descriptions of the vestibular system. In the same year, working in Edinburgh, A. Crum Brown came to similar conclusions. Of the three, Mach had the farthest reaching impact, giving a thorough description with publication of his book "Fundamentals of a Theory of Motion Perception." In what way was his approach different, thereby leading to such an impact?

Mach himself provides an answer in the summary and conclusions of his book. From psychophysical experiments he concluded that we must have detectors of angular and linear accelerations. Such a receptor organ must be located in the head. The labyrinth of the inner ear (Fig. 1) would be anatomically suitable for such a function, but the psychophysical measurements and general conclusions would be independent of such an assumption. These statements already reveal why his approach was so powerful. He differentiated between experimental observations, the logical description of the system, and its biological realization.

Experimental Observations

Originally, one of the starting points of vestibular physiology was the concept of a sensory system to detect motion with the labyrinths as specific end organs, but to which

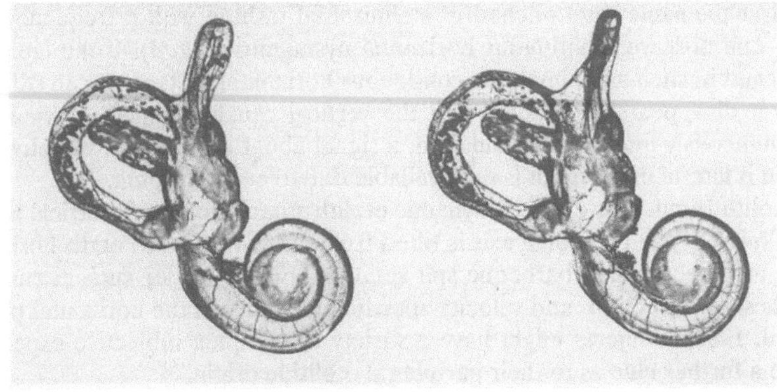

Fig. 1. Stereoscopic photograph of human membraneous labyrinth (Gray 1907)

also the visual and somatosensory system would contribute. For the physiologist and anatomist this broad concept was soon reduced to the three-neuron vestibulo-ocular reflex arc and the vestibulo-spinal reflexes. This mirrors early days of physiology when great advances were achieved by confining complex problems to simple input-output relations. This led to the discovery of basic principles of sensorimotor connections together with the necessary anatomical pathways.

Only recently was this broad concept taken up again, as several questions needed an answer. Is the three-neuron reflex sufficient to explain the phenomenon of nystagmus in a normal animal or in a patient observed clincially? The obvious answer is no. Only in the deeply comatose patient, during rapid passive head turning, might we observe the action of the three-neuron vestibulo-ocular reflex arc in isolation. Under all other circumstances, when nystagmus occurs, many more neuron populations in the brainstem and cerebellum are involved. From a physiological point of view, the following question must be considered. With adequate stimulation of horizontal canals we expect horizontal nystagmus, and we think, there is a valid physiological and anatomical explanation for it. However, if we reverse the argument, is it still correct? If we observe horizontal nystagmus, does it mean that activity responsible for it originated in the horizontal canal system? The following examples should be considered.

1. Activity Arising in Vertical Canals. Ewald (1892) introduced the method of canal plugging. The bony portion of a canal is surgically opened and sealed. If this is done for corresponding canals, no vestibular nystagmus can be elicited in this specific canal plane as no pressure gradient can build up over the cupula. Böhmer et al. (1985) report about such experiments in rhesus monkeys. After both horizontal canals had been plugged, the animals still exhibited direction-specific nystagmus over a wide range of head positions (Fig. 2). Obviously, this activity must come from the vertical canals. An additional finding is important: the canals in the monkey like in most species studied, are not exactly aligned in an orthogonal fashion (Blanks et al. 1985). In the rhesus monkey, the horizontal canals lie in a plane tilted about 15° upward relative to the stereotaxic horizontal plane, but the nullposition of the vertical canals are tilted by about 30°. Therefore the vertical null plane is by about 15° different from the plane of the lateral canals. Another paradigm explored pitching while rotating (Raphan et al. 1983). If an animal is continuously rotated about an earth-vertical axis and at the same time pitched in a sinusoidal fashion with a frequency of about 0.1 Hz, one observes continuous horizontal nystagmus (Fig. 3). If the lateral canals are plugged in such an animal, the continuous horizontal nystagmus can still be observed, but it disappears in animals with the vertical canals plugged. Obviously during simultaneously multi-axis stimulation, a signal about actual head velocity is calculated, even if lateral canal input is not available due to canalplugging.
2. Otholith Input. The classical dynamic otolith stimulation is off-vertical axis rotation in darkness. If the rotational axis is tilted from earth-vertical to earth-horizontal, then it has also been called barbeque spit rotation (Fig.4). Under such circumstances, animals show direction- and velocity-specific nystagmus in the horizontal plane (Raphan et al. 1981). Subjects might have a variety of complex subjective experiences which give a further clue as to their peripheral otolithic origin.
3. Visual Input. In the form of optokinetic stimulation visual input leads to horizontal nystagmus. This has special significance, since it was shown that such stimulation to

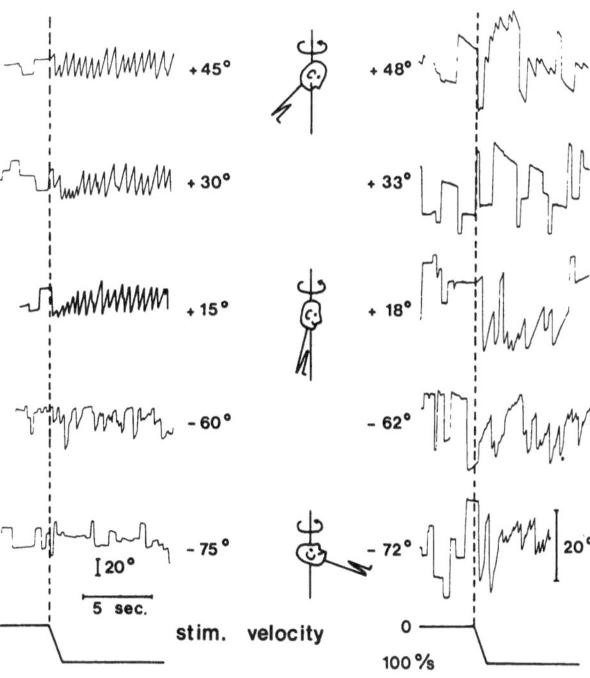

Fig. 2. Horizontal vestibular nystagmus as a function of static head position in a normal monkey (left) and in an animal with both lateral semicircular canals plugged (right). Rotation was about an earth-vertical axis with an acceleration of 100 °/s² over 1 s. With a pitch angle of -75° the axis of gaze is approximately aligned with the rotation axis, and no horizontal nystagmus is induced in the normal animal. In the canal-plugged monkey nystagmus in the compensatory direction occurs at pitch angles smaller than 33°; at pitch angles greater than 33° nystagmus is reversed and anticompensatory (Böhmer et al. 1985)

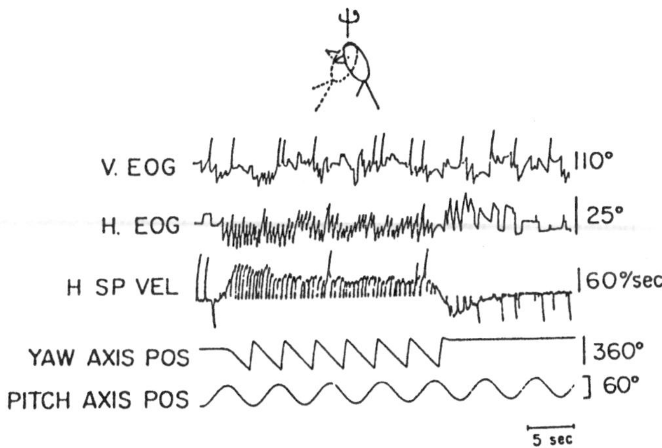

Fig. 3. Per- and postrotatory nystagmus in a monkey induced by oscillation in the fore-aft direction about a horizontal axis (pitch axis) while the animal was being rotated at constant velocity about a vertical axis in darkness. Note the continuous nystagmus during constant-velocity yaw-axis rotation and the shortened postrotatory nystagmus when the animal was decelerated (Raphan et al. 1981)

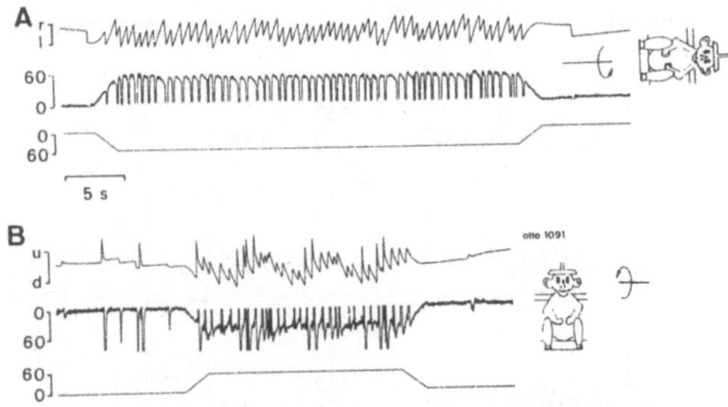

Fig. 4. Off-vertical axis rotation of a monkey in total darkness. A, Barbeque spit rotation; B, somersaulting. From top: horizontal (A) or vertical (B) eye position, eye velocity, and turntable velocity. Note that nystagmus velocity represents actual head velocity during the whole period of rotation without after nystagmus after deceleration (Henn 1982)

the nonmoving animal gives rise to direction- and velocity-specific activation of neurons in the vestibular nuclei (Fig.5) (Dichgans et al. 1973; Henn et al. 1974; Waespe and Henn 1979). Over a wide range there is even quite a close parallelism between single neuron activity in the vestibular nuclei and psychophysical phenomena (summarized in Henn et al 1980).

4. Somesthetic Input. Joint, muscle, and tendon receptors, often summarily quoted as neck receptors can induce nystagmus. While their role under normal conditions, still must be fully characterized, under pathological conditions, like after peripheral labyrinthine lesions, their gain usually increases to generate compensatory nystagmus during active or passive head movements (Dichgans et al. 1974). Further examples of somesthetic input are the nystagmus produced by continuous shoulder-arm movements (Brandt et al. 1977) termed arthrokinetic nystagmus, or the motion sensation and nystagmus produced by walking in circles. This last example was originally investigated by Purkyné (1820; reviewed in Grüsser 1984). Under more sophisticated conditions it had been reintroduced by Bles (1981). A subject walks in a small circle (diameter about 1 m) exhibiting appropriate horizontal nystagmus. If one starts to rotate the platform on which the subject walks into the opposite direction, one can adjust its velocity in such a way that the subject actually does not move, but he still believes that he is moving in circles and exhibits appropriate nystagmus. This paradigm promises interesting results after Solomon and Cohen (1987) succeeded in training rhesus monkeys for such a task.

These examples confirm that the occurrence of horizontal nystagmus does not necessarily mean that activity responsible for it arose in the lateral semicircular canals. Rather, information from many different receptor systems is utilized to reconstruct actual head velocity (Fig.6), and this information is then fed into second-order vestibular neurons which connect to the oculomotor or vestibulospinal systems. The example of increased gain of neck oculomotor reflexes after labyrinthine lesions shows that more phenomena might be revealed under conditions of pathology. Also, the

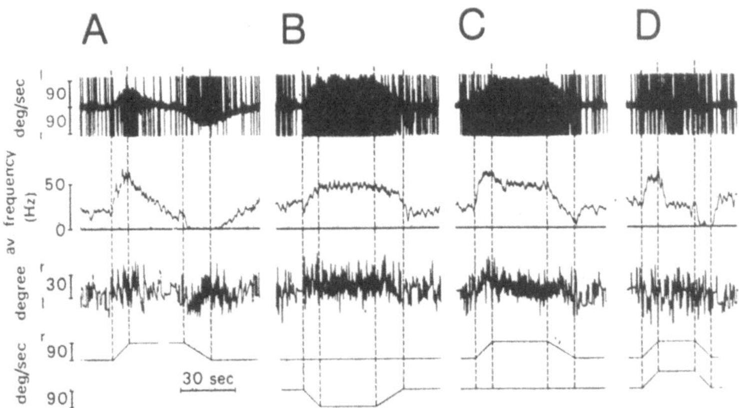

Fig. 5. Visual-vestibular interaction in the alert monkey. From above horizontal eye velocity, activity of a type I vestibular neuron, horizontal eye position, turntable and optokinetic drum position. In A, the monkey was rotated in total darkness leading to activation of the neuron and appropriate vestibular nystagmus. In B, during optokinetic stimulation while the monky is stationary, the vestibular neuron is also activated in parallel with optokinetic nystagmus. In C and D during combined or conflict stimulation, neuron activity and nystagmus velocity can be dissociated showing that the neuronal signal does not merely represent eye velocity (Waespe and Henn 1978)

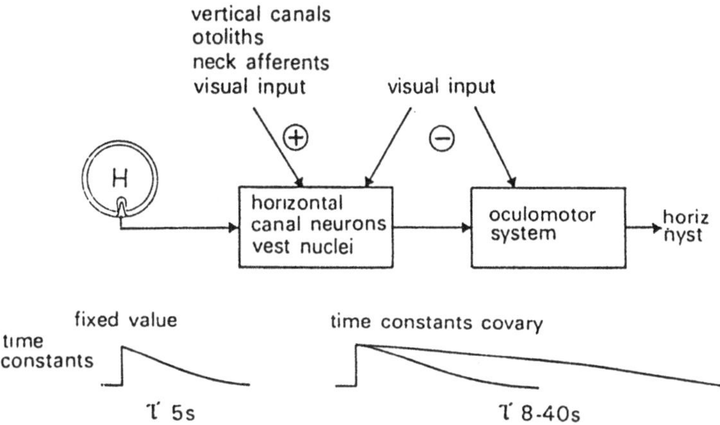

Fig. 6. Scheme of information processing in the vestibulo-ocular reflex. Neurons in the vestibular nuclei receiving an input from lateral semicircular canals, can also be activated by input from vertical canals, otoliths, neck afferents or visual input. Another step in information processing is velocity storage, i.e. the prolongation of the vestibular time constant, which, however, can vary widely depending on the state of habituation

above considerations were essentially confined to the horizontal plane. However, the labyrinth detects acceleration in three dimensions. Although in the periphery acceleration in the three planes of canals seems to be coded in the same terms, the dynamics of the vestibulo-ocular reflex is different for the horizontal, vertical, or torsional system. It is even asymmetric for up and down. This shows that even under physiological conditions gravity plays a role in modifying the response. In conclusion, we are faced with a

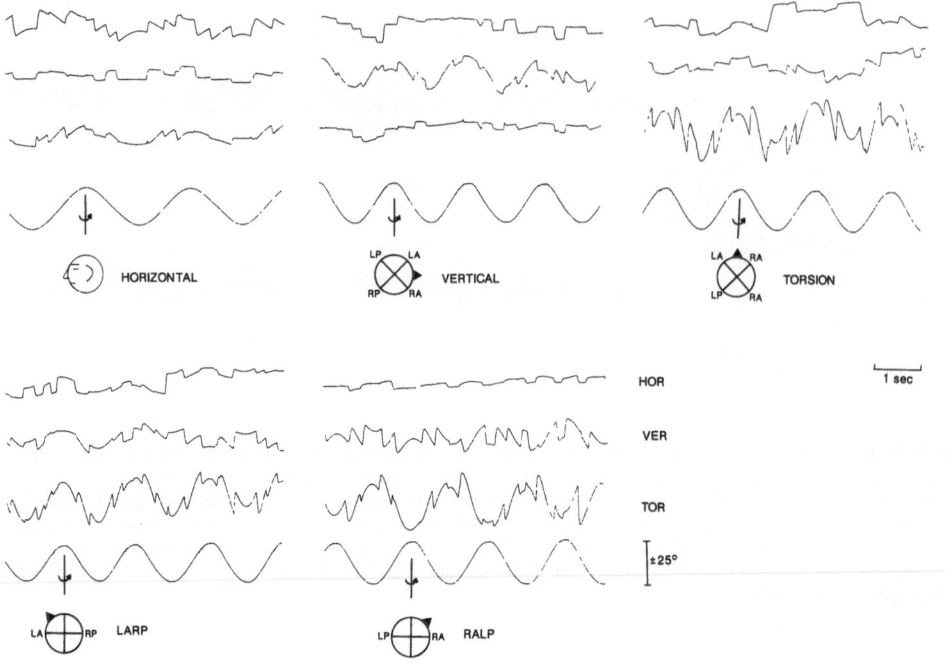

Fig. 7. Eye position recording in three dimensions in a normal monkey during rotation about different body axes. From above, horizontal, vertical, and torsional eye position, and turntable position. LARP and RALP rotations are in the planes of the left anterior - right posterior, or the right anterior - left posterior canals, respectively

wealth of observations which can certainly not be explained merely by a three-neuron reflex arc. The experimental characterization of the direct vestibulo-ocular connection was of greatest importance for a basic understanding of vestibular physiology on which to build more complicated circuits to interpret the observations quoted above.

For the experimenter, devising meaningful experiments is crucially limited by techniques. For adequate vestibular stimulation, multi-axis turntables with appriopriate optokinetic stimulators seem to depend primarily on how much money and effort one wants to spend. This practical problem was commented upon by Mach with the note that his multi-axis turntable for human experiments had cost him about one month's salary, and he gave practical advice on how to duplicate it so that critical collegues might be able to check his observations (Henn and Young 1975).

Another experimental problem is the task to make measurements with as little interference as possible for the parameter to be measured, in our case eye movements. Robinson described the magnetic search coil method in fully three dimensions in 1963. It is puzzling why it took about 20 years until this method was taken up again (Collewijn et al. 1985) to measure eye movements in three dimensions (Fig.7), especially after several models had been developed to describe the necessary three-dimensional transformations in the vestibulo-ocular reflex, but with little data to test these models. One such model had been suggested by Robinson (1982).

Further information has recently been added by detailed studies of pathophysiology. Old problems have been taken up again, such as the observations of recovery after one-sided labyrinthine lesions. A decisive new technique was added by the local administration of transmitter agonists or antagonists (e.g. muscimol or bicuculline), local anesthetics (lidocaine), or cell toxins (ibotenic or kainic acid) which inactivate or destroy cells but leave fiber systems relatively unaffected. With such techniques the velocity-to-position integrator, intercalated between the vestibular and oculomotor system, has been tentatively located to a region extending to the perihypoglossal complex and medial vestibular nuclei (Cannon and Robinson 1987; Cheron and Godaux 1987). One will probably learn much from such experiments, the idea being to characterize a local population of neurons by physiological means and then to locally interfere with their information processing. This is akin to the technique of cutting up a feedback loop in a mathematical model.

Logical Description of the System

Another important aspect of progress in understanding involves the close interaction between those who diligently observe phenomena and those who create models to describe the logical structure. Robinson (1986) and Steinman (1986) in a sort of rethorical debate each defended their respective points of view - and both are right. The wealth of observations need a structure to interconnect them in a logical scheme. Such a structural description is much too complex to be described verbally. On the other hand, such a model without the challenge of new and unexpected observations would not be particulary interesting. Robinson repeatedly stressed the importance of models (1975, 1981). One must identify a function such as gaze stabilization and describe the logical steps which are necessary to achieve such a function. As experimental data become available, these are incorporated, and the performance of the model can be tested. Often parameters need adjustment, or it becomes clear that additional functions have to be added. One aims at a functional description of the system. These anatomical pathways run in parallel, and it would be an important challenge to actually model them using parallel processing hardware.

Biological Realization

With the relatively detailed knowledge about anatomy, it seems clear how the basic circuitry of the vestibular system is realized. However, above several examples of complex three-dimensional transformations have been listed. Although off-vertical axis rotation and resultant nystagmus have been modeled, it is by no means clear how the neuronal networks actually work. Other examples are the question of how the vestibular acceleration signal is integrated and transformed to move the eyes in a compensatory fashion, or how the visual system interfaces to generate visually induced compensatory eye movements or movements towards a visual target. Further examples are habituation or plasticity. Describing these phenomena in quantitative terms has meant progress. If these descriptions can make valid predictions, then they are true, even if we do not know how they are biologically realized, or if it turns out that our hypotheses will prove wrong.

The strength of the investigation by Mach lay in the fact that he clearly separated observations and logical descriptions from hypotheses about biological realization. Observations might not be accurate enough, or they might be irrelevant, but as long as they are reproducible, they are a fact. A logical description of a function must be consistent. The one step where one might be plainly wrong is in interpretation, the question of how the observations are reduced to functions of neuronal circuitry.

When considering how our understanding of the vestibular system evolved, one is struck by the fact that Purkyně in 1820 was one of the first to make scientific observations about the vestibular system without knowing its anatomical or physiological basis. One wonders what the factors were that in 1873 led Mach, Breuer, and Crum Brown, all independently, to describing labyrinthe function in a way that is still scientifically accepted today (Henn 1984). Above, several more examples of observations or experiments have been listed which actually could have been conducted 20 or 50 years ago. This again shows that although we strive at a logical description of the world around us, the progress which leads to such a goal is by no means logical itself.

Summary

Angular acceleration in the horizontal plane leads via the three-neuron vestibulo-ocular reflex to compensatory eye movements in the horizontal plane. However, horizontal compensatory eye movements can also be elicited by stimulation of the vertical canals, otoliths, visual input, neck, and somesthetic inputs. Central vestibular structures reconstruct actual head velocity in three dimensions by using afferent information from many different receptor systems. Mathematical models were instrumental in the interpretation of experimental data, and in turn the parallel channels used in the vestibulo-ocular reflex are a challenge for modeling using parallel computing.

References

Blanks RHI, Curthoys IS, Bennett ML, Markham CH (1985) Planar relationships of the semicircular canals in rhesus and squirrel monkeys. Brain Res 340: 315-324

Bles W (1981) Stepping around: circular vection and coriolis effects. In: Long J, Baddelev A (eds) Attention and performance. Erlbaum, Hillsdale, NJ, pp 47-61

Böhmer A, Henn V, Suzuki JI (1985) Vestibulo-ocular reflexes after selective plugging of the semicircular canals in the monkey - response plane determinations. Brain Res 326: 291-298

Brandt T, Büchele W, Arnold F (1977) Arthrokinetic nystagmus and ego-motion sensation. Exp Brain Res 30: 331-338

Breuer J (1874) Ueber die Function der Bogengänge des Ohrlabyrinthes. Med Jahrbücher (Wien): 72-124

Cannon SC, Robinson DA (1987) Loss of the neural integrator of the oculomotor system from brain stem lesions in monkey. J Neurophysiol 57: 1383-1409

Cheron G, Godaux E (1987) Disabling of the oculomotor neural integrator by kainic acid injections in the prepositus vestibular complex of the cat. J Physiol (Lond) 394: 267-290

Collewijn H, Van der Steen J, Ferman L, Jansen TC (1985) Human ocular counterroll: assessment of static and dynamic properties from electromagnetic scleral coil recordings. Exp Brain Res 59: 185-196

Crum Brown A (1874) On the sense of rotation and the anatomy and physiology of the semicircular canals of the internal ear. J Anat Physiol 8: 327-331

Dichgans J, Schmidt CL, Graf W (1973) Visual input improves the speedometer function of the vestibular nuclei in the goldfish. Exp Brain Res 18: 319-322
Dichgans J, Bizzi E, Morasso P, Tagliasco V (1974) The role of vestibular and neck afferents during eye-head coordination in the monkey. Brain Res 71: 225-232
Ewald E (1892) Physiologische Untersuchungen über das Endorgan des Nervus octavus. Wiesbaden: Bergmann
Gray AA (1907-1908) The labyrinths of animals. Churchill, London
Grüsser OJ (1984) J.E. Purkyně's contributions to the physiology of the visual, the vestibular and the oculomotor systems. Hum Neurobiol 3: 129-144
Henn V (1982) E. Mach on the analysis of motion sensation. Human Neurobiol 3: 145-148
Henn V (1982) The correlation between motion, sensation, nystagmus, and activity in the vestibular nuclei. In: Honrubia V, Brazier MAB (eds) Nystagmus and Vertigo. Academic, New York, pp 115-124
Henn V, Young LR (1975) Ernst Mach on the vestibular organ 100 years ago. ORL 37: 138-146
Henn V, Young LR, Finley C (1974) Vestibular nucleus units in alert monkeys are also influenced by moving visual fields. Brain Res 71: 144-149
Henn V, Cohen B, Young LR (1980) Visual-vestibular interaction in motion perception and the generation of nystagmus. Neurosci Res Prog Bull 18: 457-651
Mach E (1873) Physikalische Versuche über den Gleichgewichtssinn des Menschen. Akad Wiss Wien (Abt 3) 68: 124-140
Mach E (1875) Grundlinien der Lehre von den Bewegungsempfindungen. Engelmann, Leipzig. Reprint (1967) Bonset, Amsterdam. English translation: Outlines of a theory of motion sensation. SLA Translation Center, John Crerar Library, Chicago
Purkyně JE (1820) Beyträge zur näheren Kenntniss des Schwindels aus heautognostischen Daten. Med Jahrb Österr Staates 6: 79-125
Raphan T, Cohen B, Henn V (1981) Effects of gravity on rotatory nystagmus in monkeys. In: Cohen B (ed) Vestibular and oculomotor physiology. Ann NY Acad Sci 374: pp 44-55
Raphan T, Cohen B, Suzuki JI, Henn V (1983) Nystagmus generated by pitch while rotating. Brain Res 276: 165-172
Robinson DA (1963) A method of measuring eye movement using a scleral search coil in a magnetic field. IEEE Trans Biomed Electron 10: 137-145
Robinson DA (1975) Oculomotor control signals. In Lennerstrand G, Bach-y-Rita P (eds): Basic mechanisms of ocular motility and their clinical implications. Pergamon, Oxford, pp 337-374
Robinson DA (1981) The use of control systems analysis in the neurophysiology of eye movements. Ann Rev Neurosci 4: 463-503
Robinson DA (1982) The use of matrices in analyzing the three-dimensional behavior of the vestibulo-ocular reflex. Biol Cybern 46: 53-66
Robinson DA (1986) The systems approach to the oculomotor system. Vision Res 26: 91-99
Solomon D, Cohen B (1987) Head and eye movements during circular locomotion. Soc Neurosci Abstr 13: 1225
Steinman RM (1986) The need for an eclective, rather than systems, approach to the study of the primate oculomotor system. Vision Res 26: 101-112
Waespe W, Henn V (1978) Visual-vestibular interaction in motion detection and generation of nystagmus in the vestibular nuclei of alert monkeys. In: Hood JD (ed) Vestibular mechanisms in health and disease. Academic, London. pp 66-72
Waespe W, Henn V (1979) The velocity response of vestibular nucleus neurons during vestibular, visual, and combined angular acceleration. Exp Brain Res 37: 337-347

Dichgans J, Schmidt CL, Graf W (1973) Visual input improves the speedometer function of the vestibular nuclei in the goldfish. Exp Brain Res 18:319–322
[illegible] (1977) The [illegible] monkey [illegible] coordination in the monkey. Brain Res 12: [illegible]
Ewald JR (1892) Physiologische Untersuchungen über das Endorgan des Nervus Octavus. Wiesbaden, Bergmann
Gray AA (1907/1908) The labyrinth of animals. Churchill, London
Griffith CR (1920) [illegible] contribution to the physiology of the [illegible] vestibular and the oculomotor system [illegible] 1:129–154
[illegible] (1974) Mach on the analysis of motion sensation. Human Factors [illegible]
Henn V (1982) The correlation between motion sensation, nystagmus, and activity in the vestibular nuclei. In: Honrubia V, Brazier MAB (eds) Nystagmus and Vertigo. Academic, New York, pp 175–189
Henn V, Young LR (1975) Ernst Mach on the vestibular organ 100 years ago. ORL [illegible]
Henn V, Young LR, Finley C (1974) Vestibular nucleus units in alert monkeys are also influenced by moving visual fields. Brain Res 71:144–149
Henn V, Cohen B, Young LR (1980) Visual-vestibular interaction in motion perception and the generation of nystagmus. Neurosci Res Program Bull 18:457–651
Mach E (1873) Physikalische Versuche über den Gleichgewichtssinn des Menschen. Akad Wiss Wien (Math-Nat) 68:124–140
Mach E (1875) Grundlinien der Lehre von den Bewegungsempfindungen. Engelmann, Leipzig. Reprint 1967 Bonset, Amsterdam. English translation: Outlines of a theory of motion sensation. [illegible] Collection Center, John Crerar Library, Chicago
Purkinje JE (1820) Beiträge zur näheren Kenntnis des Schwindels aus heautognostischen Daten. Med Jahrb Österr Staates 6:79–125
Raphan T, Cohen B, Henn V (1981) Effects of gravity on rotatory nystagmus in monkeys. In: Cohen B (ed) Vestibular and oculomotor physiology. Ann NY Acad Sci 374: pp 44–55
Raphan T, Cohen B, Suzuki JI, Henn V (1983) Nystagmus generated by sinusoidal pitch while rotating. Brain Res [illegible]
Robinson DA (1963) A method of measuring eye movement using a scleral search coil in a magnetic field. IEEE Trans Biomed Electron 10:137–145
[illegible]
[illegible]
[illegible]
[illegible] Vision Res [illegible]
[illegible] 1925
[illegible] oculomotor system. Vision Res 12:[illegible]
Waespe W, Henn V (1978) Visual-vestibular interaction in the flocculus [illegible] generation of nystagmus [illegible]

Role of Neck and Visual Afferents for Self and Object Motion Perception in Labyrinthine Defective Subjects

T. Mergner, S. Heimbrand, M. Müller, C. Siebold and W. Becker*
Abteilung Klinische Neurologie und Neurophysiologie, Universität Freiburg, Hansastr. 9, 7800 Freiburg, FRG
*Sektion Neurophysiologie, Universität Ulm, Oberer Eselsberg, 7900 Ulm, FRG

Spatial orientation is known to depend on a complex interplay of several sensory systems. Among these the vestibular system is considered to be of particular importance since it is specifically designed for sensing self-motion in space. Yet, patients with loss of vestibular functions (Ps) appear to be hardly impaired in their 'everyday life.' Does this mean that the vestibular information is redundant and can be dispensed with because other sensory channels provide equivalent information? Or do the Ps adapt to their deficit by modifying the use of the remaining cues? These questions led us to study the contribution of nonvestibular cues for motion perception in Ps. In particular, we considered two aspects:

1. Role of neck proprioceptors for the perception of passive horizontal trunk rotation in space.

In darkness, normal Ss base their perception of trunk rotation on a summation of vestibular and neck signals, and their perception of head rotation and of motion of visible objects in space in turn are based on this perception of trunk-turning (Mergner et al. 1983, 1986; Mergner and Becker 1988). In Ps, however, a neck contribution to the trunk turning sensation without the vestibular counterpart is probably detrimental, since it would create the illusion of a trunk turning during head rotation about the stationary trunk.

2. Role of visual (retinal and visuooculomotor) information on the perception of horizontal body rotation.

Previous work (Mergner and Becker 1988) had suggested that Ss are able to rely almost exclusively on the visual cue, depending on certain conditions (low stimulus frequency, initial assumptions concerning the stimulus condition, and/or choice of the visual surrounding as the reference for object motion perception). We wondered in what respect the visually induced self-motion sensation of Ps, for whom the visual input is essentially the only source of self-motion information, might differ from that of Ss.

Role of Neck Proprioceptors for the Perception of Trunk Rotation in a Dark Environment

1. In the dark, Ss perception of passive horizontal trunk rotation at frequencies/accelerations above 0.025 Hz/ 0.2 °/s^2 can be described by a linear summation of frequency-dependent vestibular (horizontal canal) and neck signals ('trunk-in-space' = 'head-in-space' + 'trunk-to-head'). In Ps both the vestibular and the neck-induced trunk turning sensations were missing up to frequencies/accelerations of 0.2 Hz/12 °/s^2. Only with stimuli of higher frequency/acceleration did the Ps perceive their trunk turning, apparently due to nonvestibular somatic inputs.

Patients. Patients were drawn from a pool of 12 students (aged 18-28 years). Loss of vestibular function was suggested by absence of vestibulo-ocular reflex upon sudden deceleration after prolonged horizontal body rotation and was due to treatment (presumably streptomycin) of meningitis during childhood (10), section of both VIII nerves because of neurofibromatosis (1), or an unknown cause (1).

Stimuli. Vestibular: Sinusoidal whole body rotation in the dark about the vertical body axis; constant amplitude, ± 8°; frequencies 0.025, 0.05, 0.1, 0.2, and 0.4 Hz. Neck: Trunk rotation about stationary head, same parameters. Combined vestibular and neck (0.2 Hz only): An 8° sinusoidal vestibular stimulus was combined with a 0°, 4°, and 8° neck stimulus, which was either in phase or in counterphase with the former.

Measure of Subjective Trunk Rotation. Concurrent indication by a rotatable pointer mounted on the chair and pivoting about the axis of body and neck rotation. Ss/Ps were to imagine a stationary point in space and to direct the pointer always at it. The estimates obtained from the compensatory pointer excursions were corrected for each individual's operator performance and expressed as gain (a gain of 1 equals physically correct compensatory pointer excursion upon body rotation).

Figure 1A shows the results of 6 Ss (two trials per S) and 9 Ps (one trial per P). The trunk-turning estimates (Ψ_{TS}) of Ss decrease at low-stimulus frequencies; this is true for both vestibular stimulation (filled curve) and neck stimulation (open curve). The decline of the vestibular curve can possibly be explained by the frequency characteristics of the vestibular (horizontal canal) system; a corresponding characteristic of the neck signal would appear to be functionally desirable, if low frequency head rotations about the stationary trunk should not elicit an illusory perception of trunk rotation.

The results obtained from the Ps are presented in an analogous form. Note that not only the vestibular trunk-turning sensation (black circles) but also the neck-induced sensation (open circles) is missing with stimuli up to frequencies of 0.2 Hz (zero estimates). Sensations suddenly appear and are close to 'normal' at 0.4 Hz. Typically, then, Ps reported that they perceived vibration, body shaking, and/or air-flow. Also, the scatter of these estimates was large, and there were overestimations and direction errors.

With combined vestibular and neck stimuli of 0.2 Hz (not shown), the Ss estimates reflected rather well the actual trunk turning, while the Ps indicated either stationarity of the trunk or small trunk excursions, the direction of which was perceived correctly in about 50% of the cases only.

Other observations further supported the notion that neither vestibular nor neck signals contribute to the trunk turning sensation of Ps in the dark:

2. Ss perception of head rotation in space can be described by the addition of a 'head-on-trunk' signal with unity gain at all stimulus frequencies tested (frequency-independent neck signal) with the 'trunk-in-space' signal that, as described above, decreases for low frequencies. In Ps, however, perception of head rotation in space depended solely on an enhanced 'head-on-trunk' signal, the trunk serving as a (subjectively stationary) reference.

Verbal estimates (method of magnitude estimation) of head rotation in space obtained from Ss during vestibular stimulation yielded a frequency-dependent estimation curve, which closely corresponded to the one depicted in Fig. 1A for the Ss trunk-turning sensation, a fact which is not surprising since the body is rotated as a whole in this stimulus condition. With neck stimulation, Ss perceived at 0.2 and 0.4 Hz stimulus frequency a small head-turning sensation in the direction of the head-to-trunk excursion. This illusion (the head actually remained stationary in space) increased considerably when stimulus frequency was lowered to 0.025 Hz. With combined neck and vestibular stimulation during head rotation about the stationary trunk, the estimates were close to unity at all the frequencies tested.

The latter results could formally be described by adding a 'high-pass' signal for the vestibular head-turning sensation with a 'low-pass' signal for the neck-induced head-turning sensation that above represented an illusion, the sum providing 'broad band-pass' information on head rotation about the stationary trunk. However, further experiments suggested a different explanation. In particular, when Ss estimated the head rotation in space during neck stimulation using the above pointer indication (see 1.), they kept the pointer more or less aligned with their stationary head; this indication was

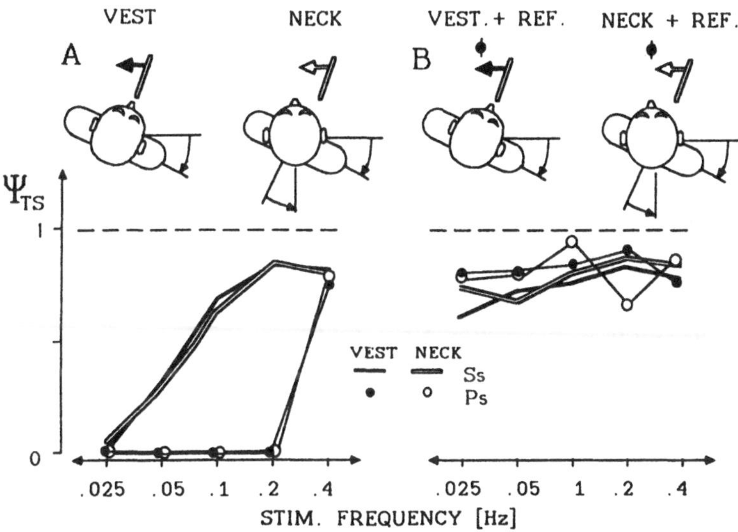

Fig. 1. Perceived trunk turning in space as a function of rotation frequency (medians of individual estimates). Insets sketch stimulus conditions (VEST, vestibular stimulation; NECK, neck stimulation). Ss, normal subjects; Ps, patients lacking labyrinthine function. **A** Stimulation in complete darkness. **B** Stimulation in presence of stationary light spot as external reference. Horizontal dashed lines show estimation of 'ideal observer'

in contrast to their verbal reports of a clear turning sensation of head in space. The discrepancy could be dissolved when two pointers were used, one to indicate head and the other to indicate space (this pointer was to be kept subjectively stationary in space); while Ss kept the head pointer roughly stationary, they moved the space pointer in the direction of trunk excursion. The deviation of subjective from objective space was small at high frequencies and increased considerably at low frequencies; the resulting curve represented essentially the reciprocal of the neck-induced sensation of trunk-turning in space (see Fig. 1A). In other words, the neck induced-turning illusion of head in space reflected the frequency deficiency of the neck-induced turning sensation of the trunk in space. This led us to assume that the head-turning illusion stems from two oppositely directed neck signals: 'head-in-space' = 'trunk-in-space' (frequency-dependent neck signal) + 'head-on-trunk' (frequency-independent neck signal). In fact, when evaluating separately the neck-induced trunk-in-space and head-on-trunk-turning sensations (the latter being indeed roughly independent from stimulus frequency), the sum of both curves closely corresponded to the actually measured curve described above for head in space. Finally, when combining vestibular and neck stimuli during head rotation on stationary trunk, the estimates were close to 'truth' at all frequencies tested.

Conceivably, this trunk-based head turning perception is advantageous as compared to one which would be derived only from the vestibular head-in-space signal; the trunk-based perception delivers a veridical message at low rotational frequencies at least in cases where the head is rotated about the stationary trunk and possibly also in cases where the trunk moves relative to ground, provided somatic receptors in the feet/legs/hips/trunk signal this trunk turning. At high frequencies, the purely vestibular sensation of head rotation in space and that derived from the trunk rotation in space can be considered equivalent.

The three Ps tested gave zero head-turning estimates with vestibular stimulation at 0.025 to 0.2 Hz. Estimates of large head excursions were given with the neck stimulation, which closely resembled the Ps estimates for 'head-on-trunk.' Their estimates during the combined stimulations again reflected the head-to-trunk excursion alone. Taken together, the data suggest that the Ps used the trunk-based 'mode' of head turning perception, taking the trunk always as stationary.

3. Ss perception of a moving light spot in the dark which they tracked with their eyes can be described by a summation of (a) a visuo-oculomotor signal of 'object-to-head', (b) a frequency-independent neck signal of 'head-on-trunk', and (c) the combined vestibular-neck information of 'trunk-in-space' rotation. Our Ps used only the first two of these signals and related them to the trunk as a subjectively stationary reference.

Ss and Ps were presented, in addition to the above vestibular and neck stimuli, with a visuo-oculomotor stimulus. They were to fixate/track with their eyes a light spot projected via a mirror galvanometer onto a surrounding cylindrical screen. Estimates of perceived object motion in space were obtained by way of a 'closed-loop nulling procedure': Ss and Ps had to move a joy stick, which coupled into the galvanometer input in such a way that the object appeared to be stationary in space. The compensatory signal from the joy stick was then considered as a measure of the perception that would have been present without the compensation. Two initial conditions were used: (a) the spot was rotated along with the body or trunk, or (b) it was kept stationary in space. The obtained data were corrected for the Ss and Ps operator perfomances obtained by 'nulling' the spot motion while head and trunk remained stationary. The signal of object-versus-head (eye-versus-head if Ss/Ps indeed fixated acurately) has a gain close to unity at all frequencies tested (Mergner et al. 1983). By subtracting this signal from the object motion perception (Ψ_{os}) indicated by the joy stick we could evaluate the vestibular and neck contributions to this perception.

Figure 2A shows the results obtained when the body and the spot were rotated together (vestibular stimulation). The continuous filled curve gives the median gains of the Ss object motion perception (Ψ_{os}) as a function of stimulus frequency. The gain is about 0.7 at 0.4 Hz stimulus frequency (a gain of 1 would indicate that Ss had effectively stabilized the spot); it declines at lower frequencies, but remains above zero (threshold) at least down to 0.025 Hz. This curve reflects the vestibular contribution to the Ψ_{os} of Ss in the dark. The interconnected filled circles in the figure give the median gains of the Ps. They obviously perceived the spot as stationary (their gain is zero) except from 0.4 Hz, where they indicated an object motion perception.

Curves that were essentially reciprocal to those just described were obtained when the spot initially was stationary, and Ss and Ps compensated, without knowing it, for an object motion illusion. In particular, the curve obtained from the Ss was close to zero above 0.1 Hz and increased at lower frequencies, whereas that of the Ps was close to unity from 0.025 to 0.2 Hz, dropping only at 0.4 Hz (not shown). These illusions, conceivably, reflect the deficit in Ss and loss in Ps, respectively, of the vestibular contribution to Ψ_{os}, i.e., the sum of 'object-to-head' (eye-to-head) and 'head-in-space' was nonzero (note that the 'head-in-space' signal may either be derived directly from the vestibular input or may result from a combination of 'head-on-trunk' and 'trunk-in-space' signals).

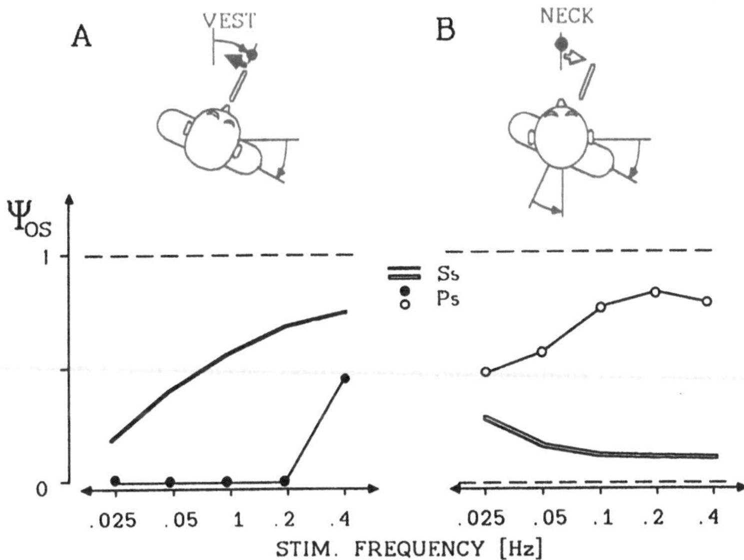

Fig. 2. Perceived object motion in space (medians of individual estimates). In a darkened room, Ss and Ps 'nulled' the actual or apparent motion of a light spot by means of a joy stick. **A** Vestibular stimulation; as long as the observer did not manipulate the joy stick, the spot rotated along with the body. **B** Neck stimulation; spot stationary in space as long as joy stick was not manipulated. Dashed lines show estimation of ideal observer

Figure 2B shows the results obtained when the trunk rotated about the stationary head (neck stimulation), while the spot was kept stationary. As indicated by the continuous open curve, the Ss moved the spot slightly in the direction of the trunk turning at 0.4 - 0.1 Hz, obviously to counteract a small object motion illusion in the direction of the head-to-trunk excursion. At 0.025 Hz, the curve shows a clear rise. The curve resembled that obtained previously for the illusion of 'head-in-space' rotation during neck stimulation, which also increased at low frequencies (see 2.). We had attributed this increase to a decrease in the 'trunk-in-space' signal at low frequencies. Since the 'object-to-head' and 'head-on-trunk' signals are frequency independent, the same argument applies to the object motion illusion in the present experiment, that is, we consider the Ss object motion sensation in the dark to be the sum of 'object-to-head,' 'head-to-trunk,' and 'trunk-in-space' signals.

This assumption received support from the data obtained from 6 Ps (Fig. 2B, interconnected open circles). During neck stimulation, their gains were high over the whole frequency range test. In other words, in order to cancel their object motion illusion they had to rotate (by means of the joy stick) the spot such that it was aligned to their trunk, which for them apparently represented 'space'.

4. By contrast to the above findings, Ss and Ps almost correctly perceived the trunk rotation when they were presented with an external reference. The external reference can be the visual scene or, in our experiments, a single light spot, which the Ss and Ps know to be stationary in space. The thus obtained trunk-turning sensation differed from the above in that it did not depend on the stimulus frequency. It could be described by a summation of a frequency-independent neck signal (trunk-to-head) with the visuo-oculomotor signals ('head-to-eye;' if eye = object = reference).

Ss and Ps were presented with a light spot which was stationary in space and on which they were to fixate. They were cued to its stationarity by rotating them first in the light and then in the dark. During the subsequent vestibular, neck, and combined stimulations they indicated trunk rotation in space by a compensatory pointer movement (see 1.).

Figure 1B shows the results obtained for Ss and 6 Ps with whole body rotation in the presence of the stationary spot (vestibular stimulus plus counterphase visuo-oculomotor stimulus). The two curves of median gain (continuous filled curve, Ss; filled circles, Ps) are very similar. They are close to unity even at low frequencies. Note that the turning sensation indicated by the Ps cannot be of vestibular origin but obviously stems from the visuo-oculomotor signal of head-to-eye (eye = object = reference). The same 'mode' of perception might have been used by the Ss. We have made such a suggestion in a previous study (Mergner and Becker 1988), when we found that Ss may switch between a self-motion sensation with the characteristic 'vestibular' decrease at low frequencies (compare Fig. 1A) and a frequency-independent self-motion sensation of visuo-oculomotor and retinal origin. The latter occurred when Ss chose, or were instructed to choose, the visual stimulus as a stationary reference (as in the present experiment). Ss then perceived the stimulus as stationary, whereas in the former case they perceived an illusory object motion (compare 3.).

Figure 1B also gives the results obtained when the trunk was rotated, and head and spot were held stationary (neck stimulus; visuo-oculomotor stimulus zero). The gain of both, Ss (continuous double lined curve) and Ps (open circles) was independent of frequency and similar to that obtained with the vestibular stimulation just described. The information about trunk rotation must stem from the neck system, since vestibular and visuo-oculomotor cues are zero in this case. Thus, Ps apparently use a frequency-independent neck input for their trunk-turning sensation in case they possess an external reference. This neck signal is complementary to the frequency-independent signal of 'head-in-space' derived from the visuo-oculomotor signal ('head-to-eye;' eye = object = reference) in our experiment (or from the oculomotor 'head-to-eye' and the retinal 'eye-to-visual object' signals when Ss/Ps do not fixate accurately). Since both Ss and Ps possess these signals, they produced essentially identical estimates in this condition.

Conclusions

In darkness, Ps have no perception of trunk rotation in space; as a consequence, also their perception of head and of object motion in space is erroneous unless the trunk is stationary. This applies to the functionally important mid-frequency (0.05 - 0.2 Hz) range, where normal Ss derive their trunk-turning sensation from frequency-dependent vestibular and neck cues. Above and below these frequencies, Ps and Ss have similar perceptions. At very low frequencies, both groups have no trunk-turning sensation and use the trunk as a (stationary) reference. At high frequencies both possess a trunk-turning sensation and use the 'physical space' (as it is reflected in the inertial momentum) as reference.

How should one interpret the fact that the Ps' loss of vestibular cues is associated with a neglect or loss of the corresponding neck cue? We consider it a functional benefit that they apparently use (and possibly try to maintain) the trunk as a stationary (zero) reference as long as there is no external reference. The certainly less advantageous alternative, the use of the head as a supposedly stationary reference, would induce a trunk-turning illusion. Another question is why Ss involve their trunk in their perception of head rotation in space in the dark, although they possess a 'direct' vestibular signal on 'head-in-space'. One reason may be that the trunk is the natural platform of the head and normally (i.e., unlike in our conditions, where Ss sat on a turning chair) contacts the ground via the legs, i.e., physical space. Possibly in the more general form the vestibularly and neck-derived 'trunk-in-space' signal is integrated with a proprioceptive 'trunk-on-feet (trunk-on-standpoint)' signal, which in contrast to the former appears to be frequency independent.

Role of Visual Information for the Perception of Self-Motion in Space

Sinusoidal rotation of a large structured visual scene elicits a visually induced sensation of self-motion ('circular vection,' CV) in Ss and Ps. In both groups it reflects the sum of a presumed visuo-oculomotor signal of 'head-to-eye' and a retinal signal of 'eye-to-scene' motion and has an approximately constant magnitude over a frequency range from below 0.025 Hz up to more than 1 Hz. Once engaged in this visual mode of self-motion perception, Ss may continue to rely on the visual cues even if a conflicting vestibular stimulus is added, that is, their sensation of self-motion resembles that of Ps, who lack the vestibular input. It is true, however, that there are 'qualitative' differences between the self-motion perceptions of the two groups. The following characteristics were observed in Ss and not in Ps: (a) With isolated visual or combined visual and vestibular stimulation, Ss are able to perceive, instead of CV, scene motion about the stationary or moving body, or when given appropriate instructions (that is, they can also rely on the vestibular instead the visual cue). (b) CV of Ss is extinguished when they track the scene with the eyes and/or the head, i.e., if retinal slip is canceled. (c) Ss perceive an 'intersensory conflict' (they report 'dizziness,' and the scatter of their estimates significantly increases) when the body-in-space rotation does not equal the body-to-scene rotation.

By contrast, Ps, who possess only the visual cue, continue to perceive CV when tracking the visual scene, and they do not register a discrepancy between body-to-scene and body-in-space rotation.

Conclusions

With body rotation in an illuminated environment, the visual system provides rather accurate information about the relative motion between observer and 'visual space', up to a frequency of about 1 Hz. However, the usefulness of this information is limited, since it is ambiguous; it can mean 'self-motion in a stationary surround' or 'scene (object) motion about the stationary body.' Thus, the brain has to decide whether the scene is an appropriate reference, and, if not, to choose an alternative. If the vestibular message is compatible with the assumption of a stationary scene, Ss can obviously take it as a 'confirmation' to use the visual reference. If the vestibular message is incompatible with the assumption, the assumption must be dropped, and the vestibular message is taken to provide the 'true' information on self-motion in space. However, the frequency deficiency of the vestibular horizontal canal system limits the use of this information. Our previous study (Mergner and Becker 1988), from which we took the Ss data, suggested that stimulus-independent factors (e.g., volition, knowledge or assumptions about stimulus conditions) also are involved in the decision process. Interestingly, in the present study our Ps, who had no vestibular cues, appeared to be unable to use the stimulus independent factors. We therefore assume that in an illuminated environment the absence of the vestibular cue is equivalent to the absence of an alternative (unless nonvestibular somatic cues provide one at high frequencies). On the other hand, the Ps data may help us to interpret the results obtained from normal Ss. The similar dynamics of the Ps' and Ss' self-motion sensations during body rotation relative to a visual scene would suggest that also Ss rely essentially on visual cues, a fact that is difficult to reconcile with a simple 'visual-vestibular convergence' theory of self-motion perception. As an alternative theory we have offered a conceptual 'multi-channel' model, according to which visual and vestibular information is conveyed separately (in parallel channels) to, or are reconstructed at, high perceptual levels (Mergner and Becker 1988).

Summary

Self- and object motion perception was investigated using psychophysical methods in patients without labyrinthine function (Ps) and compared to those of normal subjects

(Ss). The following findings were obtained for rotational stimuli in the horizontal plane.

1. In Ss, stimulation of neck afferents in the dark induces a turning sensation of the trunk in space, which has similar characteristics as the vestibular-induced sensation of head rotation in space; therefore, during rotation of the head on the stationary trunk, perceived 'trunk-in-space' = 'trunk-to-head' + 'head-in-space' = 0. In Ps, by contrast, the neck stimulus (trunk rotation versus the stationary head) evokes a sensation of head-in-space movement, whereas the trunk appears to them as stationary. The disuse or nonexistence of the neck-induced trunk turning sensation in Ps may be of functional benefit by suppressing an illusion of trunk turning when neck afferents are stimulated during isolated head rotation.

Due to the absence of vestibular and neck inputs, Ps have no trunk-turning sensation in the dark. Consequently, they refer their motion perception of head and visible objects in space to their subjectively stationary trunk; this perception is erroneous whenever the trunk is actually rotated.

2. In an illuminated environment, Ps derive their self-motion perception essentially from visual cues alone, while Ss integrate visual and vestibular information. Interestingly, also Ss may base their perception solely on visual cues, even in the presence of vestibular stimuli; their self-motion perception then closely corresponds in its quantitative aspects to that of Ps (close correspondence of the frequency characteristics). Yet, there are differences between the two groups; for example, Ss, unlike Ps, possess of a reliability measure for their perception which they apparently derive from a comparison between the visual and the vestibular cue.

References

Mergner T, Becker W (1989) Perception of horizontal self-rotation: multisensory and cognitive aspects. In: Warren R, Wertheim A (eds) Perception and control of self-motion. Erlbaum, Hillsdale (in press)

Mergner T, Nardi GL, Becker W, Deecke L (1983) The role of canal-neck interaction for the perception of horizontal trunk and head rotation. Exp Brain Res 49: 198-208

Mergner T, Rottler G, Anastasopoulos D, Becker W (1986) Influence of canal and neck inputs on visual perception of object motion in space. Behav Brain Res 20: 118-119

The Coordination Between the Lid and Eye During Vertical Saccades

W. Becker, A. F. Fuchs*
Sektion Neurophysiologie, Universität Ulm, Oberer Eselsberg, 7900 Ulm, Germany
*Regional Primate Research Center, University of Washington, Seattle, WA 98195, USA

A close coordination between vertical eye and lid movements is required to allow the lid to afford the eye maximal protection without obscuring vision. A casual observation of the vertical tracking movements of human subjects reveals a tight coupling indeed between lid and eye during slow vertical gaze changes. It is not known, however, whether this tight coupling also occurs during vertical saccades. Conceivably, a tight coupling is not required since saccadic eye movements per se already are accompanied by transient impairments of vision (Brooks and Fuchs 1975). Furthermore, the coupling for upward and downward saccades may be quite different since the levator palpebrae muscle reportedly raises the eyelid for all upward movements, but no muscle participates actively in lowering the lid during downward eye movements (Evinger et al. 1984; downward lid movements during blinks, however, are accomplished by the orbicularis oculi muscles).

To clarify these issues and to gather behavioural data for comparison with the activity of the neural apparatus controlling the position of the upper lid, we analysed the concomitant lid and eye movements during vertical saccades in both man and monkey. Vertical saccades were elicited by presenting a luminous target which stepped up and down by various amounts. Both lid and eye movements were recorded by means of magnetic induction coils. To record eye movements, we used a coil embedded in an annular suction lens (Collewijn et al. 1975). For recording movements of the upper lid, we glued an oval search coil to the lower margin of the lid near the eyelashes and recorded the rotational movement that results as it slides over the globe.

We first determined the "static gain" of the lid, i.e. the ratio of angular lid displacement to eye displacement determined from periods of fixation between saccades. It ranged from 0.89 to 1.34 in our three human subjects and was slightly larger than unity in our two monkeys.

Examples of the time course of the lid and the eye movements during vertical saccades are shown in Fig. 1 for two of our human subjects (upper panels) and one monkey (lower panel). It will be noted that the rapid changes of lid position (L) are very similar to those of eye position (E) in both the position and velocity ($\dot{E}$, $\dot{L}$) traces. Shown in isolation, the lid movements could well be taken to represent eye saccades; therefore, we call them "lid saccades". The "dynamic gain" of the lid, i.e. the ratio between the saccadic amplitudes of the lid and the eye, depends on the direction of the movement. For downward saccades it equals the static gain (except for the monkey shown in Fig. 1, lower panel), whereas for upward saccades it is generally smaller. The differences between static and dynamic gain are due to glissadic drifts (see upward lid saccade in

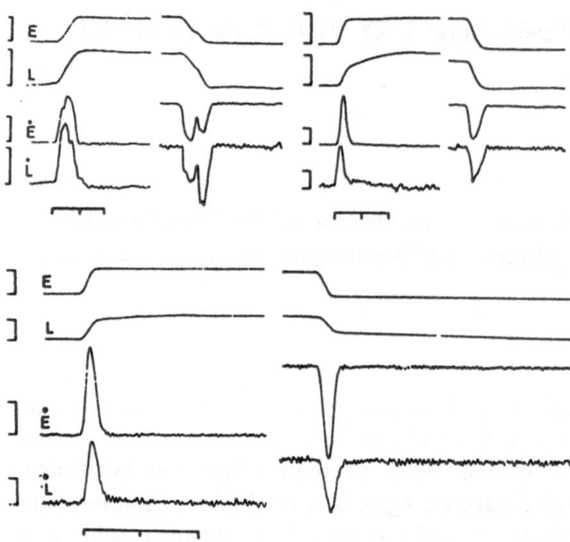

Fig. 1. Vertical saccades in two human subjects (top two panels) and one monkey (lower). E, Ė: eye position and velocity; L, L̇: lid position and velocity. Calibration bars in upper panels correspond to 10 deg and 100 deg/s, respectively, those in lower panel to 10 deg and 200 deg/s; upward deflection indicates upward movement. Time bars correspond to 200 ms (each tic = 100 ms)

Fig. 1, upper right) which were a regular feature of large upward saccades in our human subjects and in one of the monkeys. Lid saccades are closely synchronized to the eye saccades that they accompany; they start and reach peak velocity within a few milliseconds of the eye saccade, and, if not turning into a glissade, they also stop with the eye saccade. Plotted as a function of amplitude, the average peak velocity of lid saccades is similar to (for downward), or slightly slower than (for upward), that of eye saccades of similar size. Moreover, individual lid saccades exhibit the same trial-to-trial variations in peak velocity as the concomitant eye saccades and reflect also details of their velocity profiles, e.g. the double peaks in the downward saccade of the upper left panel of Fig. 1.

It is well known that a pulse of excess force is required during an eye saccade to overcome the viscous drag of the plant (Robinson 1964), and that this excess force is generated by a corresponding pulse of neural activity. The similarity of the lid and eye saccades suggests that the neural control signal for lid saccades may be similar to that for eye saccades. To examine the neural control signals for lid movements, we measured the isometric force developed by the levator palpebrae muscle during attempted vertical saccades when the lid was held in a lowered position. Also, we recorded (with the help of Dr. Kommerell, Augenklinik der Universität Freiburg) the electrical activity (EMG) in the levator palpebrae muscle during vertical lid saccades.

Figure 2 shows the force (F) exerted by the fixed lid together with the position of the contralateral, unimpeded lid (L). During upward saccades, the force clearly exceeds that required to hold the lid at its final steady state level. This level is reached slowly when the saccade ends in an elevated position (20 deg up in panel B) and more rapidly for depressed positions (10 deg down in panel A). A burst of rectified EMG activity is

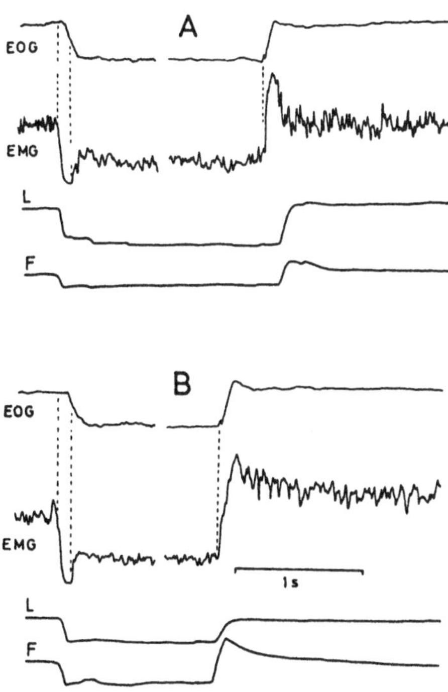

Fig. 2. Electromyogram (EMG) and isometric force of levator (F) during vertical saccades in a human subject. Note that the EMG and F recordings were obtained in different experiments and are aligned here to show general correspondence. EOG, vertical electro-oculogram recorded in EMG experiment as an indicator of lid movement. L, position of unimpeded lid recorded in isometric force experiment indicates intended movement of held lid. Eye saccades in EMG experiments were from -20 to 0 deg in A and from 0 to 20 deg in B; eye saccades in force experiment were from -20 to -10 deg (A) and +10 to +20 deg (B).

clearly seen during upward saccades (trace EOG; the vertical electro-oculogram was used as an indicator of vertical lid movement in this particular experiment); this activity leads the onset of the saccade by an average of 5 ms. The decrease in EMG activity after the burst depends on the final lid position in a way that is qualitatively similar to that of the isometric force.

We conclude from these observations that upward lid saccades are indeed generated by a "pulse-step" pattern of motoneuron activity. The activity could well be derived from that controlling the concomitant eye saccade; however, the pulse-step ratio of this activity is not exactly tailored to the dynamics of the upper lid so that upward lid saccades often are followed by glissades (Bahill et al. 1975). However, some tailoring does occur in the aftermath of saccades; the slow mechanical processes in the lid are clearly compensated by a decay of the EMG activity that is slower than that seen in oculorotatory muscles.

The generation of downward lid saccades, however, is more difficult to explain. Fig. 2 shows that such saccades are accompanied by only a slight undershoot of the isometric force below its resting level. It is true that the levator EMG activity did nearly cease, but often the EMG decreased as early as 50 ms before the onset of the saccade and resumed already 60 ms before its end (for small saccades the EMG actually resumed at saccade onset). Despite these EMG patterns, the lid movements accompanying downward saccades had a truely saccadic character and were generally faster than those in the upward direction. How downward lid saccades are achieved is particularly puzzling since during downward gaze shifts no other muscle appears to be active to overcome the surge in passive tension that must occur in the rapidly stretching levator.

Summary

The coordination between the upper lid and the eye during vertical saccades was investigated by recording their movements simultaneously with two magnetic search coils. The lid movements accompanying vertical saccades were generally a faithful replica of these saccades; hence, we call them lid saccades. To study the pattern of innervation generating lid saccades, we recorded the isometric force in the levator palpebrae muscles as well as its electromyographic activity during saccades. Upward lid saccades result from a pulse-step pattern of innervation. Downward lid saccades are difficult to explain on the basis of the observed EMG pattern.

References

Bahill AT, Clark MR, Stark L (1975) Glissades - eye movements generated by mismatched components of the saccadic motoneuronal control signal. Math Biosci 26:303-318

Brooks BA, Fuchs AF (1975) Influence of stimulus parameters on visual sensitivity during saccadic eye movement. Vision Res 15: 1389-1398

Collewijn H, van der Mark F, Jansen TC (1975) Precise recording of human eye movements. Vision Res 15: 447-450

Evinger C, Shaw MD, Peck CK, Manning KA, Baker R (1984) Blinking and associated eye movements in humans, guinea pigs, and rabbits. J Neurophysiol 52: 323-339

Robinson DA (1964) The mechanics of human saccadic eye movement. J Physiol (Lond) 174: 245-264

Ocular Pursuit of Sinusoidally Moving Targets: Is There a Sine Wave Generator in the Brain?

R.Jürgens, W.Becker and A.W. Kornhuber
Sektion Neurophysiologie, Universität Ulm, 7900 Ulm, FRG

Introduction

It is well known that periodical movements of a target are tracked by SPEM with remarkably small phase lags that appear to be incompatible with the long delay times of the visual system (which inevitably is involved in SPEM). Some authors have speculated therefore that periodical SPEM are generated by means of an 'internal function generator' which would faithfully reproduce the particular waveform of the target and which would overcome perceptual delays by adapting its phase characteristics to match the target motion (Bahill and McDonald 1983). If there are indeed such 'function generators,' one would expect SPEM (a) to have an approximately sinusoidal velocity profile during tracking of sinusoidal target movements and (b) to continue along that profile when the target movement becomes transiently invisible. We present here results which contradict these notions.

Methods

Healthy subjects (Ss) tracked a sinusoidally moving target which either (a) was continuously visible or (b) disappeared transiently after some cycles of continuous visibility, leaving the Ss in complete darkness for 0.5, 2, or 4 s. The disappearance could occur when the target reached (a) peak velocity (target crossing center position), (b) zero velocity (target at peak excursion), or (c) between these two extremes (45° before or after zero velocity). The Ss were instructed to 'try to follow the target even if it disappears for a while.'

Results and Conclusions

Figure 1 shows velocity profiles of horizontal SPEM recorded by means of a search coil while an S tracked a continuously visible target moving with an amplitude of 20° at frequencies of 0.5 Hz and 1.0 Hz. The velocity profiles are clearly triangular rather than sinusoidal suggesting that SPEM in this particular case is composed of a sequence of epochs with almost constant acceleration each lasting approximately from one velocity

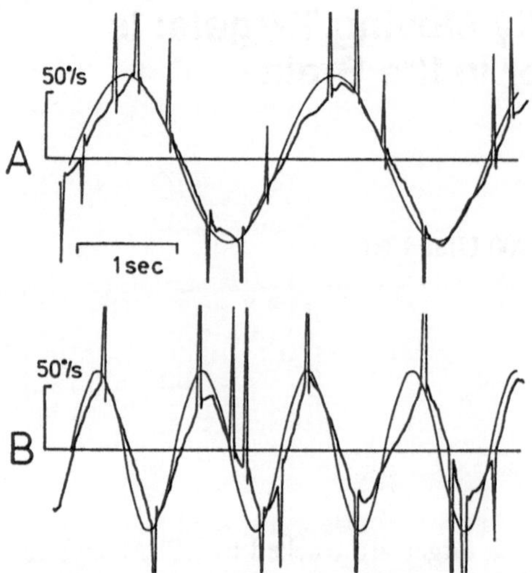

Fig. 1. Velocity profiles of sinusoidal SPEM with 20° amplitude. Target frequency 0.5 Hz (A) and 1.0 Hz (B). Thin lines, target velocity. Note the sharp breakpoints near SPEM velocity peaks

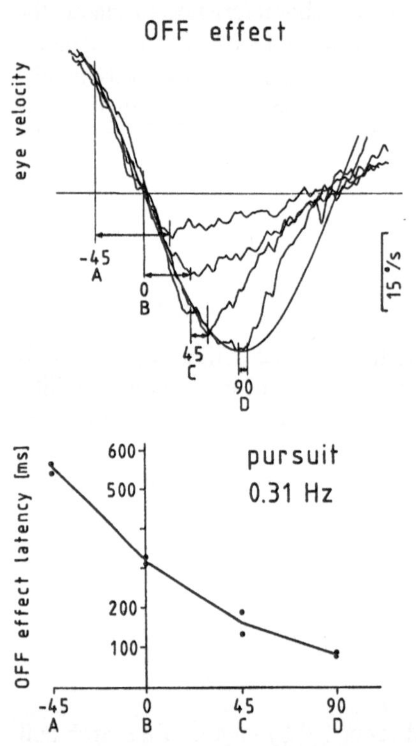

Fig. 2. Effect of sudden target blanking on SPEM. Target moved with 0.31 Hz and 20° amplitude. Grand averages across 4 Ss. Upper panel, SPEM velocity curves during blanking at four different instants. Blanking duration 2 s. Sinusoid, fit of normal SPEM (target continuously visible). Lower panel, off latency read (1) from upper panel and (2) from corresponding curves with 4 s blanking duration (not shown)

peak to the next. Triangular velocity profiles are not an exotic idiosyncrasy of a few Ss but are readily seen in many people, in particular at high frequencies (>0.5 Hz) and even with low-precision EOG recordings. Such abrupt changes in acceleration cannot be explained by assuming a continuous SPEM sine wave generator. When the target was transiently blanked, there was a latent period during which eye velocity continued to be the same, as if the target were still visible. After this 'off latency' the eye velocity decayed and entered a regime of residual sinusoidal movement in the dark at about 20% - 25% of the amplitude reached with a visible target. Interestingly, the off latency depends on when the target disappears. Figure 2 (upper panel) shows the average decay behavior corresponding to the four different instants of target disappearance (noisy curves) as well as the normal SPEM performance represented by its Fourier fundamental (sinusoidal curve). The off latencies read from these curves (arrows A, B, C, and D) are clearly a function of the instant at which the target was blanked (lower panel of Fig. 2). We speculate that the SPEM velocity is not generated as a continuous function but is 'synthesized' from segments of constant acceleration (i.e. velocity ramps) which 'aim' at successive velocity peaks. Conceivably then, it takes longer to stop such a segment just after its start (instant A) than it does during its final phase (instant D).

Summary

During tracking of sinusoidal target movements human smooth-pursuit eye movements (SPEM) often appear to be generated as discrete epochs of constant acceleration rather than in the manner of a continuously running 'sine wave generator.'

References

Bahill AT, McDonald JD (1983) Model emulates human smooth-pursuit producing zero-latency target tracking. Biol Cybern 48:213-222

The Role of Visual Feedback and Preprogramming for Smooth Pursuit Eye Movements: Experiments with Velocity Steps

A.W. Kornhuber, W. Becker, R. Jürgens

Sektion Neurophysiologie, Universität Ulm, Oberer Eselsberg, 7900 Ulm, FRG

Smooth pursuit eye movements generally subserve the task of tracking a small moving target so as to maintain it within the fovea. In order to take full advantage of the high visual acuity of the fovea, eye velocity should match target velocity precisely. This goal is complicated by the time lags of the visuo-oculomotor pathway. The smooth pursuit system therefore has developed compensatory mechanisms to overcome these lags, such as the ability to anticipate changes in target velocity if the target movement is periodic (e.g. a sine wave). On the other hand, there are many reports suggesting that the system works essentially as a velocity servo. Thus, the smooth pursuit system apparently has both a servomode and a predictive mode of operation.

At the outset of the present study we had in mind to exclude the predictive mode and to study the system's servomode in isolation. To this end, we recorded the smooth pursuit response to velocity steps of random amplitude and direction. In spite of this paradigm, however, we were not able to eliminate the interference of preprogramming. We report here on a variety of phenomena which, we feel, demonstrate an ever-present contribution of preprogramming to the initiation of smooth pursuit in responses to velocity steps.

Subjects (Ss) tracked the ramp-like movement (Fig. 1A, second trace from top) of a small luminous target starting from a center position in random direction. The ramp movement is equivalent to a velocity step (bottom trace in Fig. 1A). Target velocities were 5°/s, 10°/s, and 20°/s (randomized sequence). Eye position was recorded by a magnet search coil (top trace in Fig. 1A) and differentiated electronically to obtain eye velocity (noise $\leq$ 2°/s; see Fig. 1A). The typical eye movement response in this situation consists of a period of nearly constant acceleration with a ramp-like increase of eye velocity. The velocity pulses caused by catch-up saccades 'ride' on the ramp without affecting its linear ascent. As shown also by the examples in Fig. 1B and 1C, the velocity ramp often overshoots the target velocity; it usually is terminated by an abrupt change of acceleration which takes on a negative and again approximatively constant value (Fig. 1A, B), that is, the initial velocity ramp is followed by a sharp inflection which initiates a second ramp that finally leads to the steady state level of velocity.

A less common pattern is shown in Fig. 1C; despite the large velocity overshoot after the initial velocity ramp, the second ramp continues to accelerate the eye, albeit at a much lower rate; the steady state level, then, is reached only by means of a third, decelerating ramp.

A common feature of the three responses in Fig. 1 as well as of the majority of those observed in our experiments is that they consist of a sequence of epochs of constant acceleration. If the smooth pursuit system were a control system working with continuous negative feedback, one would expect either more oscillations after the large overshoot of eye velocity shown in Fig. 1 or, if the system is more dampened, a much smoother velocity profile. From the view of a continuous control system, a sharp turning point of eye velocity such as in Fig. 1B is a very unlikely behavior; we feel that this response is more compatible with the acceleration phase being preprogrammed. Given that this is so, the large velocity overshoot is not necessarily a flaw in preprogramming but may be a deliberate strategy to compensate for the reaction time lag and to reduce the necessity of catch-up saccades. With such a strategy, it is not surprising that the first acceleratory period, although clearly overshooting the steady-state velocity, may be followed, in some cases, by a second period of acceleration which further increases the overshoot (see Fig. 1C).

Finally, during the steady state, the velocity often oscillates slightly, and these oscillations sometimes appear to be triangular, suggesting again a sequence of epochs with different constant, albeit small, accelerations (see Fig. 1B). Note in Fig. 1C that a velocity ramp is also used to correct an inappropriate deceleration that was probably due to a wrong anticipation of the end of the target movement.

The initial acceleration program - if it exists - incorporates previous experience. Figure 2A and 2B shows the results of a quantitative analysis of the initial acceleration phase; it plots the average (across all subjects) of the mean initial acceleration and of the end velocity as a function of the target velocity. Both acceleration and end velocity obviously are functions of target velocity. However, both parameters also depend on the velocity of the preceding trial. As shown by the solid line in Fig. 2A and 2B, responses

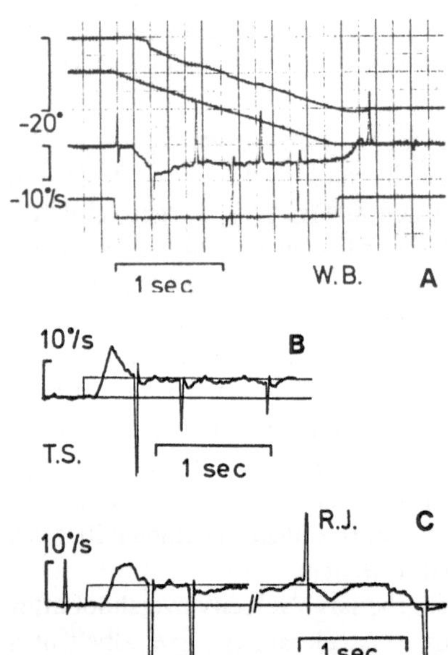

Fig. 1. A Ramp stimulus with a typical response. From top to bottom: eye position, target position, eye velocity, and target velocity. The ramp-like movement of the target is equivalent to a velocity step (bottom trace). Saccades 'ride' on the eye velocity trace and leave smooth pursuit unaffected. **B, C** Two further examples of smooth pursuit response (eye velocity: heavy curve) to a step of target velocity (thin curve)

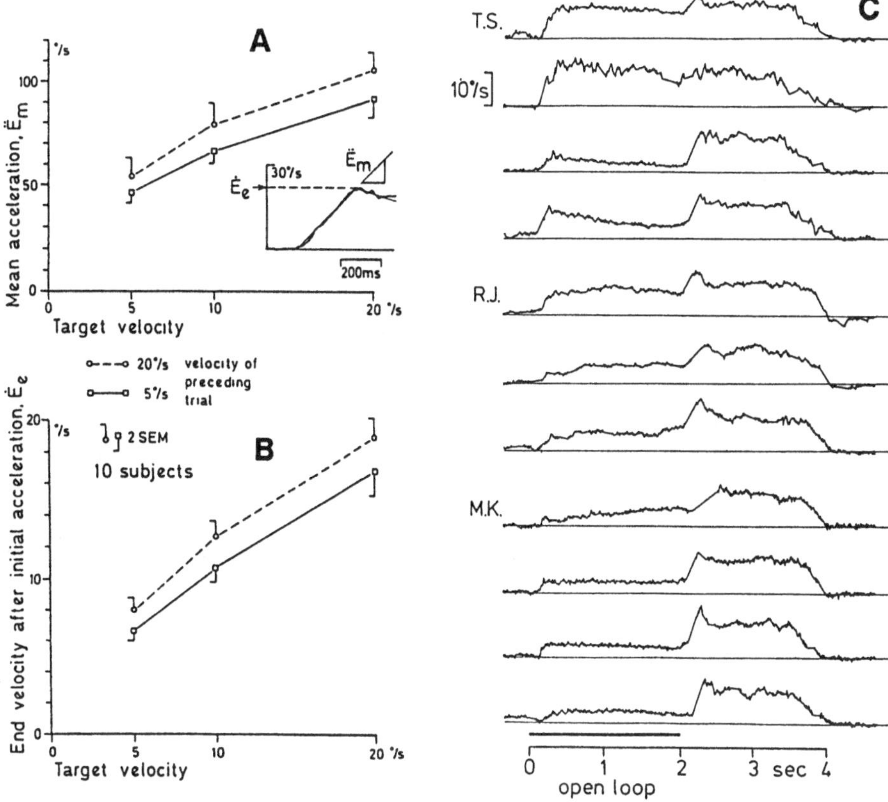

Fig. 2A,B Quantitative analysis of the initial acceleration phase. Mean eye acceleration $\ddot{E}m$ (A) and end velocity after initial acceleration $\dot{E}e$ (B) are plotted versus target velocity. For definition of $\ddot{E}m$ and $\dot{E}e$ see inset. Data from 10 Ss were averaged separately according to the velocity of the preceding trial (5°/s, continuous curves; 20°/s, discontinuous curves)

Fig. 2C. Ramp-like initiation of smooth pursuit (velocity traces of individual trials, saccades removed) without a retinal position or velocity error. The target is occasionally stabilized on the fovea (bar below bottom trace) instead of beginning the ramp movement expected by the S. Successive examples from 3 Ss

that follow trials with a velocity of 5°/s are slower than those following responses of 20°/s (dashed). We conclude that previous experience modifies the hypothesized program controlling the initial acceleration.

If the initial phase of the smooth pursuit response to velocity steps is indeed preprogrammed, one might expect that if an anticipated velocity step is unexpectedly suppressed, the program will nonetheless initiate an eye response. To test this possibility, we presented velocity steps with predictable onset and direction (to and fro target movement). In some trials we stabilized the target on the fovea (by electronic means) instead of presenting the expected stimulus, that is, we interrupted the natural visual feedback ('open loop') so that the target moved where the eye moved. Fig. 2C demonstrates successive responses from three subjects (velocity profiles, saccades removed) under this condition. During stabilization there is no position or velocity error to elicit a pursuit response, and yet we observe again a constant initial acceleration of the eye leading to an end velocity in the order of 4°-5°/s; however, there is no overshoot of the type seen in most normal, 'closed loop' responses. After this initial acceleration Ss produce a smooth movement with almost constant velocity although there is still no retinal stimulus. Interestingly, naive Ss are not aware of this situation; to them, open loop and closed loop trials have the same perceptual quality. Furthermore, there are no oscillations of eye velocity during the open loop regime. Only when we switched to the 'closed loop' condition after 2 s (right half), did we again observe a constant acceleration leading to an overshoot of eye velocity followed by oscillations around target velocity. Thus, the oscillations in the normal response are due to visual feedback. Their often triangular waveform suggests that this feedback may sometimes effect discrete adjustments of the motor output instead of continuously correcting it as a conventional servomechanism does.

Next we addressed the question of how an already initiated smooth pursuit movement continues when the visual error is suddenly nullified by stabilizing the target on the fovea ('open loop' in Fig. 3). Ss were not aware of this maneuver. As can be seen in the velocity trace in Fig. 3 (bottom), without visual error signal the eye velocity cannot be maintained at its original level; instead, eye velocity declines with an average time constant of about 4 s.

We conclude from these observations that the initiation of smooth pursuit, in a situation where the subject can expect a constant velocity stimulus, is probably of preprogrammed nature. The program typically is composed of a sequence of epochs with constant smooth acceleration. It can be called upon even in the absence of a visual error signal. However, the typical response with an overshooting eye velocity is observed only when an error signal is supplied by visual feedback. We suggest therefore that the execution of the preprogram is controlled by the visual error information. Also, the error signal is necessary to maintain eye velocity at the appropriate level.

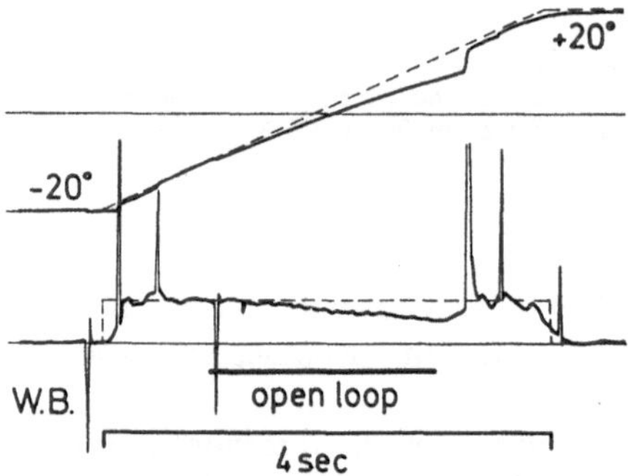

Fig. 3. Effect of unexpected stabilization of the target on the fovea ('open loop') during smooth pursuit. Top, target (dashed) and eye (continuous) position; bottom, target and eye velocity

Positional Nystagmus of Benign Paroxysmal Type (BPPN) due to Cerebellar Vermis Lesions: Pseudo-BPPN

E. Sakata, K. Ohtsu, A. Itoh and K. Teramoto
Neurological Clinic, Saitama Medical School, Saitama, Japan

Introduction

It is generally accepted that positional vertigo might result both from lesions in the central nervous system and from lesions in the peripheral vestibular organ. In 1921, Bárány described a characteristic paroxysmal vertigo and nystagmus occurring in a certain critical position, and he suggested pathogenesis of this condition attributing to otolith disease. In 1952, Dix and Hallpike described positional vertigo of the benign paroxysmal type (BPPN) due to a disorder of the otolith.

This nystagmus is very characteristic and nowadays is widely accepted among clinicians as a nystagmus due to otolith lesion. Furthermore, this characteristic nystagmus has become one of the most reliable signs for diagnosing vertigo due to disorder of peripheral vestibular organs. Although there have been numerous published papers describing positional vertigo due to lesions within the central nervous system (Bruns 1902; Alpers et al, 1944; Allen and Fernandez 1960; Fernandez and Alzate 1960; Sakata et al. 1979; Sakata et al. 1984), most of these cases are not of the benign paroxysmal type.

Recently we experienced a series of cases exhibiting typical nystagmus of benign paroxysmal type due to lesion in the cerebellar vermis.

Methods and Results

A total of 47 patients suffering from BPPN (Stenger 1955) were examined. Patients who manifested evident disorder of peripheral vestibular or auditory organs and patients with paroxysmal nystagmus of malignant type (Sakata et al. 1984) were excluded. The subjects were examined by CT scan, and some were, if necessary, examined by cerebral angiography and pneumoencephalography. In some cases, a lesion was confirmed by neurosurgical operation.

Forty-seven cases of BPPN with lesion involving the central nervous system were found. In these 47 cases we have observed evidence of cerebellar vermis infarction in 43, arachnoid cyst in the posterior fossa in 2, fourth ventricle tumor in 1, and cystic astrocytoma in the cerebellar vermis in 1. In 17 out of 47 cases, jumbling phenomenon (Dandy's symptom) was also observed. The following case is representative of these 47 cases. A 63-year-old man was admitted to our department with a 10-day history of severe vertigo attacks. For years he had been suffering from mild vertiginous attacks with occipital dull-headedness when he either lay in or got up out of bed. He had also been suffering from "jumbling of object" when he walked quickly. He visited physicians several times but nothing had been found. Ten days before the admission, he suddenly began experiencing severe vertigo every day when he either lay in or got up out of bed.

Nothing contributory was found in his past or in family histories. Neurological examination results were normal except for paroxysmal positional nystagmus. Audiometry, auditory brainstem-evoked response, caloric test, postrotatory nystagmus test, compensatory ocular counter-rolling test, Unterberger's stepping test, and gravimetry were all normal, but no suppression was observed in the visual suppression test (Takemori 1978). Although no nystagmus was elicited by positional nystagmus test, typical BPPN with a severe vertiginous sensation was precipitated by positioning nystagmus test. The nystagmus was pure rotatory, and the direction of the rotation was counterclockwise. The nystagmus increased in a rapid crescendo, and it declined rapidly. On sitting up, the nystagmus was also elicited, but the direction was reserved, i.e. clockwise. Repeated examinations did not eliminate the precipitation of the nystagmus (Fig. 1). Neuroradiological examination as well as CT-scan revealed a cystlike lesion in the dorsal part of the midline posterior fossa (Fig. 2). The cyst was removed and was diagnosed as an arachnoid cyst. Thereafter, the paroxysmal vertigo disappeared.

Discussion

In 1902, Bruns described a man with a cysticercus of the fourth ventricle who manifested periodic attacks of violent headache, vomiting, and intense vertigo with changes of head posture. He emphasized two aspects of this symptom: first, precipitation of vertiginous attack, headache, and vomiting on change of posture of the head, and second, absence of symptom during the interval of the attacks. He postulated that a fixed cyst or even a tumor might produce the symptoms and he explained the mechanism of the symptom-complex by intermittent hydrocephalus through blockage of the ventricular system. Thereafter, there were many reports (Alpers et al. 1944; Allen and Fernandez 1960; Fernandez and Alzate 1960; Sakata 1979; Sakata et al. 1984) dealing with this symptom. The symptom-complex was later called Brun's syndrome and is essentially different from BPPN mentioned below.

In 1921, Bárány described a characteristic paroxysmal vertigo and nystagmus occurring in a certain critical position of the head, and he emphasized that the factor precipitating the vertigo was not head movement but head position in space (critical position). In 1952, Dix and Hallpike described BPPN which is due to a disorder of the otolith. Nystagmus of this category is best elicited by positioning nystagmus test as originally described by them. The characteristics of this nystagmus are: there is always a marked latent period, and the nystagmus is chiefly rotatory - the direction of the rotation being towards the undermost ear. The nystagmus increases in a rapid crescendo, and thereafter it rapidly desclines. On sitting up, a slighter from of the nystagmus with reversed direction can be seen. We call this characteristic nystagmus a counterrolling positioning nystagmus. After two or three repetitions of this test the reaction declined and finally disappeared. The nystagmus could only be elicited after a period of rest, a Bárány pointed out.

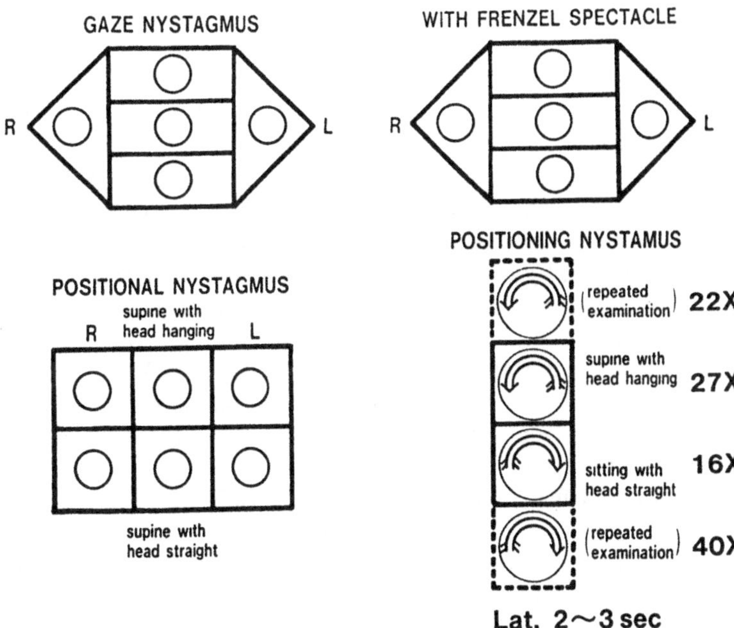

Fig. 1. A 63-year-old man. Pure rotatory, counterrolling positioning nystagmus with severe vertigo and latent period was noted

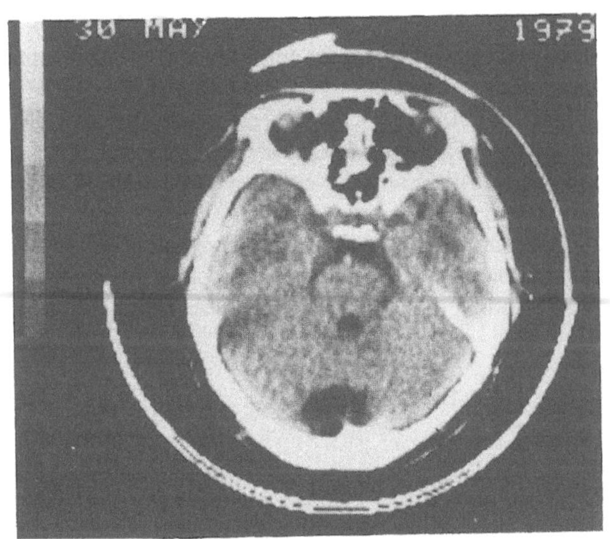

Fig. 2. A 63-year-old man. An arachnoid cyst in the dorsal part of midline posterior fossa was found (Plain CT scan, OM 15 mm)

Since then, the BPPN has been widely accepted among clinicians as a vertigo due to otolith lesion, and this characteristic nystagmus has become one of the most reliable signs for diagnosing vertigo due to disorder of peripheral vestibular organ. Of course, there have been critical experimental and clinical reports which countradict this opinion. Some reports describe the possibility that the nystagmus resulted from disorders in the central nervous system, and there is also a report indicating possible involvement of the semicircular canal. Our result suggests that the counterrolling nystagmus, best elicited by the positioning nystagmus test, cannot be a definite diagnostic sign for BPPN, since this type of nystagmus could also be found in cerebellar vermis lesion (pseudo-BPPN). Our result also suggests that the only definite difference between pseudo-BPPN and genuine BPPN is the constant appearance of the nystagmus in repeated examination. Although the positioning nystagmus test is the best way to elicite BPPN, we should always keep in mind the following two points: the procedure also stimulates semicircular canals; the existence pseudo-BPPN due to disorder of the cerebellar vermis. Concerning the pathogenesis of the pseudo-BPPN, it is a well-known fact that the flocculonodular lobe is closely related to the vestibulo-oculomotor system. We believe that the manifestation of pseudo-BPPN depends upon release of vestibular reflex from cerebellar inhibition. "Jumbling of objects" (Dandy 1968) was observed in as many as seven cases of cerebellar vermis lesion, whereas this symptom has long been believed to be a sign for bilateral disorder of peripheral vestibular organ. This fact requires further investigation concerning pathogenesis of Dandy's symptom.

Summary

Positioning nystagmus accompanied by severe vertigo has been reported in patients with partial lesions of the inner ear, especially otolith lesions. Typically this type of nystagmus represents a latent period and subsequent fatiguability. We concur with this finding and have constantly emphasized the significance of this phenomenon in clinical diagnosis.

Since we started to use CT scanning, this type of nystagmus has been noted in 47 patients, all of whom had cerebellar vermis lesions. Attention should be focused on this association. A simple coincidence could not be excluded if such a combination were seen in only 1 or 2 patients; this could be attributed to simple coincidence. But its occurrence in as many as 47 patients suggest a causal role of cerebellar vermis lesions. Its mechanism may be explained by incomplete inhibition of the vestibulo-oculomotor system including the cerebellar flocculonodular lobe or vestibulo-cerebellum.

References

Allen G, Fernandez C (1960) Experimental observation on postural nystagmus. I. Extensive lesions in posterior vermis of the cerebellum. Acta Otolarynol (Stockh) 51: 2-12

Alpers BJ,Yaskin HE (1944) The Bruns syndrome. J Nerv Ment Dis 100: 115-134

Bárány (1921) Diagnose von Krankheitserscheinungen im Bereich des Otolithenapparates. Acta Otolarynol (Stockh) 2: 434-444

Bruns H (1902) Neuropathologische Demonstration. Neurol Centralbl 21: 561-567

Dandy WE (1968) The surgical treatment of Méniére disease. Surg Gynecol Obstet 72: 421-425

Dix MR, Hallpike CS (1952) The pathology, symptomatology and diagnosis of certain common disorders of the vestibular system. Am J Otolaryngol 61: 987-1016
Fernandez C,Alzate L (1960) Experimental observation on postural nystagmus. II. Lesions of the Nodulus. Ann Otol Rhinol Laryngol 69: 94-101
Sakata E et al. (1979) Beiträge zur Pathophysiologie des Spontan- und Provokationsschwindels. HNO Praxis 4: 16-24
Sakata, E, Ohtsu, K Takahashi, K (1984) Pathophysiology of positional vertigo of the malignant paroxysmal type. ANL (Tokyo) 11: 79-90
Stenger HH (1955) Über Lagenystagmus unter der besonderen Berücksichtigung des gegenläufigen transitorischen Provokationsnystagmus bei Lagewechsel in der Sagittalebene. Arch Otorhinolaryngol 168: 220-246
Takemori S (1978) Visual suppression test. Clini Otolaryngol 3: 145-153

Two Forms of Head-Shaking Tests in Vestibular Examination

T. Kamei
Department of Otolaryngology, Gunma University, School of Medicine, Maebashi 371, Japan

Introduction

Among the various vestibular function tests clinically utilized, there are two forms of the head-shaking test. In each of these tests, to-and-fro movements are applied to the head at a rate of about two movements per second, however, the content and purpose of the tests are entirely different from each other. Although both tests have been utilized for many years, they have not received much attention in the literature. Recently, however, interest in the utility of these tests has increased, and I feel that the value of these tests should again be recognized.

Head-Shaking Test as a Method of Nystagmus Provocation

The first one is the head-shaking test as a method of nystagmus provocation. This test was designed especially for the detection of latent vestibular disequilibrium through an observation of the nystagmus occurring after head-shaking. When the head is shaken repeatedly, an increased tension occurs in the vestibular system, and this tension probably persists transiently even after the motion of the head is discontinued, giving rise to the nystagmus.

This test was initially described by Vogel in 1932. Subsequent major reports on this test include publications by Moritz (1951) and Kamei et al. (1964). Moritz (1951) conducted the head-shaking test in three separate directions, in the horizontal, frontal, and sagittal planes. In practice, however, head-shaking in the horizontal direction alone appears to be sufficient for nystagmus evaluation. Moritz further conducted the test using a series of 10 to-and-fro motions, however, a larger number of motions appears to improve the test results because of a more equal stimulation of both the right and the left inner ears.

I currently employ the following method for the head-shaking test. The patient is seated in a darkened room and given a pair of Frenzel's glasses to wear. In a normal or slightly anterior flexed position, the patient closes both eyes. The examiner, who stands in front of the patient, holds the patient's head with both hands. The head is then shaken with a to-and-fro rotational motion around a vertical axis in the horizontal plane at a rate of twice a second for 30 movements. The amplitude is approximately 90°. After the completion of the head movements, the patient immediately opens his eyes and assumes a position of frontal gaze. The resultant transiently appearing nystagmus can then be studied.

In general, head-shaking nystagmus, or to be more accurate, post-head-shaking nystagmus is a vestibular nystagmus which is never evoked in a patient with non-vestibular disturbance. I regard head-shaking nystagmus as a latent spontaneous vestibular nystagmus or latent spontaneous nystagmus in the narrow sense. As long as it is observed under Frenzel's glasses, it is always pathological and has an extremely important significance in clinical diagnosis (Kamei and Kornhuber 1974).

Head-shaking nystagmus is induced frequently in patients with peripheral and central vestibular disturbances (Kamei and Kornhuber 1974). In peripheral vestibular disturbances, recovery nystagmus is often induced in addition to deficiency nystagmus which is most easily provoked by this test (Kamei et al. 1984a). These two kinds of head-shaking nystagmus can be distinguished by the type of nystagmus. In deficiency nystagmus the head-shaking nystagmus occurs immediately after the head shaking as a strongly beating nystagmus and regresses rapidly, usually in 10 to 20 seconds. In recovery nystagmus the head-shaking nystagmus occurs after a latency period following the head shaking as a finely beating and somewhat irregular nystagmus, which generally lasts for about 30 seconds or more. In peripheral disturbances, irritative nystagmus is also induced. Generally speaking, however, this form of nystagmus is very rarely encountered. The most interesting aspect about peripheral disturbances is the appearance of nystagmus induced in a biphasic mode (Kamei 1975).

Figure 1 shows an electronystagmographic recording of the head-shaking nystagmus of this type. The patient had a right vestibular neuritis. Immediately after the head sha-

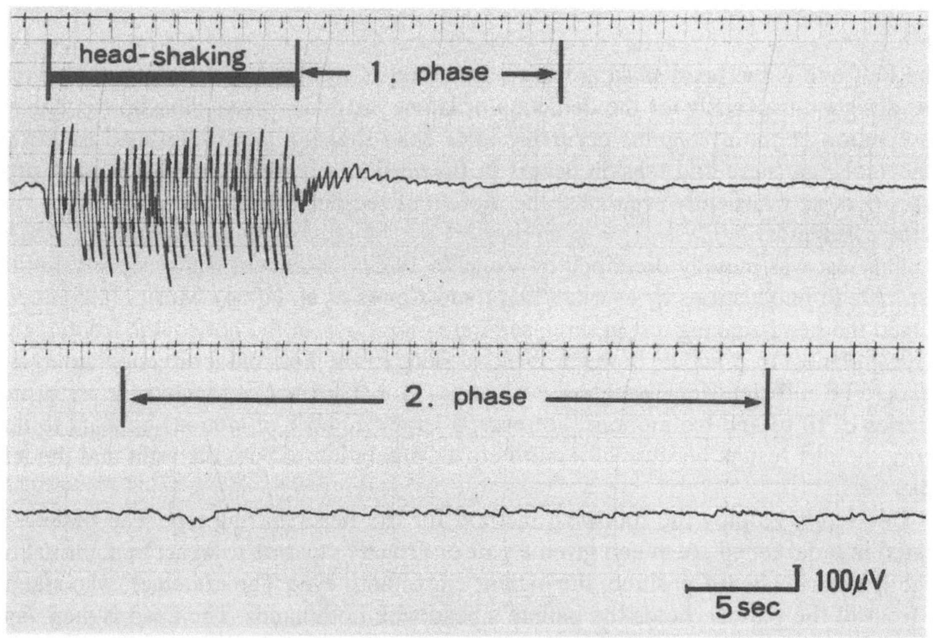

Fig. 1. Biphasic head-shaking nystagmus, observed in a patient (A.E.) with a right vestibular neuritis. Electronystagmography with horizontal lead. The lower tracing is a succession of the upper. (From Kamei 1975)

king, first phase nystagmus briskly appeared for about 12 seconds, directed to the left or healthy side. This was followed by a latent period of about 20 seconds. A small irregular phase two nystagmus then appeared toward the right or affected side for about 30 seconds. This pattern represents biphasic head-shaking nystagmus. In this type of nystagmus, phase one generally represents a deficiency nystagmus and phase two represents a recovery nystagmus. Exceptions to this rule appear to be few (Kamei 1975). This type of nystagmus, usually appears transiently at the time of transition from the stage of paralysis to the stage of recovery of the vestibular system. This phenomenon is therefore of interest from the viewpoint of vestibular pathophysiology, too.

The head-shaking test is thus a test for latent spontaneous vestibular nystagmus, and is especially useful for the localization of the affected side and for an estimation of the stage of the peripheral disturbance.

Head-Shaking Test for Evaluation of Jumbling

A second head-shaking test is the test of jumbling. Jumbling is the phenomenon seen in patients with a complete loss or marked disturbance of bilateral labyrinthine function. It consists of a loss of vestibular eye movement reflexes and resultant oscillopsia in the middle of movement of the head, especially during walking (Dandy 1941). This results in a failure of clear vision of objects during head movement. The head-shaking test described below is an almost unique method of quantification of this phenomenon.

The head-shaking test for evalutation of the jumbling phenomenon was initially suggested by Meyer zum Gottesberge in 1952 and was reconfirmed by Jatho in 1958. No remarkable reports were published subsequently until our latest study (Kamei et al. 1984b). I call this test Meyer zum Gottesberge's head-shaking test and perform it in the following manner.

The patient is seated in a lighted room. Binocular visual acuity is first measured with the head kept in a motionless state. In patients who usually require glasses, the test is performed while they are wearing them. The head of the patient is then lightly held with both hands by the examiner who stands behind the patient and the head is actively or passively shaken in a horizontal or vertical direction at a rate of two movements per second. An amplitude of 10° to 20° is utilized. The visual acuity is again measured during the head shaking. After repeating the test several times to confirm the results of the measurements, the visual acuity during head movement is compared with the visual acuity while the head is held motionless. In normal subjects, little or no decrease in visual acuity is noted due to this degree of head shaking. In subjects with a positive jumbling sign, the visual acuity during head-shaking is decreased to less than half of the visual acuity when the head is held motionless. When the visual acuity during head movement is expressed as a percentage of the value obtained while the head is motionless, a quantitative evaluation of jumbling is possible (Nakayama et al. 1986).

Figure 2 demonstrates the course of Meyer zum Gottesberge's head-shaking test results in a patient complaining of jumbling with a marked decrease in bilateral labyrinthine function due to vestibular neuritis. At the time of initial examination in July, the visual acuity during head motion was 0.2 in horizontal head shaking, or less than that in vertical head shaking which was below 20% of the acuity with the head held motionless. Recovery was noted from around September and a marked improvement

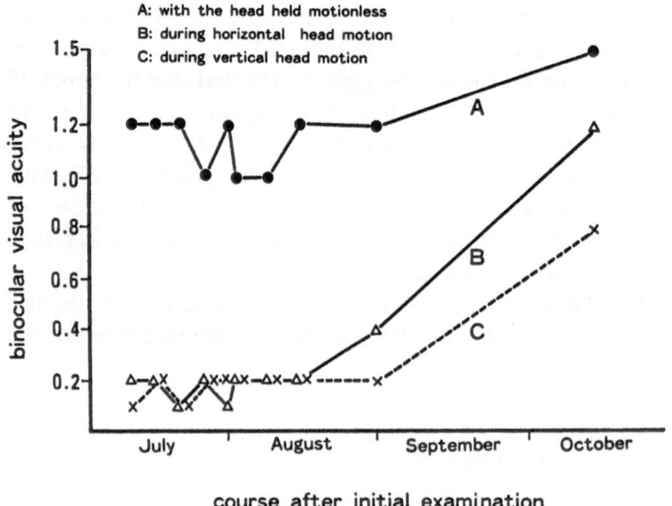

Fig. 2. The course of Meyer zum Gottesberge's head-shaking test results in a patient (T.I.) with jumbling. (From Kamei et al. 1984b)

was seen in October. Subjective jumbling improved, and followed almost the same course. Labyrinthine function was also restored. In October, a return to work was possible. In this patient, Meyer zum Gottesberge's head-shaking test well confirmed the phenomenon of jumbling.

In summary, Meyer zum Gottesberge's head-shaking test may also be seen as a vestibular function test with wide clinical applicability. It is of interest that two different vestibular function tests are mediated by similar head movements.

Summary

In clinical vestibular examination there are two forms of head-shaking tests, in each of which the patient's head is shaken in a similar way. One is the head-shaking test for the detection of latent spontaneous vestibular nystagmus. In this test, the patient's eyes are observed for nystagmus immediately after a passive rapid head-shaking around a vertical axis, using Frenzel's glasses in a dark room. The nystagmus is induced frequently in patients with peripheral and central vestibular disturbances. In peripheral vestibular disturbances, the induced nystagmus can be classified into deficiency-type nystagmus, recovery-type nystagmus, and biphasic nystagmus which is usually a mixture of the two. A second head-shaking test is the head-shaking test for the evaluation of jumbling. In this test binocular visual acuity is measured while the patient shakes his head two or three times per second, 10° - 20° horizontally or vertically, and effects compared with those when his head is still. A diagnosis of jumbling is made when the visual acuity during head shaking is less than half the visual acuity when the head is motionless.

References

Dandy WE (1941) The surgical treatment of Mènière's disease. Surg Gynecol Obstet 72: 421 - 425

Jatho K (1958) Über die Bedeutung des Dandyschen Symptoms bei einseitigem Verlust der Vestibularisfunktion. Arch Ohr Nas Kehlk Heilkd 172: 543 - 552

Kamei T (1975) Der biphasisch auftretende Kopfschüttelnystagmus. Arch Otorhinolaryngol 209: 59 - 67

Kamei T, Kornhuber HH (1974) Spontaneous and head-shaking nystagmus in normals and in patients with central lesions. Can J Otolaryngol 3: 372 - 380

Kamei T, Kimura K, Kaneko H, Noro H (1964) Revaluation of the head-shaking test as a method of nystagmus provocation. Part 1: Its nystagmus-eliciting effect. Jpn J Otol (Tokyo) 67: 1530 - 1534

Kamei T, Takahashi S, Kamada H, Muroi M (1984a) Revaluation of the head-shaking test as a method of nystagmus provocation. Part 2: Its diagnostic significance for the site of lesion. Equilibrium Res 43: 236 - 242

Kamei T, Takahashi S, Matsuzaki M, Muroi M, Yasuoka Y, Kaneko H (1984b) Jumbling, with special reference to Meyer zum Gottesberge's test. Pract Otol (Kyoto) 77 [Suppl] 2: 566 - 574.

Meyer zum Gottesberge A (1952) Störungen der visuellen Wahrnehmung nach Vestibularisausfall. Arch Ohr Nas Kehlk Heilkd 162: 62 - 66

Moritz W (1951) Auswertungen des Kopfschüttelnystagmus. Z Laryng Rhinol 30: 269 - 275

Nakayama M, Natori Y, Tachi H, Yoshizawa M, Takayama S, Miura H, Kanayama M, Kamei T (1986) Clinical investigation of vestibular damage by antituberculous drugs. Auris Nasus Larynx (Tokyo) 13 [Suppl 2]: 181 - 192

Vogel K (1932) Über den Nachweis des latenten Spontannystagmus. Z Laryng Rhinol 22: 202 - 207

Part 3

Sensory and Cognitive

The Parietal Visual System and some Aspects of Visuospatial Perception

V. B. Mountcastle, M. A. Steinmetz

Department of Neuroscience, The Johns Hopkins University School of Medicine, 725 North Wolfe Street, Baltimore, Maryland, USA 21205

Introduction

A complex function of highly developed brains is their capacity to construct and update veridical images of surrounding space and to identify accurately the spatial relations between objects in that space. These perceptual and control functions are executed ordinarily at preconscious levels, as witness the powerful but usually unnoticed role of the visual flow fields in guiding locomotion. These central images of the immediate surround can be brought immediately to conscious perception by directed attention. Studies of humans and other primates indicate that the parietal lobe system plays an important role in these aspects of visuospatial perception and in the visual guidance of pointing, reaching, and locomotion. Evidence obtained in neurophysiological studies in waking monkeys conforms with these general statements, for the functional properties of the several major classes of neurons identified in the posterior parietal homotypical cortex of the monkey mimic many of the functions attributed to the parietal system in man, and mirror the defects seen in man and other primates after parietal lobe lesions (Mountcastle et al. 1975; Lynch et al. 1977).

We limit the following description to the class of parietal neurons that is clearly visual in nature. These PVNs are of particular interest for a number of reasons. Firstly, their complex properties are not predictable from those of neurons in areas antecedent to the parietal cortex in the multistaged, transcortical systems linking the striate and parietal cortices. Those cellular properties appear to be constructed within the parietal cortex itself, in area 7a. Secondly, the pattern of neural activity in PVNs appear especially well suited to signal events in the visual flow fields during locomotion, particularly those occurring in the far visual periphery. Thirdly, it has been possible by a population analysis to show that ensembles of PVNs can provide an accurate signal of the direction of a moving stimulus, even though each single PVN provides only an imprecise signal of that direction.

The Syndrome of the Parietal Lobe System

The syndrome produced by lesions of the parietal lobe system resembles others that follow lesions of the homotypical cortex in primates, for it consists of major and disabling abnormalities of function in the perceptual and motor spheres without defects in primary sensory or motor function. There are disorders of attention, with neglect of the side of the body and of immediately surrounding space contralateral to the brain lesion. Neglect is frequently accompanied by denial that the contralateral body/world exists at all, even by denial of any illness whatsoever. There are disorders of volition, with reluctance to move the contralateral body parts or to operate within contralateral space, with full retention of primary motor function. Perhaps the most striking disorders are those of visuospatial perception. Many such patients are unable correctly to identify the spatial relations between objects in immediately extrapersonal space, such as the positions of the furniture in a room; this is the "visual disorientation" of Holmes. These patients cannot perceive the holistic aspect of scenes that they view, show defects in the topographical sense, and imperception in the periphery of the visual field. All these may occur without primary defects in vision; e.g., with normal visual acuity and full visual fields. Finally, many individuals with parietal lobe lesions show disorders of visuomotor control system, i.e., in the accurate execution of motor acts that depend upon visual feedback for guidance. The most common is optic ataxia, the inability accurately to project hand and arm to targets. Many of these disorders, originally described in humans, have been reproduced in somewhat more pallid forms in monkeys after removal of the posterior parietal cortex. These disorders are produced by lesions of a widely distributed and multiply interconnected system in which the parietal homotypical cortex is one of many nodes. This is shown by a host of anatomical studies of parietal lobe connectivity (see Fig. 1). Consonant with this is the fact that lesions in other nodes of

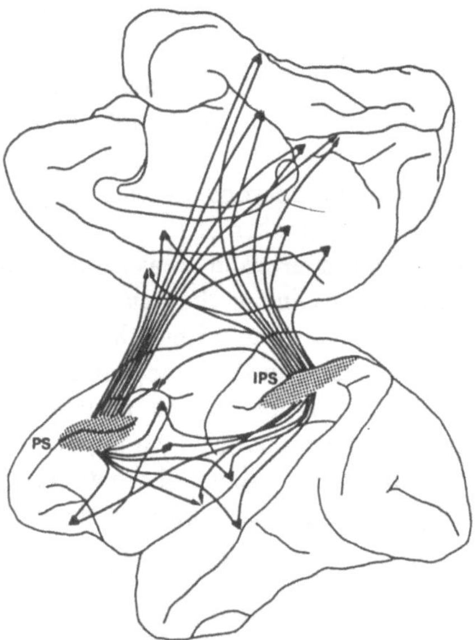

Fig. 1 A diagram to summarize the common cortical projection targets of the inferior parietal and dorsolateral frontal homotypical cortical areas. The targets shown, but unlabeled, are the orbitofrontal cortex, premotor areas exclusive of the supplementary motor area, the frontal eye fields, the anterior and posterior cingulate cortices, the frontoparietal operculum, the insula, the superior temporal sulcus, the medial parietal cortex, the parahippocampal and presubicular cortices, the caudomedial lobule, and the medial prestriate cortex. IPS, intraparietal sulcus; PS principal sulcus. (From Selemon and Goldman-Rakic 1988)

the system produce syndromes with facets that resemble those that follow lesions of the parietal lobe. For example, different but related forms of contralateral neglect may follow lesions of frontal or cingular cortices or of the superior colliculus. Moreover, the important study of Bisiach and Luzzatti (1978) shows that information concerning the spatial surround is still available within the system after parietal lobe lesion; what is lost is ready access to it.

Cytoarchitecture and Connectivity of Area 7a of the Posterior Parietal Cortex

Area 7a (or PG) of the posteromedial part of the inferior parietal lobule (IPL) of the monkey is a large region of eulaminate, homotypical cortex. There is within 7a a gradient of cytoarchitectural change in the anteroposterior direction, so that the cortex of the posterior bank of the intraparietal sulcus differs from that on the exposed surface of the IPL and the latter from cortex in the adjacent part of the upper bank of the superior temporal sulcus. Several subdivisions of 7a have been proposed, based upon these cytoarchitectural as well as connectivity differences, but no general consensus about them has yet emerged.

Area 7a is a principal target of a multichanneled transcortical system that arises in the striate cortex. This system is not simply hierarchical, for 7a receives both direct and relayed inputs from V3, V4, and V5, and the nodes of the system are themselves interconnected. Moreover, this dorsal or parietal transcortical visual system is not so completely separate from the ventral or temporal transcortical visual system as hitherto supposed, for the two systems are interconnected at several levels and receive some inputs from common sources (for reviews see Zeki and Shipp 1988; DeYoe and Van Essen 1988). Nevertheless both the detailed anatomical studies referred to and the functional properties of PVNs indicate that the dorsal system projecting to area 7a is dominated by the magnocellular stream of the visual system. The medial pulvinar and lateral dorsal thalamic nuclei project to area 7a, which thus receives a relayed input from the pretectal nuclei and the deep layers of the superior colliculus, targets of the retinocollicular component of the visual system.

This region of cortex is widely and reciprocally interconnected with other areas of homotypical cortex and with subcortical structures as well, indeed, with no less that the 15 targets shown in Fig. 1. Selemon and Goldman-Rakic (1988) have discovered that these same 15 targets receive projections from the cortex within and surrounding the principal sulcus of the frontal lobe. These convergent projections are of two types, one in which the two terminate separately in closely interdigitated columns and a second in which they terminate in different layers of the same cortical columns. The functional implications of these many convergent projections and of the two types of termination are still uncertain. In summary, its connections indicate that 7a is a node in a widely distributed system within which no hierarchical relations exist.

All the experimental results described below were obtained by recording the electrical signs of the activity of single neurons in the parietal cortex of waking monkeys as they performed one of a number of behavioral tasks. Details are given in a number of publications (Mountcastle et al. 1975; Lynch et al. 1977; Motter and Mountcastle, 1981).

Classes of Parietal Neurons and Their Modular Arrangement

A major problem in studies of area 7a is to specify exactly the functional properties of the neurons observed, for different properties are revealed in different behavioral paradigms, and because the excitability of parietal neurons is remarkably sensitive to state, particularly that of attention (Mountcastle et al. 1981). Moreover the excitability of at least three of the classes of parietal neurons, the fixation, saccade and visual neurons, is influenced by the angle of attentive gaze. Thus, the proportions of the classes of parietal neurons observed in any single behavioral paradigm is of limited value. However the results of three successive studies, each made under somewhat different circumstances, allowed us to identify the four major classes of parietal neurons indicated in Table 1.

The fixation neurons of area 7a become active when a monkey fixates attentively an object of interest, such as food in his hand, yet are inactive during his casual fixations of neutral objects in his immediate surround. Control experiments showed convincingly that these cells are insensitive to visual stimuli per se. The large majority subtend gaze fields limited to one-half or one-quarter of the field of view: only 10% are active with attentive fixation anywhere within the full range of ocular fixation (Lynch et al. 1977; Sakata et al. 1980; Yin and Mountcastle 1978). These limited gaze fields occur in either half of the field of view; 84 of 149 cells tested in detail had contralateral gaze fields. The limited gaze fields suggest that both the angle of gaze and attention influence the excitability of fixation neurons.

The oculomotor neurons of 7a include both tracking and saccade cells. Tracking neurons are inactive during steady fixation, are insensitive to visual stimuli, and become active only when a monkey tracks a moving target. They are directionally selective, but relatively insensitive to target velocities over a range of 5°/s - 50°/s. (Mountcastle et al. 1975; Sakata et al. 1983). Saccade neurons are active before and during visually evoked saccades and are inactive during spontaneous saccadic movements of the eyes. Control experiments produced convincing evidence that the saccade cells, like the fixation and

Table 1. Classification of Neurons of Area 7a

CLASS	STUDY I[a]	STUDY II[a]	STUDY III[a]	SUM	%
Fixation	155	521	266	942	28
Oculomotor[b]	88	218	163	469	14
Visual	73	NT[d]	462	535	15
Reach/Manipulation	128	136	206	470	14
Unidentified[c]	NT[d]	369	518	887	25
Special[e]	21	32	67	120	4
TOTALS	465	1276	1682	3423	100

a - Study I, Mountcastle et al. 1975; Study II, Lynch et al. 1977; Study III, Motter and Mountcastle 1981
b - Includes tracking and saccade neurons.
c - Unidentified; that is in any behavioral or stimulus sets available in the experimental series described.
d - Not tabulated.
e - Includes mainly cells with both visual receptive fields and the properties of other cells classes; e.g. fixation neurons with visual receptive fields, etc.

tracking cells, do not possess visual receptive fields and are not activated by visual stimuli. Saccade neurons are strongly directionally selective, and their discharge rates during identical saccadic movements vary with the position of the saccade within the field of gaze. This latter we regard as an angle of gaze effect (Yin and Mountcastle 1977; Mountcastle et al. 1984).

Area 7a contains a large number of visual neurons, whose functional properties are detailed in a following section.

Reach-manipulation neurons become active before and during projected movements of arm and hand towards a target. They are inactive during casual, undirected movements and are insensitive to somatic sensory stimuli. True somatic sensory neurons do not occur in area 7a; they make up large proportions of the neurons of areas 5 and 7b.

Even with extensive testing there remained in all our studies a large number of neurons whose functional properties we could not identify - 25% or more of all cells brought under observation! The most likely reason - aside from that of experimental error - is that these cells were related to sets of behavioral events not represented in the repertoire at our command.

Finally, we have observed a small number of cells with properties of more than one of the classes described: e.g., reach-manipulation or saccade, or fixation neurons with visual receptive fields. It is likely that further experiment with more complex behavioral tasks will reveal larger numbers of these special cells with complex properties.

These classes of neurons are arranged in area 7a in a modular fashion, in conformity with the general rules of columnar organization (Mountcastle et al. 1975). In general, modules containing neurons of the different classes appear to be interdigitated, but the pattern has not yet been elucidated. There is a tendency for modules of a particular cell class to be encountered more commonly in some parts of area 7a than in others; e.g., oculomotor modules are observed more commonly in the posterior bank of the intraparietal sulcus than elsewhere. However, the evidence available suggests that the modular classes are not segregated strictly to one or another of the putative cytoarchitectural subdivisions of area 7a.

The Functional Properties of Parietal Visual Neurons

General Description

The properties of PVNs of area 7a are in several ways unique among those of neurons of visual cortical areas. Their receptive fields are large, often bilateral, frequently spare the foveal and perifoveal regions, and extend into the far periphery of the visual field to include the monocular rims (Fig. 4). Two examples are given in Fig. 3, but here the peripheral limits of the fields shown were set by the 100° x 100° tangent screen upon which we back-projected testing visual stimuli. The fields were shown with other stimuli to extend to the peripheral rims of the visual fields, at least along the horizontal meridian. We emphasize that PVN receptive fields are dynamic and show a certain lability, for they may vary in size and spatial distribution with differences in the level of attention, as well as with the history of the testing stimulus within the visual field (Fig. 4).

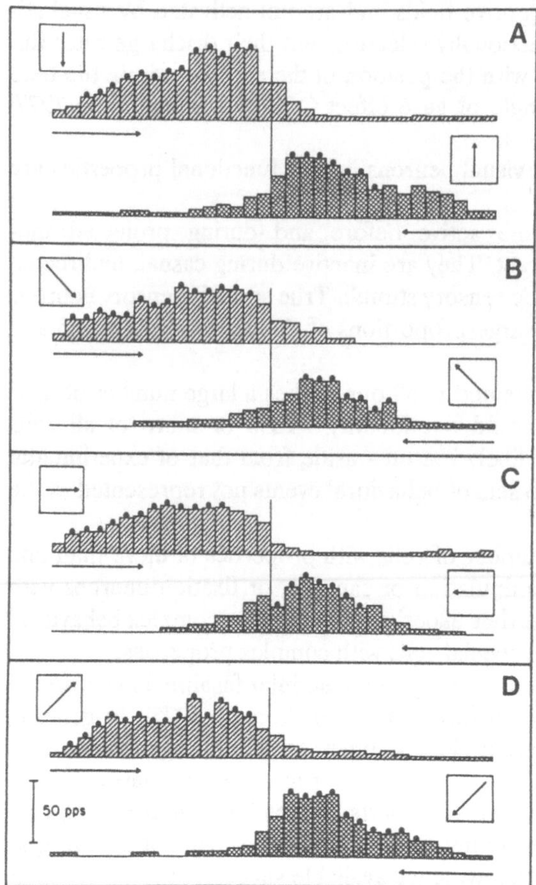

Fig. 2. Histograms of the responses of a parietal visual neuron. Pairs of histograms are shown, 1 in each pair for the responses evoked by 1 of the 2 directions of stimulus motion along each of the four meridians tested, as shown by inset boxes. Stimuli $10^{\circ} \times 10^{\circ}$ squares about 1 log above background, moving at 60°/s. Each bin of the spatial histograms = 3.125°. Directionalities significant at the 0.5 % level marked with dots. Central vertical lines = the point of fixation. The zone of transition from one significant directionality to its opposite occurred in a space of about 6° for A and D, and about 9° for B and C, bracketing the fixation point. These response patterns give an example of a strong, balanced, and radially symmetric opponent organization of directionality.(From Motter et al. 1987)

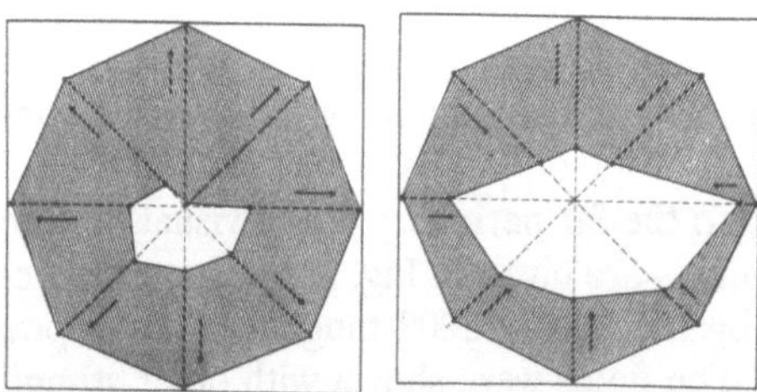

Fig. 3. Plots of receptive fields of parietal visual neurons, constructed from experiments like that of Fig. 2. To the left, field for a neuron with opponent directionality pointing outward from the point of fixation; to the right, field for a neuron with inward directionality. Peripheral edges of the fields shown mark limit of $100^{\circ} \times 100^{\circ}$ testing area; fields for these and for many parietal neurons extended into the far periphery to include, along the horizontal meridian, the monocular rims of the visual fields. (From Motter et al. 1987)

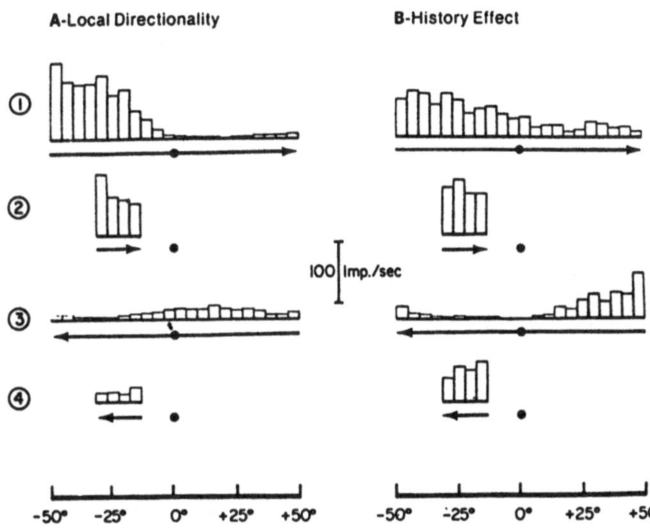

Fig. 4. A The 1st and 3rd histograms show a strong directionality of the responses of a parietal visual neuron to a 100° stimulus movement along a single meridian, centered at the point of fixation, 0°. For the 2nd and 4th histograms stimulus movements were of 20° extent, centered 20° left of the fixation point. This is one of the rare examples in which the full-field and local directionalities were congruent. **B** Experiment as in A, for another parietal visual neuron. The strong directionality shown by the responses to the full-field stimuli, histograms 1 and 3, contrasts with the lack of directionality for the responses to local stimuli, histograms 2 and 4. This is the common observation; it implies that the directional responses to 100° stimulus movement are produced by the history effect of the stimulus moving within the visual field. Stimulus speed, 90°/s; bin size 50 ms = 4.5°. (From Mountcastle et al. 1984).

PVNs are markedly sensitive to stimulus motion and to the direction of motion. Their most remarkable property is the opponent organization of this directional sensitivity (Motter and Mountcastle, 1981), described below. These cells are, however, relatively insensitive to differences in stimulus speed, for they respond well over a speed range of 20° - 400°/s, with some preference for speeds of about 100° - 120°/s. The responses of PVNs saturate when stimuli of increasing size reach about 35° - 50° square, and they are relatively insensitive to differences in stimulus luminosity, color, or orientation. Sakata et al. (1979) have shown that some PVNs are selective for depth, and quantitative studies of their candidate stereoscopic properties are badly needed.

The largest subclass of PVNs are those with large, bilateral, receptive fields. They usually respond in a sustained fashion to sustained stimuli, with onset latencies of 79 ± 6 (SD) ms. They are suppressed during a visually evoked saccade even if the stimulus remains within the receptive field. A smaller subclass of PVNs that subtend smaller and frequently contralateral receptive fields respond with onset latencies of 116 ± 26 (SD) ms. Neurons of this subclass show an increased discharge when the evoking stimulus becomes a target for a visually evoked saccade, as do many cells of the superior colliculus (Yin and Mountcastle 1978).

Opponent Directionality and Its Mechanism

The opponent arrangement of the directional sensitivity of a typical PVN is shown by the histograms of Fig. 2. The testing visual stimulus in this experiment was moved at 60°/s, for 100° along each of the eight directions of the four major meridians, centered on the point of fixation. These stimuli evoked intense responses as they moved inwardly from the periphery towards the point of fixation, but these responses rapidly dropped to background levels as the stimuli moved away from the line of fixation towards the field periphery. This neuron, like another whose receptive field is shown to the right in Fig. 3, is typical of the class of PVNs with "all-inward" patterns of directionality. An equally large class of PVNs displays the reciprocal pattern of directionality. The receptive field of such an "all-outward" neuron, determined in the manner of the experiment of Fig. 2, is shown to the right in Fig. 3

The extraordinary fact that single PVNs displayed different directional properties at different locations within their receptive fields, and particularly that the preferred directions point in opposite directions along single meridians towards or away from the point of fixation, aroused great interest as to the relevant physiological mechanisms (Motter et al. 1987). We first considered the proposition that opponent directionality is determined by local directional properties that differ systematically in different locations within the visual field, thus producing patterns of opponency such as those of Fig. 2. We compared the responses to full-field stimuli (those of 100° extent) with the responses evoked by stimuli of 20° extent, delivered in 8 directions centered at selected loci within the receptive fields, holding all other conditions constant. The result obtained was unequivocal, for only very rarely could the directionality observed when full-field stimuli were delivered along a meridian be predicted from the pattern of responses to local stimuli at points along that meridian; one of the rare examples is shown to the left in Fig. 4. Indeed, we observed that 129 of the 269 local regions tested in 92 PVNs showed no significant directional selectivity at all! These results indicated that opponent directionality is produced by mechanisms other than that of local or minute directionality.

The usual result that we obtained is shown to the right in Fig. 4. This led to the hypothesis that the directional patterns observed are determined by the history of the stimuli within the visual field. We carried out a series of condition-test experiments, in which we measured the effect of one local stimulus of 10° extent upon the response to a second such local stimulus placed at successive positions in front of the conditioning stimulus (Motter et al. 1987). The results obtained were clear and led to the model shown for inwardly directed neurons in Fig. 5. Each PVN appears to subtend a pair of superimposed receptive fields, one excitatory and one inhibitory, arranged as a double gaussian. For all-inward neurons, the excitatory field is larger than the inhibitory, leading to the net result shown to the right in Fig. 5. That is, as a stimulus moves into the visual field across its peripheral edge it first engages an excitatory field, driving the cell to high rates of discharge. With further movement of the stimulus inwardly towards the fixation point it engages the inhibitory field; the growth of inhibition quenches the response as the stimulus reaches the fixation point and suppresses it completely on the outward limb of the meridian. This feed-forward inhibition extends for 20° - 30° ahead of a moving stimulus, and lasts for several hundred milliseconds. The relative strengths of the excitatory and inhibitory influences determine whether the perifoveal region is

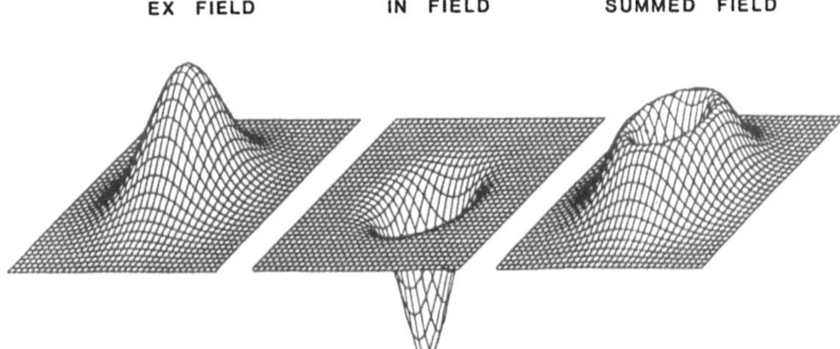

Fig. 5. Three-dimensional, double-gaussian model of the responses of a parietal visual neuron sensitive to inward stimulus motion, with radial symmetry of response intensities. The horizontal and vertical axes of the visual field are represented on the x- and y-axes, respectively, and the response amplitude on the z-axis. Left, excitatory gaussian determined by convergent excitatory inputs to the neuron. Center, inhibitory gaussian with a narrower spatial distribution than the excitatory one, thought to be generated by intrinsic cortical mechanisms. Both gaussians are centered on the point of fixation. Right, arithmetic sum of the distributions of the excitatory and inhibitory gaussians. The resulting, radially symmetric response surface is similar to that occasionally observed for parietal visual neurons, such as that of Fig. 2. Variation in the intensity of the inhibitory influence is thought to determine whether the receptive field includes or spares the perifoveal region. A reciprocal model explains in a similar way the receptive fields and directional properties of neurons with outward directionality, suche as that of Fig. 3, left. (From Motter et al. 1987)

completely spared, as for the neuron to the right in Fig. 3, or whether the response area overlaps the fixation point, as for the neuron of Fig. 2. All-outward patterns of directionality are surmised to result when the inhibitory field is larger than the excitatory field. These large receptive fields with opponent organization of directionality must result from intracortical processing within area 7a, for (a) fields with such directional patterns do not occur within prestriate areas projecting to area 7a and (b) their bilateral nature requires interhemispheric integration, which is undoubtedly executed via the heavy, reciprocal callosal connections between the areas 7a of the two hemispheres.

The Radial Distribution of Directional Sensitivities

We rarely observed PVNs whose directional sensitivity to stimuli moving along meridians were as perfectly symmetrical in the radial dimension around the fixation point in the frontoparallel plane as was that of Fig. 2, and which is implied by the model of Fig. 5. For the majority of PVNs the responses varied as a smooth and continuous function of the angular position of the meridian tested. Sinusoidal regression analysis showed this distribution to be adequately described by a sine wave function. An example is given in Fig. 6, which illustrates for an all-inward neuron a nearly perfect sinewave fit for both the means and the peak discharge rates evoked by the stimuli during the inward halves of their meridional transits. It is possible with this analysis to determine the optimal or "best" direction of stimulus movement. This analysis was made for 187 PVNs, 90 all-inward and 97 all-outward cells. Of these, the best directions of 117 were significant (R^2 greater than 0.7, sinusoidal regression analysis). These best directions are shown in Fig. 7., where those for inward and outward neurons are grouped together,

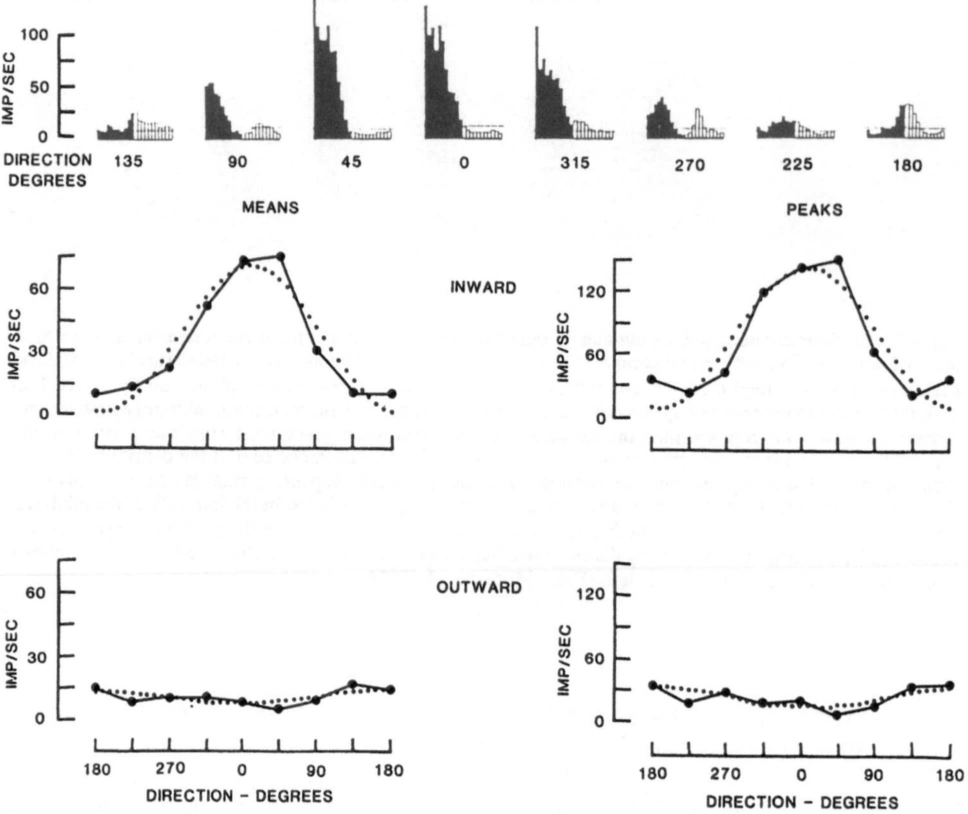

Fig. 6. Sinusoidal variation in the amplitudes of the responses of a parietal visual neuron to moving stimuli. The histograms at the top show time courses for the inward (dark shading) and the outward (unshaded) halves of 100° stimulus movements in each of the 8 directions indicated; 0° is vertical downward, angles increase counterclockwise. The mean, (left) and the peak (right) responses are plotted as functions of the direction of stimulus movements for both the inward (middle) and outward (bottom) halves of stimulus movements. Solid lines connect data values; dotted lines show sine waves fitted to the data by periodic regression. (From Steinmetz et al. 1987)

and all are shown as if occurring in right hemispheres. Seventy-five percent of the best directions point into the contralateral visual field, the remainder ipsilaterally. There is a slight preponderance (60%) for best directions in the upper half of the visual field. We show in Fig. 8 a modified model that takes into account the sinusoidal distribution of the directional responses of PVNs. It should be compared with that to the right in Fig. 5.

The properties of the parietal visual system are ideally disposed for it to play a role in the perception of self-motion, in the visual guidance of locomotion, and in the visual guidance of projected movements of the arm and hand. The relation of the properties of parietal neurons to these control mechanisms in primates has been discussed in detail in an earlier publication (Steinmetz et al. 1987).

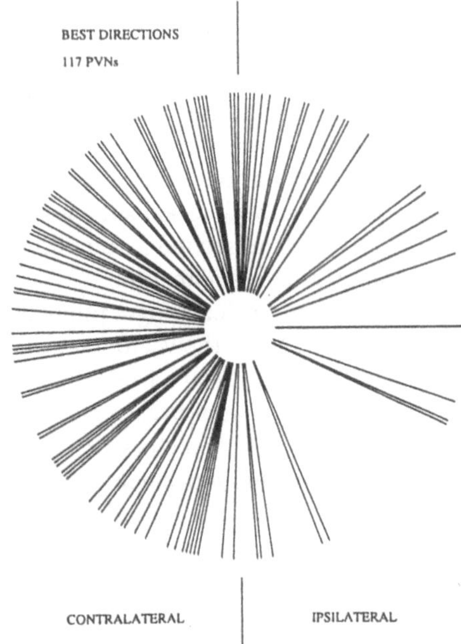

Fig. 7. Distribution of the best directions for parietal visual neurons, pointing either inward or outward with reference to the fixation point, shown as if in the frontoparallel plane and as if all neurons were located in right hemispheres. Best directions for 117 were significant, but only 97 appear in the figure; there were 20 superimpositions

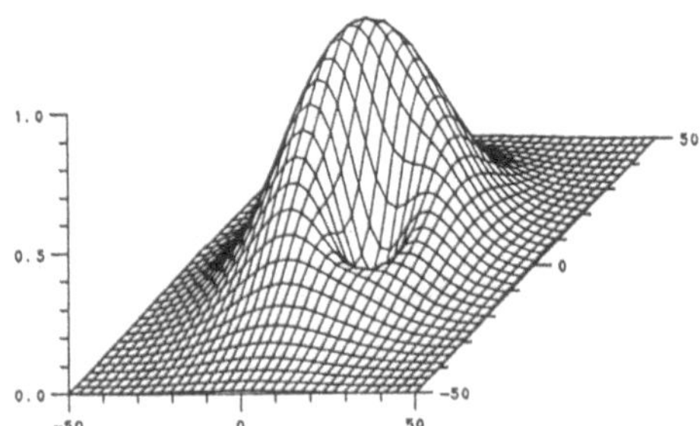

Fig. 8. Graphic of the responses of an all-inward parietal visual neuron representing the asymmetrical radial distribution of the intensities of response, fitted by a sine wave on the circle. The response surface is the result of two gaussians, one excitatory and one inhibitory. The asymmetry is produced by shifting slightly the peak of the inhibitory function relative to the excitatory one, producing the common pattern observed for parietal cells.(From Steinmetz et al. 1987)

The Population Signal of Stimulus Direction

It is obvious from the above that any single PVN provides an imprecise signal of the radial angle of a stimulus moving along a meridian into or out of the visual field. Indeed the mean index of directionality for the PVN population is obviously about 90°, given the close sinewave fits, i.e., the evoked response drops to 50% of the maximum at 45°

on either side of the optimal direction. We therefore addressed the question whether the population of PVNs considered as an ensemble might provide a more accurate signal of direction, even though it be derived from a group of imprecise elements. We chose for the analysis the simple model of linear vector summation. There are several assumptions involved. The first is that a given PVN when it discharges provides a signal of its best direction, regardless of the direction of the evoking stimulus. This is the labeled line rule, so well established for sensory systems, and particularly for the somatic afferent system. The second assumption is that there exists in the brain a computational operation for vector summation.

The method of calculating a population vector for a single direction of movement is shown in Fig. 9. Here the vectors for more than 60 neurons are represented; the length of each vector indicates the frequency of discharge evoked by a downward movement of the light, the direction of each vector being the best direction of that particular cell. The linear summation of all vectors yields a population vector pointing almost exactly in the direction of stimulus movement. Similar population vectors are shown in Fig. 10 for each direction tested, for both the inward and outward groups of cells. How accurately the population vectors predict the true stimulus directions is shown by the graph of Fig. 11; the average error is 9,3°.

Whether the particular method that we have used to analyze the ensemble signal of stimulus direction in PVNs has any relation to reality depends of course upon whether any set of central neurons receiving it has the capacity to compute and sum vectors, which is unknown. It does, however, relate to the general problem in brain physiology of how precise information about sensory events, or commands for motor action, can be embedded in populations of neurons whose individual members provide only imprecise signals of the relevant events. Certainly the precise and elegant sensory and motor performances of primates far exceed the accuracy of the signals provided for their execution by single cortical neurons. Morever, the precision of these performances appears not to be degraded by the loss of even large numbers of neurons, if those lost are ran-

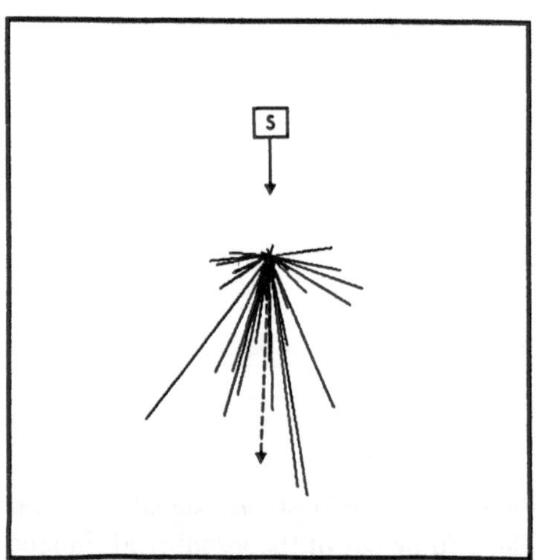

Fig.9. A model to illustrate the application of the linear vector summation analysis to the responses of parietal visual neurons to moving stimuli. See text for description

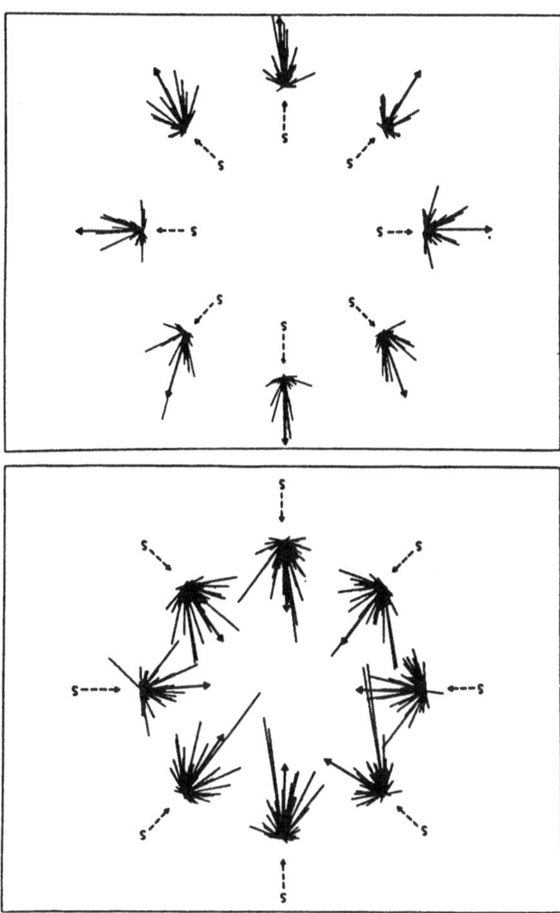

Fig. 10. Individual response vectors and the resulting population vectors (arrows) for the populations of parietal visual neurons studied in each of the 8 directions of stimulus from the periphery towards the central line of gaze; below, for cells responding to outward stimulus motion. The insets show, for each stimulus direction, the vector for each cell. It represents the intensity of the cell's response to that stimulus motion by its length, and the cell's own best direction by its direction. (From Steinmetz et al. 1987).

domly distributed in the neural populations essential for the actions. A proposition now open for study is that the interface transitions in the stages of sensory processing and through the sensory-perceptual and sensory-motor transformations occur by the flow-through of distributed signals from one population to another. Whatever the final answer to this problem, it is clear that in the neocortex neither grandmother cells nor pontifical modules exist.

State Controls and the Parietal Visual System

Obervations made under a variety of experimental circumstances show that neurons of the parietal cortex are strongly influenced by central control states. The projection and

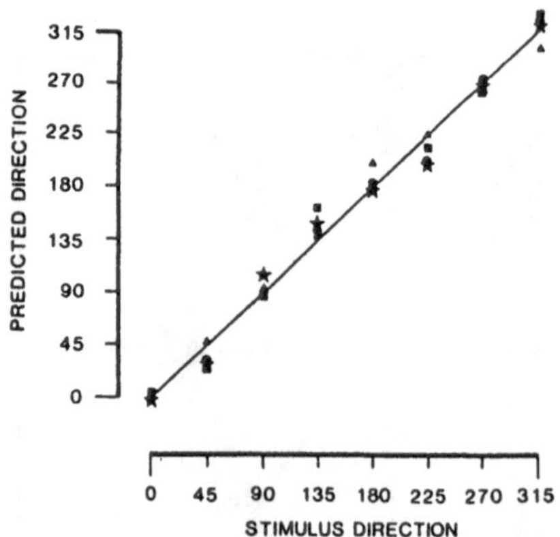

Fig. 11. Population vectors (predicted directions) shown as functions of the actual stimulus directions, 8 inward and 8 outward, for the populations of parietal visual neurons studied. The near-identity relation indicates the accuracy of the population signal of direction, on this analysis, in contrast to the imprecise signal of stimulus direction provided by any single neuron. (From Steinmetz et al. 1989)

manipulation neurons, for example, become active during movements aimed at objects of interest and not during casual manipulations executed by identical patterns of muscular contractions. Fixation neurons discharge at high rates during the visual grasp of objects of high motivational value, whether rewarding or aversive, but are inactive during casual fixations within a familiar surround. Parietal visual neurons are strongly influenced by the state of directed attention and by changes in the angle of gaze.

The Effect of Attention. The act of directed visual attention exerts a powerful influence upon the neural systems linkng the retinae and the parietal cortex (Mountcastle et al. 1981, 1987). The response of PVNs to an unattended stimulus is 3-4 times greater when the animal attentively fixates a target light that controls his behavior, as compared to the response to physically and retinotopically identical stimuli delivered in an alert but inattentive state. This facilitation is unchanged when attentive fixation is maintained in the absence of a target light. A number of control experiments made it clear that this is a specific facilitation superimposed upon the general state of arousal and alertness. The effect of attention upon parietal excitability is paradoxical in nature, for attention is well known to facilitate the behavioral response to and the central neural response evoked by attended stimuli, and simultaneously to suppress both the behavioral and the central neural responses evoked by unattended stimuli. In the present case attention produces a powerful increase in excitability in a visual system thought to operate at a preconscious level in the control of posture and locomotion and in the redirection of gaze and attention to new stimuli that appear within the visual field, particularly in its periphery.

The Effect of Gaze Angle. The excitability of several classes of parietal neurons changes with changes in the position of the eye in the orbit. Fixation neurons subtend limited gaze fields (Lynch et al. 1977), and the activity of saccade neurons changes with the spatial position of identical evoked saccades within the visual field (Yin and Mountcastle 1978). The responses of PVNs to physically and retinotopically identical stimuli are powerfully influenced by changes in the angle of gaze (Andersen and Mountcastle 1983). It is important to emphasize two further aspects of this phenomenon. Firstly, the receptive fields of PVNs are always organized in retinotopic coordinates and move with rotations of the eye in the orbit; no exception has been observed. Secondly, the angle of gaze effect is dependently related to the attentional control described above. The influence of the angle of gaze upon the responses of PVNs has evoked suggestions that concatenation of the PVN response and the gaze effect might provide a population signal serving the perceptual phenomenon of spatial constancy (Andersen et al. 1985), and a model based on this proposition has been proposed (Zipser and Andersen 1988). However, the dependency of the gaze effect upon attention makes this unlikely, for the constancy of the spatial location of objects during head and eye movements is surely independent of changes in the level of attention, over a very broad range.

The Parietal Lobe System and the Illusions of Vection

It is a common observation that optic flow in the visual periphery induces the illusory perception of egomotion in a stationary observer. This phenomenon has been studied extensively in the laboratory beginning with the observations of Mach in 1875 and Helmholtz in 1896. Several characteristics of these vection illusions suggest that PVNs may play a role in producing them. Firstly, vection illusions are facilitated when the subject is required to fixate a stationary target (Warren 1985; Wallach 1940). Similarly, attentive fixation of a target light causes a 3- to 4-fold increase in the responsiveness of PVNs to visual stimuli (Mountcastle et al. 1981, 1987). Secondly, vection illusions require moving stimuli in the visual periphery. Optic flow stimuli restricted to the central 30° of the visual field evoke the perception of motion of the visual array. In contrast, when such stimuli are presented in the visual periphery, subjects report that they themselves are moving against a stationary visual scene. This perception persists even if the central 120 deg of the visual field is blocked (Brandt et al. 1973; Dichgans and Brandt 1974; Held et al. 1975). The large bilateral fields of PVNs which often extend into the monocular rims of the visual field and spare the foveal and the perifoveal regions, the sensitivity of PVNs to moving stimuli, and the organization of their directionalities all suggest that the parietal visual system plays a role in producing vection illusions.

Visual input strongly dominates over gravitational and proprioceptive inputs. When subjects walking forward on a treadmill are presented with a scene in the periphery of their visual fields that is moving in the same direction, but at a higher velocity, they report the sensation of walking backward (Lishman and Lee 1973; Berthoz et al. 1975). This domination of optic flow over other sensory cues has also been shown to have important consequences for operators of high-speed vehicles who rely on such stimuli for accurate guidance (Gibson 1958, 1966).

Summary

Studies of primates with parietal cortical lesions suggest that the parietal system generates a neural image of immediately surrounding space and of the relation of objects within that space. Observations of parietal cortical neurons in waking monkeys reveal classes of neurons that appear as positive images of the behavioral defects produces by parietal lesions.

A major class of parietal neurons responds directly to visual stimuli. These parietal visual neurons (PVNs) subtend large, frequently bilateral receptive fields that extend to the peripheral rims of the visual field but often spare the central zone surrounding the point of fixation. PVNs are particularly sensitive to stimulus movement and direction but are relatively insensitive to stimulus velocity, orientation, form or color. The directional selectivities are commonly arranged radially with respect to the fixation point, so that directional vectors in opposite half-fields oppose one another along single meridians. Some PVNs respond preferentially to stimuli moving inwardly toward the point of fixation along any radius of the field; others respond only to outwardly moving stimuli. Such neurons respond preferentially to stimuli that arise from the surround during forward or backward locomotion. They provide strong and reciprocally antagonistic signals of shearing movements of the surround during rotation of head and eyes. In addition, their properties are those required to provide afferent feedback signals for the guidance of projection movements of the arm and hand.

The opponently distributed directional preference in the two halves of a single meridian is produced by a strong feed-forward inhibition. It is best explained on the assumption that each PVN subtends large, widely overlapping inhibitory and excitatory receptive fields. When the excitatory field is the larger, inward directionalities result, and vice versa. The response intensities evoked by stimuli moving along successive meridians in the circular dimension of a frontoparallel plane are sinusoidally distributed. On certain assumptions, a vectorial analysis of the population signal yields an accurate indication of stimulus direction.

The functional properties of PVNs appear to be synthesized within the local neuronal circuits of the parietal cortex and represent an integration of the properties of the system that projects upon the region, particularly of the multistage, transcortical system that links the striate and parietal areas, and the heavy interconnections linking the two parietal lobes via the corpus callosum. PVN properties depend critically upon the central state of attention, which exerts in this region a paradoxical effect, an increase in response to unattended stimuli, particularly those in the visual periphery.

The general properties of the parietal visual system are well suited to the role that the system is known to play in the guidance of reaching movement, posture and locomotion and, it is likely, in evoking the illusions of vection.

References

Andersen RA, Mountcastle VB (1983) The influence of the angle of gaze upon the excitability of the light sensitive neurons of the posterior parietal cortex. J Neurosc 3: 532-548

Andersen RA, Essick GK, Siegel RM (1985) Encoding of spatial location by posterio parietal neurons. Science DC 220: 456-458

Berthoz A, Pavard D, Yound LR (1975) Perception of linear horizontal self-motion induced by peripheral vision (linear vection). Exp Brain Res 23: 471-489

Bisiach E, Luzzatti C (1978) Unilateral neglect of representational space. Cortex 14: 129-133

Brandt T, Dichgans J, Koenig E (1973) Differential effects of central versus peripheral vision on egocentric and esocentric motion perception. Exp Brain Res 16: 476-491

DeYoe EA, Van Essen DC (1988) Concurrent processing streams in monkey visual cortex. TINS 11: 219-226

Dichgans J, Brandt T (1974) The psychophysics of visually induced perception of self-motion and tilt. In Schmidt FO, Worden FG (eds) The neurosciences. MIT Press, Cambridge MA, pp 123-129
Gibson JJ (1958) Visually controlled locomotion and visual orientation in animals. Br J Psychol 49: 182-194
Gibson JJ (1966) The senses as perceptual system. Hoghton Mifflin Boston
Held R, Dichgans J, Bauer J (1975) Characteristics of moving visual scenes influencing spatial orientation. Vision Res 15: 357-365
Lishman JR, Lee DN (1973) The autonomy of visual kinsesthesis. Perception 2: 287-294
Lynch JC, Mountcastle VB, Talbot WH, Yin TCT (1977) Parietal lobe mechanisms for directed visual attention. J Neurophysiol 40: 362-389
Motter BC, Mountcastle VB (1981) The functional properties of the light sensitive neurons of the posterior parietal cortex studied in waking monkeys: foveal sparing and opponent vector organization. J Neuroscience 1: 3-23
Motter BC, Steinmetz MA, Duffy CJ, Mountcastle VB (1987) Functional properties of parietal visual neurons: mechanisms of directionality along a single axis. J Neurosciene 7: 154-176
Mountcastle VB, Lynch JC, Georgopoulus AP, Skata H, Acuna C (1975) Posterior parietal association cortex of the monkey: command functions for operations within extrapersonal space. J Neurophysiol 38: 871-908
Mountcastle VB, Andersen RA, Motter BC (1981) The influence of attentive fixation upon the excitability of the light sensitive neurons of the posterior parietal cortex. J Neuroscience 1: 1218-1235
Mountcastle VB, Motter BC, Steinmetz MA, Duffy CJ (1984) Looking and seeing: the visual functions of the parietal lobe. In: Edelman GM, Gall WE, Gowan MW (eds) Dynamic aspects of neocortical functions. Wiley, New York, pp 159-193
Mountcastle VB, Motter BC, Steinmetz MA, Sestokas AK (1987) Common and differential effects of attentive fixation on the excitability of parietal and prestriate (V4) cortical visual neurons in the macaque monkey. J Neurosciene 7: 2239 - 2255
Sakata H, Shibutani H, Kawano K (1979) Depth selectivity of some visual neurons in the posterior parietal association area of the monkey. In: Ito M, Tsukahara N, Kubota K, Yagi K (eds) Integrative control function of the brain, vol 2. Kodansha, Tokyo, pp 83 -85
Sakata H, Shibutani H, Kawano K (1980) Spatial properties of visual fixation neurons in posterior parietal association cortex of the monkey. J Neurophysiol 43: 1654-1672
Sakata H, Shibutani H, Kawano K (1983) Functional properties of visual tracking neurons in posterior parietal association cortex of the monkey. J Neurophysiol 49: 1364-1380
Selemon LD, Goldman-Rakic PS (1988) Common cortical and subcortical targets of the dorsolateral prefrontal and posterior parietal cortices in the rhesus monkey: evidence for a distributed neural netword subserving spatially guided behavior. J Neuroscience 8: 4049-4068
Steinmetz MA, Motter BC, Duffy CJ, Mountcastle VB (1987) Functional properties of parietal visual neurons: radial organization of directionalities within the visual field. J Neuroscience 7: 177-191
Wallach H (1940) The role of head movements and vestibular and visual cues in sound localization. J Exp Psychol 27: 339-368
Warren HC (1985) Sensations of rotation. Psychol Rev 2: 277-278
Yin TCT, Mountcastle VB (1977) Visual input to the visumotor mechanisms of the monkey's parietal lobe. Science 197: 1381-1383
Yin TCT, Mountcastle VB (1978) Mechanisms of neural integration in the parietal lobe for visual attention. Fed, Proc 37: 2251-2257
Zeki S, Shipp S (1988) The functional logic of cortical connections. Nature 335: 311-317
Zipser D, Andersen RA (1988) A back propagation programmed network that simulates response properties of a subset of posterior parietal neurons. Nature 331: 679-684

Focal Thalamocortical Rhythms as Indicators of Attentive States in the Cat

J.J. Bouyer, P. Delagrange, M.F. Montaron, A. Rougeul-Buser, P. Buser
Institut des Neurosciences, Département de Neurophysiologie comparée, CNRS-UPMC, 9 quai St Bernard, 75005 Paris, France

Most of the recent experimental studies on the neurophysiological mechanisms of attention have utilized as a functional index either (in animal investigations) single-unit discharge patterns (e.g. Moran and Desimone 1985; Mountcastle et al. 1987; Spitzer et al. 1988) or some particular components of evoked potentials (mainly in human scalp explorations, e.g. Hillyard 1985), the expected finding being a characteristic response feature when the stimulus is attended as opposed to when it is ignored.

Our own approach has been fairly different. Our investigations of the attentive states in the cat have been based for many years now on the strategy of following the development of some specific, state-dependent electrocortical (ECoG) rhythmic activities occurring in behavioral situations that are highly suggestive of "attentiveness" of the animal, and not simply using significant stimulus sequences. After briefly describing these situations and their bioelectrical correlates, we discuss some of their underlying neurophysiological and neurochemical mechanisms as potential components of the attentional process in general.

Behavioral Paradigms

While observing behaving cats and recording their ongoing ECoG activity through electrodes implanted over various cortical areas, including motor cortex, sensory area I, posterior parietal cortex, we came across an interesting difference between two distinct behavioral situations. In common to them was a state of motionless attention of the animal, but they differed fundamentally by the specific paradigm.

The first situation was one in which the cat waited for an event to occur. A vertical opaque wall was placed in the recording room with a small hole at its bottom; behind the wall a mouse was free to move; the cat could hear it, could smell it, and at times would see it quickly appearing at the hole (Fig. 1A 1). During a standard 90 min session the animal was for most of the time (1 hour) in a condition of expectancy and behaving accordingly, immobile and watching the hole.

The second situation revealed itself (by its repercussions) to be quite different. A transparent box containing a mouse was placed in the recording room (Fig. 1B 1); the cat could now not only smell and hear its potential prey moving but could see it and follow its movements, but it could not catch it. As a rule, a standard session lasted 90 min, during which the cat usually remained immobile for about one hour, behaving as if it were in a state of focused attention toward the target.

The common advantages of the two situations were that the tested animals usually remained very quiet and immobile almost without interruption during the observation period (60 out of 90 min); adequate conditions thus existed for artifact-free recordings. However, the two situations differed markedly, based upon the ECoG used as a functional index.

Electrocortical Correlates

In pilot experiments we had noticed that some ECoG features were particularly clear and significant during these behavioral tests in the anterior part of the cortex, motor, somatosensory, and posterior parietal area. We therefore concentrated on these areas, implanting arrays of electrodes very closely placed, thus achieving an ECoG exploration with a high spatial resolution (generally an array of 12 electrodes with two neighboring electrodes being 2 mm apart).

ECoG recordings could thus reveal that both situations, that of expectancy and that of focused attention, were accompanied by the development of rhythmic activities within localized foci in the anterior cortex, in both cases contrasting with the background "desynchronized," low-voltage, fast activity which usually occurred when the animal was not in such conditions (e.g. in "active waking").

However, a major observation here has been that, in terms of ECoG activities, the two behavioral situations were essentially different: (a) during expectancy (Fig. 1A,2), a characteristic rhythmic activity developed in the somatosensory SI cortex (mainly forelimb projection area) at an average frequency of 14 Hz (Rougeul et al. 1972; Bouyer et al. 1983); (b) focused attention was concomitant with the appearance of a different rhythm (Fig. 1B,2), one at 36 Hz (again on an average, Bouyer et al. 1981), now localized upon two cortical sites (Bouyer et al. 1987), motor cortex (areas 4 gamma and 6a beta) and posterior parietal cortex (identified as area 5a; Hassler and Muhs-Clement 1964).

These observations could easily be quantified. The continously recorded ECoG activities were submitted on-line to spectral analysis based upon the FFT algorithm (Bouyer et al. 1981). Successive power spectra, each corresponding to 1 min recording (with power in μV^2 as the z-axis and frequency from 0 to 50 Hz as the horizontal x-axis) were then displayed along an oblique axis representing the total recording time within a session (which was, as a rule, programmed for 90 min). On the evolutive spectra, peaks could thus be observed at a frequency of 14 Hz (Fig. 1A,3) or 36 Hz (Fig. 1B,3).

In the following, we summarize some of the salient results obtained through systematic analysis of these two sets of rhythms. First, however, we would like to make a number of comments.

1. Regarding terminology, after some hesitations, we finally decided to use descriptive terms initially introduced by the human electroencephalographists. A 10 Hz activity was first described as wicket or mu rhythms recorded from centrally placed scalp electrodes in immobile subjects (Gastaut et al. 1957); these were as a rule suppressed through any kind of limb movement (traditionally, through clenching the fist). We

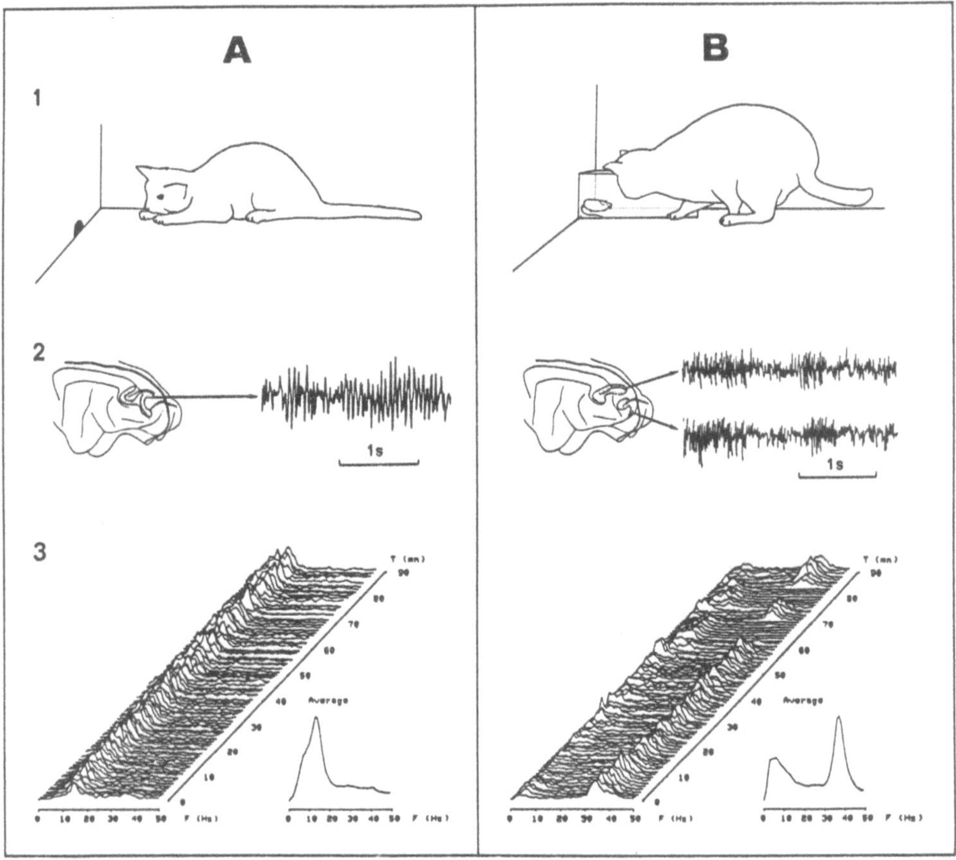

Fig. 1. Two types of waking immobility rhythms on cat cortex. **A** Expectancy: mean frequency 14 Hz. **B** Focused attention: mean frequency 36 Hz. Line 1 illustrates the most common attitude in the used experimental situation. Line 2, samples of rhythmic patterns and their localization: a single focus in SI in A and two foci, one motor and one parietal in B. Line 3, evolutive spectra taken during the 90-min recording time. Each spectrum was computed from 1-min recording. Heights of peaks indicate spectral power in μV^2 in the frequency band 0 to 50 Hz (resolution f = 0.2 Hz). Added to each set of evolutive spectra is the average spectrum computed over the 90-min recording time (From Rougeul-Buser et al. 1983)

therefore chose to use the same terminology and designated expectancy rhythms as mu. Similar activities had been recorded before our own studies by Roth et al. (1967) and Sterman and Wyrwicka (1967), who described them as "sensorimotor" rhythms. Given, however, that no mu rhythms could be recorded from the motor cortex itself, we considered this terminology as inappropriate.

Fast rhythms (16 to 45 Hz) had also been first observed in human ECoG by Jasper and Penfield (1949) and were more recently investigated through spectral analysis of the scalp EEG by Pfurtscheller (1981) and often described since as belonging to the beta rhythm band. We also adopted this term for feline fast rhythms, although their conditions of observation were not quite as specific in humans as they appeared to be in cats.

2. The conditions of observations of mu rhythms are in fact more diverse than of only expectancy. When recording from a cat placed in neutral conditions, i.e. with no particular situation of expectancy, mu rhythms can also be obtained, provided that the animal is immobile (except for eye and tail movements), but this "neutral mu" consists only in short episodes, contrasting with the long ones during expectancy; the usual description in this case is that the animal is in a state of quiet waking. There very likely exists a continuum between the neutral situation, with no particular target of interest, and that which is created when the cat is expecting a significant event to occur (in our paradigm a mouse to appear at the hole).

3. The cortical extent of each rhythm was remarkably small. It is only by using multiple, closely packed electrodes that we could identify, for each individual animal, the electrode(s) that disclosed the maximal amount of mu or beta activity, depending on the experimental conditions. As a rule, a given electrode recorded only one type of rhythm, beta or mu, except when it was implanted at the limit of areas SI and 5a.

4. Both mu and beta rhythms have appeared to be completely different from another type of synchronized activity, much more often described, namely the sleep spindles (Rougeul et al. 1972). These develop in quite distinct behavioral situations, since they are characteristic of slow sleep; it is true that their frequency is indeed very similar to that of the mu (i.e. ca. 14 Hz); their cortical extent is however much larger, with maximal amplitudes in the most medial parts of the motor and somatosensory cortices. Spindles are generally accompanied by slow delta type waves (2 to 4 Hz).

5. The rhythms described here also differ from the alpha ones which develop in the posterior cortical areas. These have in fact not yet been very well explored in the cat, the most precise information coming from observations in dogs (Lopes da Silva et al. 1978) although in conditions that were fairly different from our own. We could incidentally observe them in our cats, but in no case did they occur synchronously with the frontal rhythms; in our opinion, they should by no means be identified or confused with the latter. Other experimental paradigms should be developed to better understand the behavioral conditions for their occurrence.

6. Finally, still other (ill-defined) behavioral states are just as common in the cat as also in other species; these are suggestive of either drowsiness or withdrawal of attention. These states are also generally characterized by the development of specific rhythms, although at a lower frequency (4 to 8 Hz), appearing both in the anterior and posterior cortex, less localized than the attention rhythms, but again quite distinct from the slow sleep spindles (Rougeul et al. 1974).

Thalamic Mechanisms Involved In Mu and Beta Rhythms

Beyond this ECoG phenomenology of attentive states, we considered as a next step several problems related to the origin and underlying mechanisms of these rhythmic activities.

A first step in this analysis was to identify, for each set of rhythms, mu and beta, a specific thalamic zone (focus) that we had reasons to consider as essential for their development. Our arguments were based upon several types of observations, including: (a) gross recording from the thalamus to localize rhythms that are identical to and develop in synchrony with the cortical ones; (b) quantitative comparison of the simultaneously recorded cortical and thalamic rhythms through computation of their coherence function; (c) disappearance of one type of cortical rhythms after destroying the putative thalamic focus; (d) unit recording from the thalamus to seek for correlation between their firing and the corresponding cortical gross activity (these latter aspects are not considered further here). All these arguments lead us to conclude (as a likely hypothesis) that the thalamic area acts as a rhythmogenic zone for the corresponding cortical activity. (We do not mean thereby that rhythms are entirely generated within this thalamic area; the latter may only be the end-station of a more complex circuit possibly involving other structures.)

The rhythmogenic zone for the mu activity was thus identified within nucleus ventralis posterior (VP, forearm area). Fig. 2B illustrates some steps in this identification: 14 Hz rhythms developing in this thalamic area simultaneously with the cortical mu and coherence function illustrating a peak at about 60% around the 14 Hz band when both cortical and thalamic rhythms were compared. In addition, a complete disappearance of the mu rhythms was noticed after bilateral lesioning of the VP focus (Bouyer et al. 1983).

A parallel type of data is illustrated for the beta rhythms. In this case, however, the situation is more complex since two beta foci were identified, one lying anteriorly on the motor cortex and a second one more posteriorly in area 5a.

The rhythmogenic zone for the posterior beta focus (Fig. 2A) could be identified through the same criteria as for the mu focus, in the medial part of the posterior group of the thalamus (POm): presence of beta type activities in most of this thalamic zone, high value of their coherence with the cortical parietal beta rhythms (ca. 70%).

The rhythmogenic zone for the anterior beta focus in the motor cortex, was much more difficult to localize. We have now succeeded to characterize its presence in a rather restricted area belonging to nucleus ventralis lateralis-ventralis anterior (VL-VA).

Brainstem Control Systems For Thalamocortical Attention Rhythms

The third step in this investigation started when we could show that a variety of drugs involving the catecholaminergic, dopaminergic (DAergic), and noradrenergic (NAergic) systems induced significant and coherent modifications of either mu or beta rhythms, leading to the conclusion that the beta system was under a facilitatory (positively gating) control by the DAergic system, while the mu activity was submitted to an inhibitory (negatively gating) one from the NAergic system. We shall not go into these various detailed pharmacological studies here. Let us however summarize some of our main observations.

1. In favor of the DAergic gating, the fact that while recording units from the ventral tegmental area (VTA; containing the DAergic cell mass A10) we found that some of its cells suddenly increased firing tonically about 1 s before onset and started decreasing firing just at cessation of each train of beta rhythms (Montaron et al. 1984). According to Rogawski and Aghajanian (1982), cells with high spontaneous activity may not be DAergic. An alternative possibility, however, was recently raised by the finding that

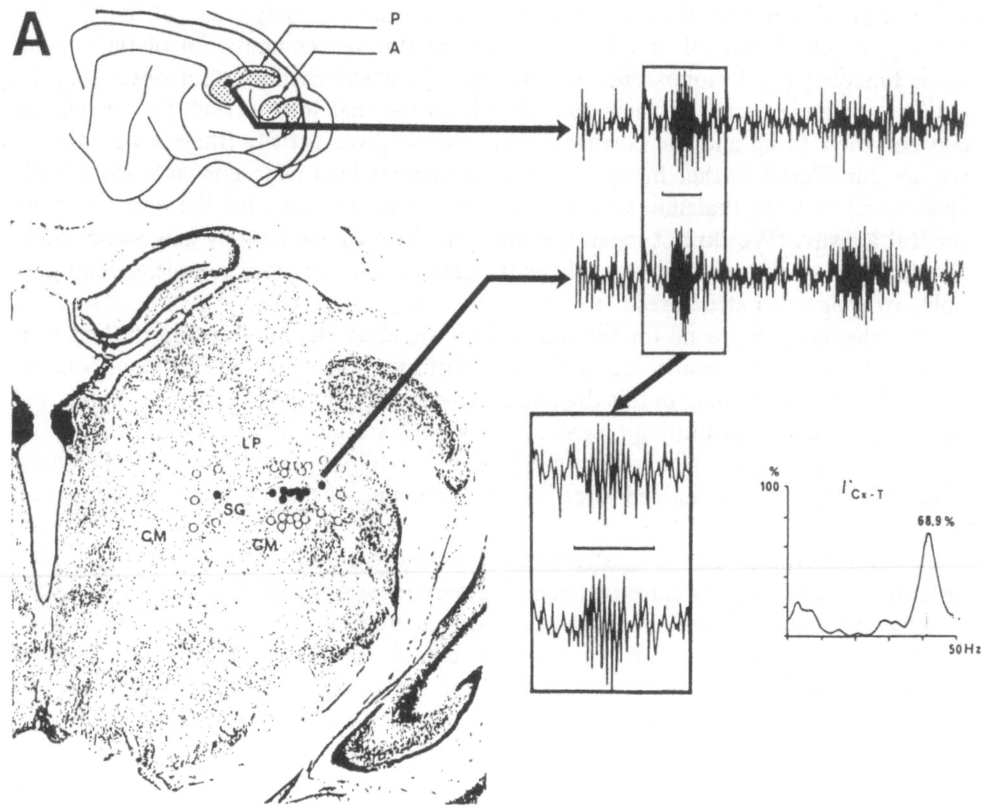

Fig. 2A. Cortical (top) and thalamic (bottom) beta activities. Rhythms during focalized vigilance (situation B on Fig.1). On top, their cortical localization. The hatched zones are data pooled from 20 cats, encompassing the extreme limits of two foci, one (A) over the cruciate sulcus, the other (P) over the ansate sulcus. At right, simultaneous recording from the posterior beta focus area 5a and from the posterior group of the thalamus. The close correspondence between the two activities is illustrated : (a) by the lower two records which are enlargements of the upper two traces in the frame, at double recording speed (calibration marks, 500 ms); (b) by the coherence function analysis between cortex and thalamus, performed over 40 records of 1.25 s each (sampling frequency 102.4 Hz; analysis performed between 0 and 51.2 Hz; resolution 0.8 Hz). Notice high coherence values in frequency band 35-45 Hz. At left, thalamic sites from which fast beta rhythms could be recorded are indicated by black dots. Open dots mark unresponsive points. All recording sites (11 cats) were located in the frontal plane A 6.5. No active points were encountered at 5 mm more anteriorly or more posteriorly from that level. Abrreviations: see Fig. 2B

some DAergic VTA cells, belonging precisely to the mesocortical pathway, lack somatodendritic autoreceptors and concomitantly display a higher frequency of firing (Roth et al. 1987). In the latter case, all cells recorded in our exploration could indeed be DAergic.

2. Another observation in the same line concerned the behavioral repercussions of VTA lesions. The animals behaved normally except when placed in a situation requiring focused attention; they were then unable to stand still and watch a target motionlessly as normal animals would do (Montaron et al. 1982), pretty much resembling the VTA syndrom described in rat (Stinus et al. 1978; Simon et al. 1980). In these lesioned animals, beta rhythms were absent (Fig. 3); they reappeared somewhat after administration of apomorphine (4 mg/kg IP).

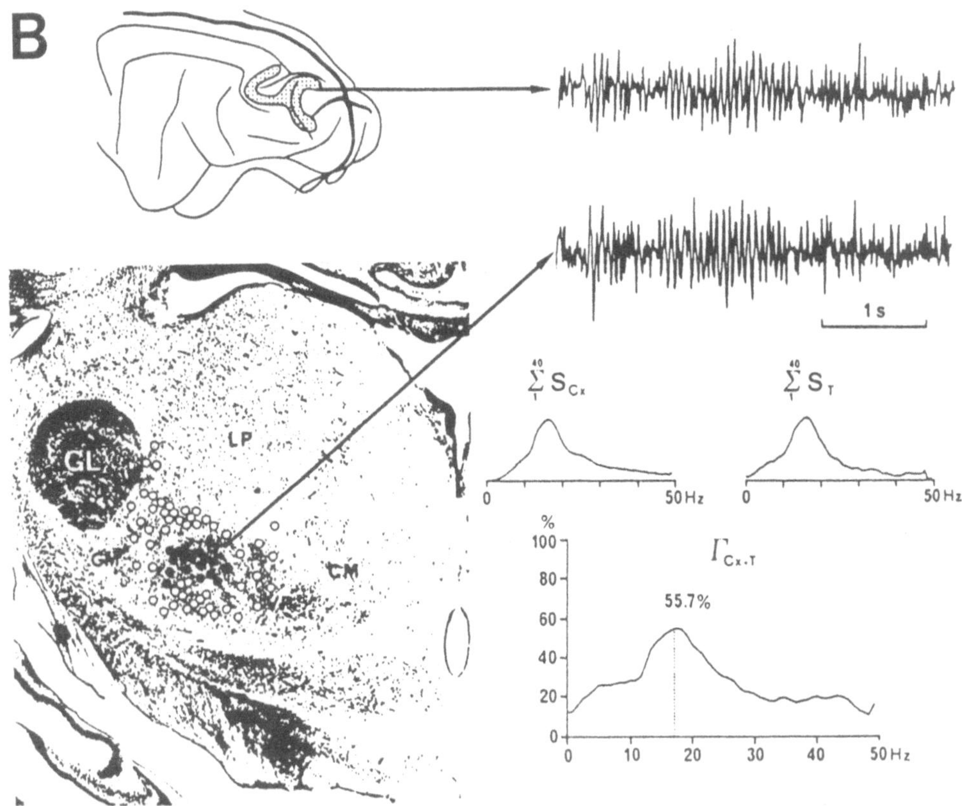

Fig. 2B. Cortical (top) and thalamic (bottom) mu activities. Rhythms during expectancy (situation A on Fig. 1). Same general arrangement as in Fig. 2A: top left, simultaneous cortical (area SI on figurine) and thalamic (nucleus ventralis posterolateralis) recordings of the mu rhythms. Below, average spectra computed over 40 episodes recorded from these derivations (maximum peak at 16 Hz). Bottom: corresponding coherence values versus frequency between cortex and thalamus, within the 0-50 Hz frequency band. Peak at ca. 16 Hz indicates important linkage between the two structures around this frequency. At left, a summary of 25 explorations in frontal plane 8.0. Positions of electrode tips that recorded mu activity are indicated by black dots; nonrhythmic derivations, by open symbols. Active points are located in nucleus ventralis posterolateralis. CM, centrum medianum; GL, geniculatum laterale; GM, nucleus geniculatus medialis; LP, nucleus lateralis posterior; SG, n. suprageniculatus

3. We then tried to determine which pathway originating from the VTA was possibly involved in the control of the beta system, the mesolimbic or the mesocortical. We therefore performed kainic lesions of nucleus accumbens (Bouyer et al. 1986), known as the first relay station on the mesolimbic route. Rather unexpectedly, the lesioned animals disclosed a syndrome that was opposite to that after VTA lesions, with a tendency to maintain fixation (on a given target or even when no significant target was present) and an apparent difficulty to switch from one fixation point to another one. Correspondingly, the amount of beta rhythms was significantly increased during these episodes (Fig. 4). These data raised a number of questions regarding the roles of the various stations of the mesolimbic and mesocortical pathways. These roles are presently under investigation, through exploring the participation of other structures that are presumably

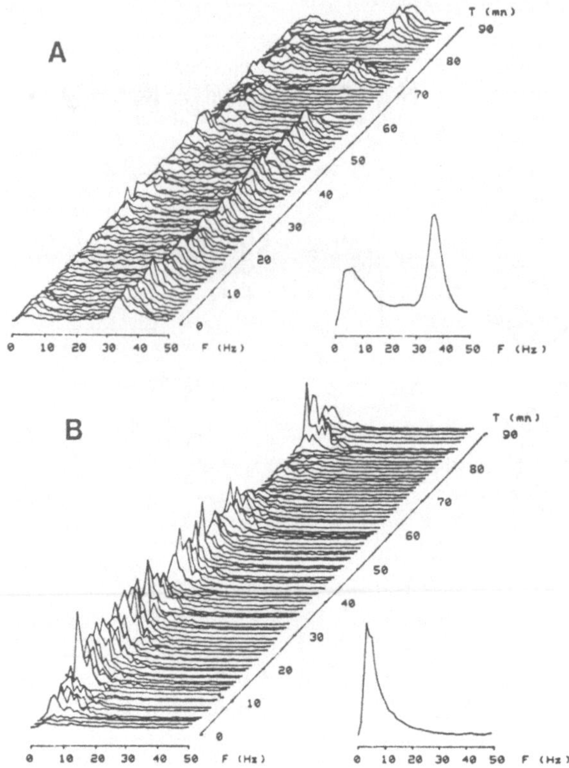

Fig. 3. Ventral mesencephalic tegmentum controls thalamocortical rhythms and accompanying behavior. Evolutive spectra taken during 90 min of continuous ECoG recording in the beta rhythm focus in the cat. Same displays as in Fig. 1. **A** Control record in normal animal in situation of Fig. 1B (mouse in transparent box); notice large amount of 36 Hz activity lasting for up to 60 min. **B** As A but after bilateral lesion of the ventral tegmental area; peaks at 36 Hz are no longer visible. The repetitive peaks at very low frequency (0 to 3 Hz) correspond to the large amount of artifacts, due to the cat's hyperactivity. (From Bouyer et al. 1983)

also involved, such as the frontal cortex, the nucleus medialis dorsalis of the thalamus and the amygdala.

4. In support of the inhibitory role of NAergic pathways, some recent behavioral and ECoG observations that were performed after treatment of the animal with DSP4, a drug that is now well-known to destroy the NAergic pathways (Jonsson et al. 1981). As a consequence, animals displayed a marked increase in the episodes of mu activitiy, accompanied by an exaggeration of their attitude of "expectancy" or "quiet waking" even in situations where a control animal (or the same subject before treatment) would have reacted with an attitude of focused attention (Delagrange et al. 1989).

Discussion and Conclusion

In brief then, we could observe the development of state-dependent ECoG synchronizations, concomitant with the appearance of two states that have in common both a bo-

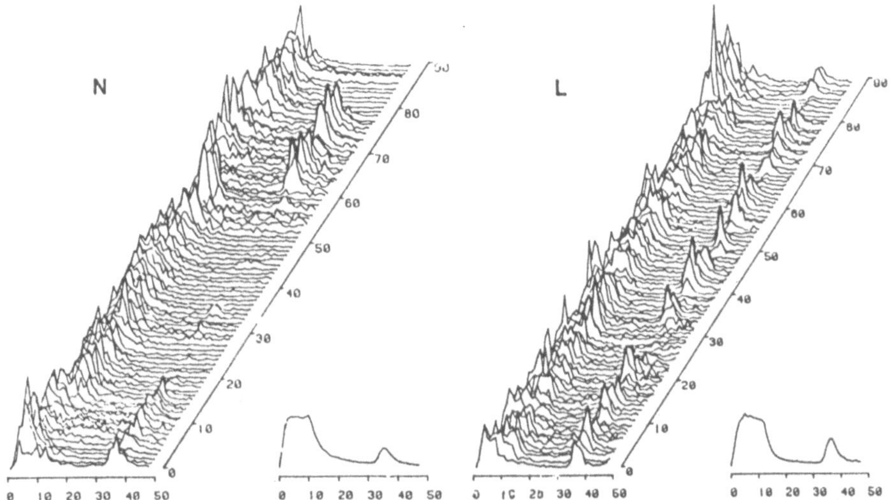

Fig. 4. Effect of lesioning nucleus accumbens. The cat was this time placed in a "neutral" situation (neither A nor B of Fig. 1). Recording from the posterior parietal beta focus. N: normal animal displays variable activities: some movements (accompanied by 0-3 Hz), some slow "drowsiness rhythms" (5-10 Hz), but only two phases with beta rhythms. Notice after bilateral lesions of nucleus accumbens (L), sustained sequences of beta rhythms. (From Bouyer et al. 1986)

dily immobilization and an attentive, i.e. a cognitive, task but differ by their particular pattern: expectancy of an invisible target to appear in one case, versus observation of a specific, visible target in the other. Several points of discussion may be raised; we concentrate upon only two of them.

1. The involved neurochemical modulatory mechanisms are twofold, noradrenergic in one case and dopaminergic in the other; these actions are not symmetrical, however, since DA favours its dependent state while NA modulation acts as an inhibitory system. The existence of some analogy to psychiatric disorders is a tantalizing issue: on the one side, the dopaminergic defect, characterized by a loss of the ability to concentrate attention upon a certain target and by repetitive shifts from one target to another; and on the other, the noradrenergic defect, which may make it difficult to pass from a diffuse or expectant type of attention to a more concentrated, focused one on a specific target.

2. The neurophysiological significance of the observed synchronizations also deserves a variety of comments.

Most recent studies on attentive processes have now emphasized that changes occur in the neuronal responsiveness of sensory channels (e.g. visual) while processing is performed of the attended stimulus (see Spitzer et al. 1988). What then can be the significance of our observations showing synchronizations of a restricted group of thalamocortical neurons as a concomitant of attention? Two (not exclusive) hypotheses can be suggested: (a) thalamocortical neuronal ensembles involved in this synchronization impair processing of nonpertinent stimuli (e.g. somatic ones); this would be the case in particular when mu develops in the VP-SI channel. Some of our observations (Delagrange et al. 1987) are indeed in favor of this issue. (b) Synchronization impairs motor integration and thus favors immobilization as part of the strategy of the animal. This might be the case, in particular, for the beta system, which involves area 5 and motor cortex, both being implicated (at least in the monkey; see Mountcastle et al. 1975; Seal and Commenges 1985) in a motor command. The best link between our data and the more traditional one on facilitation of sensory processing would be that catecholamines which modulate synchronization also modulate processing in relevant (e.g. visual) channels, such as postulated by the noradrenergic hypothesis of attention (Mason and Iversen 1979). Whether such type of relationship exists remains to be determined.

Summary

Two sets of attentive behavioral states (involving immobilization and a given pattern of cognitive activity) could be identified in the cat, correlative with the development of rhythms in specific thalamocortical channels that project upon discrete foci of the anterior cortex: one when the animal is "expecting" (waiting for an event to occur or a target to appear), accompanied by rhythms at 14 Hz (called mu), another when it is focusing its attention on a visible target, with rhythms at ca. 36 Hz (beta rhythms). The mu channel includes nucleus ventralis posterior and its cortical projection area SI; beta rhythms involve two channels, one anterior (nucleus ventralis lateralis to areas 4 and 6a), one more posterior (nucleus posterior pars medialis POm to parietal area 5a). The thalamic side of each channel seems to act as a rhythmogenic zone for the corresponding activity.

Moreover, two, in a way antagonistic systems were shown to exert a general control over these thalamocortical channels; one (partly at least DAergic) originating from the VTA favors immobile focalized attention (and the concomitant development of beta rhythms) while another, NAergic and probably governed from the locus coeruleus, restrains the attitude of motionless expectancy and blocks synchronization in the thalamocortical VP-SI channel.

References

Bouyer JJ, Montaron MF, Rougeul-Buser A, Buser P (1980) A thalamo-cortical rhythmic system accompanying high vigilance levels in the cat. In: Pfurtscheller G et al. (eds) Rhythmic EEG activities and cortical functioning. Elsevier North Holland, Amsterdam, pp 63-77

Bouyer JJ, Montaron MF, Rougeul A (1981) Fast fronto-parietal rhythms during combined focused attentive behaviour and immobility in cat: cortical and thalamic localizations. Electroenceph Clin Neurophysiol 51: 244-252

Bouyer JJ, Tilquin C, Rougeul A (1983) Thalamic rhythms in cat during quiet wakefulness and immobility. Electroenceph Clin Neurophysiol 55: 180-187

Bouyer JJ, Montaron MF, Fabre-Thorpe M, Rougeul A (1986) Compulsive attentive behavior after lesion of the ventral striatum in the cat: a behavioral and electrophysiological study. Exp Neurol 92: 698-712

Bouyer JJ, Montaron MF, Vahnee JM, Albert MP, Rougeul A (1987) Anatomical localization of cortical beta rhythms in cat. Neuroscience 22:863-869

Delagrange P, Tadjer D, Rougeul A, Buser P (1987) Activité unitaire de neurones du noyau ventral postérieur du thalamus pour divers degrés de vigilance chez le chat normal. C R Acad Sci [III] 305: 149-155

Delagrange P, Tadjer D, Bouyer JJ, Rougeul A, Conrath M (1989) Effect of DSP4, a neurotoxic agent on attentive behaviour and related electrocortical activity in cat. Behav Brain Res 33; 33-43

Gastaut H, Jus A, Jus C, Morell F, Storm van Leeuwen W, Dongier S, Naquet R, Regis H, Roger A, Bekkering D, Kamp A, Werre J (1957) Etude topographique des réactions électroencéphalographiques conditionnées chez l'homme. Electroenceph Clin Neurophysiol 9: 1-34

Hassler R, Muhs-Clement K (1964) Architektonischer Aufbau des sensomotorischen und parietalen Cortex der Katze. J Hirnforsch 6: 377-420

Hillyard SA (1985) Electrophysiology of human selective attention. Trends Neurosci 8: 400-405
Jasper HH, Penfield W (1949) Electrocorticograms in man: effect of voluntary movement upon the electrical activity of precentral gyrus. Arch Psychiat Z Neurol 183: 163-174
Jonsson G, Hallman H, Ponzio F, Ross S (1981) DSP4 (N-(2-chloroethyl)-N-ethyl-2-bromobenzylamine) - a useful denervation tool for central and peripheral noradrenaline neurons. Eur J Pharmacol 72: 173-188
Lopes da Silva FH, Storm van Leeuwen W (1978) The cortical alpha rhythm in dog: the depth and surface profile of phase. In: Brazier MAB, Petsche H (eds) Architectonics of the cerebral cortex. Raven, New York, 3:319-333
Mason ST, Iverson SD (1979) Theories of the dorsal bundle extinction effect. Brain Res Rev 1: 107-137
Montaron MF, Bouyer JJ, Rougeul A, Buser P (1982) Ventral mesencephalic tegmentum (VMT) controls electrical beta rhythms and associated attentive behaviour in the cat. Behav Brain Res 6: 129-145
Montaron MF, Bouyer JJ, Rougeul A, Buser P (1984) Activité unitaire dans l'aire tegmentale ventrale et état d'attention focalisée chez le chat normal éveillé. C R Acad Sci [III] 298: 229-236
Moran J, Desimone R (1985) Selective attention gates visual processing in the extrastriate cortex. Science 229: 782-784
Mountcastle VB, Lynch JC, Georgopoulos A, Sakata H, Acuna C (1975) Posterior parietal association cortex of the monkey: command functions for operations within extrapersonal space. J Neurophysiol 38: 871-908
Mountcastle VB, Motter BC, Steinmetz MA, Sestokasa K (1987) Common and differential effects of attentive fixation on the excitability of parietal and prestriate (V4) cortical visual neurons in the macaque monkey. J Neurosci 7: 2239-2255
Pfurtscheller G (1981) Central beta rhythm during sensorimotor activities in man. Electroenceph Clin Neurophysiol 51:253-264
Rogawski MA, Aghajanian GK (1982) Activation of lateral geniculate neurons by locus coeruleus or dorsal noradrenergic bundle stimulation: selective blockade by the alpha1-adrenoceptor antagonist prazosin. Brain Res 250: 31-39
Roth SR, Sterman MB, Clemente CD(1967) Comparison of EEG correlates of reinforcement, internal inhibition and sleep. Electroenceph Clin Neurophysiol 23: 509-520
Rougeul-Buser A, Bouyer JJ, Montaron MF, Buser P (1983) Patterns of activities in the ventrobasal thalamus and somatic cortex SI during behavioral immobility in the awake cat: focal waking rhythms. Exp Brain Res [Suppl] 7: 69-87
Rougeul A, Corvisier J, Letalle A (1974) Rythmes électrocorticaux caractéristiques de l'installation du sommeil naturel chez le chat. Leurs rapports avec le comportement moteur. Electroenceph Clin Neurophysiol 37: 41-57
Rougeul A, Letalle A, Corvisier J (1972) Activité rythmique du cortex somesthésique primaire en relation avec l'immobilité chez le chat libre éveillé. Electroenceph Clin Neurophysiol 33: 23-39
Seal J, Commenges D (1985) A quantitative analysis of stimulus- and movement-related responses in the posterior parietal cortex of the monkey. Exp Brain Res 58: 144-153
Simon H, Scatton B, Le Moal M (1980) Dopaminergic A 10 neurones are involved in cognitive functions. Nature 286: 150-151
Spitzer H, Desimone R, Moran J (1988) Increased attention enhances both behavioral and neuronal performance. Science 240: 338-340
Sterman MB, Wyrwicka W (1967) EEG correlates of sleep: evidence for separate forebrain substrates. Brain Res 6: 143-163
Stinus L, Gaffori OH, Simon H, Le Moal M (1978) Disappearance of hoarding and disorganization of eating behavior after ventral mesencephalic tegmentum lesion in rats. J Comp Physiol Psychol 92: 288-296
Roth RH, Wolf ME, Deutsch AY (1987) Neurochemistry of midbrain dopamine systems. In Meltzer HY (ed) Psychopharmacology: the third generation of progress. Raven, New York, pp 81-94

The Sensory Neuron - Where the Action Begins

A. Iggo
Department of Preclinical Veterinary Sciences, University of Edinburgh, R(D)SVS, Summerhall, Edinburgh EH9 1QH

Introduction

All the sensory information entering the central nervous system, and that leads eventually to motor action, does so in afferent fibres that are highly specialised for their particular functions. It is therefore highly germane to the subject of this symposium in honour of Hans Kornhuber to give special consideration to sensory receptors. Vernon Mountcastle has already commented on Hans Kornhuber's skills in psycho-physics, and I can extend that comment to include collaborative experiments shared with Hans in Edinburgh in 1969 when we established a remarkable ability of sensitive cutaneous non-myelinated mechanoreceptors to encode the amplitude of mechanical indentation of the skin. Figure 1 from that study shows an almost linear relation between stimulus and response, when a randomised series of 10 different stimulus amplitudes were used, and care was taken to space the stimuli in time so as to avoid inter-stimulus interactions. This result can also serve to illustrate an important feature of the cutaneous sensory receptors - when they are examined quantitatively as single units they are found to have well-defined characteristics that enable them to be categorised, into sharply delineated classes. In the broadest terms these are mechanoreceptors, thermoreceptors and nociceptors. Each class in turn can be further subdivided on the basis of what can be called their peripheral biophysical specificity.

Although the majority of the sensory innervation of the skin is by non-myelinated afferent fibres, and as Figure 1 so clearly illustrates, some of them can encode tactile stimuli with considerable precision, the C fibres are only slowly conducting and are ill-fitted to the task of providing early notice of changes in the periphery. Instead, this function is provided by the myelinated axons and their sensory receptors; the axonal conduction velocity is, at maximum, two orders of magnitude greater for the myelinated than for the un-myelinated fibres (100 m/s versus 1 m/s). The myelinated cutaneous mechanoreceptors also differentiate into several physiologically distinct categories on the basis of frequency range, persistence of response and location in the skin. Simply on the basis of the characteristics of their response to quantitatively controlled mechanical stimuli and the location and type of receptive field they fall into two groups - rapidly adapting mechanoreceptors excited by dynamic or changing indentations and slowly adapting mechanoreceptors that, in addition, sustain an afferent discharge during maintained steady indentation of the receptive field. Within each of the categories further entities can be recognised, and it is clear that the skin contains a variety of mecha-

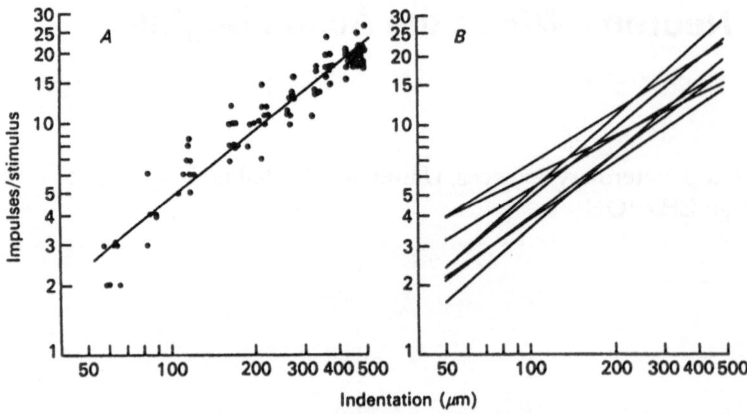

Fig. 1. a Stimulus-response relation for a C mechanoreceptor, plotting on logarithmic co-ordinates, the responses in a random intensity trial using skin indentations between 50 and 500 mu (from Iggo and Kornhuber 1977). The data were fitted with a regression line using the equation: log R = a + b log S. The regression coefficient was 0.99. **b** Regression lines for 7 C mechanoreceptors, including that in a

noreceptors with myelinated axons, an array that implies considerable encoding of peripheral stimuli at the point of entry into the nervous system.

I shall choose only examples of the specialised cutaneous mechanoreceptors. More detailed information is available in a recent review (Iggo and Andres 1985). Probably the first cutaneous mechanoreceptor to be examined electrophysiologically in detail was the Pacinian corpuscle, described in 1840 by F. Pacini. The relatively large size of the receptor makes it possible to carry out in vitro studies of the receptor in isolation (Gray and Sato 1953; Loewenstein and Rathkamp 1958; Loewenstein and Skalak 1966). Furthermore, the last-named authors showed that the typical and characteristic lamellation of the Pacinian corpuscle serves as a high-pass filter and lets only the higher frequency components of a mechanical stimulus reach the centrally located transducer elements. The combination of a relatively brief (< 10 ms) response of the transducer to a mechanical stimulus and the filtering characteristics of the lamellation result in the Pacinian corpuscle acting as a vibration detector. The receptor is thus well suited to responding selectively to relative high frequency vibrations (50-1000 Hz) and to rejecting low frequencies or steadily maintained indentation, and it can thus function very effectively to report abrupt changes in the environment.

Slowly Adapting Mechanoreceptors

The natural mechanical stimuli impinging on the body surface are not, however, all transient or vibratory, and in many situations there is a persistent, sustained or prolonged mechanical force acting on the skin. It is not surprising therefore that the skin contains mechanoreceptors capable of encoding such stimuli. Two well-differentiated slowly adapting mechanoreceptors, the SAI and SAII, exist in mammalian skin (Iggo and Muir 1969; Chambers et al. 1972). The characteristic response of the SAI receptor

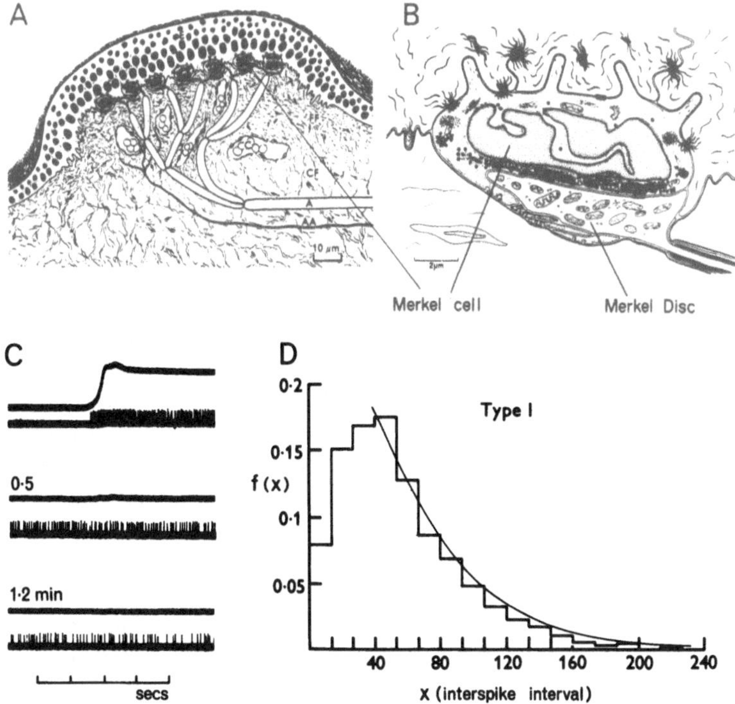

Fig. 2. The slowly adapting type I (SAI) mechanoreceptor, the Merkel 'Tastzellen' in hairy skin. **a, b** Diagrams based on electron micrographs. **c** Electrophysiological recording of response of an SAI to sustained mechanical identation, applied in upper record and continued in the lower two. **d** Frequency distribution of lengths of impulse intervals, semi-Poisson in character, during sustained indentation. (From Iggo 1977)

in hairy skin of the cat or monkey, when the skin surface is tested with punctiform stimulation, is a discharge of impulses evoked by contact of the probe with small spot-like regions of the skin. When there are several spots, the intervening skin is unresponsive to mechanical stimulation. At each sensitive spot there is a small dome-like swelling of the skin surface. The visibility of the spot is enhanced by the presence of a small tuft of capillaries at its centre. If such a spot is removed by dissection, the normal response of the afferent unit to mechanical stimulation disappears. Histological examination of a receptive spot, marked during electro-physiological recording and subsequently fixed for light and electron microscopical analysis, reveals the structure illustrated diagrammatically in Fig. 2a, b. The myelinated afferent fibre innervating a 'touch-dome' enters at the base and branches freely, each branch being myelinated up to the place that it penetrates the basement membrane of the epidermis and forms a flattened disk-like expansion that is in special relation to a cell (the Merkel cell) at the base of the epidermis. Although Merkel first described these cells, calling them 'Tastzellen' in 1875, and they were subsequently reported in man in clusters that were called 'Haarscheiben' by Pinkus (1905), their fine structure awaited electron microscopical study until Cauna (1962) first described them. The correlation of structure and function was established by Iggo and Muir (1969) in a combined electrophysiological and electron microscopical investigation. These results and conclusions have been confirmed and even rediscovered by numerous authors.

Several features of the electrophysiological characteristics of the SAI mechanoreceptor (their spot-like receptive fields, insensitivity to lateral stretch of the skin, quasi-expontential distribution of interspike intervals in the adapted afferent discharge; Fig. 2) were worked out in detail in animals such as the cat. These attributes have subsequently enabled the SAI to be identified in microneurographical studies in awake human subjects (Knibestöl and Vallbo 1980) and to be distinguished from SAII and from rapidly adapting receptors (PC and RA).

The SAI receptors can also be excited by appropriate thermal stimuli. The more effective stimulus is rapid cooling of the skin, but the ongoing discharge in response to steady mechanical pressure is also temperature dependent. The SAI are very similar in these respects to specific cold receptors. They differ, however, in having a very high sensitivity to mechanical stimulation and much faster conducting afferent fibres. Furthermore, the central destinations of the afferent fibres are different, since the SAI, in addition to sending collaterals into the dorsal horn, also send a major branch towards the brain whereas the thermoreceptor afferents terminate close to their segment of entry. The thermal receptivity of the SAI, like the SAII and other mechanoreceptors, is presumably attributable to ionic mechanisms in the nerve terminals but has not yet been satisfactorily explained. The very rapidly adapting Pacinian corpuscles, in contrast, are indifferent to thermal stimuli (Ishiko and Loewenstein 1961) that excite the SAI. Instead, there is an expected, if rather large, positive Q10 ($=2$) that enhances the size of the transducer potential of a PC at progressively higher temperatures.

Fine Structure of Merkel Cell/Merkel Disk

The fine structure of the Merkel complex has some interesting features that raise questions concerning the mechanism of transduction. In contrast to other mammalian cutaneous mechanoreceptors, the SAI afferent fibre ends in special relation to a non-neural cell, the Merkel cell. Furthermore, the normal SAI properties depend on this association with the Merkel cell (Brown and Iggo 1963; Mearow and Diamond 1983).

The notable ultrastructural features of the Merkel complex, listed in order from the epidermal side to the axon are:

1. Finger-like projections, containing longitudinally oriented parallel filaments borne on the epidermal side of the Merkel cell and inserted into invaginations in overlying special epidermal cells.
2. Absence of desmosomes attaching the projections to the epidermal cells, although hemi-desmosomes are present elsewhere.
3. Poly-lobulated nucleus with its flattened surface apposed to the side of the Merkel cell adjacent to the nerve disk.
4. Large numbers of membrane-bound dense-cored vesicles (100 nM diameter) in the cytoplasm between the nucleus and nerve ending.
5. Synapse-like junctions between the Merkel cell and Merkel disk, with an accumulation of dense-cored vesicles adjacent to this structure, but otherwise not close to the Merkel cell membrane.

These features prompt the question "Is the Merkel cell the actual transducer?" Parallels can be drawn with the hair cell of the inner ear and vestibular apparatus, where the transducer function of the hair cell has been clearly established (see Hamann and Iggo 1988, for recent reviews). In hair cells when stereocilia, borne on the apical end of the hair cell are displaced laterally in the direction of the kinocilium, they cause entry of Ca^{2+} ions through the mechanoelectrical transduction channel in the apical end of the cell (Ohmori 1988). This in turn leads to depolarisation of the hair cell and to the release of transmitter from the base of the cell, possibly by a combination of depolarisation and loading of the cell with Ca ions, with consequent depolarisation of the afferent terminals apposed to the hair cell and discharge of impulses in the afferent fibre.

The above sequence of events has not been established for the Merkel cell complex. Only recently has it been possible to test an association between any particular feature of the Merkel cell and its role in mechanotransduction. In experiments to assess the sensitivity of SAI receptors to different levels of hypoxia, Findlater et al. (1987) reported a high sensitivity of the Merkel complex compared with the relative insensitivity of the afferent fibre. Electron microscopic examination of Merkel cells fixed at the time of SAI inexcitability revealed that at the time of failure of transduction there was a statistically significant depletion of dense-cored vesicles from the Merkel cells. The degree of depletion was not simply a consequence of anoxia, although this alone did cause a reduction. A further reduction to about 30% of normal was caused by iterative mechanical stimulation and at the point of failure of afferent response in the sensory nerve fibre. In a condition of recovered mechanical sensitivity, the vesicle content in the Merkel cells was statistically indistinguishable from cells in hypoxic unstimulated domes.

Here then was evidence that the dense-cored vesicles in Merkel cells were labile whereas in previous attempts to influence the vesicle number no effect of mechanical stimulation on normal Merkel cells tissues could be found. Indeed, it is possible for normal SAI units to continue to maintain an unvarying afferent discharge to repeated mechanical stimulation, so long as the stimulus amplitude is within normal limits (Iggo and Muir 1969). Severe mechanical stimulation, on the other hand, can silence a receptor (Hunt and McIntyre 1961). Normally, then, if the vesicles are required to be discharged during mechanoelectric transduction they must be replenished very rapidly within the Merkel cell.

Exocytosis and Calcium

The release of vesicles from cells, by exocytosis, is a calcium-dependent process, and it should therefore be modifiable by manipulation of the Ca^{2+} in the environment of the SAI receptor. The perfused cat hind-limb preparation of Findlater et al. (1987) has been used by Pacitti and Findlater (1988) to test three ways to influence Ca ion permeability of Merkel cells, using cobalt (as $CoCl_2$), cadmium (as $CdCl_2$), or verapamil as calcium channel blocking agents. The concentrations chosen interfered reversibly in a dose-dependent manner with the responses of SAI receptors to standardised mechanical stimuli, and eventually both Co^{2+} and Cd^{2+} caused complete failure of the receptor to respond, even though the afferent fibres were still excitable by direct electrical stimulation. Verapamil (30 mM) also caused a progressive decline in response, to 25% of the control value.

Table 1. The vesicle density and the number of 'synaptic-like' structures in Merkel cells after saline, colbalt chloride or verapamil hydrochloride. (From Pacitti and Findlater 1988)

Drug (mM)	Granule density (number/μ)	Synapses (number/section)
Saline control	12.1 ± 1.5 (13)	0.6 ± 0.25 (13)
$CoCl_2$ (2.0)	4.3 ± 0.8** (17)	2.1 ± 0.40* (17)
Verapamil (0.03)	6.8 ± 0.8* (18)	-

Values are given as mean ± SE of mean, with number of observations in parentheses.
*P < 0.0., **P < 0.001 (t test)

These Ca^{2+} channel blockers could be acting at several places, and not necessarily in relation to their effect on mechanical sensitivity on the Merkel cell. Quantitative evaluation of effects on dense-cored vesicles in electron micrographs from saline controls, $CoCl_2$-treated and verapamil-treated SAI receptors gave values of 12.1 ± 1.5, 4.3 ± 0.8 and 6.8 ± 0.8 vesicles per μ^2 respectively - a clear indication that vesicle numbers were reduced by the Ca^{2+} channel blockers. The reduced mechanical sensitivity could be due, as in the hypoxia experiments, to this factor, especially if coupled with a reduced Ca^{2+} input to the Merkel cells because of the action of the blocking agent on the mechanoelectric transduction channel or on Ca channels elsewhere in the Merkel cell membrane. A complicating factor is the actual reduction in numbers of dense-cored vesicles, since it might be expected that a reduced Ca^{2+} intake would be expressed in reduced exocytosis and an accumulation of vesicles. It is known, however, from work on the neuro-muscular junction that Co^{2+} causes an increase in the frequency of miniature endplate potentials (Kita and van der Kloot 1973; Weakly 1973), and a similar initial discharge of dense-cored vesicles in SAI exposed to $CoCl_2$ would account for both an initial excitation with $CoCl_2$ and a depletion of vesicles. The latter effect if Co^{2+}, as elsewhere (Lavoie and Bennet 1983), inhibits vesicle transport, being due to failure of transit of vesicles from the Golgi apparatus to the storage site adjacent to the nerve terminal.

The results reviewed above clearly implicate the Merkel cell in the normal operation of the SAI as a mechanotransducer. Indeed, they may confer on it some of its distinctive properties, such as an ability to sustain high rates of discharge. It is evident that the metabolic requirements of the receptor are high, since it fails quickly once it is completely anoxic, in contrast to the afferent axon. The latter can carry tens of thousands of impulses, evoked electrically, after the receptor has failed to respond to a mechanical stimulus.

Interspike-Interval Pattern in Afferent Discharge

The individual Merkel disks and associated Merkel cells, forming as they do a terminal arborisation, feed into a single axon. There may be 50 or more Merkel cells/disks in a touch-dome, and it is not known how they co-operate in setting up an afferent discharge, although it can be predicted that each acts independently in its response to a mechanical stimulus. The presence of so many independently acting end stations and the manner of their action on the stem fibre presumably account for another distinctive feature of the SAI receptors - the quasi-exponential distribution of interspike intervals (Fig. 2) in the afferent discharge. Whether it is the independence of generator sites in the Merkel cells (Horch et al. 1974) or action potential interactions in the terminal branches (Chambers et al. 1972) that accounts for the ISI distribution, the fact is that it is the structure of the SAI receptor that is responsible.

Central Specialisation of the Sensory Neuron

The preceding section of this article has dealt with aspects of the peripheral specialisation of the sensory neuron, taking as examples cutaneous sensory receptors. Is there a corresponding central specialisation with respect to onward transmission of peripherally encoded information?

Refinements of microelectrode technique now enable intracellular recording from the larger dorsal root afferent fibres, using glass pipette microelectrodes filled with a marker substance such as HRP that can be deposited inside the axon. Orthograde and/or retrograde transport of the marker is then permitted and used to trace physiologically identified axons to their termination in the spinal cord (Brown 1981).

Careful three-dimensional reconstruction of individual indentified axons, combined on some occasions with intracellular marking of a neuron in post-synaptic contact with the marked afferent fibre, has revealed a high degree of specified order. To take an example restricted to the large myelinated mechanoreceptors of the kinds already discussed (Pacinian corpuscles, SAI, SAII and RA, i.e. hair follicle type G), the collaterals of each kind of receptor terminate in the dorsal horn in a characteristic manner, with independent saggital and rostro-caudal territories occupied by the synaptic terminals (Fig. 3). Such an organisation underlies the striking somatotopic representation of peripheral receptive fields in the spinal dorsal horn. It also offers a mechanism for the preservation of the separate identity of information encoded by the various kinds of sensory receptor.

Sensory Specificity

The SAI joins the PC as an example of a cutaneous sensory receptor with distinctive physiological properties capable of encoding in a quantitatively exact manner, certain features of the natural world. However, does the central nervous system maintain the integrity of the information? Several experimental methods can be used to seek answers to this question, for example, by electrophysiology, by psychophysics, or even a combination of the two. Hans Kornhuber was actively involved in the latter task, when together with Vernon Mountcastle and colleagues he applied strict quantitation to the psychophysics of tactile perception on one hand and to the electrophysiology of the afferent units on the other. One remarkable achievement was the analysis of flutter vibration in man and monkey (Talbot et al. 1968) that led to the attribution of vibration sense to Pacinian corpuscles and flutter sense to other rapidly adapting cutaneous receptors.

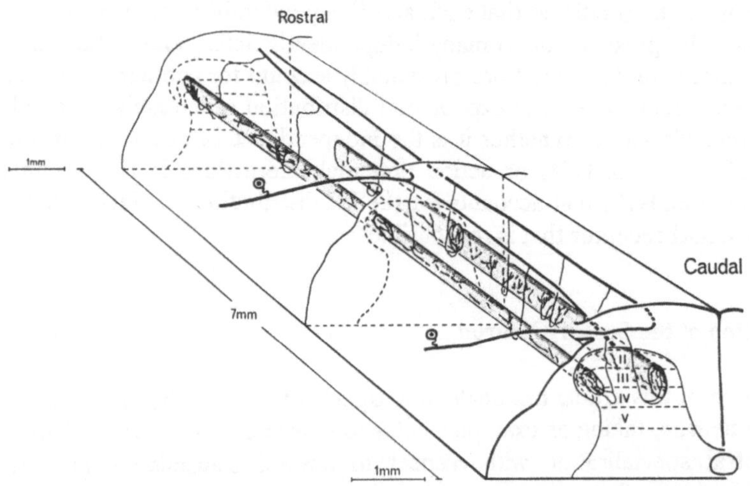

Fig. 3. Schematic representation of the arrangement of hair follicle afferent fibre collaterals in the lumbosacral spinal cord of the cat. The figure is drawn in scale and shows two afferent fibres entering the spinal cord through the dorsal roots and giving off collaterals which terminate as saggital sheets of arborizations centred on lamina III. (From Brown et al. 1977)

The use of percutaneous microelectrodes in awake human subjects has given renewed impetus to the analysis of human sensation in terms of the peripheral sensory receptors. Knibestöl and Vallbo (1980) confirmed the existence in human skin of the 4 kinds of mechanoreceptor already known in the skin of cats and monkeys. Subsequent work, reviewed by Vallbo et al. (1979) has added considerable detail. The particular interest of the results in man comes from the further advance made when electrical stimulation through the recording electrode was added as a refinement by Vallbo (1981) and Ochoa and Torebjork (1983) that allowed correlation of human sensation with the activation of an identified cutaneous mechanoreceptor. The results are truly striking: first in confirmation of the deductions by Kornhuber and colleagues in respect of flutter/vibration and, second, in the attribution of given elemental sensations to activation of a particular kind of afferent unit. For example, excitation of an SAI evoked a sense of pressure, referred to the actual receptive field of the afferent unit, whereas a PC unit evoked a sense of vibration or tickling. Another important result was that the pattern of impulse traffic in the afferent fibre did not affect the quality of the sensation. Thus activity in an SAI consistently evoked pressure, and there was no tendency for it to evoke sometimes pressure, sometimes vibration, sometimes cold, etc. The central neural machinery thus has the capacity to retain and utilise the information encoded by the peripheral receptors. These latest results also contribute to the resolution of another old problem of specificity. As mentioned earlier, the SAI, while exquisitely sensitive to tactile stimuli, also respond to temperature in a fairly consistent way. It has, in the past, been argued that they are 'spurious thermoreceptors' (Iggo 1968), a view now sustained by the new human studies that establish the SAI as pressure receptor, unable in any manner so far tested to give rise to a sense of cold or warmth.

In conclusion, I have set out in this article to assess studies on cutaneous sensory receptors, a field in which Hans Kornhuber has made distinguished contributions. The way in which his efforts have helped to illuminate the field and guide other studies has, I hope, been evident from my presentation. The final conclusion is that, even if the 'Action (in one sense) starts with the neuron', the effective neuron can with considerable conviction be claimed to be a sensory one!

Summary

Cutaneous sensory receptor studies in animals establish the existence of several clearly distinguishable categories. The slowly adapting (SAI) mechanoreceptor is described, as an example, in some detail. The mechanism of transduction and the role of the Merkel cell in this process in the SAI is discussed, especially the action of calcium ions. Sensory specificity, in relation to studies in man of single cutaneous afferent units is then discussed with the conclusion that cutaneous sensory receptors are part of a highly specific sensory mechanism.

References

Brown AG (1981) Organization in the spinal cord. Springer, Berlin Heidelberg New York, pp. 238

Brown AG, Iggo A (1963) The structure and function of cutaneous 'touch corpuscles' after nerve crush. J Physiol (Lond) 165:28-29P

Brown AG, Rose PK, Snow PJ (1977) The morphology of hair follicle afferent fibre collaterals in the spinal cord of the cat. J Physiol (Lond) 272:779-797

Cauna N (1962) Functional significance of the submicroscopal, histochemical and microscopal organization of cutaneous receptor organs. Anat Anz 111:181-197

Chambers MR, Andres KH, von Düring M, Iggo A (1972) The structure and function of the slowy-adapting type II mechanoreceptor in hairy skin. Q J Exp Physiol 57:417-445

Findlater GS, Cooksey EJ, Anand A, Paintal AS, Iggo A (1987) The effects of hypoxia on slowy adapting type I (SAI) cutaneous mechanoreceptors in the cat and rat. Somatosensory Res 5:1-17

Gray JAB, Sato M (1953) Properties of the receptor potential in pacinian corpuscles. J Physiol (Lond) 122:610-636

Hamann W, Iggo A (1988) Transduction and cellular mechanisms in sensory receptors. Prog Brain Res 74

Horch KW, Whitehorn D, Burgess PR (1974) Impulse generation in type I cutaneous mechanoreceptors. J Neurophysiol 37:267-281

Hunt CC, McIntyre AK (1960) An analysis of fibre diameter and receptor characteristics of myelineated cutaneous afferents in cat. J Physiol (Lond) 153:99-112

Iggo A (1968) Electrophysiological and histological studies of cutaneous mechanoreceptors. In: Kenshalo DR (ed) The skin senses. Thomas, Springfield Il, pp. 84-105

Iggo A (1977) Cutaneous and subcutaneous sense organs. Br Med Bull 33:97-102

Iggo A, Kornhuber HH (1977) A quantitative study of C mecha-noreceptors in hairy skin of the cat. J Physiol (Lond) 271:549-555

Iggo A, Muir AR (1969) The structure and function of a slow-ly-adapting touch corpuscle in hairy skin. J Physiol (London) 200:763-796

Ishiko N, Loewenstein WR (1961) The effect of temperature on generator and action potentials of a sense organ. J Gen Physiol 45:105-124

Kita H, van der Kloot W (1973) Action of Co and Ni at the frog neuromuscular junction. Nature 245:52-53

Knibestöl M, Vallbo AB (1980) Intensity of sensation related to activity of slowly adapting mechanoreceptive units in the human hand. J Physiol (Lond) 300:251-267

Lavoie PA, Bennet G (1983) Accumulation of $[H^3]$fucose-la-belled glycoproteins in the Golgi apparatus of dorsal root ganglion neurons during inhibition of fast axonal transport caused by exposure of the ganglion to Co^{2+}-containing or Ca^{2+}-free solution. Neuroscience 8:351-362

Loewenstein WR, Rathkamp R (1958) The sites for mechano-electric conversion in a pacinian corpuscle. J Gen Physiol 41:1245-1265

Loewenstein WR, Skalak R (1966) Mechanical transmission in a pacinian corpuscle. An analysis and a theory. J Physiol (Lond) 182:346-378

Mearow KM, Diamond J (1983) The development of mechanosensory function and synaptic morphology when regenerating axons arrive at nerve-free Merkel cells in Xenopus skin. Soc Neurosci Abstr 9:228.12

Merkel F (1875) Tastzellen und Tastkörperchen bei den Haus-tieren und beim Menschen. Arch Mikr Anat 11:636-652

Ochoa J, Torebjörk E (1983) Sensation evoked by intraneural microstimulation of single mechanoreceptor units innervating the human hand. J Physiol (Lond) 342:633-654

Ohmori H (1988) Mechanical stimulation and Fura2- flurescene in the hair bundle of dissociated hair cells of the chick. J Physiol (Lond) 399:115-137

Pacini F (1840) Nuovi organi scorperti nel corpo umano. Ciro, Pistoia, pp. 59

Pacitti EG, Findlater GS (1988) Calcium channel blockers and Merkel cells. Prog Brain Res 74:37-42

Pinkus F (1905) Über Hautsinnesorgane neben dem menschlichen Haar (Haarscheiben) und ihre verglei-chend-anatomische Be-deutung. Arch Mikr Anat 65:121-179

Talbot WH, Darian-Smith I, Kornhuber HH, Mountcastle VB (1968) The sense of flutter-vibration: comparison of the human capacity with reponse patterns of mechanoreceptive afferents from the monkey hand. J Neurophysiol 31:301-334

Vallbo AB, Hagbarth KE, Torebjörk HE, Wallin BG (1979) Somatosensory, proprioceptive and sympathetic activity in human peripheral nerves. Physiol Rev 59:919-957

Vallbo AB (1981) Sensations evoked from the glabrous skin of the human hand by the electrical stimulation of unitary mechanoreceptive afferents. Brain Res 215:359-363

Weakly JN (1973) The action of cobalt ions on neuromuscular transmission in the frog. J Physiol (Lond) 234:597-612

Parallel and Complementary Organization of Cortical Eye Movement Control and Visual Perception

O.D. Creutzfeldt

Dept. of Neurobiology, Max-Planck-Institute for Biophysical Chemistry, POB 2841, 3400 Göttingen-Nikolausberg, FRG

Introduction

In visual physiology we are faced with two complementary aspects and problems. One is concerned essentially with the anatomical and physiological representation of the retina and of visual stimuli in the brain, and the other with the involvement of various brain areas in eye movement control. The first aspect has dominated visual physiology during the last 25-30 years, leading to a model in which the visual world is layed out across the brain as on maps in which either the spatial dimensions of visual stimuli or specific aspects or features of the visual information are faithfully and orderly represented (Creutzfeldt 1983, 1985). Without going into the details of this model at this point, the question arises how this distributed information is recombined into a single and meaningful concept of the visual world. Is meaning in fact represented in feature maps of highest order and abstraction?

Visuomotor physiologists are, on the other hand, faced with the fact that eye movements are controlled and can be elicited from many parts of the brain so that many areas of the cortex are involved in the same or different aspects of eye movements. The question which arises here is that of the command structure which activates either this or that forebrain system, and of the agent or agency which selects objects in the visual environment as goals for directing the gaze (or the fovea) towards them.

While the first line of reasoning cannot avoid to use concepts of perception and cognition for interpreting its data, the second cannot help to invoke such concepts as attention, decision, or free will (e.g. voluntary eye movements). This becomes evident when studying some of the most recent reviews on the organization of cortical control of eye movements (see, for example, Andersen 1987; Fischer 1987; Heilmann et al. 1987; Robinson 1981). These are, of course, psychological concepts, and they are based essentially on an introspective analysis of cognitive and action processes, and it is obvious that they cannot do without a hierarchic organization.

Both the cognitive and the action hierarchies must be symmetrical, however. They must meet at the highest level (Jackson 1897, in 1932 p.422), where the final evaluation and analysis of the perceptual details are supposed to be combined to an idea (or *Vorstellung*), and from where the commands for appropriate action should be distributed to the various executive levels of the brain (the Wernicke scheme). We ask in the following whether the anatomofunctional organization of the brain does in fact support such a symmetrical hierarchic model, or whether we are lead by neurophysiology to consider different models. Can neurophysiology, in fact, suggest a straightforward and coherent model without referring and reverting to psychological concepts?

Cortical Control of Eye Movements Is Distributed over Many Cortical Areas.

Eye movements can be elicited by electrical stimulation from many areas and regions of the cerebral cortex (Fig. 1). This was demonstrated more than half a century ago on monkeys by Vogt and Vogt (1919) and by Foerster (1936) in men. These areas include purely sensory areas such as the primary area 17 and the prestriate visual cortex (areas 18 and 19), precentral motor areas such as the classical frontal eye field (area 8) and the supplementary motor area (SMA, area 6 a β), in which a supplementary eye field has recently been delineated (Schlag and Schlag-Rey 1987a), as well as the parietal (area 7) and the temporal association cortex (areas 21/22). These results have been essentially confirmed by recent stimulation mapping of the different eye movement related areas of the cerebral cortex. These more recent stimulation experiments (Wagmann, 1964; Robinson and Fuchs 1969; Schiller et al. 1979; Keating et al. 1982; Bruce and Goldberg 1984; Goldberg et al. 1986; Mitz and Wise 1987; Schlag and Schlag-Rey 1987b) have, in addition, revealed that the thresholds for eliciting saccades from these different areas may vary considerably and may also depend on the state of the animal. The lowest threshold regions are somewhat more restricted than suggested by the older experiments with less controlled stimulus parameters.

However, the fact that eye movements can be elicited from such a large part of the cortex concurs with the observation that eye movement related activity of single neurons can be recorded in all of these areas, except perhaps for the temporal association cortex where no saccade related activities have yet been reported to my knowledge. We

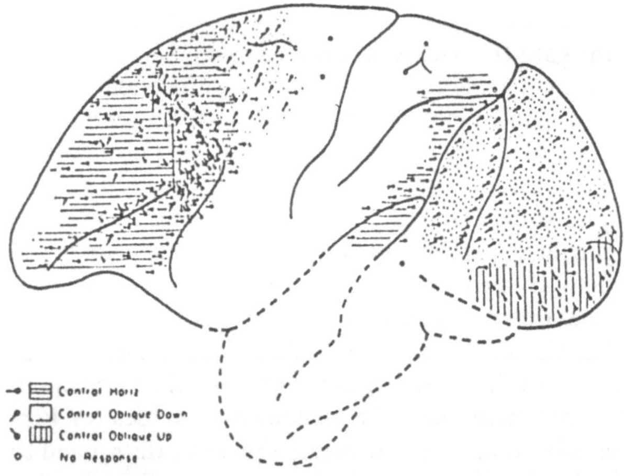

Fig. 1. Cortical map of eye movement responses to electrical stimulation of the alert, cervically transected monkey Macaca *mulatta* obtained by Wagman (1964). Locations of responses are indicated with solid points. The lines originating from the points indicate the direction of the eye movement. All responses were towards the contralateral side

shall return later to some details of eye movement related activities in these different areas. At this point, we want only to note that none of the areas where saccadic eye movements can be elicited is per se necessary for such eye movements, as can be concluded from lesion experiments in animals and from clinical observation in man.

Bilateral ablation of the frontal eye fields interferes with visual pursuit (Lynch 1987); but does not eliminate purposeful, goal-directed eye movements, as Schiller et al. (1979) have shown, nor does bilateral ablation of the supplementary frontal eye field makes them disappear completely (Guitton et al. 1985; Foerster 1936; Schiller et al. 1987). Thus, neither lesion, electrical stimulation experiments, nor, as we will see later, electrophysiological recordings suggest any clear and simple hierarchy of the cortical organization of eye movement control. Only lesion of area 17 eliminates goal-directed eye movements almost completely, because visual goals are not longer identified and represented in the cortex. However, even after area 17 lesion gaze shifts towards visual targets can still be observed under specific experimental conditions ("blind sight;" Pöppel et al. 1973; Weiskrantz et al. 1974).

On the other hand, the quality and extent of the functional deficits caused by the various lesions vary considerably. The devastating effect of area 17 lesion has already been mentioned. Even optokinetic nystagmus may be suppressed after bilateral lesions. After parietal cortex lesions, saccadic eye movements to the contralateral side are slowed down, and the latencies for eliciting eye movements into the contralateral hemifield are increased (Posner et al. 1984; Stein 1978; Lynch 1980). Visual search is impeded (Latto 1978), and after bilateral lesion tracking of fast but not of slowly moving targets is impaired, and the slow pursuit component of optokinetic nystagmus is also slowed down (Lynch and McLaren 1979). Similar effects have been observed in man. The most prominent syndrome here is, of course, Balints syndrome after bilateral lesion of the parietal cortex. This consists of what Balint called psychic paralysis of gaze, with inability to look at an object in the far visual periphery, visually guided ataxia of the hand, and difficulty in spatial attention (Balint 1909; Holmes 1919). After unilateral lesion of one parietal lobe, especially of the non-speech dominant hemisphere (because it is then not confounded with more prominent and more devastating symptoms of cognition and language), the syndrome of visual neglect is observed which has been so impressively documented by the self-portraits of the painter Räderscheidt following a right parietal infarction, collected and published by Jung (1974). It thus appears, and most authors nowadays agree, that the parietal lobe is involved in mechanisms necessary for visual attention, for focused attention in general, as well as for assessing and evaluating spatial relations between objects in the environment relative to one's own body and for appropriate visuomotor coordination (Hyvärinen 1982; Mountcastle 1976, 1981).

This does not mean, however, that the parietal cortex is sufficient for such mechanisms. Clearly, other regions of the visual cortex are also involved in visual attention and also necessary for it. Furthermore, as first observed as far back as 1895 by Bianchi, visual neglect is also observed after frontal eye field lesion, although only temporarily (Foerster 1936), but to such an extent that unilateral lesion of the frontal eye field in monkeys may be indistinguishable from a hemianopia (Latto and Cowey 1971). The often claimed hierarchic organization from the supplementary motor area to the premotor and motor cortex, in our case from the supplementary to the prearcuate frontal eye field, has also not been confirmed in recent studies; eye movements can still be elicited from the supplementary eye field after lesion of the prearcuate eye field, and spontaneous eye movements do not disappear after supplementary eye field lesions although they may be reduced. There is even a very important functional difference between the two areas in that presaccadic activation of neurons is seen in the arcuate frontal eye field only before purposeful, goal-directed saccades but not before spontaneous saccades in the dark (Goldberg and Bushnell 1981; Goldberg and Bruce 1986; Bizzi 1968). Supplementary eye field neurons, on the other hand, also discharge during or before spontaneous saccades in the dark (Schlag and Schlag-Rey 1985b, 1987a).

At this point, then, it appears that the various cortical regions which are involved in eye movement control are workingless in series than in parallel. The interface between the cortex and the oculomotor nuclei is situated in the upper brainstem with inter-, intra-, and supranuclear control loops, integrating also inputs from the vestibular system and the cerebellum (for review see Henn et al. 1982; Creutzfeldt 1988).

Complementary Contribution to Perception and Eye Movement Control of the Various Cortical Fields

The cortex uses and controls this mesencephalic machinery (see Fig. 2). It uses it because an integral part of visual perception is active exploration of the visual environment with the fovea. Thus, for visual analysis a continuous shift of gaze is necessary, and the percept is in fact a recomposition of these temporally disrupted, successive, and va-

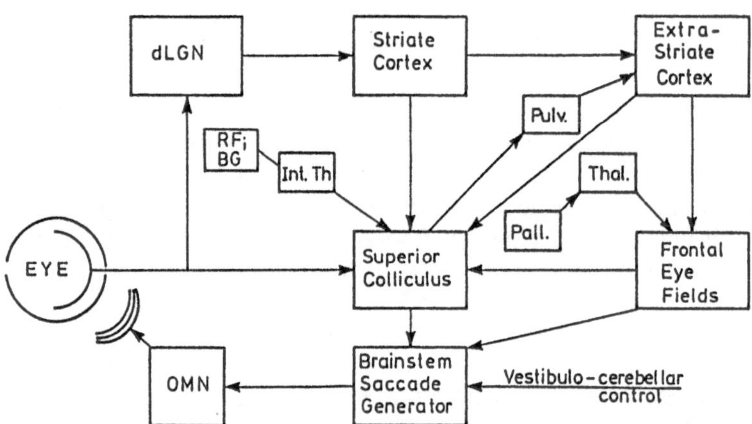

Fig. 2. Anatomical connections between the frontal eye fields and other brain structures participating in the initiation of saccadic eye movements. The frontal eye fields receive visual information via a corticocortical route originating in the striate cortex as well as directly from the thalamus. Several cortical areas, including the frontal eye fields, project to the superior colliculus, which receives a direct retinal projection through the pulvinar as well. The brainstem's saccade generator, located in the parapontine reticular formation (PPRF), receives a projection from the colliculus and an independent input from the frontal eye fields and other cortical areas. This model is minimal and primarily serves to explain how cortical areas can effect saccades, and how the frontal eye field can act without the colliculus, whereas striate, extrastriate and other cortical areas cannot. Many additional structures could be included, and structures that are included could be elaborated. dLGN = dorsal lateral geniculate nucleus; OMN = oculomotor nuclei; RF = reticular formation; Int.Th. = intralaminar nuclei of the thalamus; Pulv = pulvinar; Pall = pallidum. (From Bruce and Goldberg 1984, with some additions)

riable projections and representations of a visual scene in the various striate and prestriate visual areas. A stable and unified percept is comparable to a deconvolution of the variable representations in the retinotopic and feature-sensitive maps over the track of eye movements. Therefore, it is not suprising that eye movement related activity is seen not only in so-called oculomotor regions, and that neurons in oculomotor fields such as the frontal eye field also have visual receptive fields (see Bruce and Goldberg 1984, 1985). On the other hand, once an object of interest has been identified, it needs to be fixated and pursued not to loose it out of sight in spite of object and subject motion. Thus, the exploration program must be suppressed, the fixation and pursuit program has to be engaged, and gaze-attracting stimuli in the peripheral visual field must be suppressed. Attention must be concentrated onto the point of fixation, or, vice versa, fixation must be locked to the field of attention.

For further exploration of details, perception must be switched from the preattentive to the attentive mode, as Julesz (1985) puts it, which involves detailed comparison of adjacent structures or textures and composing them into a gestalt. This involves local and global screening of a visual scene. During reading and writing, finally, progressive eye movements are controlled by the semantic identification of one fixation frame after the other or by the whole progress of mental comprehension.

Let us now look at some cortical regions, whose neuronal activity is related to one or several aspects of visual behavior and perception. In the parietal association cortex, Mountcastle and Hyvärinen and their colleagues have found neurons which are active during fixation and pursuit and others which are active during and before saccades (for details see Hyvärinen 1982; Andersen 1987). Pursuit, fixation, and saccade neurons are clustered in small patches but mixed in the same areas and to some extent even with neurons activated in connection with arm movements. It was tempting to assign to these neurons the function of command neurons (Mountcastle 1976), commands from the highest level, so to speak, in the terminology of Jackson. It turned out, however, that these neurons were also visually excitable, and the visual receptive fields and the motion fields of individual neurons coincide to some extent (Robinson et al. 1978; Mountcastle 1981). Thus, fixation and pursuit neurons should have receptive fields in the foveal and parafoveal region of the visual field, while saccade-related neurons should have more peripherally located receptive fields. Unfortunately, exact data on this are not available to my knowledge. Furthermore, one might postulate that pursuit- and saccade-related neurons should feed into different functional subdivisions of the supranuclear oculomotor interface related to the pursuit and the saccade mode, respectively. This has not been investigated either, however.

In any case, these parietal neurons are both visual and motor. In addition, it turned out that their responses to visual stimuli may be enhanced if the subject, a monkey, is asked to pay attention to a peripheral stimulus located within the motion and receptive field, no matter whether he moves his eyes towards it or not (Lynch et al. 1977; Bushnell et al. 1981; Goldberg and Bruce 1985; Andersen 1987). This enhancement effect thus indicates that localized attention is also represented by activation of these neurons, and this may be the most important aspect of their function. It would fit with the observation that parietal lesion produces contralateral visual neglect.

Such enhancement effects have also been discovered, under various behavioral conditions, in the prelunate visual association area V4 (Fischer and Boch 1981a,b; Haenny and Schiller 1988; Moran and Desimone 1985; see also Fischer 1987) a visual area which is considered otherwise to be an area involved mainly in various aspects of visual perception (for further details see Tanaka et al. 1986). Attention-related activations do not mean, however, that such neurons or such regions select the stimulus for attention, but only that attention may make these neurons more sensitive to visual stimuli in that part of the visual field. These parietal and prestriate regions may therefore be necessary for the expression of attentive processes, which finally lead to appropriate eye movements, but they are not sufficient. An additional function of the parietal cortex must also be the change from one mode of vision (or oculomotor program) to another, as the "fixing of gaze" after parietal lesions indicates: the subject has difficulty to disengage the gaze from fixation and to jump it to another target (Holmes and Horrax 1919; Cogan 1965; Balint 1909; for further references see Hyvärinen 1982).

Attention-related activations are also found in the it dorsal prelunate area, DP. Also here, neurons with eye movement-related activities are mixed with those which are only or predominantly activated by the specific behavioral attention to a visual object (Li, Tanaka, Creutzfeldt, in preparation). Such neurons may even not be visually excitable but are visually excited only after the eye movement is terminated. These neurons in DP do not respond in connection with spontaneous eye movements in the dark or without a task but only when a fixation task is on, or when the visual environment is actively and attentively explored. The appearance of a behaviorally significant object (or person) in the field of gaze may significantly enhance their activity even without further change of predominant gaze direction (Fig. 3). Thus, the attentive fixation or visual exploration of

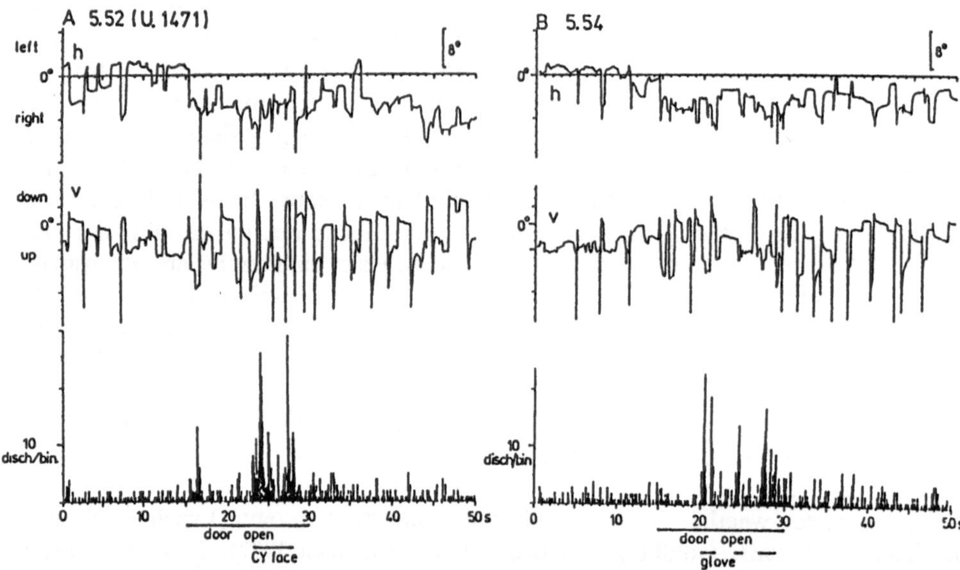

Fig. 3. Activation of a neuron in the dorsal prelunate area (DP) of a monkey. The monkey sits in a chair with his head fixed. **A** When the door to his experimental room is opened, he looks at the door (eye movement record on top, h, to the right). The unit (histogram, bottom) is shortly activated. When the face of the experimentor appears in the door, the unit is stronger activated, although the eye position does not change significantly. **B** This time, door opening hardly activates the unit, although the monkey still shifts his gaze towards the door. If a new stimulus, a glove with which the monkey is usually handled, appears in the door, the cell is again activated although the gaze does not shift significantly. (From Li, Tanaka and Creutzfeldt (1989) Brain Res 496: 307-310)

the object is represented by the activity of such neurons, but these neurons are not themselves evaluating the behavioral significance of the object, as their activation only after termination of the saccade indicates. The behavioral significance is represented by their activation, however. In this sense the prestriate areas and their neurons are necessary for object recognition, as was demonstrated recently by lesion studies of the prelunate visual association cortex in primates (Heywood and Cowey 1987) and from clinical experience (see Creutzfeldt 1983). They are not sufficient for object recognition, however, and the activities of other areas must also be added (Tanaka et al. 1986).

When regarding an object, whether exploring it or just looking at it, as in the experimental situation which Yarbus (1967) has used, our gaze jumps around (although, introspectively, we say it wanders around) with about 2-3 saccades every second. It does the same, actually, if our mind ponders (and we may ask whether our thought processes are not similarly jumpy at a similar rhythm). These continuous saccades are not spread randomly across the object but stop at resting points of the picture or object. These resting points are regions where contours cross, and the trajectories of saccadic eye movements preferentially run along contours if they exist in the picture. If the picture contains objects with fine details which constitute a gestalt, say an animal or a man, the gaze returns more frequently to it. When looking at a face, the eyes, mouth angles, and the tip of the nose are preferential resting points. Monkeys appear to explore similarly as humans although with a somewhat shorter gaze fixation period (Wurtz et al. 1982; and own observations).

The points of actual fixation and the gaze paths (contours) are represented in sufficient spatial detail only in area 17, i.e. in the primary visual cortex. From this we must conclude that the selection process for fixating a point in the visual environment takes place in area 17, and that the holding phase must be related to some extent to the activation pattern of foveal neurons.

One could then consider maximal activation of foveal cortical neurons elicited by oriented and crossing objects as hold signals for eye fixation. Decrease in such activities may unlock the fixation, so that the gaze shifts to a new contour boarder. Candidates for such a function could be "simple" cortical neurons, which are specifically sensitive to oriented contours and whose activation typically decreases after a few 100 ms. While this hold function would have to be the task mainly of simple cells in the foveal region of area 17, simple cells located in more peripheral parts of the visual field could have a function for selecting the next fixation point, i.e. another contour corner (Creutzfeldt and Nothdurft 1978). In the prelunate association field V4 one group of neurons are specifically activated by stimuli which are rich in detail (Tanaka et al. 1986). Activation of such units may also contribute to the command signals for fixation. Admittedly, this jumping of the gaze from one contrast and contour corner to the next and from one finely detailed structure to another does not represent, as such, a cognitive map or structure of the explorative (attentive) gaze shifts, but it does at least give a hint to a mechanism on which such a cognitive structure may depend.

Another aspect where area 17 neurons may be directly involved in the control of eye movements is the imperative character of small moving objects for gaze shift: we can hardly resist to turn our eyes towards a fly running across a page that we are reading or a cat running through the periphery of our visual field. Such stimuli excite predominantly the movement-sensitive complex cells in area 17, and the cells projecting from area 17 into the superior colliculus are in fact all complex cells. It is reasonable to assume that this corticocollicular pathway is involved in gaze shifts towards objects moving in the peripheral field.

A further example in which area 17 and possibly area 18 (V2) may be directly involved is the fusion of the images from the two eyes. Binocular fusion is a formidable task of the visuomotor system. It is obvious, that the fusion signals must be related to activation of binocular neurons in area 17 (and maybe 18) which are excited from homonymous points on the two retinae. When binocular innervation of cortical neurons is eliminated in early childhood due to refraction anomalies or innervation problems of the eye muscles (squint), binocular fusion may become impossible because the input to cortical neurons from the "weaker" eye is suppressed. The following experiment illustrates that binocular fusion actually depends on the presence of identical contours in the visual field of the two eyes (Fig.4). If one presents each eye with a random dot pattern in which one central field of dots is systematically displaced by a few minutes of

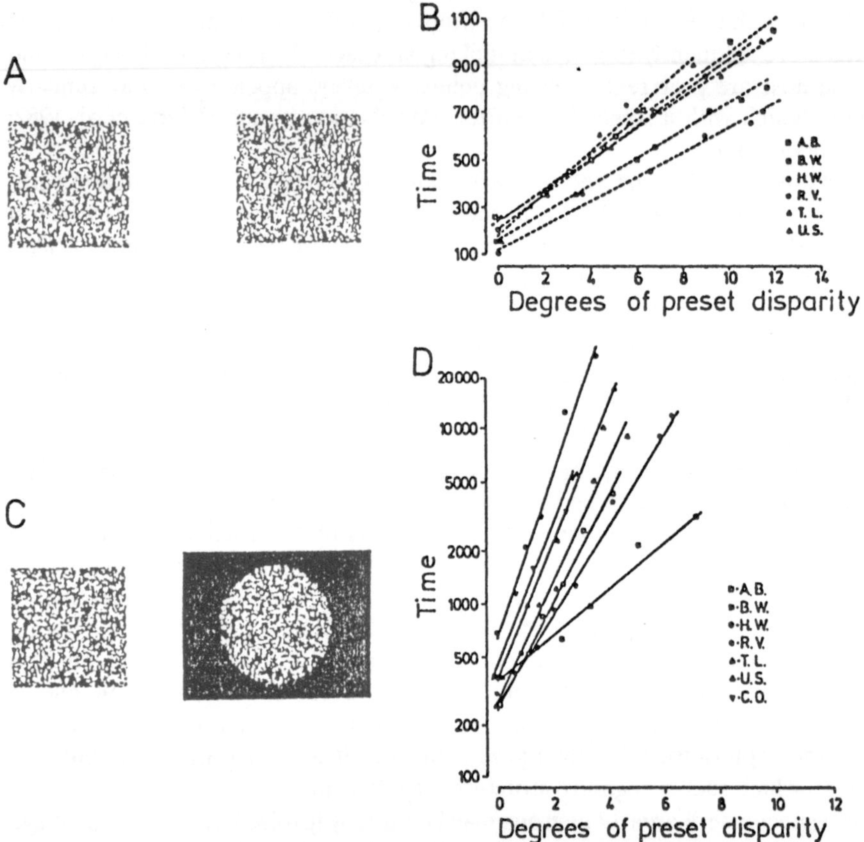

Fig. 4. Binocular fusion of random dot stereograms with identical (A) and different boarders (C). If the two dot displays are correctly fused, a small square pops out in the center. The time to fuse the two diagrams, until the stereo picture is recognized, depends on the disparity of the picture (B). When the diagrams have no common border, the time to fuse them is much longer (C). Note the different ordinates in B (linear) and D (logarithmic). Symbols and initials refer to different test subjects. (From Creutzfeldt and Vaupel, in preparation)

arc, this area appears in front of or behind the background as a stereoimage when the two images are correctly fused. We presented such dot patterns at different angles of disparity, so that the test subject had to converge his eyes (or squint) in order to superimpose the two images. Correct superimposition was reached when the stereoimage appeared. The time from the beginning of exposure of the two images to the moment at which they were correctly fused could be measured by exposing the two pictures for different lengths of time. As can be seen from Fig. 4A,B, the time needed for fusion increases linearly with the initial disparity of the two images. However, the fusion time at any disparity angle also depends critically on whether the two random dot images have some prominent contours in common. Such prominent contours are usually the edges of the images, which are identical in Fig. 4A. If one surrounds one image with a square and the other with a circular frame (Fig. 4C), the fusion time increases, at all preset disparities, by a factor of ten and more. It increases exponentially and is not possible at all at large disparity angles (Fig. 4D). By introducing identical contours into the two pictures, the fusion time can be reduced again. This experiment then suggests that fusion of the images from the two eyes depends on binocular excitation of neurons being excited by identical contours with the same orientation and in the same region of the visual field. This condition is fulfilled by the binocular simple and complex cells of areas 17 and 18 (V1 and V2).

We realize that we have now assigned command tasks for visual behavior to the typically sensory neurons in the striate and prestriate cortex. This should not mean, however, that we have thus dismissed their sensory task, but by giving them also a function for motor command we imply that their activation represents both, the presence of a sensory feature and command for appropriate visuomotor responses, or action. The fact that visual perception is a continuous exploration of the visual environment with the gaze and thus the fovea indicates, of course, that visual perception is an active process. The strategy for these explorative eye movements is laid down in the sensory properties of neurons in the primary visual cortex. It is in line with this that electrical stimulation of the points in one representation region of the visual field in the visual cortex induces a gaze shift towards that part of the visual field.

Let us now turn to the frontal eye fields. We have seen above that they are not necessary for saccadic exploration or for fixation. On the other hand, activation of neurons in the frontal eye field are, of all oculomotor neurons, most predictive for the appearance, size, and direction of a saccade. Their activation clearly precedes all saccades during visual exploration. Can we consider them, then, as the final path for cortically induced eye movement commands? But how would this be compatible with the fact that we can dismiss them without eliminating visual exploration? Let us therefore look a bit closer at the activation pattern of frontal eye field neurons during visual oculomotor tasks (Goldberg and Bushnell 1981; Bruce and Goldberg 1984; Fischer 1987). We discover different types of neurons and different responses in relation to the stimuli: pure visual responses, visuomovement responses which last from the cueing stimulus to the execution of the eye movement, and those which discharge during the saccade. These types of activity then look similar to activities found elsewhere in the premotor cortex during delayed response tasks, for example, in its hand-arm area during reaching tasks (Wise and Strick 1984; Creutzfeldt 1983). One may interpret such activities as being related to the motivation and intention to do a certain movement. These activities come closest to the actual voluntary command aspect of movements. Such commands must take into account sensory stimuli as well as the present state of the system. The frontal eye field may then be involved predominantly in eye movements depending on voluntary or instructed commands. It is in line with such an interpretation that patients with frontal eye field lesions may fail to respond to the verbal demand to move their eyes to the contralateral side (Foerster 1936; Guitton et al. 1985). The frontal eye fields are not, however, the final common path of all cortically induced eye movements and are only one of the many parallel cortical input-output loops feeding into the collicular and supranuclear mesencephalic interface in which all these commands are coordinated into purposeful movements of the eyes.

Conclusion: Cooperation of Cortical Fields in Active Visual Perception

We have seen that not only the cortical fields considered as motor fields but also those necessary for visual perception are all involved in eye movements, and that they all appear to cooperate. However, different aspects are emphasized in the different areas, and thus different emphasis may be given to the goal selection for an eye movement by each area depending on the situation and the state of the whole system. This we may call a parallel representation of active vision. Such a model does not dismiss completely the hierarchic organization but only assumes that the top level may change from one area to another according to the demand and context of a visual task. The emphasis may be on exploration, on discrimination and decision, on attention, or on volition, and the highest command would then switch accordingly from the sensory to the association areas or to the frontal eye field. The parallel representation of cortical eye movement control also does not mean that these areas are completely independent of each other and only connected through a highest command structure or agent which calls upon a specific area for a given purpose. These areas are not isolated units but connected via association fibers leading messages in both ways and, on lower levels, through convergence of outputs and feedback to the cortex. The main connection between all areas is, however, that they receive their input directly or indirectly from the same visual environment, and that any command execution, i.e. any eye movement, changes the input and thus also the condition for further commands in all of them. They are thus connected through an external, extracerebral loop (Creutzfeldt 1983). We may then consider as the main unifying condition of the various parallel cortical eye movement systems that they share the same visual environment, and that any change of the subject's relation to this visual environment either through eye or body movements affects each of these subsystems according to its access to the visual input. According to the situation induced by the demands of one system (or motivation), another system may be called upon to induce a correction or a new situation thus keeping the subject in an appropriate relation with his environment, consistent with the goal and purpose of behavior and adapted to the environment.

References

Andersen RA (1987) Inferior parietal lobule function in spatial perception and visuomotor integration. In: Mountcastle VB, Plum F, Geiger SR (eds) The nervous system. American Physiological Society, Maryland, pp 483-518 (Handbook of Physiology, vol 5)

Balint R (1909) Seelenlähmung des "Schauens", optische Ataxie, räumliche Störung der Aufmerksamkeit. Monatsschr Psychiatr Neurol 25: 25-81

Bender MB (1980) Brain control of conjugate horizontal and vertical eye movements. A survey of the structural and functional correlates. Brain 103: 23-69

Bianchi L (1895) The functions of the frontal lobes. Brain 18: 497-530

Bizzi E (1968) Discharge of frontal eye field neurons during saccadic and following eye movements in unanesthetized monkeys. Exp Brain Res 6: 69-80

Bruce CJ, Goldberg ME (1984) Physiology of the frontal eye fields. TINS 7: 436-441

Bruce CJ, Goldberg ME (1985) Primate frontal eye fields. I. Single neurons discharging before saccades. J Neurophysiol 53: 603-635

Bushnell MC, Goldberg ME, Robinson DL (1981) Behavioral enhancement of visual responses in monkey cerebral cortex. I. Modulation in posterior parietal cortex related to selective attention. J Neurophysiol 46: 755-772

Cogan DG (1965) Ophthalmic manifestations of bilateral non-occipital cerebral lesions. Br J Opthalmol 49: 281-297

Creutzfeldt OD (1983) Cortex Cerebri. Leistung, strukturelle und funktionelle Organisation der Hirnrinde. Springer, Berlin Heidelberg New York

Creutzfeldt OD (1985) Comparative aspects of representation in the visual system. Exp Brain Res [Suppl] 11: 53-81

Creutzfeldt OD (1988) Cortical mechanisms of eye movements in relation to perception and cognitive processes. In: Lüer G, Lass U, Shallo-Hoffmann J (eds) Eye movement research. Physiological and psychological aspects. Hogrefe, Toronto, pp 9-33

Creutzfeldt OD, Nothdurft HC (1978) Representation of complex visual stimuli in the brain. Naturwissenschaften 65, 307-318

Fischer B (1986) Express saccades in man and monkey. Prog Brain Res 64: 155-174

Fischer B (1987) The preparation of visually guided saccades. Baker PF et al. (eds) Reviews of physiology, biochemistry and pharmacology, vol 106. Springer, Berlin Heidelberg New York

Fischer B, Boch R (1981a) Enhanced activation of neurons in prelunate cortex before visually guided saccades of trained rhesus monkeys. Exp Brain Res 44: 129-137

Fischer B, Boch R (1981b) Selection of visual targets activates prelunate cortical cells in trained rhesus monkeys. Exp Brain Res 41: 431-433

Fischer B, Boch R (1985) Peripheral attention versus central fixation: modulation of the visual activity of prelunate cortical cells of the rhesus mokey. Brain Res 345: 111-123

Foerster O (1936) Motorische Felder und Bahnen. In: Bumke O, Foerster O (eds) Handbuch der Neurologie, vol 6. Springer, Berlin Heidelberg New York, pp 1-357

Goldberg ME, Bruce CJ (1985) Cerebral cortical activity associated with the orientation of visual attention in the rhesus monkey. Vision Res 25: 471-481

Goldberg ME, Bruce CJ (1986) The role of arcuate frontal eye fields in the generation of saccadic eye movements. Prog Brain Res 64: 143-154

Goldberg ME, Bushnell MC (1981) Behavioral enhancement of visual responses in monkey cerebral cortex. II. Modulation in frontal eye fields specifically related to saccades. J Neurophysiol. 46: 773-787

Goldberg ME, Bushnell MC, Bruce CJ (1986) The effect of attentive fixation on eye movements evoked by electrical stimulation of the frontal eye fields. Exp Brain Res 61: 579-584

Guitton D, Buchtel HA, Douglas RM (1985) Frontal lobe lesions in man cause difficulties in suppressing reflexive glances and in generating goal-directed saccades. Exp Brain Res 58: 455-472

Haenny PE, Schiller PH (1988) State dependent activity in the monkey visual cortex. I. Single cell activity in V1 and V4 on visual tasks. Exp Brain Res 69: 225-244

Heilman KM, Valenstein E, Goldberg ME (1987) Attention: behavior and neural mechanisms. In: Mountcastle VB, Plum F, Geiger SR (eds) The Nervous System, American Physiological Society, Maryland, pp 461-481 (Handbook of physiology, vol 5)

Henn V, Büttner-Ennever JA, Hepp K (1982) The primate oculomotor system. A synthesis of anatomical, physiological and clinical data. Human Neurobiol 1: 77-95

Heywood CA, Cowey A (1987) On the role of cortical area V4 in the discrimination of hue and pattern in macaque monkeys. J Neurosci, 7: 174-218

Holmes G (1919) Disturbances of visual space perception. Br Med J 2: 230-233

Holmes G, Horrax G (1919) Disturbances of spatial orientation and visual attention, with loss of stereoscopic visison. Arch Psychiatr 1: 385-407

Hyvärinen J (1982) The parietal cortex of monkey and man. Springer, Berlin Heidelberg New York, 202 pp
Jackson H (1932) Selected writings, vol II. J Taylor (ed). Hodder and Stoughton, London
Julesz B (1985) Preconscious and conscious processes in vision. Exp Brain Res [Suppl] 11: 333-359
Jung R (1974) Neuropsychologie und Neurophysiologie des Kontur- und Formsehens in Zeichnung und Malerei. In: Wieck HH (ed) Psychopathologie musischer Gestaltungen. Schattauer, Stuttgart, pp 29-88
Keating EG, Gooley SG, Pratt SE, Kelsey JE (1983) Removing the superior colliculus silences eye movements normally evoked from stimulation of the parietal and occipital eye fields. Brain Res 269: 145-148
Latto R (1978) The effects of bilateral frontal eye field, posterior parietal or superior collicular lesions on visual search in the rhesus monkey. Brain Res 146: 35-50
Latto R, Cowey A (1971) Visual field defect after frontal eye-field lesions in monkeys. Brain Res. 30: 1-24
Lynch JC (1980) The role of parieto-occipital association cortex in oculomotor control. Exp Brain Res 41: A32
Lynch JC (1987) Frontal eye field lesions in monkeys disrupt visual pursuit. Exp Brain Res 68: 437-441
Lynch JC, McLaren JW (1979) Effects of lesions of parieto-occipital association cortex upon performance of oculomotor and attention tasks in monkeys. Neurosci Abstr 5: 794
Lynch JC, Mountcastle VB, Talbot WH, Yin TCT (1977) Parietal lobe mechanisms for directed visual attention. J Neurophysiol 40: 362-389
Mitz AR, Wise SP (1987) The somatotopic organization of the supplementary motor area: intracortical microstimulation mapping. J Neurosci 7: 1010-1021
Moran J, Desimone R (1985) Selective attention gates visual processing in the extrastriate cortex. Science 229: 782-784
Mountcastle VB (1976) The world around us. Neural command functions for selective attention. Neurosci Res Prog Bull 14, [Suppl]
Mountcastle VB (1981) Functional properties of the light sensitive neurons of the posterior parietal cortex and their regulation by state controls: influence on excitability of interested fixation and the angle of gaze. In: Pompeiano O, Ajmone Marsan C (eds) Brain mechanisms of perceptual awareness and purposeful behavior. Raven, New York, pp 67-69 (IBRO monograph series, vol. 8)
Pöppel E, R Held and D Frost (1973) Residual visual functions after brain wounds involving the central visual pathway in man. Nature 243: 295-296
Posner MI, JA Walker, FJ Friedrich and RD Rafal (1984) Effects of parietal injury on covert orienting of attention. J Neurosci 7: 1863-1874
Robinson DA (1981) Control of eye movements. In: Brookhart JM, Mountcastle VB (eds) Motor control, part 2. American Physiological Society, Bethesda, pp 1275-1320 (Handbook of physiology, vol 2)
Robinson DA, Fuchs AF (1969) Eye movements evoked by stimulation of frontal eye fields. J Neurophysiol 32: 637-648
Robinson DA, Goldberg ME, Stanton GB (1978) Parietal association cortex in the primate: Sensory mechanisms and behavioral modulations. J Neurophysiol 41: 910-932
Schiller PH, True SD, Conway JL (1979) Paired stimulation of the frontal eye fields and the superior colliculus of the rhesus monkey. Brain Res 179: 162-164
Schiller PH, Sandell JH, Maunsell JHR (1987) The effect of frontal eye field and superior colliculus lesions on saccadic latencies in the rhesus monkey. J Neurophysiol 57: 1033
Schlag J, Schlag-Rey M (1985a) Unit activity-related to spontaneous saccades in frontal dorsomedial cortex of monkey. Exp Brain Res 58: 208-211
Schlag J, Schlag-Rey M (1985b) Eye fixation units in the supplementary eye field of monkey. Neurosci Abstr 25: 23-82
Schlag J, Schlag-Rey M(1987a) Evidence for a supplementary eye field. J Neurosci 57: 179
Schlag J, Schlag-Rey M (1987b) Does microstimulation evoke fixed-vector saccades by generating their vector or by specifying their goal? Exp Brain Res 68: 442-444
Stein J (1978) The effect of parietal lobe cooling on manipulative behavior in the conscious monkey. In: Gordon G (ed) Active touch. Pergamon, Oxford, pp 79-90
Tanaka M, Weber H, Creutzfeldt OD (1986) Visual properties and spatial distribution of neurons in the visual association area on the prelunate gyrus of the awake monkey. Exp Brain Res 63: 11-37
Vogt C, Vogt O (1919) Allgemeinere Ergebnisse unserer Hirnforschung. IV. Die physiologische Bedeutung der architektonischen Rindenfelderung aufgrund unserer Rindenreizungen. J Psychol Neurol 25: 401-461
Wagmann IH (1964) Eye movements induced by electric stimulation of cerebrum in monkeys and their relationship to bodily movements. In: Bender MB (ed) The oculomotor system. Hoeber, New York. pp 18-39
Weiskrantz L, Warrington EK, Sanders MD, Marshall J (1974) Visual capacity in the hemianopic field following a restricted occipital ablation. Brain 97: 709-728
Wise SP, Strick PL (1984) Anatomical and physiological organization of the non-primary motor cortex. TINS 7: 442ff

Wurtz RH, Goldberg ME, Robinson DL (1982) Brain mechanisms of visual attention. In: Thompson RF (ed) Progress in neuroscience, pp 82-90 (readings from Scientific American)
Yarbus AL (1967) Eye movements and vision. Plenum, New York

Stages of Somatosensory Processing Revealed by Mapping Event-Related Potentials

J. E. Desmedt and C. Tomberg

Brain Research Unit, University of Brussels, Boulevard de Waterloo 115, 1000 Brussels Belgium

Adapted behaviour requires the selective processing of the stream of sensory stimuli impinging on the sense organs so that distinct sensory inputs become integrated into current cognition. The perceptual evaluation of inputs involves both parallel and serial brain mechanisms whose complexity must be appreciated when designing experimental paradigms.

Our current approach involves the recording of event-related potentials (ERP) combined with bit-mapped imaging of the scalp potentials fields. Three requirements should be considered in this relation. First, the brain responses should be assessed with accurate time definition to identify critical steps of processing stages. The use of electronic averaging necessitates consistency in timing of the response items to be added algebraically.

Second, a sufficient number of recording channels should be used to ensure comprehensive topographic analysis of the ERP components that can be generated in different brain areas. Moreover, updated bit-mapping methodologies should be used to extract the information embedded in the multichannel data.

Third, the studies should be designed so as to differentiate between the two classes of brain responses that are evoked by sensory stimuli, namely the exogenous obligatory responses reflecting the physiological registration of the sensory input in cortical receiving areas, and the endogenous cognitive mechanisms whereby additional brain circuits can be activated in order to assess and compare cognitive features in conjunction with current behavioral uses of the sensory input (Desmedt and Tomberg 1989).

These two classes of responses overlap in time and are confounded in the ERP profiles elicited by the target stimuli to which the subject pays attention. Telling the subject not to pay attention to a similar sequence of sensory stimuli is not sufficient to eliminate cognitive effects and generally fails to disclose the genuine underlying exogenous baseline response. In fact, sizeable cognitive electrogeneses have been documented for ERPs to stimuli that are not in the focus of the subject's attention (Desmedt and Robertson 1977; Näätänen 1982; Hillyard and Munte 1984).

Figure 1 illustrates this problem for a four-finger paradigm in which randomly intermixed brief electric stimuli were delivered to the index and middle fingers of each hand at interstimulus intervals (ISI) varying randomly between 250 and 570 ms. In the different runs of the experiment, the subject was asked to pay attention and count mentally the shocks to one finger at a time, either the right or the left middle finger.

This task was rather difficult since the shock intensities were only about 2 mA above the subjective threshold (estimated for each finger by the method of limits). Moreover,

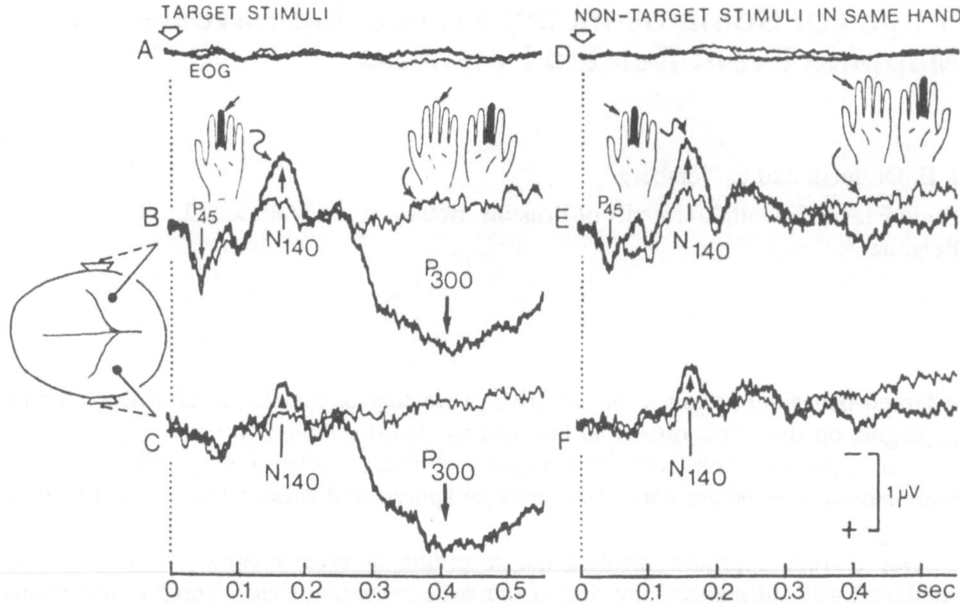

Fig. 1. ERPs to brief electric pulses delivered in randomly intermixed series to the middle and index fingers of the two hands in a normal young adult. In the figurines associated to the traces in B-C or E-F, the finger stimulated is indicated by a small oblique arrow while the target finger attended by the subject is in black. A and D, control electrooculograms. Scalp recordings from parietal areas, 3 cm behind the vertex (figurine on the left). The thinner traces superimposed as controls are ERPs to physically identical stimuli recorded in other runs of the same experiment, when the subject pays attention to the target finger in the opposite hand. N140 is larger contralaterally to the finger stimulated and is present both for the target middle finger (B-C) and for the adjacent nontarget index finger of the same hand (E-F). P300 is symmetrical and only occurs for target stimuli (B-C). (From Desmedt and Robertson 1977)

the high rate of stimulus delivery was near the normal limits of capabilities for fast sequential decision (Debecker and Desmedt 1970; Woods et al. 1980). The averaged responses recorded over the contralateral and ipsilateral parietal scalp were compared for any given finger stimuli when the subject's task involved either that hand (thicker traces in Fig. 1) or the opposite hand which was not in the focus of attention (thinner traces superimposed in Fig. 1).

These robust data revealed a hierarchical organization of the brain electrogeneses, the negative N140 component was observed in the ERP both to the target middle finger (Fig. 1 B-C) and to the adjacent non-target index finger (Fig. 1 E-F), whereas the P300 component only appeared in the ERP to the targets (B-C) (Desmedt and Robertson 1977; Desmedt and Debecker 1979). P300 was fairly symmetrical while N140 was larger contralaterally to the stimulated finger. Counterbalancing targets in the left or right hand in successive runs provided an excellent control to the effect that ERP enhancement was indeed switched by the verbal instructions directing the subject's focus of attention. The design also virtually excluded interference from shifts in vigilance levels or trend in subject's commitment or familiarity with the task.

The target shocks to the middle finger were difficult to differentiate by the subject from the non-target shocks to the adjacent index finger. In fact, only highly motivated subjects could perform with less than 5% error in the target counts of each run. The parametric set had to be carefully adjusted for each experiment, which was discontinued when fatigue started to deteriorate performance.

The results also suggested that the brain was dealing differently with the finger input at a latency of 50 to 80 ms which was considered rather early at a time, when the literature was focused on later components such as P300 or the contingent negative variation (CNV). Similar delays for the invocation of cognition-related electrogeneses were documented for other sensory modalities (see Hillyard and Kutas 1983; Donald 1983). The early contralateral P45 to finger shocks was not changed significantly with attention conditions (Fig. 1 B, E).

In these experiments, the shocks to the fingers of the opposite hand could only be neglected up to a point, since they were randomly intermixed in the very time sequence which had to be scrutinized by the subject in order to identify the targets.

Independent evidence suggests that sensory stimuli designated as irrelevant in the paradigm nevertheless involved to a variable extent cognitive electrogeneses, even though the subject followed instructions. This calls for designing a more "neutral" baseline control.

It must indeed be realized that cognitive ERP components are identified and estimated by comparing response profiles to physically identical stimuli which are recorded either under "attention" or under "neglect" conditions. Therefore, target components will be spuriously underevaluated if their profiles are titrated against "control" profiles that still include a fair amount of cognitive activities.

The issue of the earliest appearance of cognitive activities in the response to target stimuli has indeed to be reconsidered with updated experimental protocols designed to achieve a better delineation of the exogenous obligatory profiles. Desmedt et al. (1983) introduced as control the homogeneous series (100%) of (physically identical) stimuli which could be fairly thoroughly neglected by the subject.

A recent parametric study validated these conditions and made it clear that two factors can be considered necessary and sufficient to achieve "neutral" controls: First, the homogeneous series eliminates sensory mismatch in the sequence of stimuli and discourages involuntary commitment of attention resources by the subject. Second, the throughout the subject reads an interesting novel so as to be involved in an alternative cognitive activity while maintaining vigilance at a similar level.

Detailed evidence indicates that, under such conditions, the control responses fairly consistently reflect exogenous profiles that are not unduly increased by residual cognitive activities, or reduced or distorted by rate effects or habituation (Desmedt and Tomberg 1989; Tomberg et al. 1989).

When such adequate bit-mapped control data to homogeneous series of stimuli have been acquired, they can be electronically subtracted from the target ERP data evoked by physically identical stimuli under selective attention conditions. Such subtraction mapping documents the genuine endogenous cognition-related electrogeneses called upon to perform the perceptual task (Fig. 2).

Several stages of brain processing have been identified in conjunction with target finger stimuli: First, P30 and P40 cognitive components were found to disclose remarkable topographic similarity with the concomitant exogenous P27 and P45 somatosensory responses recorded in the parietal cortex contralateral to the finger stimulated.

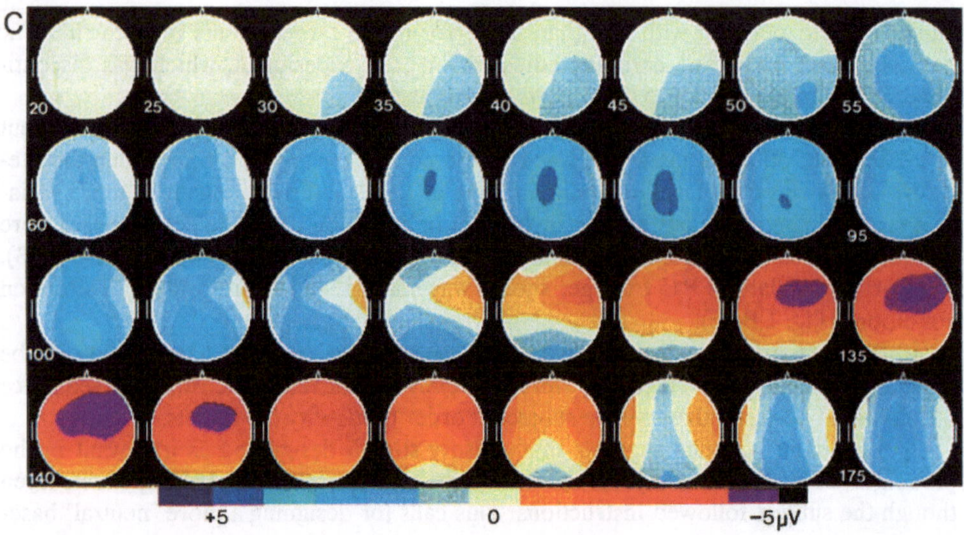

Fig. 2. Subtraction bit-mapping of grand average data from 23 normal young adults. Frozen maps based on 14 channels at 5 ms intervals between 20 and 175 ms after stimulus delivery. Color steps with hues of reds for negative and of blue for positive (scale below). Oddball paradigm with brief electric pulses delivered in random series (probabilities of 0.2 for targets and 0.8 for nontargets) to two fingers of the left hand. Control ERPs to identical finger stimuli with probability of 1.0 while subject is reading are electronically subtracted from the target ERPs. No residue is observed at 20 or 25 ms. Cognitive P30 und P40 appear at 30-45 ms, followed by P100, N140 and P200. (From Desmedt and Tomberg 1989)

These P30 and P40 are thought to reflect enhancements of the exogenous response in parietal (but not frontal) cortex receiving short-latency thalamocortical projections (Desmedt and Tomberg 1989).

Second, these components are followed from about 55 to 110 ms by complex topographic patterns of positivities labeled P100 which involve the posterior contralateral parietal areas before extending ipsilaterally and towards the frontal cortex (Fig. 2).

Thereafter, another pattern of negativities labeled N140 appears at the contralateral prefrontral areas to extend towards the midline and the ipsilateral prefrontal region, before receding on both sides (Fig. 2). N140 is followed by a posterior positivity P200 which moves towards the central regions, and by the widespread P300.

The data point to a sequential involvement of several cortical areas in posterior parietal and prefrontal regions that are known to be reciprocally related through cortico-cortical connexions (Jones and Powell 1970; Petrides and Pandya 1984; Mountcastle 1984; Goldman-Rakic 1987). They document rather precisely in space and time the parallel and sequential processing whereby the finger inputs are organized into spatial coordinates and identified as relevant objects in the sequential spatial tasks considered (Desmedt and Tomberg 1989).

References

Debecker J, Desmedt JE (1970) Maximum capacity for sequential one-bit auditory decisions. J Exp Psychol 83: 366-372

Desmedt JE, Debecker J (1979) Waveform and neural mechanism of the decision P350 elicited without prestimulus CNV or readiness potential in random sequences of near-threshold auditory clicks and finger stimuli. Electroenceph clin Neurophysiol 47: 648-670

Desmedt JE, Robertson D (1977) Differential enhancement of early and late components of the cerebral SEPs during fast sequential cognitive tasks in man. J Physiol Lond 271: 761-782

Desmedt JE, Tomberg C (1989) Mapping early somatosensory evoked potentials in selective attention: critical evaluation of control conditions used for titrating by difference the cognitive P30 P40 P100 and N140. Electroenceph clin Neurophysiol 74: 321-346

Desmedt JE, Nguyen TH, Bourguet M (1983) The cognitive P40 N60 and P100 components of SEPs and the earliest electrical signs of sensory processing in man. Electroenceph clin Neurophysiol 56: 272-282

Donald MW (1983) Neural selectivity in auditory attention: sketch of a theory. In: Gaillard AW, Ritter W (eds) Tutorials in event-related potential research: endogenous components. North Holland Amsterdam, ff 37-77

Goldman-Rakic PS (1987) Circuitry of primate prefrontal cortex and regulation of behavior by representational memory. In: Plum F (ed) Handbook of physiology. Section 1: The nervous system, vol 5, American Physiology Society, Bethesda Md, ff 373-417

Hillyard SA, Kutas M (1983) Electrophysiology of cognitive processing. Ann Rev Psychol 34:33-61

Hillyard SA, Munte TF (1984) Selective attention to color and location. Percept Psychophys 36: 185-198

Jones EG, Powell TP (1970) An anatomical study of converging sensory pathways within the cerebral cortex of the monkey. Brain 93: 793-820

Mountcastle VB (1984) Central nervous system mechanisms in mechanoreceptive sensibility In: Mountcastle VB (ed) Handbook of physiology. Section 1: The nervous system, vol 3, American Physiology Society, Bethesda, ff 789-878

Näätänen R (1982) Precessing negativity: an evoked-potential reflexion of selective attention. Psychol Bul vol 92 3: 605-640

Petrides M, Pandya DN (1984) Projections to the frontal cortex from the posterior parietal region in the rhesus monkey. J comp Neurol 228: 105-116

Tomberg C, Desmedt JE, Osaki I, Nguyen TH, Chalklin V (1989) Mapping somatosensory evoked potentials to finger stimulation at invervals of 450 to 4000 msec and the issue of habituation when assessing early cognitive components. Electroenceph clin Neurophysiol, 74:347-358

Woods D, Courchesne E, Hillyard S, Galambos R (1980) Recovery cycles of event-related potentials in multiple detection tasks. Electroenceph clin Neurophysiol 50: 335-347

On Ideation and "Ideography"

D. H. Ingvar
Department of Clinical Neurophysiology, University Hospital, S-221 85 Lund, Sweden

Current imaging techniques of functional changes in the human brain permit studies with increasing spatial and temporal resolution of cerebral events related to mental activity (Ingvar and Lassen 1975; Ingvar 1985; Risberg 1980, 1988). The ultimate aim of such studies is to describe and quantitate, as well as to picture in two and three dimensions, the cerebral biochemical and/or circulatory `structure` of thoughts and ideas, i.e. the neuronal substrate of ideation. The term cerebral "ideo-graphy" has been suggested for this type of research (Ingvar 1977).

From a simplistic neurophysiological point of view, two principally different types of ideation can be recognized.

Evoked ideation consists of either of the following :

- Cognitive events evoked by perception of actually ongoing sensory stimulation, including perception of spoken or written language
- Cognitive events which pertain to the actual performance of voluntary *motor* activity, or the production of spoken or written language

Pure ideation is defined as cognitive events which are *unrelated* to any actual sensory stimulation (including speech perception) or to any ongoing behaviour/motor performance (including speech production). Pure ideation may be divided into:

- Pure *sensory* ideation, i.e. the conceptualization of sensory messages, including the imagination of speech perception
- Pure *motor* ideation, i.e. "inner" simulation of motor acts (motor ideation) or of speech (inner speech).

Finally, a third category of pure ideation must be included here, a complex type of ideation. In this group there are indeed many hierarchial levels of complexity, from the simpler ones such as imagining a promenade in a well-known surrounding to more complex levels of problem solving, learning, recall of memories, theorizing, hypothesizing, etc., i.e. levels of cognitive activity at which simple 'sensory' and 'motor' components cannot be recognized. The neurophysiological basis for these high levels of cognition and ideation is by and large unknown. Hence, for the sake of simplicity the present review is devoted almost exclusively to evoked sensory and motor ideation and to pure sensory and motor ideation.

Events of the three types defined above may indeed be studied by electrophysiological techniques (e.g. Deecke and Freund, this volume). However, such studies lie outside the scope of the present review.

It will be claimed here that:

- Cerebral ideography provides a unique approach to study, often in a surprisingly concrete way, the functional anatomy of the highest functions of the human brain. As mentioned, however, only primary components of such functions are so far accessible to an analysis, and only some of these are dealt with below.
- Brain images of especially pure ideational events permit a mapping of brain structures which are selectively involved in the primary steps of the creation of ideas, concepts, memories, thoughts, plans, fantasies, theories, hypotheses, and cognitive events in general. Here ideation related to speech and music should be included. Cerebral ideography thus opens new approaches to the study - directly in the brain of unanaesthetized conscious humans - of the physiological basis of the highest functions, especially of cognition and language (symbol handling).
- Brain structures involved in pure ideation have a different localization than those involved in actual sensory and speech perception, or in the production of movements and speech. A selective study of brain structures involved in pure ideation therefore appears to open a new approach to understand the human psyche.
- Clinical applications of cerebral ideography, at rest and during mental activation, will reveal new aspects of the cerebral defects pertaining to cognitive and language disturbances in neurologic and psychiatric disease. Already, several rCBF studies demonstrate that this is so.

Technical Considerations

Cerebral ideography has in principle been possible since the advent of techniques to measure regional cerebral blood flow (rCBF; Lassen and Ingvar 1961; Lassen et al. 1963; Ingvar and Lassen 1975; Risberg 1980, 1988). Quantitative rCBF values obtained with 133-Xenon clearance can be considered as indirect measures of the metabolic activity of the neurons in the regions in which the flow is measured. There is normally a close correlation between the neuronal metabolism and rCBF (Roy and Sherrington 1980; Ingvar and Lassen 1975; Raichle 1979).

The first cerebral records (ideograms) of pure cognitive events in the human brain were carried out with simple rCBF techniques using only a few detectors (Ingvar and Risberg 1965, 1967; Risberg and Ingvar 1973). Currently, rCBF studies are being made with high spatial resolution using multidetector instruments and administration of Xenon 133 intravenously or by inhalation, both with two-dimensional measurements of the cortical rCBF landscape (Risberg 1980; Ryding 1986; Risberg 1987, 1988) and with three-dimensional tomographic techniques for the whole brain (SPECT; Stokeley et al. 1980; Lassen and Friberg 1988; Decety et al. 1989).

The $[^{14}C]$ deoxyglucose technique of Sokoloff et al. (1977) formed the basis for tomographic studies in man of the brain metabolism with positron-emitting isotopes (PET; Reivich et al. 1979; Raichle 1979; Phelps et al. 1979; Mazziotta et al. 1982). In recent years there has been an explosive development of quantitative PET techniques to measure regional cerebral metabolic rate (rCMR) and blood flow, as well as several molecular biological (transmitter/receptor) events in the brain. The spatial resolution of current PET techniques is only a few millimerters and the temporal resolution in some cases only a few seconds. However, so far PET techniques have only to a limited extent been used for evoked ideography (Kuschner et al. 1987; Celesia et al. 1982; Petersen et al. 1988). There has been only a few PET studies of brain events related to pure ideation (Roland et al. 1987).

The Resting Conscious State

Like in previous reviews of topics related to the present one (Ingvar 1979, 1985), we depart here from the resting conscious state, i.e. the distribution of neuronal function in the conscious awake human brain not exposed to any deliberate sensory input (covered eyes, plugged ears) or performing voluntary motor acts or solving problems. This may be considered to represent a basic type of brain activity with an "idling" type of unstructured cognition. It has been well established that normal humans in this state have a stable EEG, an overall mean cerebral oxidative metabolism of about 3 ml oxygen per 100 g brain per minute, and a mean blood flow (CBF) of about 50 ml/100 g per minute (Schmidt 1950).

Furthermore, as shown repeatedly in many laboratories, two-dimensional rCBF studies have demonstrated a stable *hyperfrontal* flow pattern in the resting state awake. The flow in premotor and frontal areas is about 10% - 20% above the hemishpere mean, and in temporal/occipital areas it is 10% - 20% below the mean (Ingvar 1979; Risberg 1980, 1988). This pattern, the resting ideogram, is also evident in some PET studies of the resting state (Mazziotta et al. 1981, 1982; Phelps et al. 1981).

As suggested by our group, the hyperfrontal resting rCBF pattern may also be principally related to the classical resting EEG alpha pattern (with closed eyes) which shows lower frequencies (alpha) in postcentral/temporal regions and higher frequencies in frontal regions (Ingvar et al. 1979). The correlation between mean EEG frequency and rCBF as well as rCMR (Ingvar et al. 1976b) corroborates this juxtaposition of the hyperfrontal resting rCBF pattern and the hyperfrontal resting EEG. Both patterns show a marked stability.

It is claimed here that the hyperfrontal rCBF (and EEG) patterns recorded in the resting state with its unstructured cognitive activity, not influenced by any deliberate incoming sensory signals or production of behavioural reactions, represents the resting cerebral ideogram of wakefulness. Elsewhere it has been proposed, in part on the basis of introspective evidence, that resting cognition implies a "simulation of behaviour", i.e. a rehearsal, refinement and optimization - as well as a formulation - of cognitive/behavioural and speech programs. Such programs are by nature serial and organized on a temporal basis and hence highly dependent upon premotor and frontal cortical regions (Fuster 1980; Ingvar 1979, 1985; Goldman-Rakic 1988, and this volume). This view is supported by much clinical evidence. Patients with a poverty of motor/behavioural reactions, as well as a lack of a goal-directed behaviour (Parkinson, schizophrenia, organic dementia, frontal lesions) show a resting rCBF pattern which is *hypofrontal* (Ingvar 1980, 1985, 1987). There are also some observations that the normal hyperfrontality disappears in sleep when the serial type of cognition pertaining to wakefulness disappears (Ingvar 1977).

All forms of conscious cognitive activity, be they of the inner "silent" type or directly related to sensory perception or to motor performance, have a temporal (serial, sequential) organization. The stream of consciousness (Davidson and Davidson 1980) has a time basis, and the "time arrow" of mental activity (*la flèche du temps*; Prigogine and Strengers 1988) is irreversible, and brain events, like biological processes in general, proceed continuously and irreversibly from the past through the present into the future.

The conscious state, e.g. the conscious awareness of the reader of these lines - in the present "now", when the lines are actually read - thus includes an awareness (now) of (a) the memorized past (text above), (b) the present visual input of these very words, and (c) an expectation (idea, concept, hope) that the argument of the present author may lead to some conclusion or consequence in the future, on the next page, at the end of the article, etc.

In view of the time paradigm used here to analyse conscious awareness, it might be apparent that cerebral ideography in principle concerns present conscious experiences of "now". These experiences may, however, be coupled to sensory perception or motor acts (evoked ideography) or be of purely "inner" nature (pure ideography). In both cases ideation contains components of the past (memories) as well as ideas about the future ("memories of the future"; Ingvar 1985).

With the use of the time paradigm above, a review will now be given of some basic findings with cerebral ideography in the four main situations outlined above.

Cerebral Ideography of Actual Sensory Perception (Evoked Ideation)

Not unexpectedly, deliberate augmentation of the sensory input, for example, by cutaneous electrical stimulation (Ingvar et al. 1976a) of by augmenting the visual input (Phelps et al. 1981, Celesia et al. 1982) augments the rCBF and rCMR in the appropriate primary cortical sensory areas. However, intense cutaneous stimuation in conscious subjects, especially involving pain sensation, also activates extensive frontal cortical regions. The resulting rCBF patterns may be interpreted as ideograms of the cognitive consequences of the sensory stimulation (Fig. 1).

The ideogram of Rest shows the unstimulated idling brain, active mainly in the two frontal lobes with its inner unstructured resting cognition. This ideogram shows the hyperfrontal pattern (see above).

The ideogram of Touch shows relative to rest a slightly more marked hyperfrontality, indicating that the cognitive consequences of the not very significant cutaneous signals are limited. The slight frontal increase suggests that the stimulus evokes only a limited mobilization of cognitive and behavioural programs since the modest stimulus can be supposed to have a low significance (meaning).

The ideogram of Pain, on the other hand, shows a markedly increased activation, mainly frontally. The experience of pain necessitates, as is well known, an acute mobilization of a great number of cognitive and behavioural action programs which are deemed necessary to deal with the pain situation. To use the time paradigm described above, the subject which experiences pain in the present mobilizes action programs to deal with pain now in order to avoid and to escape future pain. In this situation, the conscious subject indeed relates his actual experiences to previous memories and handling of pain, in the past.

This interpretation of the cerebral rest/touch/pain ideograms is not only supported by common sensical introspective evidence. There are also a number of rCBF studies

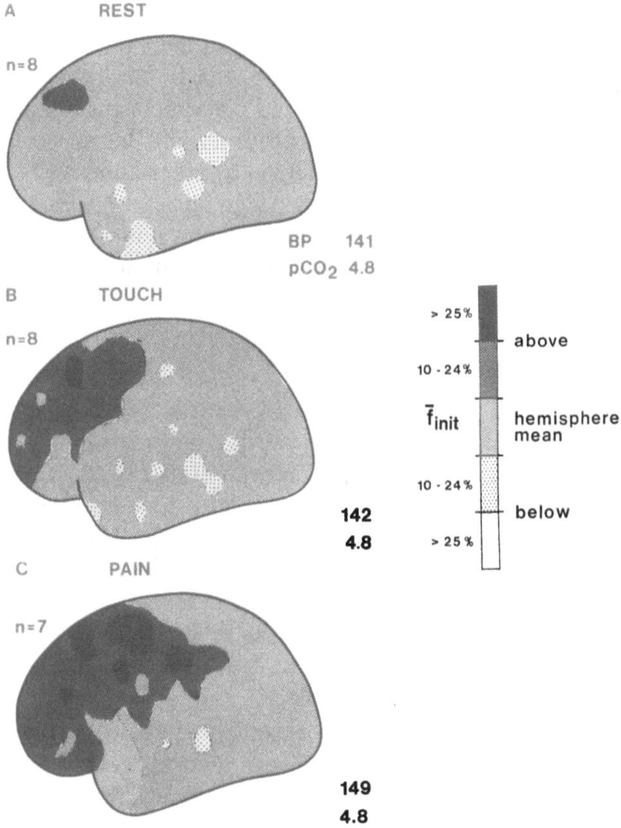

Fig. 1. Evoked ideograms during experiences of rest, touch and pain. Intraarterial multidetector xenon 133 rCBF measurements in the left hemisphere in eight subjects. Relative plots of flow changes with the resting hemisphere mean flow as a reference. **A** Rest: slight hyperfrontal flow distribution. **B** Touch: electrical stimulation of contralateral thumb region at threshold. Note increase of hyperfrontality. **C** Pain: (stimulation 3 x threshold yielding moderate pain sensation). Note: further increase of hyperfrontality and the involvement of larger, not only frontal areas. See text. (From Ingvar et al. 1976a)

which might be invoked. We found early that presentation of a problem to a subject which requires serial sequential reasoning (pure complex ideation) also augmented the resting hyperfrontality, an augmentation which appeared proportional to the mental effort (Risberg and Ingvar 1973; Risberg 1980). Here it is of interest to mention that Risberg (1974) found some personality correlation in rCBF activation studies during psychological testing. This important finding has recently obtained solid confirmation in a study by Stenberg et al. (1989). In the resting state, "ready-to-act" extroverts show a more hyperfrontal rCBF pattern than "ready-to-contemplate" introverts. Using the time paradigm above, one may look upon the hyperfrontal extroverts as having a larger inner production of action programs than the less hyperfrontal introverts. Findings of this type suggest that cerebral ideography can be used to measure not only different cognitive structures and strategies but also some of the underlying personality differences.

Further support for the temporal interpretation for the rest/touch/pain ideograms is provided by findings concerning anxiety. Anxiety implies principally an "overfuturing"

(Melges 1982), an abnormal concern for threatening future events. This state is accompanied by a marked hyperfrontality in alcoholics and demented patients which are exposed to cognitive stress (Risberg and Ingvar 1973). In more advanced cases with dementia with low performance score and no signs of anxiety in the test situation, the hyperfrontality reaction is reduced or absent (Ingvar et al. 1975). The problem of anxiety has recently been further analysed by Hagstadius and Risberg (1989) in high resolution rCBF studies. It was found that high anxiety was linked to a high degree of hyperfrontality, especially on the right side. Cerebral ideograms may thus reveal both personality traits as well as emotional aspects of cognition.

Speech Perception

Listening to sound or spoken words represents another type of evoked sensory ideation (see above). As shown by Nishizawa et al. (1982), it produces a distinct bilateral rCBF activition pattern which to a marked extent involves regions outside the auditory and the classical speech areas, especially on the left side. An activation of mainly the left frontal cortex was demonstrated. This finding signals an involvement of frontal cortical structures in the perception of spoken messages. The speech perception ideogram thus shows principal similarities with those evoked by touch and pain. In both situations the frontal activation may represent the cognitive consequences of the perceived serial sensory input of language symbols.

In principle very similar results, including frontal activations, have been recorded during perception of music (Carmon et al. 1975; Knopmann et al. 1983; Metter et al. 1981). Using again the time paradigm above, the cerebral ideograms evoked by perception of linguistic and musical symbols demonstrate an activation of frontal structures. There is thus direct ideographic support for the widely accepted "motor theory of speech perception" (Liberman at al. 1967) which requires direct access to serial speech programs in the brain for the perception of spoken words (Ingvar 1983).

Ideography of Voluntary Movements

In 1971 Olesen showed that voluntary hand movements were accompanied by an rCBF augmentation in the contralateral rolandic region (sensory motor hand area and adjacent fields). This finding was soon confirmed by Ingvar (1975) and extended by Roland who used a high-resolution multidetector rCBF instrument. He also demonstrated that the supplementary motor areas were involved in temporally structured sequential movements, and that the basal ganglia took part bilaterally, even in unilateral movements (Roland 1984).

Highly complex paradigms (maze tests, proprioceptive tests, discrimination tests, etc.) have also been used to induce rCBF changes (Roland 1984). In several of these studies, specific sensory/motor or cognitive components cannot readily be identified, and hence the interpretation of the patterns recorded is difficult.

Speech Production

Voluntary pronounciation of even very simple series of words (weekdays, series of numbers, etc.) give rise to marked bilateral cortical rCBF changes (Ingvar and Schwartz 1974; Larsen et al. 1978). The voluntary humming of a children's song also induces a bilateral activation (Ryding et al. 1987). A comparison of the rCBF patterns recorded from the right and left hemisphere during "Sunday, Monday, Tuesday ..." and those following "100, 101, 102, 103 ..." shows several regional similarities, as they do with the ideograms pertaining to humming. Further studies will be required to identify the cerebral structures responsible for the ideational and semantic contents of spoken words (or melodies hummed) as well as to differentiate in the ideograms the output motor control of the articulation and the auditory feedback effects. (See "Pure Speech Ideation",below).

Recently Petersen et al. (1988), using a rapid high-resolution PET subtraction technique, have shown that the sensory, the motor and the semantic components in simple word processing have different cortical localizations. Their important study can be viewed as an exquisite model of PET ideography of language and cognition. The regional subtraction results provided support for the existence of multiple parallel routes between the visual input (reading, speech perception), the phonological, the articulation (speech production) and the semantic coding areas. Of especial interest in the present context was the finding that the semantic processing of simple words activated frontal areas rather than posterior temporal (see "Speech perception",above).

To summarize, evoked ideation either by actual sensory stimulation (including speech perception) or by voluntary motor activity (including speech production) gives rise to cerebral ideograms in which two main components can be identified: (a) an activation of primary sensory and motor regions and (b) activation of other, mainly frontal, areas which in principle represent the cognitive consequences of the sensory input, or the volitional processes preceding the motor act. The ensuing activation patterns are often complex, and it is difficult to identify the specific regions in the brain which handle the ideational contents of a simple sensory speech input or the ideational (volitional) background of a motor/speech output. Such an identification can be made only by cerebral ideography using pure ideation.

Pure Ideation

Pure Sensory Ideation

Roland (1984; see also Roland and Friberg 1985) studied ideograms pertaining to the expection of an imagined sensory input such as a touch of a finger tip or of the mouth region. This type of selective attention induced a limited upper or lower rolandic activation and in addition a substantial premotor and prefrontal increase of rCBF. Roland and Friberg (1985) used the term "tuning" for the preparatory activation of sensory motor centres which were involved in this type of ideation, but which were not activated by any actual afferent input at the time of the rCBF measurement.

Pure Motor Ideation

Ingvar and Philipson (1977; Fig.2) were the first to record pure motor ideograms, i.e. cortical events in subjects conceptualizing a specific movement, a one-sided rhythmic hand-clenching. A contralateral activation of motor and prefrontal regions was clearly demonstrated during this type of motor ideation, and so was the apparent lack of (or at least very limited) activation of the hand area which is so evident during actual hand movements (Olesen 1971; Ingvar 1975). Later these findings have been confirmed in principle, and it has been shown that sequential movements activate the supplementary motor area bilaterally as well as the basal ganglia on both sides (Roland 1982, 1984).

A. RESTING CONDITIONS

B. Rt HAND MOVEMENTS CONCEIVED

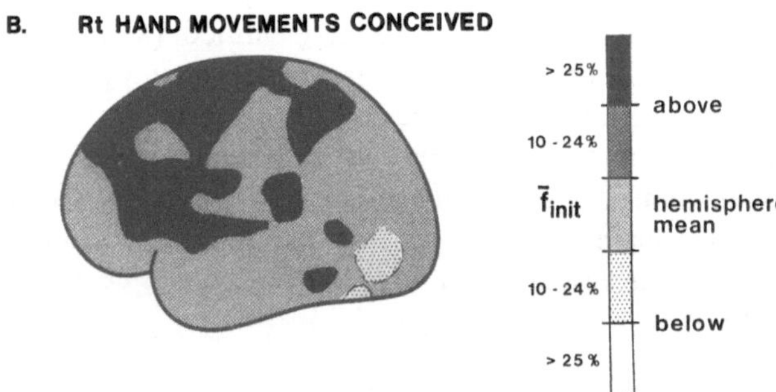

C. Rt HAND MOVEMENTS PERFORMED

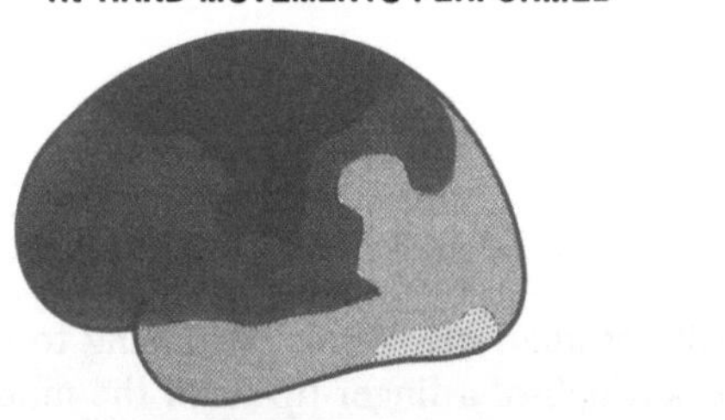

Fig. 2. Pure and evoked ideograms during motor ideation (**B**) and motor performance (**C**). Intraarterial rCBF measurements as in Figure 1. Mean of six subjects. **A** Rest: note normal hyperfrontal pattern. **B** Pure ideation: subjects thinking of rhythmic clenching movements of right hand. Note marked, mainly frontal, activation. Limited activation in rolandic area. **C** Evoked ideation: during actual performance of voluntary rhythmic clenching of right hand. Note marked activation of rolandic areas (confirming Olesen 1971) in addition to frontal areas as in B. See text. (From Ingvar and Philipson 1977)

Recently, Decety et al. (1988) have shown bilateral brain activation patterns during simulation of a graphic task (Fig.3). Not only did the ideogram of graphic simulation differ from the one pertaining to actual graphic movements, but it also showed that both paradigms, writing and simulated writing, activated the cerebellum. The cerebellar activation during motor ideation has also been confirmed during simulated tennis movements during which the absence of peripheral EMG activity was controlled (Decety et al. 1989).

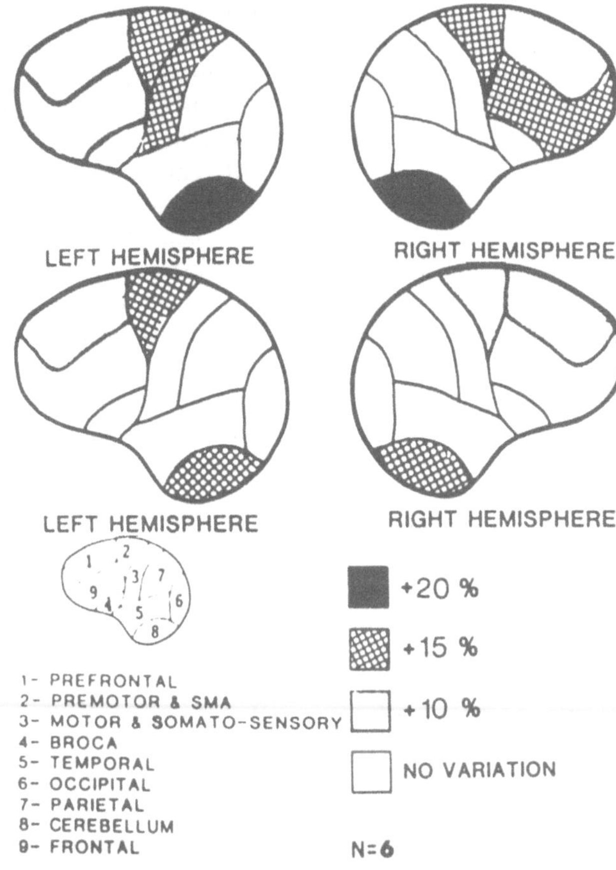

Fig. 3. Motor performance (**A**) and motor ideation (**B**). rCBF measurements with xenon-133 inhalation using a rotating gamma camera in six subjects. Mean flow changes in relation to rest in regions indicated. **A** Actual writing with right hand of figures ("one, two" etc.) on a paper. Note marked flow augmentation in the left hemisphere including the rolandic region. On the right side, premotor and prefrontal areas were activated. Note marked bilateral flow increase over the cerebellum. **B** Simulated writing with right hand of "one, two..." etc.; at the same speed as in A. Note absence of left rolandic activation but retention of bilateral premotor and prefrontal flow increase - as well as some clearcut cerebellar activation. See text. (From Decety et al. 1988)

Pure Speech Ideation (Inner Speech)

Inner speech, e.g. counting silently from 100, 101, 102 ... onwards, constitutes a pure form of a speech ideation lacking both actual sensory input and motor output. Both the auditory/proprioceptive sensory feedback and the motor/articulatory components are absent.

In an extensive study in 30 right-handed volunteers, Ryding et al. (1989) found with bilateral two-dimensional rCBF technique that silent speech activated almost exclusively the left dominant hemisphere and there in selective regions only. These regions showed a distinct relation to the Wernicke area and the upper SMA speech centre of Penfield. A lower anterior sensory/motor centre was also identified, probably related to articulatory centres (larynx, mouth, tongue) and to Broca's area. On the right side a significant frontopolar activity was also seen during silent speech. Some diffuse frontal rCBF increase was also recorded as in preliminary studies of silent speech (Lassen et al. 1978), but they did not reach significance. It should here be added that Decety et al. (1989) in a SPECT study in our laboratory has found that silent speech also activates the cerebellum, apparently in a symmetrical fashion.

The studies of silent speech illustrate clearly the value of pure ideography in analysing the symbol functions of the brain. The findings of Ryding et al. (1989) will probably be only the first of studies with different techniques, in which various paradigms are used to compare brain mechanisms responsible for the perception and the production of spoken words and the mechanisms 'behind', them, i.e. the ideas and concepts underlying sounds, words and speech. It appears of fundamental importance that this first study of silent speech ideation showed such a clear preponderance of cortical foci of activation in the left (dominant) hemisphere, and that these foci showed in so many respects a direct relation to the classical speech centres. In contrast, speaking aloud gave a widespread bilateral activation, dominanting on the right (sic) side, mainly in the temporal and rolandic areas (Ryding et al. 1987).

Clinical Considerations

So far, very few studies have been made in patients with defective cognition and language in which ideographic principles have been applied. In organic dementia and chronic schizophrenia, abnormalities have however been found with evoked ideation (psychological testing). In both disorders, a defective activation of especially frontal association cortices has been recorded (Ingvar et al. 1976; Franzén and Ingvar 1975; Risberg 1980, 1988).

It is highly inviting to suggest here that pure sensory ideography, e.g. during expectation of a (future) sensory stimulus listening to an "inner" recital or melody etc. might be used profitably in focal neurological disorders (aphasia) as well as indeed in mental disorders accompanied by cognitive and emotional disturbances. Studies with evoked ideation have already proved that specific psychological testing in schizophrenics (Wisconsin Card Sorting Test) reveals clearcut frontal abnormalities in this disease (Franzén and Ingvar 1975; Weinberger et al. 1986; Warkentin and Risberg 1989).

From the clinical point of view it may also be stressed that the ideographic approach may be used to evaluate psychiatric therapy in patients with cognitive and emotional disturbance. Kullberg and Risberg (1986; see Risberg 1988) have shown a frontal rCBF reduction following cingulotomy which alleviates anxiety. Psychopharmacological treatment (haloperidol) also lowers the frontal activity level, an effect associated with therapeutic success (Risberg 1988). It does not in fact appear far-fetched to envisage that the ideographic approach might be used the measure the effects of classical psychotherapy. Examples have been given above of how symptoms of, for example, anxiety or mental effort during problem solving are reflected in cerebral ideograms.

Right-Left Differences

Several of the studies quoted have shown important hemispheric-differences in evoked or pure ideograms. These differences are at present not well understood. Emotional factors appear to engage the right frontal region more than the left (Hagstadius and Risberg 1989). Silent speech activates, as mentioned, almost only the left (dominant) hemisphere, while speaking aloud gives rise to more pronounced activation on the right side (Ryding et al. 1987, 1989).

Much further work, especially with ideographic techniques, will have to be carried out to reconcile the many current theories on brain lateralization with solid neurophysiological including electrophysiological facts (Petersen et al. 1988).

Emotions

The large field pertaining to emotions will also merit systematic consideration in future ideographic studies. Already it has been shown that the resting ideogram (the hyperfrontal pattern) correlates to anxiety proneness. "Overfuturing" (Melges 1982) anxious persons are more hyperfrontal (more action prone) than the less anxious ones (Hagstadius and Risberg 1989). Anxious patients with brain defects show a much greater frontal activation and clearer signs of anxiety in the test situation than do normals (Ingvar and Risberg 1965, Risberg and Ingvar 1973).

Concluding Remarks

The present contribution is based upon studies from the past two decades which have used multiregional two- and three- dimensional techniques to measure mass-neuronal activity (blood flow) or metabolism in circumscribed cerebral regions. These techniques (mainly for rCBF and PET) have proved quite successful for localization and quantification of brain events related to different aspects of cognition, speech, volition, and also to some extent of emotions. The data produced so far are indeed very limited in view of the complexity of the human psyche. Nevertheless the following conclusions appear warranted:

1. There is overwhelming evidence today that mental events are coupled to an almost cosmic variety of complex metabolic and circulatory patterns in the human brain. (The important electrophysiological events are not included in this review.)

2. To distinquish multiregional rCBF and rCMR studies of mental events from other investigations of brain mechanisms, the term cerebral ideography has been introduced (Ingvar 1977).

3. Using a simplistic neurophysiological approach one may differentiate between (a) *evoked ideography*, i.e. the study of brain events (cognitive and others) directly related to sensory perception or voluntary motor activity as well as speech, and (b) *pure* ideography, i.e. brain events unrelated to sensory stimulation or motor performance, including inner speech. There are in addition a number of complex forms of ideation which have not been considered in this review.

4. Evoked ideography portrays complex cerebral functional landscape associated with perception of sensory messages and speech and their cognitive consequences. It may also picture the volitional processes which precede and accompany voluntary movements and speech.

5. Pure ideography includes selective measurements of "inner" brain events related to cognition, fantasies, expectation, planning, problem solving etc., in situations without sensory or speech input as well as motor or speech output.

6. By comparing evoked and pure ideograms such as (a) the ideograms during cutaneous electrical stimulation versus imagination of the same stimulus or (b) the ideograms during speaking aloud versus silently, it is possible to outline brain structures which are involved specifically in perception, volition and/or symbol handling.

7. In ideographic studies of silent speech it has been demonstrated for the first time that the ideas and concepts related to word symbols activate almost exclusively the left dominant hemisphere, where selectively a number of regions with clear relation to the speech centres show activity increase (Ryding et al. 1989). During aloud speech these ideational speech centres on the left side are apparently hidden by the large activations which are then induced in both hemispheres (Ryding et al. 1987).

8. Motor ideation (simulation of movements) gives clearcut cortical (rolandic) activation patterns with a prominent bifrontal component. In addition, the inner plans underlying concepts and ideas of movements activate the cerebellum. Thus for the first time cerebellum can be assigned a role in mental activity. The action plans which precede and which are a prerequisite for volitional motor acts apparently include cerebellar mechanisms.

9. Future systematic clinical applications of ideographic paradigms appear highly promising. Significant ideographic changes have already been found (at rest) in organic dementia, chronic schizophrenia and anxiety states. These changes have been of fundamental importance for the understanding of the cognitive abnormalities, including the disturbances of conscious awareness which accompany these disorders. A prominent finding appears to be a disturbance of the temporal organization of cognition which may be localized to the frontal lobes. As described elsewhere, this disturbance may be associated , for example, with the lack of abstraction in dementia, the cognitive disturbances in schizophrenia, and the abnormal concern and worry for future events in anxiety states.

Summary

The term ideography (Ingvar 1977) was introduced to denote circulatory (rCBF) and metabolic (rCMR) measurements in the human brain of events related to ideation. Evoked ideography denotes records of rCBF and rCMR changes associated with cognitive phenomena following sensory stimulation or related to voluntary motor activity (also to perception or production of speech). Pure ideography records brain events related to pure ideation which is not coupled to any ongoing sensory input or motor output. Simulated movements and silent speech are examples of pure ideational activities that give rise to specific cerebral ideograms which show changes not only in the cerebral cortex but also in the cerebellum. Ideographic paradigms appear to offer new approaches to study the physiological basis of the human psyche. Clinical ideography may be used to study the functional disturbances which underlie speech and mental abnormalities in neurological and psychiatric disease.

References

Carmon A, Lavy S, Gardon H, Portnoy Z (1975) In: Ingvar DH, Lassen NA (eds.) Brain work. Munksgaard, Copenhagen, pp 414-423

Celesia GG, Polcyn RD, Holden JE, Nickles RJ, Gatley JS and Koeppe RA (1982) EEG Clin Neurophysiol 54: 243-256

Davidson JM and Davidson RJ (1980) The psychobiology of consciousness. Plenum, New York

Decety J, Philippon B and Ingvar DH (1988) Eur Arch Psychiatr Neurol Sci 238: 33-38

Decety J et al. (1989) To be published

Franzén G and Ingvar DH (1975) J Neurol Neurosurg Psychiatr 38: 1027-1032

Fuster JM (1980) The prefrontal cortex. Raven, New York

Goldman-Rakic PS (1988) In: The nervous system, pp 373-417 (handbook of physiology, vol 5)

Hagstadius S, and Risberg J (1989) J Cereb Blood Flow Metab 8 [Suppl 1] (Abstract)

Ingvar DH (1975) In: Ingvar DH, Lassen NA (eds.) Brain work Munksgaard, Copenhagen, pp 397-413

Ingvar DH (1977) L'Encephale 3: 5-33

Ingvar DH (1979) Acta Neurol Scand 60: 12-25

Ingvar DH (1980) In: Baxter CF, Melnechuk T (eds.) Perspectives in schizophrenia research. Raven, New York, pp 107-130

Ingvar DH (1983) Hum Neurobiol 2: 177-189

Ingvar DH (1985) Hum Neurobiol 4: 127-136

Ingvar DH (1987) In: Helmchen H, Henn FA, (eds.) Biological perspectives in schizophrenia. Wiley, New York, pp 201-211

Ingvar DH, Risberg J (1965) Acta Neurol Scand 41 [Suppl 14]: 93-96

Ingvar DH, Risberg J (1967) Exp Brain Res 3: 195-211

Ingvar DH, Schwartz M (1974) Brain 97: 273-288

Ingvar DH, Lassen NA (1975) (eds.) Brain work. Munksgaard, Copenhagen

Ingvar DH, Philipson L (1977) Ann Neurol 2: 230-237

Ingvar DH, Risberg J, Schwartz M (1975) Neurology 25: 964-974

Ingvar DH, Sjölund B, Ardö A (1976) EEG Clin Neurophysiol 41: 268-276

Ingvar DH, Rosén I, Eriksson M, Elmqvist D (1976a) In: Zotterman Y (ed.) Sensory functions of the skin, Pergamon, London, pp 549-557

Ingvar DH, Rosén I, Johannesson G (1979) J Neuropsychopharm 12: 200-209

Knopman DS, Rubens AB, Selnes O, Klassen AC, Meyer MW (1983) J Cereb Blood Flow Metab 3 [Suppl 1]: 250-251

Kuschner MJ, Schwarzt R, Alavi A, Daun R, Rosen M, Silva F, Reivich M (1987) Brain Res 409: 79-87

Larsen B, Skinhöj E, Lassen NA (1978) Brain 101: 193-209

Lassen NA, Friberg L (1988) In: Olesen J, Edvinsson L (eds.) Basic mechanisms of headache. Elsevier, Amsterdam, pp 62-68

Lassen NA, Ingvar DH (1961) Experientia 17: 42-43

Lassen NA, Hoedt-Rasmussen K, Sörensen SC, Skinhöj E, Cronqvist S, Bodforss B, Ingvar DH (1963) Neurology 13: 719-727
Lassen NA, Ingvar DH, Skinhöj E (1978) Sci Am 239: 62-71
Liberman AM, Cooper FS, Shankweiter DR, Studdert-Kennedy M (1967) Psychol Rev 74: 431-461
Mazziotta JC, Phelps ME, Miller J, Kuhl DE (1981) Neurology 31: 503-516
Mazziotta JC, Phelps ME, Carson RE, Kuhl DE (1982) Neurology 32: 921-937
Melges FT (1982) Time and the inner future. Wiley, New York
Metter EJ, Wasterlain CG, Kuhl DE, et al. (1981) Ann Neurol 10: 173-183
Nishisawa Y, Skyhöj-Olesen T, Larsen B, Lassen NA (1982) J Neurophysiol 48: 458-466
Olesen J (1971) Brain 94: 635-646
Petersen SE, Fos PT, Posner MI, Mintun M, Raichle ME (1988) Nature 331: 585-589
Phelps ME, Huang SC, Hoffmann EJ (1979) Ann Neurol 6: 371-388
Phelps ME, Mazziotta JC, Kuhl DE et al. (1981) Neurology 31: 517-529
Priogogine I, Strengers I (1988) Entre le temps et l'éternité. Fayard, Paris
Raichle ME (1979) Brain Res Rev 1: 47-68
Reivich M, Alavi A, Greenberg J, Fowler J, Christman D, Wolf A, Rosenberg A, Hand P (1978) J Comput Assist Tomogr 2: 656-664
Risberg J (1974), cf. Risberg 1980
Risberg J (1980) Brain Lang 9: 9-34
Risberg J (1987) In: Wade J et al. (eds.) Impact of functional imaging in neurology and psychiatry. John Libbey, London, pp 35-43
Risberg J (1988) In: Hannay H (ed.) Experimental techniques in human neuropsychology. Oxford University Press, Oxford, pp 514-543
Risberg J, Ingvar DH (1973) Brain 96: 737-756
Roland PE (1982) J Neurophysiol 48: 1059-1078
Roland P (1984) Hum Neurobiol 2: 205-216
Roland PE, Friberg L (1985) J Neurophysiol 53: 1219-1243
Roland PE, Eriksson L, Stone-Elander S, Widén L (1987) J Neurosci 7: 2373-2389
Roy CS, Sherrington CS (1980) J Physiol 11: 85-108
Ryding E (1986) Measurement of rCBF by intravenous 133-xenon. Thesis, University of Lund, Sweden
Ryding E, Bradvik B, Ingvar DH (1987) Brain, 110: 1345-1358
Ryding E, Bradvik B, Ingvar DH (1989) J Cereb Blood Flow Metab 8 [Suppl 1] (Abstract)
Schmidt CF (1950) The cerebral circulation in health and disease, Thomas, Springfield, III
Sokoloff L, Reivich M, Kennedy C, Des Rosiers MH, Patlak CS, Pettigrew KD, Sakurada O, Shinohara M (1977) J Neurochem, 28: 897-916
Stenberg G, Rosén I, Risberg J (1989) J Cereb Blood Flow Metab 8 [Suppl 1] (Abstract)
Stokeley EM, Sveinsdottir E, Lassen NA, Rommer P (1980) J Comput Assist Tomogr 4: 230-240
Warketin L, Risberg J (1989) J Cereb Blood Flow Metab 8 [Suppl 1] (Abstract)
Weinberger DR, Berman KF, Zee RF (1986) Arch Gen Psychiatry 43: 114-125

Movement Detection and Figure-Ground Discrimination

W. Reichardt
Max-Planck-Institut für biologische Kybernetik, Spemannstr. 38, 7400 Tübingen, FRG

Introduction

Movement detectors of the so-called correlation type were proposed long ago to explain motion perception in insects (Hassenstein and Reichardt 1956; Reichardt 1957, 1961; Reichardt and Varju 1959; Varjú 1959). In the meantime, good evidence has been accumulated that this movement detection scheme can also be applied to motion detection in humans (e.g. van Doorn et al. 1982a,b; van Santen et al. 1984, 1985; Wilson 1985; Baker and Braddick 1985). More recently our interest in movement computation has focused on dynamic aspects and on the dependence of the detector output on the structure of the stimulus pattern. In addition, the properties of two-dimensional arrays of pairs of movement detectors (Reichardt and Guo 1986; Egelhaaf and Reichardt 1987; Reichardt 1987) have been investigated in detail. Individual movement detectors, however, do not provide meaningful information on a moving pattern. In addition, some spatial, physiological integration is needed, for instance in connection with a solution of the figure and ground discrimination problem (Reichardt and Poggio 1979; Reichardt 1979, 1980; Poggio et al. 1981; Reichardt et al. 1983; Egelhaaf 1985).

Movement Detector Field Theory

A simplified version of an elementary movement detector model (EMD) is shown in Fig. 1a. It is composed of two input channels spatially separated by a small interval Δx and of two subunits that are mirror images of each other. These subunits share two input channels that sample the visual field at two neighbouring points in space. The signal received by one branch of each subunit is assumed to pass through a linear temporal filter which is approximated by a delay (ε). In each subunit the delayed signal is multiplied with the instantaneous signal of the neighbouring input channel. The detector output is given by the difference between the two subunits outputs. Fig. 1b shows a pair of such EMDs, one oriented in x-, the other in y-direction. It is assumed here that a two-dimensionally contrasted brightness pattern (F) is moved parallel to the image plane of an array of pairs of EMDs. F is then described by the expression

$$F = F[x+s(t); y+r(t)] \tag{1}$$

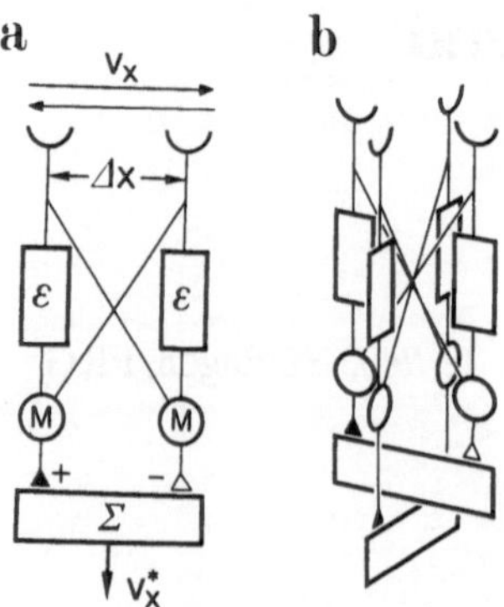

Fig. 1a,b. a Functional structure of an EMD consisting of two light receptors, signal delays ε in two channels and multiplication stages M. The functional properties of an EMD are antisymmetric since the difference between the outputs of the two branches is taken. This is why an EMD is responding to motion to the "right" (->) and to the "left" (<-) with the same amount of its output with different signs. **b** An orthogonally oriented pair of EMDs

where x, y are cartesian coordinates and s(t), r(t) are time-dependent displacements in the x- and y-direction, respectively. $ds(t)/dt = v_x(t)$ and $dr(t)/dt = v_y(t)$ represent the components of the instantaneous pattern velocity vector (v). It is assumed that Δx and accordingly Δy are small compared with the contrast changes in F. Under these conditions the inputs to an x-oriented EMD are well approximated by F and $F + \delta F/\delta x \, \Delta x$.

The ouputs of a pair of EMDs are combined to an output vector $v^* = \{v_x^* \; v_y^*\}$. The pattern velocity vector $v = \{v_x \; v_y\}$ and the output vector are related to each other by a matrix. This relation is in first approximation given by the expression

$$Z \begin{bmatrix} v_x^* \\ v_y^* \end{bmatrix} = \varepsilon e^{2q} \begin{bmatrix} q_{xx} & q_{xy} \\ q_{yx} & q_{yy} \end{bmatrix} \begin{bmatrix} v_x \\ v_y \end{bmatrix} \tag{2}$$

when making use of the substitution $F = e^q$ with $F \geq 0$. Or in short notation the relation (2) can be expressed by

$$v^* = -\varepsilon T v \tag{3}$$

The subscripts of q denote partial derivatives. It can be shown that the matrix in equation (2) represents a tensor of the order two and since v is a vector, as is v^* is a vector. Interestingly, the output of a detector pair is in the approximation given here not only proportional to the instantenous pattern velocity but is also linearly dependent on local pattern properties. These properties are represented by the tensor elements in equation (2).

An illustrative example to demonstrate some of the properties of individual detector pairs is a rotational symmetric brightness pattern as shown in Fig. 2a. When the pattern is moved in x-direction one gets for the two-dimensional array of the x-detectors the response profile plotted in Fig. 2b. Those individual x-detectors receiving inputs from the

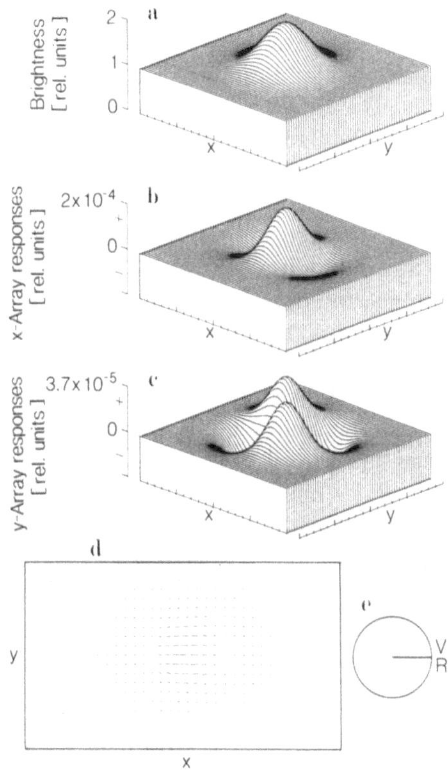

Fig. 2a-e. Computer simulation of an array (120x120) of orthogonally oriented pairs of EMDs. **a** A rotational symmetric gaussian pattern

$$F = A + Be^{-(ax)^2} e^{-(by)^2} \quad \text{with } A = B = 1 \text{ and } b = a = 0.05.$$

b Responses of the two-dimensional array of x-detectors (oriented in x-direction) to the motion of the brightness pattern in x-direction. **c** The responses of the array of y-detectors to the motion of the brightness pattern in x-direction. **d** The picture shows the central part of the array of pairs of EMDs.Most of the local vectors are elicited by the brightness pattern represented in a at the outputs of pairs of x- and y-detectors when the pattern is moved in x-direction. Most of the local vectors point into different directions. Summation in x- and y-direction, however, leads in this case to a resulting vector oriented parallel to the velocity vector as represented in **e**. (From Reichardt 1987),

periphery of the pattern produce negative responses, a consequence of the second derivatives in the tensor elements. The property means that these EMDs signal apparent motion directed opposite to real pattern motion, quite in accordance with corresponding experimental results (Reichardt 1987; Reichardt and Egelhaaf 1988). Fig. 2c contains the responses of the y-detectors. They are in general different from zero, although the pattern is moved orthogonally to the orientation of the y-detectors. A combination of Fig. 2b and Fig. 2c is shown in Fig. 2d. It contains the x-, y-dependent local vectorial responses of pairs of motion detectors. Most of the local vectorial responses point in directions different from the x-direction. Since the pattern is rotationally symmetric, it is, for trivial reasons, also symmetric with respect to the direction of pattern motion. Summation of the local vectors leads to a resulting vector that points in the same directions as the vector of pattern motion.

The typical features of the tensorial relation become even more apparent when we consider an asymmetrical pattern, as the one in Fig. 3a, rotated by 30° relative to the x, y coordinate system. The pattern is moved again in x-direction. The local vectorial responses to the motion of the brightness pattern are represented in Fig. 3b and after summation by the resulting vector in Fig. 3c. It can clearly be seen that the direction of the resulting vector deviates from the direction of the motion vector. This deviation (19.2°) is also dependent on the contrast power of the moving pattern. The examples presented here assume $v_y = 0$. The general case with $v_y \neq 0$ but equal value of $|\mathbf{v}|$ does not lead to additional information.

Since the directions of the local response vectors $\mathbf{v}^*$ usually differ from the motion vector $\mathbf{v}$, the question arises whether a one-to-one correspondence between these two vectors is granted. If so, it would in principle be possible to compute the velocity vector locally. Otherwise one is confronted with an ambiguity generally called the aperture problem. The map of $\mathbf{v}$ on $\mathbf{v}^*$ given by the transformation in equation (2) is one-to-one if the determinant of T in relation (2) does not vanish. Under these conditions the tensor T can be inverted and equation (3) solved for v. In this way the correct velocity field can in principle be calculated by using only local information on the pattern. There is only one special class of brightness patterns for which T cannot be inverted at any spatial location, and which consequently leads to ambiguous local motion measurements. In this case the determinant of T vanishes which leads to the partial differential equation

$$q_{xx}\, q_{yy} - q_{xy}\, q_{yx} = 0 \tag{4}$$

The only solutions of this differential equation are spatial brightness distributions, the logarithm of which represent so-called developable surfaces (Courant and Hilbert 1962). Except for this class of exotic patterns the map of $\mathbf{v}$ on $\mathbf{v}^*$ is one-to-one, or in

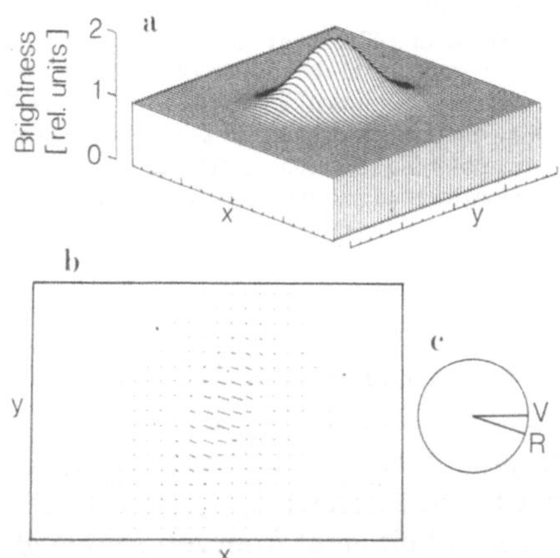

Fig. 3a-c. Computer simulation of an array of pairs of EMDs. **a** An asymmetrical gaussian pattern the long axis of which is rotated by 30° with respect to the x-axis. **b** The picture shows the central part of the array of pairs of EMDs. It represents the local vectors at the outputs of individual pairs of x- and y-oriented detectors when the pattern is moved in x-direction. The local vectors point into different directions from the motion vector. In this case the resulting vector obtained by summation of the individual responses deviates by 19.2° from the direction of the velocity vector, as shown in **c**. (From Reichardt 1987)

other words the aperture problem practically does not exist for movement detectors of the correlation type (Reichardt et al. 1988).

Figure-Ground Discrimination by Relative Motion

Much is known (see Egelhaaf et al. 1988) as to how information on large-field motion and relative motion is extracted from local motion measurements by EMDs in flies at different levels, comprising visual orientation behaviour in free (see for instance Wehrhahn et al. 1982; Wagner 1986a,b) and tethered flight (Reichardt and Poggio 1976, Reichardt et al. 1983; Wehrhahn and Hausen 1980; Egelhaaf 1985; Heisenberg and Wolf 1984;) as well as the associated response properties of visual interneurons (Egelhaaf 1985; Collett and King 1974; Hausen 1982a,b, 1984; Egelhaaf 1985). Only in female houseflies and in blowflies, however, a coherent view on how these tasks might be accomplished is now emerging at both the behavioural and the neuronal level. The investigations began with a quantitative behavioural analysis which also formed the conceptual background of the neurophysiological analysis of the fly's visual system. From the experimental data the underlying neuronal computations are abstracted, and a model circuit is formed which is sufficient to account for both information processing at the neuronal and the behavioural level.

In a typical behavioural experiment, shown in Fig. 4a, a test fly is suspended from a torque meter and positioned within a panorama of tiny random squares (ground) which also contains a small vertically oriented stripe (figure). The figure was covered with the same texture. Both figure and ground were moved (oscillated) horizontally, either together or independently (Reichardt et al. 1983). The yaw torque response of a test fly shows the same frequency as figure and ground. This is shown for two different oscillation frequencies in the initial part of Fig. 4b and 4c. When, however, figure and ground move relative to each other, the figure may in principle be distinguished; whether it is distinguished in the experiments depends on the phase relation between figure and ground motion and on the oscillation frequency of the pattern. For 90° phase shift, for instance, and high oscillation frequencies (0.5 to 8 Hz) the time course and mean values of the response profiles change considerably (Fig. 4c). A mean torque response is generated towards the figure which indicates that the fly has detected the figure and tries to fixate it. In contrast, at low frequencies (below 0.2 Hz), time course and mean value of the response differ little from those elicited by coherent motion. In conclusion, the figure is discriminated from the ground only at high oscillation frequencies.

There is good behavioural evidence that the visually induced yaw torque responses are jointly mediated by at least two parallel, bilaterally symmetrical control systems (Egelhaaf 1985, 1987; Reichardt et al. 1983). One of these control systems mediates yaw torque mainly at low oscillation frequencies and is more sensitive to motions of large stimuli than of small ones. This large-field (LF) system, thus, appears to be essentially responsible for the optomotor compensation of displacements of the entire surround and mediates course stabilization. In contrast, the other system dominates at high oscillation frequencies and is sensitive to small patterns. This small-field (SF) system mediates the detection, fixation and tracking of small moving objects.

In the fly the main projection from the retina to the brain is through three consecutive retinotopically organized visual ganglia (Strausfeld 1976) as shown in Fig. 5a. Along

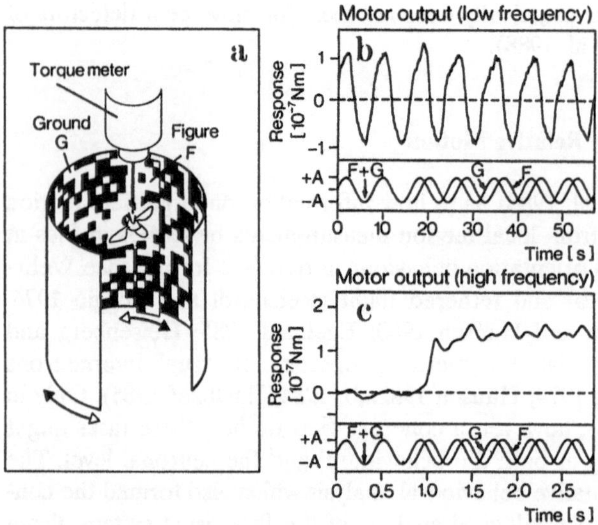

Fig. 4. Yaw torque responses to relative motion of figure and ground. **a** Experimental set-up: A test fly is fixed to a torque meter which allows measurement of the fly's turning tendency. The test fly is surrounded by a cylindrical panorama (ground, G) covered with a random texture. A vertically oriented equally textured stripe (figure, F) is placed in front of the ground. Its mean angular position was in front of the right eye 30° from the frontal midline. Its angular width amounted to 12°. Both figure and ground were oscillated either with a low (0.122 Hz) or a high (2.44 Hz) oscillation frequency. The oscillation amplitude amounted to 6°. In the experiments shown here (**b**) and (**c**) figure and ground were oscillated synchronously at the beginning and were then set to a relative phase of 90°; see stimulus traces at the bottom of the figures. The visually elicited yaw torque responses are shown in the upper diagrams. Positive and negative torques represent turning tendencies to the right and left, respectively. The responses to motion at 0° phase and at 90° phase do not differ significantly at the low oscillation frequency. At high oscillation frequency, the flies try to turn towards the position of the figure during relative motion. This is the indication that the figure is discriminated from the ground. (From Egelhaaf et al. 1988)

this pathway extensive transformation of the input information is carried out (Laughlin 1984). In the third optical ganglion, the lobula plate, the information is spatially integrated by about 50 different large interneurons. These project directly, together with input from other sensory modalities, to the motor control centres in the thoracic ganglia (Strausfeld et al. 1984; Heide 1983). The interneurons are presumed to receive input from the local EMDs. Two types of lobula plate output elements, the horizontal cells (Hausen 1982a,b) and the figure detection cells (Egelhaaf 1985), play a decisive role in the yaw torque generation. They are sensitive to horizontal pattern motion and are likely to represent the cellular analogues of the LF and SF system. The horizontal system (HS) consists of three neurons (Fig. 5a). They project into the ipsilateral part of the brain and are synaptically coupled to descending neurons. The three horizontal cells are excited by motion within their receptive fields. The figure detection (FD) cells form a group of four output cells of the lobula plate (Egelhaaf 1985). The FD4 cell scans most of the ipsilateral visual field (Fig. 5a).

The typical response patterns of both cell classes are shown in Fig. 5b and 5c. The stimulus conditions were the same as in the corresponding experiments. Fig. 5b shows the response of a horizontal cell to synchronous and relative oscillation of figure and ground. With relative motion between figure and ground the response pattern changes

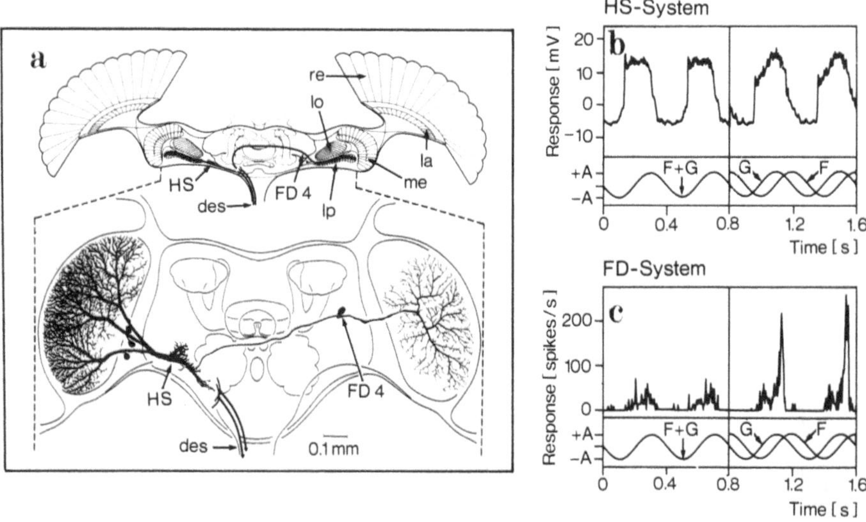

Fig. 5a-c. Neuronal components of the LF and SF systems. **a** Upper diagram: Schematics of a horizontal cross-section through the compound eyes and optic lobes of the fly. The ommatidia in the retina (re) and the corresponding columns in the visual ganglia (lamina: la; medulla: Me; lobula: lo; lobula plate: lp) are indicated. In the lobula plate the retinotopic information is spatially integrated by the large horizontal cells (HS) and figure detection cells (FD). They project into the central brain and are connected via descending neurons (des) to the thoracic-motor centres. Lower diagram: Anatomical structure of the three HS cells and the FD4 cell as revealed by cobalt and Lucifer yellow stainings. **b,c** Time-dependent responses of a HS and a FD4 cell to both 0° and 90° phase relative motion of figure and ground. The stimulus was essentially the same as in the corresponding behavioural experiments (see Fig. 4). All data were obtained with the blowfly *Calliphora*. The responses of the HS cell to stimulation with oscillation of figure and ground at 0° phase or at 90° relative phase shows only minor differences. In contrast, the FD4 cell shows only weak responses to large field motion and a characteristic sharp response peak during relative motion with 90° phase. (From Egelhaaf et al. 1988)

only slightly. This is reminiscent of the behavioural response at low oscillation frequencies. The responses of the FD cells, however, differ considerably, as shown in Fig. 5c. Relative motion of figure and ground with a phase of, for instance, 90° leads to significant response peaks. The characteristic difference between the HS and FD cells has been substantiated by measurements of the spatial integration properties of these cells (Hausen 1982a,b; Egelhaaf 1985). The HS and FD cells are selectively tuned to large-field and small-field motion and are the neuronal analogues of the LF and SF systems, respectively. The HS cells are specialized to evaluate global retinal motion fields, whereas the FD cells signal displacements of small objects against the background.

Despite considerable differences in their spatial integration properties, the responses of the HS and FD cells can be modelled by the same type of model network, provided the different model parameters are chosen appropriately (Reichardt et al. 1983; Egelhaaf 1985). A simplified version of the model circuits for the LF and SF systems are shown in Fig. 6a and 6b. They consist of the retinotopic array of EMDs (only one channel of the right eye is represented), pool cells and integrative output elements. The pool cells integrate the signals of the movement detectors and, subsequently, inhibit them via synapses of the shunting type. After the shunting operation output cells integrate the signals of the EMDs. The circuitries for the LF and SF systems differ in two

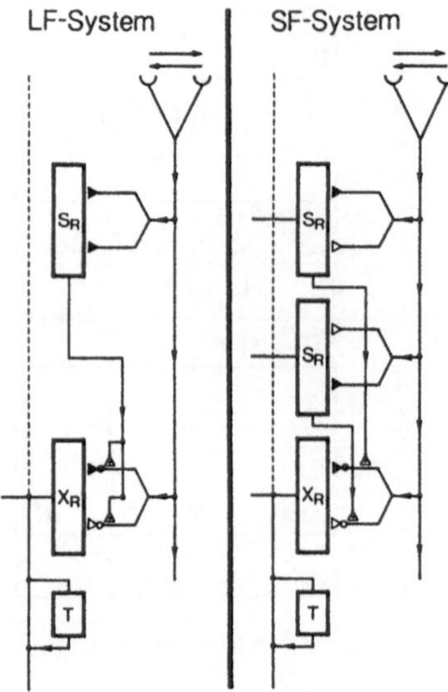

Fig. 6. Outline of the model circuitry for figure-ground discrimination. One channel of the retinotopic arrays of the LF and the SF systems, respectively, are shown for the right eye. In the LF system a pool neuron (SR) summates the EMD outputs (-◀ indicates excitatory synapses). The situation is different in the SF system. Here the pool is movement direction sensitive and consequently split into two pools with excitatory and inhibitory synaptic inputs (-◁). In addition there is experimental evidence that the pools of the SF system receive inputs from their contralateral homolog (SL; not shown in the figure). The pool outputs of the two systems are assumed to undergo saturation effects and to shunt inhibit each EMD output channel via presynaptic inhibition. The synapses involved (-◁) should therefore inhibit the output of each elementary detector channel. In both systems the cell X_R summates the outputs of the EMD channels via excitatory (-o◁) or inhibitory (-●◀) synapses. Whether an excitatory or inhibitory synapse is activated by an EMD channel depends on the velocity v and on the pattern structure. The synapses on the X output cells are assumed to operate with a non-linear characteristic, leading to postsynaptic signals that are in the LF system approximately the 1.25 and in the SF system the 3.0 power of the inputs. The motor output is controlled by the X cells via a direct channel and a channel T computing the running average of an X cell output

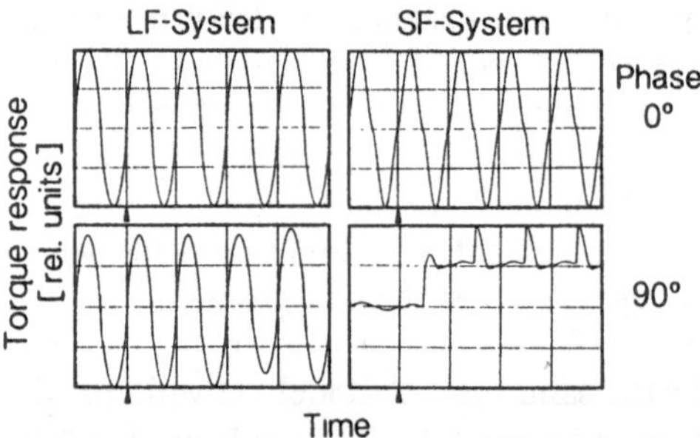

Fig. 7. Computer simulation of the LF and the SF systems. Only the result for the two experimental situations described in this paper are shown here. In the upper diagrams the oscillation frequency for figure and ground is low, whereas in the lower diagrams the oscillation frequency is high. The time scales of two computer experiments are therefore different. In the computer experiments the first cycle of figure-ground oscillation was always in phase whereas the following cycles they were 90° out of phase. As in the experiments shown before, the SF system is mainly responsible for figure-ground discrimination. The simulations are in accordance with the experimental results even in quantitative detail. They have been also tested for other phase angles and for other experimental situations

respects. Different non-linear synaptic transfer properties at the output cells are assumed. The proposed pool cells are motion direction insensitive for the LF system and motion direction sensitive for the SF system. The experiment described before have been computer simulated with the model circuits shown in Fig. 6. The results are shown in Fig. 7; they are in accordance with the experiments.

Experiments at the behavioural and at the neuronal level are mainly confined to horizontal pattern motion. In some experiments, however, horizontal and vertical pattern motions are combined. The results indicate that there seem to be no neural interactions effective between the systems responsible for motor control elicited by horizontal and by vertical pattern motions.

Summary

Movement computation and figure-ground discrimination by relative motion are fundamental for solutions of various tasks in visual orientation. A determination of the direction and speed of self-motion and of independent motion of objects from retinal activity patterns are examples for the type of problems solved by visual systems. Behavioural and electrophysiological analysis in combination with modelling has enabled us to unravel in part the process of motion computation as well as the neural mechanisms of the fly responsible for retinal large field motion and relative motion between objects (figure) and their background (ground).

References

Baker LB, Braddick OJ (1985) Temporal properties of the short-range process in apparent motion. Perception 14:181-192

Collett TS, King AJ (1974) Vision during flight. In: Horridge GA (ed) The compound eye and vision of insects. Clarendon, Oxford, pp 437-466

Courant R, Hilbert D (1962) Methods of mathematical physics, vol II. Interscience, New York

Egelhaaf M (1985) On the neuronal basis of figure-ground discrimination by relative motion in the visual system of the fly. I Behavioural constraints imposed on the neuronal network and the role of the optomotor system. Biol Cybern 52:123-140

Egelhaaf (1987) Dynamic properties of two control systems underlying visually guided turning in house-flies. J Comp Physiol [A] 161:777-783

Egelhaaf M, Reichardt W (1987) Dynamic response properties of movement detectors: theoretical analysis and electrophysiological investigation in the visual system of the fly. Biol Cybern 56:69-87

Egelhaaf M, Hausen K, Reichardt W, Wehrhahn C (1988) Visual course control in flies relies on neuronal computation of object and background motion. TINS 11: 351-358

Hassenstein B, Reichardt W (1956) Systemtheoretische Analyse der Zeit-, Reihenfolgen- und Vorzeichenauswertung bei der Bewegungsperzeption des Rüsselkäfers Chlorophanus. Z Naturforsch 11b:513-524

Hausen K (1982a) Motion sensitive interneurons in the optomotor system of the fly. I. The horizontal cells: structure and signals. Biol Cybern 45:143-156

Hausen K (1982b) Motion sensitive interneurons in the optomotor system of the fly. II. The horizontal cells: receptive field organization and response characteristics. Biol Cybern 46:67-79

Hausen K (1984) Large-field motion computations: the neural basis of visual stabilization in the flying fly. Proc Int Soc Eye Res 3:28

Heide G (1983) Neural mechanisms of flight control in diptera. In: Nachtigall W (ed) Biona report. Akademie der Wissenschaften und der Literatur zu Mainz. Fischer, Stuttgart, pp 35-52

Heisenberg M, Wolf R (eds) (1984) Vision in Drosophila. Genetics of microbehavior. In: Studies of brain functions, vol 12. Springer, Berlin Heidelberg New York

Laughlin SB (1984) The roles of parallel channels in early visual processing by the arthopod compound eye. In: Ali MA (ed) Photoreception and vision in invertebrates. Plenum, New York, pp 457-481
Poggio T, Reichardt W, Hausen K (1981) A neuronal circuitry for relative movement discrimination by the visual system of the fly. Naturwissenschaften 68:443-446
Reichardt W (1957) Autokorrelations-Auswertung als Funktionsprinzip des Zentralnervensystems (bei der optischen Wahrnehmung eines Insektes). Z Naturforsch 12b:448-457
Reichardt W (1961) Autocorrelation, a principle for evaluation of sensory information by the central nervous system. In: Rosenblith WA (ed) Sensory communication. Wiley, New York, pp 303-317
Reichardt W (1979) Figure-ground discrimination by the visual system of the fly. In: Haken H (ed) Pattern formation by dynamic systems and pattern recognition. Springer, Berlin Heidelberg New York, pp 100-121
Reichardt W (1980) Analogy between hologram formation and computation of relative movement by the visual system of the fly. Naturwissenschaften 67:411
Reichardt W (1987) Evaluation of optical motion information by movement detectors. J Comp Physiol 161:533-547
Reichardt W, Egelhaaf M (1988) Properties of individual movement detectors as derived from behavioural experiments on the visual system of the fly. Biol Cybern 58:287-294
Reichardt W, Guo A (1986) Elementary pattern discrimination (behavioural experiments with the fly Musca domestica). Biol Cybern 53:285-306
Reichardt W, Poggio T (1976) Visual control of orientation behaviour in the fly. Q Rev Biophys 9:311-375
Reichardt W, Poggio T (1979) Figure-ground discrimination by relative movement in the visual system of the fly. Part I: Experimental results. Biol Cybern 35:81-100
Reichardt W, Varjú D (1959) Übertragungseigenschaften im Auswertesystem für das Bewegungsehen. Z Naturforsch 14b:674-689
Reichardt W, Poggio T, Hausen K (1983) Figure-ground discrimination by relative movement in the visual system of the fly. Part II. Towards the neural circuitry. Biol Cybern [Suppl] 46:1-30
Reichardt W, Schlögl RW, Egelhaaf M (1988) Movement detectors provide sufficient informaiton for local computation of 2-D velocity field. Naturwissenschaften 75:313-316
Strausfeld NJ (1976) Atlas of an insect brain. Springer, Berlin Heidelberg New York
Strausfeld NJ, Bassemir U, Singh RM, Bacon JP (1984) Organizational principles of outputs from dipteran brains. J Insect Physiol 30:73-93
van Doorn AJ, Koenderink JJ (1982a) Temporal properties of the visual detectability of moving spatial white noise. Exp Brain Res 45:179-188
van Doorn AJ, Koenderink JJ (1982b) Spatial properties of the visual detectability of moving white noise. Exp Brain Res 45:189-195
van Santen JPH, Sperling G (1984) Temporal covariance model of human motion perception. J Opt Soc Am [A] 1:451-473
van Santen JPH, Sperling G (1985) Elaborated Reichardt Detectors. J Opt Soc Am [A] 2:300-321
Varjú D (1959) Optomotorische Reaktionen auf die Bewegung periodischer Helligkeitsmuster (Anwendung der Systemtheorie auf Experimente am Rüsselkäfer Chlorophanus viridis). Z Naturforsch 14b:724-735
Wagner H (1986a) Flight performance and visual control of flight of the free-flying housefly (Musca domestica L.) II. Pursuit of targets. Philos Trans R Soc Lond [Biol] 312:553-579
Wagner H (1986b) Flight performance and visual control of flight of the free-flying housefly (Musca domestica L.) III. Interacions between angular movement induced by wide- and smallfield stimuli. Philos Trans R Soc Lond [Biol] 312:581-595
Wehrhahn C, Hausen K (1980) How is tracking and fixation accomplished in the nervous system of the fly? Biol Cybern 38:179-186
Wehrhahn C, Poggio T, Bülthoff H (1982) Tracking and chasing in houseflies (Musca). An analysis of 3-D flight trajectories. Biol Cybern 45:123-130
Wilson HR (1985) A model for direction selectivity in threshold motion perception. Biol Cybern 51:213-222

Influence of Complex Visual Stimuli on the Regional Cerebral Blood Flow

B. Conrad and J. Klingelhöfer

Neurologische Klinik, Technische Universität München, Möhlstraβe 28, 8000 München 80, FRG

Measuring regional intracranial flow patterns with transcranial Doppler ultrasonography (TCD; Aaslid et al. 1982), the dynamics of local perfusion in the main parts of the occipital lobe were determined during activation of the visual cortex with specific visual stimuli of varying complexity (Klingelhöfer et al. 1988; Conrad and Klingelhöfer 1989). This is of particular importance since the conventional isotope tracer methods for measuring regional cerebral blood flow or regional brain metabolism (Lassen and Ingvar 1963; Heiss et al. 1985) do not possess a temporal resolution high enough to detect fast changes in the order of seconds corresponding to rapid alterations of the functional state and thus do not provide data on the dynamics of the adjustment of local cerebral blood flow (Raichle et al. 1984).

Using a 2-MHz pulsed Doppler device (EME TC 2-64B, FRG) the intracranial flow patterns of the P2 segment of the posterior cerebral artery (PCA) were investigated in 20 healthy subjects (18 to 32 years of age, mean age 26 years) in response to various visual stimuli as first described by Aaslid (1986, 1987). After identification of the PCA signal, the Doppler probe was adjusted and mechanically fixated with a specially developed probe holder attached to a tight headband. Flow velocity at rest (15 min) was defined as a state in which the subject was sitting relaxed in a silent, dark room. During binocular visual stimulation a slide was projected 90 cm in front of the subject with a viewing angle of 67° horizontally and 48° vertically.

White screen stimulation (600 lux) as well as checkerboard-pattern reversal stimulation (single square with an edge length of 2° of angle; reversal frequency 2 and 10 Hz) were applied in a first experiment. The subjects had to fixate a small light in the center of the screen. Simultaneously, end-tidal CO_2-concentration, thoracic respiratory excursions, and eye movements with DC-EOG (to check stable gaze control) were recorded. Additionally, blood pressure was continuously monitored noninvasively by measurements of finger pressure (Finapres, Ohmeda, USA; Boehmer 1987). Five different pictures of increasing complexity were used as visual stimuli in a second experiment. Ten subjects were requested to gaze at the picture and the other 10 to concentrate on the centrally located fixating point. The different stimulus modalities (3 in the first and 5 in the second experiment) were presented to each subject twice in a random order.

The AD-converted envelope curve of the PCA Doppler frequency spectrum, blood pressure, CO_2-concentration, and the respiratory excursions were recorded and stored on hard disk of a personal computer. Additionally, each single value of mean flow velocity (MFV) derived during one heart cycle was calculated. The prestimulus and the stimulus periods both lasted 50 s; the poststimulus phase was 100 s. In order to compare the relative amount of MFV increase in different subjects, flow velocities were normalized by dividing individual MFV values by an average MFV value calculated from the first 30 s of the prestimulus period (Fig. 1, upper part).

White screen stimulation resulted in a sudden increase in flow velocity. Thereafter, habituation occurred within 30 to 40 s following a nearly continuous gradual fall of MFV. The 10-Hz dynamic checkerboard stimulus caused a steeper onset, a higher maximum of MFV, and a more plateaulike velocity response (Fig. 1). Fifty percent of the first flow velocity maximum was reached 2.0 ± 0.8 s and 90% at 4.2 ± 1.7 s (n=20) after the flow velocity increase began. After terminating the 10-Hz checkerboard sti-

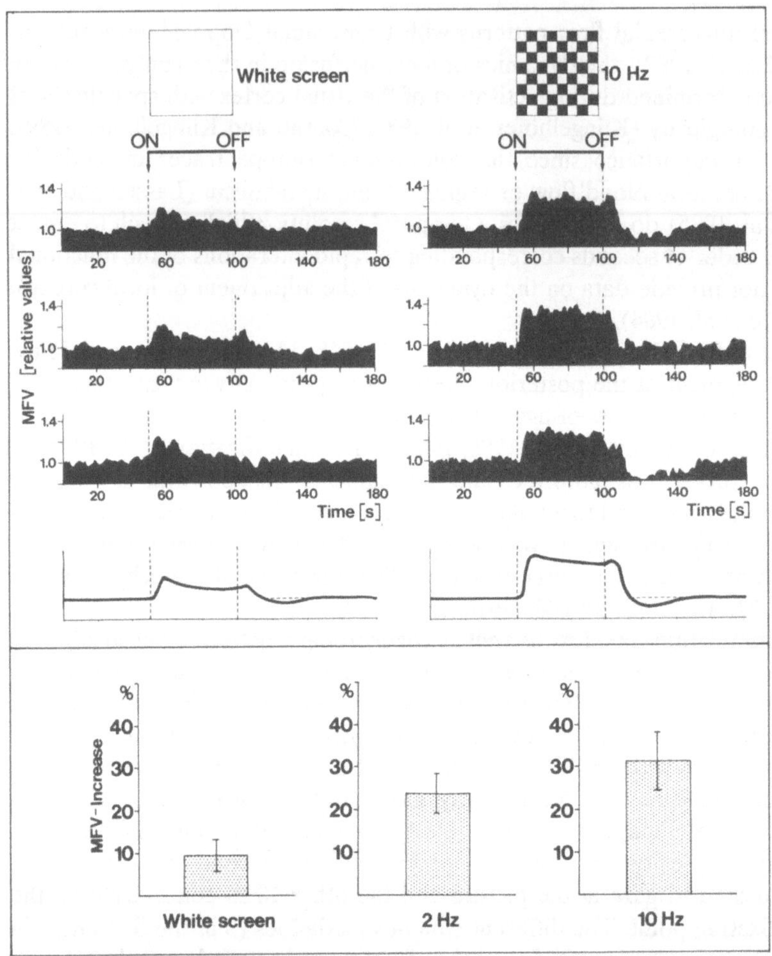

Fig. 1. Qualitative and quantitative effects of white screen and checkerboard stimulation on intracranial flow patterns of the posterior cerebral artery (P2 segment). Upper part: normalized mean flow velocity (MFV) of 3 representative subjects during white screen stimulation (left column) and 10-Hz checkerboard stimulation (right column). The schematical drawing at the bottom illustrates the general behavior of MFV changes. Lower part: percentage increase in MFV during white screen as well as 2-Hz and 10-Hz checkerboard stimulation relative to the corresponding dark phase. The subjects fixated the center point of the screen. The values given are means + SEM

mulus a second smaller increase of flow velocity occurred in 40% (n=20; "off-reaction"). In 95% (n=20) with the 10-Hz checkerboard stimulation a reduction in flow velocity occurred with a latency of 5.0 ± 2.1 s. During the poststimulus period MFV decreased below the level of the prestimulus period ("undershooting reaction") in 60% (n=20). The mean percentage increase of MFV during the stimulation phase in relation to the corresponding dark phase was highest for the 10-Hz checkerboard stimulation, reaching 30.9 ± 7.2% (n=20; Fig. 1, lower part).

With freely viewed pictures of increasing complexity (Fig. 2, I-IV) the mean relative MFV change increased continually from stimulus I to IV and reached a maximum of 38.8 ± 6.5% (n=20) with the stimulus IV. Eighty-five percent (n=20) exhibited a poststimulus "undershooting reaction." Fixating the center point of the picture IV elicited a rise of only 19.1 ± 4.2% (n=20) and generally exhibited a stronger habituation of the flow velocity comparable to the results with white screen stimulation.

The present investigation demonstrates that detailed dynamic effects of regional cerebral perfusion for different visual patterns can be detected by TCD. The strategy of stimulus perception causes distinct MFV changes in the occipital areas involved in visual information processing.

It could be argued that TCD does not record volume flow but rather flow velocity. The relationship between blood flow velocity and volume within the large basal intracranial arteries, however, is linear as long as alterations in the cerebral vascular bed are restricted to the small cortical resistance vessels. It could be confirmed by recent studies that the changes in diameter of the large basal arteries are negligible for various conditions such as hypercapnia. Intraindividual changes in blood flow velocity during TCD examination thus directly reflect changes in volume flow (Huber and Handa 1967; Bishop et al. 1986; Kirkham et al. 1986).

Viewing pictures with increasing complexity (structure, color, and depth) exhibited a further increase in flow velocity. This could be due to the fact that the visual analysis uses different processing systems independent of each other, or that the eye movements arose viewing the complex pictures increasing the variation of oculomotor projection (Creutzfeldt 1983; Livingstone and Hubel 1987). These results are in agreement with quantitative findings of PET studies (Lassen et al. 1978; Meyer et al. 1981; Phelps et al. 1982). Nearly unchanged blood pressure, heart rate, and end-tidal CO_2-concentration suggest that cardiovascular regulatory mechanisms do not influence the increase in PCA flow velocity.

The larger increase in MFV with a plateaulike velocity response during dynamic checkerboard stimulation as well as during detailed line-house stimulation as compared with white screen stimulation could be due to the increase in contrast borders in the visual field. This differential MFV behavior is similar for freely scanning versus fixating the center point of a picture.

Velocity habituation obviously occurs when the visual cortex does not receive new visual information. The rapid adaptation of MFV changes to altered stimulus conditions seems to be so sensitive that even neuronal "off effects" in the visual cortex can be observed when the stimulus is switched off. Also the delay of MFV decrease after the end of stimulation could be due to neuronal activity outlasting the stimulus, which could correspond to "visual after images" (Brown 1965). An explanation for the "undershooting reaction" occurring in the poststimulus phase could be that during visual stimulation there is an excessive blood supply causing a poststimulus counterregulation.

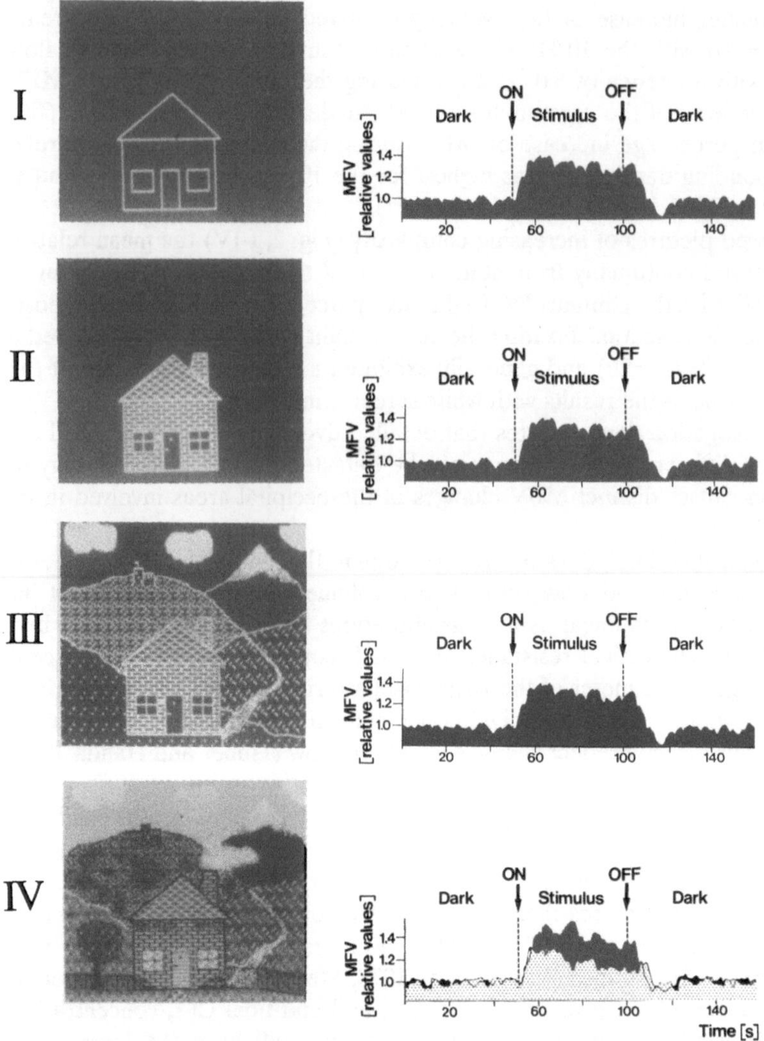

Fig. 2. Effects of visual stimulation (pictures of different complexity) on mean flow velocities (MFV) of the posterior cerebral artery. Left column: pictures of increasing complexity (I simple line-house on blue background, II detailed line-house on blue background, III detailed line-house with landscape on blue background, IV detailed line-house with landscape in color). Right column: MFV in the P2 segment of a 23-year-old subject while freely viewing stimuli I-IV. Comparing freely viewing against fixating the center point of the picture (dotted part, IV) revealed a smaller increase in MFV for the latter condition

The adaptation of MFV to the resting condition would thus reveal an oscillating regulatory mechanism.

Summary

Measuring regional cerebral blood flow or metabolism with isotope tracer methods do not provide data on the dynamics of a fast adjustment of local cerebral blood flow. Intracranial flow patterns of the posterior cerebral artery recorded by means of 2-MHz pulsed transcranial Doppler ultrasonography demonstrated detailed dynamic effects of various visual patterns on local cerebral perfusion. Visual stimuli of different complexity as well as the strategy of stimulus perception caused distinct flow velocity changes in the occipital cortex involved in information processing.

Acknowledgement. Supported by the Deutsche Forschungsgemeinschaft (SFB 330 - Organprotektion).

References

Aaslid R (1986) Transcranial Doppler examination techniques. In: Aaslid R (ed) Transcranial Doppler sonography. Springer, Wien New York, pp 39-59

Aaslid R (1987) Visually evoked dynamic blood flow response of the human cerebral circulation. Stroke 18: 771-775

Aaslid R, Markwalder TM, Nornes H (1982) Noninvasive transcranial Doppler ultrasound recording of flow velocity in basal cerebral arteries. J Neurosurg 57: 769-774

Bishop CCR, Powell S, Rutt D, Browse NL (1986) Transcranial Doppler measurement of middle cerebral artery blood flow velocity: a validation study. Stroke 17: 913-915

Boehmer RD (1987) Continuous, realtime, noninvasive monitor of blood pressure: Penaz methodology applied to the finger. J Clin Monit 3: 282-287

Brown JL (1965) Afterimages. In: Graham CH (ed) Vision and visual perception. Wiley, New York

Conrad B, Klingelhöfer J (1989) Dynamics of regional cerebral blood flow for various visual stimuli. Exp Brain Res 77:437-441

Creutzfeldt OD (1983) Cortex cerebri: Leistung, strukturelle und funktionelle Organisation der Hirnrinde. Springer, Berlin Heidelberg New York

Heiss WD, Beil C, Herholz K, Pawlik G, Wagner R, Wienhard K (1985) Atlas of positron emission tomography of the brain. Springer, Berlin Heidelberg New York

Huber P, Handa J (1967) Effect of contrast material, hypercapnia, hyperventilation, hypertonic glucose and papaverine on the diameter of the cerebral arteries - angiographic determination in man. Invest Radiol 2: 17-32

Kirkham FJ, Padayachee TS, Parsons S, Seargeant LS, House FR, Gosling RG (1986) Transcranial measurement of blood velocities in the basal cerebral arteries using pulsed Doppler ultrasound: velocity as an index of flow. Ultrasound Med Biol 12: 15-21

Klingelhöfer J, Conrad B, Frank B, Benecke R, Schneider M, Sander D (1988) Dynamics of local cerebral perfusion following different visual stimuli. J Cardiovasc Ultrasonogr 7: 99-100

Lassen NA, Ingvar DH (1963) Regional cerebral blood flow measurement in man. Arch Neurol 9: 615-622

Lassen NA, Ingvar DH, Skinhoj E (1978) Brain function and blood flow. Sci Am 239: 62-71

Livingstone MS, Hubel DH (1987) Psychophysical evidence for separate channels for the perception of form, colour, movement, and depth. J Neurosci 7 (11): 3416-3468

Meyer JS, Hayman LA, Amano T, Nakajima S, Shaw T, Lauzon P, Derman S, Karacan I, Harati Y (1981) Mapping of local blood flow of human brain by CT scanning during stable xenon inhalation. Stroke 12: 426-436

Phelps ME, Mazziotta JC, Huang SC (1982) Study of cerebral function with positron computed tomography. J Cereb Blood Flow Metab 2: 113-162

Raichle ME, Herscovitch P, Mintun MA, Martin WRW, Powers W (1984) Dynamic measurements of local blood flow and metabolism in the study of higher cortical function in humans with positron emission tomography. Ann Neurol 15: S48-S49

DC Shifts in the Human Brain: Their Relationship to the CNV and Bereitschaftspotential

W. C. McCallum, R. Cooper and P. V. Pocock
Burden Neurological Institute, Bristol BS16 1QT, England

Introduction

In this paper we present some recent developments in the study of those brain SP changes which were first discovered as a result of the pioneering studies by Kornhuber and Deecke (1965) on the Bereitschaftspotential and by Walter et al. (1964) on the contingent negative variation (CNV).

These SPs appear to reflect the progressive involvement of an individual with a particular task, whether that task is self-initiated, as in the case of the Bereitschaftspotential, or dependent on warned, external events, as in the case of the CNV. They occur in circumstances where the task concerned is at the focus of awareness, and they seem to be related to the mobilization and direction of processing resources towards the task. To the extent that they show a progressive build-up over time, they appear to be linked to a preparatory process, the end point of which is usually a decision or the initiation of a motor action.

In the four studies reported here both motor (tracking) and cognitive tasks have been used to investigate variations in workload on SPs in normal, healthy adults.

Experiment 1 - A Visual Tracking Task

In this experiment DC recording was from 4 midline electrodes (Fz, Cz, Pz and Oz), each referred to linked mastoids, in 24 subjects. Vertical and horizontal electro-oculograms were also recorded, and an on-line procedure was used to correct EEG data for eye blinks.

Subjects used a joystick to track on a video display a moving letter which began as the letter "X" but, during the course of tracking, changed to become another letter. This new letter could be either a target - to be fired at when in range - or a non-target, which had to be tracked but not fired at. The subset of letters which constituted possible targets on a given trial were displayed on the screen for 3 s prior to the start of tracking. A trial lasted for just over 28 s, the data for the whole of this period being sampled and stored for subsequent averaging at a resolution of 5 ms per point. Trials varied in difficulty. Movement could be either fast and over a relatively long distance or slow and over a short distance only; the movement could also be perturbed - i.e. erratic - or smooth and straight. Rapid selective button presses were required from the subjects when the letter first appeared, when it identified itself as a target or non-target, and when it came within range. See McCallum et al. (1988) for a more detailed description.

Figure 1 illustrates a representative single trial (no. 38) averaged across all 24 subjects. During the first second a resting baseline level was established as subjects fixated a central spot. Following the onset of an auditory warning stimulus there was a CNV-like rise of negativity in preparation for the display of letters to be memorized. This negativity remained steady or fell slightly while the letters were displayed. Thereafter it showed a further rise to the point at which the letter to be tracked appeared on the screen and tracking began. Tracking was accompanied by a further period of sustained negativity, the negative rise accelerating after the letter had revealed itself as a target or non-target. When the object (letter) was signalled as being in range, there was, in the case of targets, a further slight increase of negativity as firing took place and tracking continued to the point of impact, 2 or 4 s later. In the case of non-targets tracking was terminated, and the negativity remained relatively stable. The subject ended by fixating once more the central spot for the final 2 s, during which time the negativity was either sustained or began to fall.

The more difficult trials led to increases in tracking error and increases in scalp negativity during tracking. Over all trials there was a significant correlation between tracking error and amplitude of the negative shift. Figure 2A illustrates the significantly higher negativity at the vertex for the 12 most difficult trials (highest tracking error) compared with the 12 easiest trials (lowest tracking error).

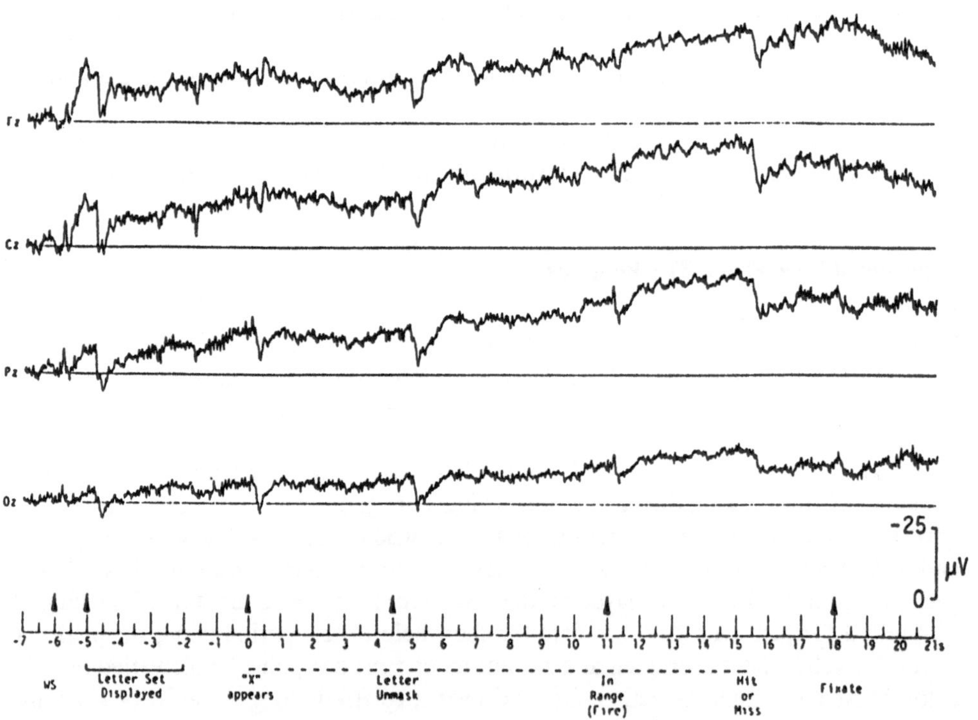

Fig. 1. Average of one representative trial (no. 38) across all subjects (experiment 1). The tracked letter was target

A comparison of trials averaged on the basis of the size of letter set (1, 3 or 6 letters) to be remembered is illustrated in Figure 2B. As memory set size increased from 1 to 6 letters so the negative rise was arrested or even reduced (the larger the set size the more marked the reduction) until with a set size of 6 the values became positive during this display and memorization phase.

Experiment 2 - Visual Tracking with an Added Secondary Task

Experiment 1 substantiated the presence of sustained negative shifts with task involvement and showed that such shifts could be sensitive to the level of workload. To increase the level of workload further and to investigate how brain resources are allocated in conditions of competing demand, a modified version of experiment 1 was repeated and further conditions were added in which a secondary task was superimposed on the tracking requirement. This task, which started with the onset of tracking, had an auditory and a visual form. Both secondary tasks were also presented in the absence of tracking. In the visual version of the secondary-task subjects were required to press a key to one of 6 colour/pattern combinations occurring in a ring of light-emitting diodes surrounding the video screen. In the auditory version of the task, subjects' key press responses were to one of 6 frequency/ear combinations of monaurally presented tones.

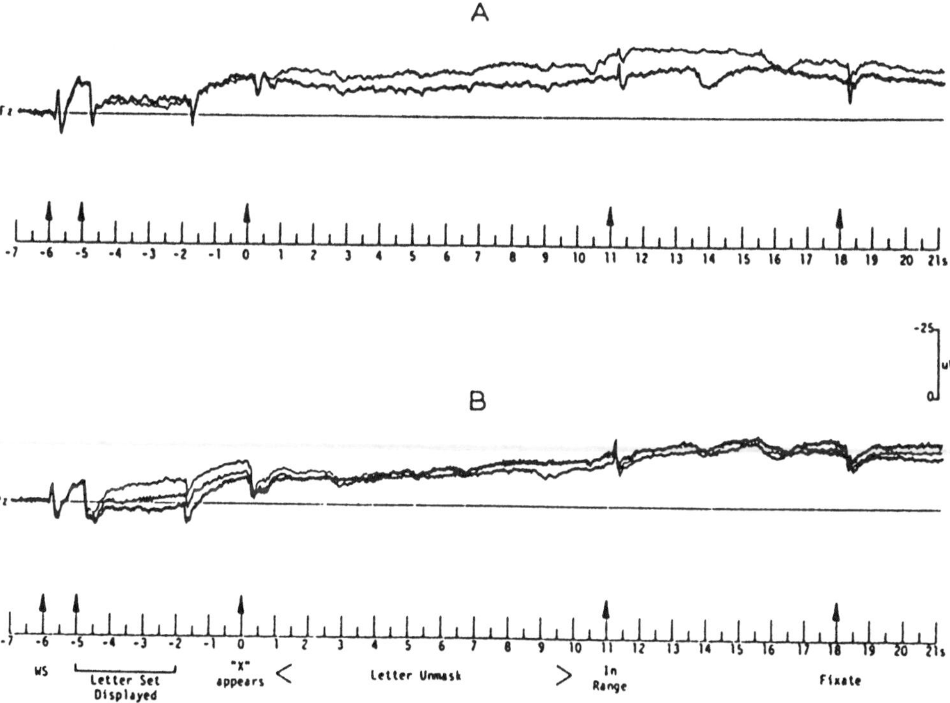

Fig. 2. Grand averages across all 24 subjects. **A.** The 12 trials with highest tracking error (thin trace) compared with the 12 trials with lowest tracking error (thick trace) at electrode Cz. **B.** Comparison of trials with set size 1 (thin trace), 3 (medium trace) and 6 (thick trace) at electrode Pz

Stimuli were presented at irregular (2-3 s) intervals. Sixteen normal adult subjects participated in this experiment.

The primary tracking task alone produced results very similar to those in experiment 1. Trials with the highest tracking error showed, at the vertex, significantly increased negativity during tracking. Memory set size showed significant decreases of negativity similar to those of the earlier experiment. The secondary tasks, presented on their own, elicited ERP patterns having a large P3 (P300) component to the designated stimulus and a smaller P3 to other stimuli.

When the primary and secondary tasks were combined, the secondary-task ERPs could be seen superimposed on the negativity of the primary task. Both secondary tasks resulted in a significant increase in negativity at electrodes Fz and Oz, but not at Cz. At electrode Pz the increase was in part significant for the auditory task but not for its visual counterpart. The auditory combination compared with the primary task alone is illustrated in Fig. 3. For both auditory and visual stimuli, the N2 component of the ERP was more negative and P3 less positive when subjects were engaged in tracking; these changes, which are illustrated in Fig. 4 for the auditory version of the task, were consistently significant at electrodes Fz and Oz, and to a lesser extent at Cz. During the course of a trial all components (N1, P2, N2 and P3) tended to decrease in amplitude during presentation of the letter set. The negative components N1 and N2 increased in amplitude at the start of tracking but thereafter steadily declined during the course of

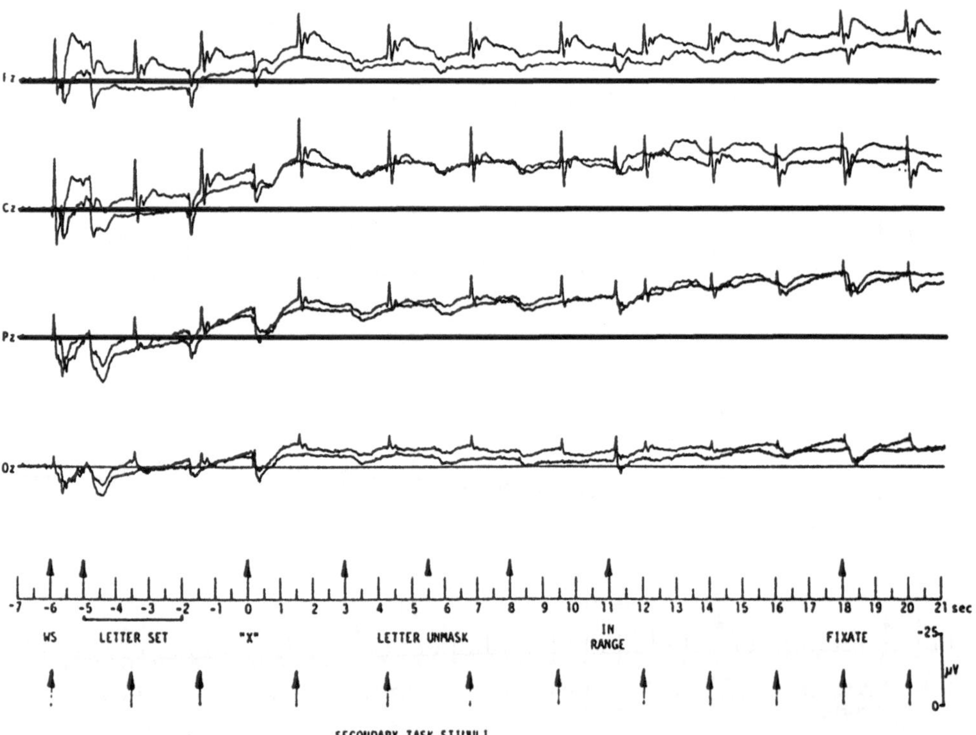

Fig. 3. Grand averages of all trials (experiment 2) for primary tracking task only (lower trace) and primary + secondary (auditory) tasks (upper trace)

tracking. Conversely, the positive components P2 and P3 declined further in amplitude at the start of tracking but then showed a steady rise as tracking proceeded.

The DC levels in the combined primary and secondary task conditions showed the same differences due to the variables governing task difficulty as those seen in the primary task alone - i.e. the more difficult the trial the higher the negativity during tracking.

The effects of workload were to be found in two distinct forms: firstly in an increasing level of negative shift as load rose, and secondly as changes in ERP components, particularly a reduction of P3 amplitude, when the secondary task was combined with the primary. This could be viewed as indexing resource reciprocity, that is, an increase

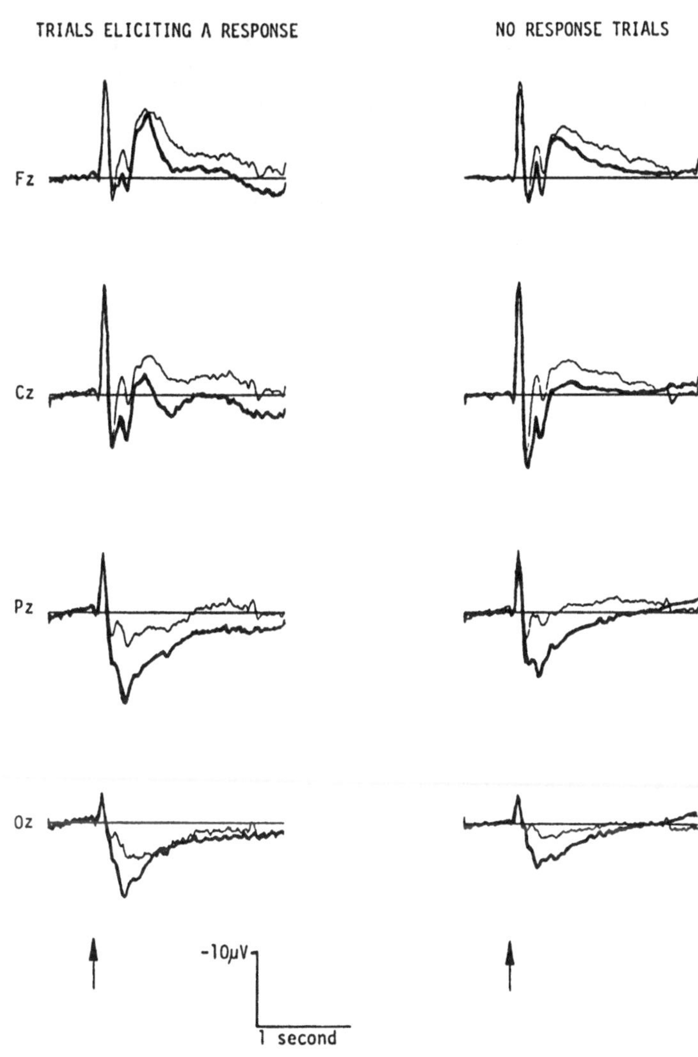

Fig. 4. ERPs to auditory secondary task stimuli (experiment 2), presented alone (thick trace) and in combination with primary tracking task (thin trace). These are grand averages across all subjects. Arrows indicate stimulus onset

of resources deployed on the primary task at the expense of a reduction in resources deployed on the secondary task.

Experiment 3 - A Cognitive Recursive Digit Summing Task

In the previous two studies relationships between operator workload and negative shifts of brain potential were clearly established. It was less clear whether the shifts represented cognitive processing load or were simply related to the maintenance of sustained motor output. We therefore designed a more cognitively based task of a similarly extended kind but without continuous motor output.

Twelve normal adult subjects undertook this task in which they watched a visual display which, during each trial, presented a sequence of single-digit numbers, preceded by a 2-s presentation of a "target" number (either 3, 6 or 9), which varied from trial to trial. Subjects pressed a button as rapidly as possible whenever the sum of the last three digits seen equalled the current target number. Target conditions could occur from 0 to 3 times in any given trial.

Recording procedures were as for the previous two experiments. Averages over 28 s were compiled for all trials on which the target was 3, 6 or 9, plus a further grand average of all trials.

Once more the task resulted in a progressively rising negative shift of potential which began with the presentation of the first digit to be summed and reached its maximum amplitude immediately following presentation of the last of the 14 digits. This widespread shift, largest at the vertex, is illustrated in Fig. 5. Compared with the general

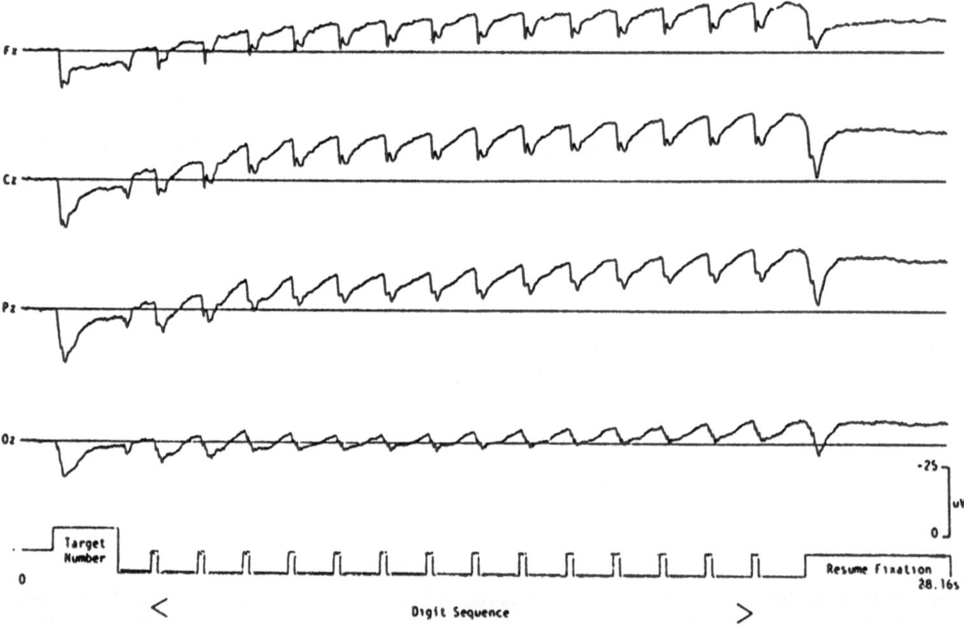

Fig. 5. Grand average of 12 subjects performing recursive digit summing task (experiment 3)

difficulty of the task, that between different target numbers was small; the three levels were not differentiated effectively either by performance or in terms of significant differences in the amplitude of the negative shift. There was, for the 2 s during which the target number was presented, a positive shift at all electrodes. Although unsuccessful in differentiating levels of cognitive load, this experiment demonstrated that a predominantly cognitive sustained task could also give rise to extended negative shifts.

Experiment 4 - Tracking with Feedback to Control the Level of Effort

One question remaining unanswered was whether the negative shifts seen were due primarily to increasing perceptual processing demands or to the application of greater "effort" by the subjects. A modified tracking experiment was therefore designed to induce subjects to expend more or less effort on the task even though the task itself remained constant in terms of the speed and pattern of movement of the object to be tracked.

Twelve normal adult subjects manually tracked on an oscilloscope screen a diamond-shaped figure which moved either in elevation only, in azimuth only or in a combination of these. Its motion was determined by sums of sinusoids. The amplitude of the motion varied through three 6 s phases on each trial; the first was low, the second high, and the third of medium amplitude. Four levels of tracking accuracy were demanded and were imposed during a trial by use of a feedback stimulus. Prior to the start of tracking a number between 1 and 4, plus arrows, appeared on the screen for 2 s, indicating the required accuracy level and the direction(s) of movement. The easiest level was 1, the most difficult 4. Feedback was conveyed by a circle which surrounded the tracked diamond. This circle expanded or contracted in size proportionate to the tracking error. As the accuracy demanded increased, it became more difficult for subjects to maintain the feedback circle at its smallest diameter; progressively smaller errors resulted in its rapid expansion. Recording conditions were as for the previous experiments; 28 s averages were computed.

As in the earlier tracking experiments, task involvement gave rise to a sustained negative shift of potential that was largest fronto-centrally. Figure 6A illustrates this shift at the vertex. However, the effect of the feedback stimulus was to worsen, not improve, tracking performance; subjects were driven to overcompensate in their efforts to maintain the smallest possible circle. Although more effort may have been expended, it was clearly not task effective, which would appear to be a prerequisite for increments in the level of negativity to occur, as the levels of negative shift were not significantly separated by this variable. Once more, the display of relevant information at the start of a trial resulted in a positive-going shift.

The study produced an additional, and unexpected, finding. Superimposed upon the sustained negative shifts was a pattern of lower amplitude, infra-slow changes, having frequencies around 0.5 Hz or slower. Further investigations, reported in full by Cooper et al. (1989), have revealed these slow oscillations to be related specifically to changes in the velocity of the tracked object (see Fig. 6B). These SPs may possibly be reflecting activity in a "feed-forward" system concerned with predicting the behaviour of the tracked object.

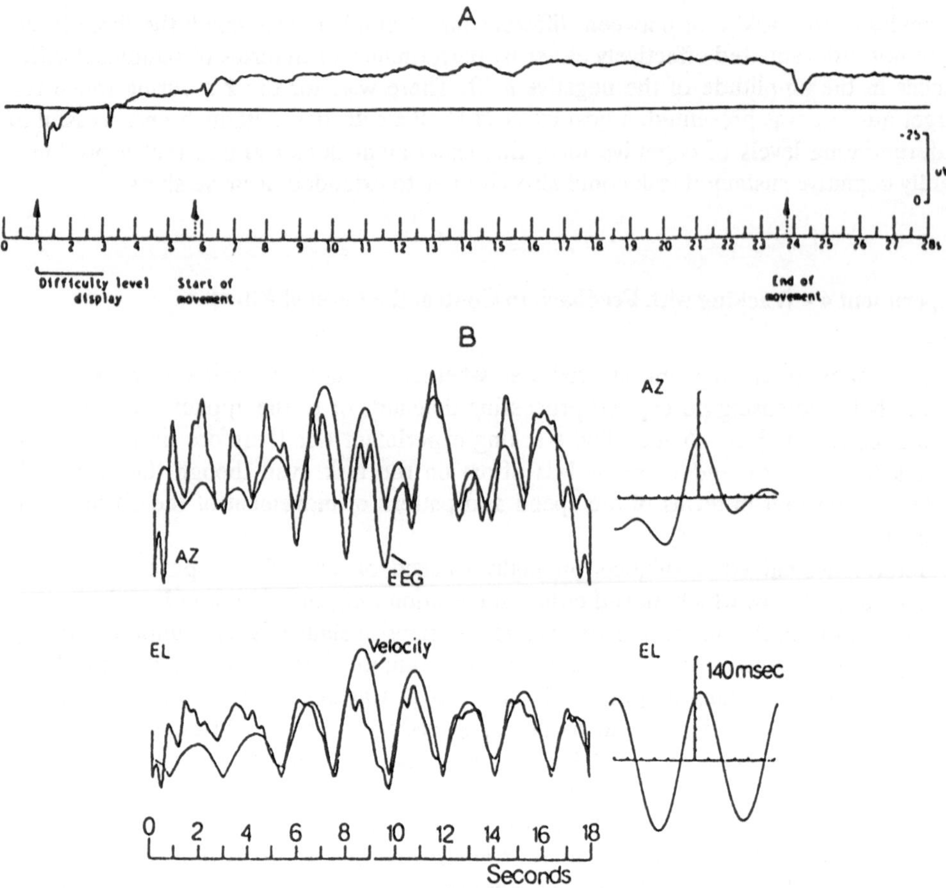

Fig. 6. A Grand average of all trials across all 12 subjects at electrode Cz (experiment 4). **B** On the left are superimposed traces of the vertex EEG and the velocity of the tracked object (rectified) during the period of tracking in experiment 4. The EEG is the grand average across subjects of all azimuth trials (upper) and all elevation trials (lower). The cross-correlograms for each pair are shown on the right. Divisions on the vertical scale are each 0.1 and on the horizontal scale 700 ms

Conclusions

Collectively the four experiments described effectively demonstrate that SP changes can occur over periods of time extending to tens of seconds. Negative shifts are not, as has been implied by some critics of the CNV, confined to late components of warning stimuli and to Bereitschaftspotentials preceding motor actions. From the studies reported it would appear that they can quite sensitively reflect an individual's level of task involvement, whether that involvement is predominantly motor or predominantly cognitive. It seems probable that several kinds of SP may be represented within the protracted shifts observed. One kind, typified by the pretracking stage of the first two experiments, is probably very similar to that observed in CNV paradigms. Almost certainly those parts of the negative shift immediately preceding anticipated motor actions contain a Bereitschaftspotential element. There is also evidence that a sustained positive-going shift is to be encountered when visual information is being absorbed and/or memorized. However, in addition to all these elements, sustained negativities are to be found which seem primarily to index the level of task involvement, that is to say, the extent to which the task concerned is demanding focalized attention, to the relative exclusion of competing demands.

Summary

Collectively the four experiments demonstrate that both cognitive and motor task involvement give rise to sustained negative slow potential (SP) shifts which can persist for tens of seconds. These shifts are linked to but distinguishable from those negative SPs which precede motor actions or prepare for events and decisions fixed in time. Factors which demonstrably increase workload tend also to increase the amplitude of the negative shifts. When secondary tasks were added to create competing demands, changes in SPs and ERPs provided evidence for reciprocity in the allocation of processing resources. Intake of visual information prior to the onset of each of the tasks resulted in changes which were, relatively, positive going. When the material had to be memorized, the magnitude of this positivity was directly related to the memory load. The final study revealed a new form of SP closely related to the velocity of object movement.

References

Cooper R, McCallum WC, Cornthwaite SP (1989) Slow potential changes related to the velocity of target movement in a tracking task. Electroencephalogr Clin Neurophysiol 72: 232-239

Kornhuber HH, Deecke L (1965) Hirnpotentialänderungen bei Willkürbewegungen und passiven Bewegungen des Menschen: Bereitschaftspotential und reafferente Potentiale. Pflugers Arch. Ges. Physiol. 284: 1-17.

McCallum W C, Cooper R, Pocock PV (1988) Brain slow potential and ERP changes associated with operator load in a visual tracking task. Electroencephalogr Clin Neurophysiol 69: 453-468

Walter WG, Cooper R, Aldridge VJ, McCallum WC, Winter AL (1964) Contingent negative variation: an electric sign of sensorimotor association and expectancy in the human brain. Nature 203: 380-384

Probability Mapping of EEG Changes due to the Perception of Music

H. Petsche, R. Rappelsberger, K. Lindner
Neurophysiologisches Institut, Universität Wien, 1090 Wien, Österreich

When our studies on the EEG during cognition were begun a few years ago, the main question was whether the spontaneous EEG contains any information about cognitive processes or, rather, would turn out as a sort of "brain garbage", which it was usually considered to be until recently. Our results taught us that the spontaneous EEG, in effect, contains a wealth of information about mental processes; therefore we have applied our method on a number of other mental tasks such as reading (Petsche et al. 1986a, 1987).

There are several reasons why we started with music as an object matter (Petsche et al. 1986b). Among them the personal interest of the authors certainly plays a leading role. Besides, looking for possible impacts in the EEG caused by listening to music seemed to us a particularly severe test for the efficiency of our method to draw information out of the ongoing EEG. It seemed more likely to obtain results in group studies when investigating special and probably more narrowed brain strategies such as may be involved in reading or mental arithmetic rather than when searching for those involved in music perception, in which far greater individual differences are expected. The evidence that these studies not only yielded significant results but even supplied insights into processing strategies differing according to sex and to the level of musical training confirms once more the usefulness of our method for the exploration of cognitive processes.

The method was developed in order to extract statistically significant differences of power and coherence from 1 min spontaneous EEG during a listening task in comparison to 1 min EEG at rest before and afterwards (a detailed description is found in Rappelsberger et al. 1988).

In the present study the subjects had to listen to the first movement of Mozart's quartett KV 458. The EEG recorded from the 19 electrodes of the 10-20 system with respect to connected ear lobe electrodes is Fourier analyzed. First the 19 EEG traces are digitized at 256/s, conspicuous artifacts are eliminated by visual inspection, and 15-30 sections of 2 s are selected from each trial for further processing. Averaged power and coherence spectra are computed (the latter between adjacent electrodes and between electrodes on homologous sites of both hemispheres). Broad-band parameters (i.e. mean values) are computed for 5 frequency bands between 4 and 32 Hz according to Gasser et al. (1982). The ln of the absolute power values is calculated for the transformation of these parameters. The mean coherence values are z-transformed (Jenkins and Watts 1986). These transformations yield approximately normal distributions of the parameters, a necessary prerequisite for the statistical tests. To be independent of statistical test variables and unknown distributions of the parameters for the determination of significant differences, a permutation procedure according to Fisher (Edgington 1980) is applied. This yields the probabilities of changes during a task with respect to the control situation before as well as after the task (significant reversible changes). The results are presented in topographic maps in colors. For this publication, however, a black and white presentation is used with squares indicating decreasing error probabilities with increasing size at different locations.

The physiological significance of the parameters power and coherence is not yet completely clear. In spite of this, there is general agreement on a few facts: at least for alpha and theta power there seems to be some correlation with the level of arousal or vigilance. When opening the eyes, there is, in addition to the theta and alpha reduction, also a beta reduction; but beta power may also increase under arousal conditions (as for instance during silent reading). Power increase and decrease thus may have different meanings with respect to topography and frequency band, and further studies are needed to arrive at sound conclusions about the significance of power changes for mental events. This is even more true for the parameter coherence, which is a measure of the correlation coefficient per frequency of two signals and thus a measure of the degree of synchrony between two areas in question. In terms of neuronal processes one may claim that the degree of coherence gives a hint of the degree of electrical coupling between the two brain regions recorded from and thus also of the amount of information exchange between them (Beaumont et al. 1978). Interhemispheric coherence thus indicates how much two corresponding cortical regions interact.

In the following, the reaction of the EEG on a group of 75 students (40 males, 23.5 ±6.3 years old, and 35 females, 22.7 ±3.8 years old) was studied when listening to the first movement of the Mozart quartett (stereophonically applied to both ears) with eyes closed. In spite of differences in the personal reactions, significant changes of power and coherence were found.

Most distinct was a decrease in alpha power, more extended in the left than in the right temporal area and also involving the precentral, frontal, and parietal areas. This decrease also concerned the theta band.

In addition, a considerable increase of interhemispheric coherence in beta 2 (18 - 24 Hz) between the posterior temporal and the parietal regions was seen. This means that during listening to the Mozart quartett these areas of the two halves of the brain are functionally linked closer together than at rest.

Fig. 1. The EEG changes caused by listening to Mozart's quartett KV 458 in men (left) and women (right) differentiated in terms of previous musical education, for at least 5 years (above) and no musical education (below). Probability maps of significant EEG power and coherence changes for 1 min listening as compared with 1 min EEG periods at rest recorded before and afterwards (significant reversible changes). The three columns in each group represent, from the left to the right: power, local and interhemispheric coherence. The five frequency bands are arranged horizontally (4 - 32 Hz). The three different sizes of the squares relate to descriptive error probabilities obtained by paired Fisher permutation test: large square 2 P = 0.01; middle sized square 2 P = 0.05; small square 2 P = 0.10. Empty squares indicate a decrease in the parameter in question, full squares an increase. Power changes caused by listening are plotted at the electrode locations, local coherence changes are indicated by squares between the two electrodes involved, and interhemispheric coherence changes are shown by two squares at the homologous electrode position on the two hemispheres, connected by a line.

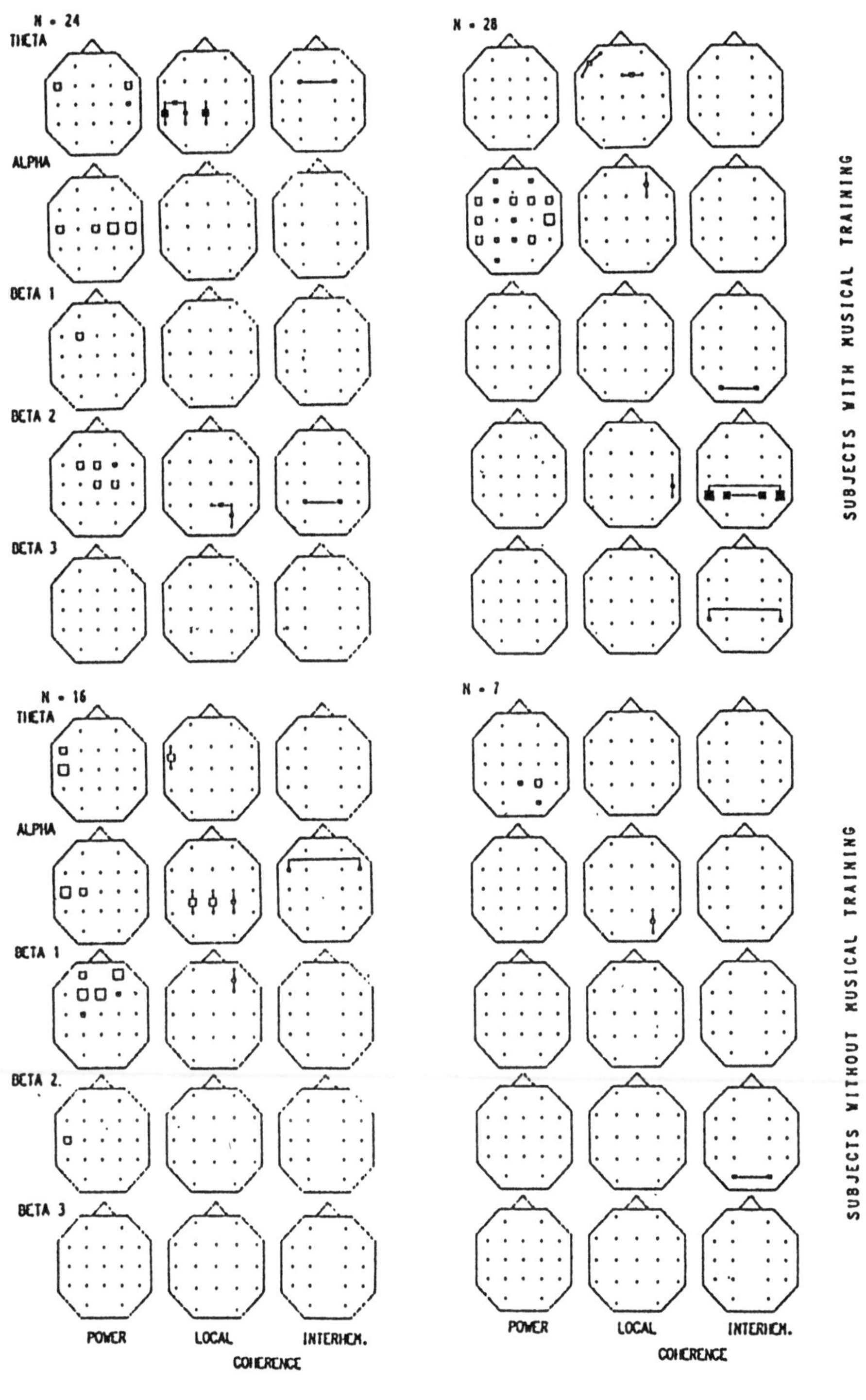
EEG CHANGES CAUSED BY LISTENING TO MOZART
N = 24
THETA
ALPHA
BETA 1
BETA 2
BETA 3
N = 28
SUBJECTS WITH MUSICAL TRAINING
N = 16
THETA
ALPHA
BETA 1
BETA 2
BETA 3
N = 7
SUBJECTS WITHOUT MUSICAL TRAINING
POWER
LOCAL
INTERHEM.
COHERENCE
MALES
POWER
LOCAL
INTERHEM.
COHERENCE
FEMALES

Local coherence changes were almost negligible in this heterogeneous group of listeners. As we hoped that the method would supply more information if this fairly multifaceted group of subjects were subdivided, we formed four subgroups of them according to sex and musical training.

Those subjects were subsumed as "musically trained" who had musical education on an instrument for at least five years, regardless their success. This seemed to be sufficient to assume that these persons would be familiar with the usual kind of Western classical music.

Figure 1 represents the findings in the four subgroups: above for subjects with and beneath for subjects without musical training, to the left for males and to the right for females. The probability maps are arranged in columns (the five frequency bands) and refer, from the left to the right, to power and to local and interhemispheric coherence changes.

As the changes of power are concerned, the following differences were found: generally, the alpha power reduction is larger in trained than in untrained subjects and also extends over larger regions of the brain; moreover, in trained subjects, this region extends over both hemispheres, whereas in untrained males alpha power reduction occurs only on the left temporal region; in untrained females there is no alpha reduction seen at all. Further sex differences concern the differences in the reduction in beta power - more beta 2 and less beta 1 in trained versus untrained subjects - which is only found in males, however.

Most impressive is the increase in interhemispheric coherence in beta 2 that occurs only in subjects with musical training and is more pronounced in females than in males and also covers a larger frequency range in the former. Furthermore, only men present increased interhemispheric coherence in the frontal areas. Increases of local coherence are only found in trained subjects.

These findings suggest that males and females, musically trained and untrained subjects, may use different basic brain strategies, when listening to music, some outcome of which may become visible in the EEG. In this context the question of the possible meaning of these findings arises.

The power decrease in the alpha and theta bands probably can be related to localized arousal effects necessary for the processing of musical information. This interpretation is also supported by farther extended zones of alpha and theta reduction in more expert subjects. The larger amount of coupling in beta 2 between the two hemispheres in trained subjects (which has not yet been found in any other test concerning music) could indicate a stronger functional interplay between regions required for the storage of musical memory (posterior part of the third temporal convolution) and the additional involvement of cortical structures, located parietally and required for recognizing spatial structures that play an important role when listening to music (Gardner 1983). The local coherence increases in the musically trained group cannot yet be interpreted satisfactorily.

Summary

The ongoing EEG was recorded from 75 students while listening to music for 1 min and compared with control periods of equal length before and afterwards. Statistically significant changes of the parameters power and coherence were calculated and represented topographically as probability maps. The differences between musically trained and untrained subjects as well as between males and females are discussed.

References

Beaumont JG, Mayer AR, Rugg MD (1978) Asymmetry in EEG alpha coherence and power: effects of task and sex. Electroencephalogr Clin Neurophysiol 45: 393-401

Edgington ES (1980) Radomization test. Marcel Dekker, New York

Gardner H (1983) Frames of mind: the theory of multiple intelligences. Basic Books, New York

Gasser T, Jennen-Steinmetz C, Verleger R (1987) EEG coherence at rest and a visual task condition for two groups of children. Electroencephalogr Clin Neurophysiol 67: 1512-1580

Jenkins GM, Watts DG (1968) Spectral analysis and its application. Holden Day, San Francisco

Petsche H, Pockberger H, Rappelsberger R (1986) EEG topography and mental performance. In: Topographic mapping of the brain. Butterworth, Stoneham 63-98

Petsche H, Pockberger H, Rappelsberger P (1986b) EEG-Studies in musical perception and performances. In: Spintge R, Droh R (eds) Music in medicine. Springer, Berlin Heidelberg New York, pp 53-80

Petsche H, Rappelsberger P, Pockberger H (1987) EEG-Veränderungen beim Lesen. In: Weinman HM (ed) Zugang zum Verständnis höherer Hirnfunktionen durch das EEG. Zuckschwerdt W, München, pp 59-74

Rappelsberger P, Krieglsteiner S, Mayerweg M, Petsche H, Pockberger H (1987) Probability mapping of EEG changes: application to spatial imagination studies. J Clin Monit 3: 320-322

Rappelsberger P, Pockberger H, Petsche H (1988) Computer supported EEG analysis in studies of cognitive processes. In: Willems JL, van Bemmel JH, Michel J (eds) Progress in computer-assisted function analysis. Elsevier, Amsterdam, pp 193-200

Program Generator Revisited: The Role of the Basal Ganglia in Language and Communication

C. W. Wallesch
Dept. of Neurology, Freiburg University, Hansastr. 9, 7800 Freiburg, FRG

Whether the basal ganglia play a role in language production has been controversial for almost a century. In 1866 Hughlings Jackson stated "that disease near the corpus striatum produces defect of expression to a great extent." During the following century only few neurologists took the presence of extracortical language representation into consideration (e.g. von Monakow 1914) until the role of the left thalamus in language functions was highlighted by the results of stereotaxic surgery (e.g. Bell 1968).

Kornhuber was probably the first who assigned a specific role in language production to the basal ganglia, namely that of a program generator (Kornhuber 1977). He repeatedly (e.g. 1974, 1977) drew attention to speech automatisms as a characteristic symptom of aphasia resulting from middle cerebral main stem occlusion with basal ganglia involvement.

This presentation of recent research of our group addresses the following questions:

1. Can basal ganglia aphasia be described as an impairment of motor programming?
2. Are speech automatisms a symptom of a program generator dysfunction?

In 1974 Kornhuber postulated that in voluntary action the area 4 and area 6 cortex integrate afferents both from the prefrontal and limbic cortex and from extracortical function generators. In his 1977 paper he extended the role of the basal ganglia in language production to that of a program generator. With respect to language function this model predicts in case of a basal ganglia lesion an aphasia syndrome with deficient programming of speech acts as its prominent symptom. Kornhuber further predicted that such patients should in their productions depend on external stimulation, and that the deficit should be transient (Kornhuber 1977).

With the advent of the CT scan in the middle 1970s the number of reports of left basal ganglia aphasia soared. Today most of these early reports seem rather worthless with respect to the question of basal ganglia participation in language functions because simple vascular anatomy was not taken into consideration (Wallesch and Papagno 1988).

Ideal subjects for the investigation of the role of the basal ganglia in language production are patients in whom cortical dysfunction is highly unlikely, namely patients with anterior choroideal infarctions, the anterior choroideal artery being a branch of the internal carotid, and patients with small hemorrhages confined to the caudate and lentiform nucleus.

Papagno and I (Wallesch and Papagno 1988) have recently reviewed the literature on basal ganglia aphasia and found altogether 20 patients with left anterior choroideal infarctions who had received language assessment. If these infarctions involve pallidum and knee of internal capsule, thus affecting the efferences from the basal ganglia, the patients, with very few exceptions, suffer from a transient aphasia. Their aphasia does not conform to one of the standard syndromes but is characterized by features of the so-called transcortical motor aphasia. This is an aphasic condition in which spontaneous speech is highly nonfluent. The patients are able only to utter a few stereotypical phrases with great effort, whereas reactive language, language in response to specific stimuli, such as repetition or naming is rapidly restored. Comprehension is only minimally affected. This aphasia syndrome obviously corresponds to Kornhuber's prediction.

The major efference of the basal ganglia projects to the ventral thalamus. A review of the literature on pathological and stereotaxic lesions (Wallesch and Papagno 1988) shows that lesions of the ventral thalamus predominate in cases of "thalamic aphasia." Aphasia resulting from ventral thalamic lesions is markedly different from the transcortical motor aphasia just described. Speech production is more fluent, with prominent semantic paraphasia (wrong choice of words), anomia, well-preserved syntax, and comprehension. Repetition is usually preserved. These patients superficially resemble Wernicke's aphasics in that they rather fluently produce errors which may render their utterances incomprehensible. The major difference is that the deficit resulting from thalamic pathology is quite focal, affecting the choice and generation of words. In Wernicke's aphasia, on the other hand, phonology, morphosyntax, and comprehension are also affected.

We tentatively explain these observations by the assumption of an integrative function of frontal cortical structures, probably area 6 in particular. Motor programs for the production of words probably do not have to be generated. Already Wernicke (1874) assumed that they were stored. Newly generated situation-related programs do not include the overlearned motor patterns of words but the selection among these. This selection seems to involve a thalamic gating which is controlled by the basal ganglia (Wallesch 1985). Furthermore, the selection processes do not affect the automatized regularization of language productions as is the case in paragrammatism - the thalamic patients' syntax is intact - but the intention-governed content, the wording of the message. In the case of basal ganglia output deficit, as demonstrated by the effects of anterior choroideal infarctions, the gating is disinhibited and transcortical motor aphasia results.

Speech automatisms/recurring utterances are inadequate stereotypical utterances which most frequently consist of neologisms (but words or phrases also occur as automatisms), and which are produced against the (presumed) intention of the patient. Typical examples of speech automatisms would be "dadada" or "zezin-zezin-zezin." Kornhuber (1974, 1977) originally viewed their presence as a symptom of a severe dysfunction of a program generator, which would have to be situated in the very periphery of linguistic processing or even at its interface with articulatory processes.

In an investigation aiming at throwing light upon the neuroanatomical, neurological, and neurolinguistic basis of automatism production Blanken, Haas, and I (Blanken et al. 1988; Haas et al. 1988; Wallesch et al. 1989) studied 59 unselected right-handed patients with a chronic (more than 3 months poststroke), single (no other vascular or nonvascular brain pathology), large (more than 2% forebrain volume) left middle cerebral infarction, 18 of whom were found to produce speech automatisms and 14 suffering from global aphasia without automatisms. Superimposed plots demonstrated a common lesion of the automatism patients in the area of supply of the lenticulostriate arteries (Fig. 1). However, most of the global aphasics without automatisms showed similar lesions.

Patient's age at the time of infarction turned out to be another important factor in the pathogenesis of speech automatisms. In patients below the age of 45, automatism production may occur in the acute phase but is rare afterwards, provided there is no diffuse or multiple pathology (Haas et al. 1988). In our patient group which consisted only of patients with middle cerebral infarctions, the infarct patterns (anterior, middle, posterior branches of the middle cerebral artery included in the infarction) did not change with patient's age. Lesions identical for size and patterns of vascular pathology frequently lead to global aphasia and speech automatisms in the elderly, but almost regularly to Broca's aphasia in younger persons. We therefore must assume that the brain changes with respect to its representation of language functions with advancing age, either in its organization or in its compensational capacity. A change in organization would be, for example an increasing focusation of representations, such as a theory of increasing lateralization with age would predict; a change in compensational capacity could be a direct result of aging or of the subclinical effects of a prolonged exposure to risk factors.

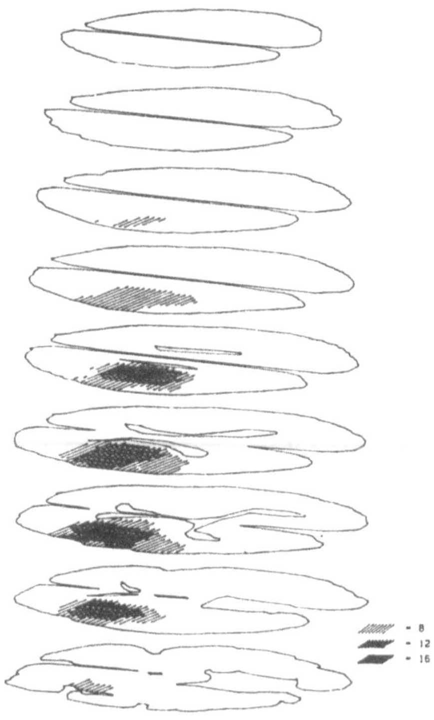

Fig. 1. Superimposed plot of the demarcated CT lesions in 18 patients with speech automatisms. Hatchings of increasing density indicate lesions shared by 8, 12, and 16 patients

If speech automatisms were a symptom in a deficit of motor programming, their presence should correlate with other motor programming impairments, such as ideomotor apraxia. We could establish a significant correlation, which was especially strong for facial apraxia, a finding which indicates a somatotopic representation of peripheral processes of the language production system (Wallesch et al. 1988).

The detailed neurolinguistic analysis of patients with speech automatisms revealed that this symptom is compatible with a rather mild degree of aphasia (Blanken et al. 1988). Patients whose oral production is confined to recurring utterances may be able to communicate by writing. These patients are very rare but extremely important for the localization of the generator of speech automatisms within the framework of a neurolinguistic language production model (Fig. 2). At present we think that there are probably three prototypical situations which can result in automatism production:

1. The periphery of the language production system - the phonological program generator (phonological segmental buffer) - is disconnected from its more central parts (for example because of severe damage to the latter). In this case the automatisms would be a symptom of the disinhibition of this component. Patients suffering from this type of disorder are severely aphasic but may be able to complete overlearned sequences, such as songs, prayings, or saying. These are automatized phonological strings which do not require linguistic assembly.
2. Phonological programming may be severely impaired and reduced to one stereotyped output, which condition is compatible with a mild degree of aphasia. These patients, then, would be able to write but not to complete overlearned material.
3. Stereotypical utterances may be a feature of a severe degree of speech apraxia, of a deficit of articulatory motor programming. These patients, too, are only mildly apha-

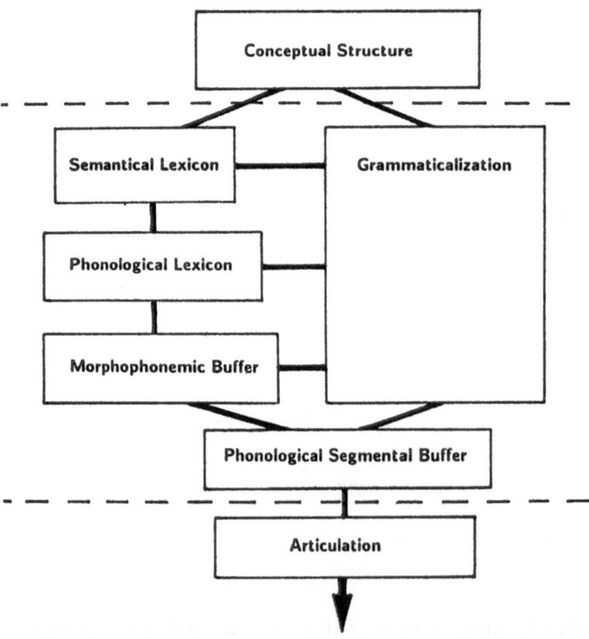

Fig. 2. Neurolinguistic model of language production. (From Blanken et al. 1988)

sic; their utterances are laborious, nonfluent, and accompanied by parapraxias until they fall into the trodden path of their automatism, which need not consist of speech sounds.

Most patients with speech automatisms, of course, present combinations or extensions of these prototypes.

What do our results mean with respect to the program generator and basal ganglia function? On the basis of our and other's data concerning aphasia syndromes with basal ganglia and thalamic lesions, we (Brunner et al. 1982, Wallesch and Papagno 1988) proposed the theory that a loop via the basal ganglia and ventral thalamus is involved in mapping intention upon linguistic functions. We believe that there are program generators involved in speech production. At present there are two in our model, namely an articulatory motor one with the defect symptom of speech apraxia and a prearticulatory phonological one which relates to the symptom of speech automatism. These are probably highly interactive, and their separation may be artificial and only due to traditional linguistic categories.

A recent theory of Alexander et al. (1986) proposes functionally distinct loops involving the basal ganglia. Our research indicates a participation of the basal ganglia in two distinct aspects of speech production: the intention-guided choice of linguistic elements and the generation of production programs. We think that these rely on different functions of the basal ganglia.

Summary

Kornhuber's (1977) proposals that (a) the basal ganglia act as a program generator in language production, and (b) that speech automatisms are a symptom of their dysfunction are reviewed in the light of recent research. The concept is supported and specified.

References

Alexander GE, DeLong MR, Strick PL (1986) Parallel organization of functionally segregated circuits linking basal ganglia and cortex. Annu Rev Neurosci 9: 357-381

Bell DS (1968) Speech functions of the thalamus inferred from the effects of thalamotomy. Brain 91: 619-638

Blanken G, Dittmann J, Haas JC, Wallesch CW (1988) Producing speech automatisms (recurring utterances) looking for what is left. Aphasiology 2: 545-556

Brunner RJ, Kornhuber HH, Seemüller E, Suger G, Wallesch CW (1982) Basal ganglia participation in language pathology. Brain Lang 16: 281-299

Haas JC, Blanken G, Mezger G, Wallesch CW (1988) Is there an anatomical basis for the production of speech automatisms. Aphasiology 2: 557-567

Jackson JH (1866) Notes on the physiology and pathology of language. Brain (1915) 38: 48-58

Kornhuber HH (1974) Cerebral cortex, cerebellum, and basal ganglia: An introduction to their motor functions. In: Schmitt FO, Worden FG (eds) The neurosciences, IIIrd study program. MIT Press, Cambridge, 11 A

Kornhuber HH (1977) A reconsideration of the cortical and subcortical mechanisms involved in speech and aphasia. In: Desmedt JE (ed) Language and hemispheric specialization in man: cerebral ERPs. Karger, Basel

von Monakow C (1914) Die Lokalisation im Grosshirn. Bergmann, Wiesbaden

Wallesch CW (1985) Two syndromes of aphasia occurring with ischemic lesions involving the left basal ganglia. Brain Lang 25: 357-361

Wallesch CW, Papagno C (1988) Subcortical aphasia. In: Rose FC, Whurr R, Wyke MA (eds) Aphasia. Cole and Whurr, London

Wallesch CW, Haas JC, Blanken G (1988) On the neurological status of speech automatisms and its significance for neurolinguistic models. Aphasiology 3:435-447

Wernicke C (1874) Der aphasische Symptomencomplex. Cohn and Weigert, Breslau

From Articular Nociception to Pain: Peripheral and Spinal Mechanisms

H.G. Schaible, R.F. Schmidt
Physiologisches Institut, Universität Würzburg, Röntgenring 9, 8700 Würzburg, FRG

Hyperalgesia and pain in joints are frequent companions of articular disorders. Pain sensations from acute or chronic arthritis are assumed to be elicited by sensory outflow from sensitized nociceptors. To obtain direct evidence on the exact changes in the sensory properties of individual nociceptive and nonnociceptive afferent units induced by inflammation we have developed a model of an acute experimental arthritis in the knee joint of the cat. With this model it is possible to study the articular sensory outflow at the single unit level but also to analyze the factors responsible for the modification of the response characteristics in the course of inflammation. Furthermore, the model provides a testing ground for peripherally acting analgesics, since it offers the interesting possibility of observing the effects of drugs on individual afferent units from inflamed joints. In addition, the impact of the afferent input both from normal and from inflamed joints on various types of spinal neurons can be studied with this approach. This brief report presents our most recent progress in this endeavor. We believe that these experiments provide useful evidence for furthering our understanding of the mechanisms of arthritic pain. It remains for future experiments to determine the pathways involved in transmitting sensory information to the thalamus and cerebral cortex and to clarify whether supraspinal centers operate in the same or a different manner from those responsible for triggering spinal reactions to painful stimuli. For details of our work the reader is referred to the references included in this communication.

Innervation of the Cat's Knee Joint

The knee joint of the cat is innervated mainly by two nerves, the medial and posterior articular nerves (MAN, PAN). The MAN is a branch of the femoral nerve. It innervates mainly the medial and anterior aspects of the joint. The PAN branches from the tibial nerve to innervate mainly the dorsal and lateral aspects. The majority of the nerve fibers in both nerves are unmyelinated (group IV or C fibers). Each of these nerves contains some 1000 unmyelinated fibers of which roughly 50% are afferent (the others are sympathetic efferents) (Heppelmann et al. 1988b; Langford and Schmidt 1983). In addition, there are about 200 myelinated fibers in each nerve of which again at least 50% are of the fine variety (group III or A delta fibers). The others are thick myelinated fibers (group II or A beta) (Heppelmann et al. 1988b). The fine afferents (group III and IV) terminate with "free nerve endings" in the joint tissue (Heppelmann et al. 1988a); those with group II afferents end in corpuscular structures of the Ruffini and Golgi-Manzoni types.

Functional Properties of Articular Receptors

Group II articular afferent units (conduction velocity >20 m/s) have no resting discharges and low mechanical thresholds to movement. They are probably not involved in joint nociception and pain. In the population of group III (conduction velocity 2.5-20 m/s) and group IV (<2.5 m/s) afferent units, resting discharges occur in no more than one-third of all fibers. The frequency of these irregular discharges is low, usually below 0.5 Hz (Schaible and Schmidt 1983a).

In their response behavior to passive movements of the knee joint, the fine afferent units (groups III, IV) can be grouped together according to their movement sensitivity ranging from units that are readily or only marginally activated by nonnoxious events through units that are activated only by noxious movements to units which despite being undoubtedly afferent are not activated by any movements, even extremely noxious ones. According to our measurements, approximately 70% of the group III and 90% of the group IV units either respond only to (potentially) noxious movements or fail to respond to movements completely (Grigg et al. 1986; Schaible and Schmidt 1983b).

The units that are excited within the motility range cannot be considered nociceptive. Fibers that are activated only by noxious motion stimuli can be referred to as nociceptors as defined by the specificity theory. The function of the afferent units which in the normal joint are not excited by movements remains obscure (but see below).

Effects of an Experimental Arthritis on Fine Articular Afferents

Method of Inflammation. The acute inflammation is induced in two steps. First, a 4% solution of kaolin suspended in distilled water is injected into the knee joint cavity using a lateral approach. Thereafter the knee joint is rhythmically bent and extended for 15 min. Finally, 0.15-0.3 ml of a 2% solution of carrageenan is injected into the joint cavity, and the joint is again flexed and extended for another 5 min. This procedure leads to a long-lasting inflammation with behavioral changes of the awake cat and histological signs of acute severe inflammation with cellular infiltration (Coggeshall et al. 1983; Grigg et al. 1986; Schaible and Schmidt 1985, 1988b).

Effects on Resting Activity. In the inflamed joint, there are clear changes in the resting and evoked discharge characteristics of fine articular afferents as observed both in single units which were observed under normal or under inflamed conditions (Coggeshall et al. 1983; Grigg et al. 1986; Schaible and Schmidt 1985) and in individual primary articular afferents which were continuously observed under normal conditions and while the joined became inflamed (Schaible and Schmidt 1988b). During inflammation resting activity is observed in 75% of group III and 83% of group IV units in the MAN. The discharges are irregular and sometimes of high frequency. Both the percentages of units with resting activity and the frequencies of their discharges are more than twice as high as in the control sample. Their resting activity might represent the neural correlate of spontaneous pain, for the large number of fibers exhibiting resting activity under inflammatory conditions also implies (and has been confirmed by direct observation (Schaible and Schmidt 1988b)) that in any case there are numerous nociceptors (usually silent in the normal joint) that exhibit resting activity under these conditions.

Effects on Movement Sensitivity. Nearly all fine afferent units from inflamed joints have low thresholds to movement, and most of them respond well to flexion and extension. The increase in the number of easily excitable afferents corresponds to the clear decrease in the number of units belonging to the other classes (Coggeshall et al. 1983; Grigg et al. 1986; Schaible and Schmidt 1988b). In experiments using continuous observation of single units our population studies were confirmed, namely that the fibers which in the normal joint respond mainly or exclusively to noxious movements largely become very sensitive to movement when the joint is inflamed. The message "noxious" is now sent to the central nervous system (CNS), even if movement takes place in the normally innocuous range (Schaible and Schmidt 1988b).

Also, it is striking that the remaining fraction of fibers that do not respond to movement is only very small in the inflamed joint (Coggeshall et al. 1983; Grigg et al. 1986; Schaible and Schmidt 1988b). It follows from this that most of these units must have become sensitive to movement, as again confirmed in continuous observations (Schaible and Schmidt 1988b). This makes clear that units which do not respond to movement in the normal joint must also be regarded as nociceptors. Thus, they respond to movement only in the case of true tissue damage. This recruitment of a large population of fibers during the inflammatory process leads to a massive amplification of afferent signals reaching the CNS.

There are interesting differences in the time course of sensitization of the various types of articular afferents which were revealed by the continuous observation of single units (Schaible and Schmidt 1988b). Low threshold units, mainly in the groups II and III fiber range, developed increased reactions mainly in the first hour after the injection of the inflammatory compounds, sometimes starting immediately after the injection (usually low-threshold group II units). In high-threshold afferents, including originally silent ones ("sleeping nociceptors") the sensitization becomes obvious within the second to third hours after induction of inflammation with a further increase later on.

Neurons in the Spinal Cord with Input from the Knee Joint

The projection of the joint afferents to the spinal cord was studied using the transganglionic transport of horseradish peroxidase as a tracer. Afferent fibers of the MAN and the PAN enter the cord via the dorsal roots L5 to S1. They form two termination fields in the gray matter of the cord: one in the superficial dorsal horn (lamina I) and a second one in the deep dorsal horn (in the laminae V, VI, and VII). The projection of these afferents is most dense in the segments adjacent to the roots of entrance, but, interestingly enough, projections of fibers are found up to the lower thoracic cord and down to the sacral cord, showing that the afferents from the knee excert synaptic connections in all caudal segments (Craig et al. 1988).

In electrophysiological studies on spinalized cats we analyzed the response properties of spinal neurons with input from the joint. We found mainly two types of neurons: (a) cells with convergent afferent input from the knee, muscles and the skin of the leg; often the receptive fields are large, extending to one or even both hindlimbs; generally these neurons respond to innocuous compression of knee and muscles and to movements in the working range of the knee, but they have a much stronger response to noxious stimuli, thus encoding the nature of stimulus (innocuous versus noxious) by their firing frequency (wide dynamic range neurons); (b) neurons which are excited only by noxious compression of knee and muscles and noxious movement of the knee joint but not by innoxious movements, light pressure, or stimulation of the skin. These latter neurons form a population of nociceptive specific spinal neurons. Both types of neurons are either ascending tract cells or nonascending interneurons (Schaible et al. 1986, 1987a).

Change of the Responses of Spinal Neurons During an Arthritis

Both types of neurons described above reacted in a typical way to the development of inflammation in the knee. Nociceptive-specific neurons without responses to innocuous movements in the normal joint started to respond to such movements when the joint was becoming inflamed. Significant reactions commenced within the second to third hours after induction of inflammation, and the size of the responses increased progressively. Wide dynamic range neurons with responses to innoxious movements of the knee already prior to inflammation showed enhanced reactions in the course of the arthritis (Schaible et al. 1987b; Schmidt et al. 1987).

Thus, an arthritis modifies significantly the discharges of different types of spinal neurons with joint input. As these neurons had ascending actions, and as the time course of the alteration of the responsiveness matched the behavioral changes seen in awake cats, they may contribute to the central mechanisms which lead to the sensation of pain and/or the corresponding "pain reactions" during an acute arthritis.

The enhanced discharges of the spinal neurons probably reflect the processing of afferent inpulses coming from the inflamed knee. But in several instances we found that in neurons with a bilateral input also the stimulation of the normal leg led to stronger reactions than prior to inflammation. This indicates that centrally originating mechanisms may contribute to the changed responsiveness of spinal neurons (Neugebauer and Schaible 1988).

Effects of a Developing Arthritis on Alpha- and Gamma-Motoneurons

From the results reported above it must be expected that the sensory inflow from the knee joint may cause reflex changes in motoneuron activity particularly in the course of articular inflammation. It has indeed been found in a series of experiments in which the responses of flexor motoneurons were tested before and during an acute inflammation that a significant fraction of both alpha- and gamma-motoneurons showed a postinflammatory increase in receptive field and an altered pattern of response to flexion and extension movements of the leg. The changes observed in the majority of motoneurons were consistent with a lowering of their reflex threshold. The findings are, in general, consistent with a flexor reflex pattern (He et al. 1988a).

Actually, inflammation-induced changes in activity of motoneurons included both excitatory and inhibitory effects. The inhibitory effects, which as the excitatory ones took 1-2 hours to develop fully, included falls in resting discharge and/or in the responses to leg movements. The inhibitory effects were particularly prominent in gamma-motoneurons: 24 of 25 units with control responses had their activity modified by the inflammation, and this included 16 units with excitatory effects and 8 units with inhibitory ones (He et al. 1988a).

A possible interpretation of these motor effects comes from incidental observations of animals with an injured or diseased limb joint: this limb is often held in a semiflexed position. This is the natural resting position of the hind limb of the cat, and for the knee joint the natural resting position is one where afferent feedback from joint receptors is at a minimum. It seems reasonable to assume that the least painful position is sought for the joint, and that it is immobilized in that position. This could be achieved by a co-contraction with some flexor bias, respresenting a balance between excitatory and inhibitory influences on motoneurons. The fact that both alpha- and gamma-motoneurons are involved suggests that the two are coactivated.

Conclusions

Relation Between Articular Nociception and Pain. In inflamed joints the sensitization and the resulting increase in activity of fine articular afferents with high mechanical thresholds (nociceptors and "sleeping" nociceptors) are mainly responsible for inducing the three major types of joint pain during an acute arthritis, namely pain at rest, pain during movement, and pain following local pressure on the joint. In contrast, units with low mechanical thresholds already in normal joints contribute little nociceptive information. Instead, they may play a role in deep (nonnoxious) pressure sensation and in signaling that the joint is about to leave its working range. Probably long before reaching consciousness these warning signals induce motor reflexes counteracting the excess of movement, thus preventing joint damage.

Mechanism of Arthritic Sensitization. Many factors are thought to contribute to the sensitization and activation of nociceptors during inflammation. But most of the evidence on which these assumptions are based is indirect and far from conclusive. For instance, it is not known which effect the edema, i.e., the increase in local tissue pressure, may have on the unit's excitability. The increase in temperature in an inflamed joint may also contribute to the increased responsiveness of articular fine afferent units. Finally, chemical factors are usually taken as most likely candidates for sensitization and inflammation. This assumption is supported by our findings that all types of fine articular units can be excited by algesic substances such as bradykinin or prostaglandins. In addition, these and other substances are able to modify the mechanosensibility of fine articular units for long periods of time both in normal and, in a much more pronounced way, in inflamed joints (He et al. 1988b; Heppelmann et al. 1986; Kanaka et al. 1985; Schaible and Schmidt 1988a).

Spinal and Supraspinal Effects of Articular Inflammation. As the afferent inflow of the spinal cord is changed in a characteristic way by inflammation, it is assumed that the spinal responses are triggered by afferent fibers, but there is also some evidence that centrally originating factors contribute to the altered responses of spinal sensory and motoneurons during inflammation. For instance, the observed altered responsiveness of spinal neurons to stimulate other parts of the legs after the inflammatory process has taken place requires further explanation. We assume that in the course of arthritis the sensitivity of the spinal neurons themselves is increased, and that now previously subthreshold afferent inputs are able to excite these neurons. As the time course of the alterations of the responses to stimuli adjacent to and remote from the joint matches the sensitization of the joint afferents, a key role of the afferent inflow from the inflamed joint in this process is likely.

References

Coggeshall RE, Hong A, Langford LA, Schaible HG, Schmidt RF (1983) Discharge characteristics of fine medial articular afferents at rest and during passive movements of inflamed knee joints. Brain Res 272: 185-188,

Craig AD, Heppelmann B, Schaible HG (1988) The projection of the medial and posterior articular nerves of cat's knee to the spinal cord. J Comp Neurol 276: 279-288,

Grigg P, Schaible HG, Schmidt RF (1986) Mechanical sensitivity of group III and IV afferents from PAN in normal and inflamed cat knee, J Neurophysiol 55: 643-653,

He X, Proske U, Schaible HG, Schmidt RF (1988a) Acute inflammation of the knee joint in the cat alters responses of flexor motoneurons to leg movements. J Neurophysiol 59: 326-340,

He X, Schmidt RF, Schmittner H (1988b) Effects of capsaicin on articular afferents of the cat's knee joint. Agents Actions 25: 221-224,

Heppelmann B, Meßlinger K, Schmidt RF (1988a) Morphological characteristics of the innervation of the cat's knee joint. Agents Actions 25: 221-224,

Heppelmann B, Heuß C, Schmidt RF (1988b) Fiber size distribution of myelinated and unmyelinated axons in the medial and posterior articular nerves of the cat's knee joint. Somatosensory Res 5: 273-281,

Heppelmann B, Pfeffer A, Schaible HG, Schmidt RF (1986) Effects of acetylsalicylic acid and indomethacin on single groups III and IV sensory units from acutely inflamed joints. Pain 260: 37-351,

Kanaka R, Schaible HG, Schmidt RF (1985) Activation of fine articular afferent units by bradykinin. Brain Res 327: 81-90,

Langford LA, Schmidt RF (1983) Afferent and efferent axons in the medial and posterior articular nerves of the cat. Anat Rec 206: 71-78,

Neugebauer V, Schaible HG (1988) Peripheral and spinal components of the sensitization of spinal neurons during an acute experimental arthritis. Agents Actions 25: 234-236,

Proske U, Schaible HG, Schmidt RF (1988) Joint receptors and kinaesthesia. Exp Brain Res 72: 219-224,

Russell NJW, Schaible HG, Schmidt RF (1987) Opiates inhibit the discharges of fine afferent units from inflamed knee joint of the cat. Neurosci Lett 76: 107-112,

Schaible HG, Schmidt RF (1983a) Activation of groups III and IV sensory units in medial articular nerve by mechanical stimulation of knee joint. J Neurophysiol 49: 35-44,

Schaible HG, Schmidt RF (1983b) Responses of fine medial articular nerve afferents to passive movements of knee joint. J Neurophysiol 49: 118-126,

Schaible HG, Schmidt RF (1985) Effects of an experimental arthritis on the sensory properties of fine articular afferent units. J Neurophysiol 54: 1109-1122,

Schaible HG, Schmidt RF (1988a) Excitation and sensitization of fine articular afferent units from cat's knee joint by prostaglandin E_2 (PG E_2). J Physiol (Lond) 403: 91-104,

Schaible HG, Schmidt RF (1988b) Time course of mechanosensitivity changes in articular afferents during a developing experimental arthritis. J Neurophysiol 60: 2180-2195,

Schaible HG, Schmidt RF, Willis WD (1986) Responses of spinal cord neurons to stimulation of articular afferent fibres in the cat. J Physiol (Lond) 372: 575-593,

Schaible HG, Schmidt RF, Willis WD (1987a) Convergent inputs from articular, cutaneous and muscle receptors into ascending tract cells in the cat spinal cord. Exp Brain Res 66: 479-488,

Schaible HG, Schmidt RF, Willis WD (1987b) Enhancement of the response of ascending tract cells in the cat spinal cord by acute inflammation of the knee. Exp Brain Res 66: 489-499,

Schmidt RF, Schaible HG (eds) Fine afferent nerve fibers and pain. VCH Verlagsgesellschaft, Weinheim, pp 1-499

Part 4

Evolution

The Evolution of Cerebral Asymmetry

J. C. Eccles

Anatomically Observed Asymmetries

There is a long history of conflicting observations on anatomical asymmetry, but the significant list is now very short. Such features as size, weight and gyral ramifications of the cerebral cortex disclose no significant asymmetry, which is amazing in the light of the asymmetries in functional properties for the left and right cerebral hemispheres.

No asymmetry has been observed for the monkey and baboon brains (Wada et al. 1975). With the ape brain there is probable asymmetry for the Sylvian fissure (f. Sylv. in Figs. 1,2) which tends to be higher in the right hemisphere (Le May and Geschwind 1975), and the left is slightly longer on average (45.7 mm) than the right (43.7 mm; Yeni-Konshian and Benson 1976).

With the human brain Rubens (1977) has reported a very careful study on the Sylvian fissure that was usually (25 out of 36 times) found to angulate upwards on the right side earlier than on the left side, where it continued on the average for another 1.6 mm before angulation. This asymmetry relates to the asymmetry of the planum temporale (Geschwind and Levitsky 1968), which is subjacent to the Sylvian fissure. It is larger on the left side in 65%, which is much less than the extreme asymmetry of speech, which has left hemisphere representation in about 95%.

Wada et al. (1975) have confirmed the usual asymmetry for adult brains and furthermore have demonstrated it in 100 brains of human babies and foetuses even as young as 29 weeks.

Functional Asymmetries

For monkeys one of the simplest indications for any cerebral asymmetry that may exist would be provided by hand preference. There have been many investigations. Hamilton (1977a, 1977b) reports very careful studies on the handedness of monkeys, which generally turn out to be symmetrical.

Unfortunately there does not seem to be a rigorous study of the handedness of chimpanzees. One can assume that normally they are ambidextrous from the long detailed account of grooming (Goodall 1986 pp. 387-408) in which there was no mention of hand preference, nor does it appear in the extensive photographic illustration.

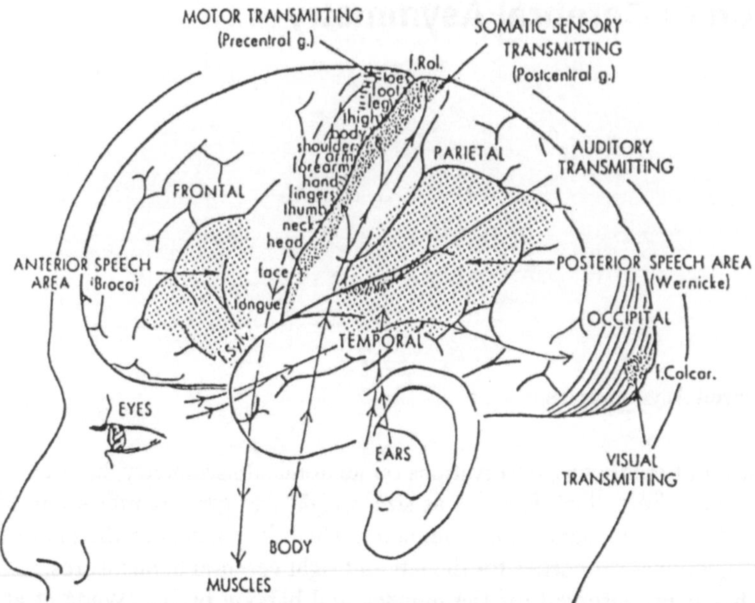

Fig. 1. The motor and sensory transmitting areas of the cerebral cortex. The approximate map of the motor transmitting areas is shown in the precentral gyrus, while the somatic sensory receiving areas are in a similar map in the postcentral gyrus. Actually the toes, foot and leg should be represented over the top on the medial surface. Other primary sensory areas shown are the visual and auditory, but they are largely in areas screened from this lateral view. Also shown are the speech areas of Broca and Wernicke

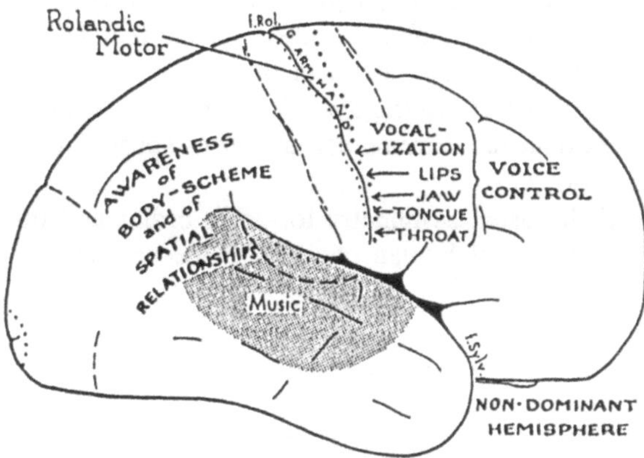

Fig. 2. Map of some functions of the right or minor hemisphere. The Rolandic (precentral) motor area matches that of the left hemisphere (Fig. 1). Proposed area for musical information processing is shown in the (right) temporal lobe. (After Penfield and Roberts 1959)

In contrast to nonhuman primates, there has been in hominid evolution an enormous development of asymmetries in the functions of anatomic symmetrical zones of the left and right hemispheres. The outstanding asymmetry is for the speech areas. In Fig. 1 it is seen that large parts of the left parietal and temporal lobes are specialized for the semantics of speech recognition and production (the Wernicke area). Yet the mirror image areas of the right hemisphere (Fig. 2) have very little functional relationship to speech. Similarly the mirror image of the Broca area in the right inferior frontal lobule appears not to be used in speech production. We may ask: what is the function of these mirror-image areas in the right hemisphere? Is there some balancing functional asymmetry? To answer these questions we turn to clinical evidence derived from local excisions of the right and left hemispheres. The parietal lobe is of particular interest.

Attention is focussed on Brodmann areas 39 and 40 (Fig. 6), which on the left side are concerned with specific speech functions. Area 39 is concerned in the conversion of visual inputs (writing or printing) into meaning, while area 40 is probably concerned with auditory inputs. Several functional disorders have been recognized in clinical studies of patients with lesions: somatagnosia, apraxia and disorders of the apprehension of spatial data.

If the right parietal lobe is damaged, the patients exhibit the most bizarre behaviour patterns, which Hecaen (1967) appropriately refers to as a "pantomime of massive neglect". The subject may neglect or deny the existence of the contralateral limbs and even neglect to clothe them. There is often failure to observe objects in the contralateral half-field of visual space. The contralateral limbs are rarely moved, yet are not paralysed. The patient tends to withdraw from and avoid the contralateral half-field of space. Yet despite this pantomime of neglect the patient may deny that he is ill at all! A remarkable case of neglect of the left side has been described by Jung (1974) in a painter whose self-portraits after the lesion were almost restricted to the right side of the face.

The right parietal lobe is specially concerned in the handling of spatial data and in a non-verbalized form of relationship between the body and space. It is especially concerned in spatial skills. Lesions result in loss of skills dependent on finely organized movements, apraxia. The disability in handling spatial data appears in writing, where the lines are wavy with the words unevenly spaced and often deformed by perseveration, e.g. the proper double letter appearing tripled, as "lettter" for letter. There seem also to be more subtle disorders of linguistic expression, with deteriorations in fluency and in vocabulary. Patients suffer from an abnormal level of linguistic fatigue.

The lesional evidence for the representation of musical appreciation in the minor hemisphere has been corroborated by the evidence derived from dichotic listening tests on normal subjects (Kimura 1973). Bever and Chiarello (1974) made a most interesting contribution to the hemispheric participation in music. Using musical inputs to the right and left ears, and so predominately to the left and right hemispheres, they discovered that musically experienced listeners recognized melodies better with the right ear, while it was the reverse with naive listeners. However, for musical appreciation, which is a gestalt, the left ear input to the right hemisphere was preferred by all listeners. Thus there is support for the hypothesis that the left hemisphere is dominant for analytic processing and the right for holistic processing.

The most direct testing for the cerebral locations of musical interpretation can be carried out by the radio-xenon testing for rCBF, the regional cerebral blood flow. The testing procedure was to present to the subject via one earphone two brief tone rhythms in quick succession. The task of the subject was to decide whether they were similar or different. The reporting was done after the rCBF period, the subject otherwise being at complete rest. The earphone input was always contralateral to the cortical study. Fig. 3 shows that the right hemisphere was much more activated, particularly in the temporal lobules and the inferior and middle parietal lobules. This total area corresponds well to the posterior speech area of Wernicke on the left hemisphere (Fig. 1). In Fig. 3 there was much less activation of the left hemisphere (Roland et al. 1976).

These observations provide an important experimental basis for the large area labelled "music" in Fig. 2. The functional asymmetry of the human cerebral cortex is a most important development in hominid evolution. The evidence from apes and monkeys suggests that our hominoid ancestors had symmetrical brains. Asymmetry is unique to hominids.

The most significant investigations on functional asymmetries of the human brain have been carried out by Sperry and associates (1979) in the study of commissurotomized patients in which there was a complete section of the corpus callosum.

Figure 4 is a diagram drawn by Sperry several years ago (1974). It is still valuable, however, as a basis of discussion of the whole split-brain story. The diagram illustrates the right and left visual fields with their highly selective projections to the crossed visual cortices, as indicated by the letters R and L. Also shown in the diagram is the strictly unilateral projection of smell and the predominantly crossed projection of hearing. The crossed representations of both motor and sensory innervation of the hands are indicated.

In general, the dominant hemisphere is specialized in respect to fine imaginative details in all descriptions and reactions, i.e. it is analytical and sequential. Also, it can add, subtract, multiply and carry out other computerlike operations. But, of course, its dominance derives from its verbal and ideational abilities and liaison to self-consciousness. Because of its deficiencies in these respects, the minor hemisphere deserves its

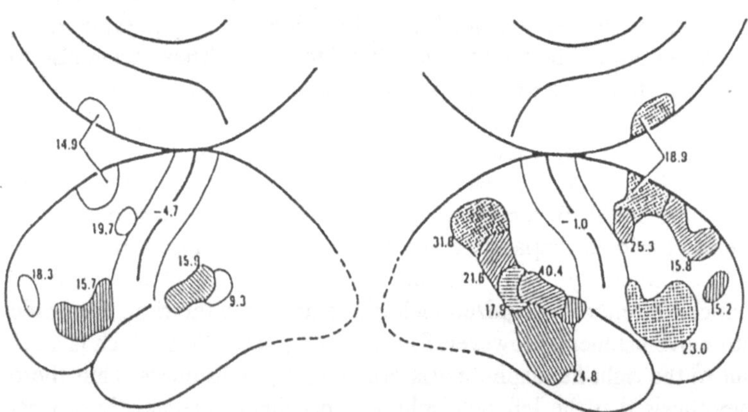

Fig. 3. Mean increases of rCBF in percent and their average distribution during auditory discrimination of temporal tone patterns. Inputs to contralateral ears. Cross-hatched areas have rCBF increases significant at the 0.0005 level (Student's test, one-sided significance level). Hatched areas: $P < 0.005$, other areas shown: $P < 0.05$. Left: left hemisphere of six subjects. Right: right hemisphere of six subjects. (From Roland and Lassen 1976)

title, but in many important properties it is preeminent, particularly in respect to its spatial abilities, with a strongly developed pictorial and pattern sense. For example, the minor hemisphere programming the left hand is greatly superior in all kinds of geometric and perspective drawings. This superiority is also evidenced by the ability to assemble coloured blocks so as to match a mosaic picture. The dominant hemisphere is unable to carry out even simple tasks of this kind and is almost illiterate in respect to pictorial and pattern sense, at least as displayed by its copying disability. It is an arithmetic hemisphere, but not a geometric hemisphere (see Fig. 5).

We think that in the light of these recent investigations by Sperry et al. (1979), there is some self-consciousness in the right hemisphere, but that it is of a limited kind and would not qualify the right hemisphere to have personhood. Thus, the commissurotomy has split a fragment off from the self-conscious mind, but the person remains apparently unscathed with mental unity intact in its now exclusive left hemisphere association. However, it would be agreed that emotional reactions stemming from the right hemisphere can involve the left hemisphere via the unsplit limbic system; hence the person also remains emotionally linked to the right hemisphere.

The structure of the speech areas was sufficiently distinctive to Brodmann, who in 1909 on subtle histological grounds constructed a map of the human cerebral cortex with separately identified areas (Fig. 6). As defined by Penfield and Roberts (1959), the

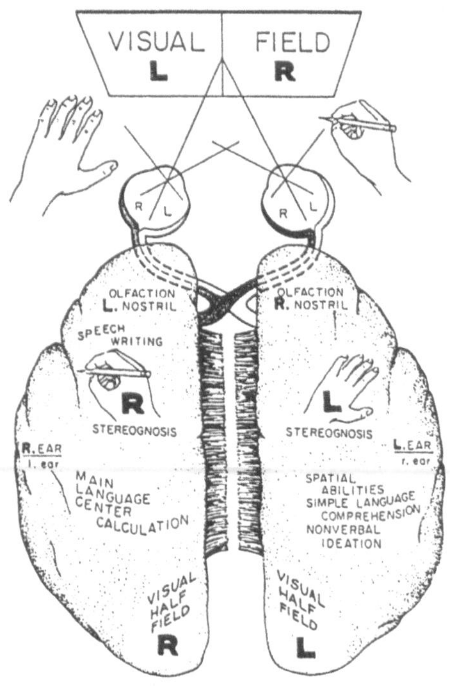

Fig. 4. Schema showing the way in which the left and right visual fields are projected onto the right and left visual cortices, respectively, due to the partial decussation in the optic chiasma. The schema also shows other sensory inputs from right limbs to the left hemisphere and that from left limbs to the right hemisphere. Similarly, hearing is largely crossed in its input, but olfaction is ipsilateral. The corpus callosum is cut. (From Sperry 1974)

DOMINANT HEMISPHERE	MINOR HEMISPHERE
Liaison to self consciousness	Liaison to consciousness
Verbal	Almost non-verbal
Linguistic description	Musical
Ideational Conceptual similarities	Pictoral and Pattern sense Visual similarities
Analysis over time	Synthesis over time
Analysis of detail	Holistic — Images
Arithmetical and computer-like	Geometrical and Spatial

Fig. 5. Various specific performances of the dominant and minor hemispheres as suggested by the new conceptual developments of Levy-Agresti and Sperry (1968) and J. Levy (1978). There are some additions to their original list

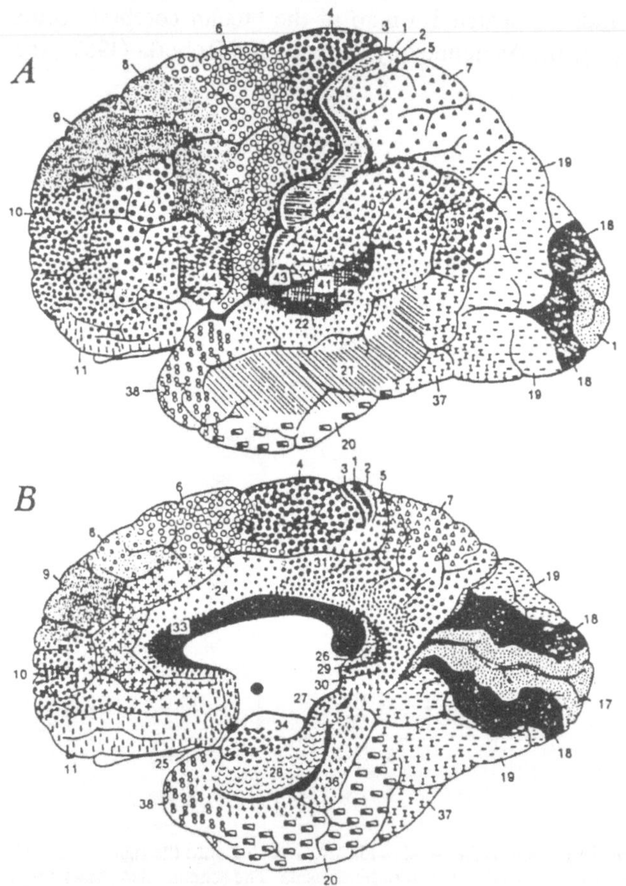

Fig. 6. Brodmann's cytoarchitectural map of the human brain. The various areas are labelled with different symbols and their number indicated by figures. **A** Lateral view of left hemisphere. **B** Medial view of right hemisphere. (From Brodmann 1909, 1912)

Wernicke area in its most extensive compass includes areas 39, 40, the posterior parts of areas 21 and 22, and part of area 37, while the Broca area includes areas 44 and 45.

The only equivalent map for an anthropoid ape was carried out by Brodmann's pupil Mauss (1911) and is shown in Fig. 7 for the orang-utang (*Pongo*). Most of the specialized speech areas of Fig. 6 are missing. At the most there are for the Wernicke area the posterior parts of areas 21 and 22 and a small area labelled 38 (39) and 40 peeping out from the Sylvian fissure, while Broca's area (44, 45) is not recognized. Unfortunately there is no equivalent cortical map for the chimpanzee brain.

It is of great interest that Flechsig (1920) found areas 39 and 40 to be the latest to myelinate among all areas on the convexity of the human cortex (Fig. 9). Myelination is delayed until after birth, and dendritic development and cellular maturation may not be completed until late childhood. These findings indicate that areas 39 and 40 developed phylogenetically as a new region of the cortex (Geschwind 1965; Tobias 1983), a conclusion that is in agreement with their virtual absence in Fig. 7. Comparison of the human brain with the ape brains indicates that this late development of areas 39 and 40 has displaced the visual areas backwards by a powerful evolutionary "force".

Hominid Evolution

The question now arises: why did functional asymmetry develop in hominid evolution? Fig. 8 illustrates the change in skulls during hominid evolution. Already *Australopithecus africanus* had a mean skull capacity of about 450 cm^3, which was a distinct advance from the hominoid ancestors, if the chimpanzee (mean 394 cm^3) can be taken as a model. On the basis of albumen dating it is assumed that the split between the pongid and hominid evolutionary lines occurred about 9.2 million years ago. The relatively small skull of *Australopithecus* did not change appreciably for 2 million years. Then came *Homo habilis* about 2.5 million years ago in a cladistical branching and with a large increase in cerebral capacity to a mean value of 646 cm^3 (Tobias 1983), a 44 % increase.

A careful study of the "brain" endocasts derived from the skulls showed that *Homo habilis* had a brain with special developments in the inferior parietal and inferior frontal lobules, which suggests the existence of speech areas (Tobias 1987). There could be no other evolutionary cause for the remarkably large development than the enormous demand for specialized mechanisms in cortical functions, as evidenced by the large speech areas of the human cortex (Fig. 1).

The Hypothesis of the Neo-neocortex

It is proposed to follow Flechsig (1920) regarding the time of myelination in the neocortex (Fig. 9) as indicating the evolutionary age. Furthermore, the delayed myelination is matched by the delayed development of the neurons and dendrites and possibly also of the synaptic inputs (Geschwind 1965, p. 273).

Areas 39 and 40 are of special interest. Comparison of the Brodmann maps for ape brain (Fig. 7) with human brain (Fig. 6) shows that rudiments of areas 39 and 40 possi-

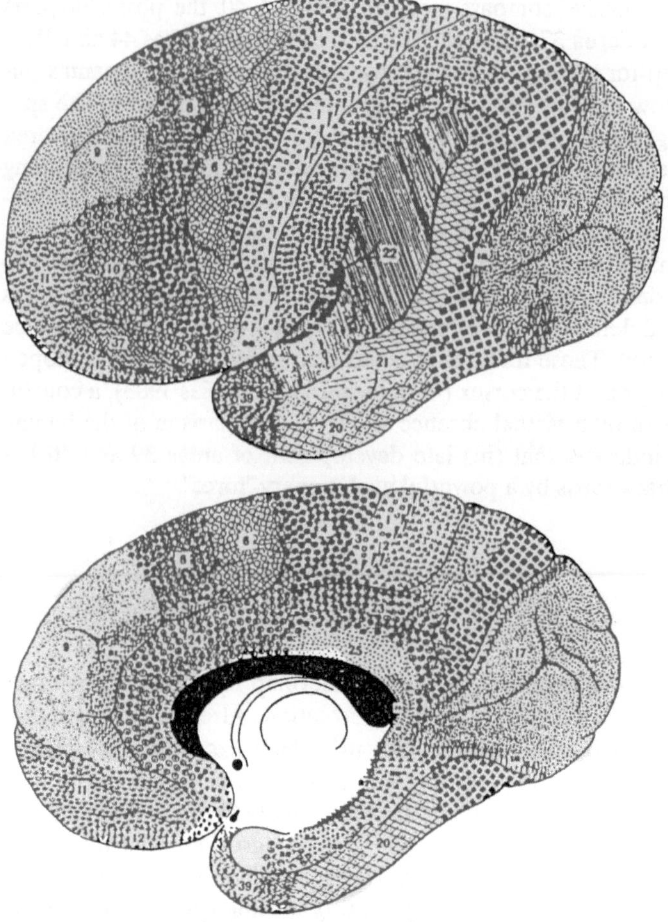

Fig. 7. Cytoarchitectonic map for the brain of an orang-utan with much the same conventions as in Fig. 6. (From Mauss 1911)

bly are recognizable in the ape. The large development in phylogenesis correlates well with the delayed myelination in ontogenesis (Fig. 9).

In Fig. 9 it can be seen that there are large prefrontal areas and inferior temporal areas of late myelination. If these areas are all late evolutionary developments, a considerable proportion of the cortical enlargement can be allotted to these new areas, which we may call neo-neocortex. The other side of the coin is the almost unchanged sizes of the primary sensory and motor areas in hominid evolution, for example area 17 in Figs. 6 and 7.

It is proposed that the neo-neocortical areas have developed in evolution for the special gnostic functions, as defined by Sperry (1982): cognitive and intellectual functions, percepts, mental images, associations and ideas. This list could be extended indefinitely. Such gnostic funtions are essential features of hominid evolution. Hitherto it was generally believed that the superior gnostic performance of the human brain was due to its magnitude. This is a crude belief with no redeeming creative idea. On the

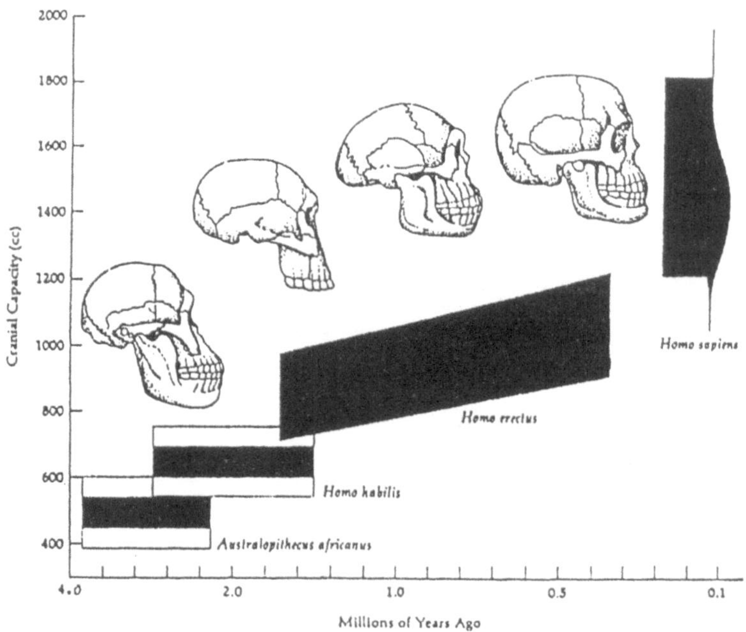

Fig. 8. Chart showing the increase in cranial capacity during the past 4 million years as shown for skulls of *Australopithecus africanus*, *Homo habilis*, *Homo erectus*, *Homo sapiens*. The shaded quadrilateral represents the time in which each species lived and its brain size. The *Homo sapiens* rectangle is for *HSN* and *HSS*. The dark spindle represents range and cranial capacity in modern *HSS* (Stebbins 1982)

contrary it is proposed that the outstanding functions of the human brain derive from the neo-neocortex which presumably would be negligible in the most advanced hominoids, on analogy with the ape model (Fig. 7). Areas 39 and 40 are the most clearly defined neo-neocortex, but the middle prefrontal and the inferior temporal lobules (Figs. 6,7) also qualify. It can be postulated that there must be some subtle microstructural features in these neo-neocortical areas. This is a great challenge to neuroanatomists. The neo-neocortical areas are undoubtedly the structural base for many of the asymmetries, for example, language on the left side and spatial construction and music on the right side.

In the symmetry of hominoids the economy of the neocortex was characterized by the duplication of all neocortical functions. In hominid evolution we can assume that there was an overwhelming demand for more neuronal networks of exquisite design in order to carry out the large evolutionary requirements, especially for the higher levels of language. Hence arose the evolutionary strategy of no longer building more neocortex with dual representation but having a single representation on one or other side with the consequent asymmetry of the neo-neocortex, and functional unification by the greatly enlarged corpus callosum. The great success of hominid evolution was assured by this economy that potentially almost doubled the cortical capacity (Levy 1977). Cortical asymmetry is the evidence of its success.

The special features in the ontogenetic development of gnostic cortical areas can be considered in relation to Fig. 10A, which depicts the neural tube of a 97-day-old mon-

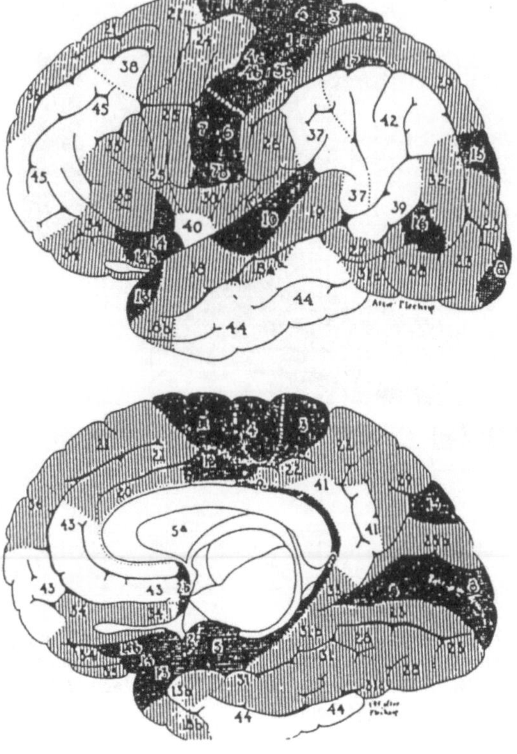

Fig. 9. Myelogenetic map of the cerebral cortex. Redrawn and altered slightly from Flechsig (1920). Cross-hatched - primordial areas; lined - intermediate areas; plain - terminal areas. Numbers refer to order of myelination; cross-hatched by birth, parallel, birth to one month; clear areas later. (From Bailey and von Bonin 1951)

key foetus (Rakic 1972) that is in process of forming the neocortex. The neural tube of the human foetus is assumed to be similar except for the important difference that in addition to the neocortex there will be special zones engaged in the neurogenesis of the neo-neocortex. In such zones the neuroepithelial cells would probably be late in mitosing to generate immature nerve cells. These nerve cells would have a wide variety of special genetic endowments, which would fit them for the task of building the distinctive neo-neocortical areas. Possibly there is clonal production of nerve cells for special functions (Meller and Tetzleff 1975). These nerve cells are arranged in a vertical minicolumn, as has been envisaged by Mountcastle (1978).

So in hominid evolution the neural vesicle of the hominoid with its neocortical generation would come to have patches for production of neo-neocortex. It can be assumed that these patches were small in australopithecines (Fig. 8), but already the greatly enlarged brain of *Homo habilis* (Fig. 8) would be produced by large neo-neocortical generating zones. With further stages of hominid evolution (Fig. 8) these zones would grow to be larger than the neocortical generating zones. Eventually with *Homo sapiens sapiens* the neural vesicle would be largely taken over by neo-neocortical generating areas. However, relative size is not the unique feature of the neo-neocortical generating

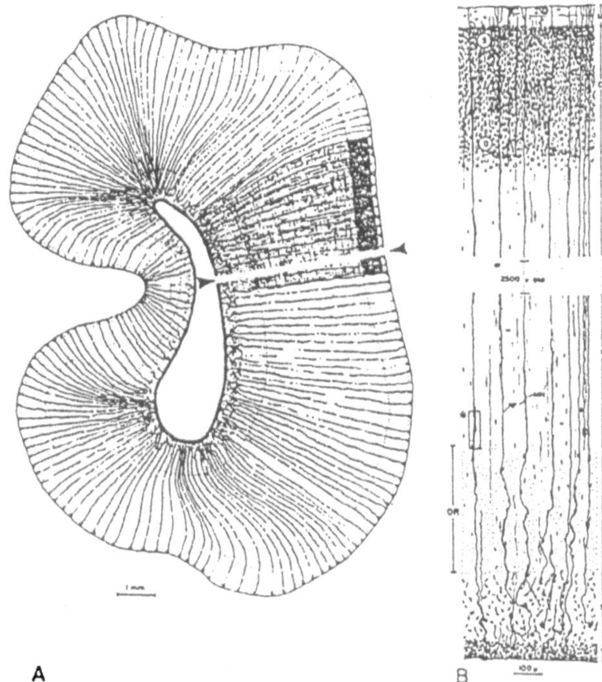

Fig. 10. A Camera lucida drawing of a Golgi-impregnated coronal section at the parieto-occipital level of the brain of a 97-day monkey foetus. The radial fibres are inscribed in slightly thicker lines than in the actual specimen in order to illustrate their arrangement at such a low magnification. (Scale equals 1 mm). The area delineated by the white strip between the arrowheads is drawn in B at higher magnification. **B** Composite camera lucida drawing of the cerebral wall in the area indicated by the white strip in A, combined from a Golgi section (black profiles) and an adjacent section stained with toluidine blue (outlined profiles). The middle 2500 μm of the intermediate zone, similar in structure to the sectors drawn, is omitted. Abbreviations: C, cortical plate; D, deep cortical cells; I, intermediate zone; M, marginal layer; MN, migrating cell; OR, optic radiation; RF, radial fibre; S, superficial cortical cells; SV, subventricular layer; V, ventricular zone (Rakic 1972)

areas. Rather it is the wide range of specialization for gnostic functions of the most diverse kinds. We can anticipate the unfolding of key problems of neurogenesis even into the distant future, problems that would relate to the special internal connectivities characterizing the unique gnostic functions, and the asymmetrical distribution of these functions. It must be realized that a limited transfer of gnostic funtions can occur at a young age, as for example occurs after destruction of the Wernicke area in infants and the transfer of speech to the right hemisphere (Milner 1974). Such plasticity is completely lost beyond puberty.

As already stated, the neo-neocortical areas are late to develop and mature in ontogenesis (Flechsig 1920; Geschwind 1965). This may provide a partial explanation of the late development of the human brain relative to that of a chimpanzee.

At birth the chimpanzee brain is 70% of the adult size, whereas with the human the ratio is only about 26%. Is the long-delayed development of the human brain attributable at least in part to its large complement of the slowly maturing neo-neocortex? It will be appreciated that this large post-partum growth became an obstetrical necessity in hominid evolution.

The hypothesis is that in some manner completely beyond scientific understanding (a mystery) the various neo-neocortical areas are specialized for transcendent functions that give gnostic performances and experiences. In the dualist-interactionist hypothesis these specific areas have intrinsic neuronal properties of a unique dynamic kind that opens them to selective interaction with the appropriate properties of the conscious mind, for example, to give perceptual experiences of light, colour, form, pattern, sound and melody, touch and structural form, odour, pain, etc. Furthermore these specific perceptual experiences are synthesized across the various sensory modalities to give the richness of sensory illumination from moment to moment. This blending and harmony occurs in the world of conscious experiences, the World 2 of Popper. Presumably this synthesis would also be occurring by interaction at the neural level of the respective neo-neocortical areas.

What is needed is an openness of mind and a humility before the challenging mysteries of the brain-mind interaction in all gnostic functions including the overarching problems of consciousness and self-consciousness.

Summary of the Neo-neocortex

By the strategy of asymmetry a large increase in neo-neocortex could be accommodated without increasing unduly obstetric hazards.

There are five special features of the neo-neocortex:

1. Phylogenetically it is the last to evolve, being a special hominid development.
2. Ontogenetically it is the last to mature, as shown by delayed myelination (Fig. 9) and by delayed dendritic and synaptic development.
3. There is functional asymmetry (Fig. 5), as for example in language, visuospatial properties and music.
4. In the young there is functional plasticity, as shown by compensation for lesions.
5. Activation of the neo-neocortex is associated with a wide variety of gnostic functions, consciousness and self-consciousness, thinking, memories, feeling, imaginings.

Summary

Initially there is consideration of the anatomical and functional asymmetries of the mammalian brain. In the evolutionary development the symmetry of the mammalian brain was almost completely preserved, even up to the apes. With hominid evolution there developed very little anatomical asymmetry, which is in striking contrast to the extreme asymmetries in higher functional performance.

This asymmetry is well displayed in the Sperry experiments with commissurotomy. For example, the function of language is almost exclusively with the left hemisphere, while spatiovisual abilities are on the right side. A generalization is that the left hemisphere is specialized for analytic performance, the right for synthetic and global.

In hominid evolution the enormous increase of cerebral cortex for *Homo habilis* through *Homo erectus* to *Homo sapiens neanderthalensis* can be attributed to the evolutionary development of communications by language of all thoughts, plans, understandings and coordinations. It is postulated that a new kind of neocortex, which may be called neo-neocortex, was evolved for these cerebral performances associated with gnostic functions: cognitive intellectual functions, mental and perceptual images, and ideas (Sperry 1982).

The great evolutionary advance from ape-like ancestors was not due to a mere increase of neocortex, but to the development of special kinds of neocortex that were not duplicated on the two sides, as had long been known for the speech areas. Thus by its asymmetrical arrangement the neo-neocortex displays an economy whereby there is a maximum of specialized cerebral performance without entailing the obstetrical hazard of a too great brain enlargement.

So in evolution there has come about mysteriously the development of large unique cortical areas for gnostic functions, which are not duplicated, but are on the one or the other side. Phylogenetically new cortex (the neo-neocortex) displays in ontogenesis a later maturation. A great experimental challenge has been revealed. A highly significant feature of the human cerebrum is its asymmetry, and furthermore that the asymmetries obtain for lobules that are intimately related to the conscious self. It is therefore proposed firstly to give an account of some cerebral asymmetries that developed in hominid evolution.

References

Bailey P and von Bonin F (1951) The isocortex of man. University of Illinois Press, Urbana IL

Bever, T.G., Chiarello, R.J. (1974) Cerebral dominance in musicians and non-musicians. Science 1985: 537-539

Brodmann, K. (1909) Vergleichende Lokalisationslehre der Großhirnrinde in den Prinzipien dargestellt auf Grund des Zellenblutes. Barth, Leipzig

Brodmann, K. (1912) Neuer Ergebnisse über die vergleichende histologische Lokalisation der Großhirnrinde. Anat Anz 41: 157-216

Flechsig, P. (1920) Anatomie des menschlichen Gehirns und Rückenmarks auf myelogenetischer Grundlage. Thieme, Leipzig

Geschwind, N. (1965) Disconnection syndromes in animal and man. Part 1. Brain 88: 237-294

Geschwind, N., Levitsky, W. (1968) Human brain: left-right asymmetries in temporal speech region. Science 161: 186-187

Goodall, J. (1986) The chimpanzees of gombe. Patterns of behaviour. Belknap Press of Harvard University Press, Cambridge, MA.

Gould, S.J. (1977) Ontogeny and phylogeny. Belknap Press, Harvard University Press, Cambridge, MA.

Hamilton, C.R. (1977a) Investigations of perceptual and mnemonic laterization in monkeys. In: Harnad S, Doty R.W., Goldstein L, Jaynes J, Krauthamer G (eds) Lateralization in the nervous system. Academic, New York

Hamilton, C.R. (1977b) An assessment of hemispheric specialization in monkeys. Ann NY Acad Sci 299: 222-232

Hecaen, H. (1967) Brain mechanisms suggested by studies of parietal lobes. In: Millikan C.H., Darley F.L. (eds) Brain mechanisms underlying speech and language. Grune and Stratton, New York, pp 146-166

Jung, R. (1974) Neuropsychologie und Neurophysiologie des Kontur- und Formsehens in Zeichnung und Malerei. In: Wieck H.H. (ed) Psychopathologie mimischer Gestaltungen. Schattauer, Stuttgart

Kimura, D. (1973) The asymmetry of the human brain. Sci Amer 228: 70-78

Le May, M., Geschwind, N. (1975) Hemispheric differences in the brains of great apes. Brain Behav Evol 11: 48-52

Levy, J. (1977) The mammalian brain and the adaptive advantage of cerebral asymmetry. Ann NY Acad Sci. 299: 264-272

Levy, J. (1978) Lateral differences in the human brain in cognition and behavioural control. In: Buser P.A., Rongeul-Buser A. (eds) Cerebral correlates of conscious experiences. North Holland Amsterdam

Levy-Agresti, J. Sperry, R.W. (1968) Differential perceptual capacities in major and minor hemispheres. Proc Nat Acad Sci 61: 1151

Mauss, T. (1911) Die faserarchitektonische Gliederung des Cortex cerebri der anthropomorphen Affen. J Psych Neurol 18: 410-467

Meller, K., Tetzlaff, W. (1975) Neuronal migration during the early development of the cerebral cortex: a scanning elctron microscopic study. Cell Tissue Res 163: 313-325

Milner, B. (1974) Hemispheric specialization: scope and limits. In: Schmitt F.O., Worden F.G., (eds) The neurosciences third study program. MIT Press, Cambridge, MA, pp 76-89

Mountcastle, V.B. (1978) An organizing principle for cerebral function: the unit module and the distributed system. In: The mindful brain. MIT Press, Cambridge, MA, pp 7-50

Penfield, W., Roberts, L. (1959) Speech and brain-mechanisms. Princeton University Press, Princeton, NJ

Rakic, P. (1972) Mode of cell migration to the superficial layers of fetal monkey neocortex. J Comp Neurol 145: 61-84

Roland, P.E., Skinhoj, E., Lassen, N.A. (1981) Focal activation of human cerebral cortex during auditory discrimination. J Neurophysiol 45: 1139-1151

Rubens, A.B. (1977) Anatomical asymmetries of the human cerebral cortex. In: Harnad, S., Doty, R.W., Goldstein, L., Jaynes J., Krauthamer, G. (eds) Lateralization in the nervous system. Academic New York, pp 503-516

Sperry, R.W. (1974) Lateral specialization in the surgically separated hemispheres. In: Schmitt, F.O., Worden, F.G. (eds) The neurosciences third study program, MIT Press, Cambridge, MA, pp 5-19

Sperry, R.W. (1982) Some effects of disconnecting the cerebral hemispheres. Science 217: 1223-1226

Sperry, R.W., Zaidel, E., Zaidel, D. (1979) Self-recognition and social awareness in the deconnected minor hemisphere. Neuropsychologia 17: 153-166

Stebbins, G.L. (1982) Darwin to DNA, molecules to humanity. Freeman , New York, pp 491

Tobias, P.V. (1983) Recent advances in the evolution of the hominids with special reference to brain and speech. In: Chagas C (ed) Recent advances in the evolution of primates. Pont Acad Scient Scripta Varia 50: 85-140

Tobias, P.V. (1987) The brain of Homo habilis: a new level of organization in cerebral evolution. J Hum Evol 16, 741-761

Wada, J.A., Clarke, R., Hamm, A. (1975) Cerebral hemispheric asymmetry in humans: cortical speech areas in 100 adults and 100 infant brains. Arch Neurol 32: 239-246

Yeni-Komshian, G., Benson, D. (1976) Anatomical study of cerebral asymmetry in the temporal lobes of humans, chimpanzees and rhesus monkeys. Science 192: 387-389

Science, Man and Meaning

V.E. Frankl
University of Vienna Medical School, Mariannengasse 1, 1010 Wien, Austria

Specialised researchers and clinicians are often confronted with people who ask: Where is that, which we would like to call a holistic medicine? Where is the unity of humanness to be seen? These people are accustomed to deploring the fact that scientists are increasingly specialised. We might define a specialist as someone who no longer sees the forest of reality for the trees of facts. However, we cannot dismiss the specialists. In an age such as ours, an age of information explosion, scientific research is not possible without teamwork and teamwork in turn is not possible without the specialists. However, I do not think that the danger really lies in increasing specialisation but rather in generalisation. Researchers sometimes produce overgeneralised conclusions and consequently overgeneralised statements.

Let me quote such a statement: "Man is nothing but a biochemical mechanism powered by a combustion system which energises computers with prodigious storage facilities for retaining encoded information." As a neurologist, I can accept it as perfectly legitimate to explain the functioning of the central nervous system in man along the lines of the analogy to a computer. The mistake really lies in the phrase "nothing but." This "nothing-but-ness" is at the root of what is usually referred to as reductionism. Man is more than a computer, man is even infinitely more than a computer, but at the same time in a certain sense he is also comparable to a computer, in the same way as a square is a part of a cube (Fig. 1). If one neglects the three-dimensionality of the cube, a two-dimensional square is left, that is if you sacrifice the third dimension, in the two-dimensional plane a square is left. Something similar holds for man: one may project his very human quality, his humanness, out of the human dimension down into a two-dimensional plane. One may even state that science is simply not possible without engaging in such projections. This would mean that science is compelled to disregard the multidimensionality of reality, to stick to the fiction that reality is something unidimensional. But the scientist should not be surprised in this case when he notices that he ends up with different, even with contradictory pictures of reality. Wherever one opens the book of reality, one is confronted with differing pictures of reality, symbolised on the left page by a square and on the right page by a circle (Fig. 2). How is it then possible to arrive at a unified concept? What happens when the left page is turned to a perpendicular or orthogonal position (Fig. 3)? Then one may well interpret the square and the circle as the projections of a three-dimensional cylinder (Fig. 4) into the horizontal plane (resulting in a circle) or, on the other hand, into the vertical plane (resulting in a square). In the light of this geometrical analogy the differences no longer contradict the unity of that object which has been projected into the two dimensions.

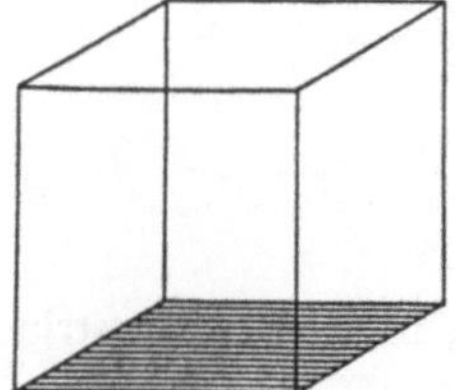

Fig. 1. The sacrifice of a dimension

Fig. 2 The book of reality

Fig. 3 Projections

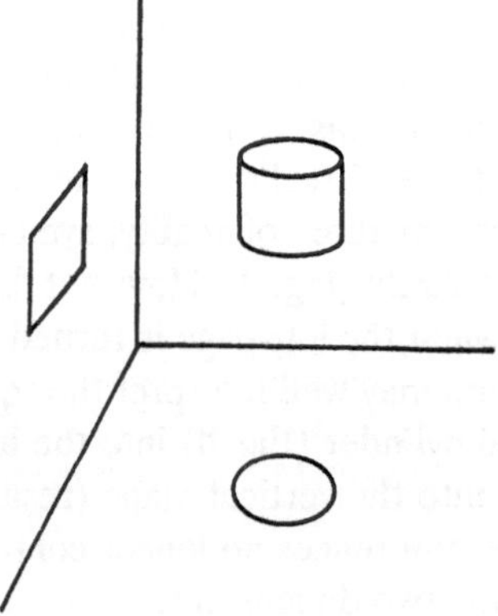

Fig. 4. Projections of a cylinder

We can proceed from this analogy to draw certain conclusions. These are that the unity becomes perceptible only if we transcend the two-dimensional planes of projections into the three-dimensional space of the object that we have projected. And this of course also happens to human beings viewed in different perspectives. I well remember many decades ago, in the 1940s, when I gave a series of three consecutive courses. Each Wednesday from 4 to 5 p.m. I lectured on neurology for the general practitioner; in this course I dealt with my patients as if they were a conglomerate of reflexes and so forth, simply mechanically organised. In the following lecture I spoke of the various schools of psychotherapy and dealt sometimes with the same patients as if they were an assembly of reactions or learning processes, or psychodynamic processes, or complexes, conflicts and so forth. But in the third hour I again called the attention of my students to the fact that man in a way is a neurological entity, in another way he follows psychological mechanisms, but ultimately man himself in his very humanness is infinitely more than that. Thus the medical profession must see the patient as a human being, to see the "homo patiens", the suffering human being behind the disease in order not to fall prey to the pitfalls and fallacies of reductionism.

Reductionism may sometimes also appear in the form of that which we might better term oxidationism; this I encountered when I was 14 years of age. In high school our science teacher claimed, quite literally, that "Life is nothing but a combustion process, an oxidation process." At that moment I jumped to my feet and threw the question at him: "Dr. Fritz, if this is so, then what meaning does our life have?" He, viewing man in reductionist terms, could of course not give an answer to the question of meaning.

What I want to convey is that this reductionistic concept of man, resulting from various overgeneralised views on man in different fields of scientific research, merely reinforces a world-wide phenomenon of our time, namely the abysmal feeling of meaninglessness, the feeling of an inner emptiness, what I have termed the existential vacuum or existential frustration. To overcome this frustration, one must understand the motivation of human beings. So, let us now assume that the different projections of man refer in the side view (Fig. 5) to the concept of man according to Freudian psychoanalysis, and in the ground plan to the concept of man according to Adlerian psychology or individual psychology. Now let us ask, what is the basic motivation within the frame of the psychoanalytic concept of man? It is the pleasure principle. In the concept that Adler offers us, the basic motivation, according to literal statements made by him, is that any human being is dominated, from the earliest childhood on, by a feeling of infe-

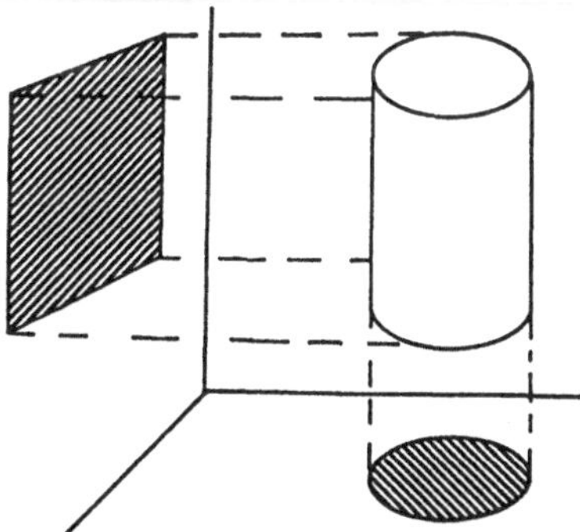

Fig. 5. The differences in the projections no longer contradict the unity of the cylinder

riority, and any human being throughout his life seeks to compensate, if not to overcompensate, this feeling of inferiority by a striving for superiority, or as Nietzsche would have called it, by the will to power. But when one looks a bit deeper, one recognises that actually the motivation theories of both psychoanalysis and Adlerian psychology pretend that man is primarily, basically, originally concerned with something within his own psyche. But actually man is not like that. In fact, man is originally out to transcend himself, man is basically open to the world, to a world full of values and meanings, of causes to serve, of persons to love. Man, to use a phrase coined by Heidegger, the great existential philosopher, is "being-in-the-world", he is directed and relating to the world and primarily, except for neurotic conditions, never concerned with himself, with his own self-actualisation. Rather, he is reaching out for meanings to fulfil, for causes to serve, for other human beings to lovingly encounter. And why does this openness, this self-transcendence quality of the human reality completely disappear and not find representation in the concepts of the current motivation theories?

If we imagine that the cylinder is not solid but is an open vessel (Fig. 6), then in the projection this openness must necessarily disappear. And this is exactly what happens in the current motivation theories. This openness, this reaching out of human beings toward meanings out there in the world, is not seen. If we regard man as someone dominated by a will to pleasure or a will to power, we totally neglect the original motivational force operant in man, that is, what I call the will to meaning. And this will to meaning is exactly that which is frustrated today in large segments of the human population all over the world. But, ironically, absolutely in contradiction to what the feeling of meaninglessness is whispering in our ears, life not only is meaningful, potentially meaningful, it not only offers to each and every individual a potential meaning to find out and fulfil; life is more, life is actually unconditionally meaningful. This means that in principle there is no situation of a human life conceivable which really can lack any sort of meaning to find and to fulfil.

But how can we understand this fact - which has been corroborated by 20 empirical studies? The answer is: simply in view of the fact that one can find meaning not only in doing a deed or creating a work. And beyond this, one not only can find another type of meaning in experiencing something or experiencing someone; this means experiencing another human being in his or her very uniqueness as a person, and this means loving him or her. Not only through work and love can life be found meaningful, but if there is

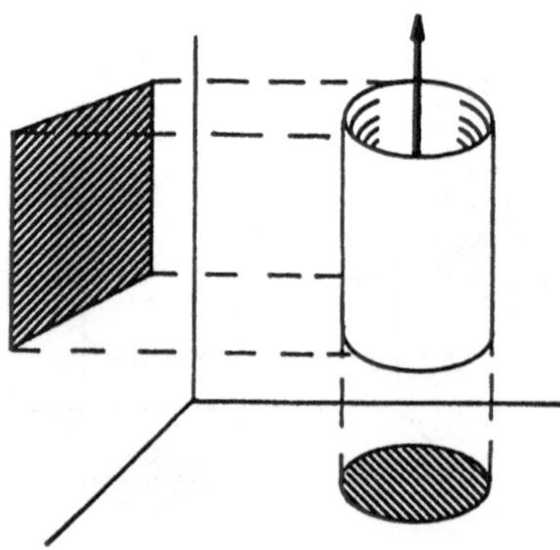

Fig. 6. The self-transcendent quality of the human reality

an unchangeable situation, for example an incurable disease, an inoperable cancer, there is still the potential to find a meaning, even in suffering, in unavoidable suffering, because exactly where you cannot change your situation, you are called upon to change your attitude to this situation, to change yourself, that is, to rise above the situation, to grow beyond the situation, to bear witness to the uniquely human potential to turn a predicament into a human achievement and accomplishment or to transform your personal tragedy into a triumph.

This sounds like abstract and idealistic preaching, but it is also very much down to earth. Let me quote something not that I have written, not that I have witnessed, but that a German Bishop, Georg Moser, has included in a little booklet. He writes the following. A few years after World War II a doctor examined a Jewish woman who wore a bracelet made of baby teeth and mounted in gold. "A beautiful bracelet", the doctor remarked. "Yes", the woman answered, "you see, Doctor, this tooth here belonged to Miriam, the other tooth here to Esther, and this one to Samuel", and she mentioned the names of her daughters and sons according to age. "Nine children", she added, "and all of them were taken to the gas-chambers." Shocked, the doctor asked: "How can you live with such a bracelet?" Quietly the Jewish woman replied: "See, Doctor, I am now in charge of an orphanage in Israel." This woman was able to turn a personal tragedy into a triumph. And from your patients, severely diseased patients, you certainly could add a whole array of similar cases, in which simple human beings have succeeded in mastering a very, very severe personal tragedy.

But there are also those people who come with the often asked question: "Why me? Why did this happen to me?" Let me recall what happened to me when I once had to supervise a group-therapeutic session in the Polyclinic Hospital. A lady had been admitted to the hospital after a suicide attempt following the death of her 7-year-old son. The group of patients discussed the problem and then asked: What is the meaning of the suffering of this lady? I then joined in the discussion and asked them: "Ladies, imagine an ape that has to serve for the development of any vaccine. And day by day this ape is punctured, blood is taken off, and the animal has to suffer. Could it ever recognise, ever understand the purpose for whose sake it has to suffer?" Unanimously the group replied, of course it could not, because an ape, an animal, cannot reach into the world of human beings, where alone such sufferings have a meaning, have a purpose. And then I pushed forward to the next question: Now what about man, are you absolutely sure that the human world, the human dimension, is something like a terminal, and that beyond this terminal there does not exist any further, any higher dimension? A dimension, that is, in which our sufferings would find an answer as to their purpose, a purpose we naturally are not capable of understanding, at least not in intellectual terms. Again the whole group replied, and as it happened, there was no religiously oriented lady among them: "Of course we cannot be sure whether or not there is a dimension beyond the human dimension in the same way as there is the higher dimension of human beings as compared with the lower dimension of animals."

Now, what holds for patients also holds for the scientist in a way, inasmuch as it is opposed to that which the late Jaques Monod declared in his famous bestseller "*Chance and Necessity*". There he contended that the whole of evolution is based on "nothing but" chance events, random events, nothing but that; there is no purpose behind it. Once I discussed this problem with Konrad Lorenz. Lorenz admitted: "Within my field of biology I cannot trace any purposefulness, I cannot see any teleology, pointing to an

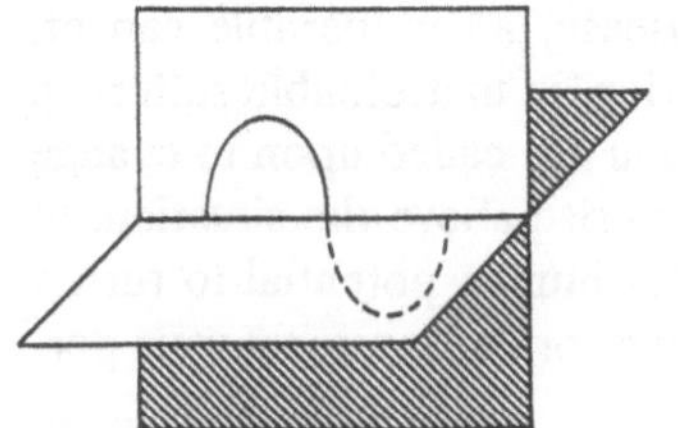

Fig. 7. Teleology does not depict itself within the plane of biology

ultimate purpose or aim. There, too, everything is only pure chance." I replied: "Dr. Lorenz, you are entitled to say that within the plane of your research as a biologist, you can only see three single, isolated, disconnected points without any meaningful connection; but have you the courage on a priori grounds to declare that no other dimension, no other plane of projection is conceivable, only this horizontal plane of biology is valid?" (Fig. 7). "There may well be a vertical plane, in which there is a sinus curve, and this sinus curve does connect the single points where it intersects the horizontal plane. Thus, there could be in another dimension - not in your biological plane but in the next higher dimension - a meaningful connection between the single points, a meaningful connection above or below them, a higher or a deeper meaning that does not depict itself in the horizontal projection of biology." And then I concluded this argument with Lorenz by telling him: "Once, dear Dr. Lorenz, once that you will have acknowledged the pure possibility of such a view, you will not only have deserved the Nobel Prize that has been bestowed upon you, but also another prize, a prize that does not yet exist, the prize for wisdom; because wisdom, I would define, wisdom is knowledge plus - that is, knowledge plus the awareness of its limitations."

And in this sense I would like to conclude my presentation by quoting Ernst von Feuchtersleben, the man for whom the first academic chair for psychiatry in the world was installed as early as in 1836. He was the author of best-selling books, among them a volume entitled *Lehrbuch der ärztlichen Seelenheilkunde* published in Vienna in 1845. And in this book I came across the following sentences that I would like to share with you, if you allow me, first in the original German: "Der echte Denker ist zufrieden, die Grenze des Denkens gefunden zu haben. Es ist eine weise Vorsehung, die diese Grenze gezeichnet hat, weil der Mensch da, wo sein Denken endet, zu handeln beginnen soll, wozu er eigentlich da ist." Now in English: "The true thinker is satisfied with having arrived at the limits of thinking, and it is a wise providence that has established those limits. Because where thinking must end, man should start acting which, after all, is his real destination."

Information and Efficiency

K. Steinbuch
Adalbert-Stifter-Straße 4, 7505 Ettlingen

Introduction

Konrad Lorenz wrote in his book *Behind the Mirror* "In the last analysis all complex structures of the entire range of organisms have arisen through the selection pressure of survival functions." What I want to show is the following: Not only physical performances have developed under selection pressure but also informational performances achieved by the sensory organs, the nervous system, information technology and cultural exchange.

I start from the following concept: all self-preserving, open systems - ranging from biological organisms to political organizations - subsist on one and the same principle: action potentials (e.g. resources of energy and material, wealth, powerful positions, etc.) are employed in the light of information about the environment in such a way that the best possible performances are achieved. Let me illustrate this basic idea by two quite different examples:

- Karl von Frisch discovered that bees returning to the beehive signal to the other bees the direction and distance of the best feeding grounds by performing a circular dance ("dance of the foraging bee").
- while I was waiting at Dallas Airport for the departure of my plane to Frankfurt, I heard the flight captain returning from the tower saying to his crew: "We shall have a strong tail wind on the northern route; this will save us flying time and fuel."

What do these two events have in common? It is the following: In both cases performance is improved by information - in one case the harvest of the bees is improved, in the other the efficiency of flight operation. Exactly this illustrates the basic idea stated above: action potentials are employed in the light of information about the environment in such a way that the best possible performances are achieved.

I would like to explain this general principle using examples and considerations taken from the daily life of highly developed societies.

Decision Making and Information

Typical reasons for the employment of information technology are:

- Production and administration are no longer competitive without the use of electonic information systems, especially the computer.

- A high-grade economy needs reliable information systems for supervising the multiple division of labour and the exchange on a large scale.
- An effective political organization with its manifold controls and checks can no longer manage without modern information and communication technology.
- Modern scientific activity must be tied into the worldwide scientific information network.

Information technology has also changed military technology, an aspect which I do not want to discuss in detail here: Modern arms systems are frequently informational systems for the discovery of targets and the guiding of missiles. Recently the "I-weapons" - the weapons of disinformation and indoctrination - have been added to the ABC weapons.

Here the question arises: Why all this expenditure for information technology? The answer is this: Correct information at the right time permits the "correct" decision. Here is a spectacular example: When in 1815 the London banker Nathan Rothschild was the first to learn by carrier-pigeons about Napoleon's defeat at Waterloo, he could make the "correct" decision and thereby make an enormous profit at the stock exchange.

Everyday work in a modern office is less spectacular, but the same sentence can be applied: The correct information at the right time permits the correct decision:

- The credit rating of the client must be ascertained before a loan is granted.
- Before engaging in a technical project we must know the state of the market, the patent situation and the production facilities.

I would like to mention here a historically interesting system of office automation: the *Informatik-System-Quelle* (an information processing system), which is remarkable for various reasons:

- Firstly, it represents the first (in 1957) public use of the concept of *Informatik* in a German-speaking country.
- Secondly, it was the first large information-processing system in the world to be realized in semiconductor technology.

The system had been developed for the automation of order processing, accounting and stock taking in a large mail-order house. Peak work loads, for which until then 1200 employees had been necessary, could now be handled with the help of the new system by only 400 people. The information-processing system found out the prices of the goods ordered - from a range of 50000 different goods - within fractions of seconds, checked whether the goods were on stock, suggested replacement deliveries if necessary, made out the bills and continually recorded the inventory. This *Informatik-System-Quelle* illustrates the fact that office work needs large amounts of information fast. It is the main task of a modern office to supply the necessary information for making the correct decisions. Recent statistics have shown that in highly developed societies more than 50 percent of all people gainfully employed work with information, either by transmitting or by processing it. These statistics include occupations which we would not normally consider to be "information professions", such as scientists, teachers and educators, employees of postal and telecommunication services.

May I mention one further example here, which is politically quite controversial in the Federal Republic of Germany, the population census, which without doubt serves the supply of information, for instance, town planning, housing construction, area planning, power supply, planning for welfare and education, labour market policy, administration, etc. No sensible decisions in these fields can be made without the information which only a census can supply.

But let us look at the problem "decision making and information" in a more basic way. If we are in an unknown area with a poor view and come to a crossroad, it is difficult to decide whether we should go to the right or to the left. Without having information as to where the roads lead, we may take the wrong way and subsequently encounter disadvantages. Having the information, for example, from a road sign - we take the correct way and avoid the disadvantages. The crossroad situation is an elementary and paradigmatic illustration of virtually all decision-making processes. Complex decisions and decision-making processes can be traced back to several such crossroad situations, to decision labyrinths. But there is always the same condition: In order to make the "correct" decision one needs not just any information but the correct information. However, not every information is "correct" in this sense:

- Information might simply be wrong. This is illustrated by a roadsign pointing in the wrong direction.
- Information may be distorted. This is illustrated by a signpost which is no longer readable.

Wrong information - or disinformation - leads to wrong decisions, be it the wrong road taken by the hiker or a wrong political or military decision. At present political conflicts are frequently conducted with intentional disinformation.

The paramount importance of information technology in medicine, for instance, becomes clear from an essay by L. Krutoff (*Schimmelpfeng-Review* no 33):

> "Every year more than a million medical publications appear worldwide ....It could have happened that a scientist had been pondering over a problem for years, had done research work only to find out by chance, perhaps after years, that somewhere a solution had already been found. The work of many years would have been pointless.
> Or: A doctor was fighting for the life of a patient suffering from a disease, which was incurable to the doctor's knowledge. The idea that somewhere in the world a remedy against this disease had already been developed was agonizing and tormenting."

These situations show quite clearly that information and information processing are of major importance in medicine. In 1969 - after several expert information systems for different areas had already been established in Europe - DIMDI was founded, an expert information system for the whole field of biosciences, especially for medicine (Deutsches Institut für Medizinische Dokumentation und Information, Postfach 42 05 80, Weißhausstraße 27, D-5000 Cologne 41). This Institute fulfils special tasks, e.g. documentation of toxins and intoxications as well as of pharmaceutical products, or the FRG version of the international classification of diseases. In order to understand the value of such medical information we must imagine ourselves in the place of a physician to whom a child is brought who has swallowed some toxic substance. The life of the child depends on whether he can find the correct therapy very fast.

For a clear comprehension of the problem "decision making and information", Figure 1 shows a diagram of the interaction between a subject system and an object system. Essential parts of the subject system are

- Evaluation.
- Internal model of the object system - which initially may be empty but forms an increasingly accurate image of the object system with increasing experience.

When a problem is to be solved, informational interaction between subject system and object system occurs:

- On the one hand, the subject system perceives the state and the changes of the object system,
- On the other, the subject system itself may perhaps change the object system by interfering.

The impulse to solve a problem results from different evaluations of different states of the object system. In any case, it is the purpose of problem solving to achieve a state of the object system which is, according to subjective evaluation, better. The internal model of the subject system creates the possibility of selecting suitable interventions and assessing whether the new situation achieved is better or worse than the former one.

A "wrong" decision may be due to two different causes:

- Wrong asessments,
- The wrong internal model

Figure 1 corresponds to the Robinson Crusoe situation, in which an individual subject system tries to solve a problem without help from other subject systems. Without prior knowledge about the object system and without the possibility of exchanging information with other subject systems, the subject system can develop an internal model of the object system only by cautiously interfering with the object system and observing the changes produced by these interventions. This procedure, called "trial and error", may

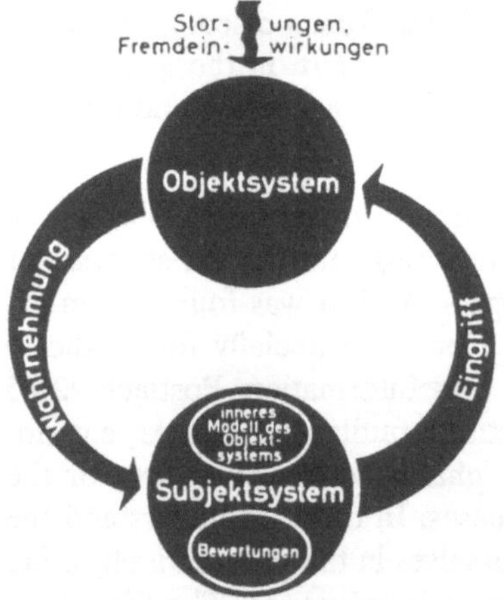

Fig. 1. Problem solving by interaction between a subjects system and a object system

help the subject systems to obtain in the beginning rough and then increasingly accurate information about the object system: its internal model forms an increasingly accurate image of the object system.

It is a trivial fact, however one with remarkable consequences, that the internal model is able to form a correct image of the object system only if its information capacity is greater than the complexity of the object system. If it is smaller, the subject system can create only an incomplete model (see below).

More interesting than the Robinson Crusoe situation shown in Figure 1 is the social situation presented in Figure 2, in which several subject systems influence each other during the process of problem solving by furnishing information; this influence may be beneficial or detrimental. According to Figure 2, one subject system may effect changes in the structure of the other subject system, in particular in its internal model, by furnishing information. A typical example is the situation when someone unfamiliar with a place shows a foreigner the way. Such changes effected by information may be

- Valuable
- Useless
- Harmful

A typical example of harmful information is the situation when the foreigner is shown the wrong way. For all decisions the question arises: Is the information on the basis of which the decision is made correct or not?

Carl von Clausewitz described these problems very impressively in his classical book *On War*:

> "War is the area of uncertainty; three-fourths of the things which action in wartime is built on lie in the fog of minor or major uncertainty. So it is primarily here that a fine and penetrating intellect is enlisted in order to ... sense the truth ....A large part of the information which one receives in wartime is contradictory, an even larger part is wrong, and by far the largest is subject to considerable uncertainty."

The conditions in our information society are - at least in part - somewhat more favourable; modern information technology has developed effective methods of transmitting information in the "correct" form to the receiver, even via impaired channels. There are, for example, encoding methods which permit the identification of distorted signs as

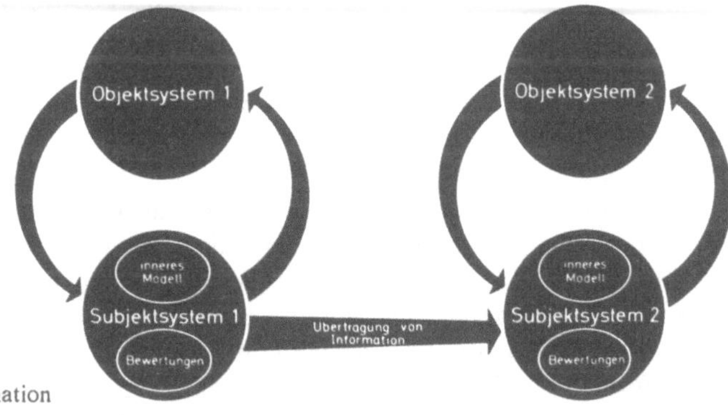

Fig. 2. Problem solving and information

"distorted" - and other methods which even correct distorted signs. Their principle is easy to understand: If one and the same sign is transmitted twice, then it is easy for the recipient to recognize whether one of the two signs arrives in a different form than the other. The recipient is alarmed; here is uncertainty! If the sign is transmitted three times, then the recipient can determine - on the strength of a majority decision - which of the signs received in a different form is likely to be the correct one. (Modern encoding methods are more sophisticated, however, this is of no importance in this connection.)

All the methods of information safeguarding use this redundancy, the roundaboutness of transmission, but none of them are absolutely sure to furnish the correct results. Nevertheless, the probability of an error remaining unrecognized has diminished by several factors of ten. However, all these methods can only safeguard the transmission; they say nothing about the correctness of the information at its source. This is less a technical problem than a cultural problem of mutual confidence and a sense of responsibility. If a recipient is to trust information, he must have experienced and received multiple confirmations of the sender's sense of responsibility. The building of confidence will probably be a central problem of our future information society. When the partners of a communication have no confidence in each other, all technical aids will be unprofitable.

In my book entitled *Schluß mit der ideologischen Verwüstung!* ("Stop the Ideological Devastation!"; Busse-Seewald, Herford, 1986) I presented some thoughts about trustworthiness and wrote that Confidence is justified by

- The proven or recognizable will to justify confidence shown, that is to act willingly, even without control
- The necessary intelligence which permits trustworthy behaviour also in unpredictable situations
- Steady behaviour, loyalty to principles

Information - Its Transmission and Processing

Much has been written about the definition of the concept of information, but there is no universally accepted definition. Most remarkable is perhaps Norbert Wiener's statement: "Information is information - it is neither matter nor energy." However this statement expresses only what "information" is *not*. Certainly the modern scientific concept of information is much wider than the colloquial one; in colloquial language we may distinguish, for example, between information and comment, but in a modern scientific sense both are information! Ignoring ontological questions, I would like to say very pragmatically that information is the matter of which efficient decisions can be made. This information can be tranmitted and processed.

Human culture began millenia ago with the invention of information storage with the help of pictorial symbols, hieroglyphics, letters on stone, clay, metal, parchment, and finally on paper. Modern information technology began with the invention of electronic information memories. Modern main-frame memories are able to store a million million signs, amounting to the content of large libraries with many thousand books. Any information addressed can be retrieved from such a master memory within a matter of seconds, which is certainly not possible in the case of large libraries. The availability of such giant memories is the prerequisite of future information systems, from which practically any information can be retrieved. One and the same unit of information may appear in various forms; the same text may appear in writing on paper, as spoken word in the air or as magnetic recording on magnetic tape.

While I shall not enter into technical details here, I would like to mention briefly the latest development in the field of telecommunication, i.e. information transmission. An enormous amount of information can be transmitted through glass fibres as thin as hair. Such hair-thin glass fibres can replace thick copper cables for the purpose of information transmission. The information contained in the whole Bible can, for instance, be transmitted via one such glass fibre within a single second. (The Bible contains approximately 5 million letters, for the encoding of which 30 million bits are needed. Modern glass fibre systems transmit 140 million bits per second, experimental systems even up to ten thousand million bit per second).

Let me comment briefly on the two concepts of information transmission and information processing:

Information transmission is the transport of messages in spaces and time. Information stored at an earlier time can, for instance, be recalled from information memories. Here we should note that the law of conservation of information is different from the laws of conservation of matter or energy: If we deliver energy or matter, we do not have it any longer, if we deliver information, we may also keep it for ourselves. Trivial, however noteworthy is the fact that information can be transmitted in space in two directions, while in time it can be transmitted in one direction only - only unilateral communication is possible; bilateral communication is possible with a human being on the moon, however not with a man of the previous generation!

Information processing produces new information out of various messages under application of defined rules. Typically the computer determines the sum by summation of addends or the product by multiplication of factors. The technique of information processing has made enormous progress in recent decades. The first computers, built by Konrad Zuse in Germany and Howard Aiken in the United States some 40 years ago, completed an arithmetical operation within a second. During the 1960s a stage was reached at which millions of operations could be completed within one second. Storage technology has developed at a similar speed. The first memories permitted only small storage capacities due to their size and energy consumption. Moreover they were very slow. In our time, many high-grade memories are available, ranging from rapid memories which reply within picoseconds (10^{-12} seconds) to magnetomotor memories which permit the storing of million million signs and the retrieval of any information addressed within seconds. Simultaneously with increasing speed of function and storage capacity, the circuits were miniaturized; VLSI (very large scale integration) places ten thousand circuit elements, - and soon it will be millions - on a chip the size of a penny. Progress in the field of hardware went hand in hand with the development of programming technique, which began with Zuse's *Plankalkül* in 1945, lead to ALGOL, the programming language developed in 1958, and has reached a wide spectrum of user-specific programming languages in our time.

What is it that computers can do? Thirty years ago it was a very common prejudice that the computer is "a complete idiot with special talents" unable to produce anything new. I warned then and I am still warning against impossibility on the strength of intuitive opinion or ideological prejudice. If asked to summarize the possibilities of artificial intelligence in a short formula I would say:

- Artificial intelligence will be able to do everything which can be traced back to a sequence of logical connections and this is more than we may assume at first thought.
- As long as computers operate exclusively on the basis of programming, they are limited by the possibilities of formalization, something which has been pointed out by Gödel.
- Where computers search for their optimum function in direct communication with the external world, they are no longer bound by the limitation of possible formalization.
- No computer can differentiate on its own between "good" and "evil"; this differentiation is the result of historical experience which the isolated computer does not have.

Recently expert systems have been playing an important role within the field of artificial intelligence. With their help, expert knowledge can be made available at any time and at any place. For this purpose expert systems employ in most cases a knowledge basis in the form of numbers, facts, relationships, rules, etc. An inference machine permits qualified conclusions in the light of this knowledge basis. Such expert systems must be kept constantly up to date; their knowledge basis must be able to accomodate a constant input of the latest experience. May I draw your attention to two typical expert systems:

- In the middle of the 1970s the expert system MYCIN was developed, which stores expert information about colour, form, way of life and environmental conditions of microorganisms. It helps to draw rather fast and correct conclusions as to the identity of pathogens (e.g. the causative organism of meningitis).
- MODIS (Motor Diagnose System) is an expert system for the diagnosis of failures in Otto engines or machines connected to them. MODIS operates with more than 2100 rules and is designed for application in automobile repair shops. It is intended to assist the mechanic in finding the failure in engines which gave trouble.

Information on these are provided *Expertensysteme im Unternehmen* edited by H. Krallmann (Erich Schmitz, Berlin, 1986). It is the purpose of expert systems:

- To prevent loss of expert knowledge (know-how securing)
- To make expert knowledge available round the clock (e.g. for shift work)
- To make expert knowledge available at all places where it is needed (e.g. in branch offices)
- To avoid accidents by rapid and reliable use of expert knowledge

I expect that such expert systems will in the future also be used to "sense" - as meant by Clausewitz - the reliability of a message. This could be achieved in the following way. With the help of adequate expert systems, information received is correlated with already established knowledge; it will thereby be possible to ascertain in many cases whether the new information is credible or not. The same will apply here as has already been observed about safeguarding information with the help of redundant encoding; no method will always supply the correct results with absolute certainty. However, the probability of unidentified errors can be reduced by many factors of ten. Therefore, I expect that with the help of expert systems we shall be able to differentiate frequently between reliable and unreliable information - just as an experienced person recognizes an untruth, where an inexperienced individual still trusts in it.

The Practice of Decision Making and Information

After these - more theoretical and speculative - considerations, I would now like to illustrate with some examples how the efficiency of highly industrialized societies is improved by information and information technology:

- Typical is the mere transfer of technology, which permits the development of high technology.
- Typical is the simulation technique, with the help of which existing and planned systems are simulated on the computer in order to permit the identification of their characteristics also in imaginary situations.
- Typical is modern automation, ranging from a simple automatic controller to modern robots with sensors, by means of which an enormous increase in efficiency was achieved.
- Typical is stock-keeping of perhaps hundred thousand different spare parts, which frequently must be dispatched all over the world within a few hours, something which would be impossible without modern computer technology.
- Typical is the economy of stock-keeping, when the best time for renewal orders is determined by the expected sales and the known delivery periods.
- Typical is the "travelling-salesman problem": If a salesman is to visit several places, his best travel route can be determined by the computer - "best" meaning that with the shortest travelling time and the lowest travelling cost.
- Typical for the interrelation between information and efficiency is the market economy as a whole, in which market data govern adaptation processes, which can never be achieved by dirigism.

All these practical examples illustrate the general principle mentioned in the beginning: action potentials are employed in the light of information about the environment in such a way that best possible performances are achieved.

The Informational Limitation of Man

The amount of information which a human being can take in through his sensory organs and store in his memory is far smaller than the amount of information necessary to give him an image of his environment. In discussing problem-solving behaviour we noted that the internal model of the subject system is able to form a correct image of the object system only if its information capacity exceeds the complexity of the object system. To speak in terms of facts we must now admit that human consciousness is no match for the complexity of man's environment and that therefore comprehensive statements about this environment are impossible. Important consequences result from this fact:

1. The inadequacy of a limited man in an enormous world necessarily means that he is constantly confronted with surprises and the workings of chance, and that the whole of "futurology" is built on sand.

2. Before complex systems, e.g. economic or political systems, intuitive predictions are frequently wrong. This "counter-intuitivity" was discovered by J.W. Forrester in 1971

while studying the redevelopment of slum areas in the USA: Contrary to the intuitive expectation that redevelopment of slums would result in social advancement of the redeveloped area, experience showed - and subsequently also theory - that redevelopment of slums leads to further social decline.

3. The awareness of our inadequacy also leads to the realization that our model of the world cannot provide understanding of the origin of our world, and that, arrogance before metaphysical interpretations and religion is therefore out of place.

The human brain has not been developed for the understanding of a complex industrial society, but for the purpose of preserving life under quite different conditions.

Decision making and mass media

The sentence "The correct information at the right time enables us to make the correct decision" holds also true in the political field - with the decision of a citizen when he casts his vote in favour of a political party, in parliamentary decisions, or for the lonely decision of the responsible politician. They all need the correct information, and disinformation will lead to wrong decisions. The mass media produce much more (and much more contradictory) information than can be processed by their consumers, and this may result in aberrations of behaviour and political decisions. Under the rule of scientific reasoning well-defined concepts are used and processed according to the principles of logic. Here intellectual discipline becomes a matter of course. Quite different is the situation under the influence of mass media. Here things are presented in such a way that they are easy to understand, and that emotional expectations are satisfied. The "illusion of understanding" and the "competence of compensating incompetence" are created. The terrible simplificators prevail on the producer side, and the "one-minute experts" on the consumer side. Arguments with them turn into mud slinging, and what results is the dictatorship of incompetence.

The main question in the case of direct communication from person to person in natural language is: Is the information correct, is there *adaequatio rei et intellectus*? In view of the flood of information produced by modern mass media we must ask the additional question: Is the information appropriate and necessary, or does it only confuse by diversion (for example, with Neil Postman's suggestion that "We are amusing ourselves to death"). While I do not want to say anything against an amusing or creative presentation of problems which have been left unheeded so far, I do want to warn against clogging up the limited channels with irrelevant matter. Again and again we hear the suspiciously unanimous complaint about the tendency towards shallowness, thrillers and pornography. However, by far more alarming is the partiality of media ideology, which leads to thoughtlessness. We always hear enthusiastic words about freedom, responsibility and emancipation, while in reality a masked ideology is instilled, the consequences of which will become evident later.

The reason and good sense of a culture governed by mass media should first of all become apparent in the reason and good sense of its media - but little appears here! The basic cultural problem of our time is mass excitement created by pretences and indifference to the real. This spirit of the time swells diabolically and creates movements which begin with enthusiasm and end in horror, or at least in disappointment. In the beginning I quoted Konrad Lorenz who wrote that performances serving the preservation of the species have been achieved under pressure of natural selection. Now, when we look at the media we see that in the absence of the pressure of selection performances prevail which destroy the species.

But let us contemplate the problem of mass media influence from a certain distance.

The following can be said of modern mass media (print media, radio and television, cable TV and satellite TV)

- They keep people occupied an average of several hours a day.
- They are perhaps the most effective means of political and commercial propaganda.
- They are even under suspicion of having decided the outcome of elections.

However their cultural effects have remained obscure. Experts state quite frankly that "Our understanding lags far behind our measurements." Research of media effects has not yet left the pre-scientific phase, in which guess-work still dominates. The theses put forward by Arnold Gehlen (*Gesamtausgabe*, vol 7, V. Klostermann, Frankfurt) are of particular interest:

> "The battle for the greatest influence fought by the intellectuals for the past two centuries has - after the popularisation of television - almost resulted in the establishment of a counter-government, which may intimidate the legal government and force it to abandon its goals ... Here lies the enormous power of the mass media whose influence can hardly be overestimated. In former times there were disputes and arguments about the issue of their influence, but since Vietnam and Watergate we may, no doubt, close the files."

Even more convincing than the examples of Vietnam and Watergate quoted by Gehlen is the fact that a single TV broadcast has brought the fishing industry to the verge of destruction within a matter of days leaving thousands of workers unemployed. The well-known TV journalist P.Voss wrote (*Herrenalber Texte 62*, Tron, Karlsruhe, 1985):

> "Political events are made, and not least with the intention of influencing the public ... Personally I am always annoyed when TV journalists pretend that our medium has no political effect, and by no means an effect which might change the political situation, i.e. the distribution of power, for instance by influencing the outcome of elections. Who passes judgement in the political trial? The voter. Who, however, pursues fact finding by the hearing of witnesses or collecting circumstantial evidence? The media. And who guarantees equality of weapons?"

Again and again it has been maintained that the mass media are completely ineffective. Walter Jens said, for instance, in 1977 on Channel II of West German TV after the murder of Hanns Martin Schleyer:

> "No sentence spoken is immune to misinterpretation ..... Rousseau is not to blame for Robespierre. This means that establishing an immediate cause-effect relationship is at least extremely problematical."

The opposite opinion was held by Heinrich Heine, who was closer than Walter Jens to the mass murder of the French Revolution:

> "Behold, you proud men of action, you are nothing but unconscious stooges of the

men of ideas, who ... traced out all your actions most precisely. Maximilian Robespierre was nothing but the hand of Jean-Jacques Rousseau."

Many opinion makers, however, seem to have or to want no understanding for the fact that they influence human behaviour and political events; many exude their aphorisms as if this were a private Glass Bead Game - and are indignant when held responsible for the consequences.

In this general confusion permit me to name quite clearly the cardinal failure in the construction of our present media culture

- Propaganda in the media is predominantly oriented to short-term interests - circulation, audience ratings, etc.
- The effect of the media becomes apparent in space and time outside of the media horizon. Therefore, there is no incentive or compulsion for those in the media to strive for reason.

In the final analysis the problem is an educational one: Do citizens possess the intellectual equipment to permit a sensible choice between:

- Credible and incredible
- Sensational and "good" information

A Critical Consideration of Kornhuber's Concept of the Brain-Mind Problem

H.-D. Henatsch
Physiologisches Institut, Georg-August-Universität, Humboldtallee 23, 3400 Göttingen, FRG

Professor Kornhuber's great merits in modern neurology and neurophysiology have already been appreciated by other participants at this meeting. My concern here is with another exciting field of his interests, namely his continuing reflection, over many years, on the philosophical consequences of nearly all aspects of the ancient psychophysical problem or in modern terms, the brain-mind problem. I had the pleasure of several private discussions with him on such topics. In these friendly exchanges our opinions were convergent in some points, but divergent in others.

I attempt here to consider briefly some of the controversial aspects of the brain-mind problem as Kornhuber has viewed it. The time available does not allow me to go deeply into details. I refer here mainly to two paradigmatic articles, both published by him in 1978: one in English with the title "A Reconsideration of the Brain-Mind Problem" (1978a), the other in German, entitled (translated) "Mind and Freedom as Biological Problems" (1978b). Numerous other, more recent papers by Kornhuber (e.g. 1984a,b; 1987a,b), in which he discusses several subquestions of our problem such as memory, recognition, will and attention, and the relative freedom of our action, must remain unconsidered here.

As a starting point, I wish to emphasize which of his statements has worried me most and evoked my principal objection: "Neuroscience shows that the body-mind problem is empirically solvable step by step" (1978b). And in another context he even claims that "in principle, the psychophysical problem has been solved ,and in a way that allows meaningful research" (1978a). These optimistic formulations irritate me in view of the historical fact that over thousands of years even the best philosophers have been unable to find a definite or at least acceptable solution to this problem. Let us try to follow the main line of Kornhuber's arguments which make him believe that "in principle" the right answer has been found.

He first reminds us that it was Descartes (1662) who sharply distinguished between the *res extensa*, that is, all matter including the living body and its brain, and the *res cogitans*, that is, the spiritual or mental aspect of our being. This was the reintroduction of dualism into philosophical thinking at the beginning of the modern age. From a dualistic viewpoint, there were, and still are, two attempts to explain the relationship between the psychic and the physical events which are commonly known: one, psychophysical parallelism, the other, psychophysical interaction. I need not report in detail how Kornhuber points out, I think rightly, that neither the one nor the other of these theories can give us a real explanation of the essence of the psychophysical relationship. I share his opinion that it is necessary to overcome the whole doctrine of dualism, which separates nature into two independent - and erroneously equalized - parts. Only in our verbal descriptions of the two kinds of phenomena we nearly cannot avoid to using terms of duality when trying to indicate their apparent diversity.

Rejecting dualistic approaches, Kornhuber then seeks some kind of monism which might offer a better way. Like many other modern researchers, he concentrates on the physiological basis of consciousness and other aspects of mind, that is, he tries to trace them back to physical events. However, he would strongly protest if we would consequently classify him as a mechanistic reductionist. He tries to avoid the two common pitfalls of monism. On one hand, he "does not intend to devaluate conscious experience as an epiphenomenon of brain processes" (1978a). Instead, he stresses that "it was an important rediscovery of the twentieth century, that the Mental or Spiritual is an independent aspect of nature" (1978b). On the other hand, he brings strong arguments against any kind of panpsychism, as can be found, for instance, in the thoughts of Spinoza (1677), Fechner (1860), and in our days by Rensch (1977): "Panpsychism, however, neglects the *evolution of order* in the history of the universe" (1978a; my italics).

What must be done, in his view, is to find a common denominator for both the psychic and the physiological events in the living organism. He finds it by stating that both processes have the same information content. So information, as defined by modern cybernetics as an expression of "order" in the world, becomes one of the key notions in his thinking. Again he notes: "The idea that information is a fundamental aspect of nature was a most important rediscovery of this century. The other important discovery was Einstein's equivalence of matter and energy, which reduced the content of the world to energy on the one hand, and order on the other" (1978a). He stresses that order and information content are not at all restricted to living beings but are also present, to various degrees, in the nonliving physical world. Here the idea of evolution as the decisive factor enters: "With the evolution of life came the need for preservation, transmission and interpretation of specific orders." The stepwise improvement of information processing in the evolution of our organisms "is a continuation on a higher level of similar processes in the physical world" (1978a). In higher order nervous systems, the combined effects of energy, order, information and evolution culminated in the appearance of an inner aspect of the surrounding world and the subject's existence in it. This symbolic internal representation of the living Self and its environment, produced by the integration of innumerable nervous activities in the brain, is the basis of consciousness and thus an essential property of mind.

So far, so good. We might accept the idea of Kornhuber and many other workers (see, for instance, the evolutionary theory of epistemology by Lorenz 1973; Vollmer 1975; Riedl 1980) that the drive to survival pushed evolution to "invent" the mind, needed for better control of the nervous system. This is a metaphoric description but not an explanation. However, Kornhuber claims that the above-mentioned combination of factors contributing to this marvellous "invention" already contains the principal solution of the brain-mind problem. He simply offers his definition of mind which he thinks is sufficiently self-explanatory: "In human, like in other higher living beings, soul or mind is the flow of information in the living nervous system" (1978a).

This is the point at which my objection begins. Of course, there is no doubt that mind uses and needs information as its vehicle, and that information processing is always essential for mental functions. But may we say that "mind is the flow of information" - and nothing more? To become meaningful, information always requires a sensing, understanding and evaluating recipient, and this is what mind provides in the brain. Here, I think, Kornhuber does just what he had promised to avoid: his postulate disqualifies mind as an epiphenomenon, this time of the cybernetic events in the brain.

Kornhuber's thesis that mind is identical with the information flow in the brain brings him in close proximity to other proponents of modern identity theories of mind and matter. They must all assume that consciousness and mentality appeared merely as a new emergent property, as a "fulguration" in the sense of Lorenz (1973), when the evolving brain matter had reached a certain high degree of complexity. But this metaphor leaves open the crucial question of how this matter could develop such a new property. Modern cybernetic models of complex neuronal networks, capable of multiple and simultaneous parallel processing of much different information, might give some help in recognizing some of the structural and functional requisites for such new tasks. But the fact that the brain-mind complex is suddenly capable of self-consciousness, of feeling as a thinking and self-acting personality, of possessing emotions and creativity, introduces a new ontological dimension of the human nature which remains unexplained.

It is easy to criticize, but much more difficult to offer an alternative suggestion. It is beyond my competence to try this. But I can at least give some helpful hints, based on recent theoretical developments, as to why this task is fundamentally so difficult, and why we cannot be perfectly sure of finding a proper answer. We have hitherto neglected an important aspect of the brain-mind problem, namely its metatheoretical character. We speak of metatheory whenever science studies, in a self-reflective way, the foundations of science itself. Thus the application of logical principles to logic itself is metalogic, the mathematical analysis of the axioms of mathematics is metamathematics, and so forth. This implies that self-reflective thinking about our thinking, any attempts at explaining the brain by the brain, also have a metatheoretical character.

An excellent account of this trend in modern science can be found in a recent book by the German molecular biologist and biophysicist Gierer (1985). He points out that metatheoretical thinking inevitably leads us to the basic limitations of our human understanding, of our epistemic capacities. We end up in uncertainties, or in undecidedness of several thinkable possibilities. I take a few examples given by Gierer: If we study the ultimate nature of matter and energy, we come to Bohr's principle of complementarity, which combines pairs of strictly contradictory properties of the microphysical elements to a unity beyond our understanding, and to Heisenberg's fundamental principle of uncertainty of the constituent physical parameters. If we take a logical system of sufficient richness, we have learned from Gödel that there are always formally correct logical statements which the system cannot decide true or false. It follows that such metatheoretical uncertainties introduce an element of basic ambiguity into our thinking. We feel that we are unable to grasp the ultimate truth, that there must be a higher dimension of reality beyond the world of observable physical objects. This dimension remains hidden to us, but its presence is indicated by apparent incongruities and paradoxes at the borderline of our thinking. We can only try to interpret them, with our imperfect language, by different metaphors, but cannot decide with certainty which one might be nearest to the hidden truth.

Applying this to the brain-mind problem, we must also take into account its metatheoretical ambiguity. This means that any serious concept of the brain-mind relationship is merely one of several possible, provisional constructs or pictures of what is going on in the brain. To take an example, the thesis of identity of mind and brain is one , which may well contain a limited element of truth; however, with the mental reservation that we do not know for sure what a "mentalized" brain really is. One can, on the other hand, reverse the identity thesis by postulating a "materialized" mind. This was attempted, for instance, by the German physicist and philosopher von Weizsäcker (1971), who some years ago said "...that man is indeed a mental being, but that the subject of natural science is nothing strange to the mind, but just mind itself, insofar as it submits to the rational operation of differentiation and objectivation" (my translation). This again seems to be a legitimate but limited approach to the truth, and it is up to you, which version you might prefer.

Finally I come back to my friend Kornhuber. I have criticized some points of his brain-mind concept, and I am sure he would like now to criticize my critique. There will be opportunities to continue our disputation. But let us not forget here to emphasize his great merits in this field of neurophilosophy, too. He has contributed many new facts about the physiological processes that underlie psychic events and their disorders in cases of brain damage or dysfunction. He has always strongly defended the real existence of mind as part of living nature, particularly in humans. Although I cannot accept his "explanation" of mind, he has rightly demonstrated that empirical neuroscience research can contribute substantially to a better understanding of the whole problem. He has given special attention to the related problem of human freedom, which is restricted but really exists, as he has always emphasized against the denials of many modern thinkers. He has given empirical hints to recognize that this freedom is not contradictory to natural laws of the physical world, but rather a consequence of the high levels of order in our brains (1984b, 1987b). And we must acknowledge his concern that our present state of insight can help us to support and advise patients with mental problems and to educate young people for better control of their living conditions.

Summing up, Professor Kornhuber has at least granted us a fresh and promising approach - though not a definite solution, as he had thought - to the psychophysical problem. Certainly there remain open philosophical questions as to the nature of mind in the brain. But we have learned from him that further progress in this approach can be reached, to use his own words, "in a way that allows meaningful research". Let us contribute, as far as our limited mental capacities allow us, to this progress.

Summary

Professor Kornhuber has claimed that, in principle, the psychophysical problem has been solved by empirical neuroscience research. His philosophical concept is critically analyzed and rejected in some points. He conceives mind as being the flow of information in highly evolved living nervous systems. His approach is not an explanation, but at its best one of several possible metaphoric pictures of the events in the brain. It is presently impossible to decide with certainty between different reasonable approaches to the problem due to its fundamental metatheoretical ambiguity.

References

Descartes R (1662) Tractatus de homine et de formatione foetus. Leiden

Fechner GT (1860) Elemente der Psychophysik. Breitkopf and Härtel, Leipzig

Gierer A (1985) Die Physik, das Leben und die Seele. Piper, München

Kornhuber HH (1978a) A reconsideration of the brain-mind problem. In: Buser P, Rougeul-Buser A (eds) Cerebral correlates of conscious experience. Elsevier/Northholland Biomedical, pp 319-334

Kornhuber HH (1978b) Geist und Freiheit als biologische Probleme. In: Die Psychologie des 20.Jahrhunderts, Vol. 6. Kindler Zürich, pp. 1122-1130

Kornhuber HH (1987b) Handlungsentschluβ, Aufmerksamkeit und Lernmotivation im Spiegel menschlicher Hirnpotentiale. Mit Bemerkungen zu Wille und Freiheit. In Heckhausen H, Gollwitzer PM, Weinert FE (eds) Jenseits des Rubikon. Der Wille in den Humanwissenschaften. Springer, Berlin Heidelberg New York, pp 376-401

Kornhuber HH (1984a) Mechanisms of voluntary movement. In: Prinz W, Sanders AF (eds) Cognition and motor processes Springer, Berlin Heidelberg New York, pp 163-172

Kornhuber HH (1984b) Von der Freiheit. In: Lindauer M, Schöpf A (eds) Wie erkennt der Mensch die Welt? Geistes- und Naturwissenschaftler im Dialog. Klett, Stuttgart pp 83-112

Kornhuber HH (1987a) Gehirn und geistige Leistung: Plastizität, Übung, Motivation. In: Rheinisch-Westfälische Akademie der Wissenschaften (ed) Vorträge N 354, 7-28. Westdeutscher Verlag, Opladen

Kornhuber HH (1987b) The human brain: from dream and cognition to fantasy, will, conscience and freedom. In: Markowitsch (ed) Information processing by the brain. Huber, Bern

Lorenz K (1973) Die Rückseite des Spiegels. Versuch einer Naturgeschichte menschlichen Erkennens. Piper, München

Rensch B (1977) Das universale Weltbild. Evolution und Naturphilosophie. Fischer, Frankfurt/Main

Riedl R (1980) Biologie des Erkenntnis. Die stammesgeschichtlichen Grundlagen der Vernunft. Parey, Berlin

Spinoza B (1677) Ethica. In: Opera Posthuma. Jan Rieuwertz, Amsterdam

Vollmer G (1975) Evolutionäre Erkenntnistheorie. Hirzel, Stuttgart

von Weizsäcker CF (ed) (1971) Das philosophische Problem der Kybernetik. In: Die Einheit der Natur. Hanser, München, pp 280-291

Brain, Mind, Freedom: Beyond Metatheory, Nearer to Reality

H.H. Kornhuber
Department of Neurology, University of Ulm, 7900 Ulm, FRG

Although this book represents the work of my friends, I have been asked to comment on Professor Henatsch's views (this volume), which I consider to be important for the education of today's students and physicians. In the contribution (34) to which Henatsch refers I said: "To talk about the relationship between brain and consciousness is an awkward task. Whatever you say, somebody will be annoyed, and most people are convinced from the outset that it is a useless problem to consider because it is unsolvable." Although Professor Henatsch seems to agree, I would like to thank him, because he is one of the few today who take the psychophysical problem seriously. Furthermore, I appreciate his and Gierer's efforts to dig out possible mistakes in existing hypotheses. Nevertheless, I would like to point out why it is better to continue research on the problem than to give up, secure in the knowledge that complete understanding is unattainable.

Self-reflection and thinking about and beyond the bases and boundaries of what he knows is a facility unique to man. Since its beginning in ancient Greece, philosophical thought was both theoretical and metatheoretical, and philosophy remained so until Hilbert (22,23), Popper (59) and Jaspers (24,25). Metatheories can, however, be either illuminating or misleading. Gierer's (15) metatheory of the psychophysical problem is not particularly helpful, for he does not look at neuroscience to find new ways for research (as for instance Schrödinger (62) looked at biology); he tries to limit scientific knowledge in order to justify the extreme variations of religions and philosophies (which he mistakenly assumes are metatheoretical in content). This attempt at delineation is proposed to end intellectual conflicts; it may be temporarily calming for some, but is in the long run counterproductive for all. Man obviously needs religion, but only that which is compatible and in fruitful communication with his knowledge (32). Isolation is counterproductive not only for the discourse between science and religion but also for cooperation among the different scientific disciplines. Intellectual isolation ultimately may tend towards murder, as has been made clear already by Socrates and underlined by Jaspers. Instead of delimitation, we should encourage communication in the search for truth. It should be remembered in this context that the success of living cells, the brain and the mind occurs because they are cooperative systems. In history, Solon (the founder of the world's first democracy) and Confucius (4) (the most influential thinker of the far east) stressed cooperation.

If an outsider looks at neurology and realizes the extraordinary number of neurons and their interconnections in the human brain, it seems hopeless to gain even a small insight in one's lifetime. Our possibilities for understanding some parts of the brain are, however, better because all organisms fulfil certain functions and therefore consist of subsystems that can be investigated in specific ways. Also we look from both without and within at the brains functions, aided in part by the lesions and diseases produced by nature. Finally, we have the cooperation of many researchers attacking a common problem.

Gierer admits that there are partial solutions to the psychophysical problem and merely states that it cannot be solved completely, but when he reaches his conclusions, he forgets that partial solutions are possible. Thus, when Henatsch ignors Gierer's statements about the existence of partial solutions and suggests that no solution is possible, he agrees with Gierer's tendency. The abundance of existing psychophysical and physiopsychic correlations, many of which are used daily in neurology, have however been the starting point for brain-mind theories since the time when Hippocrates realized that there is a natural explanation for epilepsy, which had previously been considered to be a supernatural disease. Thinkers such as Aristotle, Descartes (7) and Leibniz (49), like the 20th century authors of cybernetics, felt that we should respond to these data in the normal scientific approach, designing theories for more coherent understanding and further testing.

There have been many positive extensions of insight that have come from metatheoretical considerations. In my opinion, the largest step in the enlightenment of the human mind came when the ancient Greek philosophers, reflecting on day and night, and on land, sea and air and other kinds of order and of change came to realize that there is a deeper order in all this order and a deeper lawful change in all this change; the terms cosmos (order) and physis (growth or lawful change of order) came from this metatheoretical revolution. At the same time, researchers came to believe that man will gradually be able to understand this order. All of this seems self-evident to many of us, but not all: consider the god of the Bible, who changes sometimes his mind and does what he wants in an unpredictable manner, or the beliefs of the spiritists. Friedrich Spee von Langenfeld who was a professor of moral theology and had to serve as a pastor of "witches" (63) recognized that the terrible witch trials were the result of insufficient knowledge of nature. In our modern world of abundant information it is often forgotten that our insight into the order of the world belongs to the fundamentals of our humanity.

Kant's (30) and Nicolai Hartmann's (18) metatheoretical approaches to science produced fruitful questions. The background laid for psychopathology by Jaspers (24), Einstein's theories of relativity (11,52) and Hilbert's investigations (22,23) into the bases of logics and mathematics also employed methatheoretical considerations to produce positive extensions of insight. Thus, metatheory does not necessarily result in agnosticism or in the attitude that "anything goes", as was suggested by Gierer (he mentions Feyerabend as one of those who gave momentum to his thoughts). It is true that sceptical second thoughts are important moments in search, but we also use existing knowledge to reach constructive ends. Although complete knowledge is impossible, as was intuitively seen by Xenophanes (69), Heraclitus (20), Lessing (50) and Kant (28,30), the ancient dream of philosophy that understanding is possible has become more and more reality. It was the great metatheoretical thinker Hilbert who gave order to the inscription on his gravestone: "We must know. We will know." This was a response to Du Bois-Reymond's (8) "We don't know, we never will know" (with regard to the psychophysical problem).

When we read books from the 19th century that discuss the body-mind problem we gain a better appreciation of cybernetics. The 19th century was obsessed with matter and power to the point that it was said: If God exists, he must be a gaseous mammal. Even in our century the great scientists Hans Berger (who discovered the EEG) and Walter Rudolf Hess (the Nobel prize winner in neurophysiology) searched for the mind in the direction of a special, not yet discovered power or energy (3,21). For Plato (58), the body was mortal, the soul immortal; Descartes (7) saw the body as *res extensa*, the mind as *res cogitans*; Kant (30,31) distinguished *mundus sensibilis* from *mundus intelligibilis*; for Max Planck (57) the duality was a problem of how the brain was viewed (from outside or from inside). For the cybernetic thinkers of the 20th century, however, it is a matter of hardware and software, of the external and internal aspects of information processing and, more generally, of energy and order - especially since Einstein (10) made it clear that matter is equivalent to energy. There is still a mind-brain duality, but there is some commonality, and we can use the commonality to help patients with disorders such as psychosis or mental retardation. For example when we see that an occipital lesion may cause psychic blindness, whereas an orbito-frontal lesion causes a loss of morality (in both cases, the patient is unaware of the defect in the first weeks) we have learned something about the brain-mind problem.

Order, the main concept of the cybernetic revolution is not new but is a rediscovery from the presocratic philosophy. This idea was shared by such eminent thinkers as Pythagoras, Xenophanes (68) and Heraclitus. Heraclitus (19) saw in the cosmos, on one hand, the driving fire (we call it energy), and on the other, order and its lawful change. Aristotle (who was the most influential thinker in the western world) taught that the soul is the acting order of the body (1). Although all of us leave the results of our thinking for further testing by researchers to come, it seems unlikely that the mind will ever again be believed to be a subspecies of power or energy, although some energy is necessary to keep the mental process of information processing alive. Indeed, the mind appears to us (as to Xenophanes) a prototype of order brought into existence by a self-organizing process (the understanding of which is, of course, far from complete).

Here we find the main difference between Hans-Dieter Henatsch and the view of cybernetic authors such as Norbert Wiener (67), Karl Steinbuch (65), Konrad Lorenz (53), Manfred Eigen (9), Hermann Haken (17) and myself. Henatsch believes that information requires the mind as an understanding recipient, while the cybernetic authors see the understanding of order performed by means of already existing order of the system. This order is in the newborn infant as the result of phylogenesis and ontogenesis; it is later refined by learning and creative thinking. This order and its evolution can be recognized step by step. Demokritos (6) already realized that order may even result from chance processes such as the ordering of the pebbles on a beach. Empedocles taught that love is recognized by love (12), and Goethe (14) believed that the eye can see because it is by nature adapted to the sunlight. Also since Darwin and Vollmer (66) it has become clearer that the answer to Kant's question as to how it came that knowledge is possible is: because of the phylogenetic evolution of order. There is nothing in the data which precludes this conclusion; on the contrary: the psychophysical correlations and their reliability and validity as proven by daily clinical testing and the lack of an alternative explanation are in favor of this natural theory of the mind.

There are abundant examples from touch (55) to volition (5,41,48) and there is some understanding of the bases from the molecular (27) to the mathematical level of the system (61). I will mention just one finding which is in agreement with the theories of the distributed cerebral system (54) and the system's causality (64). There is - independent of the lobe involved - a quantitative relationship between the size of a cerebral lesion and the intellectual and behavioral deficiency (44). When we do research, all of us are looking for order by means of the order in our mind, thereby contributing to better order in both our thoughts and the "objective mind" of Hegel or Popper's "world 3" (60) or the communicative world of Herakleitos (20a) in which all minds participate when awake and collected. If the principle of the mind would be something other than order, we would not only need another type of brain, but also another world, for one of the two principles of the world is order. In the living human brain and our minds there is, however, a higher order which is able to take up, process, store and evaluate the order of the world and thus must have played a role in the biological evolution of the human brain in the past as it influences, even more, the further evolution of human culture. When we try to understand this process we have to look not only at elementary events but also on the cooperation of the parts in the system and on the systems causality. A good example is the evolution of the endocrine system (Bückmann, this book) because it is similar to the nervous system in some aspects.

Henatsch's statement that information processing requires the mind as an understanding recipient seems to indicate that in his view the principle of the mind is something other than information processing which includes communication. Is a time-honored mistake the background of this assumption: the idea of the self-sufficient mind? In reality, communication is an important point of the mind because without communication validation and improvement for functioning near to reality is impossible. Communication with the real world is, however, not possible without physical causality.

Henatsch's criticism seems, in part, a consequence of confounding the two concepts of principle and totality. One may understand the principle of memory storage without understanding all the memory content. Also, an understanding of the pictographic and alphabetic principles of writing does not insure an understanding of the whole content of a book.

Another source of uneasiness with information may be the fact that information theory considers only the quantitative aspect of order. There are, of course, many qualitative aspects of order which are reflected in our mind by impressions of beauty or sublimity and by moods, feelings and emotions like love, worry, desire for activity etc.

The central point of the problem is, however, in my opinion, freedom, a concept which is, unfortunately, even more confused, neglected and abused today than the concept of mind. The mind has sometimes been regarded as a passive, recipient being. If, however, you consider the mind not only as internal processing but as a spirit which acts out its ideas, it is clear that freedom is the central point. By freedom I mean here positive freedom which is the ability of man to make use of his hierarchically organized system, of his intelligence, conscience and good will in such a way as to live in agreement with higher values (33, 38). By will I mean the ability of man to set priorities among his needs, priorities based on long term thinking, on reason and conscience (41, 41 a). The ideas of will and freedom are nowadays more and more eliminated from the scientific literature. Unfortunately this is consistent with a metatheory that tells us that these concepts cannot be understood any way - it's the spirit of the age.

It is perhaps this apathy regarding a positive understanding of freedom, what evokes some of the arguments against a positive theory of the mind. If we only understand freedom as "freedom of...", we do not see the need for a positive theory of the mind, since independence seems to come from democracy in a similar way as electric power comes out of a plug. When we realize, however, that every "freedom of..." ultimately is based on "freedom to...", i.e., on positive freedom which consists of abilities, performance, creativity, cooperation etc. (38, 41a) we also see the need for a positive theory of the mind. Freedom from robbery, for example, is based on order maintained by a state, freedom of illusion is based on reason whereas freedom of infectious disease is accomplished by the performance of the immune system.

Freedom cannot be explained by chance, although chance may contribute to creativity. Chance is as much in the earthworm as in man, the difference between these two species, however, is mainly in the organization of their nervous systems. Nor does the division of the world in two or more worlds explain freedom. If there is no causal relation between the two worlds, they cannot interact. If, however, they interact, they ultimately belong to the same world. Freedom is not a lack of order, rather it is a higher order of behavior. One prerequisite of freedom is intelligence; this would, however, be useless if reason could not influence behavior by setting priorities from an ethical perspective; therefore, the other prerequisite is good will. Almost half of the large cerebral cortex of man, the frontal, orbital and anterior medial cortex is devoted to will (41). To seek freedom primarily in the direction of independency led to an illusionary freedom of nature. Real freedom, however, which is relative freedom, is a matter of order, information processing, decision making with reason etc. Physicians prevent degradation of freedom for instance by early treatment of hypothyroidism in newborn infants or by curing syphilitic dementia. Becoming more free is, however, usually a laborious process of developing ones abilities, using help from parents, friends and teachers, giving help to other people and by long term training in self-discipline and self-education.

The creativity of the human brain depends on an orderly mixture of deterministic functions and chance processes. Total determinism was originally not a scientific idea, but came into physics from the apocalyptic branch of theology. The founders of science, the presocratic philosophers saw on the one hand order and it's growth (physis), on the other hand chance events (6). Today it is clear that chance events occur not only in microphysics but also in macrosystems (16) including our brain. Dream content is a good example of chance events and how our brain tries to make something reasonable out of them (37). Similar chance events may contribute to our phantasy which is usually, however, under reasonable guidance by our will. But chance events alone do not explain freedom; it is only by higher organization and higher programs, by education, reason, values, good will and by feedback from the human social world, that freedom becomes a possibility: not by less determination but by higher determination, by doing service to a larger complex of facts and duties (33, 38). In ancient times, the concept of positive freedom is apparently lacking. What Nietzsche, following insights of Kant and Goethe, called "freedom to ..." (56) was in older times discussed under the concept of virtue (2). This positive freedom has to do with learning by doing, with balance, courage and authenticity. As freedom in a state is based on a stratified system with resources, social organizations and lawful decision making bodies, freedom in the human person is based on a synergetic "hierarchically" organized system of energy metabolism, hormonal regulations etc. as a basis, of numerous drives, learning, phantasy etc. and with reason, conscience and good will as responsible and unifying guides (33, 38, 41a).

Because of the incompleteness of knowledge and because of the necessity of action in life, we have to consider the practical consequences of our theories and tacit prohibitions of thinking. The modern dogma that we cannot solve such problems as mind and freedom (which was alien to such eminent thinkers as Cusanus or Leibniz) results in a kind of anti-enlightenment in which determinism and hedonism (which are extremely sceptical about good will and positive freedom) prevail. Those who push the youth into alcohol and cigarette dependency refer to a common one-sided understanding of freedom as emancipation. To cure this decadence of our mind, we must strive for a positive understanding of our task of mind and freedom and we must use positive knowledge. Because mind is order, we can help people not only by pharmacological means but also by parents' love, by education, by jobs, by logotherapy (13), etc. To counteract the tendency of affluent societies towards diminished discipline, self-destruction, towards astrology, charlatinism etc. a metatheory which encourages indifference is not helpful. What we need instead is a positive family policy (40), better education, an earlier start of creative mental work for young people, objective evaluation of research perfomance in order to encourage good work (43), departure from our self-destructive habits (35) etc.

Gierer suggests that it does not matter which type of philosophy, ideology or religion we choose: he suggests that all or at least most of them are compatible with science, peace etc. Is this true? In the Gulf war between Iraq and Iran, the Iran has indoctrinated prisoners of war with the shiite brand of moslem religion. In the Korean war the Chinese used "brain washing" methods to indoctrinate their American prisoners with communist ideology. Under communist rule in the Soviet Union scientists who did not follow the Lysenko line of anti-genetic biology were prosecuted just to mention one point of the archipel Gulag (61a, 62a). It is much longer ago that the christian church closed the platonic academy which had existed for thousand years, prosecuted heretics and "witches" and burned scientists who believed in the Kopernican heliocentric system. But as late as 1953, the pope denied freedom of religion, and only at the II. Vatican Council religious freedom was officially acknowledged by the catholic church. Luther tried to eliminate natural philosophy from religion and advised that the synagogues be burned. Under the influence of modern research, however, the protestant theology investigated the sources of its tradition. This gradual development of most religions and ideologies towards reason was since the ancient times influenced by science and science-based philosophy. Karl Jaspers (one of the few great modern philosophers who knew and appreciated the bible) pointed out that among the many religions the three biblical ones are the intolerant (26). The reason is that the other religions are based on nature while the jewish, christian and moslem religions are based on solitary historical revelation (originally, as Arthur Koestler sees it, in the jewish religion with a national touch (31a,b)). The ancient greek philosophy had natural and undogmatic religion as background as was emphasized by E. Zeller (70). The integration of the nature and science-based philosophy into the christian teaching was a great achievement of the medieval time. It was not by chance that this philosophy developed the idea of humanity and overcame ethnocentrism about 500 years before christ (47a), while the three biblical religions, based on solitary historical revelation, had - despite Jesus, Franciscus and christian patient care - a tendency towards intolerance (26). The philosopher Karl Löwith, himself of jewish origin and prosecuted by the nazi, believed that the intolerance of Marxism came from secularized biblical ideas (51).

The ancient greek philosophers, on the contrary, saw that it is more important to behave in a gentle humane way than to be a citizen of Athens or a member of a rich family. Aristippos, a student of Socrates, pointed out the importance of anthropismos, a term that was later translated to humanitas when Panaitios brought stoic philosophy to Roma. Not the limitation of knowledge but knowledge itself and, under its influence, enlightenment had a positive influence on tolerance and peace. Similarly, it is not the limitation of search that brings freedom: instead knowledge shows us the task of freedom and the means by which to help people to become or to stay more free.

Is it not so that certain types of philosophy tend to be associated with certain types of behavior? The philosophy most typical for affluent times, hedonism, tends to have destructive consequences - not only for the acting individual, but also for his relatives because everything that man does or does not do influences other people as well; e.g., alcohol embryopathy. Despite steps backwards, abuses of liberties etc., improvements of freedom have been accomplished by health policies, family policies, education policies etc. In discussions with students I realize, however, that in thinking about the problem of freedom today one must first overcome the dogma that the body-mind problem is insolvable. Any explanation of these great problems is, of course, only an approximation.

Freedom is given to us as potentiality, as a task; this was seen by Aristoteles (2) and reemphasized by Kant (29,31). My observation of wide-spread misconduct such as cigarette smoking or the daily "social" alcohol consumption which results in traumatic brain damage or early stroke (via high blood pressure, obesity and diabetes, 42), makes me believe that rather than advocate a permissive metatheory we should take serious what we know and use it to improve our habits and to advise our lawmakers. Intolerance is not a typical problem of scientists; on the contrary, we are often too tolerant regarding nonsense in our textbooks - nonsense, nota bene, which is flattering our wishes.

Two important misconceptions which are typical of our permissive society appear in current textbooks: First, it is not higher caloric intake but "social" alcohol consumption that is the most common cause of obesity , high blood pressure and type II diabetes in the human male today. The pathomechanism is hepatic insulin receptor damage by "normal" daily alcohol (which in the liver cell is oxydized to acetaldehyde, a very aggressive chemical agent), followed by hyperinsulinemia (43a, 44a) which causes down regulation of the peripheral insulin receptors thus stabilizing in a vicious circle the hyperinsulinemic state; high insulin, however, is a strong blocker of lipolysis so that daily "normal" alcohol turns us into "good users" of food by a toxic, non-caloric mechanism (46). Long standing hyperproduction of insulin may lead to a relative insuffiency of the pancreas and thus to manifest type II diabetes mellitus. But long before this happens, the hyperinsulinemia causes pathologically increased reabsorption of sodium in the kidney (leading to arterial hypertension), increased excretion of adrenalin (again leading to arterial hypertension), and stimulation of endothelial cell growth leading (in conjunction with alcohol induced hyperlipidemia) to atherosclerosis. The treatment of type II diabetes with oral insulin-liberating drugs is contraproductive because it increases the overexcretion of insulin which, as a consequence of insulin receptor damage, is invariably the main pathologic condition in the beginning of late onset of diabetes. The rational treatment of this diabetes is: zero ethanol and reduction of obesity. This treatment is also beneficial to normalize the blood lipids, for "normal" alcohol causes high plasma triglycerides and an infavorable LDL/HDL cholesterol ratio (43a).

Secondly, the majority of the livers of gentlemen who today are considered to be normal are fatty livers which have been damaged by "social" ethanol; the far reaching metabolic consequences are hyperlipemia, hyperinsulinemia, hyperadrenalinemia, arterial hypertension etc. The problem is that the normal world ranges of the liver enzymes in diagnostic use are erroneous. The most sensitive of these enzymes, gamma-glutamyltransferase (GGT) is thought to be normal up to 28 U/l (at 25° C) in the human male. Our colleagues from internal medicine, when standardizing these methods, forgot, however, that the "normal" population already is under the pathogenic influence of daily ethanol consumption. When external validation using correlations with known physiological ethanol indicators such as blood pressure or hypertriglyceridemia is applied, it turns out that almost the entire pathologic relationship is in the range which was believed to be normal; the real upper limit of normality for the GGT is 10 U/l (47), with errors at single measurements up to 12. "Normal" ethanol consumption rather than salt is the most common cause of high blood pressure in the human male in the industrialised countries and thus the usual cause of early stroke (45).

To take another example from economics, there is the dogma that consumption is good for the economy. Physicians, however, see that there is both constructive consumption (like shoes for the children), and destructive consumption (such as cocain, cigarettes and alcohol). This distinction between constructive and destructive consumption (39) is lacking in economic theory so far, although the damage caused by destructive private consumption is two orders of magnitude larger than the damage by air pollution when measured by the price of dying trees (42). Because of this deficit in economic theory, our politicians did not care when the real prices of alcohol and cigarettes went down more and more. A bottle of beer in West Germany costs now only 10% of its real cost in 1950; consequently the per capita consumptions of alcohol and cigarettes have increased to 400% of the 1950 level. Although in West Germany the costs and consequences of this destructive consumption amount to over 100 billion DM per year which is more than a third of the federal budget, the politicians reject the advice to introduce the causality principle into health economics by putting a health levy on alcohol and cigarettes (which would go to the health insurance system). - Thus, although I agree with Gierer that we cannot completely solve the brain-mind problem, it would be a mistake to follow him in his permissive conclusions. There are more practical means to improve our mind than to escape to a metatheory which tends to limit our thinking. One of the prerequisites of staying more free is to see the task of maintaining positive freedom.

Just because our knowledge of the human mind is incomplete, we should look for comparable data. Here some questions arise: Does Henatsch's view also hold for an ape or a dog? Without degrading the human genius: aren't the apes our near relatives? Isn't the physiology and psychology of a young ape, despite important differences, in many aspects similar to those of a human infant? Aren't we so similar to some of our mammalian relatives that psychopharmaca for the treatment of human psychic illnesses are usually developed in animal experiments? Doesn't the concentration on the concept of mind in only the highest mammalian form cause us to forget that a cretin is a human being? - Although the mindful human brain is the most complex organism we know in the universe, we should remember that it is a part of nature. When we emphasize man's mind, we should also underline his duty to help other life on earth.

Freedom is possibility; what we realize out of freedom depends, in part, on values we find, on persons and chances we meet and on goals we choose to work on. Real life is love, but the life's life is mind, says a verse in Goethes Divan. When concepts like mind, order, love etc. have been used in this contribution, I agree with my friend Henatsch: they are incompletely comprehensible. But concepts like these are the frame of our mental work. We feel this limitation and reject dogmatism. When we do research, it is as with hiking: behind each horizon there is another horizon. Jaspers (25) therefore named, what is called god in the bible, "das Umgreifende" (the all-encompassing, wider reality).

Summing up some points: There are many possible metatheories for a topic. Metatheories should open, not close avenues for research. The human mind is able to understand itself step by step in an empirical way. Minds live from communication and have history, to the framework of their content belong the mistakes of the spirit of the time: metatheory should try to correct them. To come closer to reality, communication, not delimitation, is the better strategy. For both theoretical and practical reasons, we have to go beyond agnostic metatheory and to use existing knowledge for solutions which cannot always be permissive. The mind recognizes the order of the world by his own higher order and influences in a synergistic way the evolution of cultural order. An important point of the mind is communication; communication with the real world is, however, not possible without physical causality. Because the mind is active, the central point of the human mind is positive freedom. This freedom, which is a task, may be improved by education, balance, courage etc. and by avoiding selfdestruction.

I thank Professor Albert Fuchs for his critical comments. The remaining mistakes of this paper are, however, not his.

References

1 Aristotle (ca. 330 BC) Peri psyches. English trans. by Hicks, 1907
2 Aristotle (ca. 330 BC) Ethika Nikomacheia. Edited by his son Nikomachos. English trans. by Rackham, 1926
3 Berger H (1921) Psychophysiologie. Gustav Fischer,Jena
4 Confucius (ca. 500 BC) Lun yü. German: Gespräche, trans by R Wilhelm.English: The sayings of Confucius.R.J.War, New York
5 Deecke L, Kornhuber HH (1978) An electrical sign of participation of the mesial "supplementary" motor cortex in human voluntary finger movement. Brain Res 159 : 473 - 476
6 Democritus (ca. 400 BC) Fragments 164 and 176. In: Die Fragmente der Vorsokratiker (Diels H, Kranz W (eds), 6th edn. Weidmann 1951
7 Descartes R (1641) Meditationes de prima philosophia. Paris. German: Meiner, Hamburg, 1956
8 Du Bois-Reymond E (1872) Über die Grenzen des Naturerkennens. 5th edn. Heit, Leipzig, 1982
9 Eigen M, Winkler R (1975) Das Spiel. Naturgesetze steuern den Zufall. Piper, München
10 Einstein A (1905) Ist die Trägheit eines Körpers von seinem Energie-Inhalt abhängig? Ann 18: 639-641
11 Einstein A (1916) Grundlagen der allgemeinen Relativitätstheorie Ann Physik 49: 769 - 827
12 Empedocles (ca. 460 BC) Fragment 109. In Diels, H, Kranz W (eds) Die Fragmente der Vorsokratiker. 6th edn., Weidmannn 1951
13 Frankl VE (1982) Der Mensch vor der Frage nach dem Sinn. 3rd edn., Piper, München
14 Goethe JW (1810) Zur Farbenlehre, Einleitung. Cotta, Stuttgart
15 Gierer A (1985) Die Physik, das Leben und die Seele. Piper,München
16 Grossmann S (1983) Chaos - Unordnung und Ordnung in nicht linearen Systemen. Phys. Bl. 39: 139-145
17 Haken H (1977) Synergetics. Springer, Berlin Heidelberg New York
18 Hartmann N (1940) Der Aufbau der realen Welt. De Gruyter, Berlin
19 Heraclitus (ca 500 BC) Fragment 30. In: Diels, H, Kranz W (eds) Die Fragmente der Vorsokratiker, 6th edn. Weidmann 1951

20 Heraclitus, Fragment 45. In: Diels, H, Kranz W (eds) Die Fragmente der Vorsokratiker, 6th edn. Weidmann 1951
20a Heraclitus, Fragment 89. In: Diels, H, Kranz W (eds) Die Fragmente der Vorsokratiker, 6th edn. Weidmann 1951
21 Hess WR (1957) Beziehungen zwischen psychischen Vorgängen und Organisation des Gehirns. II Teil. Studium Generale 10: 327-339
22 Hilbert D (1928) Grundzüge der theoretischen Logik. Springer, Berlin
23 Hilbert D (1934, 1939) Grundlagen der Mathematik, vols I and II. Springer, Berlin
24 Jaspers K (1913) Allgemeine Psychopathologie. Springer, Berlin
25 Jaspers K (1947) Von der Wahrheit. Piper, München
26 Jaspers K (1954) Die nichtchristlichen Religionen und das Abendland. In: Philosophie und Welt. Piper, München 1858
27 Kandel E (1985) Cellular mechanisms of learning and the biological basis of individuality. In Kandel, ER, Schwartz, JH (eds) Principles of neural science, 2nd edn, Elsevier, New York, pp 816-833
28 Kant I (1783) Prolegomena zu einer jeden künftigen Metaphysik, die als Wissenschaft wird auftreten können. Hartknoch, Riga
29 Kant I (1785) Grundlegung zur Metaphysik der Sitten. Hartknoch, Riga
30 Kant I (1787) Kritik der reinen Vernunft, 2nd. edn., Hartknoch, Riga
31 Kant I (1788) Kritik der praktischen Vernunft. Hartknoch, Riga
31a Koestler A (1955) Judah at the crossroads. In: The trial of the dinosaur and other essays. London, Collins, pp 106-141
31b Koestler A (1976): The thirteenth tribe. German: Der 13. Stamm. Molden, Wien
32 Kornhuber HH (1977) Über Religion. Süddeutsche Verlagsgesellschaft, Ulm
33 Kornhuber HH (1978) Geist und Freiheit als biologische Probleme. In: Die Psychologie des 20. Jahrhunderts, vol 6. Kindler, Zürich, pp 1122-1130
34 Kornhuber HH (1978) A reconsideration of the brain-mind problem. In: Buser T, Rougeul-Buser A (eds) Cerebral correlates of conscious experience. Elsevier/North Holland, Amsterdam (INSERM Symposium, vol 6)
35 Kornhuber HH (1983a) Präventive Neurologie. Nervenarzt 54: 57 - 68
36 Kornhuber H H (1983b) Wahrnehmung und Informationsverarbeitung. In: Wendt H, Loacker N, (eds) Kindlers Enzyklopädie Der Mensch, vol 3. Kindler, Zürich
37 Kornhuber HH (1984a) Neue Ansätze zu einer Theorie des Traumschlafs. Nervenarzt 54: 54
38 Kornhuber HH (1984b) Von der Freiheit. In: Lindauer M, Schöpf A (eds) Wie erkennt der Mensch die Welt? Grundlagen des Erkennens, Fühlens und Handelns. Klett, Stuttgart, pp 83-112
39 Kornhuber HH (1986a) Konsumverhalten und Gesundheit. Versicherungswirtschaft 6: 358-363
40 Kornhuber HH (1986b) Positive Familienpolitik. Diskussion. In Weigelt K (ed) Die Tagesordnung der Zukunft. Bouvier - Grundmann, Bonn, pp 107-112
41 Kornhuber HH (1987) Handlungsentschluβ, Aufmerksamkeit und Lernmotivation im Spiegel menschlicher Hirnpotentiale. Mit Bemerkungen zu Wille und Freiheit. In: Heckhausen H, Gollwitzer P, Weinert F (eds) Jenseits des Rubikon. Der Wille in den Humanwissenschaften. Springer, Berlin Heidelberg New York
41a Kornhuber HH (1988a) The human brain: from dream and cognition to fantasy, will, conscience and freedom. In Markowitsch (ed) Information processing by the brain. Huber, Bern, pp 241-258
42 Kornhuber HH (1988b) Was Blüms Reform fehlt: Kostendämpfung durch Gesundheit. Deutsches Ärzteblatt 85: 1347 - 1352
43 Kornhuber HH (1988) Mehr Forschungseffizienz durch objektivere Beurteilung von Forschungsleistungen, In Daniel HD, Fisch R (eds) Evaluation von Forschung: Methoden - Ergebnisse - Stellungnahmen. Universitätsverlag Konstanz, Konstanz, pp 361-382
43a Kornhuber HH, Backhaus B, Kornhuber J, Kornhuber AW (1989a) Risk factors and the prevention of stroke. In: Amery WK, Bousser MG, Rose FC (Eds) Clinical trial methodology in stroke. Transmedica Europe, London
44 Kornhuber HH, Bechinger D, Jung H, Sauer E (1985a) A quantitative relationship between the extent of localized cerebral lesions and the intellectual and behavioral deficiency in children. Eur Arch Psychiatry Neurol Sci 235: 129-133
44a Kornhuber HH, Kornhuber J, Wanner W, Kornhuber A, Kaiserauer C (1989b) Alcohol, smoking and body build: obesity as a result of the toxic effect of "social" alcohol consumption. Clin Physiol Biochem 7: 3-4
45 Kornhuber HH, Lisson G, Suschka-Sauermann L (1985b): Alcohol and obesity: a new look at high blood pressure and stroke. Eur Arch Psychiatry Neurol Sci 234: 357 - 362
46 Kornhuber HH, Lisson G, Suschka-Sauermann L (1985c) Adipositas und Atherosklerose als spezifisch-toxische Alkoholfolgen. Öff Gesundheitswesen 47: 488-496

47 Kornhuber J, Kornhuber HH, Backhaus B, Kornhuber A, Kaiserauer C, Wanner W (1989c) GGT-Normbereich bisher falsch definiert: Zur Diagnostik von Bluthochdruck, Adipositas und Diabetes infolge "normalen" Alkoholkonsums. Versicherungsmedizin 41: 78-81
47a Landmann M (1976) Philosophische Anthropologie. 4th edn. De Gruyter, Berlin
48 Lang W, Lang M, Kornhuber A, Deeke L, Kornhuber H H (1983) Human cerebral potentials and visuomotor learning. Pflügers Arch Eur J Physiol 399: 342-344
49 Leibniz G W (1696) Eclaircissement de systéme de la communication des substances
50 Lessing G E (1778) Anti-Goeze. Braunschweig
51 Löwith K (1953) Weltgeschichte und Heilsgeschehen. Stuttgart
52 Lorentz HA, Einstein A, Minkowski H (1958) Das Relativitätsprinzip, Darmstadt
53 Lorenz K (1973) Die Rückseite des Spiegels, Piper München
54 Mountcastle V B (1979) An organizing principle for cerebral function: The unit module and the distributed system. In F O Schmitt, F G Worden (Eds) The Neurosciences, Forth Study Program. MIT Press Cambridge/Mass., pp 21 - 42
55 Mountcastle VB, Talbot W, Darian-Smith I, Kornhuber HH (1967) Neural basis of the sense of flutter-vibration. Science 155: 597-600
56 Nietzsche F (1883) Also sprach Zarathustra. Fritzsche, Leipzig
57 Planck M (1923) Kausalgesetz und Willensfreiheit. Springer, Berlin
58 Plato (ca. 400 BC) Phaedo (The second argument: 78b)
59 Popper K (1972) Objective knowledge. Clarendon, Oxford
60 Popper K, Eccles JC (1977) The self and its brain. Springer, Berlin Heidelberg New York
61 Reichardt W (1980) Analogy between hologram formation and computation of relative movement by the visual system of the fly. Naturwissenschaften 67: 411-412
61a Sacharow AD (1980) Furcht und Hoffnung. Molden, Wien
62 Schrödinger E (1944) What is life? Cambridge University Press, Cambridge
62a Solschenizyn A (1974) Der Archipel Gulag. Scherz, Bern
63 Spee von Langenfeld F (1631) Cautio criminalis, seu de processibus contra sagas liber. (German trans. JF Ritter (1939) Cautio criminalis oder Rechtliche Bedenken wegen der Hexenprozesse. Forschungen zur Geschichte des deutschen Strafrechts, vol 1, Weimar)
64 Sperry R W (1983) Science and moral priority. Columbia University Press, New York
65 Steinbuch K (1965) Automat und Mensch, 3rd edn. Springer,ü Berlin Heidelberg New York
66 Vollmer G (1983) Evolutionäre Erkenntnistheorie 3rd edn. Hirzel, Stuttgart
67 Wiener N (1948) Cybernetics, or control and communication in the animal and the machine. Wiley, New York
68 Xenophanes (ca. 500 BC) Fragment 25. In Diels, H, Kranz W (eds) Die Fragmente der Vorsokratiker. 6th edn., Weidmannn 1951
69 Xenophanes Fragment 18. In Diels, H, Kranz W (eds) Die Fragmente der Vorsokratiker. 6th edn., Weidmannn 1951
70 Zeller E (1883) Grundriβ der Geschichte der griechischen Philosophie. Neuauflage, Magnus Verlag, Stuttgart

Neuroethological Foundations of Human Speech

D.Ploog
Max Planck Institute of Psychiatry, Munich, FRG

The ability to communicate acoustically man shares with many vertebrates; the ability to speak, however, is species-specific to *Homo sapiens*. There exists a close functional relationship between voice and speech; they operate as a tandem unit in vocal communication. Significant evolutionary transformations of the sound-producing peripheral apparatus and of the central nervous system were necessary for the emergence of speech. However, the new faculty functions within a structural framework of older functions whose biology has been adapted to the new mode of communication. The transformation of the larynx and supralaryngeal apparatus led to an enormous increase in the multiplicity of vocal patterning. This was paralleled by the increasing mobility of facial muscles, especially the lips and the tongue. Figure 1 shows the way in which man differs from the chimpanzee and other subhuman primates in regard to the vocal tract. The angulation between the mouth and the upper respiratory tract is increased, the pharyngeal space is lengthened, and the back half of the tongue has come to form the front wall of the long tract above the vocal cords. During the first months of human postnatal ontogeny a similar transformation takes place: the larynx descends, and the angulation between mouth and epiglottis increases. That this transformation is necessary for the development of intelligible speech is demonstrated by children with Down's syndrome; in more severe cases they have a supralaryngeal tract very similar to that of neonates. The genetically determined malformation makes speech impossible (Lieberman 1973).

The most important difference between voicing of animals and speech in man is that voicing is due to the activities of the vocal cords, whereas speech, especially in whispers, is due to articulatory gestures with movements of the tongue, the velum and the lips. However, the peripheral nerves and their motor nuclei in pons and medulla, which execute the movements of the larynx, on the one hand, and the articulatory apparatus, the lips, the tongue, the velum and the pharynx, on the other, are strictly homologous in the monkey and in man. This is important to remember for later conclusions.

From the peripheral sound-producing apparatus we turn to the central nervous organization of vocal behavior in nonhuman primates and in man. The squirrel monkey has been studied more thoroughly than any other primate with regard to brain structures and vocal behavior involved. Results obtained from other primate species, in particular from the rhesus monkey, are very similar if not identical (Thoms and Jürgens 1987; Ploog 1981; Jürgens and Ploog 1981).

The monkey's subcortical system yielding species-specific vocalizations when electrically stimulated, the limbic system and structures directly connected with it play an eminent role (Fig. 2). There are two regions which are of particular functional signifi-

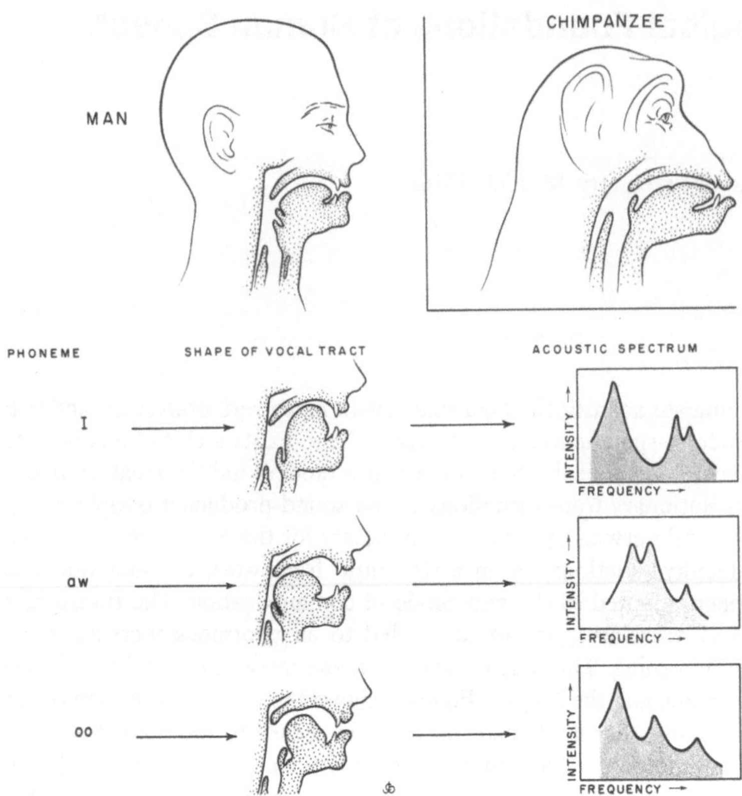

Fig. 1. The vocal apparatus of man and chimpanzee. The upper diagrams illustrate the difference in the shape of the vocal tract. The lower diagrams illustrate how movements of the tongue change the shape of the air space to generate different sounds. (After Wilson 1975)

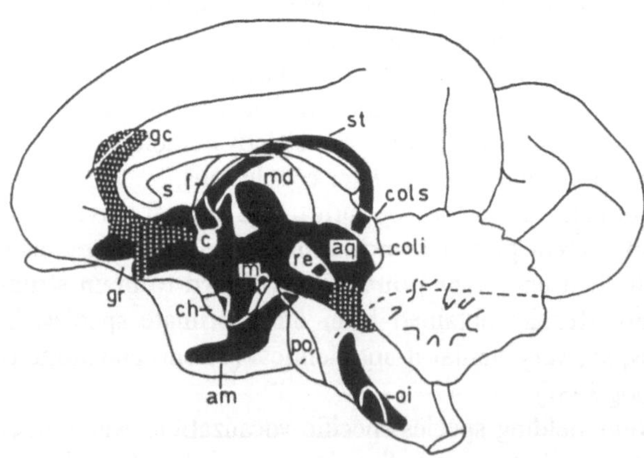

Fig. 2. Vocalization-producing brain areas in the squirrel monkey. (From Jürgens and Ploog 1976)

cance to both monkey and man: the periaqueductal gray in the midbrain and the anterior cingulate gyrus (stippled). The central gray in the caudal midbrain and laterally adjacent tegmentum is the phylogenetically oldest structure for the generation of species-specific calls. Its electrical stimulation yields vocalizations in amphibians, reptiles and mammals. Its destruction causes mutism in vertebrates (Jürgens and Ploog 1976). Also in man damage of the central gray results in mutism. The so-called traumatic midbrain syndrome leads to long-lasting serious defects of phonation and articulation.

Figure 3 shows the smallest and the largest lesions which abolished spontaneous vocalizations and vocalizations elicited by electrical brain stimulation from upper vocalization-producing areas. The lesions did not, however, abolish vocalizations from further caudal electrode positions which yielded deteriorated, abnormal calls, as Jürgens demonstrated in his findings of motor coordination of vocalizations in the lateral pontine and medullary reticular formation (Jürgens and Pratt 1979a).

In the primate the caudal midbrain area receives a direct input from, and is controlled by, the anterior cingulate gyrus. This structure is involved in the initiation and voluntary control of the voice, although to different degrees in monkey and man. Monkeys can control their vocal behavior contingent upon conditional stimuli and reward to a rather limited extent as regards duration and amplitude of a species-specific call but not in regard to the acoustic structure of the utterance. This was shown first by Sutton et al. (1974) in the macaque. After bilateral ablation of the anterior cingulate area, this instrumental response was abolished, although the monkeys were still able to vocalize spontaneously and respond to external fearful stimuli. There is, however, evidence of a reduction in the occurrence of calls.

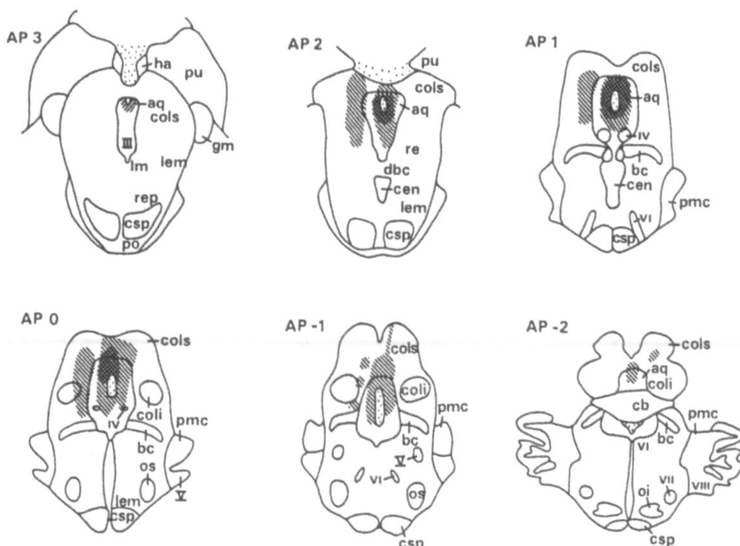

Fig. 3. Lesions which abolished "spontaneous" vocalizations elicited by electrical brain stimulation. The smaller lesion (hatching from upper right to lower left) abolished calls elicited from the nucleus striae terminalis and preoptic region. It also abolished visually induced yapping calls but not tactually provoked shrieking. The larger lesion (hatching from upper left to lower right) abolished vocalizations elicited from the gyrus rectus and spinal trigeminal nucleus. (Jürgens and Pratt 1979a)

In man the consequences of bilateral destruction of the anterior cingulate area are more severe; it results in a state of akinetic mutism similar to the traumatic midbrain syndrome. These patients may lose spontaneous speech completely for months, while emotional utterances, such as a moan in response to a painful stimulus, are still possible. Even when, after weeks, the whispered repetition of longer sentences is possible, spontaneous utterances remain reduced to a few monosyllabic words. In an intonation test, one of our patients, five years after an infarction, was not able to attach the appropriate emotion to certain exclamations, such as "shut up" or "super." The sonagrams revealed that his ability to speak with emotion was greatly reduced - measured in the range of the fundamental frequency - and that he was not able to correct this deficiency. It may be concluded then that the anterior cingulate gyrus in man serves not only the volitional initiation of the voice but also the volitional intonation of emotions. The anterior nucleus of the thalamus in man has substantially increased in volume over what would be expected in an ape thalamus of human brain size and thereby providing a larger limbic source for the cortex, especially for area 24 and the premotor cortex with area 6 and 8 (Armstrong 1986). This may explain the drastic influence of the anterior cingulate gyrus on emotional speech.

Before turning to the neocortical aspects of vocal expressions and speech, the hierarchical organization of the vocal motor system is summarized (Fig. 4).

The lowest level (I) is represented by the cranial nerve nuclei involved in phonation, the respiratory motoneurons and the lateral pontine and medullary reticular formation for motor coordination of vocalization.

The next level (II) is the caudal periaqueductal gray and laterally adjacent tegmentum. This area exerts a direct influence on call production mechanisms and can function independently of higher levels of the system, although all the higher levels are connected with level II. We can conclude that this area serves to couple the momentary motivational states and the external events to the appropriate vocal expression to be released. We may be dealing here with the neural substrate of the vocal innate releasing mechanism.

The next level (III) has not been treated in any detail. As examples, four secondary vocalization areas are represented. The structures mediate specific motivational states which induce the appropriate vocal expression; they all converge on level II.

The upper level (IV), the anterior cingulate gyrus - the limbic cortex - has been treated in more detail. Most importantly, there is a direct projection into level II which has been described by Jürgens and Pratt (1979b) as "cingular vocalization pathway." The area represents a higher order integration level that can, independently of motivational states, exert a facilitatory or inhibitory influence on level II. At the same time, it is connected with the motivational structures of the limbic system. According to its position in the circuitry of vocal behavior and its functions, it may serve as a connecting passage between limbic and neocortical mechanisms of vocal behavior (Jürgens 1986; Ploog 1988).

The next level (V) in the hierarchical organization of the vocal motor system - not shown in Fig. 4 - includes two neocortical areas which are important for vocal behavior only in man, whereas the monkey and the ape can vocalize without them. This is the supplementary speech area within area 6 which is adjacent to the anterior cingulate area 24 and the cortical representation of the larynx, pharynx, tongue and lips. First the latter will be considered. There is a direct projection from the cortical larynx area to

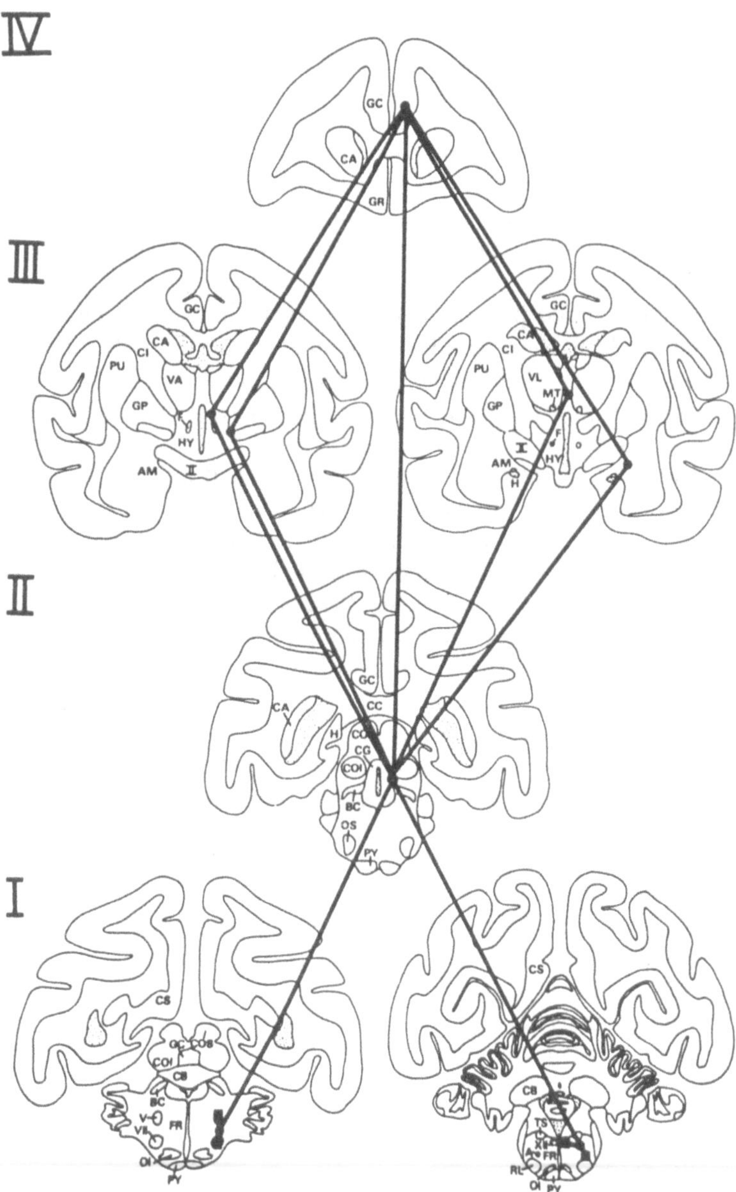

Fig. 4. Scheme of the hierarchical vocalization control in the squirrel monkey. All areas marked with a dot yield vocalization when electrically stimulated. Squares indicate cranial motor nuclei involved in phonation. The lines connecting the dots and squares represent anatomically verified direct projections in rostrocaudal direction. The dots indicate in I the reticular formation of pons and medulla, in II the periaqueductal gray and laterally bordering tegmentum, in III the mediobasal amygdala, midline thalamus and different hypothalamic areas, in IV the anterior cingulate cortex. Abbreviations: A n. ambiguus, AM amygdala, BC Brachium conjunctivum, CA caudatum, CB cerebellum, CC corpus callosum, CG periaqueductal gray, CI capsula interna, COI colliculus inferior, COS colliculus superior, CS cortex striatus, F fornix, FR formatio reticularis, GC gyrus cinguli, GP globus pallidus, GR gyrus rectus, H hippocampus, HY hypothalamus, MT tractus mamillothalamicus, OL oliva inferior, OS oliva superior, PU putamen, PY pyramidal tract, RL n. reticularis lateralis, TS n. tractus solitarius, VA n. ventralis anterior thalami, VL n. ventralis lateralis thalami, II tractus opticus, chiasma opticum, V trigeminal motor nucleus, VII facial nucleus, XII hypoglossal nucleus. (Jürgens 1986)

the cingulate gyrus shared by monkeys and man (Jürgens 1976). There is, however, a direct pathway from the laryngeal representation in the primary motor cortex to the laryngeal motoneurons in the medulla (nucleus ambiguus) which exists only in man (Kuypers 1958). This direct connection serves as the neural basis for the voluntary control of the vocal folds in man which is not possible in the monkey. In terms of evolutionary neurology, this human specialty appears to be the latest expansion of the pyramidal tract. In order to make this important difference between the human and subhuman primate obvious also on the functional level, a clinical example is presented (Fig. 5).

A right-handed patient suffered an embolic cerebral infarction in the stem of the medial cerebral artery. He had a hemiplegia on the right side with facial paresis and tongue deviation on the right. A laryngoscopical examination revealed that during respiration, as well as phonatory attempts, both vocal folds remained motionless. He could not utter a sound except that he coughed reflexively when the base of his tongue was touched. The state of complete mutism and inability to phonate was almost completely restored after 13 weeks. The patient had considerable difficulties with articulation and ability to make oral movements. The brain damage of the patient was then experimentally modeled in the monkey. The lesion was carried out bilaterally. It fitted the patient's lesion quite well, invading the homologue of Broca's area, the inferior pre- and postcentral cortex, rolandic operculum, inferior parietal cortex and large parts of the insula, claustrum, putamen and the white matter underlying the inferior frontoparietal and insular cortex.

Although the monkey's tongue, lips and masticatory muscles were completely paralyzed (as they were in the right side of the patient), so that the monkey could neither bite, chew nor lick, he could vocalize in a normal manner. The spectrographically registered vocalizations included all call types of the squirrel monkey. Consequently, and in contrast to the patient, the vocal folds were functioning, which means the central nervous control was not impaired, but exercised via the "cingular vocal pathway," whereas the patient's control, based on the primary motor cortex, was completely impaired. If the monkey would have this direct pathway for controlling his vocal folds, he would probably be able to sing.

What remains to be discussed is the function of the supplementary motor area (SMA) in monkey and man. In a comprehensive anatomical study of the efferent and afferent connections of this area, Jürgens (1984) was able to show that the SMA is more closely related to motor-coordinating than to sensory-analyzing structures. It determines the activity of the primary motor cortex to a greater extent than vice versa and participates not only in the initiation of global motor programs but also in the motor subroutines.

As far as the voice is concerned, there are several experiments which strongly support this interpretation of SMA functioning. Here is an example: squirrel monkeys, if isolated from their group, utter series of isolation peeps. Bilateral ablation of the anterior part of SMA drastically reduces the frequency of occurrence of this call but not of any one of the other call types. The ablation of the posterior part is without effect. In this situation the anterior SMA seems to be involved in the initiation of a behavior pattern not directly triggered by external stimuli. In fact, the anterior SMA is the only neocortical area involved in vocal behavior of nonhuman primates. Not only is the isolation peep the only call that can be reduced by cortical ablation, but it is also the only call

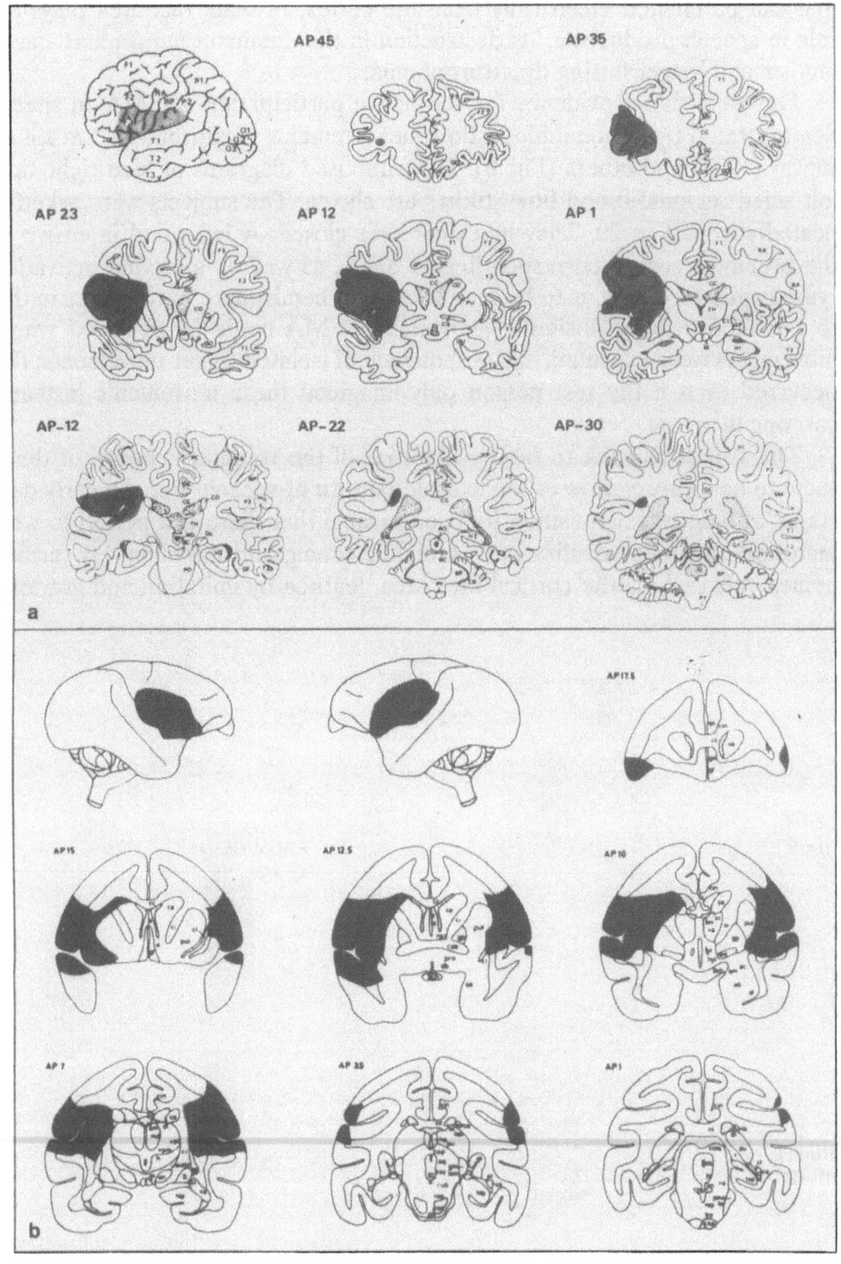

Fig. 5. Frontal sections through (a) human brain, (b) monkey brain; black zones: in a cerebral infarction area, in b cerebral lesion sites. (From Jürgens et al. 1982)

that can be elicited electrically from the cortex. In man, this area plays an important role in speech production. Its destruction in the dominant hemisphere causes transient mutism and longer lasting dysarthrophonia.

The most direct evidence for the active participation of SMA in speech has been demonstrated by regional blood flow measurements performed by Larsen et al. (1978), Ingvar (1983) and others (Fig. 6). Superimposed diagrams of nine right-sided and nine left-sided regional blood flow studies are shown. The subjects were asked to count repeatedly from 1 to 20. They had their eyes closed. A left-sided intensive activation of the premotor region corresponding to SMA, as well as a second activation of the larynx-tongue-face area, is to be seen. The right hemisphere participates in this activation to a limited degree. A similar activation of SMA could be observed when sequential movements were executed, e.g., a sequence of isolated finger movements; this activation occurred even if the test person only imagined these movements instead of actually carrying them out.

This brings us back to the mechanisms of the voluntary control of the voice which seem to be a prerogative of the learned pattern of speech. For the early developmental stages of language acquisition it is conceivable that there is a period in which the "raw material" of innate vocalization - a child's cooing and babbling - is formed into phonemes, released via the cortical face area, learned by imitation and gradually screened

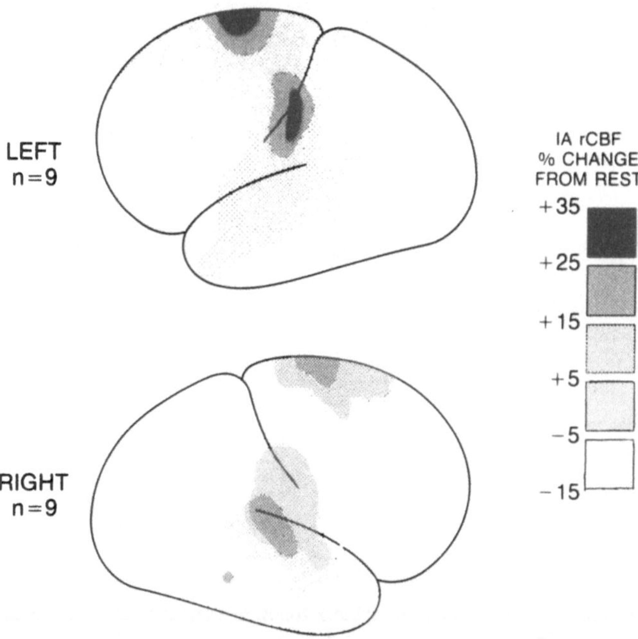

Fig. 6. Automatic speech. Superimposed diagrams of nine right-sided and nine left-sided IA rCBF studies in patients without neurological disturbances and with normal speech. The rCBF changes have been calculated in percent relative to the resting state. Scale to the right denotes magnitude of flow change. Note z-like flow change on the left side with a clear-cut flow peak in the premotor/prefrontal regions, another peak in the mouth/tongue/larynx area, and also an activation of the middle temporal region. On the right side a similar pattern was recorded, but the peaks were not as high, and less well defined, especially in the temporal region. Replotted after color TV display. (Larsen et al. 1978; from Ingvar 1983)

against interfering motorically active neighboring cortex areas and perhaps also against limbic input. The conspicuous associative movements of the extremities, especially the hands, which accompany the articulatory movements of the child learning to speak may be indicative of a functional differentiation process taking place within and between the respective areas of the motor cortex (Noterdaeme et al. 1988).

This much of voluntary phonological control has been achieved in humans only. But several preconditions and precursors of speech are already present in the nonhuman primate. Since the voluntary control of the vocal folds seems to be a major step forward in the evolution of language and speech, the tonal languages may have occurred first. They may be closest to the mechanisms of nonhuman primate vocal communication.

Summary

The use of the voice is part of motor behavior in animals and man. In an evolutionary perspective the voice is a prerequisite for the emergence of speech. In order to achieve this most advanced mode of vocal communication a gradual transformation of the peripheral sound-producing system and its central nervous control has taken place. The whole vocal motor system is organized hierarchically. Its anatomical structures and connections are homologous in monkey and man, with two exceptions. First, only in man the motor cortex, especially the face area, and parts of the premotor cortex are indispensible for vocal behavior. Second, there is a direct pathway from the laryngeal representation in the primary motor cortex to the laryngeal motoneurons in the medulla which does not exist in the monkey. It serves as the neuronal basis for the voluntary control of the vocal folds in man which is not possible in the monkey. This direct connection appears to represent the latest expansion of the pyramidal tract.

References

Armstrong E (1986) Enlarged limbic structures in the human brain: the anterior thalamus and medial mamillary body. Brain Res 362: 394-397

Ingvar DH (1983) Serial aspects of language and speech related to prefrontal cortical activity. A selective review. Hum Neurobiol 2: 177-189

Jürgens U (1976) Projections from the cortical larynx area in the squirrel monkey. Exp Brain Res 25: 401-411

Jürgens U (1984) The efferent and afferent connections of the supplementary motor area. Brain Res 300: 63-81

Jürgens U (1986) The squirrel monkey as an experimental model in the study of cerebral organization of emotional vocal utterances. Eur Arch Psychiatry Neurol Sci 236: 40-43

Jürgens U, Ploog D (1976) Zur Evolution der Stimme. Arch Psychiatr Nervenkr 222: 117-137

Jürgens U, Ploog D (1981) On the vocal control of mammalian vocalization. Trends Neurosci 4: 135-137

Jürgens U, Pratt R (1979a) Role of the periaqueductal grey in vocal expression of emotion. Brain Res 167: 367-378

Jürgens U,Pratt R (1979b) The cingular vocalization pathway in the squirrel monkey. Exp Brain Res 34: 499-510

Jürgens U, Kirzinger A, v Cramon D (1982) The effect of deep-reaching lesions in the cortical face area on phonation. A combined case report and experimental monkey study. Cortex 18: 125-140

Kypers HGJM (1958) Corticobulbar connexions to the pons and lower brain-stem in man. Brain 81: 364-388

Larsen B, Skinhöj E, Lassen NA (1978) Variation in regional/cortical blood flow in the right and left hemispheres during automatic speech. Brain 101: 193-209

Lieberman P (1973) On the evolution of language: a unified view. Cognition 2: 59-94

Noterdaeme M, Amorosa H, Ploog M, Scheimann G (1988) Quantitative and qualitative aspects of associated movements in children with specific developmental speech and language disorders and in normal pre-school children. J Human Movements Studies 15 (151-169)
Ploog D (1981) Neurobiology of primate audio-vocal behavior. Brain Res Rev 3: 35-61
Ploog D (1988) Neurobiology and pathology of subhuman vocal communication and human speech. In: Todt D, Goedeking PFD, Symmes F (eds): Primate vocal communication. Springer, Berlin Heidelberg New York
Sutton D, Larson C, Lindemann RC (1974) Neocortical and limbic lesion effects on primate phonation. Brain Res 71: 61-75
Thoms G, Jürgens U (1987) Common input of the cranial motor nuclei involved in phonation in squirrel monkey. Exp Neurol 95: 85-99
Wilson EO (1975) Sociobiology: A new synthesis. Harvard University Press, Cambridge, MA

Cricket Neuroethology: A Comparative Approach to the Nervous System

F. Huber
Max-Planck-Institut für Verhaltensphysiologie, 8130 Seewiesen, FRG

The Neuroethological Approach

Nervous systems are biocomputers designed to produce behavior. Comparative neuroethological research tries to understand how sense organs, central nervous and effector systems work to organize and control the diverse behavioral strategies of animals shaped by nature's abiotic and biotic forces to improve survival and reproductive fitness during the course of evolution.

Neuroethologists concentrate on molecular, cellular, network and higher system levels of the nervous system. They relate results of quantitative behavioral studies to data obtained with the whole instrumentarium developed in the field of neuroscience and the search for correlates and causal relationships. Neuroethologists look for computational operations within the nervous system (the software) as well as for the underlying neuronal implementations (the hardware). At present a description of behavior and nervous activities in terms of algorithms is restricted to favorable cases (Huber 1989).

The Advantage of Comparative Studies

Those neuroethologists familiar with the animal kingdom are aware that evolution is indifferent to the desires of students of animal behavior and to neurobiologists, for it is rare that all animals are equally suitable for both behavioral and nervous system analysis at all levels. Therefore, one must find the right "model organism," best suited for the questions being addressed. Crickets and frogs are favorable to elucidate the innate and phylogenetically developed machinery for acoustic communication used during intraspecific partner finding and pair formation (Huber 1983, 1988). Birds offer opportunities to approach structures and mechanisms involved in song learning (Marler 1983). Molluscs provide excellent examples to attack the molecular and cellular basis of simple forms of learning and memory (Kandel 1985). Drosophila helps to gain insight into the genetic background of behavior and nervous processing (Heisenberg and Wolf 1984), and higher mammals may help to bridge the gap to man, even in a search for steps toward cognitive functions.

There is no unique solution for selecting the best model system. My guideline has always been to look for species that execute the behavior in question even under somewhat restrained conditions which allow cellular and network analysis within the nervous system and to search for animals which exhibit behavior executed even under the strait jacket of a microelectrode. I found such animals in the group of crickets (Huber 1988).

What Can We Learn About Nervous System and Behavior by Studying Crickets?

All nervous systems from hydra to man are composed of nerve cells, connected in various ways among each other. These assemblies of nerve cells or networks which compute external sensory and proprioceptive input, set the internal state and generate patterned output giving rise to behavior. Recent comparative research in different animals has elucidated general principles mostly at the molecular and cellular levels, for instance, how ionic channels function, how calcium is involved, how connections are formed and operate among neurons at the synaptic level, and we are even able for some simpler behaviors to describe the cellular and network properties underlying them. A variety of transmitters and neuropeptides was discovered which alone or in cooperation change membrane channels and are elements to transmit information between cells or modulate connections. Thus, the term "motivation" becomes accessible to a reductionistic approach. And crickets add to the list of favorable animals (Huber 1987b).

The nervous system of crickets is segmentally and bilaterally organized, reflecting annelid ancestors. The bilateral buildup is also shown in the presence of mirror-image pairs of neurons (Fig. 1). Such a nervous system with separated ganglia is suited to study segmental integration at a single ganglion level as well as the plurisegmental cooperation among ganglia, even down to single identified nerve cells (Huber, 1983, 1989). For a neuroethologist, it is a prerequisite that behavior must guide the questions addressed to the nervous system. I therefore discuss a few aspects of cricket behavior by concentrating on acoustic communication strategies and try to bridge the gap to single neuronal operations.

Song Pattern Production

Cricket songs are characterized by their frequency spectra and their temporal organization. They consist of syllables and chirps (Fig. 2). This temporal structure is a key for species-specific song recognition, as is discussed below. The songs are broadcast by the males, and different songs are used in different behavioral contexts, for instance, the calling song for attracting conspecific and sexually responsive females; courtship songs help to direct the female at close distance to mount the male for copulation; aggressive songs are produced mainly in male-male encounters and as threatening signals to warn invaders of a male's territory.

Cricket songs are examples of rhythmic motor patterns where one forewing rubs periodically against the other. These periodical opening and closing movements are organized by a thoracic central pattern generator. Peripheral sensory feedback from the wings and near the wing joints helps to stabilize the pattern, for instance, the well-kown right over left position of the wings and the correct tooth impact between the stridulatory file and the scraper. This thoracic pattern generator is under the control of the brain (Fig. 3), which decides on the basis of sensory input, central and peripheral feed-

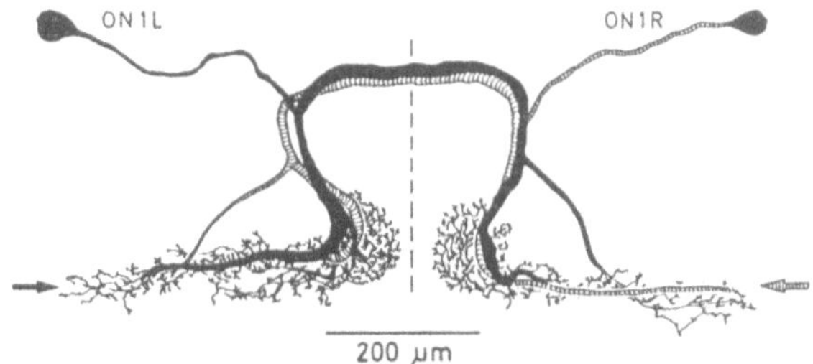

Fig. 1. A mirror-image pair of prothoracic segmental auditory interneurons in crickets as one example for the bilateral organization of the nervous system. ON1 L, ONI R, left and right omega-neurontype 1 known to sharpen the binaural contrast by reciprocal inhibition. Arrows indicate the excitatory auditory input from the ears to each ON1. Dashed line marks ganglion midline. Scale 200 µm. (Modified from Huber 1989)

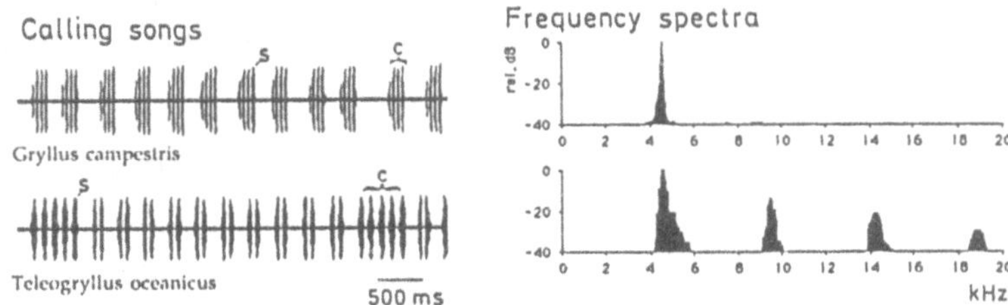

Fig. 2. Examples of calling songs of two cricket species. Left: song patterns with syllables (S) and chirps (C). Time scale 500 ms. Right: frequency spectra. Spectral energy is plotted in relative units (rel. dB). (Modified from Huber 1989)

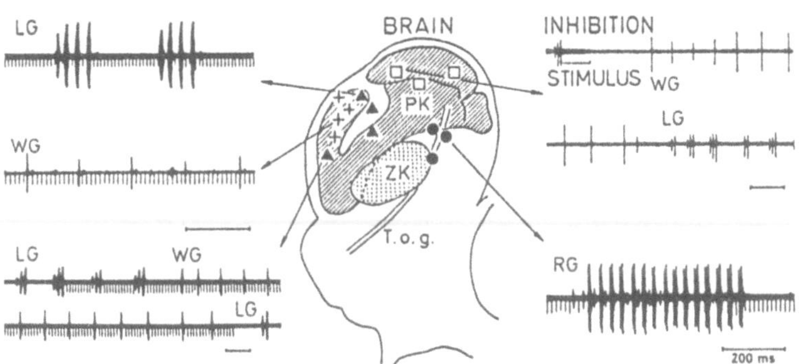

Fig. 3. Cricket songs elicited by local electrical stimulation in the brain. Middle part: Sagittal section through one half of the brain with the mushroom body (PK) and the central complex (ZK). PK represents several thousand highly ordered brain intrinsic neurons, ZK is a well-ordered network formed by an unknown number of nerve cells. T.o.g., fiber tract. Symbols mark stimulus loci and arrows indicate the effects of brain stimulation. LG, calling song; WG, courtship song; RG, aggressive song; transition between calling LG and courtship song WG; inhibition of singing and postinhibitory rebound indicated by courtschip song WG and calling song LG. Stimulus pulses are shown as regularly spaced downward deflections. Scale 200 ms. (Modified from Huber 1989)

back, and the internal state what type of song and associated behavior should be executed, as first discovered during local electrical stimulation of the brain (Huber 1960; Kutsch and Huber 1989). As in other animals and man, the cricket nervous system exhibits parallel processing. Moreover, during the performance of acoustic behavior parts of the male's nervous system cooperate. The brain remains informed about the execution of motor acts organized within the ventral nerve cord by a variety of descending and ascending neurons and loops (Otto and Weber 1982, 1985). Their is no clear sign for a strict hierarchical organization, and we have no unequivocal proof that single command neurons govern cricket behavior. Decisions are made by permanent "cross-talks" between different levels of the nervous system.

The internal state of the cricket nervous system is under the control of neuromodulator and neuroendocrine cells; the latter regulate the production and release of hormons. Only females with a high enough juvenile hormone titer become sexually responsive and are attracted by the male's call (Stout et al. 1976; Koudele et al. 1987). Ablation of cerebral neuroendocrine cells diminishes sexual responsiveness (Loher and Zaretsky 1989). Thus, vertebrate neuroethologists and even neurologists may recall features to which they are familiar when working with their backbone-spinalcord animals.

A Neuronal Correlate to Species-Specific Sound Recognition

Ethologists have formulated the concept of key sign stimuli and of central nervous filtering devives known as innate or learned releasing meachnisms (often called templates; Tinbergen 1951). Crickets offer a first insight into the cellular basis of such species-specific song filters. Female crickets in copulatory responsiveness approach a calling male only when the syllable period of its calling song lies within a restricted range, which we call the appropriate time window (Thorson et al. 1982; Fig. 4A). Such a time window

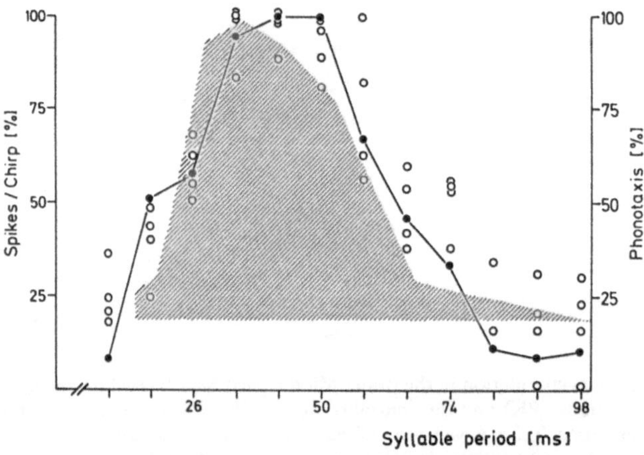

Fig. 4A. Response window for attractive syllable periods (ms) for female phonotaxis indicated by the hatched area. Phonotaxis is plotted in % (right ordinate). Activity profiles of single and identified local brain neurons, the band-pass cells, match the response window for phonotaxis (o - o, o o). The response strength is shown on the left ordinate in spikes/chirp (Modified from Huber 1989)

can be shifted by temperature since crickets belong to poikilothermic animals (Doherty 1985; Fig. 4B), where sender and receiver are temperature-coupled.

The auditory system, developed in both male and female crickets, transfers song information to the brain. At the level of the ears no single auditory sense cell and at the level of the thoracic auditory pathway no central neuron have been found to be "tuned" to the species-specific temporal demands (Wohlers and Huber 1982). But in the brain, nerve cells with band-pass properties have been identified, the activity of which correlates well with the time window for attractive syllable periods in phonotactic behavior (Schildberger 1984; Fig. 4A). Moreover, a mechanism is in sight which shapes the activity profile of the band-pass cells in the brain (Fig. 5), because high-pass and low-pass cells were also discovered which by an ANDgate converge on the band-pass cells (Huber and Thorson 1985). A similar mechanism has been proposed for frogs (Capranica and Rose 1983) indicating that evolution has used a similar neuronal strategy for song pattern recognition in two widely separated groups of animals.

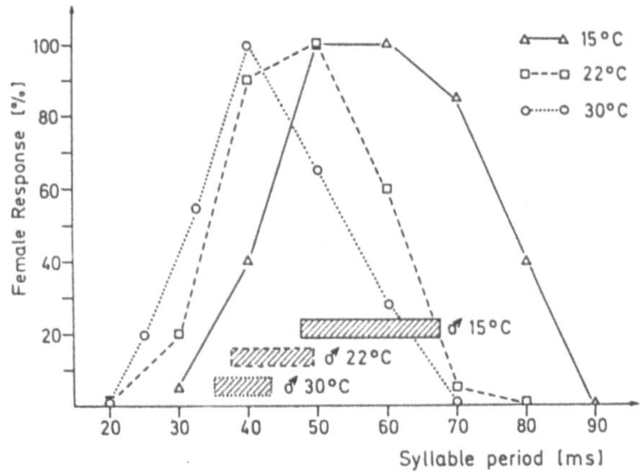

Fig 4B. Shift in the response windows for attractive syllabale periods (ms) in female phonotaxis after a change in ambient temperature. Phonotaxis is measured in % of the female's response (ordinate). Insets: Range of syllable periods in the male's calling song at the respective ambient temperatures (Modified from Doherty 1985)

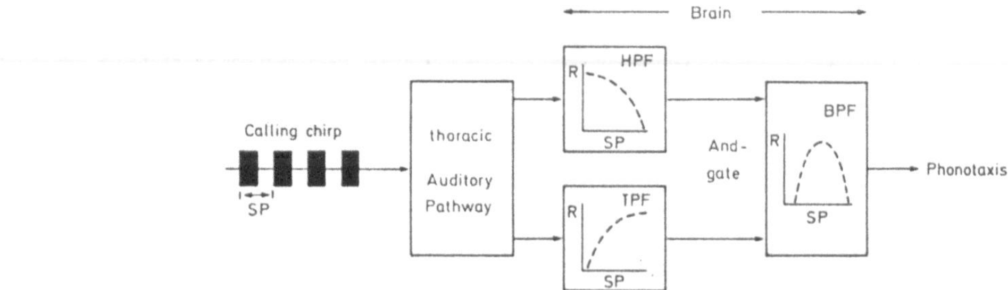

Fig. 5. Proposed functional diagram for brain-mediated song recognition. Information of the calling song pattern is transmitted to the brain via the thoracic auditory pathway. Within the brain three functional types of neurons were found, one with high-pass (HPF) -, one with low-pass (LPF) - and one with band-pass filter properties (BPF), indicated by dashed lines. High- and low-pass cells act on band-pass cells via an AND gate. R, response; SP, syllable period. (Modified from Huber 1989)

A Neural Analysis of Song Localization

Female crickets have to recognize the species-specific call of the male, and they must localize the sound source in order to approach it by walking or flying, a response called phonotaxis.

Sound localization is a binaural event based on comparison of the input from the two ears, situated within the foreleg tibiae (Huber 1987a). The tympana of each ear are driven by sound impinging on them from outside and via a tracheal tube from inside. Thus, tympana oscillations - a prerequisite for hearing - depend on sound pressure differences and phase differences of the sound waves reaching them from the outside and the inside (Kleindienst 1987). Such a pressure gradient receiver exhibits a cardoid directional characteristic as a function of sound incidence angle. At constant sound intensity each ear is most strongly excited when sound arrives from its own side, and most weakly by sound from contralateral (Boyd and Lewis 1983). The two mirror-image directional curves cross frontally and caudally, but only the frontal crossing point is stable and points to the sound source.

During phonotactic walking a cricket meanders approximately 40° to either side of this crossing point, building up a high enough intensity difference between the two ears which gives rise to a corrective turn according to the rule "turn to the side more strongly stimulated" (Wendler et al. 1980; Weber et al. 1981).

The input from auditory receptors of the two ears excites neurons on the two sides of the ventral nerve cord (Fig. 6), and a comparison is made between left and right levels of excitation according to the above-mentioned rule (Huber 1987a). This theory has been tested for the first time at the single neuronal level in crickets (Schildberger and Hörner 1988; Fig. 7).

Song and directional information is transmitted from auditory sense cells to higher order interneurons within the prothoracic ganglion, which are present there as mirror-image pairs (Wohlers and Huber 1982). Some of them, like AN1 (ascending neuron type 1), conduct the message to the brain.

In an open loop experiment phonotaxis was tested with a female cricket fixed by a holder such that it could only walk straight forward but where its legs could touch an air-suspended ball and turn it. When a calling song was broadcast from a loudspeaker in either the left or right position, the animal walked and turned the ball, respectively indicating a turning tendency to the left or right. Thus, it exhibited phonotactic behavior even under these restrained conditions. Some of the females performed this behavior even after the prothoracic ganglion had been exposed and a microelectrode inserted into one, the left AN1 cell. When the ears were stimulated by calling song from the left, this cell exhibited a pattern of excitation, copied the syllable and chirp structure of the call, and was more strongly excited than its mirror-image right partner. When stimulated with calling song from the right, the left cell was less excited, as expected by the directional characteristic of the ear. Its partner cell, however, was now more strongly excited. According to the rule "turn to the side more strongly excited" the animal turned to the right.

To test the theory further, the left AN1 cell was electrically manipulated, and by hyperpolarization its activity level was reduced even below that of the partner cell despite broadcasting the sound still from the left, its own side. The animal now turned to the right side, because its right AN1 cell was more strongly excited than the left hyperpola-

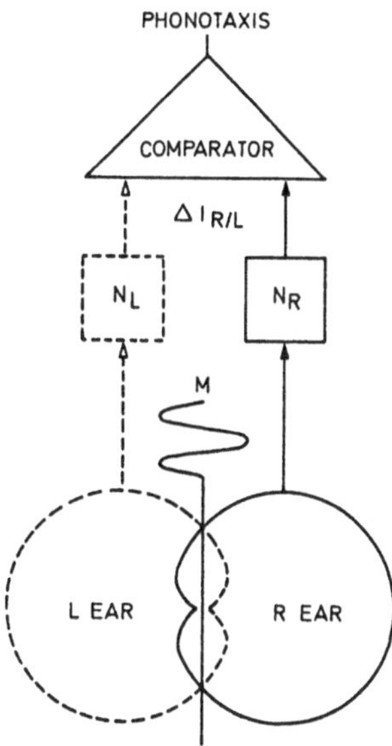

Fig. 6. Scheme to illustrate some aspects of binaural theory for directional hearing. Directional sensitivity curves of the left (l-ear, dashed circle) and the right (R-ear, solid circle) cross frontally and caudally. The meandering walking path (M) of the animal is indicated. The directional information from each ear is transmitted to central left (NL) and right (NR) neurons which conduct the message to a comparator probably situated within the brain. It calculates left and right excitation differences (ΔIR/L) and walking direction is the result of minimizing such differences.(Modified from Huber 1989)

rized cell. Thus, this experiment proved that the binaural theory has a neuronal correlate, and that a single neuron pair is sufficient to cause the predicted change in the phonotactic course. In a similar open loop set-up Stabel (1988) found that patterned information of ascending neurons which copy syllables and chirps is necessary for both recognition and localization.

Plasticity in Phonotactic Behavior

If the rule "turn to the side most strongly excited" also holds for one-eared crickets, then an animal ought to circle to the side of the remaining ear. But in some one-eared females we found phonotaxis, which indicates that such females can recognize the species-specific song by information arriving from one ear and are able to track the sound source (Huber et al. 1984; Huber 1987a; Fig. 8). Within the prothoracic ganglion, identified auditory interneurons were encountered which - when deprived long enough of excitatory input from their former ear - grew dendrites across the ganglion midline and made functional connections with the auditory fibers originating from the "wrong ear" (Schildberger et al. 1986). Their similar threshold curves and intensity response functions strongly indicate that such neurons were reconnected with perhaps identical sense cells but now originating from the other ear. Recent experiments clearly showed that such wrong connections in phonotactically active animals exhibit novel intensity re-

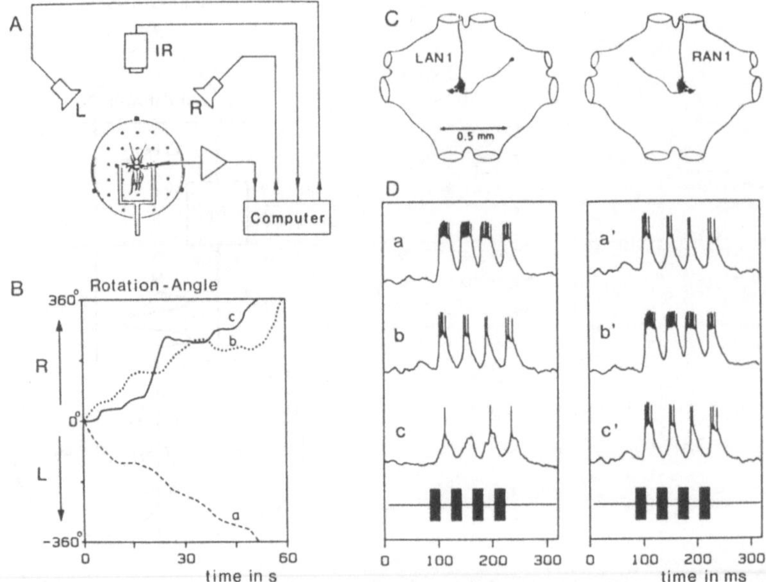

Fig. 7. Test of the binaural theory for directional hearing at the cellular level. **A** Open loop setup for studying phonotaxis. The animal is fixed by a holder and walks only straight forward, but it can turn the air-suspended ball with its legs. Turning is measured. IR - Infrared sensing and detecting system. L, left and R, right loudspeaker to broadcast the calling song. **B** Turning courses in three different situations. a (....) to the left active loudspeaker; b (....) to the right active loudspeaker; c (----) to the right with the left active loudspeaker, but during hyperpolarization of the left AN1. **C** Outlines of the prothoracic ganglion with the mirror-image AN1 cells. LAN1 left cell, RAN1 right cell. Scale 0.5 mm. **D** Responses of the LAN1 cell (a,b,c) recorded intracellularly during phonotaxis (conditions see B). a', b', c' predicted responses of the RAN1 cell if both cells have mirror-image response properties. a, a' LAN1 is stronger excited than RAN1, animal turns to the left; b, b' RAN1 is stronger excited than LAN1, animal turns to the right; c, c' LAN1 is hyperpolarized and less excited than RAN1, animal turns to the right despite sound from left. Lowest traces, calling song chirps. (Modified from Huber 1989)

sponse functions with a crossing point at a certain sound intensity. The deprived and newly connected cell is more strongly excited at low sound intensity levels and excited less strongly at higher sound intensity levels when compared with its partner cell (Schildberger and Kleindienst 1988). This crossing point of the intensity response functions of the two mirror-image cells refers to the zero-balance point for normal phonotaxis. Animals which did not exhibit phonotaxis missed such crossing points. Thus the nervous system of crickets is able to rearrange connectivities and changes functions after sensory deprivation.

Closing Remarks

Comparative neurobiology and the emerging field of neuroethology challenge us to look for species-specific solutions as manifestations of a long-lasting evolutionary process. Only by such comparison can general principles be deduced. Thus the future of comparative studies of nervous systems in relation to behavior should focus on and appreciate the diversity in form and function by looking at critical differences that are related to

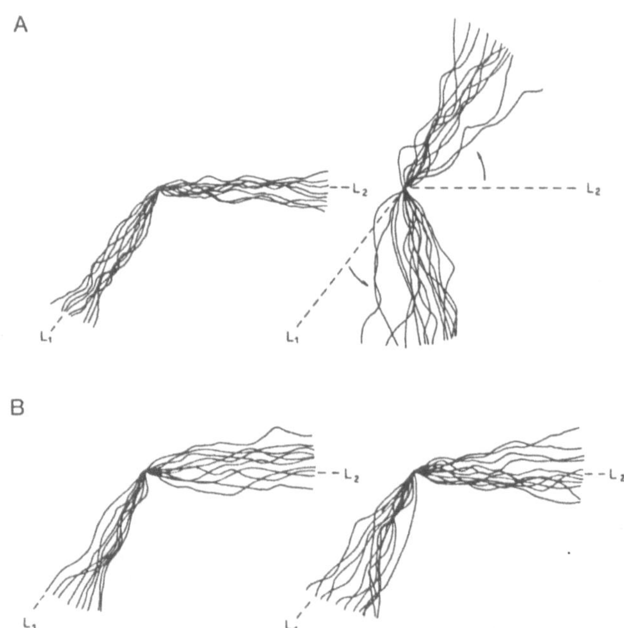

Fig. 8A Phonotactic walking courses of an intact two-eared female cricket (left) and of the same female after amputation of the right foreleg between coxa and trochanter (right). L1, L2 the two loudspeaker positions. In the animal with one ear, the courses deviate to the side of the intact (left) ear (angle 35° - 70°) and this deviation is loudspeaker dependent. **B** Left, courses of a female with the right foreleg amputated in an early larval instar and after regeneration of this leg to nearly normal length. No deviation in phonotaxis is seen (compare A, left with B, left). Right, courses of the same female immediately after amputation of the right, regenerated foreleg, again with no significant change in the walking angle. Histology of the tibial region showed absence of an auditory organ in the regenerated right foreleg. (Modified from Huber 1987a)

different capabilities in behavior, including cognitive abilities. As Bullock pointed out (see Cohen and Strumwasser 1985) "we will be on a road to understanding how the brain achieves the functions for which it evolved when we ask two - equally important - questions: (1) What are the neural correlates relevant to known behavioral differences among animals, and (2) what are the behavioral correlates relevant to known neural differences."

References

Boyd P, Lewis B (1983) Peripheral auditory directionality in the cricket (Gryllus campestris L., Teleogryllus oceanicus Le Guillon). J. Comp. Physiol. 153: 523-532

Capranica RR, Rose G (1983) Frequency and temporal processing in the auditory system of anurans. In: Huber F, Markl H (eds) Neuroethology and behavioral physiology. Springer, Berlin Heidelberg New York

Cohen MJ, Strumwasser F (1985) Comparative neurobiology: Modes of communication in the nervous system. Wiley Series in Neurobiology. Wiley, New York

Doherty JA (1985) Temperature coupling and "trade off" phenomena in the acoustic communication system of the cricket, Gryllus bimaculatus DeGeer (Gryllidae). J. Exp. Biol. 114: 17-35 (1985).

Heisenberg M, Wolf R (eds) (1984) Vision in drosophila: genetics and microbehavior. In: studies of brain function, vol 12. Springer, Berlin Heidelberg New York

Huber F (1960) Untersuchungen über die Funktion des Zentralnervensystems und insbesondere des Gehirnes bei der Fortbewegung und der Lauterzeugung der Grillen. Z Vergl Physiol 44: 60-132
Huber F (1983) Implications of insect neuroethology for studies on vertebrates. In: Ewert JP, Capranica RR, Ingle DJ (eds) Advances in vertebrate neuroethology. Plenum, New York, pp 91-138 (Nato ASI series, vol 56)
Huber F (1987a) Plasticity in the auditory system of crickets: phonotaxis with one ear and neuronal reorganization within the auditory pathway. J. Comp. Physiol. 161: 583-604
Huber F (1987b) Neuroethologie: Vom Verhalten zur einzelnen Nervenzelle. Konstanzer Universitätsreden Nr. 162: 7-39. Universitätsverlag Konstanz, Konstanz
Huber F, Thorson J (1985) Cricket auditory communication. Sci Am 253: 60-68
Huber F (1988) Invertebrate neuroethology: guiding prinicples. Experimenta 44, 428-431
Huber F (1988) Ordnungsprinzipien im Verhalten und im Nervensystem von Insekten. In: Gerok W (ed) Ordnung und Chaos in der belebten und unbelebten Natur. Wissenschaftliche Verlagsgesellschaft, Stuttgart, pp 333-357 (Verhandlungen der Deutschen Gesellschaft Naturforscher und Ärzte, 115th meeting, Freiburg 1988)
Huber F, Kleindienst HU, Weber T, Thorson J (1984) Auditory behavior in the cricket. III. Tracking of male calling song by surgically and developmentally one-eared females, and the curious role of the anterior tympanum. J. Comp. Physiol. 155: 725-738
Kandel ER (1985) Steps toward a molecular grammar for learning. Explorations into the nature of memory. In: Isselbacher KJ (ed) Medicine, science and society, symposia celebrating The Harvard Medical School Bicentennial Wiley, New York, pp 555-604
Kleindienst HU (1987) Akustische Ortung bei Grillen. Fortschr Akustik (DAGA 87): 25-46
Koudele K, Stout JF, Reichert D (1983) Factors which influence female crickets' (Acheta domesticus) phonotactic and sexual responsiveness to males. Physiol. Entomol. 12: 67-80
Kutsch W, Huber F (1989) Neural basis of song production. In: Huber F, Moore TE, Loher W (eds) Cricket behavior and neurobiology, pp.262-309. Cornell University Press, Ithaca, NY
Loher W, Zaretsky M (1989) Endocrine systems in crickets: structure and function. In: Huber F, Moore TE, Loher W (eds) Cricket behavior and neurobiology, pp.114-146. Cornell University Press,Ithaca, NY
Marler P (1983) Some ethological implication for neuroethology. The ontogeny of birdsong. In: Ewert JP, Capranica RR, Ingle DJ (eds) Advances in vertebrate neuroethology. Plenum,New York, pp 21-52 (Nato ASI series A, vol 56)
Otto D, Weber T (1982) Interneurons descending from the cricket cephalic ganglia that discharge in the pattern of two motor rhythms. J. Comp. Physiol. 148: 209-219
Otto D, Weber T (1985) Plurisegmental neurons of the cricket, Gryllus campestris L., that discharge in the rhythm of its own song. J. Insect Physiol 31: 537-548 (1985).
Schildberger K, Temporal selectivity of identified auditory neurons in the cricket brain. J. Comp. Physiol. 155: 171-185 (1984).
Schildberger K, Hörner M (1988) The function of auditory neurons in cricket phonotaxis. I. Influence of hyperpolarization of identified neurons on sound localization. J. Comp. Physiol. 163: 621-631
Schildberger K, Kleindienst HU (1988) Sound localization in one-eared crickets. In: Elsner N, Barth FG (eds) Sense organs, interfaces between evironment and behavior. Proceedings of the 16th Göttingen Neurobiology Conference. Thieme, Stuttgart, p 18
Schildberger K, Wohlers DW, Schmitz B, Kleindienst HU, Huber F (1986) Morphological and physiological changes in central auditory neurons following unilateral foreleg amputation in larval crickets. J. Comp. Physiol. 158: 291-300
Stabel J (1988) Der Mechanismus der Richtungsbestimmung und seine Beziehung zur Gesangserkennung bei der Phonotaxis der Grille (Gryllus bimaculatus De Geer). Dissertation, University of Köln, Köln
Stout JF, Gerard G, Hasso S (1976) Sexual responsiveness mediated by the corpora allata and its relationship to phonotaxis in the female cricket, Acheta domesticus L. J. Comp. Physiol. 108: 1-9
Thorson J, Weber T, Huber F (1982) Auditory behavior of the cricket. II. Simplicity of calling-song recognition in Gryllus, and anomalius phonotaxis at abnormal carrier frequencies. J. Comp. Physiol. 146: 361-378
Tinbergen N (1951) The study of instinct. Oxford University Press, Oxford
Weber T, Thorson J, Huber F (1981) Auditory behavior of the cricket. I. Dynamics of compensated walking and discrimination paradigms on the Kramer treadmill. J. Comp. Physiol. 141: 215-232
Wendler G, Dambach M, Schmitz B, Scharstein H (1980) Analysis of the acoustic orientation behavior in crickets (Gryllus campestris L.). Naturwissenschaften 67: 99
Wohlers DW, Huber F (1982) Processing of sound signals by six types of neurons in the prothoracic ganglion of the cricket Gryllus campestris L. J. Comp. Physiol. 146: 161-173

Hierarchies of Structure-Function Relationship in the Neurosciences

F. Seitelberger
Neurologisches Institut Wien, 1010 Wien, Österreich

In biology the relationship between structure and function is a central question which must be answered anew at each newly achieved level of knowledge in this science. At present we understand biological organizations in terms of the system theory as structural/functional entities, i.e. functional systems which as such in the form of hierarchic orders of net-like connections with other functional systems contribute to the maintenance of a superior unit or a system of higher complexity (Benninghoff 1935/36, 1936/37, Rothschuh 1963)

The nervous system regarding its macroscopic structure appears as a system consisting of several organ-like subsystems, the brain being the relatively most independent among them. In fact there is an enormous morphological complexity forming an abundance of innumerable discernible functional systems, e.g. sensory tracts, nuclei, areas, neural circuits, etc. I would like to mention only that in the cerebral cortex at least two different building principles are realized: one the laminar, another the modular type of neuronal geometry, and that there are together several distinct architectonics, namely the cyto-, myelo-, angio-, chemo-, transmitter and electroarchitectonics (Szentágothai 1978). Thus the brain represents a functional megasystem (Seitelberger 1982).

But, what about the function of this system? For any other organ or functional system this question can be answered by data about the physiological, physicochemical activities and their products in terms of causal definition. This is true for liver, kidneys, etc. yet is not true for the brain. Although it exhibits plenty of neurophysiogical objectivable and analyzable processes in the sense of neural mechanisms of behavioral operations, these are not the brain functions which we know and immediately experience, i.e. conscious perceptions, emotions and volition, because what we understand as the proper function of the brain is intentional behavior marshalled by ourselves, the free, self-concerned human persons (Searle 1984). These functions are nowhere present as such, neither in the EEG nor in molecular structures, neither in the transmitter traffic nor localized in neuronal structures as cortical areas, etc. In fact, this undoubtedly existing subjective domain of brain function depends on a living brain. However, it cannot be explained or defined by scientific analysis of organic activity, or by using the concept of a self-organizing system.

Thus, in the case of the brain the structure-function relationship shows a fraction or a gap. On the side of structure there is an organ of immense morphological differentiation; on the side of function, however, there is a global function comprising two radically different portions of function: namely the objectifiable neurophysiological one and another non-objectifiable one. Both parts are bound to the same organ, are performed in one process and are connected with one another by an unexplained asymmetric interdependency. Looking at conscious phenomena, one can state that they are conditioned and operated but not explained and caused by neurophysiological processes. Thus, in that portion of brain function it is the case of a categorial leap or a transition into another metaorganic dimension. The best term for this phenomenon might be categorial transformation (Oeser and Seitelberger 1988).

However, what kind of transition or transformation is this? The possible answer would be: The brain at the level of its specific functions corresponds to an information-processing system, not at all to a technical computer, but nevertheless in principale to an information-processing device in the sense of a "universal machine." Consequently it does not have to do with physicochemical entities but exclusively with self-designed equivalents or symbols of world objects and events. Reality characters appear represented or coded in certain ways in neurophysiological activities which are, however, not identical with our conscious experiences. Thus, the reality on which the brain works and which it produces in behavior does not consist of the physicochemical reality of nervous activity shown by the EEG or the imaging techniques but of the spatiotemporal patterns of excitations traveling through its subsystems, neural circuits and modules, i.e. it is not a substantial but a functional reality (Seitelberger 1982).

This functional reality could be compared only superficially with a computer program. It appears connected with the brain and its systemic processes in a not understood asymmetric interdependency. Regarding conscious performances there is nothing of parallelism or identity between neurophysiological and subjective experience, but there is obvious anisomorphia, anisochronia and bicausation, i.e. causation versus motivation, indicating a general unbridgeable difference of nature between the physical and psychic phenomena. I feel that this relation can best be characterized by using the metapher of complementarity in the sense of Nils Bohr regarding one entity of observation exhibiting two aspects which cannot be described and explained at the same time by the same methods.

The title of my contribution speaks of "hierarchies" of structure-function relationships and suggests the next point. It appears plausible to state that the brain as the central guiding instance, as it were, the citadel of the organism, represents the hierarchic summit of the pyramid of ordered complexity of functional systems which constitute the megasystem of the individual organism. But neural information processing itself exhibits a multifold stratification regarding complexity and range of products, i.e. a functional hierarchy of its own. Let me characterize roughly these levels regarding their behavioral relevance. At the basic level the regulation of the unbelievably complicated physiological processes of the living body is performed without any conscious control; yet it is experienced as an integrated report from the visceral brain on the individual's inner situation in the form of feeling or emotion, representing at the same time already a dull form of sensation of the self. Sensory perception means a higher, more complex step, because it combines information from different channels for reconstruction of outer objects, including one's own body, to abstract equivalents which are called reality models. Recognition of objects or cognition as intellectual performance means an again higher level of abstraction and synthesis because now classes of objects are conceived and group identifications of rather different single examples of objects are made by selection of characters. These functions are genetically preformed and individually learned; considering the multiplicity of reality at that level the operational instrument of language is needed. Language represents the cultural human invention of a new non-neurophysiological coding system of information processing products of the brain, i.e. of objects, concepts, relations, etc. By means of that ingenious instrument, so to say a metaorgan of brain performances comprising syntax, semantics and intentionality, the ability for purposeful simulation of existing and forthcoming but also of possible reality is given, i.e. the model-objects of thinking (a form of inner speech) which is constitutive for human intelligence (Bunge 1980; Mackay 1980). Evaluation of its results we experience as decision, and realization of results as acts of free will.

Furthermore, the abstract results of thinking processes not only control and instruct the actual and control the planned behavior, they also are verbally transmitted, stored in writing and printing and thus made available to be processed by the members of the human community.

Factually these abstract products of higher brain activity constitute their own domain of functional reality which become selforganized along proper rules and build independent, hierarchically organized systems of man-made realities which are termed knowledge, science, ethics, social systems, philosophy and religion. These products of information processing represent, as it were, the structural phase of the metabiological functional system brain whereas the process phase of this system, i.e. the totality of the performances at the level of their higher products, corresponds to human culture and tradition (Spatz 1961). Not nature but culture, or configurations of theoretical reconstruction and symbolization of reality, create the preponderant part of the world in which we are all indeed living. Technology transforms imagined possible reality into a second nature which dominates our environment and means the great and threatening challenges of mankind today (Seitelberger 1982).

I have tried to indicate that there exists not only the hierarchy of systemic order in the physiology of the living being human. Beyond the categorial gap between neurophysiology and consciousness we face another, complex hierarchy of the quasi real abstract products of higher brain functions which are encountered in progressive development, creating a more and more exact and complete representation of our environment and the world, of course, not a portrait but a fitting pragmatic reconstruction of it, in the sense of optimal evolutionary adaptation, as it were a human "idiocosmos."

This new, non-Newtonian, noncausally defined mental reality in the brain is our very biological essence, our mode of natural necessity and fate, our pride and misery, for which each individual together with all others is responsible. This selforganized metaorganic pyramid of processes, however, is not yet finished: its peak is still missing. It is our task to bridge the gap between outer and inner experience and to reduce the pseudoduality of our naive experience into the unity of our very nature. This requires not only recognizing this existential unity as integrated duality but especially realizing it in life. In other words, we must eat for a second time from the tree of knowledge in order to transform knowledge into wisdom and freedom into freelyly accepted necessity and responsibility. That would not mean the rise of a superman but would forshadow completion of genuine human nature (Seitelberger 1982).

It is the fascinating and difficult task of the neurosciences to investigate the quality and conditions of all levels of the functional systems that we have mentioned, which now are gaining increased relevance in governing our future as conscious beings. Thus, I would like to close my remarks by emphasizing the great importance of brain research as the principal transdisciplinary venture of occidental science.

References

Benninghoff, A. (1935/36) Form und Funktion. Z Gesamte Naturwiss 1: 149-160

Benninghoff A (1936/37) Form und Funktion, Teil 2. Z Gesamte Naturwiss 2: 102-114

Bohr, N. (1985) Atomphysik und menschliche Erkenntnis, vol 2. Vieweg, Braunschweig

Bunge, M. (1980) The Mind-body problem. Pergamon , Oxford

MacKay, D.M. (1980) Brains, machines and persons. Collins, London

Oeser, E., Seitelbeger, F. (1988) Gehirn, Bewuβtsein, Erkenntnis. Wissenschaftliche Buchgesellschaft, Darmstadt

Rothschuh, K.E. (1963) Theorie des Organismus. Urban and Schwarzenberg, München

Searle, J.R. (1984) Geist, Hirn und Wissenschaft. Suhrkamp, Frankfurt

Seitelberger, F. (1982) Die Evolution zur Erkenntnis. Leistungspotenzen und Leistungsprodukte des menschlichen Gehirns. In: Akten des 7. Internationalen Wittgenstein Symposiums,Erkenntnis- und Wissenschaftstheorie 22.- 29.8.1982, Kirchberg/Wechsel, pp 174-184

Seitelberger, F. (1984) Neurobiological aspects of intelligence. In: Wuketis FW (ed) concepts and approaches in evolutionary epistemology, vol 36. Reidel, Dordrecht, pp 123-148

Seitelberger, F. (1986) Informationsverarbeitung im Nervensystem. Funktionell-neuroanatomische Grundlagen.In: Biblos: Österr Z Buch Bibliothekswesen Dokumentation Bibliographie Bibliophilie 35 (1): 27-35

Seitelberger, F. (1987a) Wie geschieht Bewuβtsein? Die neurobiologischen Vorausetzungen. Psychologie Österreich 7 (1): 6-19

Seitelberger, F. (1987b) Die Raum-Zeit im Blickpunkt der Hirnforschung. In: Scharf JH (ed) Nova Acta Leopoldina vol.54 Halle/Saale, 327-344

Spatz, H. (1961) Gedanken über die Zukunft des Menschengehirns und die Idee vom Übermenschen. In: Benz E (ed) Der Übermensch. Rhein-Verlag,Zürich

Szentágothai, J. (1978) The local neuronal apparatus of the cerebral cortex. In: Buser PA, Rougeul-Buser A (eds) Cerebral correlates of conscious experience. North Holland, Amsterdam

Evolution and Phylogenetic Diversification of Chemical Messengers

D. Bückmann

Abteilung Allgemeine Zoologie, Universität Ulm, Oberer Eselsberg 24, 7900 Ulm, FRG

The program "From neuron to action" comprises all the marvellous mechanisms of the human peripheral and central nervous systems including brain and mind. What can a zoologist contribute to this field? The answer is: the history of how all this originated, and how it may have developed during the evolution of organisms.

The nervous system is the most highly developed system of cellular information transfer within a multicellular body. In all cases intercellular information transfer includes chemical information. On our way "from neuron to action" we always uncover somewhere an intercellular information transfer by a messenger substance. However, these substances are different in different animals and even in different tissues.

Presently there is some confusion about the borderlines between the groups of messenger substances such as neurotransmitters, neuromodulators, neuropeptides, peptide hormones and other hormones, tissue-specific growth factors, morphogenetic substances, chalones, poietins, prostaglandins and so on (Karlson 1984). Such confusion of our terms and concepts indicates that we may not yet have the proper understanding of the underlying principles. These might become apparent from the history of how all these systems came into existence and stemmed from each other.

The fossils of extinct ancestral organisms do not tell us anything about their intercellular information transfer. However, we can reconstruct the evolutionary history of intercellular messenger substances by comparing present-day organisms. We trace back the ancestral lines of animals with equivalent and with different messengers, and we find the branching at which a difference must have appeared in evolution.

A good example is the history of neurohypophysial hormones, as reconstructed by Acher (1984). The most primitive vertebrates, the agnatha, still possess only one such hormone, vasotocin; all the others have two. In their common ancestors, after branching off from those of the present agnatha, the gene of that peptide must have doubled. One copy remained that of vasotocin, and the other, by exchange of amino acids, became isotocin, mesotocin, or oxytocin in different vertebrates. At last, only in the common ancestors of the mammalians, also the remaining vasotocin was changed to vasopressin, and only within one subfamily, the suidae, the usual arginin-vasopressin was changed to lysin-vasopressin. Although no one has witnessed all this, we can even say at what time which changes must have occurred, how many million years ago.

An attempt to apply this method on a large scale to the whole animal kingdom (Bückmann 1984a, 1987) yielded unexpected results. (a) There is no clear relation between hormonal structure and function. This applies as well to the hormonal function in general as to the special effects of single hormones in different animals. The hormonal function is established by a functional connection through several links, which has been called a hormonal system. (b) The components of a hormonal system have not been evolved synchronously but in a sequence of different steps. The most ancient of them are the receptors. (c) In multicellular animals there are different hormonal systems of different phylogenetical ages. (d) Their distribution leads to an understanding of the conditions under which a hormonal system can be evolved. What do these four aspects mean?

1. Hormones are members of quite different groups of compounds, as amines, peptides, steroids, juvenoids and iodined thyronins, and in every case other members of the same group exist which are not hormones. Belonging to a certain type of compound does not automatically make a substance a hormone. What makes a compound a hormone is a functional system comprising the hormone-producing cells, the transport, and the target cells with specific receptors and biological responses. These are the components of what has been called a hormonal system by Karlson (1984).

As has been mentioned, there is no invariable connection of a certain hormone to a certain function. In many cases the same hormones occur in different animals but with different functions. Therefore the evolution of the messenger compounds seems to be older than their coupling to certain functions.

The insulin sequence of vertebrates forms part of the prothoracicotropic hormone in insects (Nasegawa et al. 1984). The red pigment concentrating colour change hormone of crustaceans is adipokinetic hormone in some insects. However, among insects there are "fat-flyers" and "carbohydrate-flyers." In the latter the same hormone does not mobilize lipids but serves as a hypertrehalosemic hormone (Ziegler et al 1985). Many peptide sequences which are hormones in higher animals have even been found in unicellular organisms and in higher plants, where the tasks that they fulfil in the animals do not exist (le Roith et al. 1986). The pentapeptide proctolin serves as a messenger substance in insects but as a blood-borne hormone in crustaceans (Stangier et al. 1986). Recently it has been pointed out that vertebrate gonadotropin-releasing hormone "has been recruited for diverse regulatory functions: As a neurotransmitter in the central and sympathetic nervous system, as a paracrine factor in the gonads and placenta, and as an autocrine regulator in tumor cells" (Millar and King 1988).

Even within the same animal a hormone may have different effects in different tissues and different stages. In the green caterpillar of the pussy moth, *Cerura vinula,* the ecdysteroid hormone usually causing pupation causes in the first small amounts a colour change and spinning behaviour. The chemical reasons for this colour change are even different in different parts of the integument. Later on, larger amounts of the same hormone causes pupation. The different responses cannot be ascribed to one uniform chain of biochemical effects, as each of them can be evoked independently of the other. In ligatured animals the colour change can be evoked by small doses, without automatically causing pupation, while large doses cause pupation without preceding colour change (Bückmann 1959). In other insects the same hormone may have still other functions such as calcification and decalcification (Fraenkel 1975; Bückmann 1984a) or gonadal development (de Loof 1982). Thus the ecdysteroid, which usually acts as moulting hormone, ecdysiotropic, acts in this case as a colour change hormone, melanotropic, and in adult insects it can be gonadotropic. A survey regarding prolactin by Bern (1975), even sounds somewhat reproachful at its unprincipled diversity of actions, as different as reproduction and osmoregulation. Hormones are simply "multitropic". They can be coupled to different responses. This should mean that, on the other hand, also similar functions could be controlled by different hormones. Indeed, moulting and metamorphosis are controlled by steroids in arthropods but by iodinated thyronins in vertebrates, though the vertebrates also have steroids. The hormonal control of sexual differentiation is different in every animal phylum (Bückmann 1984b).

2. Progress in receptor research (Bradshaw and Gill 1982; Csaba 1986a,b) led to the conclusion that receptors are the most generally distributed and thus presumably the oldest of the above-mentioned components of hormonal systems (Bückmann 1987). They are protein molecules which without covalent bonds fit to certain ligands and form complexes with them, which cause further responses in the cell. Such molecules exist in procaryotes as well as in eucaryotes, and there seem to be no essential differences between receptors for hormones, cell nutrients, or tissue-specific growth factors. Even cell poisons can exert their effect through forming a receptor ligand complex (Bradshaw and Gill 1982). It is still difficult to understand how receptor-ligand complexes can cause such different secondary responses within the cell.

Receptors enable the living cell to discriminate between different substances and to respond to chemical changes in its environment. This is a prerequisite for survival, and it is plausible that it must have been developed at the beginning of evolution, in the first organisms. From this first step hormonal systems must, according to their distribution among organisms, have evolved in a series of different steps.

3. A ligand may come from another cell of the same species. In unicellular organisms this would mean from another individual. Such interindividual messenger would be a pheromone. If the cells stay together as a multicellular colony of equal cells, the problem would arise for the single cell to discriminate the messenger substance of other cells from its own. In simple cell colonies this problem is solved by rhythmic intermittent release of the messenger substance (Gerisch et al. 1984).

When, however, the cells differentiate to various cell types, two decisive further steps are achieved: (a) Releaser cells forming the signal substance are separated from target cells forming the receptors. Thus a one way path of information is established. (b) At the same time, coexistence of several signal substances with different target tissues is possible. The signal substances may differ in their stability and transportation properties. Some of them act on neighbouring cells, as morphogenes, tissue-specific growth factors and prostaglandins do. The distribution of these substances among organisms is not yet known sufficiently well to draw conclusions as to their history. Only in some highly differentiated multicellular organisms have additional systems evolved.

A special development is that of phytohormones in higher plants. Their structure, with air-filled intercellular spaces, rigid cell membranes and cell connections by plasmodesms, provides special conditions for messenger transport in their body, different from that in animals.

In higher animals a special tissue of long-distance coordination has evolved, the nervous system. It works in two ways. Long extensions release transmitters immediately near the target. These transmitters are small molecules, short living and therefore acting at the site of release only. They can be unspecific without danger of acting on "wrong" cells. They transmit messages for local, quick, and short-term reactions. They are amines, amino acids and small peptides.

Other substances from the same nervous tissue are secreted into the blood for further distribution. They need more time but reach every body cell equally well. Therefore they must be long living but specifically recognized by the receptors of certain cells only. They are large and complicated peptide molecules. They transmit messages for slow and long-lasting reactions in different parts of the body.

However, there is no strict boundary between peptide hormones and transmitter substances. Both are produced by the same nervous system. Both act on receptors at the target cell surface, and typical primary actions are changes of membrane potential and the activation of secondary messengers.

There are simple multicellular animals, the Parazoa, the sponges, which have no nervous tissue. If the typical neurosecretory substances are present in them (Lentz 1968), as they are even in unicellular animals (le Roith et al. 1986), this would mean that neurosecretory substances are phylogenetically even older than the typical neurons with axons and synapses. At first large peptides transfered long-distance information in the multicellular body, and then for quick and short-lasting actions long axons and nerves evolved.

Further advance occur only in few animal groups, and they are different in each of them. This indicates that they may have evolved independently in each of these groups. All of these groups have large body sizes, high degrees of differentiation and effective blood circulations. In most cases the further achievements of hormonal systems do not occur even in an animal phylum as a whole but only in those of its subphyla which have gained large body size and complicated organization. These are, among arthropods, the insects and higher crustaceans, among molluscs the snails and cephalopods and among chordates the vertebrates.

Occurring in all of these groups and, therefore, presumably the first step in higher development of hormonal systems are neurohaemal organs. These are special structures for the release of neuropeptides into the circulation. However, in each of the mentioned groups they are constructed differently, as the vertebrate neurohypophysis, the corpora cardiaca of insects or the brain lobes of snails, with very strange and special cell types such as the "canopy cells" and their most elaborate axon connections, or the vena cava in squids, where the whole surface of large vessels forms a neurohaemal structure (Joose and Geraerts 1983).

The next step is the evolution of an entirely new type of hormonal organs, not derived from the nervous system and, at the same time, with new, non peptide hormones, again different in different animal phyla. Our thyroid hormones, iodinated thyronins, seem to be confined to vertebrates only, and juvenoid hormones are confined to insects. Steroid hormones occur in more than one animal phylum, but those serving as hormones in vertebrates belong to another type than the ecdysteroids found in arthropods and other invertebrate phyla. The receptors for these nonpeptide hormones and even the primary reactions are different from those of peptides and amines. The receptors are not located on the cell surface but are plasmatic and within the nucleus, one primary action being gene activation (Gorski 1986). The glands producing these hormones are not nervous tissue, and in every case they are typical of the special animal phylum, without homologues in other phyla, such as the thyroid gland derived from the endostyle and the branchiogenous glands in vertebrates or the prothoracic gland stemming from a maxillary gland in insects. Evidently they have evolved independently in each of the different groups, much later than transmitter and neurosecretory substances, as secondary hormonal systems.

4. How do such organs become endocrine glands? An example is the history of the thyroid gland, as reconstructed by Gorbman (1953) and Barrington (1953). The essential need of iodine makes this system susceptible to lack of iodine in fresh water and on dry land. However, it can be understood by the evolution of the thyroid from the endostyle of marine chordate ancestors. In the sea iodine is richly abundant and many animals form mucus of iodine-containing peptides. The endostyle envelops the food in such mucus. In the gut it is digested and absorbed together with the food. Thus iodinated amino acids enter the body and can serve as a characteristic signal: "food is coming". As signal they had already become indispensible when descendants of these animals conquered fresh water and dry land. In them the endostyle is separated from the foregut, and it still forms iodinated peptides at first and then cleaves them afterwards to the iodinated amino acids, which are signal substances.

The crustacean Y-organ and the insect corpora allata derive from the integument, that is, from part of their own target organ. The Y-organ even reacts to its own hormone (Kleinholz and Keller 1979). In simpler arthropods this hormone seems to be formed by the whole integument. Then part of its cells specialized in the task of forming a compound which is needed by all of them and distributing it to the others. This may be a way of better synchronizing the moulting processes in large and complicated exuviae (Bückmann 1984a).

What is common to these secondary hormonal glands? They release some typical compound into the blood in a certain biological situation be used as a signal of this situation, synchronizing all responses necessary to it. Recall that cells are capable of evolving receptors to substances in their environment and coupling the resulting complexes to quite different chains of biological responses. Unicellular animals can even be induced to form new receptors to new environmental substances by "imprinting" (Csaba 1986a,b). In cells of multicellular animals the environment of cells is the body fluid.

We find no strict relation between hormonal structure and function but rather a temporal relation. The function of hormones is the synchronizing of different responses in different places. Now we understand, why hormones *must* be multitropic. The biological problem is less to synchronize similar events than to synchronize events as different as possible, developmental, physiological and behavioural.

When the green caterpillar goes from green foliage to spin its cocoon on the brown bark for pupation, it must become red. Prolactin is at work when amphibia or fish change their surrounding from dry to wet or from sea to fresh water, or vice versa, in order to reproduce. Melatonin always causes reactions related to day-night cycles (Epple 1982) and so on.

This picture of primary and secondary hormonal systems has been complicated in the vertebrates by two exceptions. Transmitter substances are transported through the blood, as adrenaline, and several peripheral endocrine glands secrete peptide hormones. One reason may be that vertebrates are exceptional among all animals by enormous body size and warm-blooded forms with an enormous oxygen consumption, and, therefore, a high-performance circulatory system. The body fluid is moved so fast that even quickly acting substances such as adrenaline can be transported to their target tissue this way, and also peptide hormones which are formed not by nervous but by peripheral endocrine glands. These may be special acquisitions evolved only late in vertebrate phylogeny.

From this history arises a functional concept of messenger substances. All substances to which receptors can be evolved are potential messenger substances. Those which are transported over a long distance by the blood of a multicellular body are called "hormones". They must be distributed in the body fluid in a certain biological situation which requires temporal coordination of responses in different parts of the body. This distinction from other messenger substances is not a fundamental one. However, this long-distance coordination is fundamental for the evolution of large multicellular organisms.

References

Acher R (1984) Evolution of neuropeptides: neurohypophysial hormones, neurophysins and polyprecursors. Nova Acta Leopoldina NF 56. 255: 137-151.

Barrington EJW (1953) Some endocrinological aspects of protochordata In: Gorbman A (ed) Comparative endocrinology. Wiley, New York, pp 250 - 265.

Bern H (1975) On two possible primary activities of prolactins: osmoregulatory and developmental. Verh Dtsch Zool Ges 68. 40-46.

Bradshaw RA, Gill GN (eds) (1982) Evolution of hormone-receptor systems. J Cell Biochem 6: 110-185.

Bückmann D (1959) Die Auslösung der Umfärbung durch das Häutungshormon bei Cerura vinula L (Lepidoptera, Notodontidae). J Insect Physiol 3: 159-189.

Bückmann D (1984a) The phylogeny of hormones and hormonal systems. Nova Acta Leopoldina 56 Nr 255: 437-452.

Bückmann D (1984b) Vergleichende Endokrinologie und Stammesgeschichte der Sexualentwicklung. Akt Endokr Stoffw 5: 169-174.

Bückmann D (1987) Common origin and phylogenetic diversification of animal hormonal systems Experientia [Supp]l 53: 155-166.

Csaba G (1986a) Why do hormone receptors arise? Experientia 42: 715-718.

Csaba G (1986b) Receptor ontogeny and hormonal imprinting. Experientia 42: 750-759.

De Loof (1982) A new concept in endocrine control of vitellogenesis and in functioning of the ovary in insects. In: Adding ADF, Spronk N (eds) Exogenous and endogenous influences on metabolic and neural control, Pergamon Oxford, pp 165-177.

Epple A (1982) Functional principles of vertebrate endocrine systems. Verh Dtsch Zool Ges 117-126.

Fraenkel G (1975) Interactions between ecdysone, bursicon and other endocrines during puparium formation and adult emergence in flies. Am Zool 15: 29-48.

Fraenkel G, Hsiao C (1967) Calcification, tanning, and the role of ecdysone in the formation of the puparium of the facefly, Musca autumnalis. J Insect Physiol 13: 1387-1394.

Gerisch G (1984) Biochemical regulation of cell development and aggregation in Dictyostelium discoideum.Nova Acta Leopoldina NF 56, 255: 65-84.

Gorbman A (1953) Problems in the comparative morphology and physiology of the vertebrate thyroid gland. In: Gorbman A (ed) Comparative endocrinology. Wiley, New York, pp 266-282.

Gorski J (1986) The nature and development of steroid hormone receptors. Experientia 42: 744-750.

Joosse J, Geraerts WM (1983) Endocrinology. In: The mollusca. vol 4. Academic London, pp 317-406.

Karlson P (1984) The concept of hormonal systems in prospect and retrospect: Nova Acta Leopoldina NF 56: 9-20.

Kleinholz LH, Keller R (1979) Endocrine regulation in crustacea. In: Barrington EJW (ed) Hormones and evolution 1. Academic, New York, pp 159 - 213

Lentz TL (1968) Primitive nervous systems. Yale University Press, New Haven CN

Le Roith D, Roberts C Jr, Lesniak MA, Roth J (1986) Receptors for intercellular messenger molecules in microbes: similarities to vertebrate receptors and possible implications for diseases in man. Experientia 42: 782-788.

Millar P, King JA (1988) Evolution of gonadotropin-releasing hormone: multiple usage of a peptide. News Physiol Sci 3: 49-53.

Nagasawa H, Kataoka H, Isogai A, Tamura S, Suzuki A, Ishizaki H, Mizoguchi A, Fujiwara Y, Suzuki A (1984) Amino-terminal amino acid sequence of the silkworm prothoracicotropic hormone: homology with insulin. Science 226: 1344-1345.

Stangier J, Dircksen H, Keller R (1986) identification and immunocytochemical localization of proctolin in pericardial organs of the shore crab, Carcinus maenas. Peptides 7: 67-72.
Ziegler R, Eckart K, Schwarz H, Keller R (1985) Amino acid aequence of Manduca sexta adipokinetic hormone elucidated by combined fast atom bombardment (FAB)/tandem mass. Biochem Biophys Res Commun 133 (1): 337-342.

Part 5

Synaptic and Elementary Processes

Cascade-Type Reentrance: The Major Connectivity Principle of the Neocortex

J. Szentágothai
Tüzolto u. 58, Budapest 1450, Hungary

Introduction

The neocortex is indeed one of the major wonders of nature, whatever aspect is considered: (a) how it is put together during development; (b) the diversity of its major structural constituents (various neuron types) assembled according to a general blueprint of local connectivity, but with a great variation in the finer details leading to the coexistence in the same larger tissue block of various maps (or elements thereof); (c) the unbelievable richness and variety in short-, medium-, and long-range connectivity. This refers only to architectonic side. To assemble all of this information into a short coherent story is obviously impossible. The best that I can aspire for is to show a small segment of cortical connectivity (item (c) above) as it is beginning to take shape in our present understanding.

We can start with the intuitive insight,which offers itself from simply looking at well-stained Golgi specimens of cortex in any mammalian that cortico-cortical connectivity is extremely rich, as compared with connections between different parts of other major central organs. But what does this mean in real quantitative terms? In order to answer this question we must first separate excitatory connections from inhibitory ones. Since I have discussed inhibitory connectivity quite recently (Szentágothai 1987), we might discard this aspect entirely, also because it does not strictly belong to the central issue of this paper.

Theoretically, the upper limit of cortico-cortical connectivity might be a connection from any output neuron to any other part of the cortex. Such an assumption was considered for smaller vertebrates (rodents) with correspondingly smaller brains by Braitenberg (1978).

Unfortunately, we have no information whatever about the lower limit of this type of connectivity. Each pyramidal axon that leaves the cortex may have only a single stem that might be directed either to another part of the cortex or to some extracortical target. Here our knowledge stands on firmer ground. We know that the axons of pyramidal cells from lamina III (and partly from lamina II) are directed mainly towards other cortical sites. We know that the arborization space of a single cortico-cortical afferent may contain about 5000 neurons as potential synaptic targets. However, probably only a small fraction of these cells is actually directly (monosynaptically) contacted. Conversely, most axons of pyramidal (or fusiform) cells of laminae V and VI that leave the cortex are directed towards subcortical targets. But even these axons have frequently one major collateral directed - over the corpus callosum - to the cortex of the contralateral hemisphere. It has been even claimed (Krieg 1963) that the majority of the callosal fibers - also those arriving from laminae II-III - are collaterals rather than the main axons of pyramidal cells. From a few pilot studies of retrograde double labeling of pyramidal cells from injection sites in distant parts of the cortex (Schwartz and Goldman-Rakic 1984) the impression is gained that, although not abundant, pyramidal cells that project to two (or more?) distant sites of the cortex are by no means a rarity.

Short-Range (Intracortical) Connections

There are, so far, two types of identified excitatory intracortical neurons and/or connections:(a) the so called "spiny stellate" cells, preferentially located in lamina IV, and (b) local connections established by the collaterals of pyramidal cells.

We may deal very briefly with the spiny stellate neurons, one of the best studied local "interneurons" (Lund 1973; Szentágothai 1975; Somogyi 1978; Jones 1975). One might even ask oneself whether the expression interneuron is justified in these cells, because they are an important link in the neuron chain connecting input with output. In the specifically studied cases, mainly in primary sensory areas, the major role of the spiny stellate neurons is to transform the specific afferent inpulse pattern conveyed to lamina IV into a vertically ascending and descending columnar pattern of excitation (Szentágothai 1983) towards laminae III (II) and V (VI). One should not forget, however, that a considerable part of all synapses given by the axonal arbor of spiny stellate neurons is situated close to the body of the cell within lamina IV itself. As is known from a single, virtually completely reconstructed spiny stellate in the study of Martin and Whitteridge (1984), spiny stellate cells of the visual cortex (monkey) may have very considerable tangential spread (5 mm), although the preterminal and terminal branches are eventually arranged in discrete ascending columns. These connections are not considered in this paper.

Regarding connections established by pyramid cell collaterals, based on Golgi pictures from the classical period, we assumed (Scheibel and Scheibel 1970; Szentágothai 1975) that the initial collateral arborization and its connections might be rather stereotypic.And it is true that the first initial collaterals tend to ascend vertically, running parallel with the apical dendrite of the parent cell, and subsequent collaterals run in successively less steeply ascending and, finally, horizontal and even in descending courses. I have argued earlier (Szentágothai 1975, 1978b) that certain general functional conclusions might be drawn from this. Later, it turned out (Kisvarday et al. 1986) that exact reconstruction of the initial collateral arborizations and their synaptic targets disclosed specific addressing (direction towards certain targets) in many lamina III pyramidal cells. The earlier assumption (Szentágothai 1962, 1978a) that the specific targets of pyramidal cell collaterals are dendritic spines of other pyramidal cells was substantiated by the recent studies. But the connectivity between relatively neighboring pyramidal cells is certainly much more specific than was assumed initially.

Medium-Range Excitatory Connections

The two categories of short- and medium-range cortico-cortical connections cannot be rigidly separated. This is forcefully suggested by observations of Martin and Whitteridge (1984) showing a large lamina V pyramidal cell, very beautifully reconstructed, that had a very extended (several millimeters) collateral arborization, specifically located in lamina I. This arbor certainly goes beyond whatever one might accept as a cortical column or even a "hypercolumn". A single reconstructed cell does not, of course, tell us anything about the frequency of such cells in any part of the cortex. However, one of my very first observations in the visual cortex of the cat (Szentágothai 1962) later presented by Colonnier (1966) showed that in chronically isolated slabs of visual cortex in the cat,that lamina I is almost completely devoid of axons until the isolated piece contains also parts of lamina V. In such isolated preparations axons appear in large abundance, indicating that these axons must arise from cells in lamina V. At that time I interpreted this finding by assuming some specific type of "recurrent" interneuron, possibly the cells referred to in the earlier literature as Martinotti cells. Most recently Kisvarday et al. (1989) could show that pyramidal cells characterized by a very high uptake of D-[^{3}H]aspartate project massively to lamina I and as far off as 1 mm to the side of their vertical axis. So the original finding of axons surviving in lamina I of isolated cortical slabs must be reinterpreted as fibers arising from relatively local pyramidal cells of specific metabolic properties. We have here certainly a most excellent illustration of what I mean as "cascade type" connectivity. This is demonstrated by the simple diagram shown in Fig.1, which is a diagrammatic transformation of the new observation of Kisvarday et al. (1989).

Long-Range Connectivity

Here the qualification "excitatory" can safely be omitted because as far as we know all distant cortico-cortical or cortico-subcortical connections are excitatory. The pyramidal cells are excitatory in nature, no GABAergic pyramidal cells have so far been observed. The same cannot be taken granted for cortical efferents arising from fusiform neurons of lamina VI, however from the distribution and relative scarcity of GABAergic neurons in lamina VI (Gabbott and Somogyi 1986) - particularly as compared to the distribution of neurons in this layer retrogradely labeled from subcortical target sites - this would be a very unlikely assumption.

Since direct observations on cortico-cortical connections are still relatively scarce and patchy (in this context the studies of Goldman-Rakic and coworkers are the most important ; they cannot be cited here in any detail , and the reader is referred to a comprehensive summary by Goldman-Rakic 1984). We confine ourselves to general considerations. The main questions that arise here have been discussed in some detail by Braitenberg (1978), however this author leans toward the assumption that every portion - if not every efferent cell - of the cortex ought to be connected with every other. But this assumption is unlikely due to lack of space (in the white matter) to accomodate the fiber volume. My coworker, E. Lábos (see Szentágothai and Lábos, in preparation) has formalized this in the following mathematical expression:

$$k = \frac{[w - b + \sqrt{(w - b)^2 + 4 w V}]}{2 w} \tag{1}$$

where V is a volume filled with k smaller nonoverlapping bodies of equal shape and equal volume b. These bodies are interconnected by single wires with the average volume w., Assuming realistic values for both b and w the volume V of the mouse brain would be filled by k = 5700 units.

It is quite obvious that in very small brains the volume saturation (more exactly, exhaustion of available volume) would make allowance for larger cell numbers due to the small distances that must be bridged. Braitenberg (1978) has correctly realized this problem for the human brain, where specific larger cortico-cortical tracts have long been known and can be clearly demonstrated by appropriate dissection.

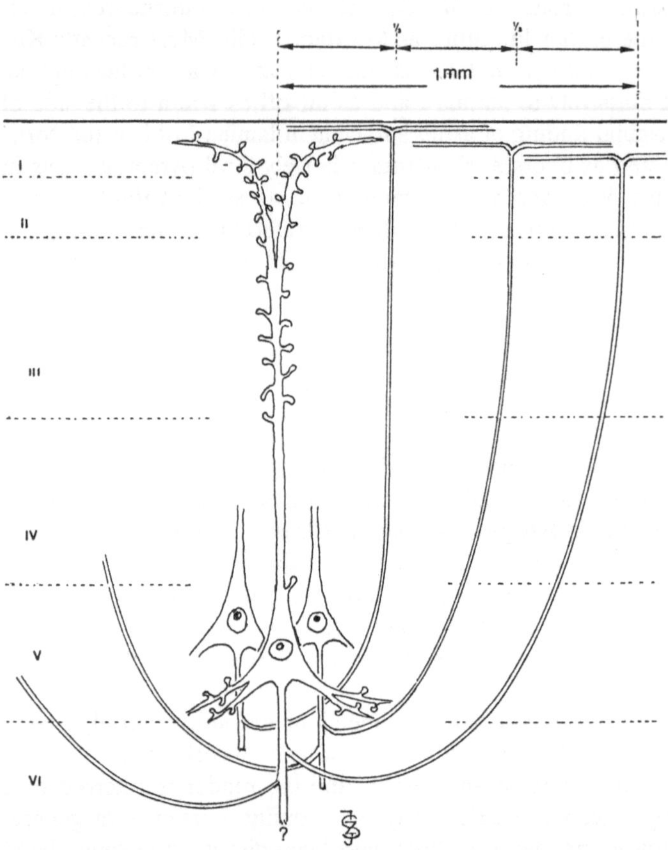

Fig. 1. Diagrammatic illustration of collateral connections of lamina V pyramidal cells, characterized by high specific uptake of D-[^{3}H]aspartate (Kisvarday et al. 1989) and projecting mainly to lamina I. The pyramidal cell drawn in some details (central) has widespread collaterals, reaching distances up to 1 mm off the vertical axis of the parent cell. Two other pyramidal cells only fragmentarily indicated are assumed to give rise to collaterals that spread only to distances of 1/3 to 2/3 of mm. This would be the simplest explanation compatible with the experimental observation of the tangential grouping of such cells, although other interpretations could also be imagined

From my own understanding of the brain I would deam it unnecessary to force the issue of "homogenety," i.e the notion that all parts of the brain are directly interconnected. This is because,first the assumption would be unrealistic for brains much larger - say, by a factor of 10 - than that of the mouse. But of almost equal importance is the consideration that a relative "homogeneity" would not require that every compartment be connected directly (monosynaptically) with every other. Such an arrangement would force unnecessary restraints upon overall connectivity. This is demonstrated with the following conceptual experiment. From the knowledge of the arborization pattern of a single cortico-cortical fiber (see, for example, Fig. 2 in Szentágothai 1978a) it becomes obvious that it terminates in such a large space of arborization that contains at least 5000 neurons (not to speak of additional dendrites entering this space from without) as potential synaptic targets. It is certain that only a small fraction of such potential targets are actual targets. Assuming that every axon of a pyramidal cell leaving the cortex had only one main branch (which is almost certainly not true, from knowing the frequent arborization in afferent axons when ascending through the white matter towards their destination in the cortex), each cortico-cortical fiber would contact at least hundreds of cells in their field of terminal arborization. From the stripelike pattern of virtually all projections into the cortex it can be inferred that the "fanning out" of preterminal branches - observed on the macrosopic scale - is repeated also on the minute scale of individual (or of small groups of) afferents. The consequence of this would be the *large-scale cascading* that would occur at every step of the cortico-cortical chain, the impulses carried by one afferent being transmitted to the cells of origin (pyramidal cells) of at least tens, if not hundreds of efferent fibers of the next neuronal link.

But let us consider the known numbers of the cortical efferent cells, the pyramidal neurons. It is generally believed ,and, probably a realistic estimate is that 60% of all cortical cells are pyramidal cells in the general type of cortices. (There is probably a larger percentage of pyramidal neurons in area 4 and possibly a smaller in the primary sensory cortices). Probably half of the 60%, i.e. 30% of the pyramidal neurons, is located in lamina III. Assuming that there are 5000 cells in such cortico-cortical column (the arborization space of the average cortico-cortical fiber), we can safely assume that we have at least 1500 pyramidal cells in each cortico-cortical column that give again rise to as many cortico-cortical fibers again. However, since cortico-cortical fibers arise also from other layers (II-VI), we may calculate almost certainly with two times this number, i.e. with 3000 cortico-cortical afferents per cortical column. (Remember that many cortico-subcortical fibers have major collaterals - mainly through the corpus callosum - that are also cortico-cortical afferents.) From the established direct cortico-cortical connections of various cortical areas, which according to the data of Krieg (1963) have connections to at least 3-5 other major cortical areas, we can safely conclude that the assumption made in Fig. 2 is exceedingly low. Even calculating with an extremely simplified cortex, stripped down to a mere mosaic of discrete columns, the overall connectivity of the cerebral cortex by chains of five successive cortico-cortical neurons would be very impressive. If we then included intracortical spread connecting over up to 10 columns, and the medium range cascading illustrated in Fig.1, the connectivity would become immediately much richer. The basic idea of this type of minimal long-range connectivity could be represented diagrammatically as is shown here in Fig.2 , with pyramidal cells having two to three cortico-cortical axon branches. By including any further link of the neuron chain the illustration would become unintelligible. This gives us

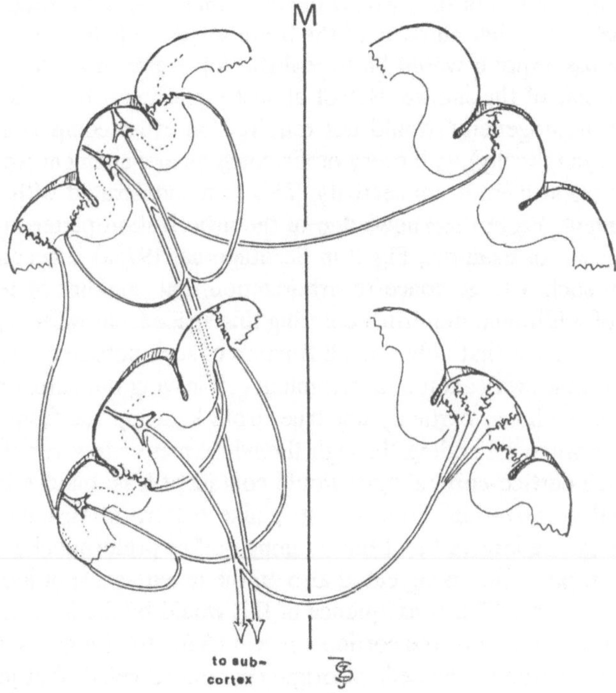

Fig. 2. Diagrammatic representation of cortico-cortical connectivity - stripped down to a base minimum - under the unrealistic assumption that pyramidal cells give rise to only one to three cortico-cortical fibers and/or collaterals. Two coronal sections of the cortex are shown on both sides of the midline (M); the upper section must be imagined as rear to the section below. Two pyramidal cells are shown in both left sections, one from lamina III (at right in the upper and at left in the lower section) and two representative pyramidal cells from laminae V and VI. The deep pyramidal cells are known to project mainly to subcortical targets, but these axons have collaterals directed towards the contralateral hemisphere. Also, supragranular pyramidal cells (from laminae II and III) are giving callosal projections, but are not represented in the diagram. The cortico-cortical fibers terminate in vertically oriented, 200-300 μm wide columns (with a larger tangential spread in lamina I); the highly probable "fanning out" of the cortico-cortical afferents by division in their final course towards the cortex (giving rise to stripes caused by a series of neighboring columns) is shown at lower right. Further neuronal links through pyramidal cells within the arborization space of the columns must to be left to the imagination of the viewer, otherwise the diagram would become too entangled

the impression that practically every compartment of the entire cerebral cortex is connected with virtually every other over 5-10 neuronal links. Thinking in terms of the known time relations in cortical processing, neuron chains of 5 to 10 links are by no means unreasonable.

However, direct, albeit multichain, cortico-cortical processing is only part of the story. There are very numerous neuron chain loops involving subcortical centers: cortico-thalamic and reverse; cortico-striato-thalamic-cortex loops; and the cortico-pontine-cerebellum-thalamo-cortical loop to mention only a few of the major large cortico-subcortico-cortical loops. Especially the so- called classical loops over the so-called "association nuclei" (a term now outdated) of the thalamus offer an exceedingly large additional variety of two-neuron chain connectivities between different cortical areas. It is therefore, hardly necessary to force the direct connection of every cortical site with every other.

If it were possible to include into a diagram of the type in Fig .2 all long-range connections of the cortex that have been traced in the literature over merely the past 20 years, the whole brain would become a network of unintelligible connections with innumerable reentrances and cascades at every single step.

Conclusion

The association made in the title of this paper is therefore no exaggeration.This leads us forcefully to the insight that the major principle of neural organization is a continuously and infinitely repeated reentrance, as emphasized already by Mountcastle (1978), of cascading neuron chain loops. In this light, the conventional concept of neural systems as reflex loops with afferent neuron chains being routed through various higher centers towards efferent chains directed towards peripheral tissues for the execution of commands, elicited or trigged by the sensory (or afferent) input, loses much of its attractiveness as a satisfactory explanation of neural activity. The vast majority of neural connections being cortico-cortical - irrespective of whether direct or indirect - attention has to be focussed increasingly upon this aspect of neural structure and function. It is quite obvious, that such a change in outlook will entail radical changes in our understanding of the "neural" at large and particularly of the phenomena listed under the concept of the "cognitive".

Summary

Excitatory cortico-cortical connectivity is discussed. Short-range (intracortical) connectivity is secured primarily over the so-called spiny stellate neurons of lamina IV and the rich initial collateral system of the pyramidal cells that are primarily addressed spines of other pyramidal cells. A new, hitherto barely known (quasi-)middle range cascading connectivity with lamina I is secured over a specific lamina V pyramidal cell system characterized by highly specific uptake of D-[^{3}H]aspartate. Long-range cortico-cortical connections may establish over five (or slightly more) neuronal links a very high degree of mutual connections between virtually all parts of the cortex.

References

Braitenberg V (1978) Cortical architectonics: general and areal. In: Brazier MB, Petsche H (eds) Architectonics of the cerebral cortex. Raven , New York, pp 443-465

Colonnier ML (1966) The structural design of the neocortex 1-23. In:Eccles JC (ed) Brain and conscious experience. Springer,Berlin Heidelberg New York, p 591

Gabbott PLA, Somogyi P (1986) Quantitative distribution of GABA-immunoreactive neurons in the visual cortex (area 17) of the cat. Exp Brain Res 61:323-331

Goldman-Rakic P. (1984) Modular organization of the prefrontal cortex. TINS 7:419-424

Jones EG (1975) Varieties and distribution of non-pyramidal cells in the somatic sensory cortex of the squirrel monkey. J Comp Neurol 160:205-268

Kisvarday ZF, Martin KAC, Freund TF, Magloczky Zs, Whitteridge D, Somogyi P (1986) Synaptic targets of HRP-filled layer III pyramidal cells in the cat striate cortex. Exp Brain Res 64:541-552

Kisvarday ZF, Cowey A, Smith AD, Somogyi P (1989) Interlaminal and lateral excitatory amino acid connections in the striate cortex of monkey. J Neurosci, 9: 647-665

Krieg WJS (1963) Connections of the cerebral cortex. USA Brain Books, Evanston, Ill
Lund JS (1973) Organiztion of neurons in the visual cortex area 17 of the monkey (Macaca mulatta). J Comp Neurol 147:455-496
Martin KAC, Whitteridge D (1984) Form, function and intracortical projections of spiny neurons in the striate visual cortex of the cat. J Physiol(Lond) 353:463-504
Mountcastle VB (1978) An organizing principle for cerebral function: the unit model and the distributed system. In: Edelman GM, Mountcastle VB (eds) The mindful brain. MIT Press, Cambridge, MA, pp 7-50
Scheibel ME, Scheibel AB (1970) Elementary processes in selected thalamic and cortical subsystems - the structural substrates. In: Schmitt FO (ed) The neurosciences second study program. Rockefeller University Press, New York, pp 443-457
Somogyi P (1978) The study of Golgi stained cells and of experimental degeneration under the electron microscope: direct method for the identification in the visual cortex of three successive links in a neuron chain. Neuroscience 3:167-180
Somogyi P, Kisvarday ZF, Martin KAC, Whitteridge D (1983) Synaptic connections of morphologically identified and physiologically characterized large basket cells in the striate cortex of cat. Neuroscience 10:261-294
Schwartz ML, Goldman-Rakic PS (1984) Callosal and intrahemispheric connectivity of the prefrontal association cortex in rhesus monkey relation between intraparietal and principal sulcal cortex. J Comp Neurol 226:403-420
Szentágothai J (1962) On the synaptology of the cerebral cortex. In: Sarkissov SA (ed) Structure and function of the nervous system. State Publishing Housefor Medical Literature , Moscow, pp 6-14
Szentágothai J (1975) The "module-concept" in cerebral cortex architecture. Brain Res 95:475-496
Szentágothai J (1978a) The neuron network of the cerebral cortex: a functional interpretation. The Ferrier Lecture. Proc R Soc Lond (Biol) 201:219-248
Szentágothai J (1978b) Specificity versus (quasi-) randomness in cortical connectivity. In: Brazier MAB, Petsche H (eds) Architectonics of the cerebral cortex. Raven , New York, pp 77-97
Szentágothai J (1983) The modular architectonic principle of neural centers. Rev Physiol Biochem Pharmacol 98:11-61
Szentágothai J (1987) The architecture of neural centers and understanding neural organization. In: McLennan H, Ledsome JR, McIntosh CHS, Jones DR (eds) Advances in physiological research. Plenum , New York , pp 111-129

Peripheral Axotomy Challenges the Central Motor Neuron and its Cellular Microenvironment

G.W. Kreutzberg
Max Planck Institute for Psychiatry, Department of Neuromorphology, Am Klopferspitz 18 A, 8033 Planegg-Martinsried, FRG

For several good reasons the nervous system is classified into a peripheral (PNS) and a central (CNS) part with distinct and in many respects different properties. One such difference is the capacity for regeneration and functional restitution after a lesion, which is present in peripheral nerves but is widely lacking in neurons of the brain and the spinal cord. An exception to this rule is the motoneuron. Motoneuron cell bodies are located in the CNS, and they have the morphology, the dendritic and synaptic organization typical of large brain stem neurons, i.e. they are essentially central neurons except for the course of their axons. Thus, they leave the CNS to innervate extrinsic target tissue, e.g. muscle or ganglia. In response to an injury to its peripheral axon the motoneuron has the full capacity to regenerate by growing a new axon which under favourable conditions is able to reach and innervate the target tissue leading to a restoration of function.

The main aim of experimental regeneration research is to recognize the principles underlying this process and to evaluate the possibilities of extending such principles to other CNS neurons thus establishing the basis for a restorative neurology. A number of changes observed so far in motoneurons as well as in primary sensory or sympathetic neurons has led to the conclusion that nerve cells have an intrinsic regeneration program which is genetically determined. The phenomenology of such a regeneration program has been studied intensively during the past two decades and has been reviewed from various points of view (see Lieberman 1971; Grafstein 1975; Bisby 1980; Kreutzberg 1982, 1986).

The reaction of the Neuron and its Microenvironment to Axotomy

The morphological changes occurring in nerve cells as a consequence of the interruption of the axon are generally known by the terms "chromatolysis", "axon reaction" or retrograde change. By electron microscopy we have learned that this phenomenon is based on changes in the organization of the granular endoplasmic reticulum. There is an enormous increase in the number of mainly free ribosomes; the cisternae of the r-ER normally arranged in Nissl bodies lose their parallel arrangement and can be found in every part of the perikaryon. This morphology has been interpreted as a hypertrophy of the cell, especially since the cell body is definitely enlarged, and the contours, normally concave, tend towards a convex shape.

These ultrastructural changes are signs of a hyperactive neuron in a metabolic sense. In fact, the increase in the organelles responsible for protein synthesis correspond to the measurements of protein changes in regenerating motoneurons. Proteins related to the cytoskeleton, such as actin, tubulin and calmodulin increased in regenerating neurons and are consequently represented to a greater extent in the slow components of axonal transport (Bisby 1980). An exception to this rule has been demonstrated for the intermediate filament protein of neurons, the neurofilament triplet. Proteins of this neuron-specific class decrease both in the neuron and in axonal transport during regeneration (Hoffman and Lasek 1980; Oblinger et al. 1987; Tetzlaff et al. 1988b). Although the significance of this is not clear, it has been speculated that the neurofilaments could play a role in the organization of the axoplasm by conferring upon it a spatial constraint which could provide stability. The decrease of neurofilament proteins in the axon could thus reflect greater fluidity and thus a facilitation of axonal flow and elongation.

In contrast to the general increase of the structural proteins mentioned, proteins related to neurotransmitter metabolism are strongly decreased in regenerating neurons of any modality. This is well known for dopamine decarboxylase, tyrosine hydroxylase, cholinergic receptors and for the cholinesterases (Frizell and Sjöstrand 1974; Hoover and Hancock 1985). In our laboratory we have studied in particular the changes in acetylcholinesterases occurring during regeneration of facial and vagal motoneurons (Kreutzberg et al. 1984; Engel and Kreutzberg 1986; Tetzlaff and Kreutzberg 1984). By studying the different molecular forms of the enzyme a rather differentiated response of regenerating motoneurons has been discovered. In the rat facial nucleus all forms of the molecule, i.e. the heavy asymmetric and the lighter globular forms, are considerably decreased and thus appear with lower activity in the cell bodies, the dendrites and the axons. The guinea pig differs highly in this respect. We found a marked increase in the A12 asymmetric form which is also the secretory form of the enzyme. The globular forms were found to decrease or to exhibit no change. An interesting aspect is that electron microscopical cytochemistry had earlier revealed evidence for an increased secretion of AChE from the dendrites of these regenerating motoneurons (Kreutzberg et al. 1975), a phenomenon for which the term dendritic secretion had been coined (Kreutzberg and Toth 1974). On the basis of the biochemical and cytochemical data it seems justified to postulate that these regenerating facial motoneurons produce a form of AChE which they normally almost completely lack. This A12 AChE is destined for export and secretion. Thus, it is found in the extracellular spaces of the facial nucleus from where it reaches the basal lamina of the local capillaries (Kreutzberg et al. 1975). It is also exported to the growing axons. In the proximal stump of the transected nerve AChE activity can be demonstrated on the surface of the newly formed axonal sprouts and in the extracellular spaces in continuity (Engel et al. 1988). The localization of the asymmetric form of AChE in regenerating facial nuclei and axons is not consistent with its function of cleaving the neurotransmitter acetylcholine at a synaptic site. A meaningful interpretation of the findings is therefore still difficult. However, the observations of AChE synthesis, location and translocation during the regeneration process have had a surprising heuristic value since they became instrumental in the discovery of dendritic secretion of proteins (Kreutzberg and Toth 1974; Greenfield 1985).

In the course of the search for a regeneration program in the neuronal genome triggered by the lesion, very early changes have been looked for. The expression of oncogene products is possibly among these early signs (Stein-Izsak et al. 1986). The enzyme ornithine decarboxylase (ODC), which catalyses the rate-limiting step in polyamine synthesis is known to react with rapid increase in response to many cellular challenges. In the case of the retrograde reaction to axotomy, ODC increases within hours and reaches a maximum of 300% over control values within 24 hours in motoneurons (Tetzlaff and Kreutzberg 1985) or over 500% in sympathetic neurons (Gilad and Gilad 1983). Since it is known that polyamines play a role in the regulation of gene activity, increased ODC activity could very well reflect the initiation of enhanced DNA, RNA and consequentely protein metabolism.

About as early as the changes in ODC activity, an increase in glucose uptake becomes apparent (Kreutzberg and Emmert, 1980; Singer and Mahler 1986). By means of the deoxyglucose technique an enhanced glucose consumption can be seen in the hypoglossal and facial nuclei in the period between one day and four weeks following nerve transection. Earlier data on the activity of oxidative metabolism, in particular the final oxidation in the Krebs cycle, have not revealed significant changes during regeneration (Lieberman 1971). It therefore seems likely that glucose is predominantly used for the production of ribose in RNA, the ribose being obtained via the hexose monophosphate shunt, identical to the pentose phosphate cycle. Enzymes involved in this pathway, e.g. glucose-6-phosphate dehydrogenase or 6-phospho-gluconate dehydrogenase in fact increase in most neurons during the regeneration process (Kreutzberg 1963; Härkönen and Kauffman 1974). From all these data the assumption seems to be justified that so-called chromatolysis represents the morphological equivalent of a metabolically hyperactive neuron with enhanced protein synthesis.

Neuron-Glial Relationship During Regeneration of Motoneurons

The observation that peripheral nerve injuries such as crush, cut or resection produce a marked "augmentation" of glial cells in the nucleus of origin dates back to the pioneering experiments by Franz Nissl (1894) who discovered the retrograde reaction of the neuron after evulsion of the facial nerve in rabbits. Today, we know that two species of glial cells are involved in the regeneration process occurring in the motor nucleus: microglia and astrocytes.

Microglia proliferation by mitosis has been documented by studying karyokinesis (Cammermeyer 1965) and by light microscopical [3H]thymidine autoradiography (Sjöstrand 1966; Watson 1965; Kreutzberg 1966) and has been recently confirmed at the electron microscopical level (Graeber et al. 1988; Streit and Kreutzberg 1987). Blinzinger and Kreutzberg (1968) discovered that the proliferated microglia are involved in a process now widely known as "synaptic stripping". This term describes the detachment and displacement of afferent synaptic terminals from the surface of regenerating neurons. It leads to a deafferentation at somatic and stem dendritic sites. These morphological changes correspond to changes in the electrophysiology of regenerating neurons reflected, for instance, in the diminuation of Ia synaptic excitation but with a compensatory increased excitability (Eccles et al. 1958; Kuno and Llinas 1970). Spontaneous miniature EPSPs with a fast rise time to peak disappear since they are produced by proximal, i.e. axo-somatic, input; EPSPs with a slow rise time to peak, however, are unchanged, suggesting the integrity of the peripheral, i.e. dendritic input (Lux and Schubert 1975). This is in agreement with the electron microscopical finding that synaptic stripping occurs mainly on the neuronal somata and not at peripheral dendrites (Blinzinger and Kreutzberg 1968; Kreutzberg and Barron 1978).

The modifiability of synapses after injury and especially in axotomized neurons has been well established in many laboratory animals and applies to many different types of peripheral, spinal and medullary neurons (see Mendell 1984). In humans, morphological data are to my knowledge lacking; however, the clinical observation of various pathological motor phenomena, such as dyskinesias, hyperexcitability or involuntary movements following a peripheral motor nerve lesion, could very likely be based on the synaptic stripping and the consecutive disarrangement of the afferents (Jankovic and Tolosa 1988). In contrast to the current view (see e.g. Eccles 1986) we have recently obtained evidence that the loss of synapses from axotomized motoneurons is not a transient phenomenon as previously thought. In a process in which astrocytes are involved, neurons can be wrapped in a way that synaptic sites become permanently occupied by astroglial processes (Graeber and Kreutzberg 1988).

Concomitant with the proliferation of satellite microglia cells, an astrocytic hypertrophy can be recognized in the facial nucleus (Cammermeyer 1955; Reisert et al. 1984). Fortunately, there is a specific marker for astrocytes which also correlates in its expression with the state of activation of astroglia. The glial fibrillary acidic protein (GFAP) is the main component of the glial filaments which belong to the intermediate filament group of cytoskeletal elements. It is cell specific for the astroglia family. After elaborating a protocol for the demonstration of GFAP at the electron microscopical level (Graeber and Kreutzberg 1985) this protein has been extensively studied in my laboratory by Graeber. A remarkable increase in GFAP immunoreactivity is demonstrated in facial nucleus astrocytes within a few days following nerve transection (Graeber and Kreutzberg 1986). The increase of GFAP antigenicity is associated with an increased appearance of glial filaments and astrocytic processes. As the resident protoplasmic astrocytes, normally very poor in GFAP, become activated by the retrograde changes occurring in the facial nucleus, they are transformed into the fibrous type. Quantitative evaluation of immunoblots have shown that GFAP increase is statistically significant as early as 24 h following the operation (Tetzlaff et al. 1988a). In electron micrographs it is apparent that relatively thick astroglial processes running through the neuropil are responsible for the strong GFAP staining. This picture differs with increasing survival times after surgery. Fine astroglia processes are now recognized covering the neuronal surface at sites where microglia cells were encountered previously (Graeber and Kreutzberg 1988). This seems to result in a wrapping of the neuronal cell bodies by double or triple lamellae of astrocytes within three weeks. Although the lamellae show typical glia filaments, GFAP immunoreactivity is decreased. We have called this process "the delayed astrocyte reaction". Unpublished long-term experiments have recently revealed that the astroglial wrapping persists as long as 300 days post operation. Since this represents more than half the life span of the laboratory rat, we assume that this change might be permanent. This leads to the conclusion that the deafferentation at the soma of facial motoneurons of the rat produced by microglia and astrocytes is an irreversible process. It remains to be demonstrated whether this is also the case in other regenerating neurons and in other species. Further investigations of this phenomenon will undoubtedly contribute to a better understanding of the apparent deficits in the functional recovery of motor nerves even following complete reinnervation of the target tissue.

References

Bisby MA (1980) Changes in the composition of labeled protein transported in motor axons during their regeneration. J Neurobiol 11: 435-445.

Blinzinger K, Kreutzberg GW (1968) Displacement of synaptic terminals from regenerating motoneurons by microglial cells. Z Zellforsch 85: 145-157.

Cammermeyer J (1955) Astroglial changes during retrograde atrophy of nucleus facialis in mice. J Comp Neurol 102:133-150.

Cammermeyer J (1965) Juxtavascular karyokinesis and microglia cell proliferation during retrograde reaction in the mouse facial nucleus. Ergeb Ant Entwicklungsgesch 38: 1-22.

Eccles JC (1986) Chromatolisis of neurones after axon section. In: Dimitrijevic MR, Kakulas BA, Vrbova G (eds) Recent achievements in restorative neurology, vol 2. Karger, Basel, pp 318-331.

Eccles JC, Libet B, Young RR (1958) The behaviour of chromatolysed motoneurones studied by intracellular recording. J. Physiol. (Lond) 143: 11-40.

Engel AK, Kreutzberg GW (1986) Changes of acetylcholinesterase molecular forms in regenerating motor neurons. Neuroscience 18: 467-473.

Engel AK, Tetzlaff W, Kreutzberg GW (1988) Axonal transport of 16S acetylcholinesterase is increased in regenerating peripheral nerve in guinea-pig, but not in rat. Neuroscience 24: 729-738.

Frizell M, Sjöstrand J (1974) Transport of proteins, glycoproteins and cholinergic enzymes in regenerating hypoglossal neurons. J Neurochem 22: 845-850.

Gilad GM, Gilad VH (1983) Early rapid, transient increase in ornithine decarboxylase activity within sympathetic neurons after axonal injury. Exp Neurol 81: 158-166.

Graeber MB, Kreutzberg GW (1985) Immuno gold staining (IGS) for electron microscopical demonstration of glial fibrillary acidic (GFA) protein in LR White embedded tissue. Histochemistry 83: 497-500.

Graeber MB, Kreutzberg GW (1986) Astrocytes increase in glial fibrillary acidic protein during retrograde changes of facial motor neurons. J Neurocytol 15: 363-373.

Graeber MB, Kreutzberg GW (1988) Delayed astrocyte reaction following facial nerve axotomy. J Neurocytol 17: 209-220.

Graeber MB, Tetzlaff W, Streit WJ, Kreutzberg GW (1988) Microglial cells but not astrocytes undergo mitosis following rat facial nerve axotomy. Neurosci Lett 85: 317-321.

Grafstein B (1975) The nerve cell body response to axotomy. Exp Neurol 48: 32-51.

Greenfield SA (1985) The significance of dendritic release of transmitter and protein in the substantia nigra. Neurochem Int 7: 887-901.

Härkönen MHA, Kauffman FC (1974) Metabolic alterations in the axotomized superior cervical ganglion of the rat. II. The pentose phosphate pathway. Brain Res 65: 141-157.

Hoffman PN, Lasek RJ (1980) Axonal transport of the cytoskeleton in regenerating motor neurons: Constancy and change. Brain Res 202:317-333.

Hoover DB, Hancock JC (1985) Effect of facial nerve transection on acetylcholinesterase, choline acetyltransferase and [^{3}H)quinuclidinyl benzilate binding in rat facial nuclei. Neuroscience 15: 481-487.

Jankovic J, Tolosa E (eds) (1988) Facial Dyskinesias. Advances in Neurology: vol.49. Raven, New York

Kreutzberg GW (1963) Changes of coenzyme (TPN) diaphorase and TPN-linked dehydrogenase during axonal reaction of the nerve cell. Nature 199: 393-394.

Kreutzberg GW (1966) Autoradiographische Untersuchung über die Beteiligung von Gliazellen an der axonalen Reaktion im Facialiskern der Ratte. Acta Neuropathol (Berl) 7: 149-161.

Kreutzberg GW (1982) Acute neural reaction to injury. In: Nicholls JG (ed) Repair and regeneration of the nervous system.Life Sciences Research Report 24, Dahlem Konferenzen 1982. Springer, Berlin Heidelberg New York pp 57-69

Kreutzberg GW (1986) Neurobiology of regeneration and degeneration. In: May M (ed) The facial nerve. Thieme, New York, pp 75-83.

Kreutzberg GW, Barron KD (1978) 5'-Nucleotidase of microglial cells in the facial nucleus during axonal reaction. J Neurocytol 7: 601-610.

Kreutzberg GW, Emmert H (1980) Glucose utilization of motor nuclei during regeneration: a ^{14}C-2-deoxyglucose study. Exp Neurol 70: 712-716.

Kreutzberg GW, Hollaender H (1983) Compatibility of horseradish peroxidase tracing with the histochemical demonstration of oxidoreductases. J Neurosci Methods 8:177-181.

Kreutzberg GW, Tóth L (1974) Dendritic secretion: a way for the neuron to communicate with the vasculature. Naturwissenschaften 61: 37

Kreutzberg GW, Tóth L, Kaiya H (1975) Acetylcholinesterase as a marker for dendritic transport and dendritic secretion.Adv Neurol 12:269-281

Kreutzberg GW, Tetzlaff W, Toth L (1984) Cytochemical changes of cholinesterases in motor neurons during regeneration. In: Brzin M, Barnard EA, Sket D (eds) Cholinesterases - fundamental and applied aspects. Walter de Gruyter, Berlin, pp 273-288.

Kuno M, Llinas R (1970) Alterations of synaptic action in chromatolysed motoneurones of the cat. J Physiol (Lond) 210: 823-838.

Lieberman AR (1971) The axon reaction: a review of the principal features of perikaryal responses to axon injury. Int Rev Neurobiol 14:49-124.

Lux HD, Schubert P (1975) Some aspects of the electroanatomy of dendrites. Adv Neurol 12: 29-44

Mendell LM (1984) Modifiability of spinal synapsis. Physiol Rev 64:260-324.

Nissl F (1894) Über eine neue Untersuchungsmethode des centralorgans speziell zur Feststellung der Lokalisation der Nervenzellen. Zentralblatt für Nervenheilkunde und Psychiatrie 17, 337-44

Oblinger MM, Brady ST, McQuarrie LG, Lasek RJ (1987) Cytotypic differences in the protein composition of the axonally transported cytoskeleton in mammalian neurons. J Neurosci 7:453-462.

Reisert I, Wildemann G, Grab D, Pilgrim C (1984) The glial reaction in the course of axon regeneration: a stereological study of the rat hypoglossal nucleus. J Comp Neurol 229: 121-128.

Singer P, Mahler S (1986) Glucose, leucine uptake in the hypoglossal nucleus after hypoglossal nerve transection with and without prevented regeneration in the Sprague-Dawley rat. Neurosci Lett 67:73-77.

Sjöstrand J (1966) Studies on glial cells in the hypoglossal nucleus of the rabbit during nerve regeneration. Acta Physiol Scand 67 (Suppl 270): 1-17.

Stein-Izsak C, Breuer O, Schwartz M (1986) Expression of the proto-oncogenes fos and myc in optic nerve regeneration. Soc Neurosci Abstr 12: 12, 7.6.

Streit WJ, Kreutzberg GW (1987) Lectin binding by resting and reactive microglia. J Neurocytol 16: 249-260.

Tetzlaff W , Kreutzberg GW (1984) Enzyme changes in the rat facial nucleus following a conditioning lesion. Exp Neurol 85: 547-564.

Tetzlaff W, Kreutzberg GW (1985) Ornithine decarboxylase in motoneurons during regeneration. Exp Neurol 89: 679-688.

Tetzlaff W, Graeber MB, Bisby MA, Kreutzberg GW (1988a) Increased glial fibrillary acidic protein synthesis in astrocytes during retrograde reaction of the rat facial nucleus. Glia 1: 90-95.

Tetzlaff W, Bisby MA, Kreutzberg GW (1988b) Changes in cytoskeletal proteins in the rat facial nucleus following axotomy. J Neurosci

Watson WE (1965) An autoradiographic study of the incorporation of nucleic acid precursors by neurones and glia during nerve regeneration. J Physiol (Lond) 180: 741-753.

Frequency and Amplitude Codes of Neuronal Signals

E. Florey
Fakultät Biologie, Universität Konstanz, 7750 Konstanz, FRG

Neurons, Signals and Information

The brain is often seen as an information-processing machine, receiving information from the sense organs delivered by sensory fibers in the form of a frequency code. Even within the brain itself, neuronal interaction is thought to be based on such a frequency code.

As far as we know, however, the sensory neurons only deliver signals. The signals are not information, nor is the frequency of nerve impulses, for this is what the signals are - coded information. The stream of nerve impulses, patterned as it may be both temporally and spatially yields information only when it is properly channeled, and when the channeled impulses can evoke responses which form temporal and spatial patterns of neuronal activity capable of creating an inner representation of the world - or of generating action which is an adequate response to the condition signaled by the senses. Information is thus generated from signals by a process which extracts meaning from impulse patterns. Such meaning arises from the anatomical arrangement, the neuronal connectivity, and the functional capacity of each part of the neuronal machine. The nervous system is an autopoietic system of almost infinite creative capacity. Its morphological structure is the result of continued interaction between its component cells, and its performance is regulated by the effects of patterned activity initiated both from within and from without via the sensory input.

We have become so accustomed to thinking of neurons as impulse-generating and impulse-conducting units, that we tend to rest satisfied with the knowledge that it is the business of neurons to generate impulses,and that neuronal activity can be measured in terms of temporal impulse patterns. If this were so, the relevant activity of the brain would be represented by the chorus of impulses. Assemblies of neurons with their collectively orchestrated impulse patterns would represent the inner as well as the outer world, and the confluent pulsed activites would prepare, initiate and execute internal and externally directed actions. Impulse patterns would represent internal reflections equivalent to perceptions, imaginations and thoughts. They would be symptomatic of reflexive actions and of more or less willfully executed behavior.

Three Ways to Study Brain Function

Neurophysiologists presently move on different levels. There are those of us who focus their attention on the impulse traffic through neuronal pathways, on the elucidation of what we like to call neuronal circuits. The nerve impulse, whatever its physical basis, is taken for granted. Synaptic transmission is mainly viewed as being either excitatory or inhibitory - the parameter of main interest being the strength of the respective synapses involved in the circuits under consideration. Neurons are regarded as integrators of synaptic inputs; "to fire or not to fire," that is the only question, and the pattern of simultaneously and successively firing neurons is then regarded as the neuronal activity proper.

There are also those of us who concern themselves with the molecular and ionic events at the cell membrane. The activation and inactivation of ionic channels become the prime concern. Synaptic events are viewed as a molecular drama in which transmitters, receptors and ionic channels are the actors, with membrane lipids and the extracellular and intracellular matrix as the supporting cast. The game is an electric one; the events are evaluated in terms of the electrical charges, potentials and currents. The sum total of the vast array of this electromolecular scene is then seen as the functioning brain.

Those of us who operate on a middle level concentrate their efforts on the elucidation of the role and function of synapses and take yet another view: the brain is seen as a vast array of synapses of many different kinds, each being activated by nerve impulses to release one or more packets of transmitter substance which, in turn, affect the respective postsynaptic cell in complex ways that include ion movements, intracellular metabolic changes, and membrane-bound events involving receptors, carriers and enzymes. Indeed, impulse generation and impulse conduction are dull events compared with the rich repertoire of synaptic processes. After all, the synapses, being biochemical machines, respond not only to arriving action currents and action potentials; their function is also affected by various chemical agents - hormones, neurohormones, neuromodulators, neurotransmitters. The result of such chemical control is an alteration of the performance of the affected synapses. On this level of physiological perspective it is not so much the biophysical as the biochemical mechanism that is of prime concern.

The first of the three physiological approaches to brain function sees the brain as a computer. The arguments are based on the morphological connectivity as described on the basis of various histological and cytological techniques and on the electrical activity recordable with extra- and intracellular electrodes from single units and larger neuronal assemblies. What is interpreted here is the temporal and spatial pattern of nerve impulses. The goal is the description and analysis of neuronal circuits in terms of the language of cybernetics and network theory.

The second approach is characterized by its biophysical outlook. The focus is on the neuronal cell membrane and on the behavior of its ion channels. The main tools of research are electrodes. The method consists of the analysis of ionic currents.

The third, intermediate level approaches brain function with the tools of biochemistry. Generation of nerve impulses and synaptic transmission are viewed as chemical events which, among other things, result in changes of electrical charge distribution across pre- and postsynaptic membrane regions but also in intracellular events, ranging from changes in energy metabolism to gene activation and induction of synthesis of special proteins which may function as enzymes, ion channels or carriers.

Needless to say, it is only when these three approaches are combined that we can hope to gain truly meaningful insights into the functioning of the real brain. The brain is a neuronal network, and it is an electrical machine, but its complex activity results from physical and chemical events, and the electrical symptoms of brain action are but one aspect which accompanies, results from - and results in biochemical processes.

The Frequency Code and the Concept of Excitation

If we look only at the pattern of nerve impulses, we note that messages are transmitted in a frequency code. It can be said that the frequency of impulses conducted along an axon is proportional to the level of excitation of the particular neuron. Such a statement is meaningful only if the term excitation is given a definite meaning. Intuitively the meaning seems clear, and yet the term excitation is by no means easy to define. Level of depolarization at the impulse-generating region of the neuron is a possible definition, but often excitation is defined in terms of the Hodgkin-Huxley model and is identified with one action potential. This second kind of excitation thus means inward current through voltage-gated ion channels.

A high frequency of such Hodgkin-Huxley excitations may thus represent a high level of neuronal excitation of the first kind. Excitability has long been considered a main characteristic of the living state. This term excitability means ability to respond to stimuli. The response in this case would be a state of excitation if excitability is the capacity to respond to a stimulus with excitation. But, clearly, excitation so defined need not be a Hodgkin-Huxley action potential; it may not even be a depolarization of the cell membrane but a hyperpolarization - as occurs when light strikes a photoreceptor in the retina of the human eye. Any kind of response, even a purely chemical one, unaccompanied by electrical changes in the cell membrane, may be regarded as an excitation. The conducted action potential, the nerve impulse, is not the only form of excitation, if by excitation we mean response of a living cell to a stimulus (whatever that may be). It would be the worst kind of tautology if we now define excitability as the ability to respond to a stimulus with excitation. But what would be the alternative? We would have to speak of "responsibility" if the term excitability would mean ability to respond rather than being able to become excited. Thus, even the state of inhibition, so well known to neurophysiologists, is a response and would thus constitute excitation!

Clearly, the meaning of the term excitation is crucial in any consideration of brain function.

Consider perception. We are convinced that the sensory neurons convey their state of excitation in a frequency code to their target sites in the brain. We also believe that after parallel processing of such frequency-coded messages, the signals converge upon neuronal assemblies or even upon specialized neurons, so-called "feature detectors" or "recognition neurons", - the term "grandmother neuron" was once coined to illustrate the meaning. Some physiologists are convinced that when this occurs, and when the final neuron or the neuronal assembly responds with excitation - that is, with one or more action potentials - this is the moment of cognition when the impulse pattern is decoded and the meaning of the messages emerges in form indeed, of patterned excitation of the final neuron(s) of the complex neuronal pathway from a sense organ or from another brain region which generates memory.

Facilitation and Posttetanic Potentiation: The Amplitude Code

Conducted action potentials may be one form of excitation, but clearly, other forms of excitation must be recognized as being physiologically significant and meaningful. When the grandmother neuron fires, it may indeed signal to other neurons that it has seen the grandmother. But what if it does not fire? Does this mean that it has no knowledge that the grandmother is there, or does this grandmother neuron know but not pass on this information to others? After all, this grandmother neuron may well receive synaptic input sufficient to generate the information "grandmother is there" even though no spikes are elicited, and the rest of the brain remains ignorant of the fact. Hodgkin-Huxley type excitation may not be the only relevant mode of excitation. Apart from any other chemical processes, the patterned synaptic input as measured in terms of postsynaptic potentials may be equivalent to perception and recognition.

It is true, however, that the grandmother neuron responds to signals brought to it through impulse-carrying channels. Nerve impulses are the signals that bridge the distance from one impulse-generating site to the next synaptic connection. And it is true that the temporal pattern of impulses is the only code that can be transmitted over the axonal pathways. The resolution of the code, however, involves another step: transformation of the frequency code into an amplitude-modulated message. This takes place at the level of the synapses. Integration of synaptic input results from amplitude-modulated signals.

This is the topic of the present contribution: the transformation of the frequency-coded neuronal messages into amplitude-modulated signals that can be integrated to result eventually in meaningful information. The processes of transformation are commonly known as facilitation and posttetanic potentiation. Synaptic transmission is not a singular event but generally occurs serially. The interval between arriving nerve impulses critically determines the quantity of transmitter released by successive impulses and thus determines the amplitude of the postsynaptic currents and of the postsynaptic potentials.

It is not sufficient simply to define a given synaptic input as weak or strong. Instead, it is more important to define the synapse in terms of its capacity for facilitation and post-tetanic potentiation. At a synapse, a single impulse may evoke a weak postsynaptic potential, but after a conditioning train of impulses the next action potential arriving after a longer interval may cause a postsynaptic potential of sufficient amplitude to trigger a spike.

It is in the context of the seriality of synaptic transmission that the phenomenon of presynaptic inhibition becomes of special interest. But before discussing this topic, let me recapitulate briefly the terms employed with regard to the general phenomenon of facilitation.

Facilitation proper is the increment of postsynaptic responses to successive presynaptic stimuli. It is commonly studied by applying trains of constant impulse frequency. Posttetanic potentiation represents a conspicuously increased postsynaptic response following a period of repetitive stimulation, the response being larger than the last facilitated response during the preceding train of stimuli.

Several mechanisms have been proposed to explain facilitation: (a) a broadening of the presynaptic action potential, (b) increasing invasion of the terminal by successive action potentials which initially decay in the proximal segment of the terminal, (c) a temporary increase in intracellular Ca activity within the terminal due to a residuum of the Ca that entered the terminal with each arriving nerve impulse. None of these hypotheses is satisfactory: a broadening of the spike, if it occurs at all, is not effective enough to explain the enormous amount of facilitation often seen. Terminals may already be fully invaded by the first action potential of a series, and yet transmitter release increases with subsequently arriving action potentials. The "residual Ca hypothesis" cannot explain why an impulse arriving shortly after one that gave rise to strong transmitter release causes depression rather than facilitation. The most serious objection to all these hypotheses is that they cannot explain posttetanic potentiation which may reach its peak several seconds after the end of the train of stimuli. These hypotheses also fail to explain antifacilitation, and they cannot account for the fact that one and the same synapse may show facilitation and antifacilitation depending on the mode of stimulation.

Recently, I have developed an alternative model of synaptic function (Fig. 1B) based on the experiments conducted by Birks and MacIntosh (1961) in the early 1960s . Using the perfused superior cervical ganglion of the cat, these authors found that the arriving nerve impulse causes not only transmitter release but also a momentarily enhanced transmitter synthesis. On the basis of observed release kinetics, they concluded that terminals contain two transmitter pools or stores. Newly synthesized transmitter would enter the first pool (store I), from which it is transferred to the second pool (store II), which then contains the releasable transmitter (presumably packaged in vesicles). The transfer represents what is commonly referred to as mobilization. It was further shown that the transmitter in either pool is subject to inactivation. The amount of transmitter in these compartments represents a steady-state condition resulting from synthesis and

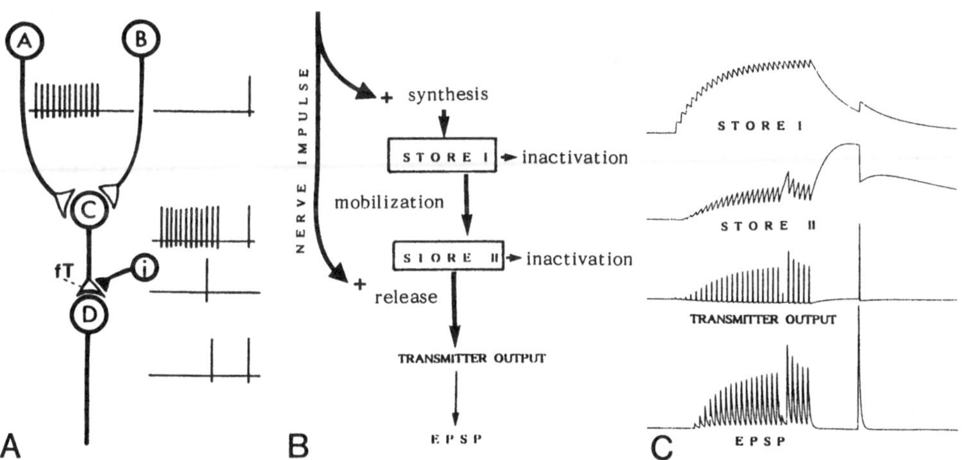

Fig.1 a-c. An alternatie model of synaptic function (for explanation see text)

inactivation, on the one hand, and mobilization and inactivation, on the other. If the arriving nerve impulse causes not only a certain fractional release of transmitter from store II but also an increment of transmitter synthesis above the resting rate of synthesis, each impulse would cause a momentary decrease of transmitter in store II and an increase of transmitter in store I. Repetitive impulses would enhance this situation, but mobilization would also be stepped up because of the increased gradient of transmitter in the two stores. If no further impulse arrives for a short time interval, store II would receive a supranormal amount of transmitter from the recently filled store I. A new impulse would now be able to elicit a potentiated transmitter release.

The model has been successfully simulated by an electronic circuit serving as an analog computer. This circuit realistically simulates all patterns of facilitation and posttetanic potentiation ever observed in any type of synapse. The model demonstrates also that by changing one or the other parameter it is possible to transform the system from one representing a certain type of synapse to that representing another. It can thus be shown how a non-facilitating synapse can become a facilitating one, and how a synapse which exhibits no sign of posttetanic potentiation can be changed to one with pronounced potentiation. In natural nervous systems, such changes may result from the action of modulator substances. They might also be effected by a particular temporal pattern of synaptic inputs.

Facilitation as such already offers interesting possibilities for the function of neuronal circuits. Assuming that only facilitated EPSPs reach the firing threshold of a particular postsynaptic neuron, rate and amount of facilitation would give rise to variable delays in a transmission line.

The phenomenon of posttetanic potentiation, however, has major consequences in neuronal circuits. A single impulse following a conditioning train of impulses which cause facilitating but still subthreshold EPSPs may now cause a suprathreshold EPSP. Assume a four-neuron system (Fig. 1A) with neurons A and B being connected to neuron C by nonfacilitating synapses capable of 1:1 transmission. Assume further that neuron C is connected with neuron D by facilitating synapses, fT, capable of posttetanic potentiation. A train of impulses in neuron A giving rise to facilitating subthreshold EPSPs in neuron D (via activation of neuron C) would prepare the C-D synapses in such a way that a single impulse in B which follows the train in A would travel the entire neuronal chain - while without the pulse train in A, the B impulse would be of no consequence.

Most intriguing is the possibility of using presynaptic inhibition (in Fig. 1A) to achieve the effect of posttetanic potentiation. Assume that a train of impulses arriving in a given facilitating terminal would cause facilitating EPSPs of subthreshold amplitude. Presynaptic inhibition of transmitter release would momentarily interrupt the train of EPSPs, but immediately afterwards the EPSPs would be potentiated to suprathreshold amplitude. The inhibition would thus be the cause of postsynaptic excitation. Fig. 1C shows the computer simulation of the events involved.

Facilitation and posttetanic potentiation provide an amplitude code which, of course, is connected with the frequency code of nerve impulses.

The interposition of an amplitude modulation in impulse-carrying chains of neurons, and especially in neuronal circuits with feedback and feed forward loops, offers an almost infinite range of possibilities for the generation of meaningful impulse patterns. The decoding of impulse frequencies appears to result from the amplitude modulation which thus becomes the key to the transformation of signals into information.

A mystery remains, of course: if amplitude is the result of impulse spacing, we still must find the determinant of the link between the two. The meaning extracted from frequency-coded neuronal messages depends on those factors which determine the various parameters underlying transmitter synthesis: inactivation and mobilization within the neuronal terminals. If we are to understand the self-referentiality of autopoietic nervous systems, we must seriously search for these determining factors.

Reference

Birks R, MacIntosh FC (1961) Acetylcholine metabolism of a sympatic ganglion. Can J Biochem Physiol 39:787-827

Balance and Imbalance of Transsynaptic Neurotransmission as Conditions for Normal and Pathological Behaviour: Examples Only

W. Birkmayer[1] and P. Riederer[2]
[1]Birkmayer Institute, Vienna, Austria
[2]Clinical Neurochemistry, Department of Psychiatry, University of Würzburg, FRG

Drugs, Action and Behaviour

For hundreds and even thousands of years drugs have been used to alter psychic behaviour, and alcohol is only one example of a drug that is still being used today. Since the beginning of our century, however, researchers have been interested in the mechanism(s) of drug action. It became clear by the early observations of Pötzl and Hess at the beginning of this century that glucose loading during melancholic phases led to changes in the response of depressed patients. Clinical, behavioural and biochemical studies after drug administration - including Freud's first trials with cocaine - led to a break-through in biochemically oriented brain research. It became evident that rauwolfia serpentina improved psychotic behaviour and decreased blood pressure by releasing biogenic amines and emptying their specific storage sites (Carlsson 1959; Brodie and Shore 1957). This loss of biogenic amines caused depressive mood in about 10 % of patients with hypertension. Generation of hypotheses that depression is caused by the loss of serotonin, noradrenaline and/or dopamine gave a first biochemical approach to the pathophysiology of psychiatric disturbances. On the other hand, research into the mechanism of action of reserpine showed its cataleptogenic effects in animal studies. This motor deficiency could be antagonized by the precursor amino acid of dopamine, L-dopa (Carlsson 1959). Later a loss of dopamine could be demonstrated in the nigrostriatal system of Parkinson's disease (Ehringer and Hornykiewicz 1960). As a consequence of all these studies, L-dopa therapy was introduced by Birkmayer and Hornykiewicz (1961) and is nowadays a therapeutic standard for treatment of Parkinson disease.

Chlorpromazine properties, e.g. sedation, were observed first in the premedication phase of surgeries. Later this effect was confirmed in schizophrenia (Delay and Deniker 1952). As pharmacological basis dopamine receptor blockade has been demonstrated.

Antidepressive effects have been described by chance for imipramine (Kuhn 1957). One of the most impressive pharmacological mechanisms of this drug is a blockade of the reuptake of noradrenaline resulting in an increase of this neurotransmitter in the synaptic cleft and at the postsynaptic receptor sites.

Iproniazid is another example for a discovery by chance, showing improved mood in patients with tuberculosis. Again, administration of this drug to depressed patients confirmed the earlier observation. Iproniazid has been found to block the deaminating enzyme monoamine oxidase (MAO). MAO-inhibiting properties were suggested to enhance aminergic neurotransmission leading to antidepressive action with elevation of mood (Crane 1957).

From these examples we learn that drugs with a rather specific action are able to provoke changes in general behaviour that are by far more complex to be described by one transmitter system only. Conversely there are so many symptoms characterizing, for example, depression that it is again unlikely to assume here a defect in one transmitter system only. Arguments for and against the noradrenaline hypothesis (Schildkraut 1965) or the serotonin hypothesis (Coppen 1972) have been established, and hypotheses including other transmitters such as dopamine (van Praag 1978), acetylcholine (Janowsky and Davis 1979) or gamma-aminobutyric acid (GABA; Scatton et al. 1986) have been evaluated. As conclusion of a critical discussion of drug action, clinical effects of drugs and biochemical post-mortem findings, it seems likely that theories focusing a specific transmitter system are not able to fulfil all criteria of a pathogenic concept of depression, to give just one example. There is, however, evidence to assume a complex interneuronal imbalance as causative factor of psychic disturbances after functional and/or morphological lesions (Birkmayer et al. 1972; Birkmayer and Riederer 1988).

Concepts of Interneuronal Dysfunction

Interaction of noradrenaline and serotonin

It is a clinical observation that patients with depression show extreme response variations when treated with antidepressants of different pharmacological profile. According to Beckmann (1982) the clinical response of patients with high concentrations of 3-methoxy-4-hydroxyphenylglycol (MHPG) *and* low 5-hydroxyindole acetic acid (5-HIAA) in the cerebrospinal fluid (CSF) is significantly better after amitriptyline, while imipramine has greater benefit in patients with low CSF MHPG *and* high 5-HIAA.

The hypothesis of a compensatory increased postsynaptic receptor function as result of a presynaptic aminergic deficit has gained much interest. In agreement with such experimental findings it could be shown that β-adrenergic adenylate cyclase coupled receptors undergo down-regulation (decrease of binding number and affinity) after long-term administration with antidepressants of varying pharmacological properties (Sulser 1983). The hypothesis of a noradrenaline-serotonin interaction predicts a functionally intact serotoninergic system in cortical areas. This assumption however does not seem to be valid in depression syndrome on the basis of post-mortem brain examinations which demonstrate a substantial decrease of serotonin and 5-HIAA (Brücke et al. 1984).

Furthermore, evidence exists that antidepressants in addition to a down-regulation of β-receptors reduce serotoninergic 5-HT-2 receptors to a similar degree. From these examples it is suggested that the direct interaction of noradrenaline and serotonin mimics only a part of normal and/or pathologic behaviour.

Aminergic-Cholinergic Interaction

Although anticholinergics do show antidepressive action, promethazine shows less impressive benefit compared to amitriptyline (Beckmann and Schmauss 1983). This clinical finding is evidence to assume that drugs with anticholinergic properties plus reuptake blockade of biogenic amines improve much more the balance of aminergic-cholinergic systems compared to drugs acting on one such system only. In the hypothalamus possible neuronal imbalance with a relative cholinergic preponderance and reduced noradrenergic transmission is evident in about 50% of depressive patients, as indicated by a pathologic dexamethasone-suppression test (Caroll et al. 1981). As noradrenaline inhibits and acetylcholine enhances corticotropin-releasing hormone the noradrenaline-cholinergic imbalance is a noteworthy factor in the pathogenesis of depression. Corticotropin-releasing hormone containing neurons are controlled by hypothalamic, limbic and other subcortical and cortical regions via monoaminergic, cholinergic and peptidergic interaction (Swanson et al. 1983). Therefore, disturbances at the level of certain endocrine parameters may indicate a cholinergic-aminergic dysfunction in depression syndrome.

Interaction of GABA and Monoamines

Recent experimental work indicates that GABA agonists and probably GABA itself modulate cerebral noradrenergic and serotoninergic neurons in an opposite manner. They stimulate noradrenergic presynaptic function and inhibit serotoninergic function. For example, progabide, a direct and centrally acting GABA agonist, in combination with a low, per se ineffectual dosis of the noradrenaline uptake blocker desipramine, enhances the release of noradrenaline and desensitizes the adenylate cyclase coupled β-receptors. It is likely that GABA agonists down-regulate β-receptors in a similar manner as classical antidepressants. This synergistic action of progabide and desipramine is shown by an increase in hypothalamic normetanephrine, an indicator of a noradrenaline release. GABA-ergic control of the noradrenergic transmission is probably mediated by GABA synapses, which are - by an as yet unknown linkage - coupled to noradrenergic cell bodies (Scatton et al. 1986).

A connection of GABA and noradrenaline is of particular interest in the treatment of anxiety. Anxiety-relieving drugs such as benzodiazepines enhance GABA-mediated inhibition of β-receptors while classical antidepressants act directly via noradrenergic receptor down-regulation.

Interaction of Dopamine with Glutamate and Acetylcholine

The most important theory of schizophrenia is the dopamine hypothesis. This hypothesis is based on pharmacological findings and suggests a hyperactive dopamine system. Clinical evidence in favour of this hypothesis is the beneficial effect of the dopamine (D2) receptor blocking neuroleptic drugs. However, there is no indication of a primary dopaminergic dysfunction in schizophrenia (Kornhuber et al. 1989). It may well be, however, that a dopamine system is linked to another transmitter system showing functional and/or morphological lesion(s). In this case the dopaminergic tonus would be enhanced. Such a possible system is represented by the glutamatergic system, which is in part under dopaminergic control (Kim et al. 1986). In agreement with this assumption recent post-mortem findings show an increase in glutamatergic NMDA binding sites in schizophrenia (Kornhuber et al. 1988), indicating both a loss of presynaptic glutamatergic neurons and a hyperactive limbic dopamine system originating in the ventral tegmental area.

It is of interest that facilitation of pharmacotoxic psychoses in Parkinson's disease is achieved by all antiparkinsonian drugs including L-dopa, dopaminergic agonists, amantadine and MAO inhibitors. By this drug-induced dopaminergic hyperactivity the glutamatergic system is beyond a threshold to adjust the interneuronal homeostasis. However, it is likely that also other transmitter systems, i.e. noradrenaline and serotonin, are involved in provoking pharmacotoxic psychoses, as indicated by post-mortem analyses (Birkmayer et al. 1974).

Similarly, biochemical and pharmacological findings led to the suggestion of a balance between dopamine and acetylcholine in the nigrostriatal system. In Parkinson's disease, this balance is changed in favour of acetylcholine, and both anticholinergics and dopamimetics improve the balance and relieve parkinsonian symptoms.

Discussion

As a conclusion from these examples it is suggested that concepts of a bi-factorial balance or imbalance of neurotransmitters is of greater clinical and therapeutic value compared to single neurotransmitter changes and their direct correlation to any type of behavior. In fact, we assume that even more than two transmitter systems must frequently be in balance to guarantee sufficient neuronal function. The nigrostriatal - strionigral loop is only one such example combining dopaminergic, cholinergic, GABA-ergic and substance P function by direct neuronal coupling while probably each of these systems are controlled by neuronal inputs from other brain areas (serotonergic, glutamatergic, etc). Another such feedback loop is the cortico-striato-thalamo-cortical loop which by combining dopaminergic, GABA-ergic, glutamatergic and probably other transmitter systems controls connections between the motor system and mental function (Carlsson 1988).

As a conclusion, we come to a working hypothesis describing the brain's cybernetic stability as responsible for our normal behaviour. The overall output of a cybernetic instability is assumed to provoke mental and motor disturbances depending on the site of functional and/or morphological lesions (Riederer and Birkmayer 1980; Birkmayer and Riederer 1989). As interindividual behaviour varies considerably, it must be assumed that balances of intra- and interneuronal feedback loops are maintained at different homeostatic levels. A wide range of cybernetic equilibria based on variations in reactivity to a stimulus (drugs may act different in healthy probands and in pathologic situations) and sensitivity of cybernetic stability (the threshold to influence a system by drugs or environment may be different in various individuals) guarantee widespread variations in mobility and psychic behaviour. This cybernetic homeostasis of brain function shows circadian and circannual fluctuations. In particular, age-related changes cause a continuous adaptation of this homeostasis leading to improved coping of all kinds of environmental stresses during life. Disorders of human behaviour seem to be based on irreversible morphological and functional lesions leading to a specific pathogenic pattern of disturbances in neurotransmitter systems which is characteristic for the loss of the normal cybernetic balance. Therefore, the hypothesis of a balance of neurotransmitters as a condition for our (normal) behaviour contributes to a better understanding of the body-mind relationship.

References

Beckmann H (1982) Biochemische Beiträge zu Klassifikation und Therapievorhersage bei endogenen Depressionen. In: Beckmann H (ed) Biologische Psychiatrie. Thieme, Stuttgart, pp 136-147

Beckmann H, Schmauss M (1983) Clinical investigations into antidepressive mechanisms. I. Antihistaminic and cholinolytic effects: amitriptyline versus promethazine. Arch Psychiatr Nervenkr 233: 59-70

Birkmayer W, Hornykiewicz O (1961) Der L-Dioxyphenylalanin (L-DOPA) Effekt bei der Parkinson-Akinese. Wien Klin Wochenschr 73: 787-788

Birkmayer W, Riederer P (1988) Depression - Biochemie, Klinik, Therapie, 4th edn. Deutscher Ärzte-Verlag, Cologne

Birkmayer W, Riederer P (1989) Understanding the neurotransmitters: key of the workings of the brain. Springer ,Vienna New York

Birkmayer W, Danielczyk W, Neumayer E Riederer P (1972) The balance of biogenic amines as condition for normal behavior. J Neural Transm 33: 163-178

Birkmayer W, Danielczyk W, Neumayer E, Riederer P (1974) Nucleus ruber und L-dopa psychosis. Biochemical post-mortem findings. J Neural Transm 35: 93-116

Brodie B B, Shore P A (1957) A concept for a role of serotonin and norepinephrine as chemical mediators in the brain. Ann N Y Acad Sci 66: 631-642

Brücke T, Sofic E, Riederer P, Gabriel E, Jellinger K, Danielczyk W (1984) Die Bedeutung der serotonergen Raphe-Kortex-Projektion für die Beeinflussung der β-adrenergen Neurotransmission durch Antidepressiva. Neuropsychiatr Clin 3: 249-255

Carlsson A (1959) The occurrence, distribution and physiological role of catecholamines in the nervous system. Pharmacol Rev 11: 490-493

Carlsson A (1988) Speculations on the control of mental and motor functions by dopamine-modulated cortico-striato-thalamo-cortical feedback loops. M Sinai J Med(N4) 55: 6-10

Caroll B J, Feinberg J, Greden, J F, Tarika J, Albala A A, Haskett R F, James N M, Kronfol Z., Lohr N, Steiner M, de Vigne J P, Young E (1981) A specific laboratory test for the diagnosis of melancholia. Arch Gen Psychiat 38: 15-22

Coppen A (1972) Indolamines and affective disorders. J Psychiatr Res 291: 163-171

Crane G E (1957) Iproniazid (Marsilid) phosphate, a therapeutic agent for mental disorders and debilitating diseases. Psychiatr Res Rep Am Psychiatr Assoc 8: 142-152

Delay J, Deniker P (1952) Le traitement des psychoses par une methode neurolytique derivée de l'hibernotherapie. In: Cossa Maison P (ed) Congrès des médicins aliènistes et neurologistes de France, vol 50. Librairie de l'Academie de Médicine, Paris, pp 497-502

Ehringer H, Hornykiewicz O (1960) Verteilung von Noradrenalin und Dopamin (3-Hydroxytyramin) im Gehirn des Menschen und ihr Verhalten bei Erkrankungen des extrapyramidalen Systems. Klin Wochenschr 38: 1236-1239

Janowsky D S, Davis J M (1979) Psychological effects of cholinomimetic agents. In: Davis K L, Berger P A (eds) Brain acetylcholine and neuropsychiatric disease. Plenum, New York

Kim J S, Kornhuber H H, Kornhuber J, Kornhuber M E (1986) Glutamic acid and the dopamine hypothesis of schizophrenia. In: Chagass C, Josiassen R C, Bridger W H, Weiss H J, Stoff D, Simpson G S (eds) Biological psychiatry 1985. Elsevier, Amsterdam, pp 1109-1111

Kornhuber J, Mack-Burkhardt F, Riederer P, Hebenstreit G F, Reynolds G P, Andrews H B, Beckmann H (1989) (^{3}H)MK-801 binding sites in post-mortem brain regions of schizophrenic patients. J Neural Transm 77:231-236

Kornhuber J, Riederer P, Reynolds G P, Beckmann H, Jellinger K, Gabriel E (1989) 3H-Spiperone binding sites in post-mortem brains from schizophrenic patients: relationship to neuroleptic drug treatment, abnormal movements, and positive symptoms. J Neural Transm 75: 1-10

Kuhn R (1957) Über die Behandlung depressiver Zustände mit einem Iminobenzyl-Derivat (G 22355). Schweiz Med Wochenschr 87: 1135-1140

Riederer P, Birkmayer W (1980) A new concept: brain area specific imbalance of neurotransmitters in depression syndrome - human brain studies. In: Usdin E, Sourkes T L, Youdim M B H (eds) Enzymes and neurotransmitters in mental disease. Wiley, New York, pp 261-280

Scatton B, Nishikawa T, Dennis ., Dedek J, Curet O, Zivkovic B, Bartholini G (1986) GABA-ergic modulation of central noradrenergic and serotonergic neuronal activity. In: Bartholini G, Lloyd K G, Morselli P L (eds) GABA and mood disorders. Experimental and clinical research. Raven, New York, pp 67-75

Schildkraut J J (1965) The catecholamine hypothesis of affective disorders: a review of supporting evidence. Am J Psychiatry 122: 509-522

Sulser F (1983) Deamplification of noradrenergic signal transfer by antidepressants: a unified catecholamine-serotonin hypothesis of affective disorders. Psychopharmacol Bull 19: 300-304

Swanson L W, Sawchenko P E, Rivier J, Vale W W (1983) Organization of ovine corticotropin-releasing factor immunoreactive cells and fibres in the rat brain: an immunohistochemical study. Neuroendocrinology 36: 165-186

Van Praag H M (1978) Amine hypotheses of affective disorders. In: Iversen L L, Iversen S D, Snyder S G (eds) Handbook of psychopharmacology, vol 13. Plenum, New York, pp 187-275

Behavioural Pharmacology of Brain Glutamate

W. J. Schmidt
Abt. Neuropharmakologie, Universität Tübingen, Mohlstr. 54/1, 7400 Tübingen, FRG

Introduction

Excitatory amino acids, most probably glutamate or aspartate, are common transmitters in the brain. Many interneurons in the cortex and most, perhaps all, corticofugal neurons use glutamate as their transmitter. Receptors for excitatory amino acids are characterized according to their preferred agonists, i.e. N-methyl-D-aspartate (NMDA), kainate and quisqualate. Presynaptically an alpha-amino-phosphono-butyrate preferring receptor may be localized (for review see Cotman and Iversen 1987).

In behavioural pharmacology we ask how a drug changes animal behaviour. The present paper deals with the functions of the NMDA receptor since for this subclass, most specific antagonists have been developed, e.g. MK-801 and 2-amino-5-phosphonovalerate (AP-5). MK-801 is able to cross the blood-brain barrier. At the NMDA receptor it does not compete with L-glutamate but blocks the receptor-gated cation channel; therefore it is called a use-dependent noncompetitive antagonist. In contrast, AP-5 does not readily cross the blood-brain barrier (it was therefore injected intracerebrally) and blocks the NMDA receptor competitively. MK-801 and AP-5 have anticonvulsive and anxiolytic effects and protect against ischemia- or NMDA-induced neurodegeneration (Clineschmidt et al. 1982).

Behavioural Changes Induced by Blockade of Glutamatergic Transmission

Systemic administation of the NMDA receptor antagonist MK-801 produces the following behavioural changes in rats:

Locomotion: In a circular runway equiped with photocells MK-801 doses of 0.1 and 0.3 mg/kg p.o. increased activity (Clineschmidt et al. 1982). Rats moving freely in an 8-arm radial maze (spontaneous alternation) also exhibited enhanced locomotor activity (dose dependently 0.08, 0.16, 0.33 mg/kg i.p.; Bischoff et al. 1988).

Sniffing was enhanced by MK-801 to such an extend that at the higher doses of 0.16 and 0.33 mg/kg it was performed in a sterotypic manner, at 0.33 mg/kg from 15 to 145 min after administration (Bischoff et al. 1988).

Delayed alternation in the T-maze was impaired by MK-801. Rats were rewarded for alternating between two arms of a T-maze; between each arm visit a delay of 30 s was interposed. At 0.16 mg/kg the rats made more errors, i.e. they reentered the last visited arm.

Amnesia was induced at the highest dose of MK-801 (0.33 mg/kg). In an 8-arm radial maze, the rats lost their ability to remember the spatial location of food. They reentered arms from which food had previously been collected. We were able to show that only working memory (short-term memory that holds information within a trial) was impaired, but that reference memory (long-term memory that holds information across trials) was intact (Bischoff et al. 1988).

The angle between two consecutive arm entries in the 8-arm radial maze was changed by MK-801. While control rats preferred an angle of 45° or 90°, MK-801 treated rats preferred an angle of 90° or 135°. Preliminary evidence indicates that this is not due to NMDA- or opiate receptors but may be mediated by sigma receptors.

Feeding was inhibited by MK-801 (0.33 mg/kg); however, this does not imply an effect of this drug upon a regulation centre but is most probably due to response competition, i.e. the animal cannot eat because it sniffs in a stereotypic manner.

Glutamate in the Basal Ganglia

Glutamate in the Substantia Nigra

The competitive NMDA-antagonist AP-5 (5 nmole/0.5 μl) was stereotaxically injected into the pars compacta of the substantia nigra (AP 5.0/5.3 behind bregma; L±2; V 2.2/2.0 above I.A.; incisor bar 3.3 mm below I.A.).

AP-5 induced stereotyped sniffing and antagonized catalepsy induced by 0.5 mg/kg haloperidol i.p.. With regard to these variables it mimics the effects of systemically administered MK-801. Tentatively, this may be explained by an excitatory projection of corticofugal glutamatergic neurons upon dopaminergic neurons in the substantia nigra, releasing dopamine from dendritic varicosities. The nigral dopamine is supposed to act on somatic autoreceptors, inhibiting the nigral output to the striatum. The blockade of the excitatory glutamatergic influence upon these dopaminergic neurones by AP-5 may result in a hyperactive nigrostriatal dopaminergic system that in turn may produce stereotyped sniffing as well as the observed anticataleptic effect (Schuster and Schmidt 1988).

Glutamate in the Striatum (N. Caudatus and Putamen)

The role of striatal glutamate was investigated by stereotaxic injections of AP-5 (50 nmol/0.5 μl) to the anterodorsal striatum of rats (AP 1.7; L±2; V 5; incisor bar 3.3 mm below I.A.). AP-5 in the striatum induced locomotion, stereotyped sniffing (Schmidt 1986) and impairments in the delayed alternation task in the T maze (Hauber 1988) thus mimicking the effects of systemically administered MK-801. However unlike systemic MK-801, AP-5 in the striatum does not change 8-arm maze performance (Hauber 1988). Thus, it can be concluded that amnesia is not due to NMDA receptor antagonism in the striatum. From various studies it seems that NMDA receptors in the limbic system (hippocampus septum and cingulum) are responsible for amnesia.

In another approach to study the involvement of striatal NMDA receptors, we destroyed the cell bodies of neurons that bear NMDA receptors within the striatum. The neurotoxin quinolinic acid (30 nmol/1 μl) was injected at the same site as - in other rats - AP-5 had been. Quinolinic acid destroys striatal output neurons using GABA and substance P as transmitters (Beal et al. 1986). The lesioned animals showed enhanced locomotion, sterotyped sniffing and were less liable to the cataleptic effects of 0.5 mg/kg haloperidol i.p. . These data corroborate the results obtained with AP-5 injections and confirm the involvement of NMDA receptors in the control of these behavioural variables.

An Outline of the Functions of the Striatum (N. Caudatus and Putamen)

The general anatomical connections of the striatum are now fairly well understood. The striatum receives dopaminergic afferents ascending from the midbrain and glutamatergic inputs from the whole cortex; most inputs derive from the frontal and prefrontal cortex. The main output of the striatum goes, via pallidum or substantia nigra and the thalamus, to the cerebral cortex - in the case of the motor circuit to the supplementary motor area (Alexander et al. 1986). The functions of the striatum are less clear; one reason for this are the poor results obtained with lesion-, stimulation- and single-cell recording procedures. (These techniques were extremely useful in the analysis of cortical and cerebellar functions.)

Glutamate is considered to be an excitatory transmitter. While glutamate may transmit specific information, most probably the nigrostriatal dopaminergic neurons do not transmit specific sensory or motor information. The dopamine released by these neurons in the striatum may act as a neuromodulator. Depending on the activity of glutamatergic neurons, or of the cell onto which dopaminergic neurons impinge, dopamine exerts a facilitatory or an inhibitory influence.

In functional terms, dopamine may be regarded as inhibiting striatal activity. The spontaneous firing rate (about 4.2 spikes/s) of populations of striatal neurons and the frequency of caudate spindles (about 15/s), which may give a measure of striatal excitability, are reduced by application of dopamine, apomorphine, amphetamine or by electrical stimulation of the substantia nigra. Also, spreading depression using KCl decreases striatal excitability, i.e. reduces the frequency of caudate spindles.

A dopamine deficit has opposite effects in producing signs of enhanced striatal excitability. Dopamine receptor blockers such as neuroleptic drugs (haloperidol, pimozide) and a dopamine deficit produced by 6-OH-dopamine lesion in the substanitia nigra enhance firing rate and spindle activity.

Enhanced striatal activity seems to exert inhibitory control over cortical neurons and behaviour. This may produce parkinsonian symptoms in humans or behavioural depression like catalepsy in animals.

Our findings show that striatal dopamine and glutamate may exert opposite effects on behaviour. The actions of the NMDAantagonist AP-5 are similar to those of dopamine agonists at least to some extent. Like AP-5, dopamine agonists impair delayed alternation, increase locomotion (however only when injected into the ventral striatum, the nucleus accumbens), induce stereotyped sniffing, decrease grooming activities and increase the frequency of switching from one behavioural activity to another (Schmidt 1986, Bury 1987).

While dopamine antagonists produce the opposite effects or antagonize the dopamine-induced stereotypies, the effects of NMDA in the striatum are to be interpreted with caution. Agonists of excitatory amino acids, including NMDA or L-glutamate, excite all or nearly all central neurons (axons of passage are spared). Nevertheless NMDA (3.3 and 6.6 nmol/0.5 µl) injected into the striatum produced some behavioural changes just opposite to those produced by AP-5 or dopamine agonists. NMDA decreased sniffing, locomotion and the frequency of switching from one behavioural activity to another (Schmidt and Bury 1988).

A Speculation About the Functions of the Striatum

The striatum may be involved in response selection, i.e. it selects which behaviours will be carried out and which will be not. With increasing dose response of dopamine-agonists or NMDA antagonists more and more behaviours are executed collateral to an ongoing behaviour; an increased switching from one behaviour to another results. At high doses stereotypies occur that are performed with high frequency and without guidance by external stimuli. On the contrary, dopamine-antagonists at modest dose levels suppress spontaneous behaviour, but the reactions to external stimuli are intact. This shows that the animal has become dependent on exteroceptive guidance; a phenomenon also reported from parkinsonian patients. Behaviours occurring collateral to an ongoing behaviour are suppressed, and the animal cannot switch from one behavioural activity to another in the absence of an external stimulus (Schmidt 1984). High-dose levels of dopamine antagonists produce behavioural depression and catalepsy. The animals become akinetic and show muscular rigidity. Only strong sensory stimuli (cold water, tail pinch) can elicit responses. Thus, the striatal transmitters dopamine and glutamate control (a) the balance between spontaneous switching and maintaining ongoing behaviour and(b) the balance between the use of exogenous and endogenous information. It is a consequence of these effects that only at a sufficient degree of dopamine activity the organism is able to initiate or to switch to another behaviour in the absence of external guidance.

Animal Models of Human Diseases

Parkinson's Disease

The antidopaminergic drugs reserpine or haloperidol induce behavioural symptoms that resemble those of Parkinson's disease in humans. Rigor and akinesia induced by reserpine (2.5 mg/kg) or haloperidol (0.5 mg/kg) were antagonized by MK-801 dose dependently (0.08; 0.16; 0.33 mg/kg). Thus, tentatively, MK-801 or other NMDA antagonists may have a therapeutic potential as antiparkinsonian agents (Bubser and

Schmidt 1988). In this connection it is of interest that commonly used antiparkinsonian drugs have recently been shown to be NMDA antagonists as well (Olney et al. 1987).

The Glutamate Hypothesis of Schizophrenia

It was first proposed by Kim et al. (1980) that - alternatively to the classical dopamine hypothesis of schizophrenia - an impaired glutamatergic transmission may also be taken into consideration to underlie the disease. Up to now there are no reasons that argue against this view, for example, MK-801 produces excitation, insomnia, light-headedness and psychotic episodes in humans. Also our present animal experimentation favours this new hypothesis: systemic and intrastriatally administered NMDA antagonists produce behavioural stimulation, increased switching from one behavioural activity to another and stereotypies. These symptoms are also seen in schizophrenic patients (Lyon et al. 1986).

A New Animal Model to Detect Antipsychotics

In animals, dopamine-agonists induce an abnormal and characteristic stereotyped behaviour related to stimulation of dopamine receptors in the striatum and n. accumbens. These stereotypies are widely recognized to represent an animal model for psychosis. Its antagonism is one of the most useful screening models for neuroleptic drugs. However, dopamine-induced stereotypy is antagonized only by the "classical" neuroleptics, not by "atypical" ones; clozapine, for instance, even enhances amphetamine-induced stereotypy.

However, we have shown that sniffing stereotypy, induced by intrastriatal injection of AP-5, is antagonized by classical (e.g. haloperidol 0.1 mg/kg i.p.) and by atypical neuroleptics (e.g. clozapine 10, 5, 2 mg/kg s.c.; Schmidt 1986). This test gives no false positive results for buspirone. Also MK-801 induced stereotypies are specifically antagonized by neuroleptics (haloperidol, clozapine) but not by other drugs, for example benzodiazepines (midazolam 0.3 mg/kg i.p.). If further studies confirm these findings, this test may prove useful to detect new antipsychotic drugs on the basis of their ability to antagonize the behavioural effects of impaired glutamatergic transmission.

Summary

Inhibition of glutamatergic transmission at the N-methyl-D-aspartate (NMDA) receptor in the striatum produces behavioural stimulation and stereotypies similar to those produced by dopamine agonists. This behavioural syndrome is considered to represent an animal model for psychosis. While dopamine-induced stereotypies are antagonized only by classical antipsychotics, stereotypies induced by inhibition of glutamatergic transmission are antagonized by classical and atypical antipsychotics. These results support the glutamate hypothesis of schizophrenia.

References

Alexander GE, De Long MR, Strick PL (1986) Parallel organization of functionally segregated circuits linking basal ganglia and cortex. Annu Rev Neurosci 9: 357-381

Beal MF, Kowall NW, Ellison DW, Mazurek MF, Swartz KJ (1986) Replication of the neurochemical characteristics of Huntington`s disease by quinolinic acid. Nature 321: 168-171

Bischoff C, Tiedtke PI, Schmidt WJ (1988) Learning in an 8-arm-radial-maze: effects of dopamine- and NMDA-receptor-antagonists. In: Elsner M, Barth FG (eds) Proceedings of the 16th Göttingen neurobiology conference. Thieme, Stuttgart, p 358

Bubser M, Schmidt WJ (1988) The NMDA antagonist MK-801 counteracts haloperidol-induced catalepsy. In: Elsner M, Barth FG (eds) Proceedings of the 16th Göttingen neurobiology conference. Thieme, Stuttgart, p 356

Bury D (1987) Steuerung von Verhaltensweisen der Ratte im "open field" durch Dopamin und Glutamat im Striatum. In: Elsner N, Creutzfeldt O (eds) Proceedings of the 15th Göttingen neurobiology cConference. Thieme, Stuttgart

Clineschmidt BV, Martin GE, Bunting PR, Papp NL (1982) Central sympathomimetic activity of (+)-5-methyl-10, 11-dihydro-5H-dibenzocyclohepten-5,10-imine (MK-801), a substance with potent anticonvulsant, central sympathomimetic, and apparent anxiolytic properties. Drug Dev Res 2: 135-145

Cotman CW, Iversen LL (1987) Excitatory amino acids in the brain - focus on NMDA receptors. Trends in Neurosci 10: 263-265

Hauber W (1988) Striatal glutamate and maze performance. In: Elsner N, Barth FG (eds) Proceedings of the 16th Göttingen neurobiology conference. Thieme, Stuttgart

Kim JS, Kornhuber HH, Schmid-Burgk W, Holzmüller B (1980) Low cerebrospinal fluid glutamate in schizophrenic patients and a new hypthesis on schizophrenia. Neurosci Lett 20: 379-382

Lyon N, Mejsholm B, Lyon M (1986) Stereotyped responding by schizophrenic outpatients: cross-cultural confirmation of perseverative switching on a two-choice task. J Psychiatr Res 20: 137-150

Olney JW, Price MT, Labruyere J, Salles KS, Frierdich G, Müller M, Silverman E (1987) Anti-parkinsonian agents are phencyclidine agonists and N-methyl-aspartate antagonists. Eur J Pharmacol 142: 319-320

Schmidt WJ (1984) L-dopa and apomorphine disrupt long- but not short-behavioural chains. Physiol Behav 33: 671-680

Schmidt WJ (1986) Intrastriatal injection of DL-2-amino-5-phosphonovaleric acid (AP-5) induces sniffing stereotypy that is antagonized by haloperidol and clozapine. Psychopharmacology (Berlin) 90: 123-130

Schmidt WJ, Bury D (1988) Behavioural effects of N-methyl-D-asparate in the anterodorsal striatum of the rat. Life Sci 43: 545-549

Schuster G, Schmidt WJ (1988) Glutamatergic control of behavior in the midbrain of rats: effects of the NMDA-antagonist AP-5. In: Elsner N, Barth FG (eds) Proceedings of the 16th Göttingen neurobiology conference. Thieme, Stuttgart

Effects of CO_2 on Neural Functions

H. Caspers and E.-J. Speckmann
Institute of Physiology, University of Münster, Robert-Koch-Str. 27a, 4400 Münster, FRG

Introduction

Since the classical studies of Lorente de No (1947) on peripheral nerves numerous investigations have been devoted to the effects of carbon dioxide on other neural structures and functions (Speckmann and Caspers 1974; Caspers et al. 1987). One general aspect in the variety of specific results obtained in these studies is that the actions of CO_2 turn out to be partly excitatory, partly depressive in nature. The following presentation focuses on this particular problem. In a first section some examples of different CO_2 effects on EEG waves as well as on the membrane potential and discharge frequency of single neurons are described. A final section then deals with the mechanisms possibly involved in generating opposite responses.

Effects of CO_2 on the EEG and on the Membrane Potential and Discharge Frequency of Single Neurons

1. The prominent feature of CO_2 effects on spontaneous cortical activity is the well-known depression of EEG waves in the conventional frequency range (see Fig. 1). Similar actions have been observed with a variety of cortical potentials evoked by direct, afferent, and sensory stimulations (see Caspers and Speckmann 1972, 1974). Finally, also high-voltage fast potential fluctuations such as seizure discharges are extinguished when the pCO_2 is raised to a critical level. These changes in bioelectrical activity reflect the wellknown anticonvulsive properties of hypercapnia. All effects of CO_2 reported so far show adaptation and a distinct rebound if the elevated pCO_2 is kept constant for some minutes and then reduced to normal values.

Besides the depression of both spontaneous and evoked EEG potentials which clearly predominates in the spectrum of cortical CO_2 effects, there appear, however, some further activity changes possibly due to excitatory actions of CO_2 on special generator structures (Lehmenkühler et al. 1977; Caspers et al. 1979). This finding is illustrated in Fig. 1. As the power spectra of EEG waves indicate, islands of high-frequency waves in the range of 60-70 c/s emerge from the background of reduced normal activity as soon as the arterial pCO_2 reaches a level of about 70 mmHg. The fast potential fluctuations occur on both hemispheres and show a high coherence in

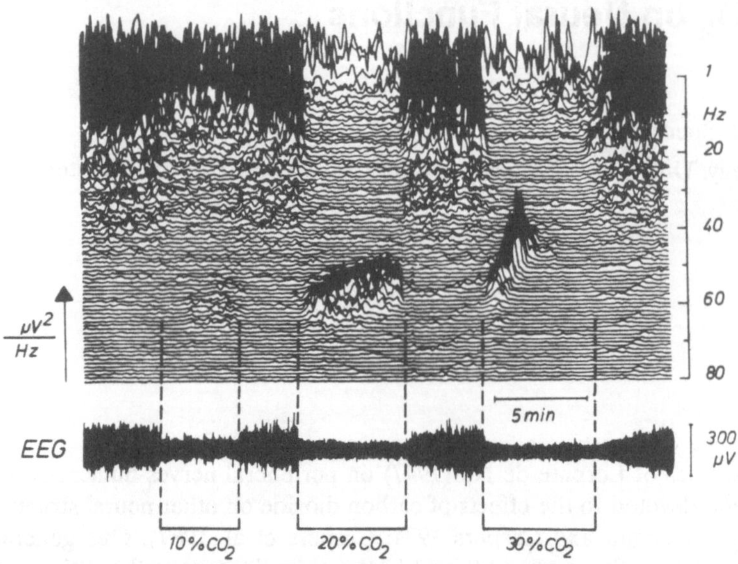

Fig. 1. Effects of CO_2 on spontaneous EEG activity in an anesthetized rat ventilated artificially at a constant rate to adjust the initial level of arterial pCO_2 to 40-50 mmHg. Besides the original EEG waves power spectra of the curves before, during, and after administration of 10%, 20% and 30% CO_2 in the inspired gas mixture are displayed. Depression of conventional EEG is associated with provocation of high-frequency waves. (From Caspers et al. 1979)

bilateral as well as in fronto-occipital derivations. With a further rise in pCO_2 they initially increase in amplitude and slightly decrease in frequency until they finally disappear again, when the pCO_2 exceeds approximately 130 mmHg.

The occurrence of high-frequency EEG waves, which tend to appear in spindlelike groups of some 100 ms duration, seems to be related to another excitatory CO_2 effect sometimes observed during focal interictal seizure discharges. Whereas generalized seizure activity is progressively suppressed, focal interictal spikes often show a transient increase in frequency, if the pCO_2 is raised to higher values (see Zschocke and Heyn 1971). Further studies revealed that single spikes occurring during hypercapnia appear with maximum probability in the rising phase of a spindle group (Caspers et al. 1979). The finding suggests that focal seizure discharges may be triggered and accelerated to some extent by the same neuronal processes giving rise to the high frequency waves.

In summary, the spectrum of CO_2 effects on bioelectrical cortical activity is governed by the depression of both spontaneous and evoked potentials in the conventional EEG. However, also opposite activity changes can be observed that seem to be excitatory in nature.

2. The described effects of CO_2 on cortical field potentials are paralleled by opposite actions on single neurons. In the majority of nerve cells of vertebrates studied in situ CO_2 causes the RMP to increase (Washizu 1960; Caspers and Speckmann 1972; Carpenter et al. 1974; Speckmann and Caspers 1974). A representative finding obtained in a motoneuron of the rat spinal cord is displayed in Fig. 2. In these cells the hyperpolarization is associated with an abrupt suppression of antidromically evoked action potentials when the pCO_2 is continuously raised. At the same time repetitive neuronal

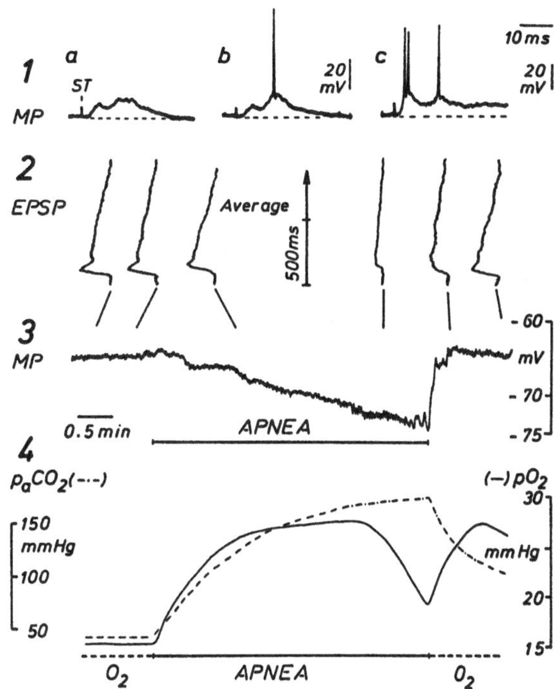

Fig. 2. Increase in membrane potential (MP) of a rat spinal motoneuron (3) during a period of apnea following the administration of 100% O_2. The changes in arterial pCO_2 and in tissue pO_2 at the recording site are displayed in 4. In addition to the MP the excitatory postsynaptic potentials (EPSP) evoked by stimulation of the motor cortex with single electrical pulses were measured and averaged throughout the experiment (2). The curves in 1 represent original oscilloscope recordings of the postsynaptic responses to stimuli (ST) of increasing intensity (a, b, c). (From Caspers et al. 1984)

discharges elicited by intracellular current injections become progressively reduced, thus turning from a tonic to a phasic type of response. This phenomenon fits with the damping effects of CO_2 known from studies on peripheral nerves (Monnier 1955). Besides these actions pointing partly to a direct influence of CO_2 on neuronal membranes, interferences with synaptic transmission processes are evident. During hypercapnia excitatory postsynaptic potentials (EPSP) decrease in amplitude with the membrane resistance often being enhanced. This statement holds true for both mono- and polysynaptic EPSP (see Fig. 2). As a result, impulse transmission is rapidly depressed, and the discharge frequency of spontaneously active units declines (Fig. 3B,1). Neurons showing this type of response have been labeled as I neurons (see Speckmann and Caspers 1974).

In contrast to the majority of vertebrate neurons in situ, snail neurons usually depolarize when the pCO_2 in the conventional bath fluid is elevated (Chalazonitis 1963; Brown and Berman 1970; Thomas 1976; Zidek et al. 1978; Thomas and Meech 1982). Similar observations have been made with cultured mammalian cells (Bingmann et al. 1981; Caspers et al. 1987). Samples of this type of response are given in Figs. 4 and 5. Apart from such reactions encountered in isolated cells of both invertebrates and vertebrates, excitatory effects of CO_2 have been found to occur also in single units in situ. Besides the well-known activation of peripheral and central chemoreceptors (see Fig.

3A), increases in discharge frequency appear, for instance, in spinal interneurons (Fig. 3B,2). However, the percentage of such elements, classified as E neurons, is rather small. In their investigations on the rat spinal cord Speckmann et al. (1970) estimated this type of cell not to exceed 5%-10% of the whole population studied.

All effects of CO_2 reported above tend to adapt when the administration is maintained or repeated at shorter intervals. This fact is particularly demonstrable using the discharge frequency of a chemoreceptor-fiber unit of the carotid body as an indicator (Fig. 3A). In summary, opposite effects of CO_2 are demonstrable also in single neurons. They raise the question as to the mechanisms involved in generating the different responses.

Influence of Microenvironment on CO_2 Effects

The different effects of CO_2 on various neural functions may have several explanations. At first it might be supposed that they are determined mainly by special membrane

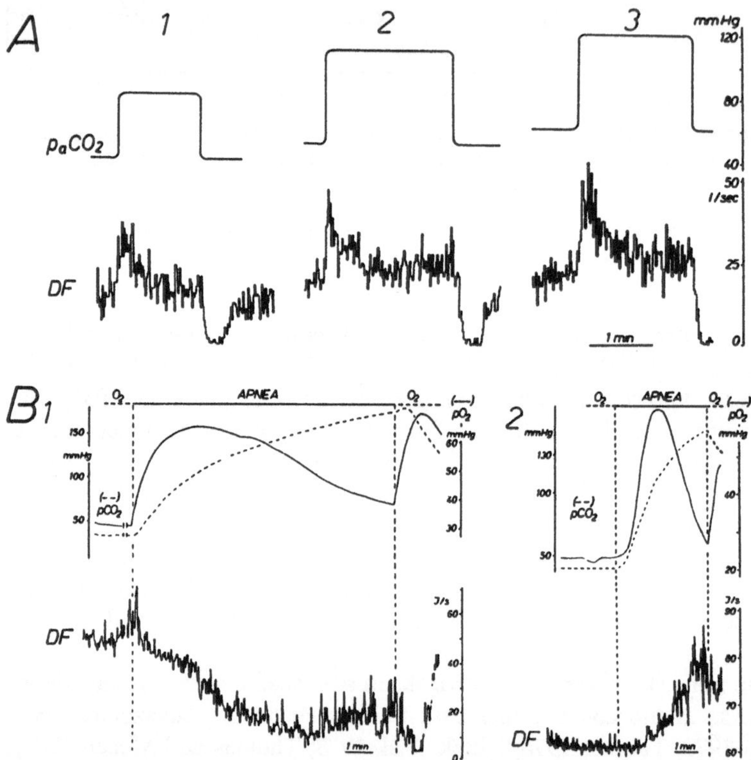

Fig. 3. Different effects of CO_2 on the discharge frequency (DF) of spontaneously active neurons. **A** Increase in DF of a single chemoreceptor-fiber unit in the rat carotid body during a stepwise rise of arterial pCO_2 ($paCO_2$). The effect shows adaptation and postexcitatory depression. **B** Decrease (1) and increase (2) in DF of two interneurons in the rat lumbar cord during periods of apnea following administration of 100% O_2. The evoked changes in arterial pCO_2 and in tissue pO_2 are displayed in the upper tracings. (Part A in collaboration with D. Bingmann, part B from Caspers et al. 1979)

properties of the neurons concerned. However, such an interpretation is at variance with the fact that a rise in pCO_2 can cause both de- and hyperpolarizations in the same type of cells, depending on whether these are studied in their natural milieu in situ or, for instance, in a cell culture superfused by a conventional bath fluid. These observations point to the microenvironment playing an essential role in influencing the direction of the CO_2 response. In this context the buffer capacity in the extracellular space obviously represents an important factor (Fenn and Asano 1956; Thomas 1974, 1976). The assumption has been studied in more detail (Zidek et al. 1978; Bingmann et al. 1981). The results of such an experiment performed on an isolated neuron of the buccal ganglion of the snail *Helix pomatia* are illustrated in Fig. 4. At first the tracings confirm that

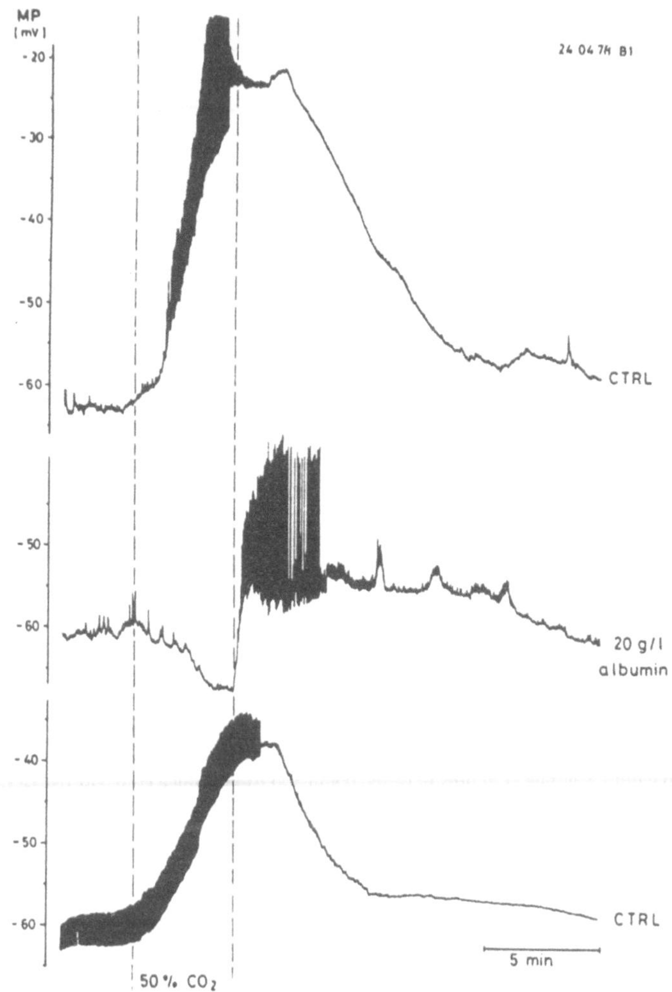

Fig. 4. Changes of the membrane potential (MP) of neuron B1 in the buccal ganglion of the snail *Helix pomatia* during a rise of the pCO_2 (dashed vertical lines) in the bath fluid. The depolarization induced in the upper and lower control recordings (CTRL) reverses to hyperpolarization after 20 g/l albumin have been added to the bath solution. (Experiments in collaboration with W. Zidek, A. Lehmenkühler and H. Lange-Asschenfeldt)

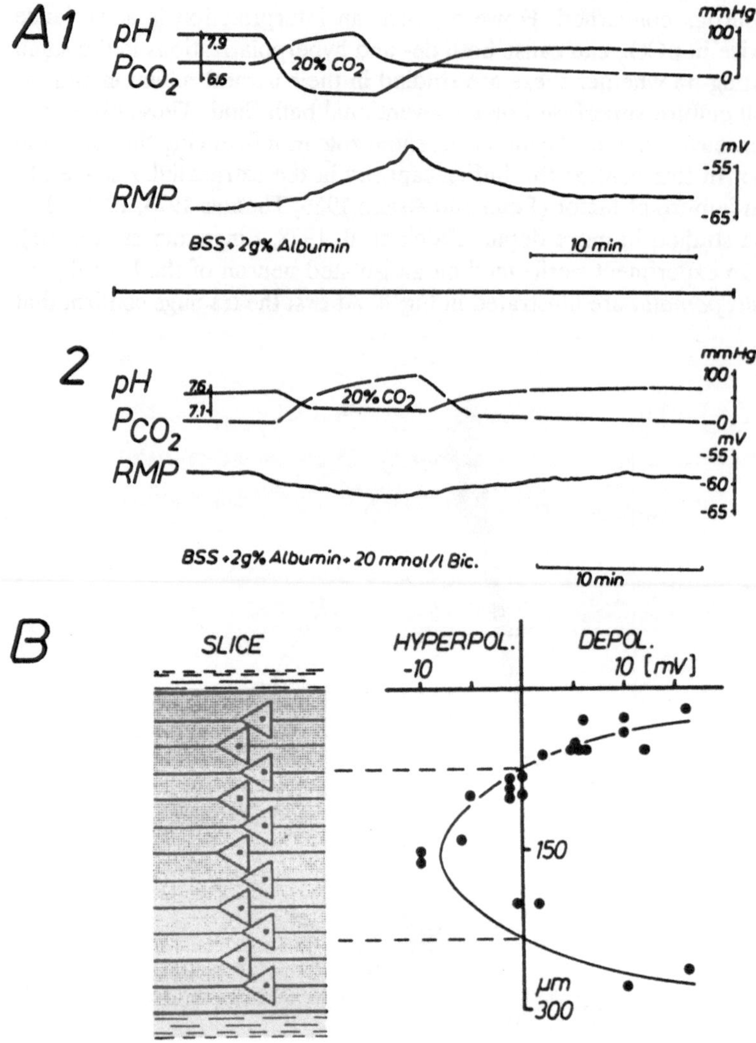

Fig. 5. Effects of CO_2 on the resting membrane potential (RMP) of mammalian neurons in vitro. **A** Changes in the RMP of a cultured dorsal root ganglion cell of the rat spinal cord during a rise of the pCO_2 and a lowering of the pH in the bath, respectively. The usual depolarization (1) reverses to hyperpolarization (2) when 20 mmol/l bicarbonate are added to the fluid. **B** Changes in the RMP of neurons in a slice of the guinea pig hippocampus at depths between 0 and 300 μm during a constant rise of the pCO_2 in the superfusate (10% CO_2). Cells in the outer layers show depolarizations, those in the inner ones hyperpolarizations. In a border zone between the two, measurable effects can be lacking. (Part A from Bingmann et al. 1981, part B from Caspers et al. 1986)

a rise of the pCO_2 in the bath solution usually causes these cells to depolarize. After having applied a macromolecular buffer (albumin) to the superfusate, the response of the RMP to the same increase in pCO_2 is reversed in sign. With prolonged or repeated administrations of CO_2 the effect tends to adapt. Corresponding results are obtained with cultured mammalian neurons. Representative curves recorded from a dorsal root ganglion cell of the rat spinal cord are displayed in Fig. 5A. The tracings indicate that the usual depolarization of the cell evoked by a rise in pCO_2 turns to hyperpolarization, when the buffering power of the bath fluid is enhanced. In these cases the bicarbonate buffer proved more effective than albumin and hemoglobin.

The influence of the micromilieu on the effects of CO_2 becomes evident, moreover, from another experimental observation illustrated in Fig. 5B. In this diagram the RMP of neurons in a slice of the guinea pig hippocampus, determined during a constant rise of the pCO_2 in the superfusate, is plotted against the recording depth between 0 and 300 μm. The measurements show neuronal depolarizations to prevail in the outer zone of the slice in which cells are in an immediate contact with the bath fluid, whereas hyperpolarizations predominate in inner layers, where a more or less natural micro-environment is preserved. This profile of CO_2 effects flattens with increasing duration of an experiment.

In summary, the findings indicate that the direction of CO_2 effects on neural functions is influenced by the buffer capacity of the extracellular medium. The question arises as to the elementary mechanisms involved.

Discussion

The results of numerous studies performed on elementary actions of CO_2 suggest that the polarity of the evoked changes in membrane potentials during hypercapnia depends essentially on shifts of the pH gradient across the membranes. It has been shown that lowerings of the pH in the extracellular space, on the one hand, induce neuronal depolarizations (Brown and Berman 1970; Lux and Müller 1987). On the other hand, a decline in the intracellular pH elicited, for example, by H^+ injections gives rise to hyperpolarizations (Thomas 1974). This effect is possibly attributable to the liberation of calcium ions from intracellular binding sites, which in turn activates calcium dependent potassium currents. The concentration of extracellular K^+ has actually been found to rise (Lehmenkühler 1979; Caspers et al. 1987). With a high buffer capacity of the extracellular fluid the initial fall in intracellular pH induced by the rapid diffusion of CO_2 seems to exceed the acidic shift in the surrounding medium and thus to cause the RMP of the cells to increase. At a lower external buffering power the evoked initial fall in pH would prevail in the extracellular medium and consequently give rise to neuronal depolarizations. Such mechanisms would explain, moreover, the occurrence of adaptation and rebound phenomena.

Independent of a further clarification of such problems, the described findings have shown that opposite effects of CO_2 on neural functions can arise, in principle, from different buffer capacities of the surrounding media. It requires further investigations to find out whether the opposite responses of I and E units in situ can actually be attributed to such a mechanism. Another problem is whether the depolarizing effects of CO_2

usually found in neurons of invertebrates can be shown to persist if the cells are studied in their natural environment.

Summary

The effects of CO_2 on various neural functions are marked by both depression and excitation:

1. In the EEG the prominent feature of CO_2 effects is the depression of waves in the conventional frequency range. This action includes highly synchronized potential fluctuations such as seizure discharges, thus reflecting the anticonvulsive properties of the metabolite. At higher CO_2 pressure levels, however, groups of 60-70 c/s waves appear, clearly distinguishable against the background of reduced normal activity. These waves can trigger interictal spikes.

2. In the majority of spinal and cortical neurons studied in situ the resting membrane potential (RMP) increases when the pCO_2 is raised. The hyperpolarization is accompanied by a reduction and final extinction of postsynaptic potentials, with the membrane resistance usually being enhanced. In spontaneously active cells of this type, labeled as I neurons, hypercapnia evokes a reduction in the discharge frequency. In contrast to this behaviour, the RMP of other units decreases when the pCO_2 is elevated. The groups of cells showing such a response comprise isolated neurons of invertebrates, cultured mammalian cells studied in vitro as well as chemoreceptors and a small population of interneurons labeled as E neurons. In spontaneously active units of this type hypercapnia causes a rise of the discharge rate.

3. A depolarization of neuronal elements evoked by a rise of the pCO_2 can be reversed to hyperpolarization when the buffer capacity of the extracellular medium is enhanced, and vice versa. The mechanisms and consequences of this effect are discussed.

References

Bingmann D, Kienecker EW, Caspers H, Knoche H (1981) Chemoreceptor activity of sinus nerve fibres after their implantation into the wall of the external carotid artery. In: Belmonte C, Pallot DJ, Acker H, Fidone S (eds) Arterial chemoreceptors. Leicester University Press, Leicester pp 92-104

Brown AM, Berman PR (1970) Mechanism of excitation of aplysia neurons by carbon dioxide. J Gen Physiol 56: 543-558

Carpenter DO, Hubbard JH, Humphrey DR, Thompson HK, Marshall WH (1974) Carbon dioxide effects on nerve cell function. In: Nahas G, Schaefer KE (eds) Topics in environmental physiology and medicine. Springer, Berlin, Heidelberg, New York, pp 49-62

Caspers H, Speckmann EJ (1972) Cerebral pO_2, pCO_2 and pH: changes during convulsive activity and their significance for spontaneous arrest of seizures. Epilepsia 13: 699-725

Caspers H, Speckmann EJ (1974) Cortical DC shifts associated with changes of gas tensions in blood and tissue. In: Remond A (ed) Handbook of electroencephalography and clinical neurophysiology, vol 10A. Elsevier, Amsterdam, pp 41-65

Caspers H, Speckmann EJ, Lehmenkühler A (1979) Effects of CO_2 on cortical field potentials in relation to neuronal activity. In: Speckmann EJ, Caspers H (eds) Origin of cerebral field potentials. Thieme, Stuttgart, pp 151-163

Caspers H, Speckmann EJ, Lehmenkühler A (1984) Electrogenesis of slow potentials of the brain. In: Elbert T, Rockstroh B, Lutzenberger W, Birbaumer N (eds) Self-regulation of the brain and behaviour. Springer, Berlin Heidelberg New York, pp 26-41

Caspers H, Speckmann EJ, Bingmann D, Lehmenkühler A (1986) Wirkungen von CO_2 auf das Membranpotential einzelner Neurone. In: Grote J, Thews G (eds) Funktionsanalyse biologischer Systeme, 15. Steiner, Stuttgart, pp 185-195
Caspers H, Speckmann EJ, Lehmenkühler A (1987) DC potentials of the cerebral cortex. Seizure activity and changes in gas pressures. Rev Physiol Biochem Pharmacol 106: 127-178
Chalazonitis N (1963) Effects of changes in pCO_2 and pO_2 on rhythmic potentials from giant neurons. Ann NY Acad Sci 109: 451-479
Fenn OW, Asano T (1956) Effects of carbon dioxide inhalation on potassium liberation from the liver. Am J Physiol 185: 567-576
Lehmenkühler A (1979) Interrelationships between DC potentials, potassium activity, pO_2 and pCO_2 in the cerebral cortex of the rat. In: Speckmann EJ, Caspers H (eds) Origin of cerebral field potentials. Thieme, Stuttgart, pp 49-59
Lehmenkühler A, Speckmann EJ, Caspers H (1977) Synchronous 60-70 c/s EEG waves evoked by an increase in pCO_2. Electroencephalogr Clin Neurophysiol 43: 131
Lorente de No R (1947) Carbon dioxide and nerve function. Stud Rockefeller Inst Med Res 131: 148-193
Lux HD, Müller TH (1987) Calzium-abhängige Schrittmacherprozesse an der neuronalen Membran mit dem Zeitbedarf paroxysmaler Vorgänge. In: Speckmann EJ (ed) Epilepsie 86. Einhorn, Reinbek, pp 16-26
Monnier AM (1955) Die funktionale Bedeutung der Dämpfung in der Nervenfaser. Ergeb Physiol 48: 230-285
Speckmann EJ, Caspers H (1974) The effect of O_2 and CO_2 tensions in the nervous tissue on neuronal activity and DC potential. In: Remond A (ed) Handbook of electroencephalography and clinical neurophysiology, vol 2C. Elsevier, Amsterdam, pp 71-89
Speckmann EJ, Caspers H, Sokolov W (1970) Aktivitätsänderungen spinaler Neurone während und nach einer Asphyxie. Pflügers Arch 319: 122-138
Thomas RC (1974) Intracellular pH of snail neurones measured with a new pH-sensitive glass micro-electrode. J Physiol (Lond) 238: 159-180
Thomas RC (1976) The effect of carbon dioxide on the intracellular pH and buffering power of snail neurones. J Physiol (Lond) 255: 715-735
Thomas RC, Meech RW (1982) Hydrogen ion currents and intracellular pH in depolarized voltage-clamped snail neurones. Nature 299: 826-828
Washizu Y (1960) Effect of CO_2 and pH on the responses of spinal motoneurons. Brain Nerve 12: 757-766
Zidek W, Lehmenkühler A, Caspers H, Lange-Asschenfeldt H (1978) Macromolecular buffering reverses the CO_2 effect on the membrane potential in snail neurons. Pflügers Arch 377: 43
Zschocke S, Heyn D (1971) Influence of the pCO_2 on macropotentials and single cell discharges in cortical seizure activity. Electroenceph Clin Neurophysiol 30: 265

The R-Wave Biography of a Brain Potential

J. Kriebel[1], B. Grözinger[2] and H.H. Kornhuber[3]
[1]Abteilung Neurologie und Psychiatrie, Bundeswehrkrankenhaus, Akademisches Krankenhaus der Universität, Oberer Eselsberg 40, 7900 Ulm, FRG
[2]Sektion Neurophysiologie, Universität Ulm, Oberer Eselsberg, 7900 Ulm, FRG
[3]Abteilung Neurologie, Universitätsklinik, Steinhövelstr. 9, 7900 Ulm, FRG

Bereitschaftspotential-Paradigm

Since 1971 we have investigated prespeech potentials (Grözinger et al. 1972) using the standard BP recording and analysis conditions (Kornhuber and Deecke 1964). In summary, the experiments showed prespeech activity to be distributed widely over the head (Grözinger et al. 1974, 1976, 1979, 1980), with some differences between homologues bifrontal electrode placements. The vertex recordings having the highest amplitudes and longest latences of BP before speech onset emphasized the importance of the supplementary motor area (SMA) in speech production (Grözinger et al. 1979, 1980). The SMA may have some role in the translation of motivation into action, for it is juxtaposed to and connected with the limbic cortex (Grözinger et al. 1979; Kornhuber 1980; Kriebel et al. 1988). Initiation and timing (Kornhuber 1984) might be its task also for speech activity. The right temporo-parietal recordings were significantly lower in BP amplitudes (Grözinger et al. 1979, 1980).

Artifact Controls and Animal Experiments

The interpretation of the data presupposed detailed observation of artifacts. The multiple artifact influences on brain potentials had to be controlled with consistency and accuracy from the very beginning (Grözinger et al. 1972, 1974, 1980). Control of artifact contaminations included, for example, changing positions of reference electrodes including extracranial references, recordings via source derivations, controls of eye and lid movements, head and lead movements, manifold myographic recordings, controls of galvanic skin restistance, glossokinetic potentials, and electrical field artifacts (Grözinger et al. 1974, 1980). The elimination and/or control of these artifacts supported the intracranial origin of the prespeech potentials. However, a sinusoidal-like intermingling potential often could not be excluded. It proved to be respiration correlated. We therefore named this potential respiration wave or R-wave (Grözinger et al. 1972). Fig. 1 shows the superposition of R-waves and BP. Due to the intermingling of interhemispheric asymmetrical R-waves no clear determination of amplitudes and latences of BP is possible. The R-waves are sometimes dominant in a way that BP is not obvious at all. The lower part of Fig. 2 represents such a situation prior to articulatory movements without phonation. After activity onset all potentials, e.g. evoked potentials,

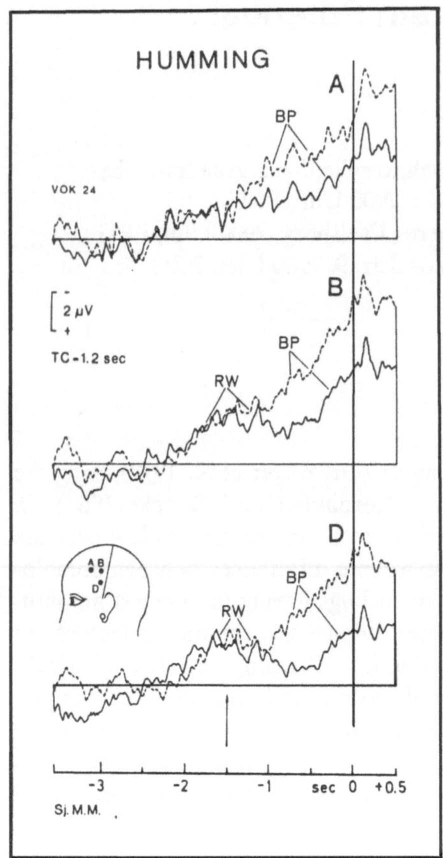

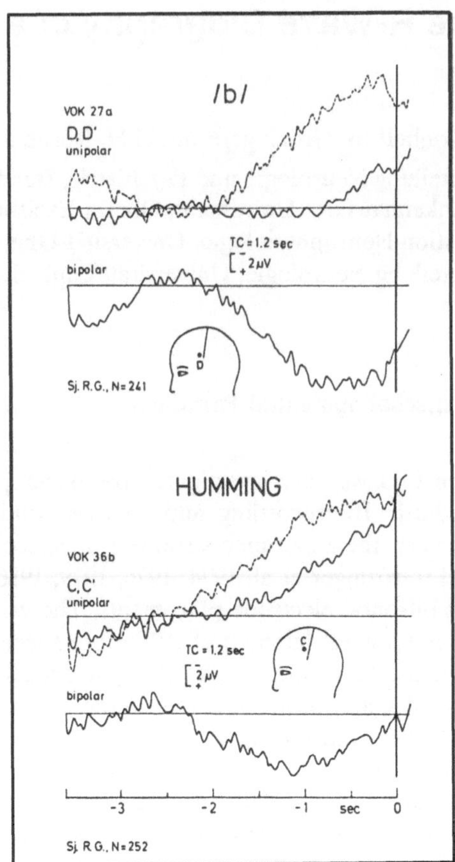

Fig. 1. Superposition of Bereitschaftspotentials (BP) and R-waves (RW) preceding humming and preceding repetitive speaking on the phoneme (b). Dashed lines: left hemisphere; full lines: right hemisphere. Analysis time four seconds. Vertical lines indicate onset of activity. Monopolar recordings versus linked ears; right half shows additionally bipolar recordings clarifying the superposition of R-waves

are masked by glossokinetic potentials (Grözinger et al. 1974, 1980). Expressive speech activities depend on exact coordination with respiration. This is obvious if respiration under resting conditions is compared with respiration during continuous speech (Fig. 2, upper part), the differences being caused by the task of maintaining relatively constant subglottic pressure for phonation.

To exclude some intracranial but not cerebral sources of the R-wave, such as respiratory changes in CSF pressure or blood pressure, we were forced to continue the experiments with animals (cats and pigs). The pigs were anesthetized (halothane) and immobilized in complete muscular relaxation (Imbretil) for undisturbed artificial ventilation. Arterial blood pressure (Statham gauge; femoral artery) and blood gas concentrations (O_2, CO_2) also under hypoxia and hypercapnia (14% O_2; 2%, 4% and 6% CO_2) were recorded. Neurophysiological conditions were as follows: bilateral vagotomy was done to exclude afferent input of various receptors from the lungs and tracheobronchial tree, etc. Phrenic nerve recordings (bipolar silver hook electrodes) were

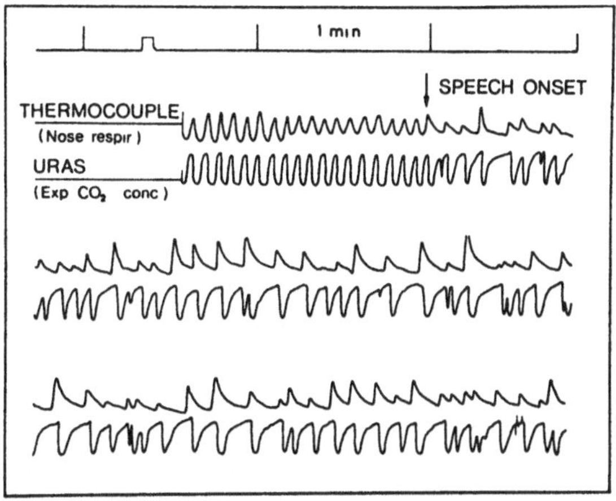

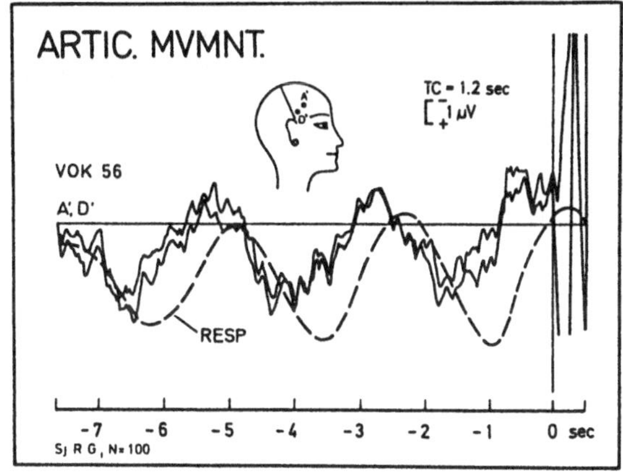

Fig. 2. Upper part: comparison of respiration during rest, and respiration during speech activity. Respiration was recorded by a thermocouple in the nose (upper line) and from expiratory CO_2 concentration (lower line). The time scale is given in minutes. The clear differences in respiration during speaking indicate cerebral respiratory control.Lower part, average of brain potentials (predominating R-waves) and nose respiration (dashed line) preceding articulatory movements. Analysis time eight seconds. N=100. Vertical line indicates activity onset

used to measure the efferent output of the cerebral respiratory centers. EEG recordings were taken from the intact skull (Diekmann et al. 1980; Kriebel et al. 1988). The autocorrelations of the artificial ventilation and phrenic nerve activity differed. Of course the autocorrelations functions (ACF) of phrenic nerve activity was less expressed. The biological signal (phrenic nerve activity, e.g. respiratory brain stem activity) varied more than that of the mechanically performed ventilation. Nevertheless, cross-correlation functions (CCF) and cross-correlation coefficients (CCC) between phrenic nerve activity and EEG, e.g. R-waves, were significantly larger than the CCF and CCC between artificial ventilation and R-wave. This clear result demonstrates that efferent respiratory neural activity is the main source of the R-wave (Diekmann et al. 1980). This is compatible with the hypothesis that the R-wave may be a bioelectric correlate of the cerebral coordination of respiration and expressive speech activity.

Some similarities of the topographical distribution of the BP and R-waves prior to speech activities as well as the necessity to investigate the R-wave under continuous speaking conditions encouraged us to leave the BP paradigm for further experiments. There was a need for topographical experiments under conversational speech conditions. Therefore the BP paradigm was replaced by the analyses of CCC and CCF gained during continuos speaking of the subjects (Diekmann et al. 1980). In the first step the CCF and CCC of the R-wave and respiration were compared with normal breathing under resting conditions and under voluntarily slowed respiration. We thus looked for differences between autonomic and voluntary control of respiration. Recording of continuous speaking, counting and verbal thinking followed. The timing of the respiration pattern was more precise under autonomic control. Inspite of this, the CCF and CCC of respiration and R-wave are under better voluntary control. This confirms the involvement of the cerebral cortex. An overview of all topographical comparisons of the CCC of respiration and R-waves during continuous speaking, counting, repeating different syllables and during verbal thinking is given in Fig. 3 (lower parts). With verbal thinking and normal breathing we have test conditions without involvement of motoric speech activities, resulting in rather low values. Greater CCC are found under conditions including motoric speech activities and under voluntarily controlled slowed breathing. Interhemispheric differences can be observed during counting only. The results together suggest that brain areas involved in expressive speech activity may be also projection areas for the R-wave.

TOPOGRAPHICAL COMPARISON
OF CORRELATION FUNCTIONS

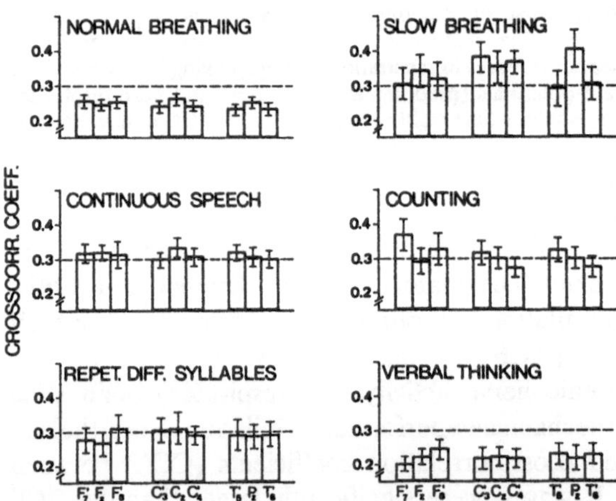

Fig. 3. Topographical comparison of cross-correlation coefficients (CCC) of respiration and R-wave during normal breathing, slow breathing, continuous speech, counting, repeating syllables and verbal thinking. All colums represent mean values of seven subjects and the 95% confidence intervals. The horizontal dashed line marks for reference reasons the cross-correlation coefficient of 0.3. The electrode placements are indicated below

Summary

This publication describes the biography of the R-wave from an "artifact" to a respiration-correlated brain potential with clear neurophysiological functions. During our investigations with the typical Bereitschaftspotential (BP) paradigm the R-wave intermingled with the prespeech potentials and disturbed (like an artifact) their exact measuring (amplitudes and latences). Manifold artifact controls and animal experiments proved the cerebral origin and the functional interpretation as a bioelectric correlate of the coordination of motoric speech activities with respiration. New approaches using cross-correlation techniques instead of the BPparadigm allow the investigation of respiration and speech activities under more complex test conditions (e.g. continuous speaking, voluntarily influenced respiration, etc).

References

Deecke L, Grözinger B, Kornhuber HH, Kriebel J, Kristeva R (1979) Cerebral potentials related to voluntary movements, speech production and respiration in man. In: Speckmann EJ, Caspers E (eds) Origin of cerebral field potentials. Thieme, Stuttgart , pp 132-140

Diekmann V, Grözinger B,Kornhuber HH, Kriebel J, Bock KH (1980) Evidence for the cerebral origin of the R-wave in the pig. Prog Brain Res 54: 103-108

Grözinger B, Kornhuber HH, Kriebel J, Murata K (1972) Menschliche Hirnpotentiale vor dem Sprechen. Pflugers Arch Physiol 332: 100

Grözinger B, Kornhuber HH, Kriebel J (1974) Methodological problems in the investigation of cerebral potentials preceding speech: determining the onset and suppressing artifacts caused by speech. Neuropsychology 13: 263-270

Grözinger B, Kornhuber HH, Kriebel J (1976) EEG-investigation of hemispheric asymmetries preceding speech. The R-wave. In: McCallum, Knott (eds) The responsive brain. Wright, Bristol, pp 103-107

Grözinger B, Kornhuber HH, Kriebel J (1979) Participation of mesial cortex in speech. Exp Brain Res [Suppl] 2: 189-192

Grözinger B, Kornhuber HH, Kriebel J, Szirtes J, Westphal KTP (1980) The Bereitschaftspotential before speaking of words and sentences. Prog Brain Res 53: 798-804

Kornhuber HH (1980) Introduction. Prog Brain Res 54

Kornhuber HH (1984) Attention, readiness for action, and the stages of voluntary decision - some electrophysiological correlates in man. Exp Brain Res [Suppl] 9: 420-429

Kornhuber HH, Deecke L (1964) Hirnpotentialänderungen beim Menschen vor und nach Willkürbewegungen, dargestellt mit Magnetbandspeicherung und Rückwärtsanlyse. Pflugers Arch Gesamte Physiol 281: 52

Kriebel J, Grözinger B, Haag C, Kornhuber HH (1988) Topography of the R-wave during continuos speaking, counting, verbal thinking, normal respiration and voluntarily controlled respiration. Electroencephalogr Clin Neurophysiol 69: 17

Summary

This publication describes the beginnings of the R-wave from an "artifact" to a respiration-related brain potential with clear neurophysiological functions. During our investigations with the typical Bereitschaftspotential (BP) paradigm the R-wave interfered with the premotor potentials and distorted (as an artifact) their exact measuring (amplitudes and latencies). Manifold artifact controls and animal experiments proved the cerebral origin and the functional interpretation as a bioelectric correlate of the coordination of motor (speech) activities with respiration. New approaches using non-averaging techniques instead of the BP paradigm allow the investigation of vocalization and speech activities under more complex test conditions (e.g. continuous speaking, voluntarily influenced expirations, etc.).

References

Deecke L, Grözinger B, Kornhuber HH, Kriebel J, Kotchoubey B (1979) Cerebral potentials related to voluntary movements, speech production and vegetative ... [illegible] ... (eds) Origin of cerebral field potentials. Thieme, Stuttgart, pp 125–[illegible]

Grözinger B, Kornhuber HH, Kriebel J, ... (1984) Evidence for the extracranial origin of the ... [illegible] ... Exp Brain Res 54:[illegible]

[illegible]

The Proton-Activated Sodium Current: Activation Conditions in Mammalian Central Neurons

H.D. Lux and R. Grantyn

Abteilung Neurophysiologie, Max-Planck-Institut für Psychiatrie, Am Klopferspitz 18a, 8033 Planegg, FRG

Neurons are known to be extremely sensitive to changes in pH levels. Small alkaline shifts in the neuronal environment, as observed during stimulation-induced repetitive activity (Kraig et al. 1983), tend to produce an increase in neuronal excitability. Acidic shifts could be a consequence of an excess of strongly dissociated anions by release of metabolically generated lactate (Kraig et al. 1983) or by a loss of cations (Na^+, Ca^{2+}) due to activated neuronal inward currents.

The consequence of a rapid decrease in extracellular pH is in marked contrast with the effect of slow acidification which depresses neuronal activity. Experiments with dissociated cell cultures (Krishtal and Pidoplichko 1980; Konnerth et al. 1987; Grantyn and Lux 1988) showed that neurons react to a rapid increase of $[H^+]_o$ by strong depolarization (Fig. 1) due to a transient Na current (Fig. 2A). This Na current in response to a step change of $[H^+]_o$, termed $I_{Na(H)}$, is unaffected by tetrodotoxin (TTX) but blocked by inorganic Ca^{2+} channel blockers. In chick dorsal root ganglion (DRG) neurons, voltage-activated Ca^{2+} currents were abolished during $I_{Na(H)}$ (Konnerth et al. 1987) within milliseconds (Davies et al. 1988).

In most neurons, $I_{Na(H)}$ is a large current as compared with other cationic inward currents. The conductance increase in response to elevation of $[H^+]_o$ lasts for several seconds. Depending on the amount of involved channels, this conductance may largely shunt the neuronal membrane, or cause a more restricted depolarization in specific dendritic compartments of the neuron. The former blocks the normal patterned activity,

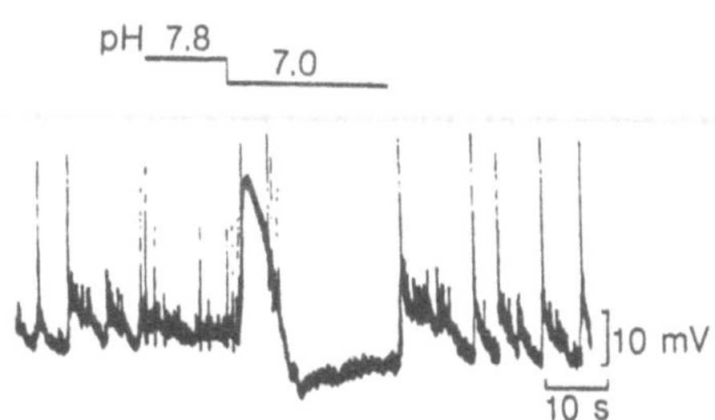

Fig. 1. Transient depolarization with subsequent suppression of spontaneous synaptic activity in a cultured neuron from the rat hypothalamus. After two weeks in vitro, these neuronal cultures form a synaptic network and generate spontaneous burst-like activity (Misgeld and Swandulla 1989). The record was taken in a HEPES-buffered salt solution containing 0.2 mM Ca^{2+} and 4 mM Mg^{2+}. A rapid pH step can easily be applied to cultured neurons by switching the solutions which superfuse the investigated cell via a closely placed multibarrel micropipette

as is the case during epileptic discharges, while the latter might be an effective means of enhancing the action of excitatory neurotransmitters.

In order to assess further the functional significance of INa(H), the following questions ought to find an answer: (a) Is INa(H) neuron specific? (b) What are the conditions for activation and inactivation of INa(H) in brain tissue? (c)) How is INa(H) regulated? These questions could be adressed partially by investigating different cells from the vertebrate central and peripheral nervous system.

INa(H) Represents a Specific Property of Neurons and Appears Early in Development

Our studies on INa(H) were carried out on mature (Konnerth et al. 1987; Davies et al. 1988) and developing (Gottmann et al. 1989) DRG neurons and on a variety of cells from the mammalian central nervous system (retinal ganglion cells, neurons from the rat tectum, hippocampus and neocortex). INa(H) can be elicited in any of these neurons, but not in rat cardiac myocytes and skeletal muscle fibers (Davies et al. 1988, and unpublished data). INa(H) of central neurons decays more slowly and also has a significantly higher sensitivity to the blocking action of divalent cations as compared with INa(H) of DRG neurons. However, the pH dependency of the activation and inactivation processes appears to be very similar in peripheral and central neurons. In developing neurons, INa(H) channels appear to be expressed ahead of Ca and Na channels (Gottmann et al. 1989; Grantyn et al. 1989) and also of channels activated by amino acids (Grantyn et al. 1989).

Activation of INa(H) Is Facilitated by Slow Alkalinization and Decreased Levels of Extracellular $[Ca^{2+}]o$ and $[Mg^{2+}]o$

In rat tectal neurons, half-maximal responses to $[H^+]o$ were generated by increasing the extracellular pH to 6.8 after applying a holding pH of 7.9. Decreasing the holding pH leads to progressive inactivation of INa(H), the latter being half complete at pH 7.7. Alkalinization enhances, thus, the sensitivity of the neuronal membrane to a sudden increase in $[H^+]o$.

The effect of $[Ca^{2+}]o$ on INa(H) is complex. By promoting alkalinization in situ (Chester et al. 1986) and by counteracting inactivation of INa(H) in cultured DRG neurons (Konnerth et al. 1987) extracellular Ca^{2+} may sensitize neurons to sudden elevation of $[H^+]o$. At the same time, Ca^{2+} acts as a channel blocker, the half-blocking concentration of Ca^{2+} being 1.25 mM for INa(H) of tectal neurons. In the latter cells, INa(H) was always markedly increased by reducing both $[Ca^{2+}]o$ and $[Mg^{2+}]o$ below the "physiological" level (Grantyn et al. 1988).

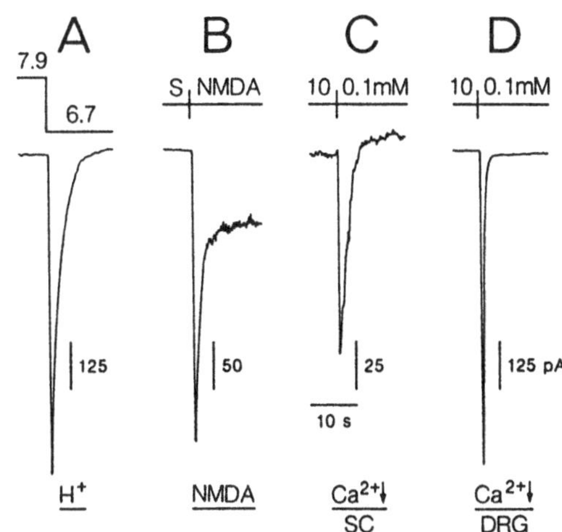

Fig. 2. Configuration of transient sodium inward currents, as induced by a rapid pH decrease (A), application of 200μM NMDA (B) and low Ca^{2+} solution (C, D). Giga-seal voltage-clamp recording from a cultured neuron of the rat superior colliculus (A-C). D, record from a cultured chick DRG neuron

Rapid Application of low Ca^{2+} Solutions or NMDA Activates a Current Similar to INa(H)

In view of the double effect of $[Ca^{2+}]_o$ on INa(H), it can be expected that the largest current amplitudes should be elicited by combining a step change from high to low $[Ca^{2+}]_o$ with an stepwise increase in $[H^+]_o$. This is, indeed, the case. Moreover, a sudden transition from high to low $[Ca^{2+}]_o$ in the absence of pH changes may activate a current similar to INa(H) (Hablitz et al. 1986; Fig. 2C).

In search for other factors which might activate currents similar to INa(H) in the absence of larger shifts in $[H^+]_o$ and $[Ca^{2+}]_o$, we tested a number of transmitter substances, including excitatory amino acids (EAA). It was found that *N*-methyl-D-aspartate (NMDA) at a concentration of 100- 200 μM induced transient sodium currents which decayed with the same time constant as INa(H) (Fig. 2B). This transient NMDA-activated sodium current INa(NMDA)T is produced by channels which are different from those generating the classical persistent current response to NMDA. In the presence of 1.5 mM $[Ca^{2+}]_o$ a voltage-dependent block was seen with the persistent component of INa(H) while the transient NMDA-activated conductance was not voltage-dependent, i.e. behaved as INa(H) (Grantyn and Lux 1988).

INa(NMDA)T also shares another feature with INa(H): it fully inactivates during exposure to low pH. Low pH reduces, to some extent, all ionic conductances. However, the effect on INa(NMDA)T is by far more pronounced than that on other currents, including the persistent NMDA-activated current (Fig. 3). Finally, it was shown that during INa(H) NMDA failed to elicit a transient sodium current, a finding which could be explained by convergent actions on proton and NMDA receptor sites at one channel type (Grantyn and Lux 1988). By contrast, the persistent NMDA-activated current displayed in rat tectal neurons a perfect summation with INa(H).

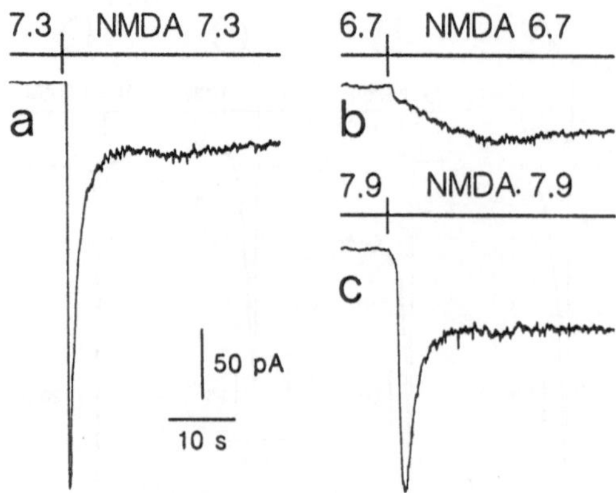

Fig. 3. Influence of pH on NMDA-activated currents in rat tectal neurons. Variation of holding pH together with the pH of NMDA-containing solution. Note complete inactivation of INa(NMDA)T in (b)

INa(H) Is Antagonized by Ca^{2+} and NMDA Antagonists

A pharmacological investigation of INa(H) was undertaken to test the effect of known selective blockers of low-voltage activated Ca^{2+} currents (nickel, amiloride), high-voltage-activated Ca^{2+} currents (cadmium, ω-conotoxin) and NMDA-activated currents (D-2-amino-5-phosphonovaleric acid). It was found that all these substances reduced INa(H), the most effective agent being the amiloride derivative 5-aminodiethylamiloride (ADEA). ADEA is a blocker of the Na^+/H^+ exchanger in rat brain synaptosomes (Jean et al. 1985) and a selective antagonist of low-voltage activated Ca^{2+} currents (Tang et al. 1988). So far, no substance has been found which would block INa(H) without affecting any other ionic conductance.

Our demonstration of a proton-activated sodium current in mammalian central neurons raises the question about the functional consequences that such conductance may have in the brain. INa(H) is likely to contribute to epileptic activity. It is in this regard remarkable that tonic ictal episodes, as induced in the hippocampal slice by superfusion with Ca^{2+}-free solution, were associated with ionic currents that had the typical time course of INa(H) (Yaari and Konnerth, personal communication).

At present, a physiological role of INa(H) is more difficult to elucidate. We showed that NMDA is capable of activating a current with the characteristics of INa(H) in the absence of large deviations from physiological levels of $[H^+]_o$ or $[Ca^{2+}]_o$. This finding suggests that colocated NMDA and proton receptors could mediate a coupled gating process which first leads to strong depolarization by activation of INa(H) or INa(NMDA)T and then to Ca^{2+} inflow by opening voltage-dependent NMDA channels that underly the persistent component. This coupled activation process may be crucial for long-term modification of neuronal release sites and thereby determine the efficacy of synaptic transmission in developing and mature brain circuits.

References

Chester M, Chan CY, Nicholson C (1986) Stimulus evoked alkaline shifts in the in vitro turtle cerebellum. Abstr Neurosci Soc 12:696

Davies NW, Lux HD, Morad M (1988) Site and mechanism of activation of proton-induced sodium current in chick dorsal root ganglion neurones. J Physiol (Lond) 400:159-187

Gottmann K, Dietzel ID, Lux HD, Ruedel C (1989) Proton-induced Na^+ current is present in neuronal precursor cells prior to the development of voltage-dependent Na^+ current and high voltage-activated Ca^{2+} current. Neurosci Lett (in press)

Grantyn R, Lux HD (1988) Similarity and mutual exclusion of NMDA- and proton-activated transient Na^+ currents in rat tectal neurons. Neurosci Lett 89:198-203

Grantyn R, Perouansky M, Lux HD (1988) Proton-gated sodium currents in mammalian central neurons. Abstr Neurosci Soc 14:597

Grantyn R, Perouansky M, Rodriguez-Tebar A, Lux HD (1989) Expression of depolarizing voltage- and transmitter-actived currents in neuronal precursor cells from the rat brain is preceded by a proton-activated sodium current. Dev Brain Res 49:150-155

Hablitz JJ, Heinemann U, Lux HD (1986) Step reductions in extracellular Ca^{2+} activate a transient inward current in chick dorsal root ganglion cells. Biophys J 50:753-757

Jean T, Frelin C, Vigne P, Barbry P, Lazdunski M (1985) Biochemical properties of the Na^+/H^+ exchange system in rat brain synaptosomes. J Biol Chem 260:9678-9684

Konnerth A, Lux HD, Morad M (1987) Proton-induced transformation of calcium channel in chick dorsal root ganglion cells. J Physiol (Lond) 386:603- 633

Kraig RP, Ferreira-Filho CR, Nicholson C (1983) Alkaline and acid transients in cerebellar microenvironment. J Neurophysiol 49:831-850

Krishtal OA, Pidoplichko VI (1980) A receptor for protons in the nerve cell membrane. Neuroscience 5:2325-2327

Misgeld U, Swandulla D (1989) Quisqualate receptor-mediated rhytmic bursting of rat hypothalamic neurons in dissociated cell culture. Neurosci Lett 98:291-296

Tang CM, Presser F, Morad M (1988) Amiloride selectively blocks the low threshold (T) calcium channel. Science 240:213-215

The Impact of Advanced Computing on Medicine

W. Witschel
Department of Theoretical Chemistry, University of Ulm, Oberer Eselsberg, 7900 Ulm, FRG

Introduction: Advanced Computing in the Biosciences in the USA and Deficits in German Medical Education

Computers are used in the medical clinic mainly in connection with laboratory automatization or clinic administration. Examples for the application of advanced computing in medical research are isolated. One reason is the conservative computer policy, which allows only difficult access to a central main-frame for medical students and research workers. This is in contrast to both education and research in the USA, where widely distributed high-level computing with good graphics facilities encourages program use and development in the life sciences. It is no surprise that both hardware and software for in the computational life sciences were developed in the USA, and to a lesser extent in Oxford, Cambridge, Groningen and Brussels. In the USA, BIOCOMP is strongly supported by the NIH and NIH grants, as well as by private foundations. BIOCOMP represents the application of powerful computers with excellent graphics and user-friendly, self-explanatory software to the life sciences. Let me list briefly some examples from medical computations in relation to drugs and clinical pharmacology which I found in connection with the preparation of a technical report to the European Community.

Database and molecular properties for protein crystallography include the following:

- Determination of macromolecular structures by X-ray crystallography and two-dimensional electron density maps
- Analysis of primary sequence data and their correlation with structure and function (receptor modelling)
- Molecular simulation and molecular graphics.

Mathematical modelling of biological systems for neurochemistry and neurophysics of the brain can be used in the following analysis:

- Three-dimensional analysis of dynamic processes such as blood flow, cell locomotion and tissue deformation
- Diffusion of ions and transport through cell membranes
- Transport through porous media, filtration
- Fate of toxic materials (e.g. drugs) in the body or organ compartments

- Solution of fluid flow or diffusion equations in three dimensions with boundary conditions from absorption, breakdown and transport
- Brown-motion dynamics, motion of ions in channels

As some of these applications of BIOCOMP need a strong theoretical background which cannot be sketched here, I shall concentrate on the following topics: computers and local area networks; suitable software for the modelling of drugs and drug receptor interactions; and some suggestions for the education of medical students in BIOCOMP on a voluntary basis.

Survey of Computers

What Is Needed from the Computer?

Various techniques of computation are linked to the nature of the experiment, which is also strongly related to various disciplines (Figure 1); this include:

- Chemo- and biometrics - techniques for laboratory automatization
- Experiments - sketched in Figure 1b
- Computer Experiments - calculations of receptor-drug interactions, transport and metabolism of drugs
- Number crunching - large-scale numerical calculations of molecular properties, molecular dynamics of proteins and dynamics in vivo processes
- Graphics - the graphical representation of the results on high-speed high-resolution multicolour workstations

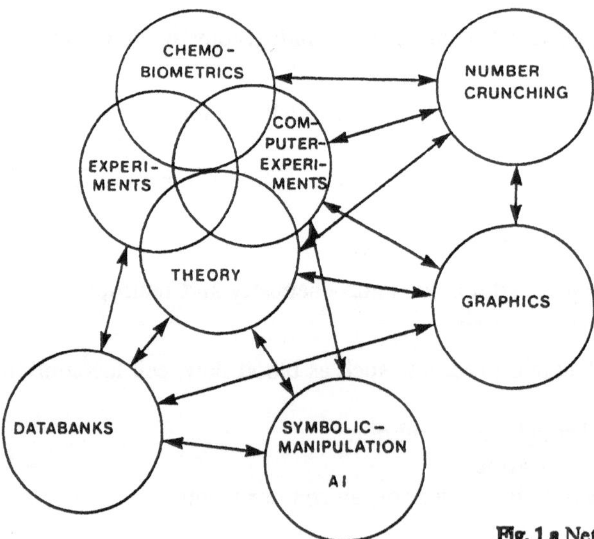

Fig. 1 a Network between computing and experiment

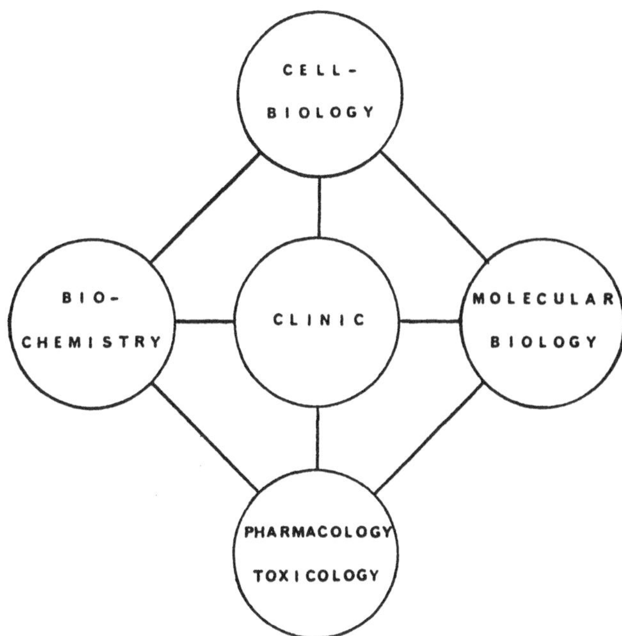

Fig. 1 b Experiments

- Symbolic manipulation or artificial intelligence (AI) - of increasing important not only for diagnostics such as in the program MYCIN but also for pharmatherapy; based on heuristic medical rules
- Databanks - storing not only the relevant literature, but also the numerical and diagnostic data and observations; made "intelligent" by methods of AI

Computer Organization

Modern computing is characterized by networking of distributed computers. As an example, we sketch our own department and university network (Figure 2), which of course is only an approximation to ideal plans, restricted by budget problems. Though only two years old, the hardware described in Figure 2 has already become outdated by new micro supercomputers or personal supercomputers such as following:

- APOLLO 10000: high performance, good graphics, no software available.
- ARDENT: high performance, no software available.
- HP 9000-835SRX: medium performance, excellent graphics, only two program packages available.
- SILICON GRAPHICS: several models ranging from a personal workstation 4P-20 to the multiprocessor high-performance 4D-120; excellent graphics and an excellent software catalogue for molecular modelling and protein design. This system represents at present the state of the art.
- STELLAR: High performance, good graphics. Though only two months in the market, this system already offers some standard program packages.

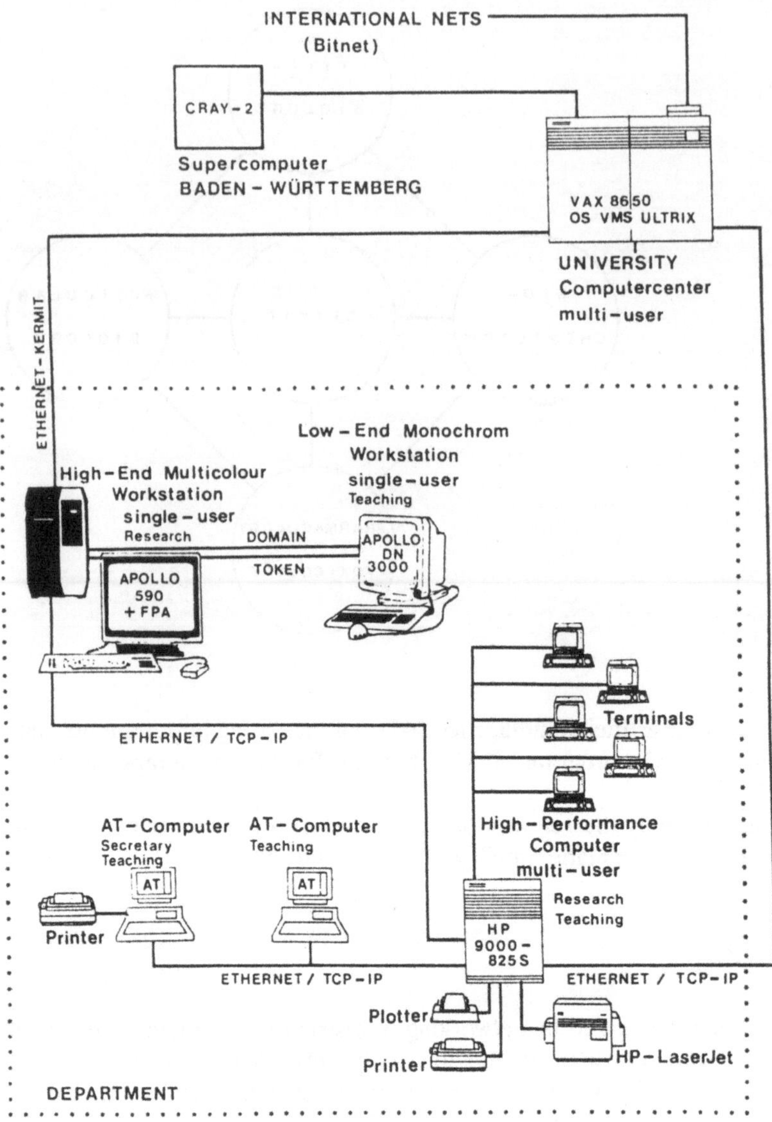

Fig. 2 Department network at the university

In buying a workstation one should consider:

- Available software
- Reliability
- Maintenance and maintenance costs
- Price/performance ratio
- Possibility of upgrading (which can be crucial because of the hard competition in the market)
- Probability that the company will be in business in the forseeable future

Software

For the effective application of BIOCOMP one needs a toolbox with various methods (Figure 3). We cannot go into details of the toolbox as it depends strongly on the level of BIOCOMP. There are a number of self-explanatory programs which can be understood and applied immediately, whereas others on the research level need considerable theoretical background which should be taught jointly by professors of science and medicine. The contents of the toolbox are explained in detail in the references given on molecular modelling. As the printing of colour plates is very expensive, we omit here detailed examples and refer to the references given below. We shall mention two examples: quantum chemistry and databanks.

Quantum chemistry is the numerical solution of the molecular Schrödinger equation in various approximations. It allows calculation without experimental measurements (at various levels of accuracy) of molecular properties relevant for BIOCOMP:

- Molecular structure in ground and excited states
- Electronic structure
- The kinetics of chemical reactions
- Molecule-solvent interaction
- Electrical and magnetic properties
- Thermodynamic properties
- Charge density and molecular interactions
- Conformations by seeking the potential energy minimum for internal rotations
- Vibration properties

Databanks contain the ordered information of structure, physical and biological properties and references. Methods of artificial intelligence using heuristic rules can be used to combine the properties of various substance classes. Examples of databanks used in BIOCOMP include

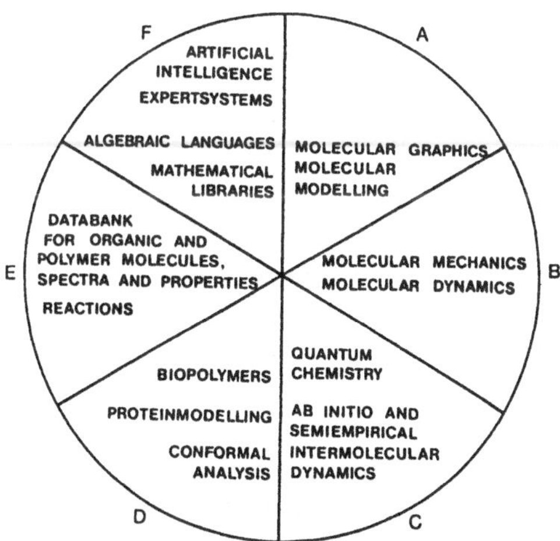

Fig. 3 Software toolbox

- Brookhaven Protein Database: cartesian coordinates of 274 macromolecules determined by X-ray diffraction.
- Cambridge Structural Database: structural and crystallographic information on 45000 organic and metal organic compounds. A set of search, retrieval, analysis, graphics and plotting programs is available.
- Genbank: a database which contains complete nucleotide sequences of some 6500 RNA and DNA sequences.

References

As the field of computer applications to the life sciences is new, I shall give a number of references to introduce the physician to the problem of understanding drug action and BIOCOMP. DuBois et al. (1985) give a survey on chemical ideograms and molecular graphics. McGammon and Harvey (1987), Hol (1986), Kollman (1987), Frühbeis et al. (1987) and Richards (1988) reviewed molecular modelling. Radunz (1988), Mager (1988) and Klebe (1988) give introductions to the understanding and optimization of drug action. Richards' (1983) monograph is an excellent introduction to quantum pharmacology, which is supplemented from the technical point by Clark (1985). Molecular mechanisms of drug action by BIOCOMP are reviewed by Fletterick and Zoller (1986) and surveyed by Coulson (1988).

Some Suggestions for the Introduction of BIOCOMP in Medicine

Progress in the introduction of BIOCOMP in the area of medicine can be made along two lines:

- A surprisingly large number of medical students is interested in modern methods, both experimental and computational/theoretical, which they want to link to experiences and observations in the clinic. Unfortunately, in the FRG a joint MD-Ph.D. program is not possible. However, it should be possible that professors from medical departments and science departments join in research and education. Joint courses in theoretical and practical aspects of BIOCOMP should be given for interested students on a voluntary (or "with honours") basis.
- Medical students should be given similar opportunities as science or engineering students to use high-power graphic workstations with user-friendly software at their working places. They would learn to use these new tools by doing. As the gap between certain fields in chemical research has widened between the USA and Europe, strong efforts in this field are necessary.

Summary

Molecular modelling and molecular graphic, together with X-ray crystallography, two-dimensional NMR and protein chemistry, are important tools in drug design. It is suggested that the same techniques can also be used as tools in the understanding of drug action in the medical clinic. Hardware and software together with applications are briefly surveyed.

Acknowledgements. I wish to thank Prof. Kornhuber for the many stimulating discussions we have had on neurochemistry and drug design. Part of this work was supported by the Merckle GmbH, Blaubeuren, FRG.

References

Clark, T (1985) A handbook of computational chemistry. Wiley, New York

Coulson, C. J (1988) Molecular mechanisms of drug action. Taylor and Francis, London

DuBois J E, Laurent D and Weber J (1985) Chemical ideograms and molecular computer graphics. visual comput 1:49

Fletterick R and Zoller M (1986) eds Computer graphics and molecular modelling, Cold Spring Habor Laboratories, Cold Spring Harbor

Frühbeis H, Klein R, Wallmeier H. (1987) Computergestütztes Moleküldesign - ein Überblick. Angew Chem 99:413

Hol W G (1986) Proteinkristallographie und Computergraphik auf dem Weg zu einer planvollen Arzneimittelentwicklung, Angew Chem 98:765

Klebe G (1988) Neue Strategien bei der Wirkstoffsuche. Arzeimittelforschung 38:484

Kollman P (1987) Molecular modelling. Ann Rev Phys Chem 38:303

Mager P P (1988) Zur Optimierung von bioaktiven Leitstrukturen I-III. Pharma Unserer Zeit 17:106,129,177

McGammon J A Harvey S C (1987) Dynamics of proteins and nucleic acids. Cambridge University Press, Cambridge

Radunz H E (1988) Vom Screening zum Drug Design. Moderne Methoden der Wirkstoff-Findung. Pharm unserer Zeit 17:161

Richards W G (1983) Quantum pharmacology. Butterworth, London

Richards W G (1988) Computer-aided molecular design. Sci Prog 72:481

Biomagnetic Measuring Technique: State of the Art and Prospects for the Future

U. Eckener
Dornier GmbH, POB 1420, 7990 Friedrichshafen, FRG

Biomagnetism

Humans, and other living beings, show magnetic effects by two principles:

- Magnetic flux as a cophenomenon of electrical currents due to electrical activity of neurons, ion flow in the muscles, etc.
- Magnetic induction, due to varying magnetic susceptibilities of different tissues or components in the organism

Table 1 shows the most essential of these effects. The figures in parentheses give the order of magnitude of the relative magnetic field. For comparison, the magnetic field of earth magnetism is shown. The signals differ by five orders of magnitude.

The detection and separation of these signals afford accordingly high sensitivity and dynamic range of the measuring system. This is realized by the SQUID magnetometry, which is based on quantum mechanical effects (SQUID = superconductive quantum interference device).

SQUID magnetometry

The principle of the SQUID is shown in Figure 1. The heart of the device is a superconductive ring. Due to a quantum mechanical effect, the electrons in the material, for example Niob, come into a collective state when the material is cooled to very low temperature (approximately 9° K). In this state, they can form an electrical current without any ohmic losses, hence no voltage is necessary to keep the current running. This state is called superconductivity.

When the magnetic field around the superconductive ring changes, a current starts by itself. The magnitude of this current is such that it compensates the external field changes. Hence the magnetic flux in the area enclosed by the ring stays equal, for example zero. This is called the Meissner effect.

When the ring is prepared with a "weak link", i.e. a thin contact which is not superconductive, the electrons can pass this link by tunneling. (That is a quantum mechanical effect: even if there is an insurmountable potential barrier, the electron has a certain probability of existence beyond the barrier.)

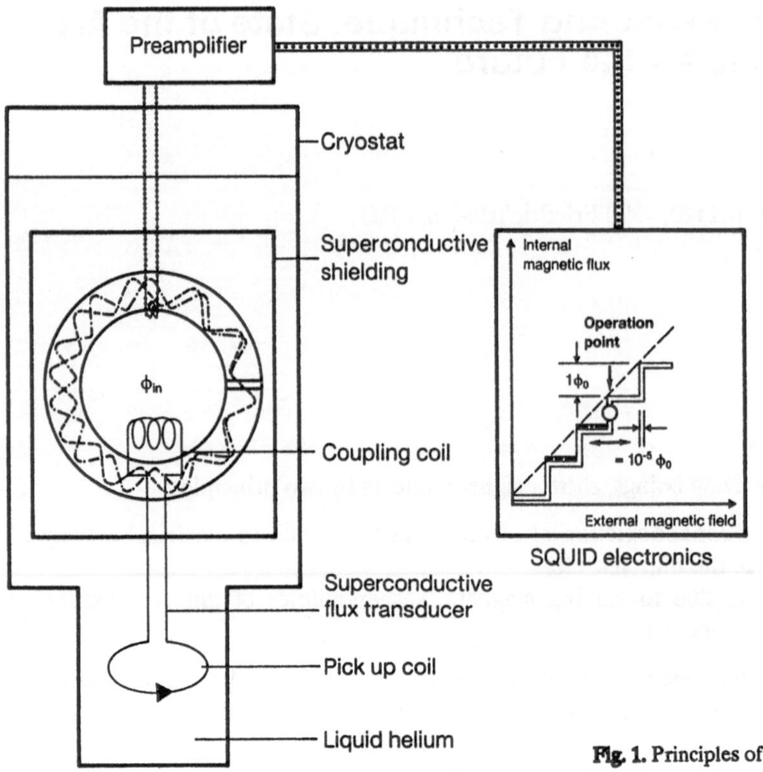

Fig. 1. Principles of SQUID magnetometry

This works until a certain maximum current which cannot be exceeded if the contact is reached. Thus, when the magnetic field increases further, the current breaks down for a moment, admits the entering of one flux quantum in the inner area of the ring and continues magnetic shielding at a lower current rate. This is called the Josephson effect.

The short normal conductive state can be detected by an external, inductively coupled electronic circuit, by which a single magnetic flux quantum is counted. This is the only known quantum mechanical phenomenon which can be detected directly by a macroscopic effect.

Since a flux quantum is a very small portion, the device has the high resolution of about 10^{-11} T. By a further electronic trick an interpolation of the actual shielding current between two counting steps is possible, thus improving the resolution to about 10^{-16} T.

Up to now this has remained a theoretical value because a major problem is the discrimination between the signal to be measured and the background noise. One way for reducing magnetic noise is the use of a shielding chamber, which to a certain extent shields external magnetism. A second way is constructing the pickup coils which couple the magnetic signal from the source to the SQUID such that they do not pick up the magnetic field itself but the first or the second spatial derivative of the magnetic field. Magnetic noise from distant sources is eliminated by this measure, but the sensitivity decreases by a factor of 10 or 100.

Magnetoencephalography

The SQUID magnetometry is a highly promising tool for studying the brain activity in research and in diagnosis.

The additional information and better localization of regional brain activity supplied by the magnetoencephalography is to be used in three important fields of neurology and to improve medical diagnoses and contribute to disease prevention by early recognition of dangerous states. These are the early recognition and localization of epileptic activity, early recognition of endogenous psychoses, and the investigation of cerebral activity in the context of deliberate motor performances and learning processes (Deecke and Kornhuber 1978; Kornhuber 1985).

The common denominator of these fields, with regards to magnetoencephalography, is the expectancy that the source of normal and pathological activities, especially in deep-seated brain structures, can be more easily found and localized than has been possible with the EEG. When compared to other localization methods being developed, magnetoencephalography offers one distinct advantage: It has an extremely high time resolution (in the range of 1 ms), whereas other diagnostic tools based on metabolism (measurement of the regional blood circulation in the brain) or positron-emission tomography require more than ten seconds.

The advantage of magnetoencephalography in comparison to electroencephalography lies in the fact that the magnetic field is not shielded or distorted by the organic tissue of the head in contrast to the electric field. Thus the location of active neurons also from inner areas of the brain seems to be possible up to a spatial resolution of a few millimeters (Fig. 2).

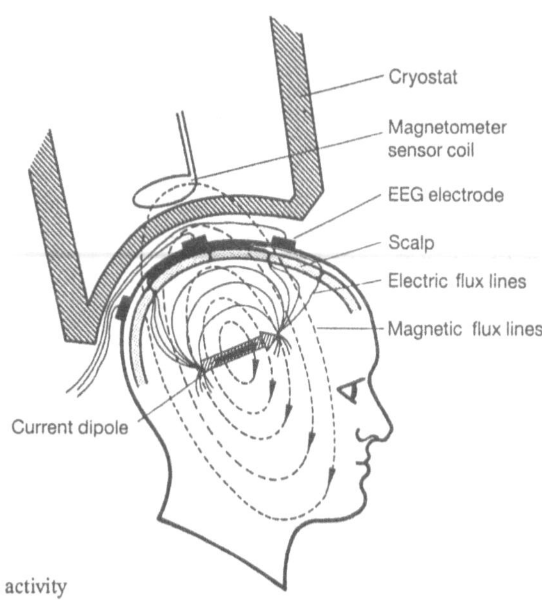

Fig. 2. Electric and magnetic field due to brain activity

Since many sources are active at a time, measurings from 20 or more different spots must be made simultaneously and correlated in a mathematical procedure in order to find the locations of the individual sources. This is a rather complicated mathematical problem which leads in principle to ambiguous sets of solutions. Additional boundary conditions must be included in order to decide which set of solutions represents the physical reality. Joint theoretical and experimental research on this topic is taking place in New York, Los Angeles, Rome, Berlin and at Dornier as well. One device, representing today's state of the art is at the Neurological University Clinic in Vienna.

Present Work and Prospects

The device that Dornier is developing for Professor Kornhuber's institute is to expand the present state of the art by the following features:

- 28 SQUID channels [2 x (11 + 3)].
- Over 190 dB dynamic range.
- Tangential and normal components of the magnetic field are picked up (up to now normal only).
- The SQUID channels can be distributed optionally to 96 + 6 pickup coils placed at 32 + 2 different spots. This enables the user to ascertain the most suitable configuration for a specific task.
- The coils are designed as magnetometers. Spatial derivatives are determined mathematically, not by gradiometers. Thus more flexibility is achieved.
- The data processing includes simultaneously measured EEG values.
- The data processing includes individual spheric shell approximation of the head and in a next step a near-real geometric approximation of the head.

Up to now, SQUID systems involve a relatively complicated apparatus, mostly designed for one special diagnostic purpose. On-line imaging of measurements, in a three-dimensional or multiplane presentation is not yet possible due to the high calculation effort. The need of liquid helium for cooling of the SQUID is practicable but a bit troublesome.

Development of future equipment towards miniaturized SQUID systems, allowing for 100 or more channels in smaller and more flexible cryocontainers with less cooling effort and at lower cost. Faster computers, time-saving algorithms and direct transfer of digitized geometric data into the processing loop are further goals for a future SQUID system.

The anticipation of the new perowskite-type high-temperature superconductive materials might be too euphoric because of the noise problems at higher temperature and because of their very early state of development at present. Nevertheless, we are working on these problems and keeping SQUID application in mind.

The Ulm project will yield further scientific, interpretational and empirical knowledge, subtly differentiated requirements and new incentives for biomagnetic diagnostic methods. Although we no longer have the "Ulmer Schachteln" carrying humans and human ideas down the river from Ulm, we have Hans Helmut Kornhuber as a compass for present and future efforts.

Table 1. Biomagnetic phenomena

Neuromagnetism		
	- Spontaneous activity	($10^{-12}/10^{-14}$T)
	- Schizophrenia, endog. psych.	
	- Epilepsy	
	- Motor/lingual diseases	
	evoked potentials	(10^{-14}T)
	- Visual	
	- Auditory	
Cardiomagnetism		
	- Adult	(10^{-11}T)
	- Fetal (FHR)	(10^{-12}T)
Myomagnetism		(10^{-11}T)
Liver iron load		
	- Hemosiderosis	(10^{-9}T)*
	- Thalassemia	
Lung	- Contaminants	($10^{-9}/10^{-11}$T)*
	- Pneumography	
	- Clearance	
Magnetic field of earth		(10^{-5}T)

* Depends on external stimulating field

Summary

Magnetic signals of an organism contain significant information which can be used in diagnosis. Since these signals are weak and overlayed and disturbed by the "magnetic noise" of the environment, sophisticated tools are required to make use of them. The SQUID magnetometry is an adequate measuring method for this purpose. For a current project of the University of Ulm, Department of Neurology (Prof. Kornhuber), equipment for neuromagnetic measurements based on an advanced SQUID technology is being developed.

References

Cohen D (1968) Magnetoencephalography: evidence of magnetic fields produced by alpha rhythms currents; Science 161:784

Williamson S J, Kaufmann L (1981) Biomagnetism. J. Magn Mater 22:129

Deecke L, Kornhuber H H (1978) An electrical sign of participation of the mesial supplementary motor cortex in human voluntary finger movement. Brain Res 159:473-476

Kornhuber H H (1985) Zur Pathophysiologie und Therapie der Schizophrenie. In: Huber G (ed) Basisstadien endogener Psychosen und das Borderline-Problem. Schattauer, Stuttgart

Part 6

Neurological Sciences I

(Psychiatry)

Advances in Schizophrenia Research

G. Huber
Psychiatrische Klinik und Poliklinik der Universität Bonn, 5300 Bonn 1, FRG

Schizophrenia is mainly genetic. But environmental factors, both biological and psychosocial, play a role in manifestation and even more in course and outcome (Huber 1980). No overrepresentation of lower social classes was found in the Bonn study (Huber et al. 1979) with respect to the social class of the parents and the highest social class achieved by the patient before the onset of the disease. Only at the time of catamnestic follow-up does there occur an unequal distribution in favor of the lower social class. This is a consequence of the illness in the sense of the "drift hypothesis".

Most frequently the disease begins in the third decade of life, followed in frequency by the fourth and second decades. The age at onset is lower in men than in women: 70% of men but only 47% of women fell ill before the age of 30. The outcome in males is poorer than in females, but without statistical significance. Females fell ill more frequently during the post-partum period and more rarely during pregnancy. The life-time risk is the same for both sexes. However some findings of the Bonn study suggest that schizophrenia is more common among women.

Precipitating factors seem to represent necessary but not sufficient conditions for the manifestation of psychotic episodes in about a quarter of the patients. The Bonn findings support the assumption that psychological precipitants may be regarded as favorable with regard to long-term outcome. The intra- and interindividual variability of schizophrenia is very marked in most measures, especially with respect to course and outcome. Patients who became ill before the psychopharmacological era had a poorer prognosis than patients who became ill after 1951. Patients who were somatically treated within one year after onset, including that of prodromes, had a highly significant better long-term prognosis than those treated only later or never at all (Gross et al. 1983).

Long-Term Course and Outcome

Until recently there have been no results of life-long studies that approach general applicability. The three European long-term investigations, the Zurich (Bleuler 1972), Lausanne (Ciompi and Müller 1976) and Bonn studies, agree on the following points. Diagnosis must be made independently of outcome because investigations of long-term outcome would be meaningless if an unfavorable outcome were used as an obligatory criterion *for* and a favorable outcome *against* the diagnosis of schizophrenia. Several factors were found in the European long-term studies as prognostically favorable. These factors are identical with criteria used to classify schizoaffective, schizophreniform or cycloid psychoses, e.g. acute onset, depressive syndromes, precipitating life events and not abnormal primary personality (Gross et al. 1986b).

The average duration of illness in the Bonn study was 22.4 years. In the Bonn and Zurich study 22% showed complete psychopathological remissions; 43% noncharacteristic types of remission, mainly slight pure deficiency syndromes ("pure defect") determined by dynamic and cognitive basic symptoms; only 35% revealed characteristic or typical schizophrenic psychoses. Approximately 90% were living permanently at home at the last follow-up; two-thirds of them had for many years no longer been under professional care. 56% were socially recovered, i.e. fully employed, yet only two-thirds of them were at previous occupational level and one-third below the premorbid level (Huber et al. 1979).

Decades after the onset more than half are fullly employed and show full remission or slight pure residues. There are four groups of course types, each accounting for one quarter of all schizophrenics: a favorable group with nearly 100% social remissions, a relatively favorable group, a relatively unfavorable group, and finally an unfavorable group in which the social recovery rate drops to 2% in course type XII. This course type includes the so-called schizophrenic catastrophes, developing within the first 3 years of the illness - 4% in the Bonn study, a rate lower than that found in earlier times (e.g. Bleuler 1941).

The results of the European long-term studies lead to a revision of three classical opinions about schizophrenia: the doctrines of incurability, of incessant progression and of the fundamental heterogeneity and numinous singularity of schizophrenia. The long-term outcome is largely independent of the duration of the illness. There is no increasing deterioration in the third, fourth and fifth decades of the disease. Only the shortest courses from 10 to 14 years tend to be more favorable because of neuroleptic treatment in this group, as the data of the Bonn study indicate. The illness often shows the trend towards improvement in the sense of the so-called "second, positive bend" still 20 to 40 years after onset. The results of the Bonn study allow the statement that most of the schizophrenics are most of time not schizophrenic, i.e. guided by the psychopathological cross-sectional picture one cannot diagnose schizophrenia.

According to Zubin (1987) it is an important result of the European long-term studies that the outcome is far more favorable than psychiatrists had assumed previously. Zubin thinks the results of these studies mean a revolution in our knowledge about schizophrenia. They have provided, as Zubin wrote in 1987, a proclamation of emancipation for schizophrenia from the yoke of inevitable chronicity not unlike Abraham Lincoln's proclamation of emancipation for the American slaves. In the United States these results of the Zurich and Bonn studies were confirmed by the Vermont study (Harding and Strauss 1985).

Concept of Basic Symptoms and Basic Stages

The concept of basic symptoms has been developed in the 1950s starting with the disclosure and description of cenesthetic, central-vegetative and perception disorders

(Huber 1957; Gross and Huber 1972), and has led to a new doctrine regarding symptoms in schizophrenias (Huber 1983, 1986a; Süllwold and Huber 1986). The dynamic and cognitive basic symptoms are deficiencies, subjectively experienced as deficiencies and impairments, missed before the disease in intraindividual comparison. The prepsychotic prodromes and outpost syndromes and the postpsychotic reversible or irreversible basic stages are determined by basic symptoms. Because of the far-reaching overlapping of the phenomenological picture of the pre- and postpsychotic basic stages, it was possible to construct the Bonn Scale for the Assessment of Basis Symptoms (BSABS) for a standardized documentation of all types of basic stages (Gross et al. 1987). For theory and etiology but also for therapy it is very important that the basic deficiencies really form the basis of the schizophrenic first-rank symptoms. We could demonstrate transitions between productive-psychotic phenomena and episodes and the preceding and following basic symptoms that constitute the pre- and postpsychotic basic stages. The symptoms can pass over in the same patient from uncharacteristic level 1 to more or less characteristic level 2 into typical schizophrenic symptomatology of level 3, including Schneider's first-rank symptoms, and vice versa. As long as no fixation takes place in level 3, this development is reversible. Follow-ups proved the existence of psychopathological transitions between basic symptoms and psychosis or, in the usual but misleading terminology, between negative and positive schizophrenia (Huber 1966, 1985; Huber et al. 1979; Gross et al. 1986; Klosterkötter 1988; for overview see : Süllwold and Huber 1986).

Each basic symptom is in the Bonn schedule operationally defined, commented upon and illustrated by characteristic statements. The Bonn scale rests upon the complaints and deficiencies experienced and related by the patients as deficiencies, whereas, for instance, the Iowa Scale (SANS; Andreasen and Olsen 1982) is based almost completely on the observation of the patients by the investigator. The cognitive level 2 basic symptoms show a very marked intraindividual fluctuation and thus very different degrees of "process activity," a term defined according to actual phenomenological criteria in 1968 and modified recently (Huber and Penin 1968; Gross et al. 1988). Occurrence and increase in basic symptoms depend on precipitating factors, e.g. mental or physical strain, certain everyday social situations and emotionally affecting minimal occasions. Here precipitating environmental factors are more or less "normal stressors," overcharging the morbogenic diminished capacity of working-up information. The objection that basic symptoms are not specific for schizophrenia overlooks, among others, three points: First, a specific psychopathological symptomatology does not exist at all, for example first-rank symptoms and also the basic symptoms themselves occur also in well-known definable brain diseases and not exclusively in schizophrenias. Second, descriptions and definitions given in the BSABS prove the more or less characteristic quality for the majority of basic symptoms, especially for cognitive basic deficiencies, phenomena which occur also in organic brain diseases but not in neurotic or psychopathic personality disorders. Third, we could prove by follow-up that and in what manner subjectively experienced basic symptoms pass over into productive-psychotic phenomena (first-rank symptoms), i.e. in schizophrenic symptoms in the conventional sense.

The basic symptoms were assumed to be closer to the hypothesized somatic substrate than the most more complicated psychotic "end-phenomena" (Huber 1966). Many data support the assumption that basic symptoms are the consequence of disorders of information processing and - genetically determined - biochemical abnormalities in the limbic system (Gross et al. 1973).

Disorder of Information Processing?

Hints for this hypothesis are, among others, the results of psychological, neurophysiological, EEG, neurochemical and morphological investigations and the observation that the basic symptoms occur - even if rarely - in organic brain diseases, preferentially of the limbic system (Huber 1982; for overview see: Süllwold and Huber 1986). Patients with pure deficiency syndromes show a decrease in psychological tests which correlates with the subjectively experienced basic symptoms confirmable with the Bonn Scale or the Frankfurt Questionnaire. EEG deviations, for example, abnormal rhythmicity (theta- or alpha-parenrhythmia), related to the limbic system were found in process-active basic stages (Penin et al. 1982). Quantitative EEG studies concerned with the reactivity of the brain to stimuli argue in favor of a disorder in the later phases of information processing (Koukkou 1984; Huber 1986a). Six EEG signs of schizophrenia were described by Kornhuber's group, for instance, that the Bereitschaftspotential lasts twice as long as in normals (Kornhuber et al. 1984). The higher degree of intraindividual variability of the electric states may be due to diminished stability of attention and unsteadiness of dynamic drive, correlating with the variance in reaction time and attention tests and the dysmetria of saccadic eye movements (Diekmann et al. 1985; Kornhuber 1985). According to the basic symptom concept, positive symptoms cannot be clearly separated from so-called negative symptoms. Patients with basic symptoms, i.e. with "negative schizophrenia", can develop positive schizophrenia and vice versa. There is a very marked intraindividual fluctuation between the three levels of basic symptoms. Therefore, "negative" basic symptoms of level 2, especially most of the cognitive basic deficiencies, are strictly speaking "positive" symptoms in statu nascendi ("microproductivity").

Neurochemical Findings

Some neurochemical findings may also speak for alterations in limbic areas. We found that untreated patients with high process activity compared with inactive patients and normal probands, showed significantly higher levels of DA and NA and at the same time significantly lower levels of TSH and triiodothyronine (Gross et al. 1988). Drugs inducing an inactivation of hippocampus, e.g. amphetamine and phencyclidine, may cause symptomatic schizophrenia. Because PCP acts at glutamate receptors it can be valuable in producing a model for schizophrenia. In this context the hypothesis that the primary cause of the schizophrenia is not a hyperfunction of dopaminergic but a hypofunction of glutamatergic systems, as suggested by Kornhuber, is an area worthy of more research.

Symptomatic schizophrenias are observed in temporal lobe epilepsy. Paroxysmal transition syndromes, such as the aura prolongata may be another phenomenological model for schizophrenia. The list of aura symptoms described by Wieser in stereoelectroencephalographic studies is largely identical with basic symptoms described in the Bonn Scale (Gross et al. 1987).

Morphological findings

Starting in 1953 we identified in psychopathological-pneumoencephalographic studies a subgroup of schizophrenias with irreversible pure deficiency stages ("pure defect") and a slight ventricular enlargement. Our CAT scan investigations confirmed these findings (Gross et al. 1982). Meanwhile about 50 studies have shown CT brain abnormalities. With quantitative-morphometric methods Bogerts (1985) could show a significant decrease in volume of amygdala and hippocampus and of the wall structures of the third ventricle. These neuropathological findings, corresponding the neuroradiological deviations, point to an atrophy or hypoplasia in the mentioned areas of limbic system (see Huber 1985).

Final remark

The assumption of a genetically based brain disorder acting via neurobiochemical mechanisms and susceptible to certain kinds of nonspecific everyday stress is quite compatible with the results of the cited and other findings. Here, of course, we could not organize and summarize the mass of information and discuss and appreciate the diversity of approaches and findings about schizophrenias. Perhaps only today are many of us old and wise enough to understand fully Kornhuber's message of 1971 (in Huber 1971) as he asked us to bear in mind even in the search for the causes of schizophrenia that Psychiatry is much too difficult for the psychiatrists.

Summary

Many thousands of papers and books have been published on schizophrenia. There are not very many facts consistently found. In the following I comment upon only a few points, some facts rather consistently found and some results belonging in our opinion to the advances in research about "the schizophrenias" in recent years.

References

Andreasen NC, Olsen S (1982) Negative versus positive schizophrenia. Arch Gen Psychiatry 39: 789-94

Bleuler M (1941) Krankheitsverlauf, Persönlichkeit und Verwandschaft Schizophrener und ihrer gegenseitigen Beziehungen. Thieme, Leipzig

Bleuler M (1972) Die schizophrenen Geistesstörungen im Lichte langjähriger Kranken- und Familiengeschichten. Thieme, Stuttgart

Bogerts B (1985) Schizophrenie als Erkrankung des limbischen Systems. In: Huber G (ed) Basisstadien endogener Psychosen und das Borderline-Problem. Schattauer, Stuttgart.

Ciompi L, Müller C (1976) Lebensweg und Alter der Schizophrenen. Eine katamnestische Langzeitstudie bis ins Senium. Springer, Berlin Heidelberg New York (Monographien aus dem Gesamtgebiet der Psychiatrie, vol 12)

Diekmann V, Reinke W, Grötzinger B, Westphal KP, Kornhuber HH (1985) Diminished order in the EEG of schizophrenic patients. Naturwissenschaften 72: 541-542

Gross G, Huber G (1972) Sensorische Störungen bei Schizophrenien. Arch Psychiatr Nervenkr 216: 116-130

Gross G, Huber G, Schüttler R (1973) Verlaufsuntersuchungen bei Schizophrenen. In: Huber G (eds) Verlauf und Ausgang schizophrener Erkrankungen. Schattauer, Stuttgart

Gross G, Huber G, Schüttler R (1982) Computerized tomography studies on schizophrenic diseases. Arch Psychiatr Nervenkr 231: 519-526

Gross G, Huber G, Schüttler R (1983) Verlauf schizophrener Erkrankungen unter den gegenwärtigen Behandlungsmöglichkeiten. In: Hippius H, Klein K E (eds) Therapie mit Neuroleptika. perimed, Erlangen

Gross G, Huber G, Schüttler R (1986a) Long-term course of Schneiderian schizophrenia. In: Marneros A, Tsuang M T (eds) The schizoaffective psychoses. Springer, Berlin Heidelberg New York

Gross G, Huber G, Armbruster B (1986b) Schizoaffective psychoses long-term prognosis and symptomatology. In: Marneros A, Tsuang M T (eds) The schizoaffective psychoses. Springer, Berlin Heidelberg New York

Gross G, Huber G, Klosterkötter J, Linz M (1987) BSABS. Bonner Skala für die Beurteilung von Basissymptomen (Bonn Scale for the Assessment of Basic Symptoms). Springer, Berlin Heidelberg New York

Gross G, Huber G, Klosterkötter J, Rao ML (1988) Klinisch-neurochemische Korrelationsuntersuchungen bei schizophrenen Erkrankungen. In: Gross G, Huber G (eds) Neuere pharmakopsychiatrische und neurochemische Ergebnisse der Psychosenforschung. Tropon, Cologne (Das ärztliche Gespräch 44)

Harding CM, Strauss JS (1985) The course of schizophrenia: an evolving concept. In: Alpert M (ed) Controversies in schizophrenia. Changes and constancies. Guilford, New York

Huber G (1957) Pneumencephalographische und psychopathologische Bilder bei endogenen Psychosen. Springer, Berlin Göttingen Heidelberg (Monographien aus dem Gesamtgebiete der Psychiatrie und Neurologie, vol 79)

Huber G (1966) Reine Defektsyndrome und Basisstadien endogener Psychosen. Fortschr Neurol Psychiatr 34: 409-426

Huber G (ed) (1971) Ätiologie der Schizophrenien. Bestandsaufnahme und Zukunftsperspektiven. Schattauer, Stuttgart New York (1. Weißenauer Schizophrenie-Symposion 23. u. 24.4.1971)

Huber G (1980) Hauptströme der gegenwärtigen ätiologischen Diskussion der Schizophrenie. In: Peters U H (ed) Die Psychologie des 20. Jahrhunderts, vol 10. Kindler, Zurich

Huber G (ed) (1982) Endogene Psychosen: Diagnostik, Basissymptome und biologische Parameter. Schattauer, Stuttgart New York (5. Weißenauer Schizophrenie-Symposion)

Huber G (1983) Das Konzept substratnaher Basissymptome und seine Bedeutung für Theorie und Therapie schizophrener Erkrankungen. Nervenarzt 54: 23-32

Huber G (ed) (1985) Basisstadien endogener Psychosen und das Borderline-Problem. Schattauer, Stuttgart New York (6. Weißenauer Schizophrenie-Symposion)

Huber G (1986a) Negative or basic symptoms in schizophrenia and affective illness: introduction. In: Shagass C, Josiassen R C, Bridger W H, Weiss K J, Stoff D, Simpson G M (eds) Biological psychiatry 1985. Elsevier, Amsterdam (Proceedings of the IVth World Congress of Biological Psychiatry, Philadelphia 1985)

Huber G (1986b) Laudatio. Verleihung des Hans-Jörg-Weitbrecht-Preises 1985. In: Huber G (ed) Zyklothymie - offene Fragen. (Das ärztliche Gespräch 41). Tropon, Cologne

Huber G, Penin H (1968) Klinisch-elektroencephalographische Korrelationsuntersuchungen bei Schizophrenen. Fortschr Neurol Psychiatr 36: 641-659.

Huber G, Gross G, Schüttler R (1979) Schizophrenie. Verlaufs- und sozialpsychiatrische Langzeituntersuchungen an den 1945-1959 in Bonn hospitalisierten schizophrenen Kranken. Springer, Berlin Heidelberg New York (Monographien aus dem Gesamtgebiet der Psychiatrie, vol 21)

Klosterkötter J (1988) Basissymptome und Endphänomene der Schizophrenie. Springer, Berlin Heidelberg New York (Monographien aus dem Gesamtgebiete der Psychiatrie, vol 52)

Kornhuber HH, Kornhuber J, Kim JS, Kornhuber ME (1984) Zur biochemischen Theorie der Schizophrenie. Ergebnisse und Kasuistik. Nervenarzt 55: 602-606

Kornhuber HH (1985) Zur Pathophysiologie und Therapie der Schizophrenien. In: Huber G (ed) Basisstadien endogener Psychosen und das Borderline-Problem. Schattauer, Stuttgart New York

Koukkou M (1984) Elektroenzephalographische Studien der Informationsverarbeitung bei akuten und ehemaligen schizophrenen Patienten, Neurotikern und psychisch Gesunden. In: Hopf A, Beckmann H (Hrsg) Forschungen zur Biologischen Psychiatrie. Springer, Berlin Heidelber New York

Penin H, Gross G, Huber G (1982) Elektroenzephalographisch-psychopathologische Untersuchungen in Basisstadien endogener Psychosen. In: Huber G (ed) Endogene Psychosen: Diagnostik, Basissymptome und biologische Parameter. Schattauer, Stuttgart New York

Schneider K (1987) Klinische Psychopathologie. 13. unveränd. Aufl. Mit einem Kommentar von G. Huber und G. Gross. Thieme, Stuttgart

Süllwold L, Huber G (1986) Schizophrene Basisstörungen. Springer, Berlin Heidelberg New York (Monographien aus dem Gesamtgebiet der Psychiatrie, vol 42)

Wieser HG (1982) Zur Frage der lokalisatorischen Bedeutung epileptischer Halluzinationen. In: Karbowski K (ed) Halluzinationen bei Epilepsien ihre Differentialdiagnose. Huber, Bern

Zubin J (1987) Closing comments. In: Häfner H, Gattaz W F, Janzarik W (eds) Research for the causes of schizophrenia. Springer, Berlin Heidelberg New York

Cognitive Basic Symptoms of Thought, Perception and Action in Idiopathic Psychoses and Limbic System

Gisela Gross

Psychiatrische Klinik und Poliklinik der Universität Bonn, 5300 Bonn 1, FRG

By developing the concept of basic disorders and basic symptoms and by concerning himself with the apparently uncharacteristic stages in the course of schizophrenia Huber has drawn attention to the subjectively experienced basic symptomatology. In his long-term follow-up investigations of Heidelberg and Bonn outpatients he has shown that uncharacteristic basic stages prevail in the long-term course of schizophrenic diseases when compared to typical schizophrenic periods (Huber 1957, 1961, 1966). The psychopathological picture of the basic stages is characterized by manifold dynamic and cognitive deficiencies, experienced and communicated by the patients themselves as disorders and impairments. The subjectively experienced basic symptoms offer a chance to analyse more closely what so-called schizophrenic symptomatology may truly be. In 1966 Huber described them as substrate-close basic symptoms. They are called "substrate-close" because they are nearer to the hypothesized somatic substrate than the typical schizophrenic phenomena, formed and modified by secondary working-up processes. And we call them "basic" symptoms, because they represent in our opinion the real primary symptoms of schizophrenia and constitute the basis of the fluctuating florid symptomatology (Huber 1976, 1983). Hypothetically, basic symptoms are ascribed to impairments of selective information processing, of lack of hierarchies of habituation and to genetically determined biochemical disturbances in the limbic system. Thus it would be justified to replace the term "schizophrenia" by "limbopathy" (Huber 1971, 1983; Gross and Huber 1985). The assumption of a limbopathy and a connection between limbic structures and basic symptoms is supported by the results of neurophysiological, electroencephalographic and psychological investigations, e.g. by Broen (1968), Hasse-Sander et al. (1982), Huber et al. (1979), Poljakov (1973) and Kornhuber (1983). Morphological alterations in the temporal and diencephalic area of the brain, e.g. the slight dilatation of the 3rd ventricle in a subgroup of schizophrenic patients with pure defect syndromes (Huber 1957; Gross et al. 1982) and a distinct abnormal rhythmicity in EEG point to mesolimbic structures (Huber and Penin 1968). Uncharacteristic basic symptoms occur not only initially but also in later stages and especially in residual states in the sense of pure defect (*reiner Defekt;*- Huber 1961, 1966; Huber et al. 1979). These pure deficiency syndromes or postpsychotic irreversible basic stages are often rather an inclination to failure, remaining almost completely compensated under favourable environmental conditions, as Huber stated 20 years ago.

For theory and etiology but also for therapy it is very important that the more or less uncharacteristic basic deficiencies really form the basis of the schizophrenic first-rank symptoms. The question is whether there exist connections between reversible psychotic manifestations and the preceding and following dynamic and cognitive deficiencies, and whether the pre- and postpsychotic basic stages are constituted by them. By follow-up investigations it was possible to prove the existence of a psychopathological transition state between basic symptoms and psychosis, or in another terminology between negative and positive symptoms, and to disprove the objection that basic symptoms occur in a number of different psychic disturbances in the same manner. This objection, namely that the basic symptomatology cannot be specific for schizophrenia, overlooks three points: (a) A specific psychopathologic symptomatology does not exist at all, and schizophrenic first-rank symptoms do not occur exclusively in schizophrenias but also in definable brain diseases. (b) Descriptions and definitions of the Bonn Scale show the more or less characteristic quality for the majority of basic symptoms especially for cognitive basic deficiencies, phenomena which can occur also in organic brain diseases and schizoaffective psychoses but not in neurotic or psychopathic personality disorders. (c) We could prove by follow-up investigations that, and in what manner subjectively experienced basic deficiencies or minus symptoms can pass over into positive or productive psychotic phenomena; these are schizophrenic symptoms in the conventional sense. For the clinical use of the concept of basic symptoms and basic disorders it is important that all basic symptoms display extreme fluctuations, both in their manifestation and degree, depending on given situations or strains or on an endogenous level, i.e. without a discernable cause. Therefore it is possible only by continuous observation to recognize and document the more or less uncharacteristic and often only transitory level 2 basic symptoms before going on to florid (level 3) symptomatology, i.e. to a higher degree of process activity. Because of the substantial overlapping of the psychopathological picture and the phenomenological aspects of pre- and postpsychotic reversible and irreversible basic stages we have grouped these unpsychotic manifestations of schizophrenic diseases, not typical in the sense of classical concepts of schizophrenia, under the term of basic stages in a broader sense (Huber 1976, 1983). Furthermore, it was possible to construct the Bonn Scale for the Assessment of Basic Symptoms (BSABS) for the standardized documentation of basic symptomatology of all pre- and postpsychotic basic stages (see Huber, this volume). This Bonn rating scale is based on the subjectively experienced and reported complaints of 750 patients; the basic symptoms are registered in 5 main and 1 additional category in 98 single items (Gross et al. 1987).

We would now like to consider cognitive basic symptoms of thought, perception and action in detail. The cognitive thought disorders are one of the most frequent basic symptoms and reported in approximately 70% by patients with pre- and postpsychotic basic stages. They are based on disturbances of intake and processing of information, probably localized in the limbic system and hypothesized to be caused by disturbances in different neurotransmitter systems. These cognitive (level 1 and level 2) basic symptoms, not yet typical for schizophrenia, can be summarized under the heading of "losing control of thought processes", described by Huber in 1966. The thought disorders are differentiated in the Bonn Scale in 17 subcategories. In the sub-type interference of thought (BSABS C.1.1), patients notice the intrusion of notions that have nothing to do with the current train of thought, often tied to external impressions. They cause increased distractability and disturbances of selective attentiveness. Patients in postpsychotic irreversible basic stages, i.e. pure defect syndromes, report, for example, about obsessional perseveration of certain thoughts referring to preceding unimportant events (C.1.2) or about disturbances of concentration (C.1.5) and of memory (C.1.8 - C.1.11) most frequently. The patients try to counter these disturbances and the subjectively experienced vacancy of mind or blocking of the current train of thought (C.1.4) by shielding themselves, and organizing their work, by reducing the speed of work, by finishing tasks immediately, by writing everything down, or by trying to improve lost abilities through training them. Patients with disorders of receptive speech (C.1.6) either do not grasp the meaning of written or spoken words, sequences of words or phrases or only with great effort and incompletely (Süllwold 1983). In case of disturbances of expressive speech (C.1.7), the self-perceived complication of speech due to deficient actualization of fitting words, the patient may try to compensate this deficit by having excessive use of ingrained speech patterns (Süllwold 1983). With an increase of such stereotypes, a patient's vocabulary can become destitute. If this symptom is markedly intensified, it can lead to poverty of speech or poverty of contents of speech, as Andreasen and Olsen (1982) described in the Iowa Scale.

The cognitive perception disorders are divided into 11 subtypes. The perception disorders are observed mainly in the beginning of schizophrenic diseases and in post-psychotic reversible and/or irreversible basic stages. They occur mainly paroxysmally or phasically in short phases persisting only seconds or minutes, seldom hours, days or weeks. Most frequent are perception disorders on optic field, e.g., blurred vision (C.2.1), hypersensitivity to light or visual stimuli in general and photopsias (C.2.2). Further, patients complain about seeing things as nearer or more distant than normally, about micro- and macropsy, changes of perception on face and shape of other people or of ones own face or body, uncertainty in estimation of distances, experiences of pseudomovements of objects, alternations of intensity and/or quality of colours, dysmegalopsia and an abnormally long persistence of optic stimuli (C.2.3). Sometimes the optic perception is focused only on unimportant details of objects which are experienced as isolated from the whole field of perception (C.2.9), or the comprehension of the meaning of perceptions is disturbed, or patients suffer from an aroused state of perceptual awareness (C.2.8). Other perception disorders concern a change of intensity and quality of acoustic perception, e.g., hypersensitivity to noises or acoasms (C.2.4) and disturbances of the perception on olfactory, gustatory or tactile field (C.2.6). All the mentioned perception disturbances do not fulfil the criteria of hallucinations or delusions. Similar disturbances of perception were observed in organic brain diseases in the region of diencephalon, described, for example, by Beringer, Schuster, Pötzl and Ewald (see Huber 1957) in patients with brain stem encephalitis or tumours in the thalamic region.

Disturbances of experienced motor behaviour can be seen as an inability to suppress concurrent reaction tendencies or as loss of hierarchies of habituation; phenomenologically there exist similarities and correspondencies to symptoms occurring in circumscribed diseases of the extrapyramidal system. Like all basic symptoms these phenomena may also pass over from uncharacteristic (level 1) to more or less characteristic (level 2) and finally into typical schizophrenic symptoms (level 3) or vice versa. In 1957 Huber described these as neurological-psychopathological transitional phenomena, to which also belong cenesthesias, which are qualitatively abnormal bodily sensations, central-vegetative disturbances (Gross 1987) and the just mentioned cognitive disorders of perception. The motor symptoms comprise, among others, motor interference (C.3.1), where intended movements suddenly are interrupted by competitive impulses of motions. Pseudospontaneous movements such as visual spasms or the so-called automaton syndrome (Huber 1957) occurring without intention of the patient and without the criterion of being influenced. They are also observed mainly in the beginning of schizophrenic diseases. If the patients complain about motor blockades (C.3.2), they can perform their intended movements only with great difficulty, or they are blocked totally. The phenomenon of being spellbound (Huber 1957) belongs to this subtype, too. Here the patients are unable to move or to speak for seconds or minutes without disturbances of consciousness. This state is the counterpart of the automaton syndrome. The loss of automatic abilities (C.3.3) requires an increased level of attention to carry out properly everyday motions and actions, usually done automatically, for example a good secretary is unable to type fluently, or one cannot dress or comb oneself, and the tasks require full attention, in contrast to the healthy state. Further, patients describe the subjectively realized retardation of psychomotor functions and of speaking and self-percepted motor disturbances in the sense of extrapyramidal or tic-like hyperkinesias, (e.g. grimacing, pseudoexpressive movements such as permanent kneading of hands or snapping with the fingers).

As all the other basic symptoms, the motor phenomena are not specific for schizophrenic diseases. They can occur also in defined and known brain diseases, e.g. in brain stem affections, phenomenologically identically and not distinguishably (Kleist, see Huber 1957). This phenomenological relationship or identity can support the hypothesis of the substrate-close character of the basic symptomatology and make it obvious that the basic symptoms are caused by disturbances of brain functioning in the area of the limbic system. Like all basic symptoms, cognitive thought, perception and motor disturbances can manifest or deteriorate without discernable cause and can also be precipitated by environmental factors, for example by conflicts subjectively not important but more or less overcharging the capacity of working-up information diminished morbogenically. The tolerance threshold to such normal stressors is lowered, and the "ceiling" is reached earlier than in healthy persons and in patients before onset of the disease. Cognitive basic symptoms and also the dynamic deficiencies are probably caused by disorders of information processing. The phenomenal similarity of cognitive thought, perception and motor disturbances to symptoms caused by affections of limbothalamic regions (reported by Huber 1957) and other findings, for example, by Kornhuber (1985) mentioned in several contributions to this volume point to functional disorders in the limbic system as a biological basis for the psychopathological symptomatology in schizophrenia.

References

Andreasen N C, Olsen S (1982) Negative versus positive schizophrenia. Arch Gen Psychiatry 39: 789-794

Broen W E (1968) Schizophrenia. Theory and research. Academic, London

Gross G (1987) Zur Frage zentral-vegetativer Störungen bei idiopathischen Psychosen. Fundam Psychiatr 1: 123-133

Gross G, Huber G (1985) Das Konzept der Basissymptome in der klinischen Anwendung. In: Janzarik W (ed) Psychopathologie und Praxis. Enke, Stuttgart

Gross G, Huber G, Schüttler R (1982) Computerized tomography studies on schizophrenic diseases. Arch Psychiatr Nervenkr 231: 519-526

Gross G, Huber G, Klosterkötter J, Linz M (1987) 2BSABS. Bonner Skala für die Beurteilung von Basissymptomen (Bonn Scale for the Assessment of Basic Symptoms). Springer, Berlin Heidelberg New York

Hasse-Sander I, Gross G, Huber G, Peters S, Schüttler R (1982) Testpsychologische Untersuchungen in Basisstadien und reinen Residualzuständen schizophrener Erkrankungen. Arch Psychiatr Nervenkr 231: 235-249

Huber G (1957) Pneumencephalographische und psychopathologische Bilder bei endogenen Psychosen. Springer, Berlin Göttingen Heidelberg (Monographien aus dem Gesamtgebiete der Psychiatrie und Neurologie, vol 79)

Huber G (1961) Chronische Schizophrenie. Hüthig, Heidelberg

Huber G (1966) Reine Defektsyndrome und Basisstadien endogener Psychosen. Fortschr Neurol Psychiatr 34: 409-426

Huber G (ed) (1971) Ätiologie der Schizophrenien. Bestandsaufnahme und Zukunftsperspektiven. Schattauer, Stuttgart (1. Weißenauer Schizophrenie-Symposion am 23. und 24.4.1971)

Huber G (1976) Indizien für die Somatosehypothese bei den Schizophrenien. Fortschr Neurol Psychiatr 44: 77-94

Huber G (1983) Das Konzept substratnaher Basissymptome und seine Bedeutung für Theorie und Therapie schizophrener Erkrankungen. Nervenarzt 54: 23-32

Huber G, Penin H (1968) Klinisch-elektroencephalographische Korrelationsuntersuchungen bei Schizophrenen. Fortschr Neurol Psychiatr 36: 641-659

Huber G, Gross G, Schüttler R (1979) Schizophrenie. Verlaufs- und sozialpsychiatrische Langzeituntersuchungen an den 1945 - 1959 in Bonn hospitalisierten schizophrenen Kranken. Springer, Berlin Heidelberg New York (Monographien aus dem Gesamtgebiete der Psychiatrie, vol. 21)

Kornhuber H H (1983) Chemistry, physiology and neuropsychology of schizophrenia: towards an earlier diagnosis of schizophrenia I Arch Psychiatr Neurol Sci 233: 415-422

Kornhuber H H (1985) Zur Pathologie und Therapie der Schizophrenie. In: Huber G (ed) Basisstadien endogener Psychosen und das Borderline-Problem. Schattauer, Stuttgart

Poljakov J (1973) Schizophrenie und Erkenntnistätigkeit. Thieme, Stuttgart

Süllwold L (1977) Symptome schizophrener Erkrankungen. Springer, Berlin Heidelberg New York (Monographien aus dem Gesamtgebiete der Psychiatrie, vol 13)

Süllwold L (1983) Schizophrenie. Kohlhammer, Stuttgart

Phenomenological Aspects and the Measurement of Negative or Basic Symptoms in Schizophrenia

R. Schüttler
Bezirkskrankenhaus Günzburg, Department of Psychiatry II, University of Ulm, 8870 Günzburg, FRG

The Problem

Basic symptoms in schizophrenia are defined as noncharacteristic symptoms which are experienced and communicated by the patients themselves. These symptoms may be amazingly similar to the symptomatology of certain organic brain syndromes. Evaluating basic symptoms qualitatively and quantitatively in schizophrenic disorders and organic brain syndromes we need a valid operational documentional system. None of the known documentional systems (AMDP, BPRS, PSE) seemed to record these special symptoms sufficiently (Schüttler et al. 1982).

The Günzburg Self-Rating-Scale for Basic Symptoms

Items

In constructing the Günzberg Self-Rating Scale for Basic Symptons (GSBS) we took as a basis the finding of the Bonn group (Huber et al. 1984), that schizophrenic patients complain not only about the well-known positive symptoms but also about numerous noncharacteristic symptoms in the long-term course of the disorder (Schüttler 1985). Huber has called these disturbances "basic symptoms." They may be present for only short times , may fluctuate or may remain constant for the whole duration of the disorder. The occurrence of basic symptoms is also possible in prepsychotic stages (prepsychotic isolated outpost syndromes, prodromes); (Huber et al. 1980). We consider the basic symptoms to be the consequence of disturbances in information processing and genetically determined biochemical abnormalities in the limbic system (Schüttler et al. 1981).

We took the battery of items used by the Bonn group in constructing our self-rating scale for basic symptoms (Schüttler et al. 1985). The scale includes three categories of items: dynamic deficiencies, cognitive disturbances of thought and cognitive disturbances of perception. These are described in Tables 1 - 3.

Table 1. Dynamic deficiencies

- Diminished vigor, energy, endurance, patience
- Diminished drive, activity, impulse, initiative
- Decreased emotional resonance
- Feeling of unfeelingness
- Incapacity to enjoy and to be pleased
- Increased exhaustibility
- Increased impressionability

Table 2. Cognitive disturbances of thought (leack of control of thought processes)

- Interference of thought
- Disturbances of concentration and memory
- Pressure of thought
- Obsessional perseveration of thought
- Impairment of selective attention
- Thought blocking of the current train of thought
- Disturbances of thought initiative

Table 3. Cognitive disturbances of perception (sensorial disturbances)

- Sensation of numbness, stiftness and feeling strange
- Sensation of motor weakness
- Circumscribed sensation of pain
- Vestibular sensation
- Sensation of movement, pulling or pressure inside or on the surface of the body
- Alternation of quality of hearing and seeing (including diminished or increased intensity)
- Pseudomovements
- Dysmorphopsia

Validation

We tested the preliminary form of the GSBS in a pilot study with 127 patients of various diagnoses (organic brain syndromes, 16; schizophrenic disorders, 29; affective psychoses, 13; neurotic disorders, 37; addiction, 32).Of these 127 patients, 28 were inpatients and 99 had been released after inpatient treatment within 5 years. Their mean age was 33.8 years.

We evaluated statistically whether there were significant differences between the groups. The most important result was that we were able to differentiate patients with organic brain syndromes or schizophrenic disorders from those with neurotic disorders. This result contradicts the opinion that residual syndromes and basic stages of schizophrenic disorders are psychopathologically heterogeneous and clearly distinct from organic brain syndromes. However, we must note the preliminary character of these results because at this time we are lacking an analysis of items and of reliability of results. These results were encouraging for further efforts at constructing a self-rating scale for basic symptoms (Schüttler et al. 1985).

We valuated these categories of basic symptoms using factor analysis. Stepwise extraction of factors demonstrated a relatively good operationality of the hypothetical categories. Further statistical and conceptual evaluation led us to a settlement of 3 factors accounting, respectively, 41.9%, 33.3% and 24.8% of explained variance. Criteria for this settlement were the intrinsic values of the factors, the practicability of interpretation, the increase of variance by other additional factors and the optimal correspondence of items with singular factors.

Furthermore, we found the complexity of most items to be 1 (78%). Settlement of factors used items with factorloadings of at least 0.40. Most items showed the highest loading on the intended factor. Factor 1 represents only items for dynamic deficiencies, factor 2 represents to 80% items for cognitive disturbances of thought, and factor 3 represents to 92% items for cognitive disturbances of perception. Further statistical analyses (selectivity; factor analysis with 4 factors; intercorrelation analyses) revealed a correlation among the factors and hence the possibility of determining a sum score of basic symptoms. These results confirm the operational basis for a set of empirical hypothetical categories of basic symptoms (Schüttler 1987).

A problem in validity of the GSBS is the similarity of basic symptoms to certain side effects of neuroleptics. Therefore, we had to differentiate the basic symptoms from similar symptoms. We attempted to do this by examining the correlation between neuroleptic dose and values on GSBS of schizophrenic patients. A high correlation between neuroleptic dose and GSBS value would mean a poor distinction between basic symptoms and neuroleptic side effects; a low or no correlation would mean a good distinction. We analyzed our sample with 2 different tests: first, the total neuroleptic dose of each patient was registered in chlorpromazine equivalents (procedure A) secondly, an individual dose evaluation was made independently by 3 psychiatrists (procedure B). Analysis by procedure A revealed a low, by procedure B no correlation. We consider procedure B really to be the more valid. The results of both procedures strengthen our opinion that the GSBS registers genuinely basic symptoms independent of neuroleptics.

Discussion

Lacking previously a proper instrument for operational documentation of the basis symptoms of schizophrenic patients, we constructed the GSBS to measure noncharacteristic complaints in the long-term course of schizophrenic disorders as are typically described by the patients themselves. Specific categories are dynamic deficiencies, cognitive disturbances of thought and cognitive disturbances of perception.

Construction of the scale involved specifying and selecting items and evaluating their reliability and validity. We demonstrated a good reliability of the new test instrument. A pretest demonstrated also a valid differentiation between schizophrenic and neurotic patients; a further study is being constructed to replicate these preliminary results. The factor analysis showed a good differentiation of the 3 hypothetical categories of basic symptomatology.

Providing a reliable and valid test instrument for rating basic symptoms makes it possible for other investigators to replicate results regarding the theoretically and practically important basic symptoms. Surely, the occurrence of noncharacteristic symptoms and syndromes is not questioned in the long-term course of schizophrenic disorders.

However, there are many discussions about the formation and the "relative specificity" of "substrate-close" basic symptoms. It is argued that these complaints are seen with neuroses, organic brain syndromes, "reactive psychosis", systemic disorders in general medicine and with aging, futhermore the similarity with hospitalism was been pointed out and it is argued, that basic symptoms may represent a "social artifact".

To confront such objections we plan to test a group of surgical and internal inpatients of a general hospital without psychiatric illness using the GSBS. Furthermore, we are investigating a group of neurotic patients (in terms of DSM III classification) who for the first time in their lives are inpatients of a psychiatric hospital. We are paying particular attention to compare the frequencies of symptoms and symptom clusters in GSBS between the patients with neurotic disorders and these with schizophrenic disorders. Further findings, such as results of other test methods or neuroradiological examination, are being collected.

References

Huber G, Gross G, Schüttler R, Linz M (1980) Longitudinal studies of schizophrenic patients. SchizophrBull 6 : 592-605

Huber G, Gross G, Schüttler R (1984) Schizophrenie. Eine verlaufs- und sozialpsychiatrische Langzeitstudie. Springer, Berlin Heidelberg New York

Schüttler R (1985) Phenomenological aspects of pure deficiency syndromes in schizophrenia. Psychiatry 1

Schüttler R (1987) Zur stützenden psychotherapeutischen Langzeitbehandlung bei Schizophrenien. In: Huber G (ed) Fortschritte in der Psychosenforschung? Schattauer Stuttgart

Schüttler R, G Huber, Gross G (1981) Reine Defektsyndrome und Basisstadien schizophrener Erkrankungen. Bedeutung für die nosologische und ätiologische Hypothesenbildung. In: Huber G (ed) Schizophrenie. Stand und Entwicklungstendenzen der Forschung. Schattauer Stuttgart

Schüttler R, Gross G, G Huber (1982) Zum Problem der operationalisierten Dokumentation der Potentialeinbuße bei reinen Defizienzsyndromen. In: Huber G (ed) Endogene Psychosen: Diagnostik, Basissymptome und biologische Parameter. Schattauer Stuttgart

Schüttler R, Bell V, Blumenthal S, Neumann N-U, Vogel R (1985) Zur Potentialeinbuße in idiopathischen Basisstadien bei organischen Psychosyndromen und neurotischen Symptombildungen. In: Huber G (ed) Basisstadien endogener Psychosen und das Borderline-Problem. Schattauer, Stuttgart

Evoked Brain Potentials and Psychometric Data in Children at Risk for Schizophrenia

H. Schreiber[1], G. Stolz-Born[1], J. Born[2], J. Rothmeier[3], A.W. Kornhuber[1], and H.H. Kornhuber[1]

[1]Abteilung Neurologie, Universitätsklinik, Steinhövelstr.9, 7900 Ulm, FRG
[2]Abteilung für angewandte Physiologie, Universität Ulm, Oberer Eselsberg, 7900 Ulm, FRG
[3]Abteilung Neurologie, Psychiatrisches Landeskrankenhaus Weissenau, 7980 Ravensburg, FRG

Introduction

While we have learnt to manage psychotic schizophrenic symptomatology with hallucinations fairly well, most schizophrenic patients with only negative symptoms are not diagnosed and treated - many of them are socially maladjuste (Kornhuber 1983, 1985). We still lack early diagnostic parameters that allow an earlier diagnosis. This is especially true for cases in which no typical "positive "symptoms, e.g. paranoia and hallucinations, are prevailing, but vague "negative" symptoms, like social retreat or lack of will. For the latter impaired information processing due to an attentional deficit is considered a basic dysfunction (Kraepelin 1909-1915; E.Bleuler 1911; Mc Ghie and Chapman 1961; Huber et al. 1979; Kornhuber 1983; Kornhuber 1985, Schmid-Burgk et al 1982). Besides in clinical observations (McGhie and Chapman 1961) this attention deficit could be objectified by various kinds of psychometric and cognitive tests, such as reaction times (Shakov 1963) and recording of slow evoked potentials, especially the P3 component (Roth et al. 1980). On the assumption that there is an important hereditary factor in schizophrenia, these methods have been used in groups genetically at risk for the disease (Friedman et al. 1986; Stolz and Kornhuber 1984, unpublished data; Rothmeier and Kornhuber 1985). The aim of the present study is to clarify further whether there can be found early diagnostic signs for a schizophrenic disposition by focusing on a comparison of slow event-related potentials with performance on mental and psychomotor tasks and saccadic eye movements. For this purpose a high-risk group and a control group of children were tested.

Methods

Twelve high-risk and 12 control children aged between 9 and 16 years (mean, 12.6 years for both groups) were tested (details in Table 1). The high-risk children were offsprings of at least one schizophrenic parent according to DSM-III; three children had two schizophrenic parents. The children were matched by age, sex, education and environmental factors, thus forming 12 comparable pairs.

The children performed an oddball paradigm while late auditory evoked potentials (AERPs) were registered. By binaural stimulation the subjects were offered two different types of tones, a frequent (80% probable) standard tone pip (1200 Hz) without any

Table 1. Personal data for high-risk and matched control children. Age, sex (male: m, female: f); MI indicates practice in playing a musical instrument; Educ indicates educational background: lo - refers to level of high school, elementary school or lower, me - intermediate level (German *Realschule*), hi college. Env means environmental background: ho - child lives at home, fop - child stays with foster parents or relatives, fho - child is in a foster home. Psy indicates which parent (F: father, M: mother) suffers from schizophrenia

CONTROLS						HIGH-RISK						
No	Age	Sex	MI	Educ	Env	No	Age	Sex	MI	Educ	Env	Psy
01	13.1	M	yes	hi	ho	16	12.7	M	yes	me	fop	M
02	13.2	M	no	lo	ho	18	13.1	M	no	lo	ho	F/M
03	16.1	F	no	me	ho	14	15.9	F	yes	hi	ho	M
04	11.4	M	yes	lo	ho	24	10.2	M	no	hi	ho	M
05	11.6	F	yes	lo	ho	19	11.8	F	yes	lo	ho	F/M
06	9.0	M	no	lo	ho	22	9.0	M	yes	lo	ho	F
07	12.6	F	no	lo	ho	15	12.7	F	yes	me	ho	M
08	14.5	F	no	me	ho	17	14.11	F	yes	hi	fop	M
09	14.3	M	yes	lo	ho	21	15.4	M	no	lo	fho	F/M
10	9.8	M	yes	lo	ho	25	11.5	M	no	lo	ho	M
11	11.3	F	yes	lo	ho	20	11.4	F	yes	hi	fop	M
12	14.4	F	yes	me	ho	23	13.6	F	yes	hi	ho	F

task relevance and a rare (20% probable) target tone pip (800 Hz) which had to be counted silently. Recordings were obtained from EEG over the frontal (Fz), central (Cz) and parietal (Pz) midline. For analysis EEG epochs (660 ms) were amplified, digitized and stored on magnetic tape for off-line averaging. Baseline-to-peak amplitudes and latencies were determined for maxima and minima within latency bins accounting for N1 (60-140 ms post-stimulus onset), P2 (140-240 ms), N2 (180-360 ms), and P3 (300-660 ms). N2 and P3 were assessed in the target condition only. Sweeps with artifacts (e.g. eyeblinks) were excluded by a computer program. The children were also submitted to an extensive test battery including performance on the Wechsler Intelligence Scale (WIS), the d2 attention test of Brickenkamp, and a visual-acoustic reaction time paradigm with regular and irregular preparatory intervals. Assessment of saccadic eye movements comprised dysmetrias and nonfixations in a random step stimulus pattern.

Results

Figure 1 depicts typical individual evoked potentials averaged for matched pairs. The evoked responses to standard pips are synchronous, while the waveforms to target pips are more complex and differ between subjects. There are no significant differences for latency or for amplitude of the evoked components in the standard condition. The same is true for amplitudes of N2 and P3 in the target condition. N2 and P3 latencies to target tones, however, are on average substantially longer in the high-risk children. This is both an individual and a group effect. Concerning N2, 10 of 12 high-risk children display longer latencies than the corresponding matched controls; for P3 this is true in 9 of 12 cases. Grand average data for N2 and P3 latencies indicate 260 ms (SE 9 ms) and 528 ms (SE 33 ms) in the high-risk group and 218 ms (SE 6 ms) and 429 ms (SE 32 ms), respectively, in the control group; this effect is statistically significant (Fig. 2). As a stri-

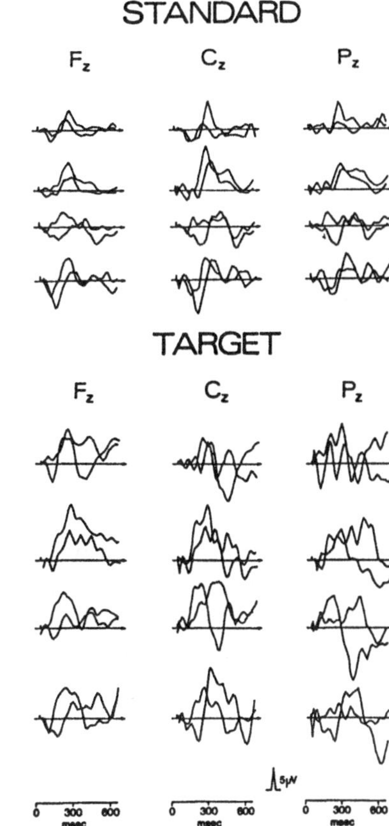

Fig. 1. Representative averaged AERPs to standard and to target tone pips from four high-risk subjects (dotted lines) and their matched controls (solid lines) recorded from frontal (Fz), central (Cz) and parietal (Pz) midline electrodes

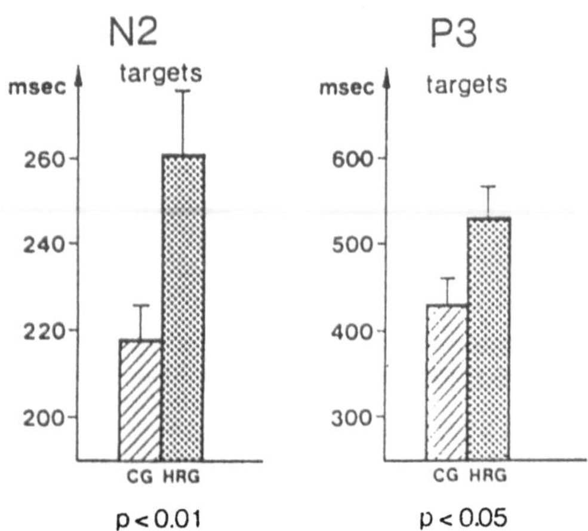

Fig. 2. Left panel: N2 latencies to target tone pips in the high-risk group (HRG) and in the control group (CG). Right panel: latencies of P3 to target tone pips (at Pz) for both experimental groups. Significant ($p < 0.05$) differences between groups are indicated

king observation, in the target condition, the high-risk children do not show the homogenously distributed N2 latencies of control children but a double-peaked distribution, with five children having very long latencies (among them are two very high-risk children with two schizophrenic parents). For a summary of psychometric data see Table 2. High-risk children as a group performed poorer, but not statistically significant, in all tests. Since we were in search of variables with reliable group specificity as a possible diagnostic tool, a stepwise discriminant analysis on striking ERP and psychometric variables was performed. The most reliable group specificity was found for N2 latency, reaching 75% of correct classifications in the classification matrix. Saccadic eye movements, cognitive data, d2 test, and reaction times did not reliably distinguish between groups.

Discussion

The essential finding of the present study is that children not manifestly ill but at risk for schizophrenia can be distinguished from normal controls by processing of rare, task-relevant acoustic stimuli. This is most reliably true for latency of the N2 and P3 components of the evoked response. Amplitudes did not yield any significant differences, probably because of a slurring effect due to interindividual variability. Subject sampling does not seem to be responsible for the differences in latency as the children were matched according to age, sex, education and environmental factors. Cognitive performance was quite comparable between groups (Table 2). Moreover, no significant correlation was found between IQ and latency. Latency effects were manifest in the comparison of individual matched pairs, excluding averaging errors. The fact that the delay in latency in high-risk children occurs only in the target condition could be due to a qualitative difference in controlled versus automatic processing of information or to a reflection of unstable attentional sets. The results agree with repeated findings of a delayed P3 and with recent data on a significant delay of the N2 in schizophrenic patients (Brecher et al. 1987). Whether the extent of slowed latencies could be a diagnostic criterion remains to be investigated. Fortunately, only a minority of the high-risk children will later develop schizophrenia (Rosenthal and Kety 1968). But, unfortunately, the biological conditions for the development of the disease are still not known. Perhaps new etiological aspects (Kornhuber and Kornhuber 1987) will soon give us more insight into the causal factors of the disease and suggest better possibilities of treatment.

Addendum. In the meantime we have been able to confirm our data about delayed P3 and N2 latencies in high-risk children. In a second study under way with 24 matched high-risk and control children the delay proves to be statistically significant ($p<0.01$) for P3 and showed a trend ($p<0.2$) for N2. Prolonged P3 marked ten and N2 seven high-risk children in 12 matched pairs.

Summary

The present study reports on event-related potentials (ERPs) and psychometric performance in children genetically at risk for schizophrenia. The high-risk children belon-

Table 2. Psychomotor and mental performances, saccadic eye movements

VARIABLE	CONTROLS mean	(SD)	HIGH-RISK mean	(SD)
Warned RTs/ms (acoustic/visual) with constant foreperiod				
0.5 s	175,0	(21,2)	194,5	(35,0)
1.0 s	198,5	(34,2)	209,6	(27,0)
2.0 s	201,8	(31,6)	229,7	(31,8)
4.0 s	222,0	(37,5)	248,4	(36,9)
8.0 s	248,2	(42,9)	279,3	(35,8)
16.0 s	273,1	(44,5)	298,4	(53,5)
Warned RTs/ms (acoustic/visual) with inconstant foreperiod				
0.5 s	240,1	(32,9)	244,9	(25,6)
1.0 s	240,2	(34,3)	243,0	(24,4)
2.0 s	254,7	(38,0)	257,4	(29,5)
4.0 s	255,5	(39,6)	264,7	(32,5)
8.0 s	266,5	(47,2)	281,0	(27,4)
16.0 s	275,9	(48,0)	287,1	(41,3)
Verb.-IQ	108,7	(10,4)	101,6	(13,6)
Perf.-IQ	117,5	(15,1)	116,4	(15,0)
Full.-IQ	114,9	(11.6)	110,0	(13,7)
d2-Attention Test of Brickenkamp				
Full-score	320,1	(63,0)	319,9	(89,6)
Error-score	16,2	(11,0)	31,1	(42,1)
Saccadic eye movements				
Hypometrias (≤50%)	2,9	(2,0)	4,4	(2,7)
Hypometrias (≥60%)	16,1	(7,2)	18,1	(6,7)
Non-Fixations (≥5°)	13,8	(10,0)	19,3	(15,0)

ged to families with at least one schizoprenic parent. Twelve high-risk and 12 matched control children (aged 9-16 years) were tested. Late ERP components were registered in an auditory oddball task. In addition, various kinds of psychometric tests and saccadic eye movements were assessed. Results revealed prolonged latencies of the N2 and P3 components of the ERP in the high-risk children. They also performed poorer in the psychometric test battery and in saccadic eye movements. A reliable group difference, however, was yielded only by the N2 and P3 components. By these parameters the high-risk group could be divided into two subgroups which might be of prognostic relevance.

References

Bleuler E (1911) Dementia praecox oder die Gruppe der Schizophrenien. In: Aschaffenburg G (ed) Handbuch der Psychiatrie. Leipzig

Brecher M, Porjesz B, Begleiter H (1987) The N2 component of the event-related potential in schizophrenic patients. Electroencephalogr Clin Neurophysiol 66:369-375

Friedman D, Cornblatt B, Vaughan H Jr, Erlenmeyer-Kimling L (1986) Event-related potentials in children at risk for schizophrenia during two versions of the continuous performance test. Psychiatr Res 18:161-177

Huber G, Gross G, Schüttler R (1979) Schizophrenie. Verlaufs- und sozialpsychiatrische Langzeituntersuchungen an den 1945 bis 1959 in Bonn hospitalisierten schizophrenen Kranken. Springer, Berlin Heidelberg New York (Monographien aus dem Gesamtgebiet der Psychiatrie, vol 21)

Kornhuber HH (1983) Chemistry, physiology and neuropsychology of schizophrenia: towards an earlier diagnosis of schizophrenia I. Arch Psychiatr Neurol Sci 233:415-422.

Kornhuber HH (1985) Zur Pathophysiologie und Therapie der Schizophrenie. In: Huber G (ed) Basisstadien endogener Psychosen und das Borderline-Problem. Schattauer, Stuttgart pp 129-144
Kornhuber HH, Kornhuber J (1987) A neuroimmunological challenge: schizophrenia as an autoimmune disease. Arch Ital Biol 125:271-272
Kraepelin E (1909-1915) Psychiatrie. Ein Lehrbuch für Studierende und Ärzte, vols 1-4, 8th edn. Barth, Leipzig
McGhie A, Chapman J (1961) Disorders of attention and perception in early schizophrenia. Br J Med Psychol 34:103
Rosenthal D, Kety SS (eds) (1968) The transmission of schizophrenia. Pergamon, Oxford
Roth WT, Pfefferbaum A, Horvath TB, Berger PA, Kopell B (1980) P3 reduction in auditory evoked potentials of schizophrenics. Electroencephalogr Clin Neurophysiol, 49:497-505
Rothmeier J, Kornhuber HH (1985) Ähnlichkeiten und Unterschiede psychischer Leistungen bei schizophrenen Patienten und ihren Angehörigen. Dtsch Med Wochenschr 110 (4):157
Schmid-Burgk W, Becker W, Diekmann V, Jürgens R, Kornhuber HH (1982) Disturbed smooth pursuit and saccadic eye movements in schizophrenia. Arch Psychiatr Nervenkr 232: 381-389
Shakov D (1963) Psychological deficit in schizophrenia. Behav Sci 8: 275-305

The Order of EEG Activity of Schizophrenic Patients and the Influence of Haloperidol and Biperidene on the EEG Order of Healthy Subjects

V. Diekmann, B. Grözinger, K.P. Westphal, W. Reinke, and H.H. Kornhuber
Neurologische Klinik, Universität Ulm, 7900 Ulm, FRG

Introduction

It has been demonstrated previously that in a Bereitschaftspotential paradigm schizophrenic patients showed characteristic changes in the EEG compared to normal control groups (e.g. Grözinger et al. 1984; Diekmann et al. 1985). Often patients with confirmed diagnosis were medicated with neuroleptic and additionally with anticholinergic drugs. Therefore the question arises whether the reported EEG changes were due to the medication or to disease. We tried to solve this problem by investigating (a) a group of never treated schizophrenics and (b) a group of healthy volunteers treated with haloperidol. Some results of these studies using classical statistical methods have been published recently (Westphal et al. 1989,and this volume). In this paper we present some results of these studies using uncertainty analysis (UA; Reinke and Diekmann 1987), a method based on Shannon's information calculus (Shannon and Weaver 1949). UA extracts the information contained in the variability of the measured values and is especially useful to detect similarities in time space-processes.

Methods

In study I we compared 9 never treated schizophrenics with 13 normal subjects who were matched in terms of age, sex and education to the patients. In study II we analyzed the influence of haloperidol on a group of 15 healthy students. The EEG of the patients or normals was recorded with eyes closed from frontal, central and parietal midline and lateral regions. The registration was unipolar, and the averaged ear registrations served as reference. The right-handed subjects performed self-paced voluntary fast movements of the right fingers (study I) or of the right index finger (study II). Smoothed FFT power spectra were calculated by averaging about 50 movements. Three time epochs, each of 1 s duration were analyzed: REST, starting 2.5 s before movement onset, Bereitschaftspotential period (BP; 1 s before movement), and motor period (MOT). As measures of uncertainty we computed entropy values for the mean power densities (MPD) of the classical frequency bands (delta, theta, alpha and beta) and for the total spectrum (Diekmann et al. 1985; Reinke and Diekmann 1987). In our case the entropy values served as order numbers of the intra- and interindividual variabilities of the

MPDs. Low entropy values mean low variability or high order and vice versa. In study II we recorded the EEGs in four sessions; details are given in Table 1.

Table 1. Experimental sessions

Session I	Premedication
Session II	After 14 days of 0.04 mg/kg haloperidol per day
Session III	After 21 days haloperidol combined with 6 mg biperiden per day during the last seven days
Session IV	Postmedication

Results

The greatest differences between normals and schizophrenics (study I) were found in the theta band (Fig. 1). The entropies in Fig. 1 are scaled in percentages of the maximal possible values. 0% means perfect order or no variability and 100% means no similarity between the measured values, or total chaos.

In normals the changes in entropy of the vertex and the central and parietal lateral electrode positions showed a clear connection to movement. In the lateral positions the entropies increased from REST to the BP period and decreased from BP to MOT. At the vertex (Cz) these changes are reversed. The entropy of Fz and Pz remained nearly stable in all epochs. Remarkable were the low entropy values of the vertex during the BP period and of C4 during the movement period. In contrast to the normals in the group of never treated schizophrenics almost no changes were found on the contralateral (left) hemisphere (C3, P3) and at the vertex, and instead of a decrease on the ipsilateral (right) hemisphere an increase of the entropy was found over the central area (C4) from the BP period to the MOT period. However, the most remarkable difference was the low order (high entropy) of the vertex MPD of the schizophrenics compared to that of the normal subjects especially in the BP epoch and the low order of the MPD of C4 during the movement epoch. These differences between the schizophrenics and the normals were significant with $p=0.002$ (Cz difference) and $p<0.001$ (C4 difference; difference test of χ^2 variables after Pearson et al. 1932 using the methods described by Reinke and Diekmann 1987; and a correction of clustering effects, Brier 1980). These results of the theta band contrast with those of the other frequency bands which showed in both groups nearly the same behavior of movement-related changes for electrode position entropies. However, as in the theta band, in the other frequency bands most of the entropy values in the group of schizophrenics were much higher than the corresponding values in the group of normals, i.e. the normals showed less variability than the schizophrenics.

Figures 2 and 3 show some results from study II. In Fig. 2 the beta band joint entropies of the three time epochs (REST, BP and MOT) are plotted as a function of the medication for the frontal, central and parietal lateral and medial areas. It should be pointed out that the joint entropies are not an average over the measured values but are multivariate measures. In our case the so-called maximum entropy was calculated for the triad of measured time epoch values, which includes all the finger movements of all subjects, and it served as measure of the intra- and interindividual similarity of this triad. Strong differences were found between the contralateral (left) and ipsilateral

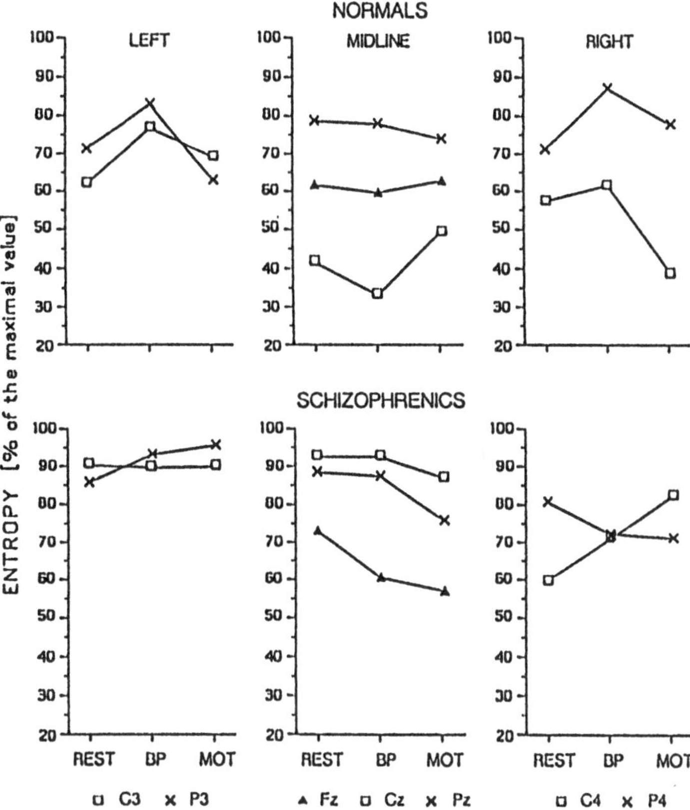

Fig. 1. Entropy of the theta band mean power density of the electrode positions C3, P3, Fz, Cz, Pz, C4 and P4 as function of the three time periods REST, Bereitschaftspotential period (BP) and movement period (MOT) for normal subjects (top) and never treated schizophrenic patients (bottom). Entropies are given as percentages of their maximal possible values. Schizophrenics showed a diminished order of EEG activity (higher values of entropy) compared to normals, most evident at the Cz and C4 electrodes

(right) hemisphere and the midline. Remarkable were the pronounced increase in disorder for F3 and C3 during haloperidol and the increase in order during the additional administration of biperiden, which exceeds the premedication values. C4 showed the same behavior but on a lower level of order. In contrast to these results in the midline, Cz and Pz showed an increase in order during haloperidol and a decrease in order with the additional administration of biperiden.

Figure 3 shows a topographic representation of the alpha band entropies of the electrodes (univariate entropies) and some joint entropies of pairs of electrodes (bivariate entropies). Entropies are drawn as a function of the time epochs (horizontal) and of the medication (vertical). In the figure the univariate entropies (boxes) are plotted in five different shades from "white" (very low order) to "black" (very high order). The bivariate entropies are represented in five classes of lines (thickness) connecting the electrode pairs. Very thick lines mark high similarities, and a missing line means very low similarities. In the REST epoch and before any medication we found a pattern

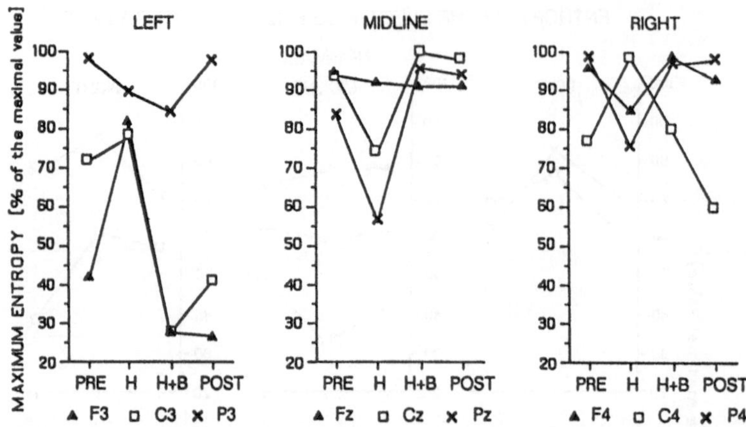

Fig. 2. Beta band maximum joint entropy of the three time epochs of the electrode positions F3, C3, P3, Fz, Cz, Pz, F4, C4 and P4 as function of the medication. PRE (premedication), H (haloperidol administration), H + B (haloperidol + biperiden administration) and POST (postmedication). Entropies are given as percentages of their maximal possible values. Haloperidol diminished the order (increased entropy) of the contralateral frontal and the central areas. Additionally administered biperiden restored the order towards the premedication state

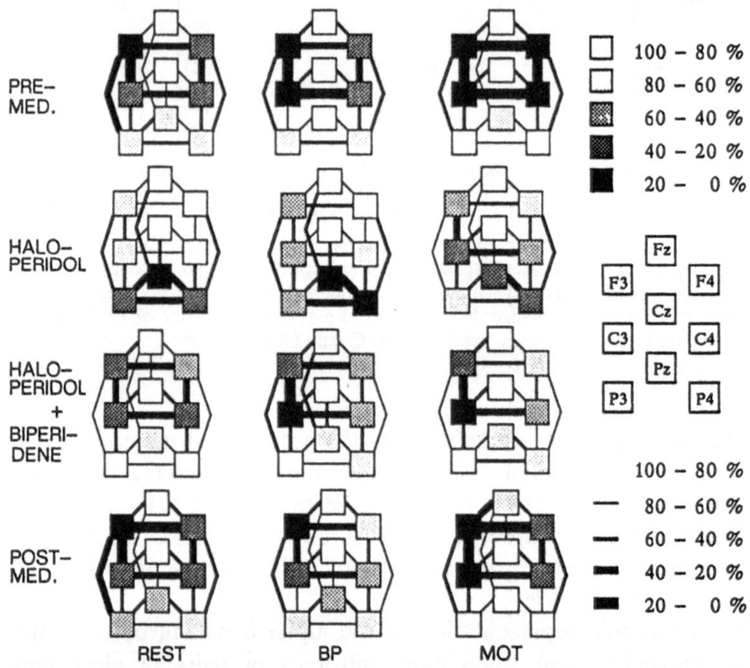

Fig. 3. Topographic representation of univariate and bivariate entropies for the electrodes F3, Fz, F4, C3, Cz, C4, P3, Pz and P4 of the mean power density of the alpha band. Entropies are drawn as a function of the medication and of the three time periods REST, Bereitschaftspotential period (BP) and movement period (MOT) and are given as percentages of their maximal possible values. The univariate entropies (boxes) are drawn in five different shades. The bivariate joint entropies (combinations of two electrodes) are represented in five classes of lines (thickness). The pattern of high order over frontal and central lateral areas in the premedication state is disturbed during haloperidol and changed towards premedication during the additional administration of biperiden.

of high order at the frontal and central lateral areas (FCL pattern). During the administration of haloperidol this pattern was changed very much. Additional administration of biperiden restored the pattern towards the premedication state. In the post-medication session the pattern looked nearly the same as in the premedication session. In the BP and MOT periods quite similar changes in the FCL pattern were found.

Discussion

It is known from previous studies (Westphal et al., this volume) that normal subjects enhance theta MPD during the BP period (i.e., the preparatory process) over parietal and central cortex, and that schizophrenics (treated as well as never treated) do not show this characteristic. We found that schizophrenics differ not only in the medians of the theta MPD (Westphal et al., this volume) but also in the variability of the theta MPD values. This is valid for all three time epochs (most pronounced over the vertex). The disturbance in the EEG order in the preparatory epoch may be related to the disturbed volition in schizophrenics, as discussed by Kraepelin (1905). That this disturbance in the EEG order is most prominent at the vertex in the group of never treated schizophrenic patients accords with the known important part of the supplementary motor area in the preparation of voluntary movements (Deecke and Kornhuber 1978; Roland et al. 1980; Kornhuber 1984).

In addition, we found in the schizophrenic group compared to the normal group a difference in the changes in entropy values from the preparatory (BP) period to the movement period (MOT) which is most prominent over the ipsilateral motor cortex (C4). This correlates to the more variable reaction time of the schizophrenics, to their diminished speed in attention tests, and to their dysmetria of saccadic eye movements (Schmid-Burgk et al. 1982). This result demonstrates that UA yields additional information beyond that which may be extracted from the median values of the EEG spectral parameters.

It is known that haloperidol in normals causes an increase in the alpha band MPD (Reesz 1989), but nothing is known about the variability of the MPD. The disturbance in the frontal and central pattern of EEG order during haloperidol administration indicates that haloperidol increases the variability of the MPD values. It is known that haloperidol affects the dopamine receptors in the basal ganglia, which play a major part in the initiation of voluntary movements (Kornhuber 1984). However, biperiden compensated these receptor changes to some degree, and this to a higher degree of order in the EEG activity.

Summary

In two studies we demonstrated the advantage of UA, a multivariate nonparametric statistical method for the analysis of the EEG power spectrum. In addition to the information which may be extracted from the medians of the measured values, the information included in the variability of measured values or in sets of measured values (e.g.triads) could be extracted.

In a population of never treated schizophrenics we found compared to a population of normal subjects, an increased variability of the mean power density in the theta band EEG power spectrum during a voluntary motor task. This diminished order of EEG activity is most pronounced over the central midline cortex. Over the ipsilateral central cortex schizophrenics showed a decrease in order from the preparatory epoch (BP epoch) to the motor epoch, whereas normals showed an increase. In the other frequency bands no such differences were found.

In normal subjects haloperidol diminished the order in the beta band EEG activity over the contralateral frontal and over the central areas in a voluntary right index movement task. In the alpha band haloperidol caused a disturbance in a pattern of high order over the frontal and central lateral cortex. Both changes returned to nearly the premedication state with the additional administration of biperiden.

References

Brier S S (1980) Analysis of contingency tables under cluster sampling. Biometrika 67:591-596

Deecke L, Kornhuber H H (1978) An electrical sign of participation of the mesial "supplementary" motor cortex in human voluntary finger movement. Brain Res 159:473-474

Diekmann V (1985) Kurzzeit-Spektral-Analyse von ereignisbezogenen EEG-Abschnitten. In: Gänshirt H, Berlit P, Haack G (eds) Kardiovaskuläre Erkrankungen und Nervensystem. Neurotoxikologie. Probleme des Hirntodes. Springer, Berlin Heidelberg New York, pp 881-885 (Verhandlungen der Deutschen Gesellschaft für Neurologie, Vol 3)

Diekmann V, Reinke W, Grözinger B, Westphal K P, Kornhuber H H (1985) Diminished order in the EEG of schizophrenic patients. Naturwissenschaften 72:541-542

Grözinger B, Neher K D, Westphal K P, Diekmann V and Kornhuber H H (1984) The EEG in schizophrenia. Changes in relation to voluntary movement in the theta and delta band. Naturwissenschaften 17:320-321

Kornhuber H H (1984) Mechanism of voluntary movement. In: Prinz W, Sanders A F (eds) Cognition and motor process. Springer, Berlin Heidelberg New York, pp 163-173

Kornhuber H H, Deecke L (1965) Hirnpotentialänderungen bei Willkürbewegungen und passiven Bewegungen des Menschen: Bereitschaftspotential und reafferente Potentiale. Pflügers Arch 284:1-17

Kraepelin E (1905) Einführung in die Psychiatrische Klinik, 2nd edn. Barth, Leipzig

Pearson K, Stouffer S A, David F N (1932) Further applications in the statistics of the Tm(x) Bessel function. Biometrika 24:293-350

Reesz J (1989) Studie mit Langzeitmedikation zur Wirkung von Haloperidol und der Kombination von Haloperidol und Biperiden auf den EEG-Frequenzgehalt von gesunden Normalpersonen vor und während einer Willkürbewegung. Thesis, University of Ulm

Reinke W, Diekmann V (1987) Uncertainty analysis of human EEG spectra: a multivariate information theoretical method for the analysis of brain activity. Biol Cybern 57:379-387

Roland P E, Larsen B, Lassen N A, Skinhoj E (1980) Supplementary motor area and other areas in organization of voluntary movements in man. J Neurophysiol 43:118-136

Schmid-Burgk W, Becker W, Diekmann V, Jürgens R, Kornhuber H H (1982) Disturbed smooth pursuit and saccadic eye movements in schizophrenia. Arch Psychiatr Nervenkr 232:381-389

Shannon C, Weaver W (1949) The mathematical theory of communication. University of Illinois Press, Urbana

Westphal K P, Grözinger B, Diekmann V, Scherb W, Reesz J, Leibing U, Kornhuber H H (1989) Comparison of untreated and treated schizophrenic patients, normals and neuroleptic treated normals: "hypofrontality" and different EEG spectra during voluntary movement. Psychiatry Res (in press)

Progress in Neuropsychopharmacology: The Use of Thymosthenic Substances in Schizophrenia

P.A.J. Janssen
Janssen Pharmaceutica, Turnhoutseweg 30, 2340 Beerse, Belgium

Introduction

The pharmacotherapy of psychotic disorders that started with chlorpromazine (Delay et al. 1952) has been revolutionary for the treatment of mental diseases of non-organic origin. The development of haloperidol meant further progress in this field as this agent was more potent and had less adverse effects than chlorpromazine. It nevertheless took several years before dopamine antagonism was put forward as a possible mechanism of action for neuroleptic drugs (Carlsson and Lindqvist 1963).

From the early 1970s, the advent of radioligand binding studies provided new tools for studying neurotransmittor receptors in the brain and for investigating drug interaction with various receptors. The technique allowed identification, classification and refined definition of receptors, and this was a new impetus for investigating the role of serotonin in the brain. Of the alleged serotonin receptor binding sites (from 5-HT_1 and subtypes to 5-HT_3) that have been described, the pharmacological and clinical importance of the 5-HT_2 receptor system is discussed here with particular attention to substances characterized as potent 5-HT_2 receptor blockers.

There are several reasons for studying the potential psychotropic effects of compounds that interact with brain serotonin receptors: there is a vast quantity of 5-HT_2 receptors in the mammalian brain, down-regulation of 5-HT_2 receptor sites following chronic administration of certain tricyclic antidepressants has been described (Peroutka and Snyder 1980), elevated serotonin blood concentrations have been found in chronic schizophrenic patients (Delisi et al. 1981), and finally in drug discrimination studies the subjective feelings induced by LSD, a purported serotonin agonist, are selectively antagonized by 5-HT_2 antagonists.

Pipamperone

In the late 1960s, with only a limited number of pharmacological tests related to serotonin antagonism as background but relying on adequate clinical observation, it was possible to draw a clear profile of pipamperone (Lambert et al. 1969). Pipamperone was found to normalize remarkably emotional tone and behaviour and was therefore considered to be particularly useful in the treatment of relational disorders. Its antipsychotic activity at higher doses was characterized by antiautistic, disinhibiting and resocializing effects that were most helpful in the treatment of chronic psychoses, while displaying a very low liability to develop extrapyramidal symptoms (EPS). Moreover, pipamperone was found to be a good sleep inducer, capable of regulating sleep-wakefulness rhythms that are often disturbed in a wide variety of psychic and mental disorders.

Setoperone

Many years later, in 1983 and 1984, pharmacological and clinical studies with the potent serotonin and moderate dopamine antagonist setoperone further evidenced the new possibilities to treat negative symptom behaviour in chronic schizophrenic patients (Ceulemans et al. 1985). However, very poor bioavailability (<1%) of setoperone prompted us to hypothesize that similar clinical therapeutic results could be obtained by using the potent, selective and bioavailable serotonin antagonist ritanserin.

Ritanserin

Although very important indications for the characterization of ritanserin could be obtained from its receptor binding profile, because this substance shows a very potent and selective binding to the 5-HT_2 receptors in rat frontal cortex, the major property of this compound was discovered in clinical pharmacology studies and more specifically in polysomnographic EEG studies (De Clerck et al. 1987; Idzikowski et al. 1986). In a placebo-controlled crossover sleep study in 9 subjects in Great Britain, a single dose of 10 mg ritanserin in the morning was shown to nearly double the amount of slow wave sleep in healthy volunteers, while the benzodiazepine nitrazepam caused light sleep to increase at the expense of deep sleep. This effect of ritanserin is dose dependent and consistent after repeated intake.

As slow-wave or deep sleep, sometimes called the "restorative phase" of human sleep, is often impaired in schizophrenic patients (Hiatt et al. 1985), this predominant property of ritanserin further stimulated our interest in the study of a very selective serotonin antagonist in schizophrenia.

For clinical testing, a double-blind design was chosen in which ritanserin could be compared with placebo in parallel groups of predominantly type II schizophrenic patients. The double-blind medication was given as an add-on to the existing neuroleptic therapy, and the evaluation was based on a Clinical Global Impression (CGI), the Brief Psychiatric Rating Scale (BPRS) and the Simpson and Angus Scale for EPS.

In comparison with placebo, a significant improvement was observed with ritanserin for the total BPRS, where the largest contribution to the clinical efficacy was a reduction in negative and affective symptoms such as anergia, anxiety/depression and activation.

It is remarkable that the EPS also improved significantly during ritanserin therapy, while they remained unchanged under placebo. Striking among these observations was the significant reduction of tremor, a finding that at a later stage was confirmed in Parkinson patients treated with ritanserin (Riederer et al. 1986; Hildebrand and Delecluse 1987). Although this pharmacological association of a serotonin antagonist and a dopamine antagonist results in a substantial clinical benefit, the practical usefulness of a monotherapy for the treatment of psychotic patients remains undeniable.

Risperidone

Therefore we selected R 64 766 (risperidone), which, on the basis of its in vitro receptor binding profile can be qualified as a potent 5-HT_2 receptor blocker with concomitant binding properties to the dopamine D_2, histamine and noradrenergic receptors but not to the cholinergic ones. In vivo, risperidone was shown to be a very potent serotonin antagonist with, in second place, a dopamine antagonistic activity as evidenced by amphetamine antagonism in rats. At slightly higher doses an interaction with the noradrenergic and histaminergic system was also shown in animal pharmacology.

After completion of the acute and chronic toxicological experiments, pilot clinical studies with risperidone were initiated using an oral solution. Psychotic patients from broad diagnostic categories but clearly presenting florid symptoms such as hallucinations and delusions or severe contact disturbances were selected to enter a dose-finding pilot study with risperidone. During the four weeks administration of risperidone that followed the arrest of the previous medication and a one-week placebo wash-out, a significant improvement on the CGI and on BPRS severity scores (Figure 1) was observed. These results are equally distributed over the different clusters of the BPRS, suggesting an activity on both positive and negative symptoms of schizophrenia.

Interestingly, and in agreement with earlier results with other serotonin antagonists, a consistent reduction in extrapyramidal symptomatology with respect to previous therapy was described during treatment with risperidone, whilst no new or acute EPS were reported. In this trial low median doses of 3 to 4 mg per day produced significant improvement and were very well tolerated.

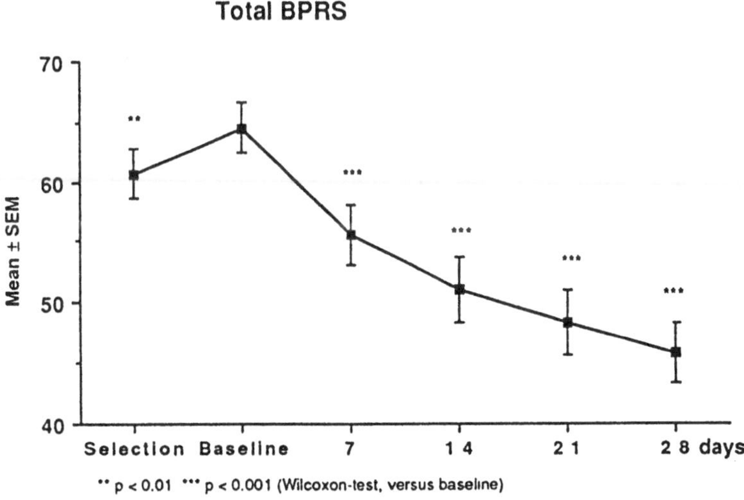

Fig. 1. Risperidone in psychotic patients: BPRS total score

In a subsequent study with rising doses of risperidone from 10 to 25 mg daily, using weekly increments of 5 mg, the general clinical efficacy in chronic schizophrenic patients was evaluated whilst adverse experience and safety of the drug administration were closely monitored.

The CGI was rated weekly on a scale from 0 (poor) to 3 (very good) overall therapeutic effect. There is a significant increase from the first week of treatment (Figure 2). Increasing sedation at higher doses is probably responsible for the decline in improvement of the global impression at the end of the study. The mean total BPRS score of these patients shows a trend to deterioration of the general psychopathology during the placebo wash-out and significant improvement during the 4 subsequent weeks on active risperidone therapy. The cluster of negative symptoms (emotional withdrawal, motor retardation, blunted affect) which are usually very difficult to treat with conventional neuroleptic therapy, also show consistent improvement during the first half of the study. The maximum benefit is obtained at a daily dose of 15 mg. Above this dose, increased psychomotor retardation and bradykinesia are responsible for a gradual neutralization of the original response. These effects influence the total EPS ratings as measured by the Simpson and Angus scale, where after 3 and 4 weeks the pretrial values are slightly exceeded. These results confirm our earlier findings that efficient antipsychotic therapy with risperidone can be expected at daily doses of 10 mg or less.

Discussion and Conclusion

Clinical studies with specific 5-HT_2 antagonists like ritanserin and mixed serotonin-dopamine antagonists like setoperone and risperidone, may reduce rather than widen the gap between different views on amine imbalances in chronic schizophrenia.

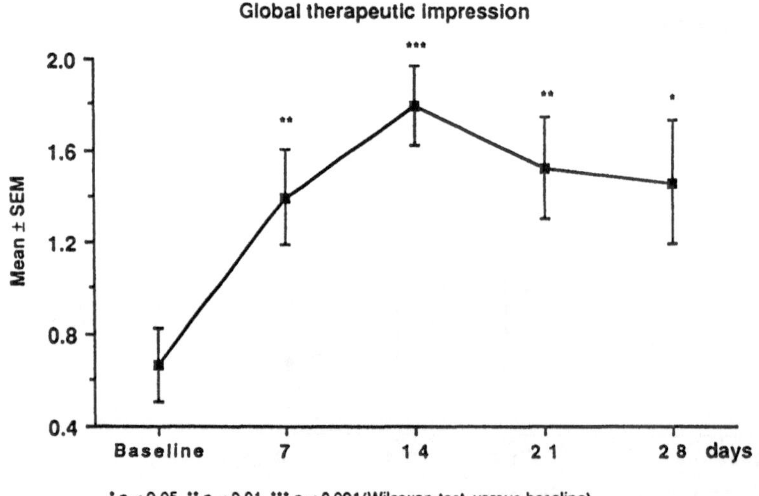

Fig. 2. Dose-rising study of risperidone in psychotic patients: global therapeutic impression

After Crow (1980) had made a distinction between type I schizophrenic patients exhibiting paranoid psychotic behaviour purportedly due to abnormally high dopamine receptor density and type II patients primarily showing negative symptoms, the successful application of serotonin antagonists especially in the latter group supports those who claim a substantial role for both dopamine and serotonin in chronic psychosis.

All studied agents with serotonin antagonistic properties produced an improvement in all BPRS parameters related to mood. These thymosthenic properties seem related to their ability to block 5-HT_2 receptors because they appear also with the most specific 5-HT_2 antagonist ritanserin. The major improvement on drive and mood may be strongly associated with their properties to increase human slow-wave sleep (Reyntjens et al. 1986).

Concomitantly, the reduction of EPS observed with all serotonin antagonists warrants special attention as none of these agents has any anticholinergic activity.

In conclusion, the advent of thymosthenic agents seems to offer two new strategies in the future management of chronic psychoses:

1. A pharmacological monotherapy with either a selective dopamine D_2 antagonist or selective 5-HT_2 antagonist, depending on the polarity of the pattern of complaints (positive versus negative symptoms).

2. A pharmacologically combined dopamine/serotonin antagonist treatment by either dosing 2 selective compounds (e.g. haloperidol plus ritanserin) to the needs of the individual patient or giving monotherapy with combined serotonin/dopamine antagonistic properties (risperidone).

As the international trend is not in favour of combination therapies, a substance with a double profile would have the best probability to success in modern pharmacotherapy. According to the results obtained so far, risperidone fully complies with the requirements of psychiatric practice.

Summary

Central dopamine antagonist properties are the common denominator of antipsychotic drugs. After the discovery of pipamperone in the 1960s, other 5-HT_2 antagonists such as setoperone and ritanserin have proven to be very useful in the treatment of psychotic patients especially in their negative and affective symptoms. The availability of ritanserin as selective serotonin antagonist and risperidone as combined serotonin-dopamine antagonist offers the possibility of maintenance therapies in chronic psychoses based on the energy and mood improving (thymosthenic) properties of these compounds.

References

Carlsson A, Lindqvist M (1963) Effect of chlorpromazine or haloperidol on formation of 3-methoxytyramine and normetanephrine in mouse brain. Acta Toxicol Pharmacol (Copenh) 20: 140-144

Ceulemans DLS, Gelders YG, Hoppenbrouwers M-LJA, Reyntjens AJM, Janssen PAJ (1985) Effect of serotonin antagonism in schizophrenia: a pilot study with setoperone. Psychopharmacology (Berlin) 85: 329-332

Crow TJ (1980) Molecular pathology of schizophrenia: more than one disease process? Br Med J 280: 66-68

Declerck AC, Wauquier A, van der Ham-Veltman PHM, Gelders Y (1987) Increase in slow-wave sleep in humans with the serotonin-S_2 antagonist ritanserin (the first exploratory polygraphic sleep study). Curr Ther Res 41: 427-432

Delay J, Deniker P, Haal JN (1952) Utilisation en thérapeutique psychiatrique d'une phénothiazine d'action centrale élective (4650 RP). Ann Med Psychol 110: 112-116

Delisi LE, Neckers LM, Weinberger DR, Wyatt RJ (1981) Increased whole blood serotonin concentrations in chronic schizophrenic patients. Arch Gen Psychiatry 38: 647-650

Hiatt JF, Floyd TC, Katz PH, Feinberg I (1985) Further evidence of abnormal non-rapid-eye-movement sleep in schizophrenia. Arch Gen Psychiatry 42: 797-802

Hildebrand J, Delecluse F (1987) Effect of ritanserin, a selective serotonin-S2 antagonist, on parkinsonian rest tremor. Curr Ther Res 41: 298-300

Idzikowski C, Mills FJ, Glennard R (1986) 5-Hydroxytryptamine-2 antagonist increases human slow-wave sleep. Brain Res 378: 164-168

Lambert PA, Bouchardy M, Marcou G, Gradel F (1969) Principales indications du pipampèrone. 67ième Congrès Psychiatrique et Neurologique de Langue Francaise, Bruxelles

Peroutka SJ, Snyder SH (1980) Regulation of serotonin-2 ($5HT_2$) receptors labelled with (3H)-spiroperidol by chronic treatment with the antidepressant amitriptyline J Pharmacol Exp Ther 215: 582-587

Reyntjens AJM, Gelders YG, Hoppenbrouwers M-LJA, vanden Bussche G (1986) Thymosthenic effects of ritanserin (R 55667), a centrally acting serotonin-S_2 receptor blocker. Drug Dev Res 8: 205-211

Riederer P, Auff E, Birkmayer W, Brücke T, Goldenberg G, Maly J, Müller C, Pötzl G, Sofic E, Schnaberth G (1986) Ritanserin in the treatment of tremor-dominant Parkinson's disease. Proceedings of the 4th European Workshop on Clinical Pharmacology, Pamplona, p 40

Dopamine D_2-Receptors in Post-mortem Human Brains from Schizophrenic Patients

J. Kornhuber[1], P. Riederer[1], G. P. Reynolds[2], H. Beckmann[1], K. Jellinger[3], and E. Gabriel[4]

[1] Department of Psychiatry, University of Würzburg, FRG
[2] Department of Pathology, University of Nottingham, UK
[3] Ludwig Boltzmann Institute for Clinical Neurobiology, Vienna, Austria
[4] Psychiatric Hospital Baumgartner, Vienna, Austria

The currently predominant biological hypothesis of schizophrenia is the dopamine hypothesis. It is based mainly on the fact that most antipsychotic drugs possess a common ability to block central dopamine D_2 receptors, implying a hyperactive dopaminergic system in schizophrenia. There is little direct evidence for overactive dopaminergic neurons in schizophrenia. Thus in recent years the alternative hypothesis of changed postsynaptic dopamine receptors has been investigated. Dopamine receptors, as defined by the binding of various ligands, e.g. [^{3}H]spiperone (Owen et al. 1978; Mackay et al. 1982; Seeman et al. 1984; Pimoule et al. 1985; Mita et al. 1986) and [^{3}H]flupenthixol (Cross et al. 1981), were measured at either a single concentration or a range of concentrations to determine maximum number of binding sites (Bmax) and apparent equilibrium dissociation constant (K_D). Elevated D_2 receptor densities in schizophrenics have been reported in most of these studies, while the D_1 receptors were found to be unchanged by most investigators using either a single concentration of [^{3}H] ligand (Cross et al. 1981; Pimoule et al. 1985) or a range of concentrations to determine D_1 receptor densities (Seeman et al. 1987). The interpretation of these results, however, remains difficult, since most patients included in these studies had been treated with neuroleptic drugs. Long-term neuroleptic administration results in an increase in D_2 (Owen et al. 1980; Mackenzie and Zigmond 1985) but not D_1 receptors (Mackenzie and Zigmond 1985) in animal experiments. It is possible, therefore, that changes in D_2 receptors in schizophrenia are due mainly to chronic neuroleptic treatment.

Recently, new evidence for the dopamine hypothesis of schizophrenia has emerged from in vivo positron emission tomography studies (Wong et al. 1986), the finding of higher dopamine levels in left compared to right amygdala in post-mortem schizophrenic brains (Reynolds 1983) and elevated synaptosomal high-affinity uptake of dopamine (Haberland and Hetey 1987). Therefore, we have reexamined dopamine D_2 receptors in post-mortem brains of schizophrenic patients paying attention to neuroleptic treatment, abnormal movements and psychopathology prior to death. Some of these results were published previously (Kornhuber et al. 1989).

Putamen samples were obtained at autopsy from 27 control patients and 27 schizophrenic patients. Histopathological examination was performed on all brains to exclude out other abnormalities such as tumour, infarction, anoxia, brain atrophy or Alzheimer's disease. Post-mortem delay time was less than 10 hours in all but 2 cases (14 and 20 hours, controls). Control patients had no history of neurological or psychiatric disorder. Patients were included in the schizophrenic group only if they met the diagnostic criteria of Feighner et al. (1972). Furthermore, most psychotic patients (n = 25) conformed to the diagnosis of schizophrenia according to the International Classification of Diseases (ICD-9; Table 2). Patients were classified as drugfree when there was no neuroleptic treatment at least 3 months prior to death ($n=9$; in 3 schizophrenic patients there was no neuroleptic treatment for at least 1 year before death). If a mean neuroleptic drug-free time was calculated for a group of patients, 90 days was used for 90 or more drug-free days. The presence or absence of tardive dyskinesia prior to death was recorded from case notes. Furthermore, two groups of patients were selected by retrospective analysis of case reports. One group exhibited more positive schizophrenic symptoms at time of death, such as hallucinations and delusions (corresponding to ICD-9, 295.3), and one group exhibited more negative symptoms, e.g. loss of volition and flattening of affect (corresponding to ICD-9 295.6; Table 2).

For each brain sample, a saturation analysis was performed to determine apparent K_D and Bmax values, using $[^3H]$spiperone. In order to check the method, tissue samples from another subarea of the putamen of 7 of our schizophrenic patients were analysed in parallel by Seeman and his group in Toronto (Seeman et al. 1984). There was a good correlation between these and our own Bmax values ($r=0.82$, $p<0.05$). Nonparametric statistics (Spearman's rank correlation coefficient, Wilcoxon Mann Whitney U-test, Pearson's chi-square test, Kolmogoroff-Smirnoff test) were applied throughout using the two-tailed approach. P-values higher than 0.05 were regarded as not significant.

Figure 1 and Table 1 demonstrate the dependence of Bmax and K_D values for $[^3H]$spiperone binding on prior neuroleptic treatment. Bmax values of patients treated with neuroleptics within a short period prior to death are elevated compared to controls ($p<0.05$; Table 1) and decrease as a function of drug-free time ($r=0.51$, $p<0.01$; Fig. 1). Bmax values in off-drug schizophrenic patients are lower than control values (not significant). The K_D values scatter widely and are greatly elevated in brain tissues from schizophrenics treated with neuroleptic drugs in the days before death. Compared to the decrease in Bmax with neuroleptic-free time, the K_D values decrease more rapidly (within two weeks) and are not different from control values thereafter (Fig. 1).

Table 1. Case data and Bmax and K_D values: correlations between Bmax and K_D values are indicated (r)

	Controls	Schizophrenics		
		All	On drug	Off drug
	(n=27)	(n=27)	(n=18)	(n=9)
M/F	7/20	9/18	7/11	2/7
Age (years)	72.8±15.4	70.6±8.4	69.3±8.5	73.3±7.9
Bmax (pmol/g tissue)	19.3±7.1	23.0±11.1	26.8±11.2 *	15.3±5.8
K_D (nM)	0.125±0.062	0.475±0.671 **	0.640±0.775 **	0.144±0.036
r	0.27	0.43 *	0.48 *	-0.48

*$p<0.05$, **$p<0.001$ compared to controls

Table 2. Mean neuroleptic-free time and binding characteristics in post-mortem putamen from patients with more positive (ICD-9 295.3) versus more negative (ICD-9 295.6) symptoms

ICD-9 subtype	295.3 (n=6)	295.6 (n=17)	
Bmax (pmol/g tissue)	29.15±10.53	20.93±10.45	ns
K_D (nM)	0.539± 0.350	0.488±0.818	ns
Drug-free time (days)	4.2±5.1	50.65±41.26	*

ns = not significant, * $p<0.05$

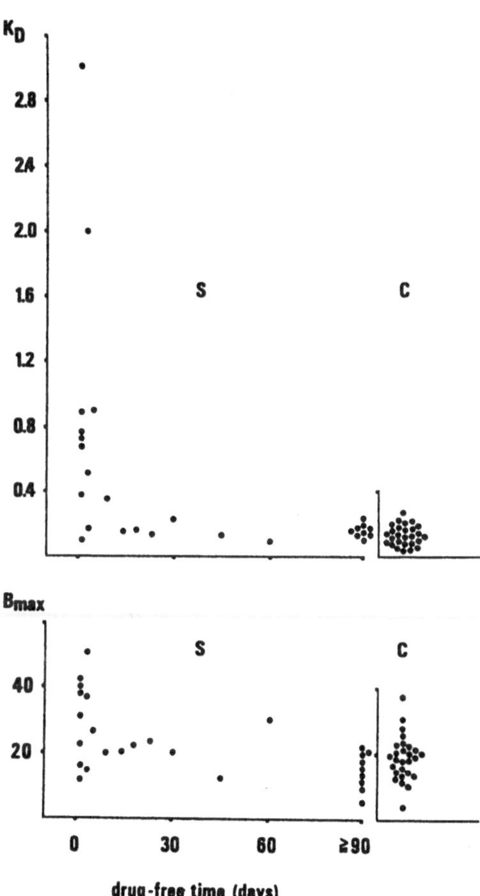

Fig. 1. Individual Bmax (pmol/g tissue) and K_D (nM) values in post-mortem putamen from schizophrenics (S) and controls (C). The neuroleptic-free time interval prior to death (day 0) is indicated

Enhanced K_D values reflect an apparent decrease in receptor affinity, which is thought to be the result of residual neuroleptic drug in post-mortem brain tissue competing with the tritiated ligand (Owen et al. 1978, 1979; Mackay et al. 1982; Seeman et al. 1984). The rapid normalization of K_D values with increasing drug-free time found in the present study is compatible with this assumption. The faster drop in K_D values following drug withdrawal, compared to Bmax values, is in agreement with animal experiments (Owen et al. 1980; Murugaiah et al. 1982). We found a significant positive correlation between K_D values (enhanced K_D values represent residual neuroleptic activity) and Bmax values in on-drug schizophrenics, while no significant correlation was observed in off-drug patients and controls (Table 1). These results are compatible with the view that enhanced Bmax values are mainly iatrogenic and are in agreement with the findings of Mackay et al. (1982) who found unchanged Bmax values for [^{3}H]spiperone binding in the caudate nucleus and nucleus accumbens in drug-free schizophrenic patients. Furthermore, the results are in agreement with in vivo positron emission tomography measurements using either [^{11}C]methylspiperone (Herold et al. 1985) or the highly selective D_2 antagonist [^{11}C]raclopride (Farde et al. 1987; but see also Crawley et al. 1986; Wong et al. 1986).

The course of tardive dyskinesia is widely ascribed to long-term administration of neuroleptic drugs, which are dopamine receptor antagonists. The prevailing dopamine receptor hypersensitivity hypothesis (Klawans and Rubovits 1972) was derived from behavioural experiments. The up-regulation in the number of D_2 receptors following long-term administration of neuroleptic drugs would explain the hypersensitivity seen in behavioural experiments and offers a neurobiological substrate of tardive dyskinesia. However, this hypothesis does not explain various aspects of the syndrome (Fibiger and Lloyd 1984). In agreement with this criticism we found nearly identical dopamine receptor densities in putamen samples from schizophrenic patients who suffered from tardive dyskinesia prior to death compared to those who did not (Table 3). Since all clinical data was obtained from retrospective analysis, it appears possible that some of the schizophrenic patients did show tardive dyskinesia which was not documented in

Table 3. Mean neuroleptic-free time and binding characteristics in post-mortem putamen from patients with the absence or presence of tardive dyskinesia prior to death

	Tardive Dyskinesia		
	Yes (n=5)	No (n=22)	
Bmax (pmol/g tissue)	22.97±10.22	22.69±11.52	ns
K_D (nM)	0.274±0.261	0.521±0.729	ns
Drug-free time (days)	51.20±38.79	35.18±40.39	ns

ns = not significant

the case reports. Our results, however, are in agreement both with other direct D_2 receptor measurements (Crow et al. 1981, 1982; Cross et al. 1985) and with indirect neuroendocrinological data (Tamminga et al. 1977).

Table 2 shows a nonsignificant increase in Bmax and K_D values in patients with more positive schizophrenic symptoms. However, this seems to be related to the drug-free time interval, which was shorter in these patients ($p < 0.05$). Recently, elevated D_2 receptor densities were found in patients with mainly positive schizophrenic symptoms (Mita et al. 1986), and there was a good correlation between [^{3}H]spiperone binding and severity of positive symptoms (Crow et al. 1981). Usually, neuroleptic drugs are more effective in the treatment of positive compared to negative symptoms. Therefore, it is likely that the neuroleptic-free time interval prior to death is shorter in patients with positive symptoms compared to those with negative symptoms. This hypothesis is substantiated by our data. The dosage of neuroleptic treatment probably correlates with the severity of positive symptoms. Both factors could result in D_2 receptor densities correlating with positive symptoms.

It has been claimed that Bmax values of D_2 receptors in schizophrenic patients have a bimodal distribution (Seeman et al. 1984). In the present study the D_2 receptor densities in the putamen of neither controls nor schizophrenics differed statistically significantly from a normal distribution ($p = 0.99$ and $p = 0.33$ respectively, Kolmogoroff-Smirnoff test; Fig. 2). This may, however, be due to the smaller number of cases in the present investigation.

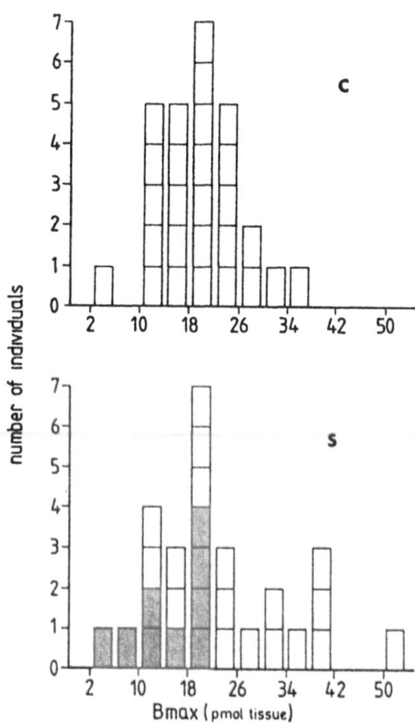

Fig. 2. Distribution of dopamine D_2 receptor densities in the putamen of control (C) and schizophrenic (S) patients. The hatched rectangles indicate patients who were drug free for at least 3 months

From our data we conclude that changes in Bmax and K_D values in schizophrenics are almost entirely iatrogenic, and that there is no obvious relationship to abnormal movements or positive schizophrenic symptoms prior to death.

References

Crawley JCW, Crow TJ, Johnstone EC, Oldland SRD, Owen F, Owens DGC, Smith T, Veall N, Zanelli GD (1986) Uptake of ^{77}Br-spiperone in the striata of schizophrenic patients and controls. Nucl Med Commun 7: 599-607

Cross AJ, Crow TJ, Owen F (1981) ^{3}H-Flupenthixol binding in post-mortem brains of schizophrenics: evidence for a selective increase in dopamine D_2 receptors. Psychopharmacology (Berlin) 74: 122-124

Cross AJ, Crow TJ, Ferrier IN, Johnson JA, Johnstone EC, Owen F, Owens DGC, Poulter M (1985) Chemical and structural changes in the brain in patients with movement disorders. In: Casey DE, Chase TN, Christensen AV, Gerlach J (eds) Dyskinesia: research and treatment. Berlin Heidelberg New York, Springer pp 104-110

Crow TJ, Owen F, Cross AJ, Ferrier N, Johnstone EC, McCreadie RM, Owens DGC, Poulter M (1981) Neurotransmitter enzymes and receptors in post-mortem brain in schizophrenia: evidence that an increase in D_2 dopamine receptors is associated with the type I syndrome. In: Riederer P, Usdin E (eds) Transmitter biochemistry of human brain tissue. London, Macmillan, pp 85-96

Crow TJ, Cross AJ, Johnstone EC, Owen F, Owens DGC, Waddington JL (1982) Abnormal involuntary movements in schizophrenia: are they related to the disease process or its treatment? Are they associated with changes in dopamine receptors? J Clin Psychopharmacol 2: 336-340

Farde L, Wiesel FA, Hall H, Halldin C, Stone-Elander S, Sedvall G (1987) No D_2 receptor increase in PET study of schizophrenia. Arch Gen Psychiatry 44: 671-672

Feighner JP, Robins E, Guze SB, Woodruff RA, Winokur G, Munoz R (1972) Diagnostic criteria for use in psychiatric research. Arch Gen Psychiatry 26: 57-63

Fibiger HC, Lloyd KG (1984) Neurobiological substrates of tardive dyskinesia: the GABA hypthesis. Trends Neurosci 7: 462-464

Haberland N, Hetey L (1987) Studies in postmortem dopamine uptake. II. Alterations of the synaptosomal catecholamine uptake in postmortem brain regions in schizophrenia. J Neural Transm 68: 303-313

Herold S, Leenders KL, Turton DR, Kensett MJ, Pike VW, Clark JC, Brooks DJ, Crow TJ, Owen F, Cooper S, Johnstone EC (1985) Dopamine receptor binding in schizophrenic patients as measured with ^{11}C-methylspiperone and PET. J Cereb Blood Flow Metab 5: S191-S192

Klawans HL, Rubovits R (1972) An experimental model of tardive dyskinesia. J Neural Transm 33: 235-246

Kornhuber J, Riederer P, Reynolds GP, Beckmann H, Jellinger K, Gabriel E (1989) ^{3}H-Spiperone binding sites in post-mortem brains from schizophrenic patients. Relationship to neuroleptic drug treatment, abnormal movements and positive symptoms. J Neural Transm75:1-10

Mackay AVP, Iversen LL, Rossor M, Spokes E, Bird E, Arregui A, Creese I, Snyder SH (1982) Increased brain dopamine and dopamine receptors in schizophrenia. Arch Gen Psychiatry 39: 991-997

Mackenzie RG, Zigmond MJ (1985) Chronic neuroleptic treatment increases D-2 but not D-1 receptors in rat striatum. Eur J Pharmacol 113: 159-165

Mita T, Hanada S, Nishino N, Kuno T, Nakai H, Yamadori T, Mizoi Y, Tanaka C (1986) Decreased serotonin S_2 and increased dopamine D_2 receptors in chronic schizophrenics. Biol Psychiatry 21: 1407-1414

Murugaiah K, Theodorou A, Mann S, Clow A, Jenner P, Marsden CD (1982) Chronic continuous administration of neuroleptic drugs alters cerebral dopamine receptors and increases spontaneous dopaminergic action in the striatum. Nature 296: 570-572

Owen F, Cross AJ, Crow TJ, Longden A, Poulter M, Riley GJ (1978) Inreased dopamine-receptor sensitivity in schizophrenia. Lancet 2: 223-226

Owen F, Cross AJ, Poulter M, Waddington JL (1979) Change in the characteristics of ^{3}H-spiperone binding to rat striatal membranes after acute chlorpromazine administration: effects of buffer washing of membranes. Life Sci 25: 385-390

Owen F, Cross AJ, Waddington JL, Poulter M, Gamble SJ, Crow TJ (1980) Dopamine-mediated behaviour and ^{3}H-spiperone binding to striatal membranes in rats after nine months haloperidol administration. Life Sci 26: 55-59

Pimoule C, Schoemaker H, Reynolds GP, Langer SZ (1985) [^{3}H] SCH 23390 labeled D_1 dopamine receptors are unchanged in schizophrenia and Parkinson's disease. Eur J Pharmacol 114: 235-237

Reynolds GP (1983) Increased concentrations and lateral asymmetry of amygdala dopamine in schizophrenia. Nature 305: 527-529

Seeman P, Ulpian C, Bergeron C, Riederer P, Jellinger K, Gabriel E, Reynolds GP, Tourtellotte WW (1984) Bimodal distribution of dopamine receptor densities in brains of schizophrenics. Science 225: 728-731

Seeman P, Bzowej NH, Guan HC, Bergeron C, Reynolds GP, Bird ED, Riederer P, Jellinger K, Tourtellotte WW (1987) Human brain D_1 and D_2 dopamine receptors in schizophrenia, Alzheimer's, Parkinson's and Huntington's diseases. Neuropsychopharmacology 1:5-15

Tamminga CA, Smith RC, Pandey G, Frohman LA, Davis JM (1977) A neuroendocrine study of supersenitivity in tardive dyskinesia. Arch Gen Psychiatry 34: 1199-1203

Wong DF, Wagner HN, Tune LE, Dannals RF, Pearlson GD, Links JM, Tamminga CA, Broussolle EP, Ravert HT, Wilson AA, Toung JKT, Malat J, Williams JA, O'Tuama LA, Snyder SH, Kuhar MJ, Gjedde A (1986) Positron emission tomography reveals elevated D_2 dopamine receptors in drug-naive schizophrenics. Science 234: 1558-1563

Phencyclidine - A Challenge to Schizophrenia Research

M. E. Kornhuber
Abteilung Neurologie und Psychiatrie, Bundeswehrkrankenhaus, Oberer Eselsberg 40, 7900 Ulm, FRG

Based on the antidopaminergic action of antipsychotic drugs, the prevailing biochemical hypothesis on schizophrenia suggests a hyperfunction of central dopaminergic systems to be of pathogenetic relevance in this disorder. Despite numerous attempts to verify this hypothesis no convincing evidence has as yet appeared in support (see J.Kornhuber et al., this volume). Therefore the pathogenetic substrate underlying schizophrenia may well be in a different neurotransmitter system linked to dopaminergic neurons. Such relations to the dopaminergic system exist for example, for cholinergic, GABAergic and glutamatergic neurons (Kim et al. 1981; J.Kornhuber et al. 1984 a, b; Kornhuber and Kornhuber 1986). Among these the glutamatergic neurons seem to play an extraordinary role.

Glutamate is the predominant excitatory transmitter in the vertebrate central nervous system (Iversen 1984), especially that of corticofugal fibres (Cotman et al. 1987; Kim et al. 1977; J.Kornhuber et al. 1984 b). Mesencephalic dopaminergic neurons receive glutamatergic input from the prefrontal cortex (J.Kornhuber et al. 1984 b), a brain region which may be of relevance in schizophrenia. The glutamatergic input to the striatum is subject to dopaminergic mesencephalic modulation (see Kornhuber and Kornhuber 1986 for review). Dopamine agonists and antagonists may change striatal glutamate levels (Kim et al. 1981; J.Kornhuber et al. 1984 a, 1985). The finding of reduced cerebrospinal fluid glutamate in schizophrenic patients led to the glutamate hypothesis of schizophrenia (Kim et al. 1981, 1985; H.H.Kornhuber et al. 1984).

Some hope for insight into the pathogenesis of schizophrenia comes from the investigation of model psychoses. The best model psychosis for schizophrenia at the present time seems to be the phencyclidine (PCP) psychosis (Javitt 1987; Luby et al. 1959; Luisada 1978; Snyder 1980). PCP is a frequently used halucinogen in the USA and may cause psychotic episodes that were misdiagnosed as schizophrenia when the drug was introduced. Compared to the amphetamine psychosis the psychotomimetic effect is achieved more rapidly, and there may occur negative symptoms in addition to productive psychotic symptoms thus resulting in the full clinical schizophrenia-like syndrome.

Several lines of evidence (coming from pharmacological, physiological and behavioural studies) indicate that the central effects of PCP are at least in part mediated via the NMDA receptor complex (Anis et al. 1983; Cotman and Iversen 1987; Jones et al. 1987; M.E.Kornhuber et al. 1985) which is one of three known glutamate receptor subtypes (Watkins and Olverman 1987). This PCP-interaction with the NMDA receptor complex may account, for example, for analgesia, diminished storage of memory and behavioural stereotypies (Collinridge and Bliss 1987; Jones et al 1987; M.E.Kornhuber et al. 1985; Schmidt 1986). The hypothesis that PCP produces its psychotomimetic effects via the glutamatergic system (H.H.Kornhuber et al. 1984; M.E.Kornhuber et al. 1985) has since then received much interest (Etienne and Baudry 1987; Javitt 1987).

The aim of this study was to characterize the effect of PCP in a glutamatergic model system (locust neuromuscular junction) that might also serve the purpose of screening in future pharmacological studies with PCP. Distal muscle fibres of the extensor tibiae muscle of the locust *Schistocerca gregaria* were kept in slowly flowing locust saline (NaCl 150, KCl 10, $CaCl_2$, Tris-maleate 2 and sucrose 90 mM; pH 6.8). By antidromic stimulation of the axon of the slow excitatory neuron (SETi) excitatory junction potentials (EJPs) not leading to muscular contractions were recorded with a single intracellular glass microelectrode. PCP (range $5x10^{-6}$ to $5x10^{-5}$ M) did not affect the resting membrane potential or the membrane resistance, as estimated from the time constant of decay of EJPs (ME Kornhuber und Walther 1987). Bath applied PCP ($2.5x10^{-5}$ M) reduced the EJP amplitude by about 65%, indicating a blockade of the neuromuscular transmission (Fig. 1). The PCP derivative ketamine was about 10 times less effective in blocking the neuromuscular transmission than PCP (results not shown). Since a report on the postsynaptic effect of PCP in this preparation has already appeared (Idriss et al. 1985) we investigated the presynaptic properties under PCP.

We therefore studied

1. The coefficient of variation of the EJP amplitudes (CV)
2. The failure rate, i.e. the rate of non-following EJPs to nervous stimulation
3. The rate of spontaneously occurring transmitter release, i.e. miniature excitatory junction potentials (MEJPs) under PCP.

The results were as follows (summarized in Table 1).

1. PCP leads to an increase in CV by roughly 20% at $2.5x10^{-5}$ M, indicating a reduced number of transmitter quanta per EJP since CV depends on the number of trans-

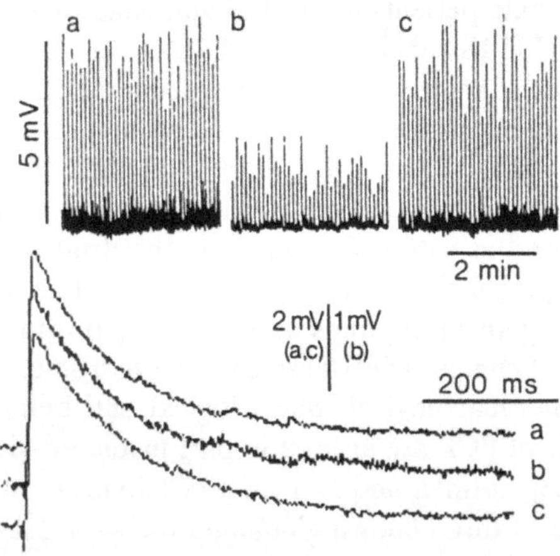

Fig. 1. Effect of $2.5x10^{-5}$ M PCP on amplitude and time course of excitatory junction potentials (EJPs). **A** Before. **B** In the presence of. **C** After PCP. The lower part of the figure shows single records from the upper part at an extended time scale. Gain at B in the lower part is twice that A and C to facilitate comparison of the time course of the EJPs

Table 1. Effect of 2.5×10^{-5} M PCP on various parameters of the neuromuscular transmission of the locust. Mean values (± SEM) of 4 experiments are given. *$P < 0.05$; n.s., not significant. The failure rate has been omitted here because at 2.5×10^{-5} results of but one experiment have been available

	Control	PCP	%Change
EJP amplitude (mV)	6.30±0.9	2.30±0.4	- 64 *
CV of EJP amplitude	0.21±0.05	0.24±0.05	+19 *
MEJPamplitude (mV)	0.21±0.01	0.10±0.01	- 53 *
Number of MEJP/s	4.50±0.6	4.40±0.6	n.s.
Resting membrane potential (mV)	61.40±6.1	61.50±6.4	n.s.
Time constant of EJP (ms) 86.00±21	86.00±20	n.s.	

mitter quanta contributing to the EJPs (Del Castillo and Katz 1954). Increasing CV means decreasing quantal content since the relative difference between single EJP amplitudes and the mean EJP amplitude increase relative to the absolute value of the EJP amplitude.

2. Since the transmitter release depends on the Ca^{2+} concentration in the external solution, the failure rate (see above) was studied at a reduced Ca^{2+} concentration. PCP led to an increased failure rate (about 25% at 2.5×10^{-5} M), again indicating a diminution in the quantal content.
3. The number of MEJP/s, however, did not show a significant change under PCP (2.5×10^{-5} M), indicating an unaltered spontaneous transmitter release under PCP.

The mechanism of reduction of quantal transmitter release by PCP may be explained by a blockade of Na^+ and/or Ca^{2+} channels. A blockade of K^+ channels cannot account for these results since this would rather diminish transmitter release, as was shown in the vertebrate CNS under PCP. This difference does not, however, rule out a possible role for this preparation in future pharmacological screening of drugs that may alter glutamatergic neurotransmission, thus serving the possible development of new antipsychotic drugs.

References

Anis N A, Berry S C, Burton N R, Lodge D (1983) The dissociative anaesthetics, ketamine and phencyclidine, selectively reduce excitation of central mammalian neurons by N-methyl-aspartate. Br J Pharmacol 79 : 565-575

Blaustein M P, Bartschat D K, Sorensen R G (1986) Phencyclidine (PCP) selectively blocks certain presynaptic potassium channels. NIDA Res Monogr 64 : 37-51

Collingridge G L, Bliss T V P, (1987) NMDA receptors - their role in long-term potentiation. Top Neurol Sci 10: 288-293

Cotman C W,Iversen L L, (1987) Excitatory amino acids in the brain - focus on NMDA receptors. Top Neurol Sci 10: 263-265

Cotman C W, Monaghan D T, Ottersen O P, Storm-Mathisen J (1987) Anatomical organization of excitatory amino acid receptors and their pathways. Top Neurol Sci 10: 273-280

Del Castillo J, Katz B (1954) Quantal components of the end plate potential. J Physiol 124: 560-573

Etienne P, Baudry M (1987) Calcium dependent aspects of synaptic plasticity, excitatory amino acid neurotransmission, brain aging and schizophrenia: a unifying hypothesis. Neurobiol Aging 8: 362-366

Idriss M, Albuquerque E X (1985) Phencyclidine (PCP) blocks glutamate activated postsynaptic currents. FEBS Lett 189: 150-156

Iversen L L (1984) Amino acids and peptides: fast and slow chemical signals in the nervous system? Proc R Soc Lond Bio 221: 245-260
Javitt D C (1987) Negative schizophrenic symptomatology and the PCP (phencyclidine) model of schizophrenia. Hillside J Clin Psychiatry 9: 12-35
Jones S M, Snell L D, Johnson K M (1987) Inhibition by phencyclidine of excitatory amino acid stimulated release of neurotransmitter in the nucleus accumbens. Neuropharmacology 26: 173-179
Kim J S, Hassler R, Haug P, Paik K S (1977) Effect of frontal cortex ablation on striatal glutamic acid level in the rat. Brain Res 132: 370-374
Kim J S, Kornhuber H H, Schmid-Burgk W, Holzmüller B (1980) Low cerebrospinal fluid glutamate in schizophrenic patients and a new hypothesis on schizophrenia. Neurosci Lett 20: 379-382
Kim J S, Kornhuber H H, Brandt U, Menge H G (1981) Effects of chronic amphetamine treatment on the glutamate concentration in cerebrospinal fluid and brain: implications for a theory of schizophrenia. Neurosci Lett 24: 93-96
Kim J S, Kornhuber H H, Kornhuber J, Kornhuber M E (1985) Glutamic acid and the dopamine hypothesis of schizophrenia. Biol Psychiatry 1109-1111
Kornhuber H H, Kornhuber J, Kim J S, Kornhuber M E (1984) Zur biochemischen Theorie der Schizophrenie. Nervenarzt 55: 602-606
Kornhuber J, Kornhuber M E (1986) Presynaptic dopaminergic modulation of cortical input to the striatum. Life Sci 39: 669-674
Kornhuber J, Kim J S, Kornhuber M E, Kornhuber H H (1984 a) The interaction of the dopaminergic and glutamatergic systems in vivo: experiments with L-dopa and bromocriptine. CINP Congress p 189
Kornhuber J, Kim J S, Kornhuber M E, Kornhuber H H (1984 b) The cortico-nigral projection: reduced glutamate content in the substantia nigra following frontal cortex ablation in the rat. Brain Res 322: 124-126
Kornhuber J, Kim J S, Kornhuber M E (1985) Substantia nigra glutamate under L-dopa and fluphenazine. J Neurochem 44S: 108A
Kornhuber M E, Walther C (1987 a) The electrical constants of the fibres from two leg muscles of the locust Schistocerca gregaria. J Exp Biol 127: 173-189
Kornhuber M E, Kornhuber J, Zettlmeißl H, Kornhuber H H (1985) Phencyclidin und das glutamaterge System. In: Keup W (ed) Biologische Psychiatrie. Springer, Berlin Heidelberg New York
Kornhuber M E, Walther C, Kornhuber J (1987 b) Phencyclidine depresses transmitter release at the neuromuscular synapse of the locust. Neurosci Lett 83: 185-189
Luby E D, Cohen B D, Rosenbaum G, Gottlieb J S, Kelley R (1959) Study of a new schizophrenomimetic drug - sernyl. Arch Neurol Psychiatry 81: 363-369
Luisada P V (1978) The phencyclidine psychosis: phenomenology and treatment. NIDA Res Monogr 21: 241-254
Schmidt W J (1980) Intrastriatal injection of DL-2-amino-5-phosphonovaleric acid (AP5) induces sniffing stereotypy that is antagonized by haloperidol and clozapine. Pharmacopsychiatry 90: 123-130
Snyder S H (1980) Phencyclidine. Nature 285: 355-356
Watkins J C, Olverman H J (1987) Agonists and antagonists for excitatory amino acid receptors. Top Neurol Sci 10: 265-272

Neurotoxic Metabolites of Tyrosine/Dopamine in Cerebrospinal Fluid and Serum of Normal Men and Neurological Patients. A Sign of the Activity of free Oxygen Radicals?

H. Zettlmeißl, S. Häusermann, H. Maurer, H.H. Kornhuber
Neurologische Universitätsklinik Ulm, 7900 Ulm,FRG

Introduction

In order to look more closely at neurotransmitters and their metabolites in cerebrospinal fluid, a reliable, highly selective HPLC method was developed (Zettlmeißl et al.1986). With this method several unknown peaks appeared. One was identified as o-tyrosine, another as m-tyrosine. These substances were present in trace amounts in healthy persons and in higher concentrations in some neurological patients, especially patients with amyotrophic lateral sclerosis and with multiple sclerosis in exacerbations of this disease. The physiological form of tyrosine in the metabolism is p-tyrosine.

For the hydroxylation of phenylalanine the enzyme phenylalanine hydroxylase uses small amounts of hydroxylradicals. Besides this physiological use of radicals, free oxygen radicals may be generated in living beings if protective mechanisms are deficient. Therefore we looked for polyhydroxylated compounds of the tyrosine/dopamine metabolism.

Methods

For this purpose we developed a new HPLC method which is able to separate many aromatic compounds at once, using their "eigenfluorescence" (excitation 290 nm, emission 360 nm).

Results

In this way we were able to determine the number of hydroxylated substances separated from tyrosine to 6-hydroxydopamine without disturbance by other substances. The compounds were separated without interference (Fig. 1). When analyzing biological samples in a short time under protection from oxygen, we found 6-hydroxydopamine and 5-hydroxydopamine (Fig. 2). The identification of these substances in human cerebrospinal fluid and serum was done by identical samples given to the biological matrix. 6-Hydroxydopamine is a neurotoxic substance. 5-Hydroxydopamine is a false neurotransmitter. 6-Hydroxydopamine is a very unstable compound which can be easily oxidized

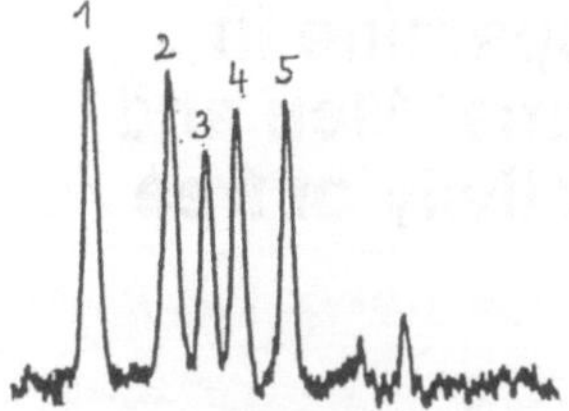

Fig. 1. Chromatogram of the pure substances, showing the ability of the HPLC method to separate, among others, 6-hydroxydopamine (2), 5-hydroxydopamine (3), dopa (4), and 6-hydroxydopa (5). (1) is para-tyrosine

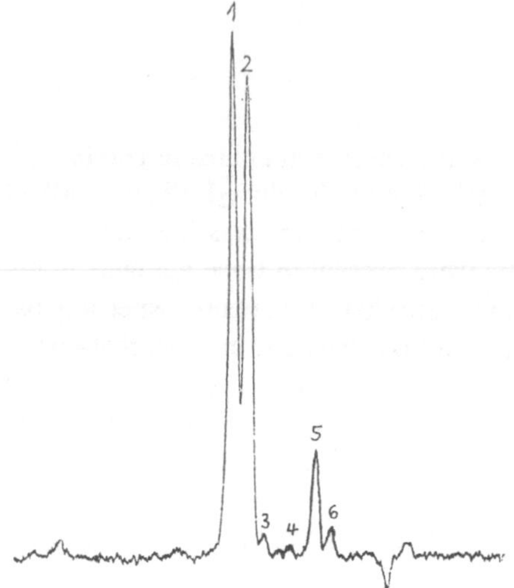

Fig. 2. Chromatogram from human serum with p-tyrosine (1), the chinone form of 6-hydroxydopamine (2), 6-hydroxydopamine (3), 5-hydroxydopamine (4), dopa (5), and 6-hydroxydopa (6). O-tyrosine, m-tyrosine and dopamine are so far separated that they appear much later on the same chromatogram

to its chinone form. This chinone is highly reactive and adds amino, mercapto, and hydroxylgroups in a short time. One of these reaction products increases when we add 6-hydroxydopamine to the biological matrix.

The amount of the neurotoxic substances and their reaction products seems to depend on factors such as infectious disease, alcohol consumption and cigarette smoking as far as we can tell from our pilot studies.

We started experiments in vitro using oxidating substances and compounds like phenylalanine, tyrosine . Under these conditions, a large number of hydroxylized reaction products appear. This system is useful to test the effects of radical scavengers. The efficiency of substances such as vitamin E, β-carotine, glutathione, *N*-acetyl-cysteine and others has been demonstrated with this system.

Interestingly, an oscillating reaction was found in this in vitro redoxsystem with p-tyrosine and 5-hydroxydopamine.

Discussion

Since 5-hydroxydopamine and 6-hydroxydopamine have not been identified before in humans or in animals, and since no enzymes are known for their biosynthesis, it may be that their appearance is a sign of the activity of free oxygen radicals in unphysiological amounts.

Summary

Using a new HPLC method, neurotoxic metabolites of the tyrosine/dopamine pathway have been identified in human cerebrospinal fluid and serum. The appearance of these substances is perhaps a sign of unphysiological activity of free oxygen radicals.

References

Zettlmeißl H, Blome J, Kornhuber H H (1986) A sensitive, fast durable, highly selective method for the determination of amino acids and biogenic amines in the cerebrospinal fluid and other body-fluids and tissues. Arch Ital Biol 124: 129 - 132

Searching for New Antiischemic Compounds: Theoretical and Practical Aspects

B. Wilffert and T. Peters

Janssen Research Foundation, Raiffeisenstraβe 8, 4040 Neuss 21, FRG

Many substances act in an antiischemic manner, e.g. membrane-stabilizing drugs, cationic amphiphilic compounds,such as β-sympathicolytics, local anesthetics and all agents that suppress cardiac function, e.g. calcium entry blockers. However, high concentrations are necessary, and the side effects are pronounced. It would be desirable to find agents which affect only the pathophysiological processes observed during ischemia without influencing physiological processes. The optimal stage of interaction would be early in the ischemic cascade.

What happens during ischemia? Oxygen shortage results in activation of the anaerobic glycolysis which leads to an increased lactate production and a decrease in the intracellular pH. The inner leaflet of the plasmalemma contains acidic lipids such as phosphatidylserine (Post et al. 1988), a binding partner for calcium which seems to contribute to physiological regulation of intracellular release and sequestration of calcium ions (Lüllmann and Peters 1977). Increasing the proton concentration, as observed during ischemia, releases calcium from this pool by decreasing the binding affinity of phosphatidylserine for calcium (Peters 1986a). Another important intracellular calcium store, the sarcoplasmic reticulum also releases calcium upon acidification, at least in the heart (Fabiato 1985). Both processes contribute to an increase in the calcium concentration in the cytosol, which opens a putative nonspecific cation channel (Smith 1988). This results in an influx of calcium and sodium and an efflux of potassium according to the respective concentration gradients. The increased calcium concentration activates phospholipase causing a release of free fatty acids, prostaglandin and leukotriene formation and membrane disruption in concert with the calcium-induced protease activation. Furthermore, as soon as the extracellular potassium concentration exceeds a threshold of about 12 mmol/l, potential dependent calcium entry processes are activated in blood vessels and neurons causing vasoconstriction and neurotransmitter release (Akerman and Heinonen 1983). The action potential is shortened during ischemia (Sperelakis 1988) and delayed after-depolarizations could be expected during reperfusion of the ischemic tissue, which may be the basis of some arrhythmias associated with reperfusion (Coetzee et al. 1987).

For screening purposes it would be advantagous if the ischemic cascade could be mimicked in isolated tissues. Cardiac glycoside intoxication shows many similarities with ischemia. In therapeutic concentrations of cardiac glycosides the calcium concentration in the cytosol and therefore inotropy increases as a result of the interaction with Na-K-ATPase, and a shortening of the action potential duration is observed (Peters 1986b). During cardiac glycoside intoxication the intracellular calcium concentration rises, further activating the transient inward current, thereby further increasing the sodium concentration already enhanced by Na-K-ATPase inhibition and provoking delayed after-depolarizations (Borchard and Ravens 1986; Coetzee et al. 1987). Transient inward current and Na-K-ATPase inhibition both contribute to a cellular sodium overload which enhances the calcium overload via the Na-Ca exchange, and the same cascade as observed during ischemia is observed. Therefore the major difference between the development of ischemia and cardiac glycoside intoxication seems to be that ischemia starts with a decrease in cellular pH, and in cardiac glycoside intoxication the first toxic change is an excessive increase in the calcium concentration in the cytosol, which during ischemia is subsequent to the pH-drop.

Therefore we decided to use cardiac glycoside intoxication as a screening model for antiischemic compounds. We use the following experimental protocol. Electrically paced rat left atria (1 Hz) are treated with the compound under investigation (Fig. 1). Then a therapeutic positive inotropic concentration of ouabain (8.5 x 10^{-6} mol/l) is administered. This allows an impression of the activity of the new compound on physiological contractile processes (electrical stimulation and therapeutic ouabain concentrations). Cardiac glycoside intoxication is elicited by gradually increasing the stimulation frequency (2 and 3 Hz) without changing the ouabain concentration (Peters 1986b). Both an increase in diastolic and a decrease in systolic tension of the paced atria as characteristics of cardiac glycoside intoxication are observed. Therefore information about the effectivenes of the agents under investigation on these pathologic processes is obtained. At the end of the experiment the atria are analyzed for their ion content by

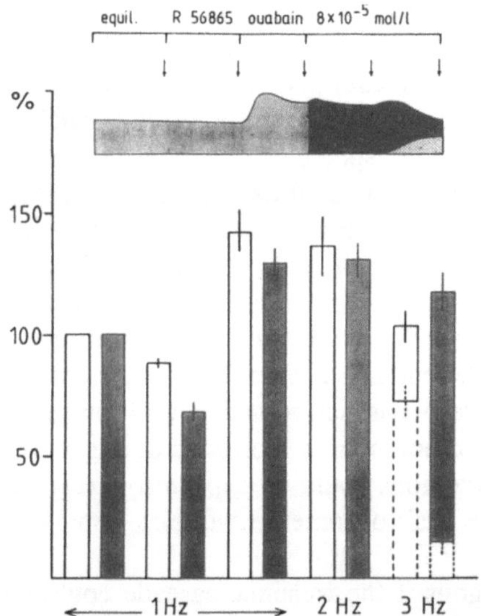

Fig. 1. The contractile force (systolic and diastolic) in % of initial value of rat left atria. After a 40 minute equilibration period atria were treated with R 56865 (10^{-6} mol/l) (control: open columns; treated: filled columns); 45 minutes later ouabain (8 x 10^{-5} mol/l) was added. The inset shows the experimental protocol. The arrows indicate the contraction represented by the columns. Data are presented as means ± S.E.M (n = 6-9)

means of atomic absorption spectrophotometry to estimate sodium and calcium overload.

One of the agents found with this test is R 56865 (*N*-[1-[4-(4-fluorophenoxy)butyl]-4-piperidinyl]-*N*-methyl-2-benzothiazolamine). At a concentration of 10^{-6} mol/l R 56865 acts in a slightly negatively inotropic manner and slightly affects the positive inotropic action of ouabain (Fig. 1), both measures for the interaction with physiological processes. The inhibition of the increase in diastolic tension elicited by ouabain intoxication, however, is strongly reduced. The ouabain-induced calcium overload is completely blocked, the sodium overload by about 50%. R 56865 appeared to be very effective against delayed after-depolarizations caused by ouabain (Vollmer et al. 1987). This agent will be tested further for its antiischemic properties. Flunarizine is also active in this ouabain-intoxication model and shows clear antiischemic effects. In rats respirated with N_2 the ionic events in the extracellular space of the brain can be measured with ion-selective electrodes (Fig. 2). Under these ischemic conditions pH declines, and the potassium concentration rises slowly up a threshold of about 13 mmol/l and then suddenly increases up to about 80 mmol/l. Concomitantly the cells depolarize, and a calcium influx represented by a decrease in the extracellular calcium concentration takes place (Höller et al. 1986). Flunarizine (50 mg/kg, i.p.) increases the threshold for the terminal depolarization and thereby prolongs anoxia tolerance times. Therefore flunarizine is an effective antiischemic agent which may play a role in its mechanism of action in stroke and migraine.

In conclusion, ischemia and cardiac glycoside intoxication have many similarities. We use cardiac glycoside intoxication in a test for screening for antiischemic substances, and R 56865 and flunarizine are examples of compounds active in this model. For a number of these compounds antiischemic properties could be demonstrated in a model for brain ischemia, and side effects pointing towards an interaction with physiological contractile processes of the heart are not observed in the necessary concentration range.

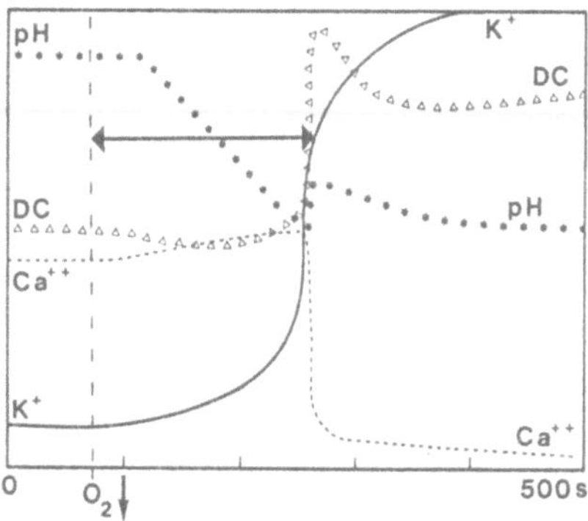

Fig. 2. The pH, DC potential, Ca^{2+} and K^+ concentration in the extracellular space in the brain of rats subjected to respiration with N_2.

Summary

Ischemia and cardiac glycoside intoxication have much in common. The major difference is that ischemia starts with a pH drop and cardiac glycoside intoxication with an excessive increase in the calcium concentration in the cytosol, which is, however, subsequent to the change in pH during ischemia. The accompanying cascade characterized by calcium and sodium overload, delayed after-depolarizations, etc. is comparable in ischemia and cardiac glycoside intoxication. Therefore we apply cardiac glycoside intoxication as a screening model for antiischemic compounds in isolated tissues. For a number of compounds active in this model antiischemic properties were observed.

References

Akerman KEO, Heinonen E (1983) Qualitative measurements of cytosolic calcium ion concentration within isolated guinea-pig nerve endings entrapped arsenazo III. Biochim Biophys Acta 732: 117-121

Borchard M, Ravens U (1986) Intracellularly applied sodium mimics the effects of ouabain in single cardiac myocytes. Eur J Pharmacol 131: 269-272

Coetzee WA, Opie LH, Saman S (1987) Proposed role of energy supply in the genesis of delayed afterdepolarizations implications for ischemic or reperfusion arrhythmias. J Mol Cell Cardiol [Supp 5] 19 : 13-21

Fabiato A (1985) Use of aequorin for the appraisal of the hypothesis of the release of calcium from the sarcoplasmic reticulum induced by a change of pH in skinned cardiac cells. Cell Calcium 6: 95-108

Höller H, Dierking H, Dengler K, Tegtmeier F, Peters T (1986) Effect of flunarizine on extracellular ion concentration in the rat brain under hypoxia and ischemia. In: Battistini N, Courbier E, et al. (eds) Acute brain ischemia medical and surgical therapy. Raven , pp 229-236

Lüllmann H, Peters T (1977) Plasmalemmal calcium in cardiac excitation-contraction coupling. Clin Exp Pharmacol Physiol 4: 49-57

Peters T (1986a) Calcium in physiological and pathological cell function. Eur Neurol[Suppl 1] 25 : 27-44

Peters T (1986b) Glycoside receptors in the heart. Prog Pharmacol 6: 65-80

Post JA, Langer GA, Op den Kamp J, Verkleij AJ (1988) Phospholipid assymetry in cardiac sarcolemma. Analysis of intact cells and 'gas dissected' membranes. Biochim Biophys Acta 943: 256-266

Smith TW (1988) Digitalis. Mechanism of action and clinical use. N Engl J Med 318: 358-365

Sperelakis N (1988) Regulation of calcium slow channels of cardiac muscle by cyclic nucleotides and phosphorylation. J Mol Cell Cardiol[Suppl 2] 2O : 75-105

Vollmer B, Meuter C, Janssen PAJ (1987) R 56865 prevents electrical and mechanical signs of ouabain intoxication in guinea-pig papillary muscle. Eur J Pharmacol 142: 137-14O

Experimental Intracerebral Hematoma and Treatment with Flunarizine in Rats

B. Kleiser[1], J. Van Reempts[2], E. Horn[1] and H.H. Kornhuber[1]

[1]Department of Neurology, University of Ulm, 7900 Ulm, FRG

[2]Department of Cardiovascular Pathophysiology, Janssen Research Foundation, 2340 Beerse, Belgium

The risk of morbidity and mortality from intracerebral hemorrhages which occur after head trauma, birth injury, ruptures of aneurysms or intracranial vessels is still rather high. Unfortunately, there is no convincing pharmacotherapy available so far (Johnson 1987). In this study we describe an animal model for an intracerebral hematoma, its effects on locomotor behavior and the influence of flunarizine, a calcium overload blocker (Borgers et al. 1983; Van Nueten et al. 1978) on recovery from the hematoma-induced neurological deficits.

Adult male Wistar rats weighing about 300 g were operated under pentobarbital anesthesia (60 mg/kg intraperitoneal). A trepanation was performed 3.5 mm lateral to the midline and 1.0 mm anterior to bregma on the left side of the skull. Two weeks later, 50 µl arterial blood of a donor rat was injected into the foreleg area of the motor cortex at the stereotactic coordinates 1 mm posterior to bregma, 3.5 mm lateral and 2 mm below the dura by holding the canula under an angle of 45 degrees. After this treatment, some animals developed a hemiparesis which was characterized by the failure of lifting a hanging leg on a platform within 10 seconds (Fig. 1).

Locomotor behavior was investigated by means of a rotating board (size 21.5 x 4.5 cm; 0.5 cm thickness) turning around its longitudinal axis (Fig. 1). The initial velocity was 2 rounds per minute; every 30 seconds it was increased up to 18/minute in seven steps. The rats were trained to stay by running and climbing on the rotating platform one week after the operation. The time was determined when the animal fell from the rotator at one level of velocity for the third time. If the animal did not fall down even at the highest speed, the experiment was stopped after 20 minutes. A short running time means a reduced locomotory performance. Control investigations in four groups of untreated animals demonstrated a high stability of mean running times (Table 1).

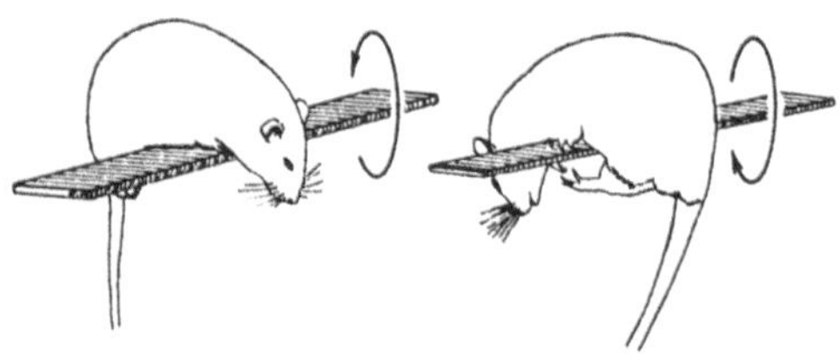

Fig. 1. Two rats walking in the rotator. Left: Rat without neurological deficits. Right: Rat with a paresis in the left hindleg. This animal has a hematoma in the motor cortex of the right hemisphere

Table 1. Running times (s) of 4 groups of untreated rats on the rotator. The animals were chosen randomly out of a group of 25 rats. SD = standard deviation, SEM = standard error mean, n = number of rats

	Mean	SD	SEM	n
Group 1	682	297	121	6
Group 2	676	216	82	7
Group 3	670	431	216	4
Group 4	671	351	124	8

Pharmacological treatment was started five minutes after the stereotactic application of the hematoma. 13 animals were treated with flunarizine and 12 with the solvent. First, flunarizine was given at a dose of 0.1 mg/kg by slow i.v. injection into a tail vein (1 ml/kg). Three and six hours later, flunarizine 20 mg/kg, dissolved in vegetable oil, was given orally. The placebo group received the vehicles.

In the flunarizine group 7 animals developed a hemiparesis at the right side of the body, mainly in the foreleg, while the other 6 animals showed no neurological deficits. In the control group, 8 animals developed a hemiparesis; the other 4 revealed no deficits. The running time until falling down of animals without neurological deficits calculated from all recordings obtained in the course of this investigation was 722 ± 88 s, while in the animals with hemiparesis it was 257 ± 31 s (means $\pm$ 2 x SEM). This reduction of motor performance was significant ($p < 0.002$, U test).

The animals with hemiparesis showed a significant improvement if they were treated with flunarizine. In particular, 4 days after creation of the hematoma the running time of the hemiparetic animals treated with flunarizine was 296 ± 89 s, while in the hemiparetic animals of the solvent group the running time was 170 ± 70 s on the rotator. Six days after the blood injection the corresponding values were 386 ± 71 s for the hemiparetic rats with flunarizine and 202 ± 98 s for the controls. The increase of the running time in the flunarizine group was significant at $p = 5\%$. No significant differences were found five hours and two days after creation of the cerebral lesion (Fig. 2). Animals without neurological deficits showed no significant differences in their running times.

Several studies with ischemic brain damage in animals have shown a protective effect of flunarizine with a reduction in the size of cell damage (Van Reempts et al. 1983, 1987; Wauquier et al. 1985). Probably the mechanisms of neuronal damage caused by ischemia and by intracerebral hemorrhage are similar. Penetration of calcium into the neurons may be regarded as an important factor initiating cellular damage (Borgers et al. 1983; Siesjö 1981). Compatible with this interpretation is the fact that the motor performance of the animals with hemiparesis two days after the hematoma is lower than four hours afterwards (suggesting a mechanism of delayed cell damage not due just to a mechanical compression by the blood), and that the therapeutic action becomes apparent in this late stage.

From the clinical point of view it was important to choose a model with behavioral parameters for the evalution of neurological deficits. The running times on the rotating platform showed significant differences between animals with and without hemiparesis so that it can be regarded as a reliable test for a damage in the motor cortex and in the white matter below the motor cortex. Comparing the groups with and without medication, the investigation showed a similar impairment of performance directly after the

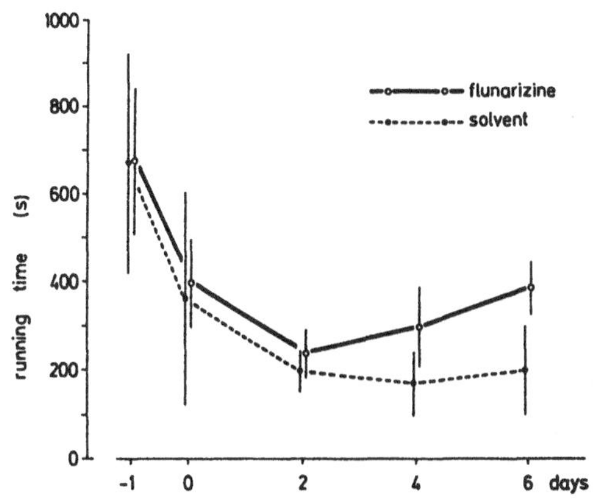

Fig. 2. Influence of flunarizine on the running time (s) in rats with hemiparesis. The abscissa gives the day of neurological testing before (-1) and after creation of the intracerebral hematoma which was performed at day 0. The values for day 0 refer to the results 4 hours after the lesion (hematoma). Mean values ± 2 x standard error of mean are given. n = 7 for the flunarizine group, n = 8 for the control group

creation of the hematoma, but later a significantly better recovery from hemiparesis was found in the group with flunarizine. No significant differences occurred in the animals without neurological deficits. Therefore, the effect of flunarizine is not due to a nonspecific activation. The drug may partially cure neurological deficits caused by cell damage secondary to an intracerebral hematoma.

Acknowledgement. We would like to thank Dr. A. Kornhuber for statistical evaluation of the data.

References

Borgers M, Thone F, Van Reempts J, Verheyen F (1983) The role of calcium in cellular dysfunction. Am J Emerg Med 1: 154-161

Johnson R T (1987) Current therapy in neurologic disease. Decker, Toronto

Siegel S (1956) Non parametric statistics for the behavioural sciences. McGraw-Hill, New York, pp 116-127

Siesjö B K (1981) Cell damage in the brain: a speculative synthesis. J Cereb Blood Flow Metab 1: 155-185

Van Nueten J M, Van Beek J, Janssen P A J (1978) Effect of flunarizine on calcium-induced responses of peripheral vascular smooth muscle. Arch Int Pharmacodyn Ther 232: 45-52

Van Reempts J, Borgers M, Van Dael L, Van Eyndhoven J, Van de Ven M (1983) Protection with flunarizine against hypoxic-ischemic damage of the rat cerebral cortex. A quantitative morphologic assessment. Arch Int Pharmacodyn Ther 262: 76-88

Van Reempts J, Van Deuren B, Van de Ven M, Cornelissen F, Borgers M DSc (1987) Flunarizine reduces cerebral infarct size after photochemically induced thrombosis in spontaneously hypertensive rats. Stroke 18: 1113-1119

Wauquier A, Fransen J, Clincke G, Ashton D, Edmonds H L (1985) Calcium entry blockers as cerebral protecting agents. In: Godfraind T, Vanhoutte P M, Govoni S, Paoletti R (eds) Calcium entry blockers and tissue protection. Raven, New York, pp 163-172

"Normal" Alcohol Consumption and Well-Being: Relationship Between Reduction of Alcohol Consumption and Changes in Well-Being

J. C. Aschoff and A. Knaps
Department of Neurology, University of Ulm, 7900 Ulm, FRG

Introduction

On several occasions Kornhuber (Kornhuber 1982, 1984, 1986a,b,c; Kornhuber et al. 1985a,b; Scheben et al. 1987) has pointed out that the usual low-dose alcohol consumption is one of the main reasons for high blood pressure, the risk of cardio- and cerebrovascular infarction, but also for traffic accidents (Richter and Hobi 1975) and many other problems. As a consequence, low-dose alcohol consumption is also one of the main factors for increasing costs in our public health service. Neither these problems nor those, which we deal with in this paper relate to the problems of "alcoholism"; they relate exclusively to problems connected with so-called normal drinking behaviour, that is, they apply to most of us.

West Germany now ranks fourth in alcohol consumption with 12 liters per year per person (*Jahrbuch zur Frage der Suchtgefahren 1988*). Thirty years ago average yearly consumption was only around 3 liters. It is also well known that in West Germany the large majority of adult males above the age of 16, and close to 30% of adult women, consume alcoholic beverages nearly every day. Only 6% of our population never uses alcohol (Feuerlein and Küfner 1977; Trojan 1980). Between 8% and 25%; depending upon their profession, drink alcohol while working. The percentage of teetotallers has remained constant over the years while that of daily or occasional drinkers has increased significantly over the past 20 years.

Kornhuber (1986b) has also pointed out that a large number of somatic complaints is due to this low-dose dependency on alcohol. Many of our medical colleagues deny this; they treat stomach problems, headaches, high blood pressure and all kinds of general complaints by giving all kinds of medical advice, and especially by prescribing medication, but not by simply asking the patient for a strict alcohol reduction or, even better, total abstinence. Given the facts collected by Kornhuber and his group, we have considered the drinking behaviour of students using the following questions:

1. To what extent are our students able to reduce their alcohol consumption?
2. Is there a relationship between alcohol consumption and subjective well-being and general somatic complaints, and to what extent do well-being and somatic complaints change in relation to alcohol reduction?

We should note a statement by Burish et al. (1981) from their studies at the Vanderbilt University in the USA: "Studies of college students suggest that most of them drink, many drink excessively and drinking is a major part of social activities on and off campus".

Methods

Questionnaires were administered to 320 students of the University of Ulm (206 males, 114 females). More than 50% belonged to the age group 22 to 25 years; only 8% were in the age group 32 to 36 years.

We asked our students, who remained anonymous, about their drinking behaviour. We also asked them to stop alcoholic beverages of all kinds over a period of three weeks.

At this first contact as well as three weeks later we applied two self-rating scales, the Adjective Mood Scale (*Befindlichkeits-Skala*) for the assessment of fluctuations in well-being and the Complaint List (*Beschwerden-Liste*) which reflects the degree of somatization and describes somatic or general complaints (von Zerssen 1976a,b). These scales have been translated into many languages (von Zerssen 1986). Normative values have been provided from data collected from well over 4000 subjects from the general population, investigated in epidemiological studies as well as from all kinds of clinical patients. In these self-rating scales the higher values represents less well-being and lower values more well-being.

Results

Alcohol consumption by students of the University of Ulm is shown in Fig.1. Only 8% of male students but 27% of female students use no alcohol at all (or only very little). High alcohol consumption is found in only 3% of female students, but in 13% of male students. Compared to the general West German population female students show rather similar values, whereas male students have a somewhat lower alcohol consumption rate.

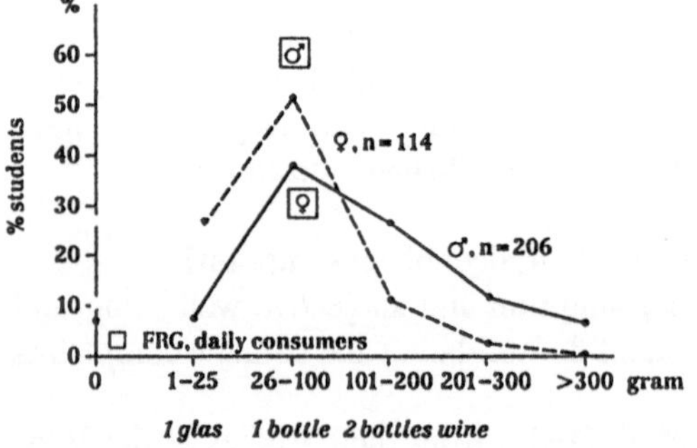

Fig. 1. Amount of alcohol consumed per week by 320 students at the University of Ulm

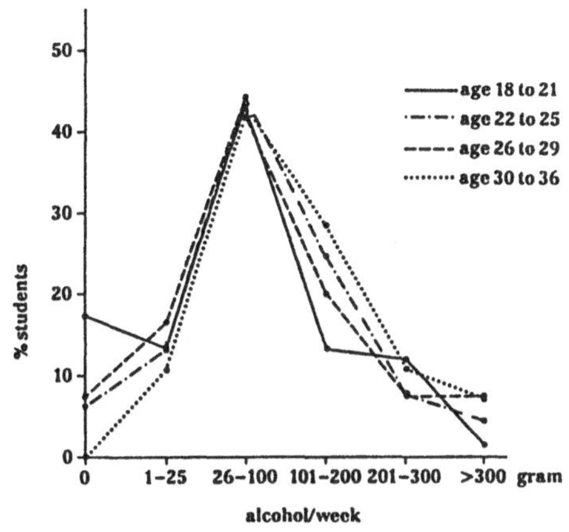

Fig. 2. Amount of alcohol consumed per week, according to age, by 320 students at the University of Ulm

If we group our students into sub-groups according to age (Fig. 2), we detect no differences with respect to the most usual amount of alcohol consumption, that is, one to two bottles of wine per week (or comparable amount of beer). Only in very low and in very high alcohol consumption do we find - as expected - that hardly a student above the age of 30 drinks no alcohol at all, while under the age of 21, that is, in our youngest students, around 15% do not drink any alcohol (or only very occasionally). On the other hand, we find no heavy drinkers among the youngest students, but around 10% of our students aged 26 to 36 are heavy drinkers.

The main task in this investigation was not, however, to determine amounts of alcohol consumed but the correlations between the feeling of well-being and the amount of somatic complaints and the amount of alcohol consumed. As shown in Table I, lowest scores (that is adequate well-being and no somatic complaints) were found in those students who drink some (i.e. very moderate) alcohol, which means only in the evening and not daily (between one glass and one bottle of wine per week or one to five bottles

Table I. Assessment of well-being and of somatic complaints with two self-rating scales. Correlation to amount of alcohol consumption

Alcohol/week	n	Bfs	BL
no alcohol consumption	27	15,8	14,4
1- 25 g	43	17,3	14,4
26-100 g	139	14,0	13,0
101-200 g	69	17,7	17,6
201-300 g	28	16,2	16,0
>300 g	14	19,2	17,2

Bfs = Adjective mood scale (Befindlichkeitsskala)

BL = Complaint list (Beschwerdenliste)

of beer per week). In both directions, that is, with a lower or non-existent alcohol consumption, on the one hand, and with higher alcohol consumption, on the other, the state of well-being is reduced and somatic complaints increase. This takes place only to a small extent when no or little alcohol is consumed (probably due to alcohol intolerance in connection with somatic diseases), but with a sudden increase with higher amounts of alcohol consumption. None of these students, however, can be called "alcoholic".

All students were asked to stop or at least to reduce their alcohol consumption over a period of three weeks. Fig. 3 shows the outcome: the lower the alcohol consumption had been prior to testing, the larger the percentage of students who were able to give up alcohol or at least to reduce alcohol consumption. In the higher consumption group, some 40% to 50% of the students could not, or were unwilling to, reduce their alcohol consumption.

What happened to the feeling of well-being and to the somatic complaints when alcohol consumption was reduced, and what kind of changes occurred in those who could not reduce their consumtion? Figure 4 gives the results for those with no alcohol re-

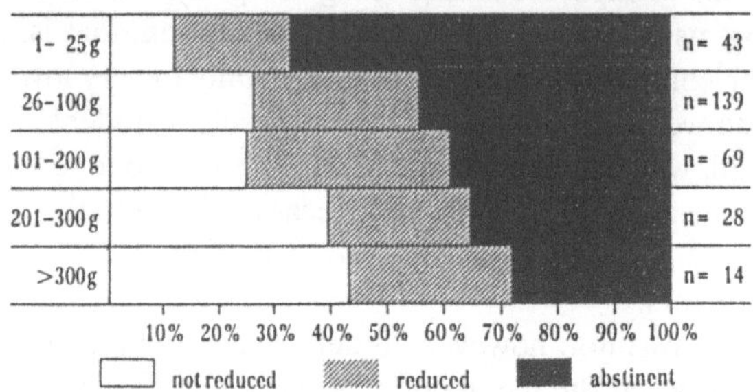

Fig. 3. Ability to reduce or abstain from alcohol, in correlation to amount of alcohol consumed prior to test

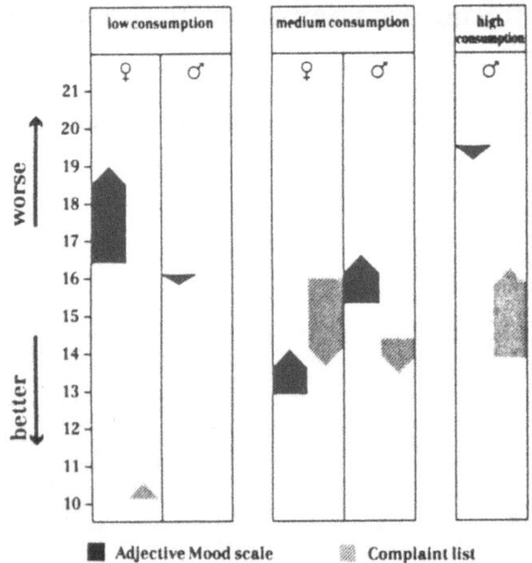

Fig. 4. Changes in mood and somatic complaints within three weeks on two self-rating scales, with no reduction of alcohol consumption

duction. There are some changes for the better and some for the worse, especially changes towards depressive tendencies in those with low alcohol consumption and no reduction. Otherwise changes occurred in both directions, mostly to a rather small degree. In Fig. 5 changes are shown for the group of students who stopped their alcohol consumption totally. All changes were for the better, i.e. all scores were lower on the Adjective Mood Scale and on the Complaint List. In low and medium consumption it was especially the improvement in somatic complaints which impressed us. A high alcohol consumption prior to cutting it out completely was especially effective in improving depression.

Comparing all those students who did not reduce their alcohol intake at all with those who reduced or stopped, we find a remarkable change for the better on both of the self-rating scales among those in whom alcohol was reduced and no real changes in those who did not reduce their alcohol intake (Fig. 6).

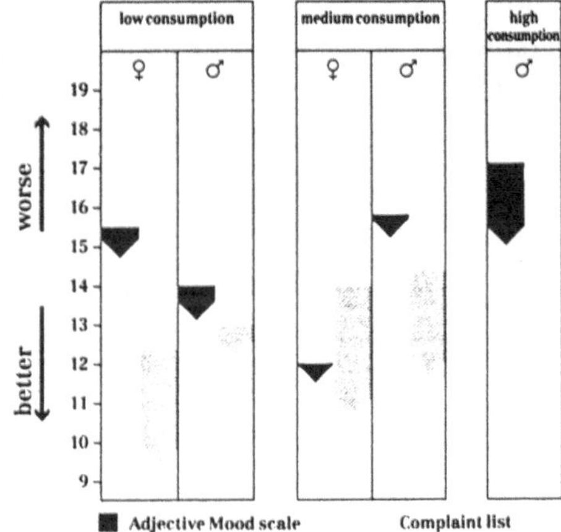

Fig. 5. Changes in mood and somatic complaints within three weeks on two self-rating scales after total cessation of alcohol consumption

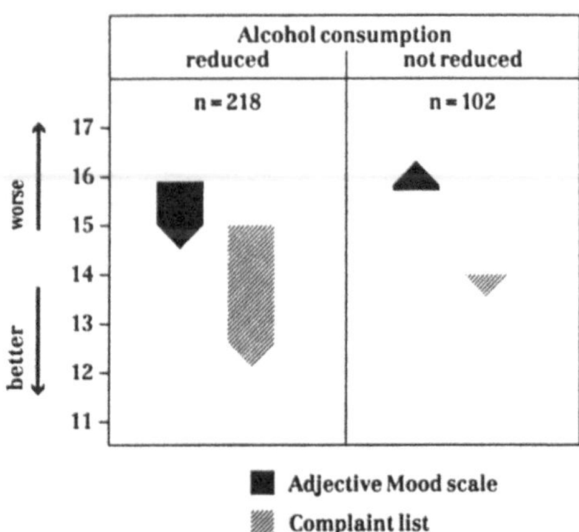

Fig. 6. Changes in mood and somatic complaints within three weeks on two self-rating scales. Comparison between reduction and non-reduction group

Discussion

Although we investigated the problem in a population subgroup, the results can easily be applied to all of us. A previous investigation (Aschoff 1986) showed an astonishing correlation between alcohol consumption and the feeling of well-being and degree of somatic complaints (Fig. 7). For the first time, to our knowledge, we could now show that reduction of or total abstinence from alcohol consumption for a given period of time reduces the number of somatic complaints and improves the general state of health, i.e. it leads to a more balanced mood.

These findings do not refer to alcoholics but to an average population subgroup with a so-called low-dose alcohol dependency. The results lead to the conclusion that it is worthwhile in every respect for the individual to reduce alcohol consumption. In social drinkers, i.e. those with low-dose dependency, the reversibility of many of so-called general complaints after abstinence has been demonstrated with other methods by Cala et al. (1983). Alcohol consumption therefore should be reduced to the point that alcoholic beverages are not consumed every day and if at all, then only in the evening and preferably only at weekends or on festive occasions. Although this recommendation seems ridiculous with respect to the well-known high alcohol consumption in the general population, following it would still be the best thing that we could do for our health.

Summary

Daily "social drinking", i.e. the usual low-dose dependency on alcohol, although different from alcoholism, leads to a number of somatic and psychic complaints. Using two

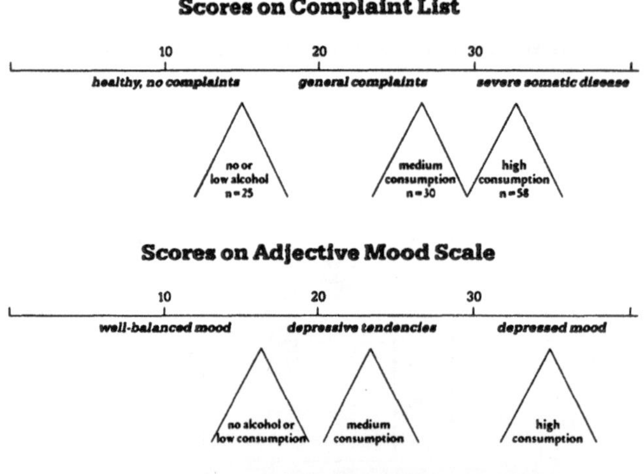

Fig. 7. Scores on the two self-rating scales, according to amount of alcohol consumption. (From Aschoff 1986)

self-rating scales, we found that reduced alcohol drinking (or even better total abstinence) over a period of three weeks (320 students at the University of Ulm) leads to a more balanced mood, to an increased well-being and to a reduction of somatic complaints.

References

Aschoff J C (1986) Körperliche Beschwerden und Befindlichkeit in Abhängigkeit von Alkoholkonsum und Substitution mit Magnesium. Med Welt 37: 449 - 453

Burish T G, Maisto S A, Mitch Cooper A, Sobell M B (1981) Effects of voluntary short-term abstinence from alcohol on subsequent drinking patterns of college students. J Stud Alcohol 42 (11): 1013 - 1020

Cala L et al. (1983) Results of computerized tomography, psychometric testing and dietary studies in social drinkers, with emphasis on reversibility after abstinence. Med J Aust 2: 264 - 269

Feuerlein W, Küfner Z (1977) Alkoholkonsum, Alkoholmiβbrauch und subjektives Befinden, eine Repräsentativerhebung in der Bundesrepublik Deutschland. Arch Psychiat. Nervenkr. 224: 89 - 106

Kornhuber H H (1982) Neue Ansätze zur Therapie alkoholischer Leiden. In: Der Alkoholkranke. Tübingen pp 67-78 (Schriften der Bezirksärztekammer Südwürttemberg, Vol 2)

Kornhuber H H (1984) Bluthochdruck und Alkoholkonsum. In: Rosenthal J (ed) Arterielle Hypertonie, Springer, Berlin Heidelberg New York

Kornhuber H H (1986a) Gesundheitsschäden durch "normalen" Alkoholkonsum. In: Deutsche Hauptstelle gegen Suchtgefahren (ed) Jahrbuch zur Frage der Suchtgefahren 1986. Neuland Hamburg

Kornhuber H H (1986b) "Normaler" Alkoholkonsum - eine der Ursachen von Bluthochdruck, Adipositas und Atherosklerose. Schweiz Rundsch Med [Prax] 75: 1577-1579

Kornhuber H H (1986c) Konsumverhalten und Gesundheit. Versicherungswirtschaft 6: 358 - 363

Kornhuber H H, Lisson G, Suschka-Sauermann L (1985a) Alcohol and obesity: a new look at high blood pressure and stroke. Eur Arch Psychiatr Neurol Sci 243: 257-362

Kornhuber H H, Lisson G, Suschka-Sauermann L (1985b) Adipositas und Atherosklerose als spezifisch-toxische Alkoholfolgen. Öff Gesundheitswes 47: 488 - 496

Richter R, Hobi V (1975) Die Beeinträchtigung der Fahrtüchtigkeit bei Blutalkoholkonzentrationen um 0,50%. Schweiz Med Wochenschr 105 (27): 884-890

Scheben B, Henkler C, Kornhuber A, Kornhuber H H, Maier V, Molz K H, Swobodnik W, Wechsler JG (1987) On the road to stroke: hepatic steatosis and hyperinsulinaemia associated with "normal" alcohol use in young males. Verh Dtsch Ges Neurol 4: 674 - 675

Trojan A (1980) Epidemiologie des Alkoholismus und der Alkoholkrankheit in der Bundesrepublik Deutschland. Intern Welt 7: 241-250

Zerssen D von (1976a) Die Beschwerden-Liste. Manual. Beltz Weinheim.

Zerssen D von (1976b) Die Befindlichkeits-Skala. Manual. Beltz Weinheim.

Zerssen D von (1986) Clinical self-rating scales (CSRS) of the Munich Psychiatric Information System (Psychis München). In: Sartorius N, Ban TA (eds) Assessment of depression. Springer Berlin Heidelberg New York

"Normal" ("Social") Daily Alcohol Consumption, Arterial Hypertension, Glucose Tolerance, Plasma Insulin, C Peptide, GGT, S-Adenosylmethionine, Plasma Lipids and Obesity: Insulin Receptor Damage and Mediators Versus Repair Mechanism of Toxic Alcohol Effects

B. Backhaus, A.W. Kornhuber and H.H. Kornhuber

Neurologische Universitätsklinik Ulm, Oberer Eselsberg 45, 7900 Ulm, FRG

Introduction

Arterial hypertension is the most important risk factor for stroke (Kannel and Wolf 1983). Recently it has become evident that a large part of "essential" hypertension, especially in the human male, is due to daily alcohol consumption in moderate amounts (Kornhuber 1984). It has been hypothesized that hepatic insulin receptor damage resulting in hyperinsulinemia may be one mediating mechanism (De Fronzo 1981; H.H. Kornhuber et al. 1985; Oehler et al 1982; Rower et al. 1981; Scheben 1987; Stout 1979). The obesity which is correlated with "essential" hypertension is also due to the toxic effects of daily alcohol consumption rather than to increased caloric intake (H.H. Kornhuber et al. 1985b, 1989b). It would be important for the patients to have easily accessable signs of the risk for alcohol-mediated hypertension, obesity and type II diabetes. Unfortunately there are up to now no simple laboratory tests available for the lighter stages of alcoholic liver damage (hepatic steatosis). For instance, the most important enzyme in this respect, GGT, is in cases of fatty liver usually in limits nowadays generally (though erroneously) considered to be "normal." The purpose of the present investigation was to look with several methods including HPLC for changes that are related to daily "social" alcohol consumption and the related elevation of blood pressure and obesity.

Methods

Hospital inpatients, mainly from the orthopedic clinic, were investigated immediately after their admission. All persons were given an oral glucose tolerance test. Besides glucose, the plasma insulin and the C peptide levels were measured. Other parameters included blood pressure, body weight (Broca Index) plasma lipids, GGT and the serum amino acids (measured with HPLC; Zettlmeißl et al. 1986).

Results

Remarkable was that the patients with "social" ("normal") daily alcohol consumption and with a GGT of 9-15 U/l showed in the glucose tolerance test an increased level in C peptide, glucose and plasma insulin both fasting and 60 minutes (Fig.1.) and 120 minutes postprandial, although the "normal" range of GGT in West Germany is believed to be up to 28 U/l (at 25°C), in the United Kingdom 51 U/l in the male (at 37°C). The patients with daily alcohol consumption (independent of the quantity) had, contrary to patients with only occasional consumption, a higher level in arterial blood pressure (both systolic and diastolic; Fig.2.) and in the serum S-adenosylmethionine (Fig.3.). From about 30 amino acids measured by HPLC, mainly S-adenosylmethionine was correlated with the other alcohol effects, such as elevated blood pressure, increased plasma insulin, GGT, serum lipids and the Broca Index. Those persons with a higher S-adenosylmethionine level also had higher plasma insuline. The patients with high plasma lipids (cholesterol and triglyceride) exhibited an elevation of plasma insulin. The plasma C peptide level was increased in correlation with the insulin and all the alcohol parameters, indicating that the "normal" alcohol consumption causes true hyperinsulinemia due to insulin receptor damage and not just a pass effect.

Discussion

The elevation of plasma insulin under daily "normal" alcohol is due to a regulation compensating for the insulin receptor damage; this appears earlier and is more sensitive than glucose elevation. GGT is also a sensitive enzyme indicating hepatic steatosis, but only if the normal range limit is merely up to 10 U/l, thus much lower than believed until now (for details see H.H. Kornhuber et al. 1989a; J. Kornhuber et al. 1989). S-adeno-

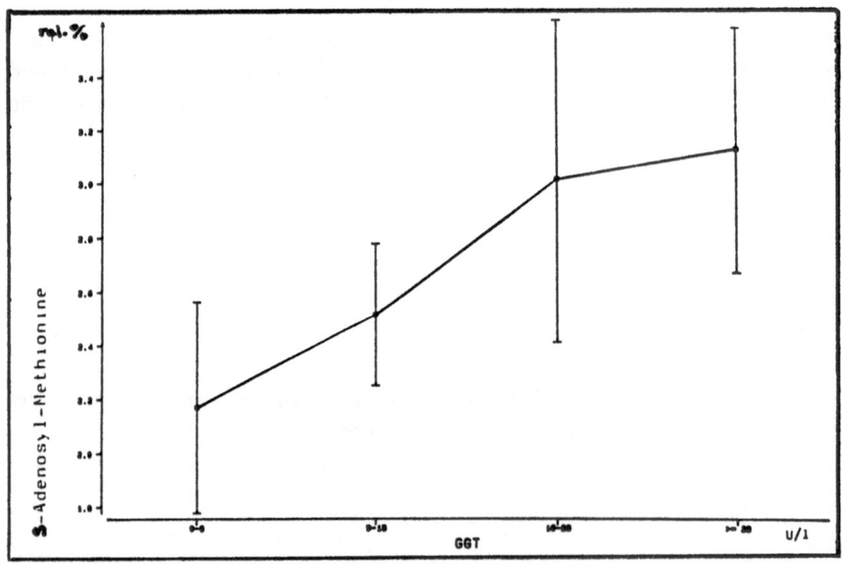

S-Adenosyl-Methionine as related to GGT

Fig.1. Plasma insulin after 60 minutes (OGTT) as related to GGT

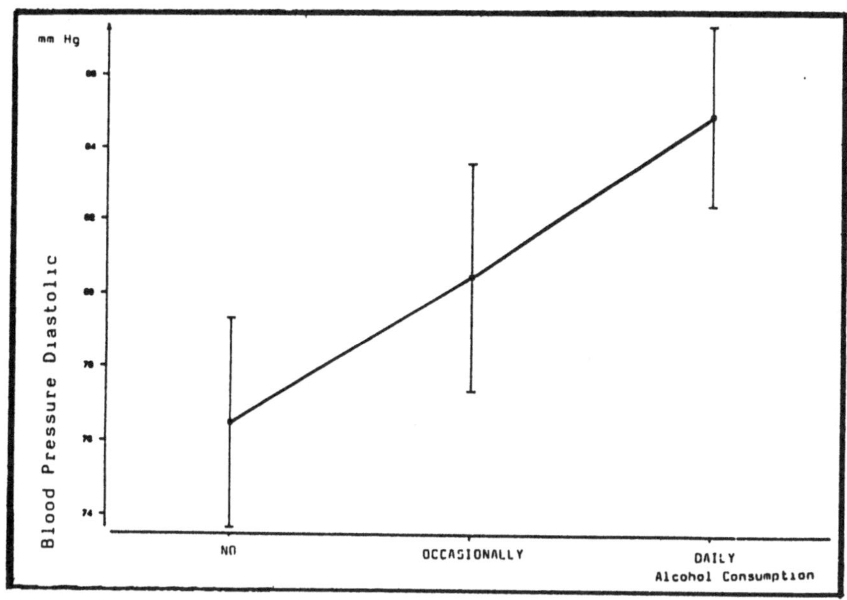

Diastolic blood pressure as related to alcohol consumption

Fig.2. Diastolic blood pressure as related to alcohol consumption

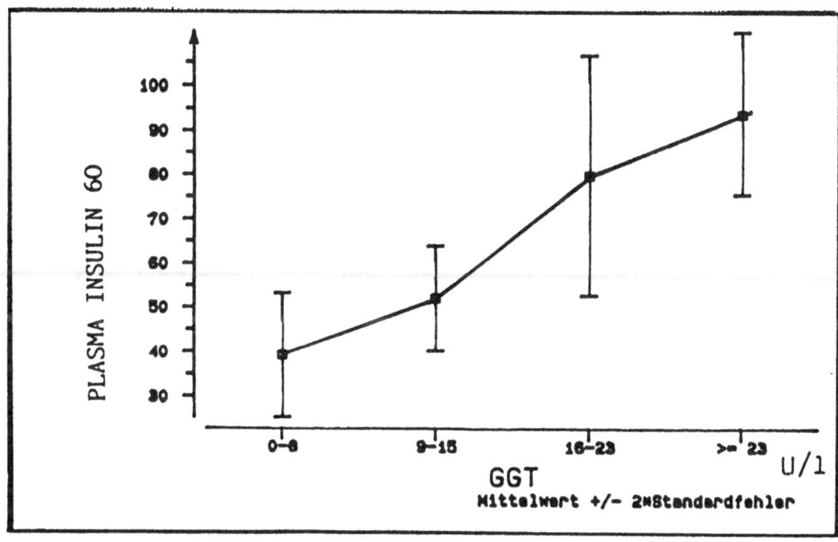

Plasma insulin after 60 minutes (OGTT) as related to GGT

Fig.3. S-Adenosylmethionine as related to GGT

sylmethionine is probably related to glutathione, which is an important substance for the detoxication of ethanol and its even more toxic metabolites including acetaldehyde, free radicals, etc. Thus, the elevated level of S-adenosylmethionine under "social" alcohol may represent a repair mechanism.

Summmary

Arterial hypertension, the most important risk factor for stroke, is caused mainly by daily "normal" alcohol consumption, at least in the human male. The purpose of this investigation was to look with several methods (among them HPLC and glucose tolerance test) for changes that are related to daily "normal" alcohol and the associated hypertension. Elevations of plasma insulin, gamma-glutamyltransferase (GGT), plasma lipids and S-adenosylmethionine were related to daily "normal" alcohol. The "normal" range of GGT as assumed is erroneous.

References

DeFronzo RA (1981) Insulin and renal sodium handling: clinical implications. Int J Obes [Suppl 1] 5: 93-104

Kannel WB, Wolf PA (1983) Epidemiology of cerebrovascular disease. In Ross Russell RW (ed) Vascular disease of the central nervous systems Livingstone, Edinburgh

Kornhuber HH (1984) Bluthochdruck und Alkoholkonsum. In: Rosenthal J (ed) Arterielle Hypertonie, 2nd edn. Springer Berlin Heidelberg New York, pp 149-162

Kornhuber HH, Lisson G, Suschka-Sauermann L (1985a) Alcohol and obesity: a new look at high blood pressure and stroke. Eur Arch Psychiatr Neurol Sci 234: 357-362

Kornhuber HH., Lisson G, Suschka-Sauermann L (1985b) Adipositas und Artherosklerose als spezifischtoxische Alkoholfolge. Öff Gesundheitswes 47: 488-496

Kornhuber HH , Backhaus B, Kornhuber J, Kornhuber AW (1989a) Risk faktors and the prevention of stroke. In: Amery W K, Bousser MG, Rose FC (eds) Clinical trial methodology in stroke. Transmedica Europe, Tunbridge Wells

Kornhuber HH , Kornhuber J, Wanner W, Kornhuber A, Kaiserauer C (1989b) Alcohol, smoking and body build: obesity as a result of the toxic effect of "social" alcohol consumption. Clin Physiol Biochem 7:203-216

Kornhuber J , Kornhuber HH, Backhaus B., Kornhuber A, Kaiserauer C, Wanner W (1989) GGT-Normbereich bisher falsch definiert: Zur Diagnostik von Bluthochdruck, Adipositas und Diabetes infolge "normalen" Alkohol Konsums. Versicherungsmedizin 41: 78-81

Oehler G, Bleyl H, Matthes K J (1982) Hyperinsulinemia in hepatic steatosis. Int J Obes [Suppl 1] 6: 137-144

Rower JW, Young JB, Minaker KL, Stevens AL, Palotta J, Landsberg L(1981) Effect of insulin and glucose infusions on sympathetic nervous system activity in normal man. Diabetes 30: 219-225

Scheben B, Henkler C, Kornhuber A, Kornhuber HH, Maier V, Molz KH, Swobodnik W, Wechsler JG (1987) On the road to stroke: hepatic steatosis and hyperinsulinaemia associated with normal alcohol use in young males. Verh Dtsch Ges Neurol. 4: 674-675

Stout RW (1979) Diabetes and artherosclerosis - the role of insulin. Diabetologica 16: 141-150

Zettlmeißl H, Blome J, Kornhuber HH (1986) A sensitive, fast, durable, higly selective method for the determination of amino acids and biogenic amines in the cerebrospinal fluid and other body fluids and tissues. Arch Ital Biol 124: 129-132

EEG Signs of a Disturbed Voluntary Process Prior to Voluntary Movements in Schizophrenia

K.P. Westphal, B. Grözinger, V. Diekmann, and H.H. Kornhuber
Abteilung Neurologie, Universität Ulm, 7900 Ulm, FRG

Introduction

Attenuation of alpha activity has been found before and during voluntary movement over the activated motor areas (Kornhuber and Deecke 1965). As schizophrenics suffer from voluntary motor disturbances (Manschreck 1986), an abnormality in the attenuation pattern in voluntary movements had been hypothesied. Since the time of Kraepelin (1905) disturbed motor behaviour in schizophrenics has been considered to represent a disturbance of will. Therefore we expected different attenuation effects prior to voluntary movement onset because the voluntary process takes place before the movement onset and may be altered by the disturbance of will in schizophrenics. Moreover, we looked for changes in theta activity as higher theta activity is reported in relation to task engagement (Rugg and Dickens 1982) or to learning (Lang et al. 1988) and may have some aspects of volition. To compare the different functional states we analysed the EEG (segments of 1 s duration) during rest and before and during voluntary movements. To describe the influence of neuroleptics, the data of treated and untreated schizophrenics and matched controls were compared with that of normals treated with haloperidol for 14 days.

Methods and Patients

Subjects with closed eyes performed a voluntary self-paced flexion movement with their right fingers. Three-movement related EEG periods, each of 1 s duration were analysed by fast Fourier transformation: a resting period beginning 2.5 s before movement onset, a Bereitschaftspotential period (BP period or premovement period) beginning 1.0 s before the movement, and a movement period starting with the movement onset. The mean power density (MPD) of the movement-related spectra for each period were compared. Nonparametric statistical tests and medians were used. In order to exclude interindividual variability the medians were normalized. The 31 treated patients (23 males, 8 females) contained different diagnostic subtypes: ICD 295.3, n = 17; ICD 295.1, n = 10; ICD 295.6, n = 3, ICD 295.2, n = 1; they had a mean illness duration of 8.5 years. A control group of 21 subjects was matched in age, sex and education. A second group of 13 untreated schizophrenics were compared with a matched subgroup of 13 of the 21 normals. Nine of the 13 untreated schizophrenics had never been treated; 4 had been without medication for a minimum of 6 months. The mean illness duration in the untreated group was 6.2 years. Subtypes were: ICD 295.3, n=7; ICD 295.1, n=3; ICD 295.6, n=3. Finally, 15 healthy male paid students were treated for 14 days with haloperidol (0.04 mg/kg per day), the EEG was recorded on the 14th day and before and after the drug treatment.

Results

Theta MPD. Unlike all normal groups the 13 untreated schizophrenics did not enhance theta MPD in the premovement period. The difference in medians between premovement period and movement period in each subject differed significantly between untreated patients and matched controls (Utest, $p<0.05$ at Pz, $p<0.10$ at P4). Furthermore, the enhancement of theta MPD in normals was altered by neuroleptics. In all neuroleptic-treated groups highest values were seen in the rest period followed by pronounced attenuation of theta MPD during the movement period.

Alpha MPD. Premovement attenuation of alpha MPD differed between 31 treated patients and matched controls over the motor cortex at C3' (Fig. 1). The difference in medians of the rest period minus the premovement period calculated for each subject was significantly higher in the controls than in the 31 treated schizophrenics ($p<0.05$, Utest). The 13 untreated schizophrenics showed the same tendency. Premovement attenuation over the motor cortex (C3') was even more pronounced in the normals treated with haloperidol. Furthermore, haloperidol caused significantly more attenuation during the movement period over central and parietal areas.

Discussion

In our studies normal subjects demonstrate the same functional behaviour in the EEGspectra prior to voluntary movements of the right finger. They enhance theta MPD during the premovement period (i.e. the preparatory process) over the central and parietal cortex and attenuate alpha MPD over the activated motor cortex. Neither sign is found in untreated and treated schizophrenics. This disturbance of the premovement EEGspectra patterns may be related to a disturbed volition in schizophrenics, as suggested already by Kraepelin (1905). Especially the lack of enhancement of theta MPD could correspond to an alteration of motivational impulses since theta activity is correlated with task engagement (Rugg and Dickens 1982) and with learning (Lang et al. 1988). Lang et al. (1988) suspect a relationship between the frontomedial limbic system and the theta activity registered in the EEG over the frontomedial cortex. Thus, less theta activation during the voluntary process in schizophrenics may be related to a limbic dysfunction in schizophrenia (Bogerts et al. 1985; Zec and Weinberger 1986). The lack of attenuation of alpha activity before the voluntary movement over the motor cortex in schizophrenics may be a sign of disturbed motor activation during the voluntary process prior to movement onset. This may also be related to a disturbance of the basal ganglia which probably participate in the generation of voluntary movements

MEAN POWER DENSITY OF THE ALPHA-BAND

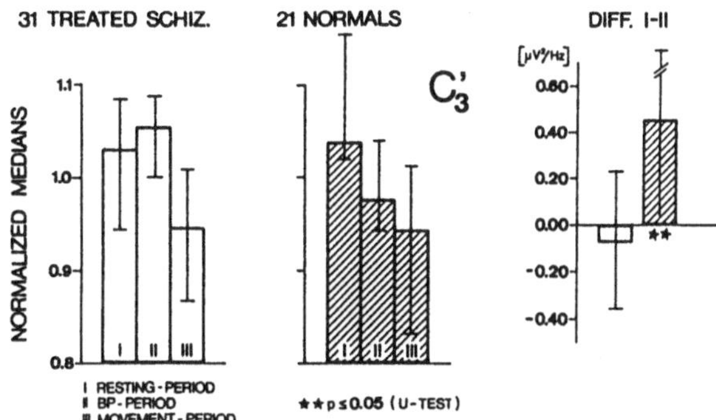

Fig. 1. Left: Comparison of 3 movement-related EEG periods over the motorcortex. Treated schizophrenics show no attenuation during the BP period. Right: Difference of rest minus BP period for schizophrenics and normals; significant difference of attenuation between both groups

(Kornhuber 1984). Moreover attenuation of alpha activity is related to activation of the thalamo-cortical projections (Buser 1987). Less attenuation before the movement may be influenced by altered impulses of the basal ganglia on this thalamo-cortical projection.

Summary

We compared movement-related EEG spectra at rest, during the Bereitschaftspotential period and during movement in untreated and treated schizophrenics. In contrast to normals schizophrenic populations exhibited neither an enhanced theta power during the voluntary process nor an attenuated alpha power over the motor cortex prior to voluntary movement onset. The lack of theta enhancement reflects a disturbance of volition in schizophrenia.

References

Bogerts B, Meertz E, Schonfeldt-Bausch R (1985) Basal ganglia and limbic system pathology in schizophrenia: a morphometric study. Arch Gen Psychiatry 42:784-791

Buser P (1987) Thalamocortical mechanisms underlying synchronized EEG activity. In: Halliday AM, Butler SR, Paul R (eds) A textbook of clinical neurophysiology Wiley, Chichester, pp 595-621

Kornhuber HH (1984) Mechanisms of voluntary movement. In: Prinz W, Sanders AF (eds) Cognition and motor process. Springer, Berlin Heidelberg New York, pp 163-173

Kornhuber HH, Deecke L (1965) Hirnpotentialänderungen bei Willkürbewegungen und passiven Bewegungen des Menschen: Bereitschaftspotential und reafferente Potentiale. Pflugers Arch Physiol 248:1-17

Kraepelin E (1905) Einführung in die Psychiatrische Klinik, 2nd edn. Barth, Leipzig

Lang W, Lang M, Kornhuber A, Diekmann V, Kornhuber HH (1988) Event related EEG-spectra in a concept formation task. Hum Neurobiol 6:295-301

Manschreck TC (1986) Motor abnormalities in schizophrenic disorders. In: Nasrallah HA, Weinberger DR (eds) The neurology of schizophrenia. Elsevier, Amsterdam, pp 65-96 (Handbook of schizophrenia, vol 1)

Rugg MD, Dickens AM (1982) Dissociation of alpha- and theta activity as a function of verbal and visuospatial tasks. Electroencephalogr Clin Neurophysiol 53:201-207
Zec RF, Weinberger DR (1986) Brain areas implicated in schizophrenia: a selective overview. In: Nasrallah HA, Weinberger DR (eds) The neurology of schizophrenia. Elsevier, Amsterdam, pp 175-206 (Handbook of schizophrenia, vol 1)

Part 7

Neurological Sciences II

(Neurology and Neurosurgery)

The Role of the Magnification Factor in the Recovery Process of Visual Field Defects After Retrogeniculate Lesions

B. Messing and H. Gänshirt

Neurologische Universitätsklinik Heidelberg,Im Neuenheimer Feld, 6900 Heidelberg, FRG

In 1987 in a follow-up study we reported on visual field defects with vascular damage of the geniculostriate pathway. This study included 37 cases with infarctions or hematomas and homonymous field defects or cortical blindness. Investigations were carried out in the first week after the event, and repeated 6 months, 1 year and 3 years after the stroke.

The extent of the spontaneous functional restoration of the visual field was the greatest in cortical blindness, more extensive in hemianopia than in quadrantanopia. The ratio of restoration was approximatively .25 for quadrantanopia, .50 for hemianopia, and 1.00 for cortical blindness. These values would also have been found if the respective visual field defect had been completely restored, complying with Wilders "law of initial value."

With regard to the functional restoration of the visual field and the anatomical substrate, lesions in the occipital lobe and the occipital pole had the best prognosis; lesions in the striate cortex had the poorest. The process of restoration came to an end within about six months at the latest (Messing and Gänshirt 1987 a, b).

The major aspect here is that vision components return after hemi- or quadrantanopic defects in special regular patterns. In hemianopias the restoration starts in the peripheral visual field from the adjacent quadrant of the contralateral side. Cone- or finger-shaped extensions proceed either symmetrically or asymmetrically in the blind area. In quadrant deficits, vision returns first in the peripheral part of the blind quadrant which is adjacent to the functioning quadrant of the same side (Fig. 1).

Our present study began the investigation at a very early stage, if possible on the first day, and calls for follow-up at short time intervals. The assigning of priority to the visual fields of the first days and weeks arises from the results of our initial study which demonstrated that the decisive events in the recovery process take place within that time.

Table 1. Localization and percentage of visual field improvement after three years (n = 37)

Lateral geniculate body	10.9%
Optic radiation	13.4%
Striate cortex	8.3%
Occipital pole	44.0%
Lateral occipital cortex	23.0%

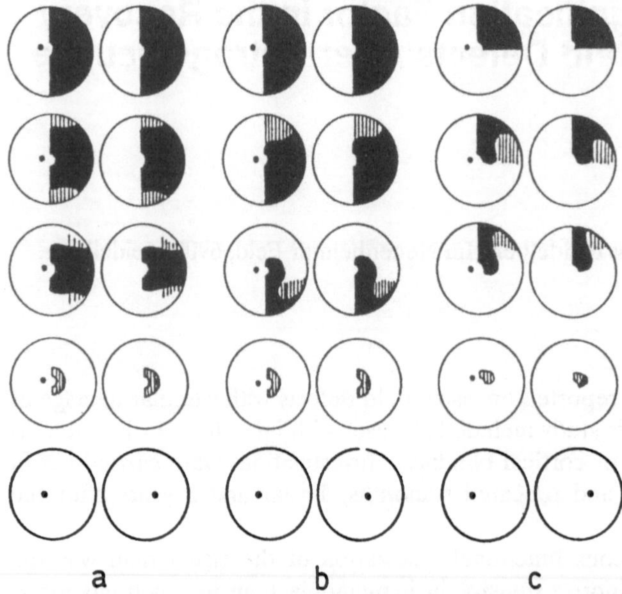

Fig. 1. Diagram of the restoration process after hemianopic and quadrantanopic visual field defects. In hemianopias cone-shaped extensions propagate toward the blind area in the periphery, mostly in a rather symmetrical way (a), occasionally emanating from only one quadrant (b). In quadrantanopia the seeing cone is observed to start without exception in the peripheral field adjacent to the functioning quadrant of the same side. Movement is always perceived first (dashed areas). The process of restitution can end at any stage. Further recovery is unlikely after a period of 6 months

The first case (Fig. 2) shows an incomplete restoration after a left-sided hemianopia with the finger-shaped extensions into the blind part of the visual field. This patient regained the vision in the lower left quadrant and developed a small Riddoch cone directed to the periphery of the upper left quadrant. On the 40th day of restoration the visual field could not be distinguished from a quadrantanopia.

In the next case (Fig. 3) of a hemianopia the visual field was restored in a symmetrical way. In the extensions, movements are seen first. After complete closing of the peripheral cones a Riddoch field remained in the central area on the 13th day, giving way to normal vision within the following 30 days. In less than 2 months the field defect had recovered entirely.

A similar course of restoration is shown in Fig. 4, a complete recovery of a hemianopia within 6 months. The restoration proceeded from the upper quadrant to the lower one. Such asymmetric restoration is responsible for the relatively long recovery time of half a year.

The last patient with cortical blindness (Fig. 5), investigated on the first day and followed up for 170 days, showed the finger-shaped extensions in the same way as the hemi- and quadrantanopic defects did.

These are four examples of a recent series. Going into the literature, numerous comparable visual fields can be found. In 1917 Poppelreuter published several dozen, and Riddoch reported 10 cases in the same year (Riddoch 1917), without exception soldiers of World War I with occipital bullet- or shell wounds. In Riddoch's contribu-

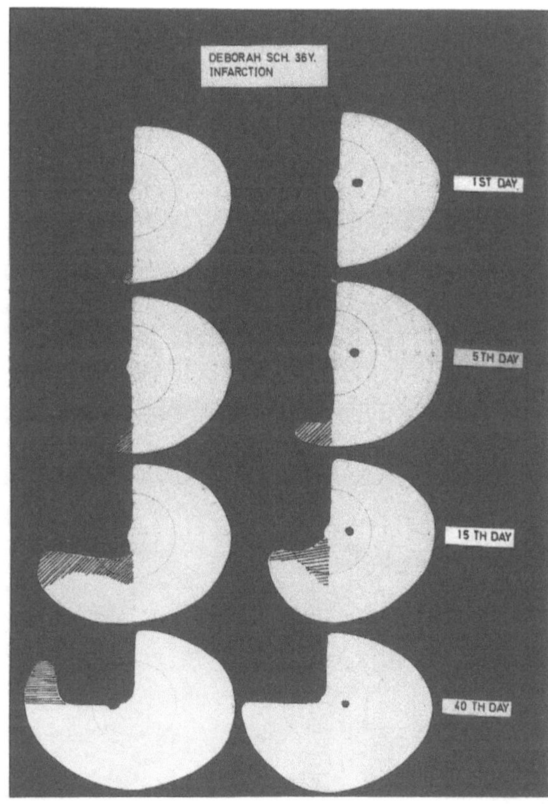

Fig. 2. Partial restoration after a left-sided hemianopia. Vision is regained only in the lower left quadrant with a small peripheral Riddochcone in the left upper quadrant (dashed areas: perception of movement only)

tion the later named Riddoch phenomenon is described for the first time. The patients of Bender and Kanzer (1939), Koerner and Teuber (1973), Zihl and von Cramon (1986), and Zihl et al. (1977) were also traumatic and vascular cases. It therefore comes as a surprise that nowhere is a statement made concerning these particular patterns.

The apparently obvious explanation for this phenomenon, the localization of the visual field periphery in the rostral striate cortex with the advantage of collateral blood supply via the middle cerebral artery, cannot be consistent since brain injury cases also yield these patterns. Neuronal hypotheses include sprouting, reactivation of remnants of nerve fibers from the embryonic period, and an extension of the bilateral retinal projection over 5° on both sides beyond the vertical axis. However, these are either not sufficiently supported or do not correspond to the time course of the process.

In a report by Baumgartner (1977) on the spread of fortification spectra it can be recognized that the single line elements are magnified where the fortification band is developing to the periphery of the visual field. A migraine patient in our department with a visual aura drew a very similar pattern of the scintillations on request (Fig. 6).

The background of this phenomenon is the magnification factor, i.e. the linear extent of the cortex concerned with each degree of visual field, as first defined by Daniel

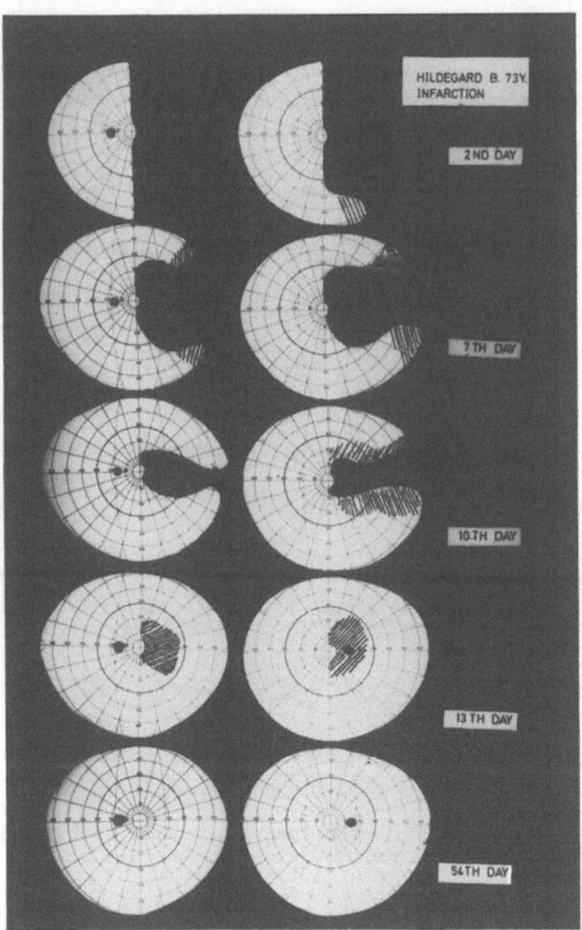

Fig. 3. Complete restoration of the visual hemifield in a symmetrical way within less than 2 months

and Whitteridge in 1961. This factor falls off from the center to the periphery, which is most readily discernible in an initially equivalent lesion and constant rate of recovery in the visual cortex. Progressive cortical recovery sets a gain of the visual field in motion, becoming larger in area units or degree from the center to the periphery. Recovery of equal areas in the overall disturbed cortical area entails 10 to 20 times greater visual field region in the periphery as compared to the center because the number of cortical columns in the visual cortex decreases per area unit of visual field toward the periphery. This can also be shown by the increasing size of the retinal receptive fields (Fig. 7).

Finally, we offer a comment to Riddoch's outstanding work concerning the recovery process. He says that this process for the appreciation of movement begins in the periphery of the field and extends inwards toward central vision, and he remarks in addition that in cases in which movement is perceived in the affected field there is some return of vision, whereas in those in which no movement is perceived after an interval of some months has elapsed the affected field probably remains permanently blind. Our results

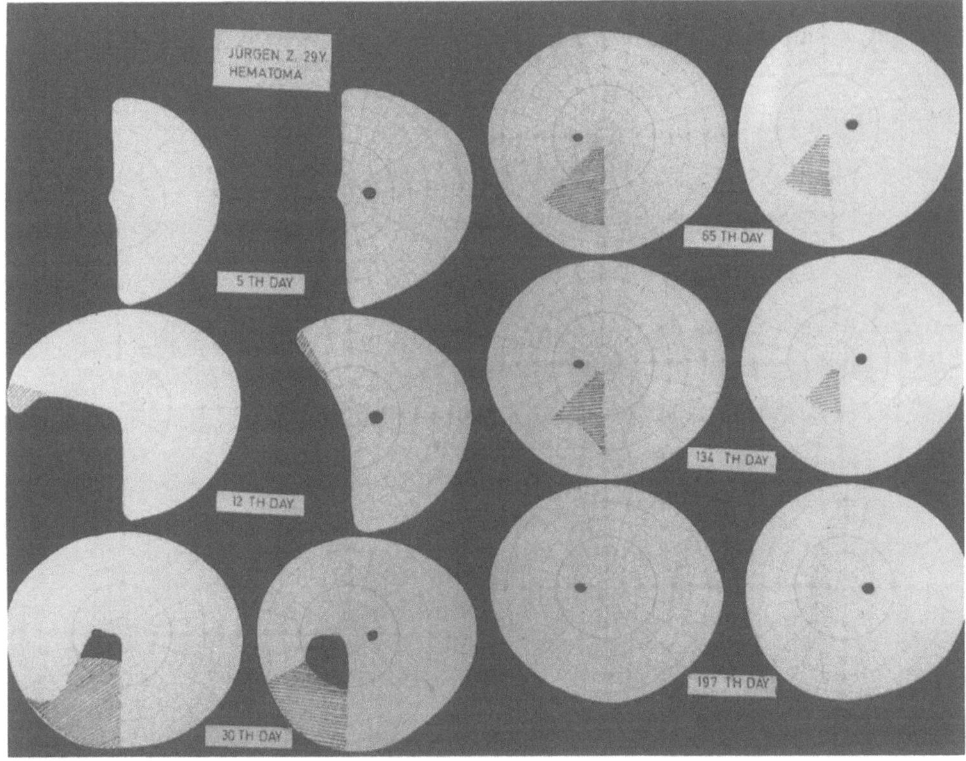

Fig. 4. Complete restoration of the visual hemifield in an asymmetrical way within about 6 months

agree completely with those of Riddoch. Actually the restoration begins in the periphery; the perception of movement is restored first. The prognosis is poor when movement perception has not been restored within some months.

Introducing the magnification factor of Daniel and Whitteridge (1961) and the concept of the columnar arrangement of the visual cortex of Mountcastle (1978), we are able to understand why the restoration of the visual field after retrogeniculate damage is initiated in the periphery, and that this highly dynamic process is a systematic one, which means that it becomes predictable. We venture to propose that our topic, which also could be called "From the arrangement of neurons to the understanding of visual perception recovery" comes close to the aim of this symposium.

Summary

The spontaneous restoration of hemianopias, quadrantanopias and cortical blindness after vascular damage of the geniculostriate pathway follows systematic principles and is predictable. The process is initiated in the periphery by development of cone-shaped areas of vision in the blind areas, with the perception of movement restored first. The central parts of the visual field recover last. The neuronal background of this phenome-

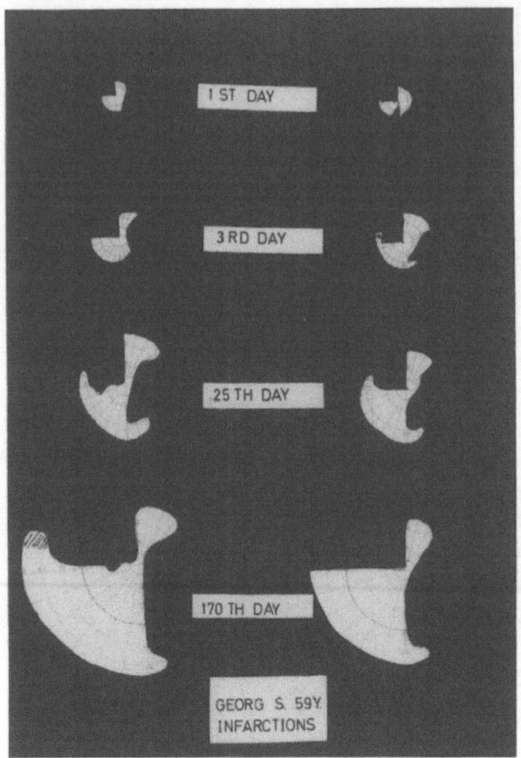

Fig. 5. Cortical blindness in which the development of the cone-shaped extensions can be shown even within the first half year after bilateral occipital infarctions

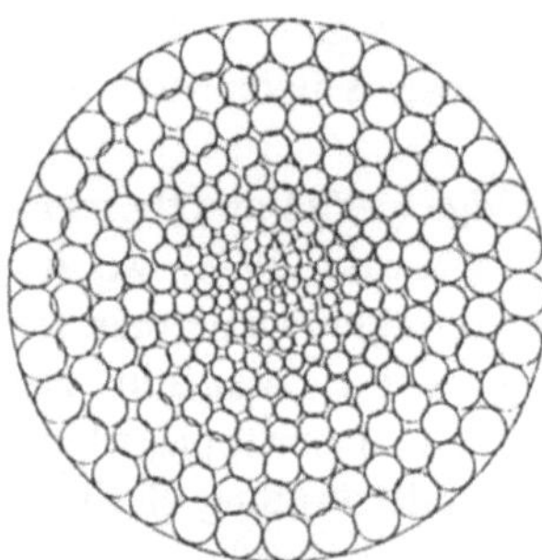

Fig. 6. Spread of fortification spectra in a patient with migraine. The single line elements are magnified where the fortification band develops to the periphery of the visual field

non is explained by the magnification factor, i.e. the linear extent of the visual cortex concerned with each degree of visual field, setting in motion a gain of visual field enlarging from the center to the periphery. This phenomenon is in accordance with Riddoch's observation that the perception of movement begins in the periphery and moves to the center.

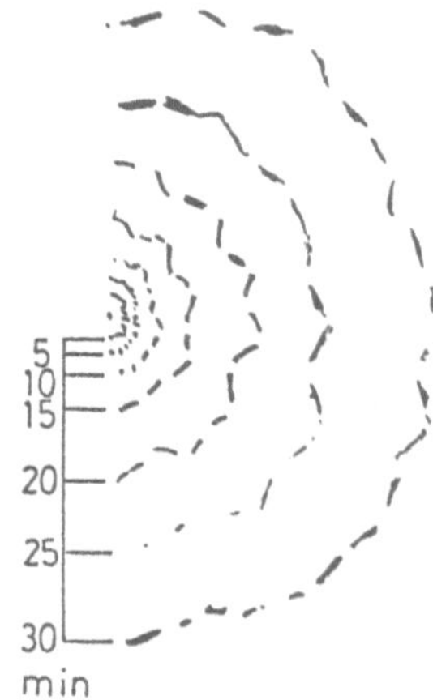

Fig. 7. The size of the retinal receptive fields increases from the center to the periphery because the number of cortical columns in the visual cortex decreases per area unit of visual field toward the periphery

References

Baumgartner G (1977) Neuronal mechanisms of the migrainous visual aura. In: Rose CF (ed.) Physiological aspects of clinical neurology. Blackwell, Oxford, pp 111-121

Bender MB, Kanzer MG (1939) Dynamics of homonymous hemianopias and preservation of central vision. Brain 62: 404-421

Daniel PM, Whitebridge D (1961) The representation of the visual fields on the cerebral cortex in monkeys. J Physiol (Lond), 159: 203-221

Koerner F, Teuber H-L (1973) Visual field defects after missile injuries to the geniculo-striate pathway in man. Exp Brain Res 18: 88-113

Messing B, Gänshirt H (1987a) Spontanverlauf vaskulärer, retrogenikulärer Gesichtsfeldstörungen. In: Poeck K, Hacke W, Schneider R (eds) Verhandlungen der Deutschen Gesellschaft für Neurologie. Springer, Berlin Heidelberg New York, pp 228-236

Messing B, Gänshirt H (1987b) Follow-up of visual field defects with vascular damage of the geniculostriate visual pathway. Neuroophthalmology 7: 231-242

Mountcastle VB (1978) An organizing principile for cerebral function: the unit module and the distribution system. In: Edelman GM, Mountcastle VB (eds) The mindful brain. MIT Press, Cambridge / MA, pp 7-50

Poppelreuter L (1917) Die Störungen der niederen und höheren Sehleistungen durch Verletzung des Okzipitalhirns. Voss, Leipzig Die psychischen Schädigungen durch Kopfschuβ im Kriege 1914-16 vol.1

Riddoch G (1917) Dissociation of visual preceptions due to occipital injuries, with special reference to appreciation of movement. Brain 40: 15-57

Uhthoff L (1915) Beiträge zu den hemianopischen Gesichtsfeldstörungen nach Schädelschüssen, besonders solchen im Bereich des Hinterhauptes. Klin Mbl Augenheilkd 55: 104-125

Zihl J, von Cramon D (1986) Recovery of visual field in patients with postgeniculate damage. In: Poeck K, Freund H-J, Gänshirt H (eds) Neurology. Springer, Berlin-Heidelberg New York, pp 188-194

Zihl J, von Cramon D, Brinkmann R, Backmund H (1977) Verlaufskontrolle und Prognose bei Gesichtsfeldausfällen von Patienten mit cerebrovaskulären Störungen. Nervenarzt 48: 219-224

Oxygen Free Radicals and Radical Scavengers in Neurology

H.E.Kaeser and H.Langemann

Neurological Clinic, University of Basle, Section of Neurobiology, Department of Research, Cantonal Hospital, 4031 Basle, Switzerland

Free radicals have become fashionable in recent years. Hundreds of articles and many regional and national meetings have been devoted to them. The first international symposium on free radicals will take place this year in Vienna. Free radicals seem to be involved in one way or another in many diseases, such as cancer, ischemia, circulatory shock, inflammation and liver cirrhosis (Table 1).

Some authors speculate that they play a role in brain cell death and in aging (Ames et al. 1981; Siesjö 1984). Radical scavengers have been administered in many experimental and clinical conditions. The results are on the whole positive, though still controversial. In neurology free radicals are not yet a popular topic, but this may change in the near future.

Aerobic organisms cannot live without oxygen, but this means that there is a permanent danger of autoxidation. O_2 itself is a diradical with two unpaired electrons spinning in the same direction in two different orbitals. It is curiously inactive, in contrast to its partially reduced products such as the superoxide radical ($O_2^{\cdot -}$), hydrogen peroxide (H_2O_2) and the hydroxyl radical ('OH Table 2). Most of these species are extremely short-lived but highly aggressive, and they attack both polyunsaturated fatty acids of membranes and DNA, thereby initiating chain reaction (Halliwell and Gutteridge 1985; McCord 1988; Pryor 1986). Oxygen radicals are produced in the mitochondria of all aerobic cells (Table 3).

Hemoglobin is oxidized to methemoglobin in considerable amounts and generates peroxides (Halliwell and Gutteridge 1985; Pryor 1986). Activated macrophages make

Table 1. Diseases with possible radical-mediated reactions

Disease	
Emphysema	+ + +
Cancer	+ + +
Stroke	+
Myocardial infarction	+
Arthritis	+ +
Atherosclerosis	+
Cirrhosis	+
Retrolental fibrosis	+ + +
Cataract	+ +
Aging	+
Parkinson	+ +

Table 2. Oxygen-derived free radicals

O_2 + e-	-------->	$O_2^{\cdot -}$
$2\,O_2^{\cdot -} + 2H+$	-------->	$H_2O_2 + O_2$
$H_2O_2 + Fe^{2+}$	-------->	Fe^{3+} + 'OH + OH^-
$H_2O_2 + Cu^+$	-------->	Cu^{2+} + 'OH + OH^-

Table 3. Sources of radicals

Hemoglobin	-------->	Methemoglobin
Mitochondria		
Phagocytes		
Hyperoxia		
Ischemia - Reperfusion		
Trauma - Edema		
X rays		
Burns		
Toxins		
Redox active drugs		
Catecholamines and others amines		
O_2-producing enzymes:	Xanthine oxidase	
	Tryptophan dioxygenase	
	Indoleamine dioxygenase	

use of superoxide and hydroxyl radicals to fight bacteria (Forman 1986). Hyperoxia induces increased amounts of oxy-radicals (Fridovich and Freeman 1986). Reperfusion of tissues after ischemia converts xanthine dehydrogenase to xanthine oxidase, which generates radicals (McCord 1985, 1988). Inflammation, burns and X rays liberate oxyradicals and activate macrophages. Some oxyradical like drugs, e.g. phenobarbital, clozapin, nitrofurantoin, certain antibiotic and antiparasitic drugs, clofibrate and anti-malaria medications generate radicals (Mitchell and Russo 1987). Chemical and natural products in the air are other sources of radicals. The danger of autoxidation is so high that only a very elaborate defense mechanism can cope with it.

Defense Mechanisms Against O_2 Damage

A first safeguard against autoxidation is the spatial separation of radicals and target molecules in organelles and cytosol (Fridovich and Freeman 1986). In addition, a complicated system of defense mechanisms protects organs from damage by activated forms of O_2 (Table 4). O_2 is reduced to water by 4 successive electron transfers in the presence of cytochrome oxidase. When some partially reduced oxygen escapes, it is picked up by the second defense line, the metalloenzymes superoxide dismutases (SOD). Some of these enzymes, for instance the Cu^{II}-Zn^{II} SOD, are distributed in the cytosol, whe-

Table 4. Antioxidant defense mechanisms

1. Cytochrome oxidase
2. Superoxide dismutase (SOD) Cu^{II} Zn^{II}, Mn
3. Catalases, glutathione peroxidase
4. α-tocopherol, ascorbic acid
5. Repair mechanisms

reas the Mn-containing SOD is bound to the matrix of the mitochondria. Mn SOD transforms the superoxide radicals to hydrogen peroxide ($2\ O_2^{-} + 2H^{+} = H_2O_2 + O_2$). Hydrogen peroxide is oxidized by the scavengers of the third defense line, i.e. the catalases (CA) in the peroxisomes and glutathione peroxidase in the cytosol. The Se-containing glutathione peroxidase seems to be of particular importance for the elimination of H_2O_2 by reduction to H_2O (Fridovich and Freeman 1986; Pryor 1986). Two vitamins constitute the fourth defense line: vitamins C and E. Alpha-tocopherol (vitamin E) arrests the chain reactions of polyunsaturated fatty acid in biological membranes. Ascorbic acid regenerates the oxidized form to vitamin E (Fridovich and Freeman 1986). In spite of all these defense strategies damage to biological membranes and to DNA may occur. DNA may be repaired without a loss of genetic code as a part of the normal turnover or by other special mechanisms (Fridovich and Freeman 1986).

Besides these defense mechanisms other radical scavengers play an important role. Uric acid, ceruloplasmin, beta-carotene, metal chelators and MAO inhibitors must be mentioned here. Other strong antioxidants in the CSF have not yet been identified. (Ames et al. 1981; Fridovich and Freeman 1986; Table 5).

There is a complicated interaction between the oxidizing enzymes, their substrates and energy metabolism (Fig. 1). It has been compared to a house of cards, and it seems almost impossible to study single events separately (Mitchell and Russo 1987). In spite of this we shall try to describe what happens in ischemia and in inflammation.

Table 5. Other radical scavengers

Uric acid
Allopurinol
Cysteine
Ceruloplasmin
Beta-Carotene

Xanthine Oxidase

After ligation of a coronary artery several biochemical changes take place. High-energy phosphatase is degraded to hypoxanthine and xanthine, and at the same time xanthine dehydrogenase is converted to xanthine oxidase. When the ischemic tissue is reperfused, a burst of superoxide is generated by the oxidation of xanthine (McCord 1985; Fig.

2). A second source of superoxide is the disruption of mitochondria (McCord 1988). Under ischemic conditions cytochrome oxidase is no longer able to reduce O_2 to water, so that partially reduced oxygen may leak into the cytosol (Burton 1988; McCord 1988). In addition, inflammatory cells migrate into the ischemic tissue. These cells (polymorphonuclear leukocytes, macrophages and monocytes) are secondary sources of free radicals (Forman 1986). These free radicals, particularly OH radicals, have a direct toxic effect on myocardial cells (Burton 1988). At the same time free radical scavengers such as SOD, catalase, glutathione peroxidase and membrane-bound alpha-tocopherol are decreased (Gardner 1988; McCord 1988). When preparations of catalase and SOD were applied before and after the ligation, the size of myocardial infarct and the amount of creatine phosphokinase liberated by the damaged myocardium were both

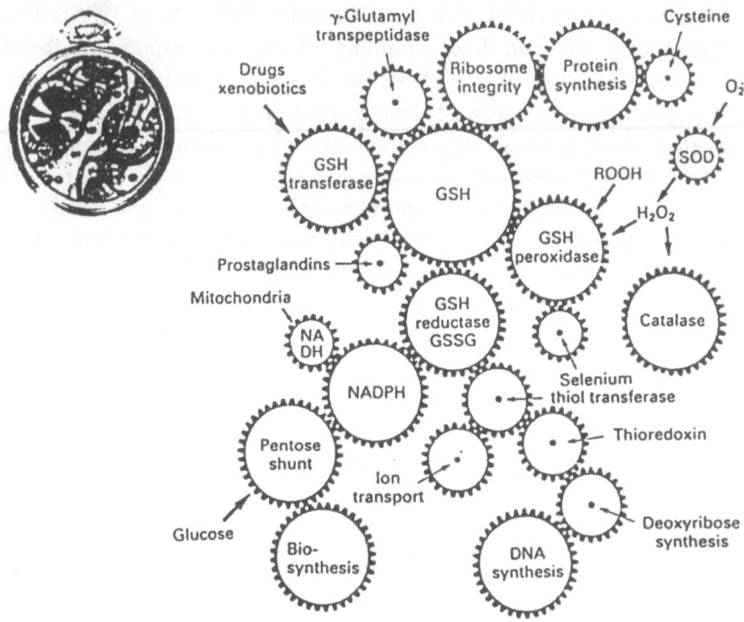

Fig. 1. Interrelationsship of different radical-scavenging and radical-generating systems (From Mitchell and Russo 1987)

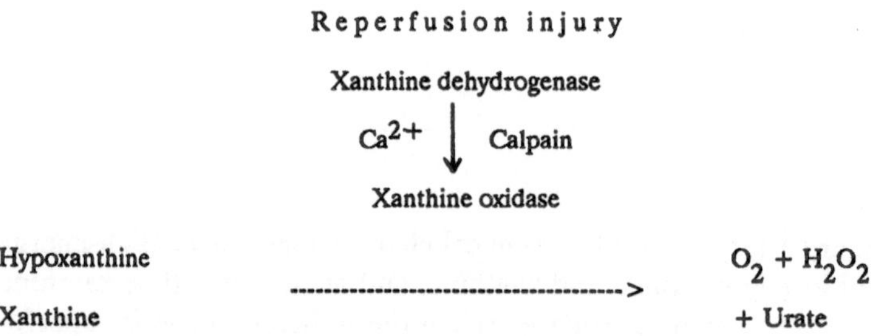

Fig. 2. Superoxide generated by xanthine dehydrogenase

reduced (McCord 1985, 1988). An important species difference regarding xanthine oxidase was observed. In the rat and dog the amount of xanthine dehydrogenase/oxidase in the heart is high, in rabbits and in man very low. For this reason allopurinol, which is a specific xanthine oxidase inhibitor, was effective in protecting the heart from ischemia in some species, but not in rabbit and man (Miura et al. 1988).

Inflammatory processes can also lead via a series of events to ischemia. Neutrophil cells are activated by an extracellular chemoattractant and stick to the capillary endothelium causing extravasates. They produce superoxide and additional enzymes such as myeloperoxidase and plug the capillary beds. The microcirculation is impaired causing edema and increased interstitial pressure, resulting in ischemia (McCord 1985, 1988).

As mentioned above, xanthine oxidase is not involved in the generation of reperfusion injuries in the myocardium of rabbits and man. In the central nervous system, however, this enzyme seems to play a role in all species. After ligation of the middle cerebral artery, the tissue concentration of uric acid (UA) increases greatly up to a maximum after 24 hours (Kanemitsu et al. 1986). This increase is likely to be due to transformation of xanthine dehydrogenase to xanthine oxidase in ischemic tissues. We had the opportunity of studying the changes of uric acid in experimental autoimmune encephalomyelitis (EAE). UA levels were determined in the spinal cord of Lewis rats with chronic relapsing EAE and seemed to correlate with symptomology (Honegger et al. 1986). During the first attack tissue concentration rose to high levels. It was 10 - 20 times higher than in control animals and decreased during recovery, though still remaining at a level 1 - 3 times that of controls. In animals with only weight loss but without overt clinical deficit UA increased to 2 - 3 times control values, and in animals which did not react to the inoculation the values of UA stayed within normal limits. A common pathophysiological explanation for ischemia and EAE could be related to the finding that in both conditions the blood-brain barrier (BBB) is damaged, as can be seen by the contrast medium accumulation in computer tomography. This may not, however, be the only factor involved, as a breakdown of the BBB is not sufficient to cause a tissue concentration which is higher than the normal serum level. There is also an active transport system for hypoxanthine and UA through the endothelium of brain capillaries. One wonders whether the high tissue concentrations of UA are the result only of the brain damage, or whether this substance, which is an inhibitor of xanthine oxidase and a radical scavenger, might serve some useful purpose. We have no answer to this question at the present time, although our recent experiments provide additional information. Preliminary results show that allopurinol administered from the tenth day after inoculation partially protects Lewis rats from EAE. This study is still in progress, and it is not yet clear what the necessary dose is, or which day after inoculation is the critical one for supression of the EAE reaction.

Glutathione

The glutathione peroxidase system is of particular importance in Parkinson's disease and Parkinson syndrome of toxic origin and in radiation and drug-induced cytotoxicity (Mitchell and Russo 1987). The oxidative metabolism of dopamine and DOPA by MAO-B is associated with the production of H_2O_2 and with conversion of glutathione

(GSH) to the oxidized form (GSSG). Normally, the natural defense mechanism includes SOD and glutathione peroxidase, which inactivate O_2^- and H_2O_2. The levels of GSH are believed to be limiting in this process. In Parkinson's disease the progressive loss of nigrostriatal neurons is much more severe than the normal age- dependent decrease. An explanation of this could be an impairment of the natural defense mechanism, leading to damage of nigrostriatal neurons by oxygen radicals. Indeed, severe deficiency of GSH has been found at autopsy in the sustantia nigra of parkinsonian patients (Perry et al. 1982; Perry and Voon Wee Yong 1986). In this case, treatment with large quantities of L-DOPA and decarboxylase inhibitors would acclerate the degeneration of nigrostriatal neurons (Riederer and Przuntek 1987). This was the clinical impression and led to the recommendation that the doses administered should be as low as possible. In this context an epidemiological survey by Rajput (1984) in Rochester/Minnesota is of special interest. He found that dementia was almost 3 times more frequent in patients treated with L-DOPA than in Parkinson patients without DOPA or in the normal population of corresponding age . Although this study may be open to some questions, we wonder whether DOPA could also damage other neuronal populations, particularly the cholinergic system.

6-OH-Dopamine is a dopamine derivative which destroys adrenergic and serotoninergic neurons. If applied locally in animals, it induces a selective degeneration of dopaminergic neurons. This effect can be prevented by a selective MAO-B inhibitor and possibly also by sulfhydryl compounds. The selective MAO-B inhibitor selegiline (deprenyl) also prevents the destructive effects of MPTP (*N*-methyl-4-phenyl-1, 2, 3, 6-tetrahydropyridine; Riederer and Przuntek 1987). This substance is converted by MAO-B to the charged molecule MPP^+ which accumulates in the zona compacta of the substantia nigra. Further metabolism of MPP^+ releases free radicals which destroy dopaminergic neurons. A single injection of MPTP in mice lowers the GSH content of the substantia nigra, although this can be prevented by selegiline (deprenyl) or by the antioxidants alpha-tocopherol and beta-carotene (Riederer and Przuntek 1987; Voon Wee Yong et al. 1986).

At the present time Parkinson's disease is not considered to be of hereditary origin in the majority of cases. Chronic exposure to a toxic substance which selectively damages nigrostriatal neurons, such as 6-OH-DOPA or MPTP, is an attractive hypothesis. Inhibition of MAO oxidase by selegiline and treatment with antioxidants such as vitamin E, ascorbic acid or glutathione modulators might be effective in slowing the progressive degeneration of nigrostriatal neurons and prolonging the life expectancy of Parkinson patients (Perry and Voon Wee Yong 1986; Riederer and Przuntek 1987).

Much more research will have to be carried out before the role of free radicals and radical scavengers is elucidated. Even if many observations are still controversial, no one doubts the importance of free radicals for the life and death of cells, including neurons.

Summary

Highly active oxygen derivatives are produced in every aerobic cell. A series of defense mechanisms with antioxidants and radical scavengers protect normal tissues from autoxidation. Reperfusion after organ ischemia generates large amounts of oxygen and hydroxyl radicals which cannot be inactivated by the endogenous scavengers. In these cases the administration of water-soluble and lipid-soluble scavengers may protect the tissues from damage. Allopurinol, for instance, seems to have such an effect, not only in heart and brain ischemia but also in experimental autoimmune encephalomyelitis. It is possible that autoxidation plays a role in Parkinson's syndrome and in the Parkinson dementia complex. Oxygen free radicals and radical scavengers are an important new field of investigation, particularly in ischemic, inflammatory, degenerative and neoplastic diseases of the nervous system.

References

Ames BN, Cathcart R, Schwiers E, Hochstein P (1981) Uric acid provides an antioxidant defense in humans against oxidant- and radical-caused aging and cancer: A hypothesis. Proc Natl Acad Sci USA 78: 6858-6862

Burton KP (1988) Evidence of direct toxic effects of free radicals on the myocardium. Free Radic Biol Med 4: 15-24

Forman HJ (1986) Oxidant production and bactericidal activity of phagocytes. Annu Rev Physiol 48: 669-680

Fridovich I, Freeman B (1986) Antioxidant defenses in the lung. Annu Rev Physiol 48: 693-702

Gardner TJ (1988) Oxygen radicals in cardiac surgery. Free Radic Biol Med 4: 45-50

Halliwell B, Gutteridge JMC (1985) Oxygen radicals and the nervous system. Top Neurol Sci 8: 22-26

Honegger CG, Krenger W, Langemann H (1986) Increased concentrations of uric acid in the spinal cord of rats with chronic relapsing experimental allergic encephalomyelitis. Neurosci Lett 69: 109-114

Kanemitsu H, Tamura A, Sano K, Iwamoto T, Yoshiura M, Iriyama K (1986) Changes of uric acid levels in rat brain after focal ischemia. J Neurochem 46: 851-853

McCord JM (1985) Oxygen-derived free radicals in postischemic tissue injury. N Engl J Med 312: 159-163

McCord JM (1988) Free radicals and myocardial ischemia: overview and outlook. Free Radic Biol Med 4: 9-14

Miura T, Yellon DM, Kingma J, Downey JM (1988) Protection afforded by allopurinol in the first 24 hours of coronary occlusion is diminished after 48 hours. Free Radic Biol Med 4: 25-30

Mitchell JB, Russo A (1987) The role of glutathione in radiation and drug induced cytotoxicity. Br J Cancer 55: 96-104

Perry TL, Voon Wee Yong (1986) Idiopathic Parkinson's disease, progressive supranuclear palsy and glutathione metabolism in the substantia nigra of patients. Neurosci Lett 67: 269-274

Perry TL, Godin DV, Hansen S (1982) Parkinson's disease: a disorder due to nigral glutathione deficiency? Neurosci Lett 33: 305-310

Pryor WA (1986) Oxy-radicals and related species: their formation, lifetimes and reactions. Annu Rev Physiol 48: 657-667

Rajput AH (1984) Epidemiology of Parkinson's disease. Can J Neurol Sci 11: 156-159

Riederer P, Przuntek H (eds) (1987) MAO-B-inhibitor selegiline (R-deprenyl). J Neural Transm [Suppl] 25

Siesjö BK (1984) Brain cell death in ischemia and aging: are free radicals involved? Monogr Neural Sci 11: 1-7

Voon Wee Yong TL, Perry L, Krisman AA (1986) Depletion of glutathione in brainstem of mice caused by N-methyl-4-phenyl-1,2,3,6-tetrahydroypridine prevented by antioxidant pretreatment. Neurosci Lett 63: 56-60

Immunotherapy in Multiple Sclerosis: Current Status and Future Prospects

R.E. Gonsette
Belgian National Center for MS, 1910 Melsbroek, Belgium

The monitoring of lymphocytes subsets in treated patients yields interesting information about different effects of various immune treatments and allows immunotherapy for multiple sclerosis (MS) to be conducted under safer conditions. MS selectively interrupts functional transmission "from neuron to action" and it is therefore a good opportunity to present this paper as a tribute to Prof. Kornhuber, who has pursued invaluable research in the field of MS.

It is now accepted that MS is an immune disease. As a consequence, more than 25 different techniques of immunotherapy have been applied but it is still difficult to evaluate the usefulness of most of them. However well controlled recent clinical studies have been completed and definite conclusions can be drawn for two substances: AZA and CsA.

The results of the English-Dutch double-blind trial with AZA (Hughes 1988) clearly show that compared with a placebo group, a mild but not significant tendency to a slower progression is observed in treated patients. Nevertheless, the risks associated with chronic administration of AZA in MS definitely outweigh its mild benefit. This is also true for the German-Dutch study with CsA (Kappos et al. 1988), the conclusion of which is that CsA cannot be the drug of final choice in MS because of unwanted side-effects and the absence of a significant efficacy.

Another recent technique, TLI (Cook et al. 1987) appears more interesting. Indeed, patients experiencing a sustained lymphocytes count below 900/mm^3 after TLI, show a significantly less rapid progression compared with sham irradiated patients, but this single treatment does not induce a permanent remission, and the potential morbidity associated with retreatments remains a serious drawback.

We must thus admit today that AZA and CsA are ineffective in preventing MS progression, and that TLI is probably efficacious but of limited value for practical reasons.

The publication by the group of Hommes et al. (1975) concerning CY treatment in progressive MS and that by our group (Gonsette et al. 1977) about remitting-progressing patients have rekindled the interest of other investigators for this potent immunosuppressive agent.

To summarize briefly our first publication, a short-term intense immunosuppression for 2 - 3 weeks was induced with iv CY doses of 1 - 2 g every 2 or 3 days, until there was a lymphopenia of 1000/mm^3. In this preliminary open trial, a marked reduction in the annual relapse rate (ARR) calculated over a period of 2 years before and after treatment was observed after a mean follow-up for 4.5 years. This reduction apparently did not result from a spontaneous decrease due to the evolution of the disease since it was observed not only when the patients were used as their own control but also when treated patients were compared with a retrospectively matched historical control group.

Recently, Killian et al. (1988) have also reported a marked decrease in the ARR in a single-blinded trial after monthly iv CY for 1 year in relapsing-remitting forms. When the placebo group was given CY at the end of the study, they also experienced a significant decrease in episodes. No changes in the Kurtzke scores were observed in either group during this short observation period.

The ARR is a poor indicator of disease evolution. Nevertheless in our experience, a transient stabilization was observed in more than 60% of the patients. It has been confirmed later by Hauser et al. (1983) in a controlled study that CY is able to halt transiently the progression of the disease in progressive MS.

Of note is that at that time, Huston (1983) already reported about the advantages of monthly treatments with iv CY, avoiding alopecia and prolonged hospitalization. This has been confirmed by Kornhuber and Mauch (1986) after weekly administration.

Recently, in a single-blinded study with a short-term intense CY immunosuppression in progressive MS (Likosky 1988) no difference was found in clinical stabilization in individual patients between the CY and the placebo groups. However, when considering the mean increase in EDSS from baseline values, a lower augmentation was observed in the group of treated patients.

In 1977 we came to the conclusion that a short-term intense iv CY immunosuppression was effective in transiently reducing exacerbation and progression rates in most remittent-progressive forms, but that some form of maintenance therapy was required to prolonge the beneficial effects.

In a second protocol (Gonsette 1985), maintenance therapy was begun after an acute iv immunosuppression with oral doses of 50 mg CY every day, later tapered to 50 mg every other day. Sixty patients were enrolled and matched to sixty "nonmaintained" patients (single iv immunosuppression). After a mean follow-up of 6 years, it appeared that the augmentation of the mean Kurtzke disability scale (KDS) was definitely lower in patients with maintenance therapy (KDS 2.44 after versus 2.27 at entry) than in patients without maintenance therapy (KDS 4.08 after versus 2.05 at entry).Interestingly, the same has been found in a recent study by Goodkin et al. (1987) with subsequent iv retreatment every other month.

However, even if the risk of cancer apparently was not increased in our patients treated with CY for years, there is no doubt that a prolonged administration of this drug in other immunologic diseases has been associated with a higher incidence of malignancies. In a study concerning rheumatoid arthritis, it has been clearly demonstrated by Baker et al. (1987) that the risk of cancer was significantly increased in CY-treated patients compared with nonimmunosuppressed patients, and that both the cumulative dose and the mean duration of CY treatment were the most important risk factors.

Thus a third protocol was designed in 1985 as an alternative mean of administration that would keep the efficacy of CY and reduce immediate side effects as well as long-term adverse reactions related to cumulative doses.

CY is given as a single iv dose of 650 mg/mm^3 once a week for 6 weeks on an outpatient basis. As it is not possible to predict for how long a given dosage of CY remains effective on the immune system, an interval of 6 months was arbitrarily chosen for further periodic retreatments (1 g iv).

It has been shown since then by Goodkin et al. (1987) that weekly CY induction, which does not provoke hair loss, has the same favorable clinical effects as daily infusions that provoke alopecia in all patients. Thirty-two patients have been treated so far with this technique, with a mean follow-up of 18 months. As far as the clinical tolerance is concerned, gastrointestinal side effects with nausea and vomiting were frequently experienced but were markedly reduced in most patients with oral or iv usual antiemetics. No patient had alopecia and hair loss was limited to mild thinning in a few patients. Fatigue lasting for several days was observed in some cases. As a rule, this mode of administration was much better accepted by the patients and only two dropouts were related to immediate side effects.

The first clinical trial directing CY immunotherapy by monitoring immune parameters has been recently published by Myers et al. (1987). CY was given at monthly intervals, increasing the dose to achieve a reduction in B and CD4 cells below the 5th percentile of a normal control population.

Lymphocytes subsets typing is performed as a routine in our MS patients, and changes in CD4 and B cells observed after weekly CY administration according to our third protocol were compared with those produced by the technique of Myers et al. (1987). It appears that at the end of the 6-week treatment, the mean reduction in percentages from baseline values is the same as or even higher than that obtained during the maximum immunosuppressive period after monthly CY administration (Table 1). It must be stressed that weekly administration limited to 6 weeks does not provoke hair loss whereas monthly infusion for 10 to 12 months systematically causes alopecia.

If we refer to the corresponding percentiles in our normal population, it appears that weekly administration of CY for 6 weeks in MS patients provoked a mean reduction in all lymphocyte subsets to the 5th percentile after 6 weeks and/or 6 months and to the 10th percentile after 18 months, except for B cells that are reduced to a lesser extent (to the 25th percentile; Table 2).

If the immune effects of periodic intermittent CY injections after 2 years are compared to those observed after maintained oral administration for years, it appears that the most pronounced immunosuppression is obtained with the intermittent technique, but that they do not reach the 5th percentile values however (Table 3).

Nevertheless, CY is not the definite treatment of MS and more effective and less toxic substances must be found.

Table 1. Intermittent bolus iv CY immunotherapy: lymphocytes subsets changes (absolute values)

Mean reduction (in %) from pretreatment values	Lymphocytes	CD4	CD8	B cells
6 weeks treatment	40	42	33	56
+ 6 months	40	49	41	58
Myers et al.		47	22	50
+ 1 year	29	35	25	53
+ 1.5 year	34	38	30	55

Table 2. Effects of intermittent bolus injection of CY on immunologic parameters (absolute values)

	Lymphocytes	CD4	CD8	B cells
Healthy subjects	2310 (50)*	931 (50)	602 (50)	227 (50)
MS before CY	2390 (50)	1134 (70)	562 (40)	336 (85)
CY 6 weeks	1413 (< 5)	661 (10)	381 (10)	151 (25)
CY 6 months	1398 (< 5)	586 (5)	333 (5)	142 (20)
CY 12 months	1695 (10)	736 (20)	424 (15)	161 (25)
CY 18 months	1597 (10)	710 (15)	397 (10)	152 (25)

* Corresponding percentiles in a normal population within parentheses

Table 3. Intermittent injections versus oral maintenance: long-term immune effects (2 years) (absolute values)

	Lymphocytes	CD4	CD8	B cells
Healthy subjects	1521 (5)*	579 (5)	349 (5)	102 (5)
Intermittent injection	1597 (10)	710 (15)	397 (10)	152 (25)
Oral maintenance	2044 (30)	833 (35)	740 (75)	180 (35)

* Corresponding percentiles in a normal population within parentheses.

Ridge et al. (1985) have shown that mitoxantrone (MX) suppresses clinical signs in rats with a developing or an established EAE. This has been confirmed recently by Lublin et al. (1987) in mice and in guinea pigs in our laboratory (Gonsette et al. 1988). In our experience, both clinical signs and histological lesions were markedly reduced after administration of MX.

According to Fidler et al. (1986) MX exerts a potent suppressive influence on the humoral response through a marked reduction in B cell number. They have also shown that MX inhibits helper function and enhances suppressor activity (Fidler et al. 1985). MX seems therefore the drug of choice for MS therapy especially as its immediate and long-term toxicity is markedly lower than that of CY.

We are presently testing the tolerance and the effects of MX on lymphocytes subsets in a preliminary clinical study. Immediate side effects (nausea, vomiting) appear somewhat milder than those observed after CY infusions, and none of the first ten patients treated with MX had alopecia. Lymphocytes serial determinations clearly show that MX produces more marked effects on B and HLA-DR cells than CY, with the result of a reduction in these two populations below the 1st percentile. In opposition to CY, a differential effect of MX on CD4 and CD8 cells or marked changes in CD4/CD8 ratio have not been observed so far.

In conclusion, weekly intermittent administration of CY followed by single-dose retreatments every 6 months currently appears to be the safest technique in MS patients in whom an immunotherapy is required. There is a reasonable hope that in the future long-term immunosuppression with MX in MS will yield similar or even better clinical benefits with markedly reduced immediate side effects and long-term adverse reactions.

Summary

Recent blinded controlled trials have demonstrated that azathioprine (AZA) and cyclosporine A (CsA) are of no benefit in multiple sclerosis. Total lymphoid irradiation (TLI) and cyclophosphamide (CY) appear to influence both the relapse rate and the progression of the disease. The authors report briefly about clinical results with different techniques of CY administration, as well as about their respective effects on lymphocytes subsets. It appears that controlled intermittent injections do not provoke alopecia, produce a marked immunosuppression and reduce the risks of long-term severe adverse reactions.

References

Baker GL, Kahl LE, Zee BC, Stolzer BL, Agarwal AK, Medsger TA (1987) Malignancy following treatment of rheumatoid arthritis with cyclophosphamide. Long-term case-control follow-up study. Am J Med 83: 1-9

Bisteau M, Devos G, Brucher JM, Gonsette RE (1989) Prevention of experimental allergic encephalomyelitis in guinea pigs with desferrioxamine, isoprihosine and mitoxantrone. In Gonsette RE and Delmotte P (eds) Exerpta Medica. International Congress Series 863, pp 299 - 300

Cook SD, Devereux C, Troiano R, Zito G, Hafstein M, Lavenhar M, Hernandez E, Dowling PC (1987) Total lymphoid irradiation in multiple sclerosis: blood lymphocytes and clinical course. Ann Neurol 22: 634-638

Fidler JM, Smith F, Gibbons J (1985) Mitoxantrone inhibits helper function and enhances suppressor activity. Agents and Actions 16: 607-608

Fidler JM, Quin DeJoy S, Gibbons JJ (1986) Selective immunomodulation by the antineoplastic agent mitoxantrone. I. Suppression of B lymphocyte function. J Immunol 137: 727-732

Gonsette RE (1985) Combined acute and chronic immunotherapy with cyclophosphamide (Endoxan) in MS. Eur Neurol 24: 437

Gonsette RE, Demonty L, Delmotte P (1977) Intensive immunosuppression with cyclophosphamide in multiple sclerosis. Follow up of 110 patients for 2 - 6 years. J Neurol 214: 173-181

Goodkin DE, Plencner S, Palmer-Saxerud J, Teetzen M, Hertsgaard D (1987) Cyclophosphamide in progressive multiple sclerosis. Maintenance versus nonmaintenance therapy. Arch Neurol 44: 823-827

Hauser SL, Dawson DM, Lehrich JR, Beal MF, Kevy SV, Propper RD, Mills JA, Weiner HL (1983) Intensive immunosuppression in progressive multiple sclerosis. A randomized, three-arm study of high-dose intravenous cyclophosphamide, plasma exchange, and ACTH. N Engl J Med 308: 173-180

Hommes OR, Prick JJG, Lamers KJB (1975) Treatment of the chronic progressive form of multiple sclerosis with a combination of cyclophosphamide and prednisone. Clin Neurol Neurosurg 78: 59-72

Hughes RAC, Mertin J (1988) Double-masked trial of azathioprine in multiple sclerosis. Lancet 179-183.

Huston DP (1983) Immunosuppression in multiple sclerosis. N Engl J Med 309: 240-241

Kappos L, Patzold U, Dommasch D, Poser S, Haas P, Krauseneck P, Malin JP, Fierz W, Graffenried BU, Gugerli US (1988) Cyclosporine versus azathioprine in the long-term treatment of multiple sclerosis - Results of the German multicenter Study. Ann Neurol 23: 56-63

Killian JM, Bressler RB, Armstrong RM, Huston DP (1988) Controlled pilot trial of monthly intravenous cyclophosphamide in multiple sclerosis. Arch Neurol 45: 27-30

Kornhuber HH, Mauch E (1986) Immunosuppressive Cyclophosphamide-Therapie der multiplen Sklerose mit wenig Nebenwirkungen. Dtsch Med Wochenschr 46: 1778

Likosky WH (1988) Experience with cyclophosphamide in multiple sclerosis: the cons. Neurology 38: 14-17

Lublin FD, Lavasa M, Viti C, Knobler RL (1987) Suppression of acute and relapsing experimental allergic encephalo-myelitis with mitoxantrone. Clin Immun Immunopathol 45: 122-128

Myers LW, Fahey JL, Moody DJ, Mickey MR, Frane MV, Ellison GW (1987) Cyclophosphamide "pulses" in chronic progressive multiple sclerosis. A preliminary clinical trial. Arch Neurol 44: 828-832

Ridge SC, Sloboda AE, McReynolds RA, Levine S, Oronsky AL, Kerwar SS (1985) Suppression of experimental allergic encephalomyelitis by mitoxantrone. Clin Immun Immunopathol 35: 35-42

Effective Treatment of Multiple Sclerosis with Cyclophosphamide with Little Side Effects

E. Mauch, H.H. Kornhuber, U. Pfrommer, A. Hähnel, H. Laufen, and H. Krapf
Fachklinik für Neurologie, Dietenbronn GmbH, 7959 Schwendi 1, FRG

Introduction

Gonsette et al. (1977), Hommes et al. (1980), Hauser et al. (1983) and Carter et al. (1986) demonstrated the efficacy of intensive immunosuppression with CY in MS. The side effects, however, were severe, for example alopecia in 100% of the patients. Furthermore, the total CY dose per life should not be higher than about 50 g (Kornhuber et al. 1987).

Patients and Methods

Only patients with a chronic progressive course were included. For all patients the diagnosis of MS was clinically definite according to the criteria of McDonald and Halliday (1977). Both groups consisted of 21 cases. There was no significant difference between the two groups as regards sex, age or degree of disability at the start of the investigation. However, the progression of the disease was significantly faster in the CY group than in the control group ($p < 0.05$). A dose of 8 mg/kg CY was given intravenously at intervals of 4 days until the lymphocyte count was down to half the initial value (however, not below 1000/µl). The total dose averaged 1.9 g CY per patient. In order to avoid hemorrhagic cystitis, the neurogenic bladder was carefully trained (Kornhuber 1986; Kornhuber and Riebler 1984) and cystitis was cured; furthermore mesna (20% of the CY dose) was given before the CY infusion as well as 4 and 8 hours after. To counteract nausea and vomiting patients received antiemetics the day of the CY infusion and the day after. Furthermore at least 3 l fluids were given the day of the CY infusion. The patients in the CY group did not receive any cortisone or ACTH. To take care of the mutagenic and teratogenic risk due to CY, the patients were advised to practise contraception over 1 year. The clinical course was evaluated (a) by the Kurtzke disability scale, (b) by a quantitative neurological examination and (c) according to the abilities of the patient in daily life activities. All patients were reinvestigated 1 year after the treatment.

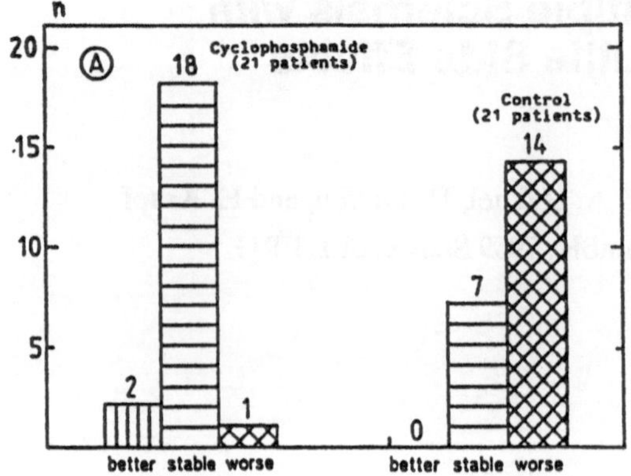

Fig. 1A. Changes in Kurtzke disability status scale

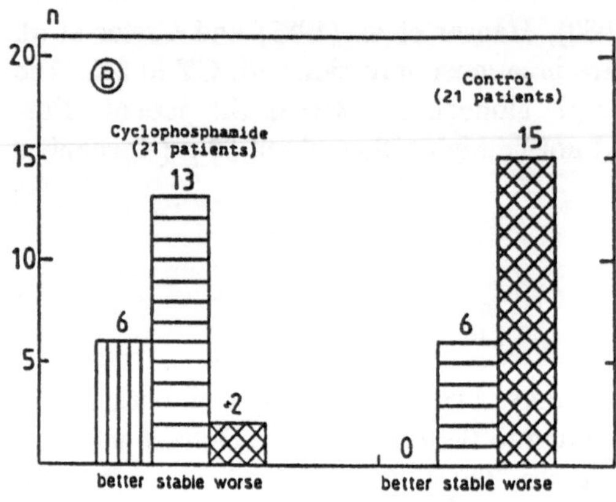

Fig. 1B. Changes in quantitative neurological examination

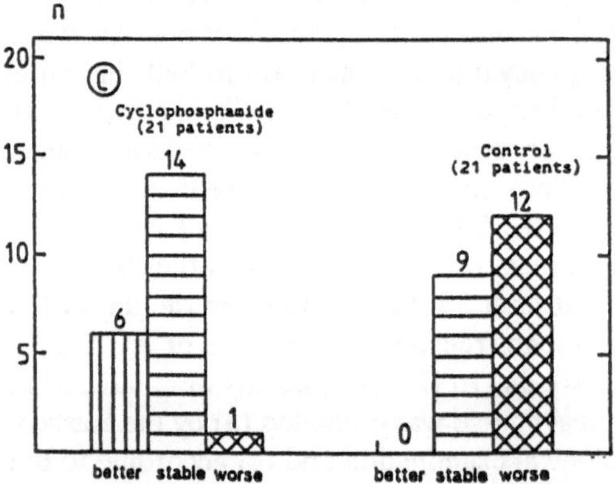

Fig. 1C. Changes of the abilities in daily life activities

Results

As shown in figure 1A-C there is a highly significant difference between the efficacy of CY treatment and the standard therapy in chronically progressive MS ($p < 0.001$, chi square test). The main side effect of CY treatment was loss of appetite in 4 patients, 1 of whom also had vomiting. Two patients complained of faintness for the two days they received the antiemetic drugs. There was no loss of hair in the CY group.

Conclusions

Low-dose CY interval therapy is effective in MS for a period of nearly 1 year. The side effects in this interval scheme are tolerable. The long-term risk of subsequent malignancy is negligible if the urinary bladder is trained, the cystitis is cured before the CY treatment, and the total life time dose of 50 g CY is not exceeded (Kornhuber and Mauch 1986; Kornhuber et al. 1987). Thus the benefit versus risk ratio is favorable. To make the treatment more effective, we now combine short high dose cortisone with low-dose CY treatment.

Summary

Twenty-one multiple sclerosis (MS) patients with a chronically progressive course were treated with low dose of cyclophosphamide (CY). The control group consisted of 21 MS patients with a chronically progressive course who received the standard treatment (ACTH or cortisone). For 20 of the 21 patients in the CY group the degree of disability (Kurtzke scale) was stable over 1 year. In the standard therapy group, 7 out of 21 patients were stable over 1 year, while 14 showed progressive disability. Improvements occurred only after CY treatment, although the patients in the CY group were on average the more severe cases with a faster progression of the disease. A quantitative neurological examination and a score of the abilities of patients in daily life activities showed nearly identical results. The beneficial effect of CY in chronic progressive MS was thus highly significant ($p < 0.001$).

References

Carter JL, Dawson DM, Hafler DA, Fallis RJ, Stazzone L, Hauser SL, Weiner HL (1986) Five-year experience with intensive immunosuppression in progressive multiple sclerosis using high-dose IV cyclophosphamide plus ACTH. Neurology [Suppl. 1]:284

Gonsette RE, Demonty L, Delmotte P (1977) Intensive immunosuppression with cyclophosphamide in multiple sclerosis. J of Neurol 214:173-81

Hauser SL, Dawson DM, Lehrich JR, Beal MF, Kevy SV, Propper RD, Mills JA, Weiner HL (1983) Intensive immunosuppression in progressive multiple sclerosis. A randomized, three-arm study of high-dose intravenous cyclophosphamide, plasma exchange, and ACTH. N Engl J Med 308:173-80

Hommes OR, Lamers KJB, Reekers P (1980) Effect of Intensive immunosuppression on the course of chronic progressive multiple sclerosis. J Neurol 223:177-90

Kornhuber HH (1986) Symptomatische Therapie der Multiplen Sklerose. Kassenarzt 8:35-41

Kornhuber HH, Mauch E (1986) Immunsuppressive Cyclophosphamid-Therapie der multiplen Sklerose mit wenig Nebenwirkungen. Dtsch Med Wschr 111:1778

Kornhuber HH, Riebler R (1984) Mit der Multiplen Sklerose leben. Die häusliche Behandlung der MS. Schattauer, Stuttgart

Kornhuber HH, Mauch E, Petru E, Schmähl D (1987) Zum Malignitätsrisiko bei Cyclophosphamid-Therapie der multiplen Sklerose. Dtsch Med Wschr 112:530

McDonald WI, Halliday AM (1977) Diagnosis and classification of multiple sclerosis. Br Med Bull 33:4-8

Neurogenesis and Pathogenesis of Glia: Immunological Studies

K. Warecka
Neurological Department, University of Lübeck, 2400 Lübeck, FRG

Introduction

At present, it seems to be generally agreed that proteins play a major part in the growth, differentiation and dedifferentiation of a cell. Nevertheless, it is still not known what determines the differentiation of a cell in a specific direction, and what controls the normal growth of the cell, although some protein factors regulating mammalian cell growth are now well established. Furthermore, we do not know why the cell dedifferentiates. It is almost generally accepted that proteins bring it about, and that proteins play a key role in all of the biological processes, physiological as well as pathological, including cell dedifferentiation.

This paper contains data on a glycoprotein isolated from white substance of human brain in our laboratory by affinity chromatography on Sepharose B (Warecka et al. 1972) and ConA columns (Brunngraber et al. 1975), which by immunological criteria is CNS specific (brain, spinal cord, optic nerve; Warecka and Bauer 1967), it could be found in very early stages of cell differentiation and disappears in the process of dedifferentiation (von Schweinitz und Krain 1976) or cell damage (Strache 1983). The glycoprotein has a MW of 45000 (Weber and Osborn 1969; Brunngraber et al. 1974). Analytical data indicated a high content of carbohydrates: on the average 0.25 mg hexose per 1 mg protein. Estimation of the sugar composition indicated the presence of mannose, glucose, galactose, fucose, glucosamine, galactosamine and NANA (Brunngraber et al. 1975).

Studies on neuronal-enriched and glial-enriched brain fractions (Rose 1967; Vogel 1971) have indicated an essentially glial localization of this glycoprotein (Warecka and Müller 1969).

Phylogenetic studies show an identity of the brain-specific glycoprotein with an adequate component of *Macaca Mulatta* brain, a semiidentity with a corresponding component of rat brain and no identity with a brain-specific glycoprotein of a mouse brain. However, a semiidentity exists between rat and mouse brain-specific proteins (Warecka 1980). From the point of view of the molecular biology it may be assumed that the primary structure of these brain-specific proteins is homologous in all vertebrates, and that the number of identical antigenic determinants increases with the closer phylogenetic relationship of various species. All these findings are in agreement with general laws of development.

Table 1. Biochemical properties of the brain-specific antigenic α2-glycoprotein

It is water soluble,

it is an acidic glycoprotein - (average 0.25 mg hexose/1.0 mg protein)

and it contains carbohydrate residues in molar ratio:

Galactose	Mannose	Glucose	Fucose	Glucosamine	Galactosamine	NANA
1.0	1.7	1.4	0.4	2.3	0.3	0.7

It binds to Concanavalin A

It gives upon SDS-polyacrylamide gel electrophoresis one prominent periodic acid - Schiff staining band at apparent - Mw 45000

Table 2. Biological properties of human brain-specific α2-glycoprotein

specifity:	presence in:		not found in:
organs	brain hemispheres spinal cord n. opticus	white substance	peripheral nervous system
cells	glia fraction		neurons oligodendrocytes myelin

Studies on brain of mutant mice such as NMRI, jimpy and quaking and healthy mice of the mutant strains showed a semiidentity with each other (Lange 1976).

Immunological studies performed on four embryonic, 19 fetal and 12 brains from infants show the appearance of α2-GP in the human ontogenesis simultaneous with the occurrence of myelination gliosis (Warecka and Müller 1969).

The concentration of the α2-GP increases during the period of myelination and remains constant thereafter. However, the fetal brain-specific α2-GP differs in its composition from an adequate component of adult. The fetal α2-GP remains in the adult brain the whole life through. In addition, the fetal brain contains some amount of adult protein depending upon age. These and other data suggested that this glycoprotein may play an important role in the formation and maintenance of the myelin sheath. This also suggested the presence of the glycoprotein in oligodendroglia, since it is these cells that are responsible for the formation of the myelin sheath. However, immunological studies in vitro with isolated oligodendroglia cells, isolated myelin fraction and with MBP have not shown any reaction with anti-α2-GP serum. Subsequent studies on the in vivo loca-

Table 3. Phylogenesis

PHYLOGENESIS:

COMPLETE OR ALMOST COMPLETE IDENTITY WITH AN EQUIVALENT PROTEIN OF MACACA MULATTA.

SEMIIDENTICAL WITH A BRAIN-SPECIFIC PROTEIN OF RAT:

NO IDENTITY WITH MOUSE BRAIN-SPECIFIC PROTEIN.

VARIOUS STRAINS OF MICE AS NMRI (jp/y), QUAKING (qk/qk) AND CONTROLS (jp/x; x/x; x/y) (qk/+; +/+) SHOW "GENETIC DIFFERENCES"

Table 4. Results of immunological and histological studies in 4 embryonic, 19 fetal and 12 infantile human brains

length of embryo and fetus (cm)	weight of embryo and fetus (g)	weight of brain (g)	age	antibodies against brain-specific protein	histological examination	number of examined cases
0 8	1 5		3 - 4 weeks	no material	no material	1 +)
7 0	21 8	3 80	8 - 9 weeks	–	neuroblast glioblast (neurons)	1
9 0	23 6	4 02	10-11 weeks	–		1
12 0	35 0	7 70	12-13 weeks	not tested		1
23 0	103 2	15 45	14-18 weeks	+	(myelination glia) neuroblast glioblast (neurons) (macroglia	2
37 0	1082 0	174 00	24-28 weeks	+	myelination glia in the cerebrum thin myelin sheats in the pons	9 ++)
46 0	1700 0	250 00	32-36 weeks	+	myelination glia in the cerebrum thin myelin sheats in the midbrain	8 +++)
54 0	3445 0	330 00	40-42 weeks birth	+	myelination glia myelin and differentation of glia	2
		750 00	3 - 4 months	+	myelination glia myelin and differentation of glia	2
		1200 00	1 - 5 years	+	brain full developed	8

+) antibodies against serum proteins
++) 5 of these survived up to 3 days
+++) 5 of these survived up to 17 days
In those collectives, in which more than one brain was examined, the length and weights are concerned as an average

lization of this antigen by immunocytochemical as well as immunoelectron microscopic techniques clearly showed that astrocytes rather than oligodendroglia were labeled with specific antibodies directed against α2-GP (Ghandour et al. 1982; Fig.1).

These studies ruled out a direct association of this glycoprotein with the process of myelination. It was concluded that the antigen represents an index of astrocyte matura-

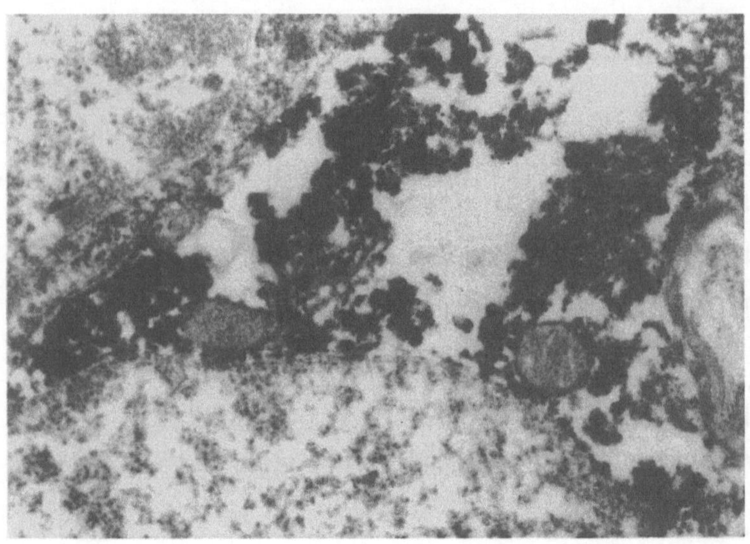

Fig. 1. Labeled astrocyte in the granular layer of cerebellum (immunoperoxidase stained)

tion or its degree of differentiation (Langley et al. 1982). With regard to human brain pathology, two results have been obtained. Firstly, studies on 137 human primary brain tumors obtained during operation showed a progressive loss of α2-GP during the process of dedifferentiation of brain cells, i.e. during malignant transformation from astrocytoma to glioblastoma multiforme. It was observed that transformation of tumor cells run contemporary with a loss of some sugars, especially of NANA. The α2-GP found here then resembles the fetal form and the α2-GP of lower mammals (Warecka 1975). Secondly, investigations on MS brain tissue showed that this glial marker disappears in MS plaques. On the junction between brain tissue and plaques the amount of α2-GP is decreased as well in the so-called normal tissue of MS brain (see Table 6).

We do not know why glioblasts have no access to the genetic information needed for their development into mature glial cells. In any case, either they do not have the capacity to synthesize adequate proteins for building up mature astrocytes or oligodendrocytes, or some enzymes destroy the proteins, or perhaps other mechanism(s) are in play. We do not know what the role of α2-GP is. In cultures of brain cells of newborn rats (immunocytochemical techniques) an expression of α2-GP could be shown in precursors of glial cells. These precursor cells developed into type II astrocytes or oligodendrocytes (Bhat et al. 1986; Fig. 2).

A new hypothesis based on recent studies with tissue cultures suggests that the primary lesion in MS could be related to damage to oligodendroglial progenitor cells (Elias 1987). It could be assumed that the absence of α2-GP alters the course of normal development of astrocytes and/or oligodendrocytes to maintain myelin throughout the life span. With regard to primary brain tumors we do not know at the present time enough about α2-GP to say whether the splitting of sugars promotes the disintegration of the astrocyte cells or is an expression of damage which has already occured.

Table 5. Glia-specific glycoprotein in percentage of total water-soluble brain protein in brain tissue and glial brain tumor

	localization	% of total brain protein [1]
white substance [2]:		
frontal lobe	f.	54.24
occipital lobe	o.	35.81
temporal lobe	t.	29.89
parietal lobe	p.	26.25
tumor:		
astrocytoma [3] I–II	t.	9.60
" I–II	t.–p.	9.00
astrocytoma II	p.	7.55
" II	p.–o.	9.82
" II–III	t.	2.76
astrocytoma III	t.	2.08
" III	p.	2.60
astrocytoma IV (= glioblastoma multiforme), 28 tumors	various	0
oligodendroglioma 5 tumors	various	0

1) *all values multiplied by 10^3*
2) *mean value from 5 human adult brains*
3) *grades after KERNOHAN*

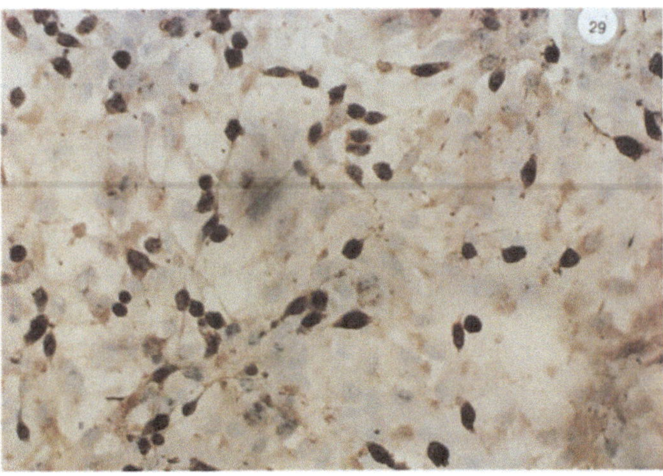

Fig. 2 Immunocytochemical localization of α2-glycoprotein in precursor glial cells of rat brain culture (17-day old)

Table 6. Quantity of α2-glycoprotein in MS brain and MS plaques in comparison to normal brain

	tissue specifity	$\left[\frac{\alpha_2\text{-glycoprotein (g/l)}}{\text{total protein (g/l)}} \times 10\right]$	
brain area	material from normal brain *	macroscopic normal material from MS-brain	plaque
frontal lobe (w.s.)	2.62	0.39 0.89 0.61 0.71	0.15 - - -
parietal lobe (w.s.)	2.60	0.63 0.98	- -
occipital lobe (w.s.)	2.13	0.26 0.82 -	0.00 0.63 0.00
temporal lobe (w.s.)	2.33	0.35 -	0.30 0.00
cerebellum	1.84	0.65	-
midbrain	0.71	0.42 -	0.00** 0.00
aqueductus cerebri region	0.34	-	0.00
olive (sup.)	0.81	0.77 0.43	0.00 -
n. opticus	1.40	- -	0.00 0.00

(w.s.) = white substance

* = average of 5 brains

** = as well at the edge of plaques no α_2-glycoprotein was found

\- = no material

Summary

A CNS-specific glycoprotein (α2-GP) of MW 45000 and a relatively high content of NANA was isolated from white matter of human brain by affinity chromatography on Sepharose B and ConA columns. Many studies indicated an essentially glial localization of this glycoprotein. α2-GP appears at a very early stage of human ontogenesis and remains for the rest of life. However, the "adult" protein differs in quality and quantity from the "fetal" α2-GP, as well as from an adequate component of lower mammals, excluding the rhesus monkey. In cultures of brain cells from newborn rats, an expression of α2-GP by glial precursor cells could be shown. These precursor cells develop into type II astrocytes or to oligodendrocytes. With regard to brain pathology in humans two results have been obtained. Firstly, the α2-GP decreases progressively during malignant transformation from astrocytoma to glioblastoma multiforme; secondly, this glial precursor marker disappears in multiple sclerosis (MS) plaques. Whether normal glial cells could not be generated in these pathological processes, or wether other mechanism(s) may provoke the cell damage remains unknown.

References

Bhat N R, Arimoto K, Warecka K, Brunngraber E G (1986) Expression of α2-glycoprotein by glial precursor cells: an immunocytochemical study with glial cultures. Dev Brain Res 29: 31-36

Brunngraber E G, Susz J P, Warecka K (1974) Electrophoretic analysis of human brain-specific proteins obtained by affinity chromatography. J Neurochem 22: 181-182

Brunngraber E G, Susz J P, Javaid J, Aro A, Warecka K (1975) Binding of Concanavalin A to the brain specific proteins obtained from human white matter by affinity chromatography. J Neurochem 24: (805-806)

Elias SB (1987) Oligodendrocyte development and the natural history of multiple sclerosis. new hypothesis for the pathogenesis of the disease. Arch Neurol 44: 1294-1299

Ghandour M S, Langley O K, Gombos G, Vincendon G, Warecka K (1982) Cellular localization of the brain specific α2-glycoprotein in rat cerebellum: an immunohistochemical study. Neuroscience 7: 231-237

Lange U (1976) Gehirnspezifische Proteine bei Mäusemutanten "Jimpy" und anderen Mammalia. Thesis, Medical College of Lübeck

Langley O K, Ghandour M S, Vincendon G, Gombos G, Warecka K (1982) Immunoelectron microscopy of α2-glycoprotein. J Neuroimmunol 2: 131-143

Rose S P R (1967) Preparation of enriched fractions from cerebral cortex containing isolated, metabolically active neuronal and glial cells. Biochem J 102: 33-43

Strache J (1983) Das Vorkommen von gehirnspezifischem Alpha2-Glykoprotein in Hirnen verstorbener Patienten, die an Multipler Sklerose gelitten haben. Thesis, Medical College of Lübeck

Vogel H M (1971) Die Lokalisation des gehirnspezifischem Alpha2-Glykoproteins in Gehirnzellen. Thesis, Medical College of Lübeck

von Schweinitz und Krain J C (1976) Das Vorkommen von gehirnspezifischem Alpha2-Glykoprotein in Hirntumoren. Thesis, Medical College of Lübeck

Warecka K (1975) Immunological differential diagnosis of human brain tumors. J Neurol Sci 26: 511-516

Warecka K (1980) Onto- and phylogenesis, maturation and genetic differences of glia specific glycoprotein. In: Baumann N (ed) Neurological mutations affecting myelination. Elsevier/North-Holland, Amsterdam (INSERM symposium no 14)

Warecka K, Bauer H (1967) Studies on "brain-specific" proteins in aqueous extracts of brain tissue. J Neurochem 14: 783-787

Warecka K, Müller D (1969) The appearance of human "brain-specific" glycoprotein in ontogenesis. J Neurol Sci 8: 329-345

Warecka K, Möller H J, Vogel H M, Tripatzis I (1972) Human brain-specific alpha2-glycoprotein: purification by affinity chromatography and detection of a new component; localization in nervous cells. J Neurochem 19: 719-725

Weber K, Osborn M (1969) The reliability of molecular weight determinations by dodecyl sulfate-polyacrylamide gel electrophoresis. J Biol Chem 244: 4406-4412

Paired Stimuli in the Diagnosis of Peripheral and Central Nervous Diseases

H.J. Lehmann and H. Gerhard
Neurologische Klinik, Universitätsklinikum Essen, 4300 Essen, FRG

Introduction

If at various intervals double electric stimuli are applied to the nervous system we find the well known absolute refractoriness followed by a relative refractory period. This may give some information on the functional state of the nervous system.

Peripheral Nervous System

In acute animal experiments segmental demyelination of peripheral nerves was induced by diphteria toxin, experimental allergic neuritis and circumscribed neuritis (Fig. 1). The relative refractory period was delayed regarding amplitude and latency of responses to the second stimulus. In contrast, in primary axonal lesions (Wallerian degeneration, thallium neuropathy) no change in the relative refractoriness could be observed (Fig. 2). Identical results have been obtained applying trains of stimuli. With rising frequency of stimuli transmission stopped earlier in segmental demyelination but remained unaltered in primary axonal lesion.

Investigation in human neuropathy comprised polyneuritis, local entrapment like carpal tunnel syndrome and polyneuropathies such as in uremia, alcoholics and diabetics (Fig. 3). Originally we hoped that differentiation between primary axonal lesion such as in uremia and segmental demyelination would be possible, as found before in animal experiments. However, the results were disappointing (Fig. 4). Polyneuritis and neuropathies in human beings showed a rather uniform change of refractory period, apparently independent of the primary axonal or segmental origin of lesion.

The following conclusions can be drawn from these results. In acute animal experiments changes in the response to repetitive stimuli were obvious only in demyelinating diseases and were missing in primary axonal neuropathies such as Wallerian degeneration or thallium neuropathy.

In contrast, in human diseases changes have even been observed in primary axonal atrophy such as in alcoholics and in uremia (Kaeser 1965; Dyck et al. 1971). The reason may be that longer lasting human diseases lead to changes in fiber spectra which are not observed in acute axonal neuropathies in animals. Longer lasting diseases may cause atrophy of single fibers and consequently loss of fiber density, changes in the fiber spectra due to different involvement of larger and smaller fibers, fiber hypoplasia with

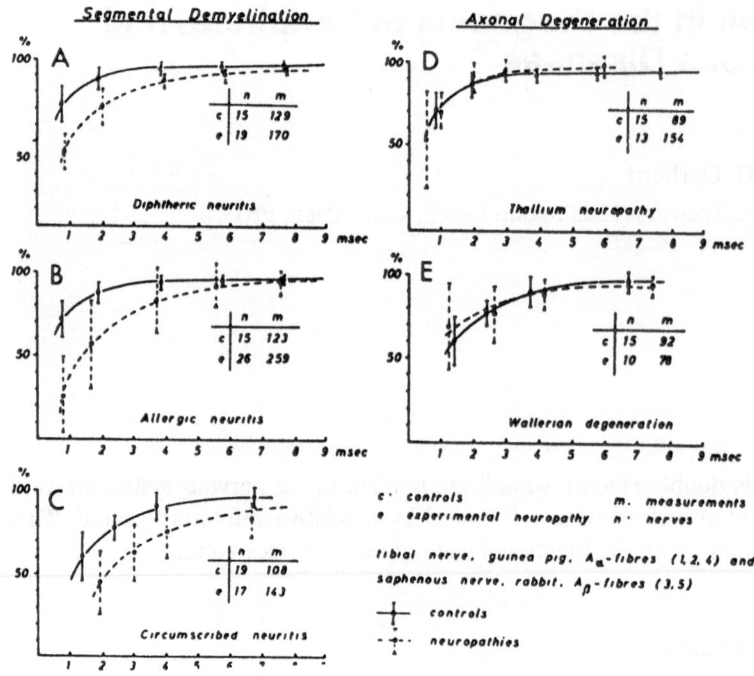

Fig. 1. Relative refractory period of peripheral nerves. Animal experiments: Ordinate: amplitude in percentage of response to single shock. Abscissa: interval of paired stimuli. Regarding the restoration of amplitude of responses to single shock in segmental demyelination (A,B,C) the relative refractory period is prolonged whereas in acute axonal degeneration (D,E) the results do not differ from controls. (From Lehmann 1973)

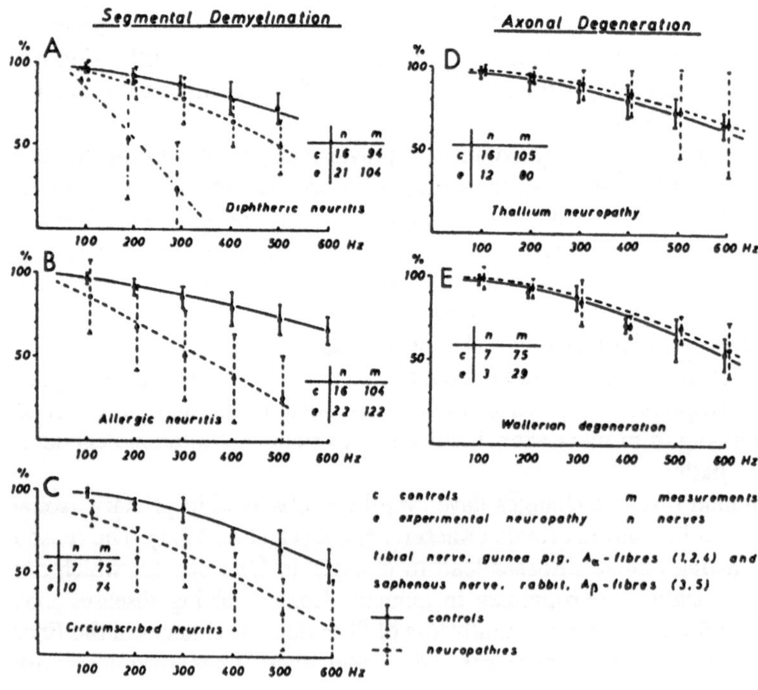

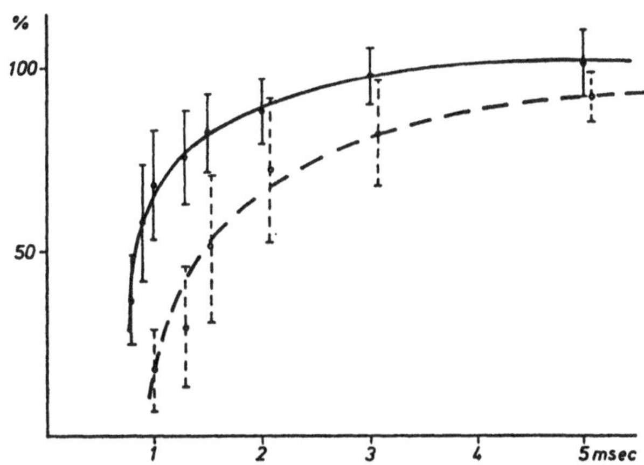

Fig. 3. Refractory period in uremic neuropathy. N. suralis: - = controls (17 nerves) --- = patients with uremia (7 nerves). In this chronic disease, which is due to a primary axonal lesion (Dyck et al. 1971) there is a marked decrease of amplitudes as compared with control nerves. (From Tackmann et al. 1974)

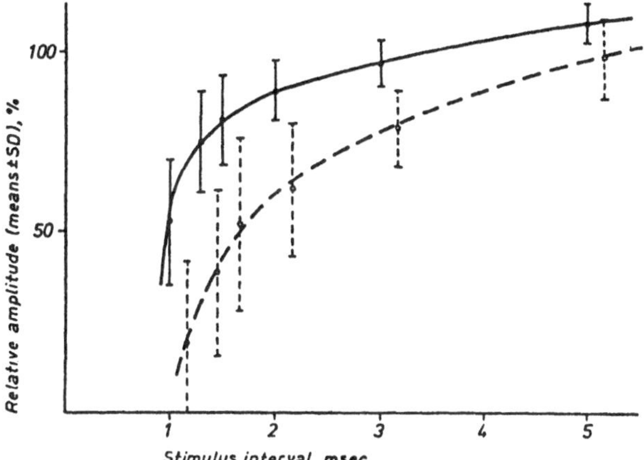

Fig. 4. Relative refractory period in carpal tunnel syndrome. - = controls (12 median nerves) --- = patients with CTS (11 nerves). In this chronic, primarily segmental demyelination there is an equal prolongation of relative refractory times as in chronic primary axonal lesion. (From Tackmann and Lehmann 1974)

reduction of axon calibers and myelin layers, due to fiber dystrophy. Regenerating processes, leading to relative predominance of smaller fibers in the nerve transection may even occur. All these changes can influence the fiber spectra of diseased nerves and lead to a predominance of smaller fibers having other membrane properties and a lower safety factor (Tasaki 1955; Mc Leod et al. 1973]. In conclusion, alteration of the re-

Fig. 2. Transmission of trains of stimuli. Peripheral nerves, animal experiments. In segmental demyelination (A,B,C) amplitudes decrease significantly with rising frequency whereas in axonal degeneration (D,E) there is no significant difference from controls. (From Lehmann 1973)

sponse to repetitive stimuli in human neuropathy must not mean primary demyelinating disease. Double stimulation of peripheral nerves has only a limited value in early detection of beginning neuropathies.

Central Nervous System

In spite of this somewhat disappointing result of our search in the peripheral nervous system we started anew with double stimuli evoked potentials in the central nervous system.

In animal experiments on rabbits two models of spinal cord lesions were investigated: ischemia and experimental allergic encephalomyelitis. Spinal cord ischemia was induced by infrarenal ligation of the abdominal aorta. Due to anatomical conditions this is effective in rabbits, whereas, for instance, in dogs and cats a suprarenal ligation would be necessary which prevents longer lasting follow-up studies (Fig. 5). Investiga-

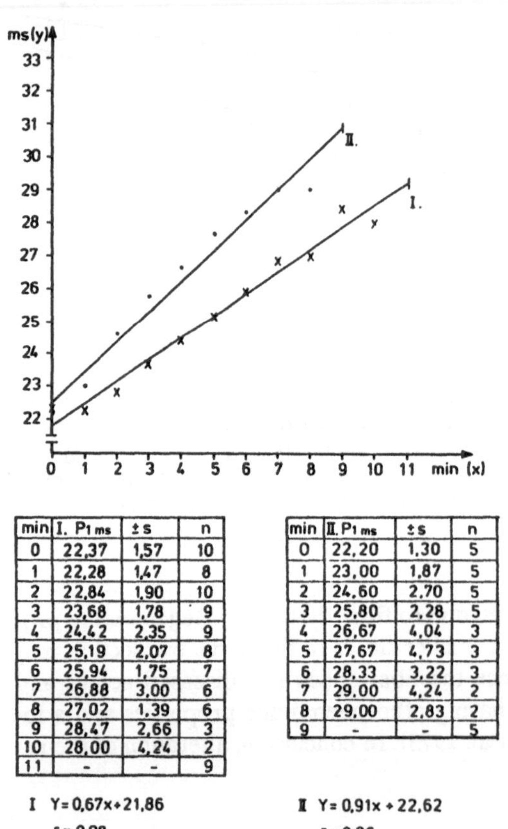

min	I. P1 ms	±s	n
0	22,37	1,57	10
1	22,28	1,47	8
2	22,84	1,90	10
3	23,68	1,78	9
4	24,42	2,35	9
5	25,19	2,07	8
6	25,94	1,75	7
7	26,88	3,00	6
8	27,02	1,39	6
9	28,47	2,66	3
10	28,00	4,24	2
11	-	-	9

min	II. P1 ms	±s	n
0	22,20	1,30	5
1	23,00	1,87	5
2	24,60	2,70	5
3	25,80	2,28	5
4	26,67	4,04	3
5	27,67	4,73	3
6	28,33	3,22	3
7	29,00	4,24	2
8	29,00	2,83	2
9	-	-	5

I $Y = 0{,}67x + 21{,}86$, $r = 0{,}98$

II $Y = 0{,}91x + 22{,}62$, $r = 0{,}96$

Fig. 5. Evoked scalp potentials after spinal cord ischemia. Paired stimulation of tibial nerves in 10 rabbits, 100 ms stimulus interval. I: response to first stimulus, II: response to second stimulus. Ordinate: Latency of response. Abscissa: minutes after end of aortic ligation. For further explanation see text

tion of tibial single and double stimuli (somatosensory evoked potentials, SEP) with recording from the scalp was done before surgery and then during the first minutes of an aortic ligation lasting one hour. Fig. 5 demonstrates the changes in single (I) and double (II) stimuli evoked P1 potentials during the first minutes after ligation. P1 on second stimulus (II) ceases after 8 minutes, P1 on first stimulus (I) after 10 minutes of ligation. Latency of first stimulus response (I) reacts more slowly than latency of second response (II). This shows that double stimuli SEP demonstrates a spinal cord ischemia earlier than single stimulus SEP. This is of interest in monitoring spinal cord function in aortic aneurysm surgery.

Figure 6 shows a human aortic aneurysm. The aorta had to be clamped for 15 minutes during surgery. Double stimulated tibial nerve SEP shows prolongation of latency of the first response which vanishes after 12 minutes (Fig. 7). Latency of the second response increases earlier; response ceases already after 9 minutes.

Monophasic experimental allergic encephalomyelitis (EAE) was induced in 18 rabbits by sensitizing with 100 mg homologous spinal cord tissue emulsified with complete Freund's adjuvant. Three weeks after immunization the animals started to develop ascendent tetraparesis followed by brainstem signs. Histopathology revealed preponderance of the lesions at lumbar spinal cord without any involvement of the peripheral nervous system. Somatosensory evoked potentials elicited by single stimuli to the tibial nerve revealed no significant change in latency of P1 after three weeks. P1 latencies were delayed from 23.6 ± 4.1 ms to 26.6 ± 4.9 ms after six weeks with an additional drop of amplitude.

Recording of double stimulus SEP proved to be the most sensitive (Fig. 8). They showed a marked increase in relative refractory time already within the first week of EAE disease when single stimulus SEP was only slightly changed in a few animals. The relative refractory time is the period during which the response to the second stimulus

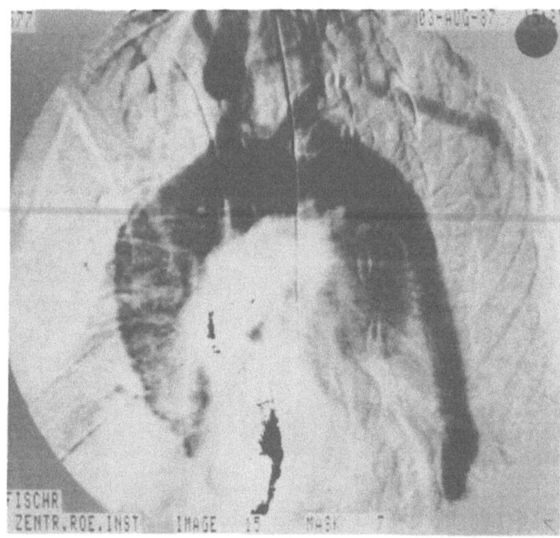

Fig. 6. Aortic aneurysm in a 57-years-old patient. State before surgery. (Courtesy of Prof. E. Löhr, Essen)

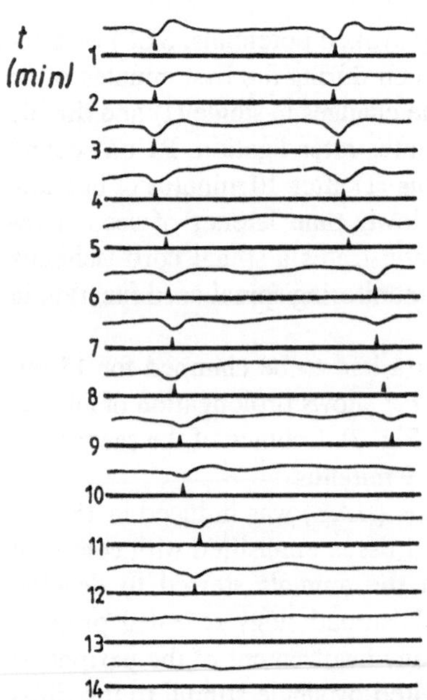

Fig. 7. Paired stimuli: Tibial nerve SEP during aortic clamping. Same patient as in Figure 6. P1 potential to first stimulus (left row) lasts longer (12 min.) and is less delayed than P1 response to second stimulus (right row), which vanishes already after 9 minutes

Double stimuli	before EAE			after 3weeks			after 6weeks		
	P1/N2	±s	n	P1/N2	±s	n	P1/N2	±s	n
100	2,0	0,8	18	1,5	0,9	18	0,8	0,3	8
90	2,1	0,9	18	1,2	0,9	18	0,6	0,4	8
80	1,8	0,7	17	1,3	0,4	17	0,5	0,4	7
70	1,8	0,6	17	1,0	0,3	16	0,4	0,3	6
60	1,6	0,7	16	0,8	0,3	18	0,4	0,4	5
50	1,0	0,3	16	0,75	0,7	15	0,3	0,4	5
40	0,8	0,6	15	0,7	0,3	14	0,3	0,4	5
30	0,7	0,5	12	0,8	0,3	12	0,3	0,2	4
25	0,6	0,4	12	0,5	0,2	10	0,3	0,2	3
20	0,5	0,3	3	-	-	1	-	-	1

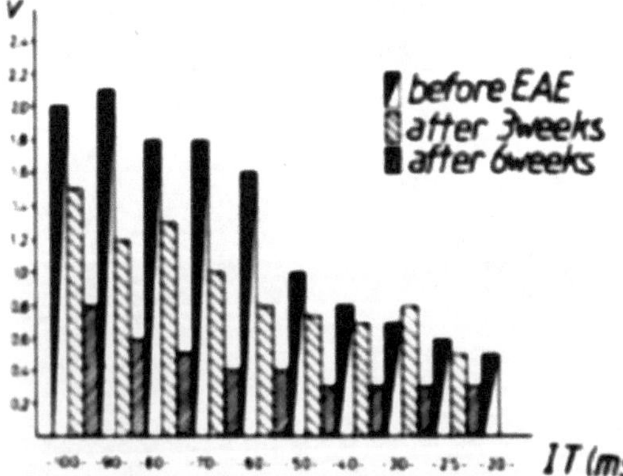

Fig. 8. Tibial nerve SEP in EAE. Paired stimuli. 18 rabbits. Reduction of the amplitude of second P1 response is related to interstimulus time (IT= 20-100 ms) and to duration of EAE (before onset of disease and after 3 and 6 weeks)

is significantly delayed and/or the amplitude is significantly reduced (<50%). This refractory time was increased from 33.5 ± 5 ms before immunization to 58.4 ± 4.9 ms (74.3 %) after three weeks and to 63.4 ± 5.1 ms (89.3 %) after six weeks. The amplitude of the second P1 response dropped from 1.8 μV before immunization to 1.0 after three weeks and 0.4 after six weeks at 70 ms stimulus interval (Table 1). The electrophysiological changes were paralleled by clinical impairment. According to the IKEDA grading, the rate of animals suffering from grade 3 increased from 22 % (3 weeks) to 50 % (6 weeks). The absolute refractory time remained essentially unchanged during the whole period. In conclusion, SEP with double stimuli is superior to that with single stimulus in the detection of early lesions in these acute demyelinating processes of CNS. This may be of interest even in acute bouts of multiple sclerosis.

In human beings, double stimuli SEP was investigated in the 1960s by Allison (1962) and Shagass and Schwartz (1964) and by others in the 1970s. A major advance for investigation was the possibility electronically to subtract the response to the first stimulus. This renders an isolated record of the second stimulus response for investigation. Twenty-five healthy subjects served as controls for the following studies of double stimulus SEP in multiple sclerosis, alcoholics and diabetics. Single and double stimuli were delivered at the tibial nerve. P1 of the obtained scalp SEPs were regarded as pathologic if latency exceeded three times the standard deviation of controls.

In 15 patients with definitive multiple sclerosis but without signs of spinal cord involvement, single stimulus tibial nerve SEP was within normal limits (Fig. 9). However in nearly one-third of these cases, at stimulus intervals between 40 and 60 ms six out of fifteen had an abnormally prolonged latency and in 60 % of these cases the cerebral refractory time was significantly prolonged.

Similar results have been obtained in 17 diabetics, 12 without and 5 with polyneuropathy (Fig. 10). Although single stimulus tibial nerve SEP was within normal limits in nearly half of the cases refractory period was prolonged, prolongation of the latency exceeding double standard deviation of controls in one-third of these cases.

Even in alcoholics with normal response to single stimuli there was a considerable number with pathologically prolonged latency to the second stimulus (Fig. 11), refractory time being prolonged in 30 % of the patients. None of these patients had signs of peripheral nerve involvement. Although an anatomical clue is still lacking, these data indicate that further search for spinal cord involvement in diabetics and in alcoholics may be worthwhile.

Table 1. Absolute and relative refractory time of tibial nerve evoked scalp potential and clinical state (IKEDA grading) in 18 rabbits before and after 3 and 6 weeks EAE

Refractory time					
before EAE		after 3 weeks		after 6 weeks	
absolute	relative	absolute	relative	absolute	relative
20.8 ± 4.9 n = 18	33.5 ± 5 n = 18	22.3 ± 4.8 n = 18	58.4 ± 4.9 n = 18	23.4 ± 5.9 n = 8	63.4 ± 5.1 n = 8
Clinical findings (acc. to IKEDA)					
GRADE 0	18	GRADE 0	-	GRADE 0	-
GRADE 1	-	GRADE 1	6	GRADE 1	-
GRADE 2	-	GRADE 2	8	GRADE 2	3
GRADE 3	-	GRADE 3	4	GRADE 3	4
GRADE 4	-	GRADE 4	-	GRADE 4	1
GRADE 5	-	GRADE 5	-	GRADE 5	-

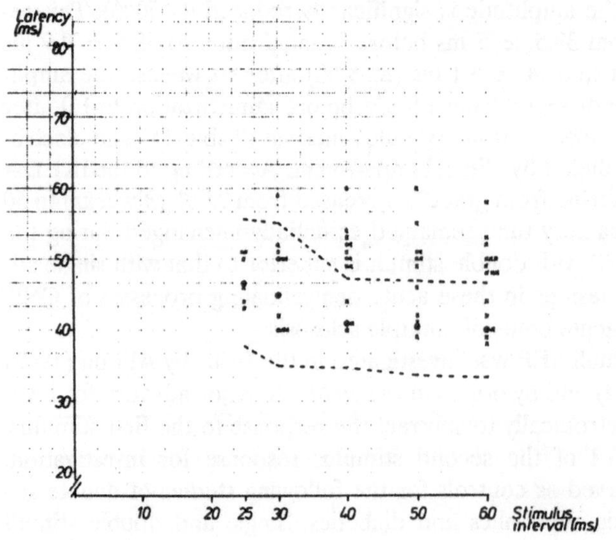

Fig. 9. Paired stimuli. Tibial nerve SEP in multiple sclerosis. Prolongation of latency of response to second stimulus in 15 patients related to double standard deviation (= = =) of controls (25 subjects). For further information see text

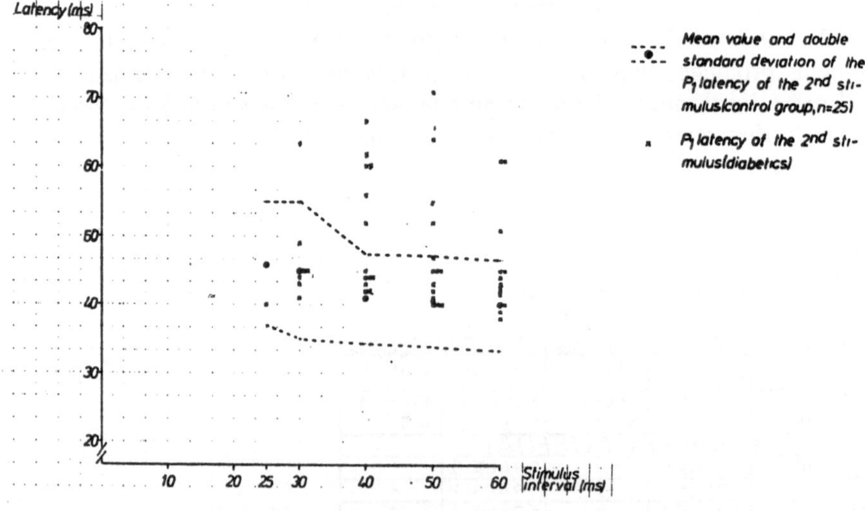

Fig. 10. Paired stimuli. Tibial nerve SEP. Prolongation of latency of the response to second stimulus in diabetics related to controls. For further information see text

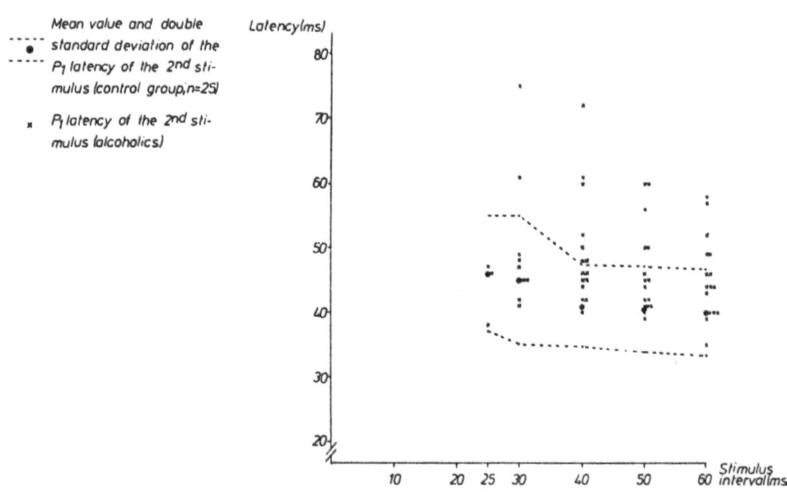

Fig. 11. Paired stimuli. tibial nerve SEP. Prolongation of latency of the response to second stimulus in alcoholics. For further information see text

Conclusions

It can be concluded, that application of paired stimuli and evaluation of the second response are of no major clinical interest in diseases of the peripheral nervous system. However, it seems that evoked potentials elicited by paired stimuli in CNS diseases can give answers to some questions concerning the functional state. This may be true for the somatosensory system as well as for the magnetic stimulated pyramidal tract. From recent investigations we have also gained the impression that in certain cases double stimulus responses may contribute to the generator research in somatosensory evoked potentials.

Summary

Responses evoked by paired electric stimuli can give some information on the functional state of peripheral nerves in acute neuropathy. This may be of some interest in animal experiments. But paired stimulation is of no major diagnostic value in more chronic courses of peripheral nervous diseases, especially in human neuropathies. However, in diseases of the central nervous system responses evoked by double stimulation of somatosensory or pyramidal pathways can give some valuable insight into the functional state if lesions are elicited by ischemia or inflammation. In these CNS lesions the findings in human diseases mirror the results of animal experiments. Paired stimuli may therefore offer an interesting tool for the diagnostic procedure in human CNS diseases.

References

Allison T (1962) Recovery function of somatosensory peripheral nerve and cerebral evoked responses in man. Electroencephalogr Clin Neurophysiol 17: 128 -138

Dyck PJ, Johnson WJ, Lambert EH, O'Brien PC (1971) Segmental demyelination secondary to axonal degeneration in uremic neuropathy. Mayo Clin Proc 46: 400 - 429

Kaeser HE (1965) Veränderungen der Leitgeschwindigkeit bei Neuropathien und Neuritiden. Fortschr Neurol Psychiatr 33: 221 - 250

Lehmann HJ (1973) Segmental demyelination and changes in nerve conduction in experimental circumscribed neuropathy. In: Desmedt JE (ed) New developments in electromyography and clinical neurophysiology vol2. Karger, Basel, pp 145 - 157

McLeod JG, Prineas JW, Walsh JC (1973) The relationship of conduction velocity to pathology in peripheral nerves. In: Desmedt IE (ed) New developments in electromyography and clinical neurophysiology vol2. Karger, Basel, pp 248 - 258

Shagass C, Schwartz M (1964) Recovery functions of somato-sensory peripheral nerve and cerebral evoked responses in man. Electrencephalogr Clin Neurophysiol 17: 128 - 133

Tackmann W, Lehmann HJ (1974) Relative refractory period of median nerve sensory fibres in the carpal tunnel syndrome. Eur Neurol 12: 309 - 316

Tackmann W, Ullerich D, Cremer W, Lehmann HJ (1974) Nerve conduction studies during the relative refractory period in sural nerves of patients with uremia. Eur Neurol 12: 331 - 339

Tasaki I (1955) Measurement of the capacity and the resistance of the myelin sheath and the nodal membrane of the isolated frog nerve fibre. Am J Physiol 181: 630 - 650

Early Speech Education: An Epidemiologic Study

D. Bechinger, H.H. Kornhuber, and W. Schmidt
Department of Neurology, University of Ulm, 7900 Ulm, FRG

Introduction

Early treatment of motor handicaps has been generally accepted. Early treatment of speech and language difficulties has until recently not been considered as a necessary method for achieving better development and education. This study was undertaken to investigate whether early education for speech and language favors a better state of language development and therefore should be applied as a general procedure.

With early treatment children are evaluated for speech and language difficulties in kindergarten. If there are slight or moderate difficulties in speech, parents are instructed how to overcome these, and regular ambulatory supervision by teachers is offered. With severe disturbances special preschool classes with more intense and daily treatment by speech therapists are recommended. Thus most of these children mȧy show no more speech difficulties when they enter primary school. In the district where there was no early treatment at the time of investigation children received treatment for overt speech difficulties only if the parents noticed these, and if the attending physician consented and referred them to speech therapy.

Methods

A total of 2312 children of a single age class in two country districts of comparable socioeconomic level were examined during their second grade of primary school. Of these 1273 children were of a district with early treatment and 1039 from a district where early treatment was not yet established. Because of the necessary consent of parents only 82% of this age class were examined on average, not significantly different in the two districts.

The examination of the children had to be short because teaching in the classroom should not be disturbed too much. The tests and possible difficulties in speech to be found are listed in Table 1.

With the articulation test, objects in a series of pictures had to be named. Furthermore, children had to reproduce two sentences and describe a picture story with a series of five complex pictures.

Table 1. Examinations conducted

Test used	Disturbance searched for
Articulation test	Dyslalia
Description of a picture story	Dyslalia Stuttering Dysgrammatism
Reproduction of sentences	Dyslalia Dysgrammatism Comprehension (Attention deficit)
Reading	Reading disability
Spontaneous speech	Dyslalia Dysgrammatism Stuttering
Rating by the classroom teacher	All speech disturbance

Results

With the articulation test we found in both districts many children with sigmatism, somewhat more children in the district with no early treatment; the difference, however, was not significant. Sigmatism is regarded by many parents and teachers to be no disturbance. With other dyslalias there was a significant difference between the two districts (Fig. 1). Another difference between the two districts was that in the one with no early treatment children with speech difficulties were found mostly in special classes for mentally retarded children, while in the district with early treatment the children were attending special classes for speech difficulties. Stuttering was slightly more frequent in the district with no early treatment; the difference, however, was significant only on one sided testing.

In the test with reproduction of two sentences (Fig. 2) we found in both districts children with a high frequency of errors in semantics in both districts but many fewer children with errors in grammar. In the district with early education, children made signifcantly fewer errors in grammar.

In the description of the picture story (Fig. 3) children in both districts produced about the same quantity of errors in semantics and grammar. The difference was not significant. If we compare, however, children with dyslalia and children without dyslalia, those with dyslalia produced significantly more errors in grammar. Also in the reading test (Fig. 4) there was no significant difference between the districts, but the children with dyslalia were developing difficulties in reading especially dyslexia significantly more often.

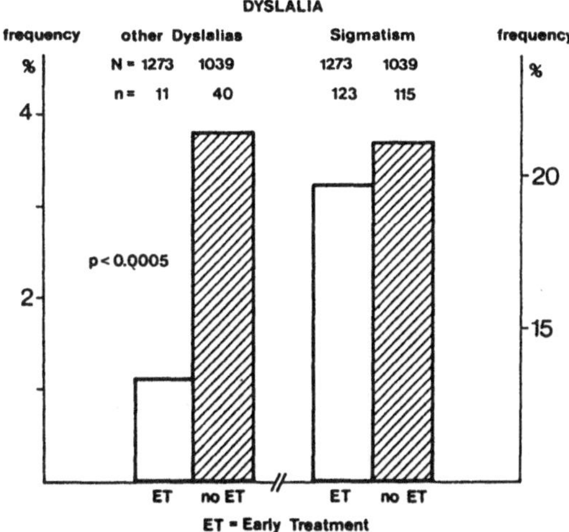

Fig. 1. Articulation test. Children in the district with early speech therapy show slightly fewer sigmatism but significantly less other articulation disorders (dyslalia) than children with no early treatment

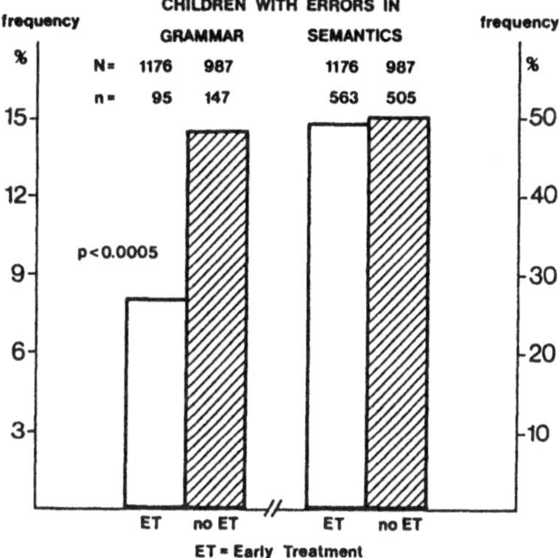

Fig. 2. Reproduction of sentences. Children with early treatment have the same number of semantic errors but significantly fewer errors in speech morphology (grammar). Children with German as a second language are excluded

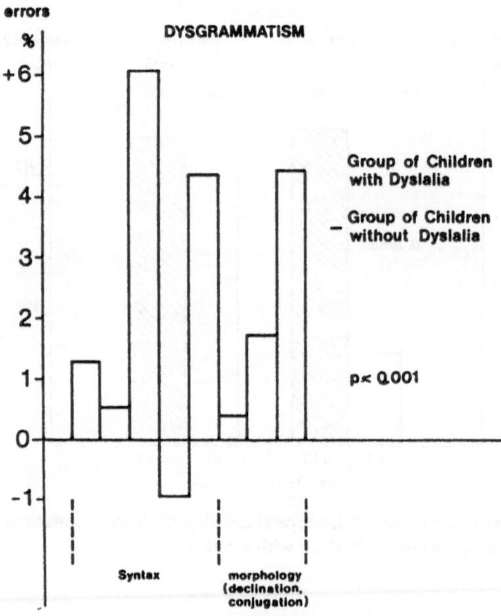

Fig. 3. Description of a picture story. Errors of syntax and morphology are significantly more frequent in the group of children with dyslalia

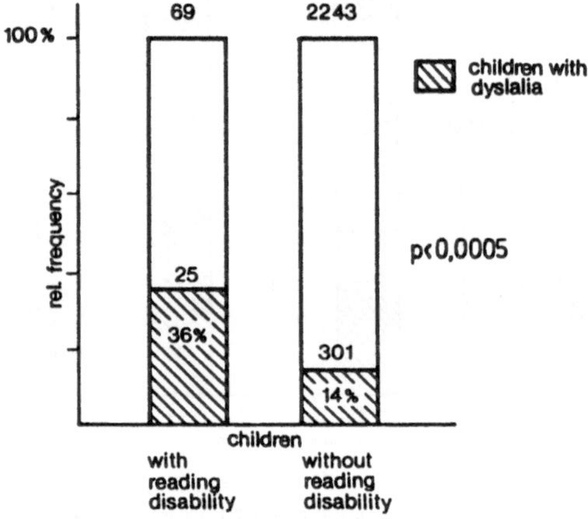

Fig. 4. Reading test. Children with reading difficulties had a dyslalia significantly more often

Discussion

These results show that early speech education is useful. There are fewer children with dyslalia in the district with early treatment. Children with dyslalia more often have difficulties with grammar, i.e. they make more errors in syntax and grammatic morphology, although they do not show more difficulties than other children in semantic speech abilities. These findings correspond to the opinions of pedagogues (Zuckrigl 1964) and child neurologists (Rapin and Allen 1982). Children with dyslalia are more inclined to devolop dyslexia, i.e. reading or writing difficulties. Therefore speech education should not be delayed until school but must be started in kindergarten to cure speech retardation before school starts. At this early stage and if the lack in speech development is not too large, mothers may act as therapists (Bechinger et al. 1984) with daily exercise under a weekly supervision of a speech therapist. This has also been emphasized for the treatment of stuttering (Cave 1977). In 1984 an article in *Lancet* was published (Lincoln et al. 1984) in which usefulness of the treatment of aphasia was doubted because after half a year of treatment aphasic patients with treatment were not better than patients without treatment. The reason for this may be that treatment in these cases was given only two hours per week. Obviously treatment should be more intensive. With professional, personal, treatment, however, this is not possible in most cases; it requires the help of the family.

Summary

A total of 2312 children belonging to the same age group were examined for speech and language difficulties during their second grade of school; 1273 came from a district where early treatment of speech and language disturbances had been an established procedure for years and 1039 from a district with no early treatment. The children in the district with no early treatment showed significantly more severe dyslalias. Furthermore, children with dyslalia develop significantly more dysgrammatism and later dyslexia. Thus early treatment is superior in preventing speech difficulties, but it should be done intensively and daily with the help of the family.

References

Bechinger D, Kornhuber HH, Schmidt W (1984) Ein epidemiologischer Beweis für die Wirksamkeit der Frühbehandlung zerebraler Sprachstörungen mit Hilfe der Mütter. Verh Dtsch Ges Neurol 3: 999-1002

Cave D (1977) Assessment and treatment of stuttering in children. Dev Med Child Neurol 19: 410-412

Lincoln N B, Mulley G P, Jonas A C McQuirk E, Lendrem W, Mitchell I R A (1984) Effectiveness of speech therapy for aphasic stroke patients. A randomized controlled trial. Lancet 1: 1197-1200

Rapin I, Allen D A (1982) Progress towards a nosology of developmental dysphasia. Experta Med Int Congr Ser 579: 25-35

Zuckrigl A (1964) Sprachschwächen. Neckar, Villingen

Hope for a Drug Treatment in Acute Stroke

P.-J. Hülser, H. Bernhart, C. Marbach, A. W. Kornhuber, and H. H. Kornhuber
Department of Neurology, University of Ulm, 7900 Ulm, FRG

The situation concerning drug treatment in acute stroke has been disappointing up to now. Besides prevention in embolic stroke (Canadian Cooperative Study Group 1978; Cerebral Embolism Task Force 1986; UK-TIA Study Group 1988) none of the specific drug regimens has been proven to be successful. Especially the widely used hemodilution has failed to exhibit beneficial effects in recently conducted studies (Frei et al. 1987; Scandinavian Stroke Study Group 1987; Italian Acute Stroke Study Group 1988).

Numerous other drug regimens have been tested,however up to now no reliable advice has been possible. Most probably the main reason for this situation is that therapy in most patients is started too late, regarding the survival time of cerebral neurons in ischemically damaged areas. Therefore we need a drug which can be given as early as possible, that, is by the practitioner prior to transfer to the clinic. This means that the drug must be safe and without risk in those cases that have causes other than ischemic infarction, especially cerebral hemorrhage. The recent development of new fibrinolytic drugs will probably not provide such drugs because it is unlikely that fibrinolysis will be possible prior to CT or NMR to exclude hemorrhage.

Therefore clinical studies about the efficacy of calcium antagonists which can cross the blood-brain barrier through their lipophilic properties are mandatory. Nimodipine was the first drug tested in this field. A multicenter double-blind placebo-controlled study found that there is some favorable effect (Gelmers et al. 1988), but this study has been critisized because the mortality in the placebo group was uncommonly high especially due to bronchopneumonia, and the beneficial effect was restricted to a group of patients with a medium handicap at the beginning of the study. Further, the neurologic outcome was demonstrated for only the first 28 days after the acute event; data on the final outcome were missing. Nimodipine still has considerable blocking effects on the physiologic voltage - dependent calcium channels; thus it depresses blood pressure, which may be contraproductive in stroke.

Flunarizine, a calcium overload blocker, in contrast, has been shown to be effective in preventing the pathologic influx of calcium ions in cellular damage; thus it has a cerebroprotective effect, but it practically does not influence the physiologic calcium channel (Peters 1984). This property explains why flunarizine does not exert significant negative inotropic effects on normal heart muscle and does not influence the myogenic tone of blood vessels. In experimental settings flunarizine reduces postischemic hypoperfusion (Van Nueten 1982), scavanges free radicals (Phillis et al. 1985), and has a cellular cerebroprotective effect. In various animal models of hypoxia and local or global brain ischemia a favorable effect of flunarizine on functional and morphologic parameters was found when the drug was given before (Silverstein et al. 1986; Van Reempts et al. 1987) or after the injury (Deshpande and Wieloch 1985). Even more important, recently a beneficial effect of flunarizine on cerebral hemorrhagic lesions was demonstrated in animal experiments (Kleiser et al. 1989). The drug is now available for intravenous application from Janssen. Therefore we decided to conduct a multicenter double-blind controlled study; this is now under way. However, we first conducted a small study with healthy volunteers receiving 50 mg flunarizine in one dose with careful control of side effects; we then conducted an open pilot study in stroke patients to investigate the tolerance of a high-dose intravenous application of this drug to find a suitable dose. The main side effect was transient weariness. The velocity of saccadic eye movements dropped slightly, in parallel with the subjective weariness. Blood pressure and heart rate were not affected. No pharmacogenic depression or extrapyramidal side effects were detected although 2 x 25 mg flunarizine was given to the stroke patients daily over 7 days and thereafter 30 mg daily for 21 days (Hülser et al. 1988). The quantitatively measured clinical outcome showed a favorable course. The recovery curve (Mathew score; Mathew et al. 1972) resembled an exponential course in all subscores including sensorimotor, mental functions and activities in daily life (Fig. 1). The individual scores in the examined population exhibited a broad dispersion at the beginning which decreased during the period of observation. This means that even patients with severe deficits had a favorable outcome on average. Although it is at present not yet possible to give a recommendation on the basis of clinical improvement, the results of animal experiments together with the low level of side effects in clinical studies give hope that immediate treatment with calcium overload antagonists such as flunarizine could be a therapy which can be started already by the general practitioner and may yield at least a small benefit for the patient. More could be expected from prevention, the possibilities of which have not at all been exploited yet (Kornhuber 1983; Kornhuber et al. 1989).

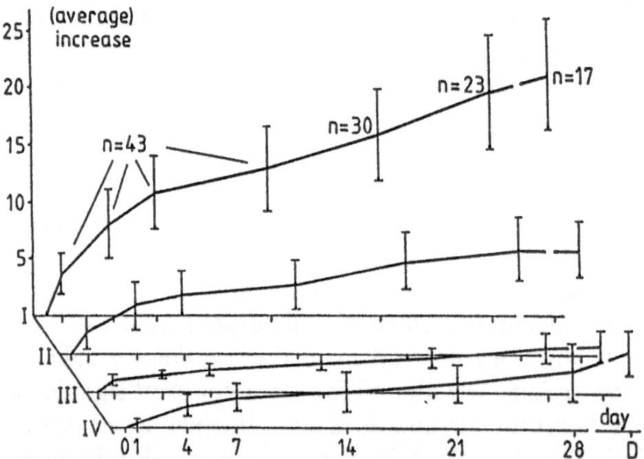

Fig. 1. Increase in the average Mathew score compared with the initial value. The bar indicates twice standard deviation. The curves depict the average values for 43 patients during the first 2 weeks. Thereafter the number of patients decreases because more and more patients were discharged from hospital. I: total score, II: mental status score, III: sensorimotor score, IV: ability of daily life score

References

Canadian Cooperative Study Group (1978) A randomized trial of aspirin and sulfinpyrazone in threatened stroke. N Engl J Med 299:53-59

Cerebral Embolism Task Force (1986) Cardiogenic brain embolism. Arch Neurol 43:1-84

Deshpande JK, Wieloch T (1985) Amelioration of ischaemic brain damage following postischaemic treatment with flunarizine. Neurol Res 7:27-29

Frei A, Cottier C, Wunderlich P, Lüdin E (1987) Glycerol and dextran combined in the therapy of acute stroke. A placebo-controlled, double-blind trial with a planned interim analysis. Stroke 18:373-379

Gelmers HJ, Gorter K, de Weerdt CJ, Wiezer HJA (1988) A controlled trial of nimodipine in acute ischemic stroke. N Engl J Med 318:203-207

Hülser P-J, Bernhart H, Marbach C, Kornhuber HH (1988) Treatment with an i. v. calcium overload blocker (flunarizine) in acute stroke. Eur Arch Psychiatry Neurol Sci 237:253-257

Italian Acute Stroke Study Group (1988) Haemodilution in acute stroke: results of the Italian haemodilution trial. Lancet 1:318-321

Kleiser B, van Reempts J, van Deuren B, Haseldonckx M, Borgers M, Horn E, Eβeling K, Widder B, Kornhuber HH (1989) Favourable effect of flunarizine on the recovery from hemiparesis in rats with intracerebral hematomas. Neurosci Lett 103: 225-228

Kornhuber HH (1983) Präventive Neurologie. Nervenarzt 54:57-68

Kornhuber HH, Backhaus B, Kornhuber J, Kornhuber AW (1989) Risk factors and the prevention of stroke. In: Amery WK, Bousser MG, Rose FC (eds) Clinical trial methodology in stroke. Transmedica Europe,Tunbridge Wells

Mathew NT, Meyer JS, Rivera VM, Charney JZ, Hartmann A (1972) Double-blind evaluation of glycerol therapy in acute cerebral infarction. Lancet 2:1327-1333

Peters T (1984) Wirkung von Kalziumantagonisten auf zelluläre Kalziumbewegungen. Therapiewoche 34:6011-6029

Phillis JW, Delong RE, Towner JK (1985) The effect of lidoflazine and flunarizine on cerebral reactive hyperemia. Eur J Pharmacol 112:323-329

Scandinavian Stroke Study Group (1987) Multicenter trial of hemodilution in acute ischemic stroke. I. Results in the total patient population. Stroke 18:691-699

Silverstein FS, Buchanan K, Hudson C, Johnston MV (1986) Flunarizine limits hypoxia-ischemia induced morphologic injury in immature rat brain. Stroke 17:477-482

UK-TIA Study Group (1988) United Kingdom transient attack (UK-TIA) aspirin trial: interim results. Br Med J 296:316

Van Nueten JM (1982) Selectivity of calcium entry blockers. In: Albertini A, Paoletti R (eds) Calcium modulators. Elsevier, Amsterdam, pp 199-208

Van Reempts J, van Deuren B, van de Ven M, Cornelissen F, Borgers M (1987) Flunarizine reduces cerebral infarct size after photochemically induced thrombosis in spontaneously hypertensive rats. Stroke 18:1113-1119

The Essence of Aphasia - Disturbed Control of Language Production, e.g. in Phonemic Paraphasia: A Quantitative Comparison of Spontaneous Speech in Aphasia and Dementia

C. Marbach, R.J. Brunner, H.H. Kornhuber, B. Müller, C.W. Wallesch, and J.M. Hufnagl

Department of Neurology, University of Ulm, 7900 Ulm, FRG

Introduction

The instruction for the Aachener Aphasie Test (AAT) makes a point of the fact that in cerebral atrophies the aphasic syndromes are blurred, but no details for the possible differentiation of one from the other are mentioned. In clinical practice and especially in speech therapy frequently a most crucial distinction is to be made between aphasia-like disturbances in demented patients and aphasias due to focal lesions.

Patients and Methods

Nineteen demented patients (11 senile, 4 alcoholic, 4 multi-infarction), 34 aphasics (15 Broca, 15 Wernicke, 2 amnesic, 2 transcortical-motor), 13 patients with right hemispheric lesions and 10 healthy aged subjects were investigated by means of a series of tests, including (a) analysis of spontaneous speech (semistandardized interview, evaluation criteria according to Brunner et al. 1982), (b) Token Test (de Renzi and Vignolo 1962; modified by Orgass 1976), and (c) Colored Progressive Matrices (Raven 1956). All verbal responses were tape-recorded and quantitatively analyzed regarding fluency, word categories, tenses, syntactical structure, phonemic errors and other neurolinguistic pathology. The flow of information was investigated by counting units of information/minute. These units were defined by the following three features: (a) no further segmentation possible, (b) relevance, (c) no redundance. The cerebral lesions/atrophy were verified by means of CT scans with a matrix of 256 x 256.

Results

The main difference between demented and aphasic patients was a significantly higher production of phonemic paraphasias by the aphasic patients in their spontaneous utterances (Fig. 1). On the average there were four times more words with phonemic errors in aphasics. Upon ESPA analysis (Peuser 1978) there were twice as many elisions, substitutions, permutations and additions per phonemically distorted word in the aphasic productions as compared to the demented patients. Although the demented patients produced a greater number of stereotypies than aphasics, there were no non-meaningful recurring utterances in dementia as they occur in global aphasia (Brunner et al. 1982). There was no significant difference between aphasic and demented patients on the Token Test (Fig. 2). The groups were also comparable for fluency, paragram-

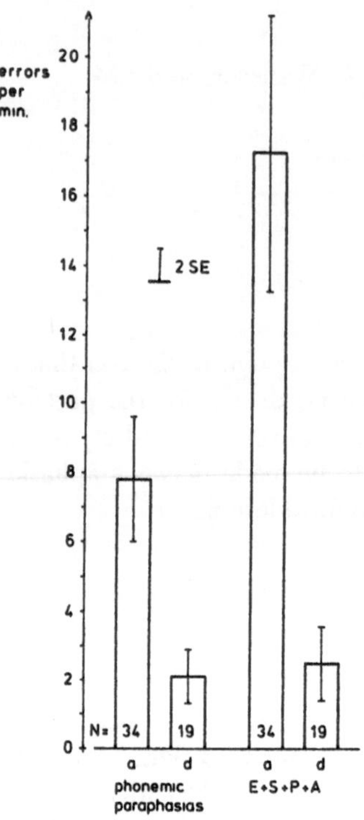

Fig. 1. Phonemic paraphasias and ESPA in aphasic (a) and demented (d) patients. E: Elision; S: Substitution; P: Permutation; A: Addition

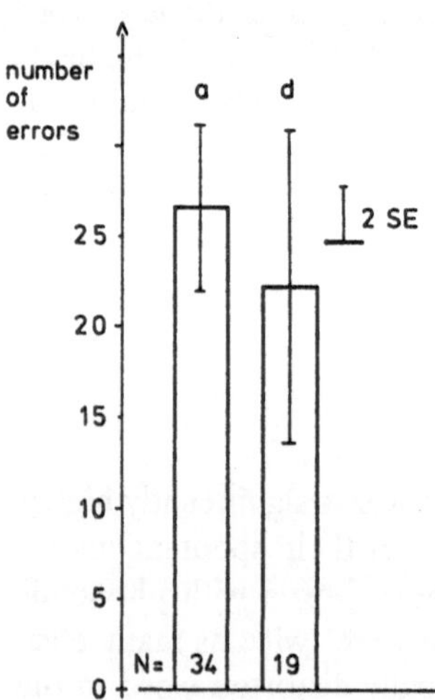

Fig. 2. Token Test results (Orgass modification) in aphasic (a) and demented (d) patients

matism, stereotypies, use of pronouns and many other aspects. Both aphasics and demented patients differed from patients with right hemisphere lesions by a significant reduction in information flow per minute. The speech production in demented patients was characterized by circumlocutions, adherence and reduction in coherence. The number of words needed for one unit of information was twice as high in the demented patients as compared to the aphasics.

Discussion

Phonemic paraphasia is obviously an essential feature of aphasia and is prominent both in the Broca and Wernicke varieties. We interpret their occurrence as a result of defective checking for and perception of errors. The maintenance of correct phonemic structure in dementia serves as a distinction to aphasia. While there is a specific disturbance of language in aphasia, the essence of dementia is a disturbance of prelinguistic cognitive processes. The results of the patients with diffuse and left hemisphere lesions in the Token Test are poor in contrast to those of patients with lesions of the right hemisphere. This finding emphasizes that the Token Test measures a function of the left hemisphere. The language disturbances in dementia are obviously due to the bilaterality of the underlying cerebral pathology.

Conclusion

Aphasia and dementia may be differentiated on the basis of verbal behavior by linguistic analysis of spontaneous speech with special emphasis on phonemic paraphasia.

Summary

We investigated 34 aphasics, 19 demented patients, 13 patients with right hemisphere lesions and 10 healthy aged controls using computer tomography (CT) scan and quantitatively evaluated linguistic and neuropsychological tests. The main difference between the aphasics and the demented patients was a much higher rate of production of phonemic paraphasias in the aphasics. Phonemic paraphasia is regarded as a prominent and diagnostic element of aphasic language disturbance and interpreted as the result of defective checking for errors. Otherwise, there were many similarities between aphasic and demented patients, including in the level of Token Test performance.

References

Brunner R J , Kornhuber H H , Seemüller E , Suger, G , Wallesch C W (1982) Basal ganglia participation in language pathology. Brain Lang 16: 281-299

De Renzi E, Vignolo L (1962) The Token Test: a sensitive test to detect receptive disturbances in aphasics. Brain 85: 665-678

Orgass B (1976) Eine Revision des Token Tests. Diagnostika 22: 7087

Peuser G (1978) Aphasie. Eine Einführung in die Patholinguistik. Fink (Patholinguistika 3)
Raven C J (1956) Standard progressive matrices. Lewis London

Planning Strategies of Intracranial Microsurgery

W. Seeger
Neurochirurgische Universitätsklinik der Albert-Ludwigs-Universität, 7800 Freiburg,FRG

Summary

Operative procedures in deep-seated areas of the brain by microsurgery must consider anatomical variations in the brain and variations in the topographical relationships of the lesions. Most of these problems are to be recognized in the preoperative CT, angio-

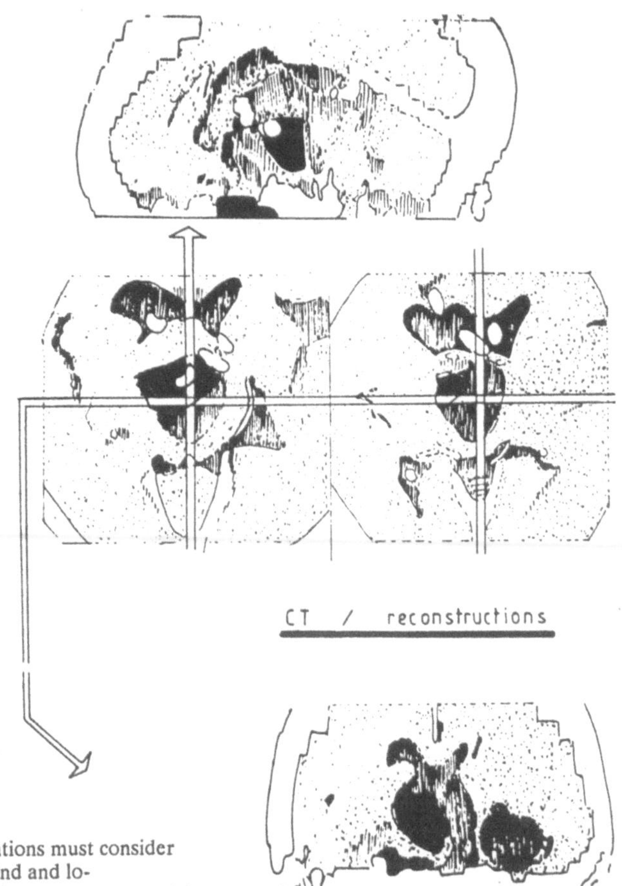

Fig. 1. Planning strategies before operations must consider 10 points.The first is diagnosis of the kind and location of the lesion.Here a cystic teratoma of the 3rd ventricle

grams and NMR image and allow definition of operative approach and operative technique. Most important is preservation of limbic structures.(For literature see W. Seeger, *Planning Strategies of Intracranial Microsurgery*, Springer Vienna, New York, 1986.)

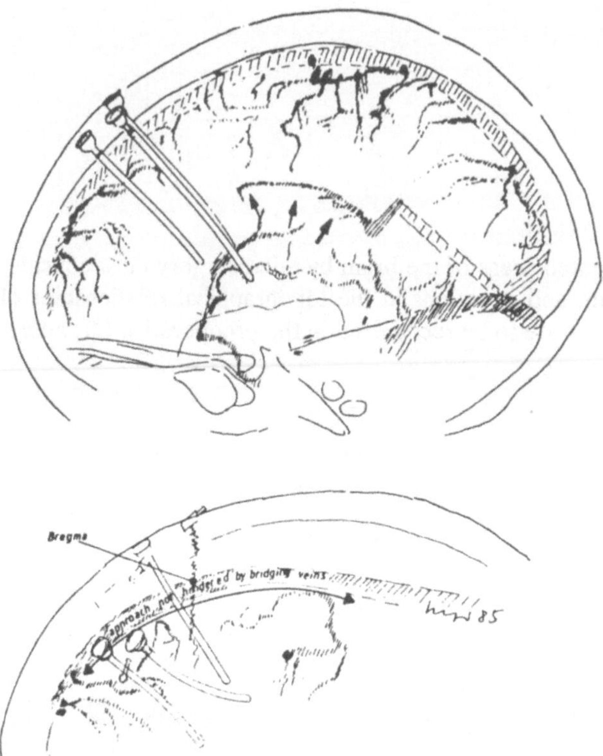

Fig. 2. Veins lie mostly on the surface of the brain. At an operative approach we must know first the course and variants of veins, in this case of the veins of fissura longitudinalis. Then the usual analysis of arteries close to the lesion is important

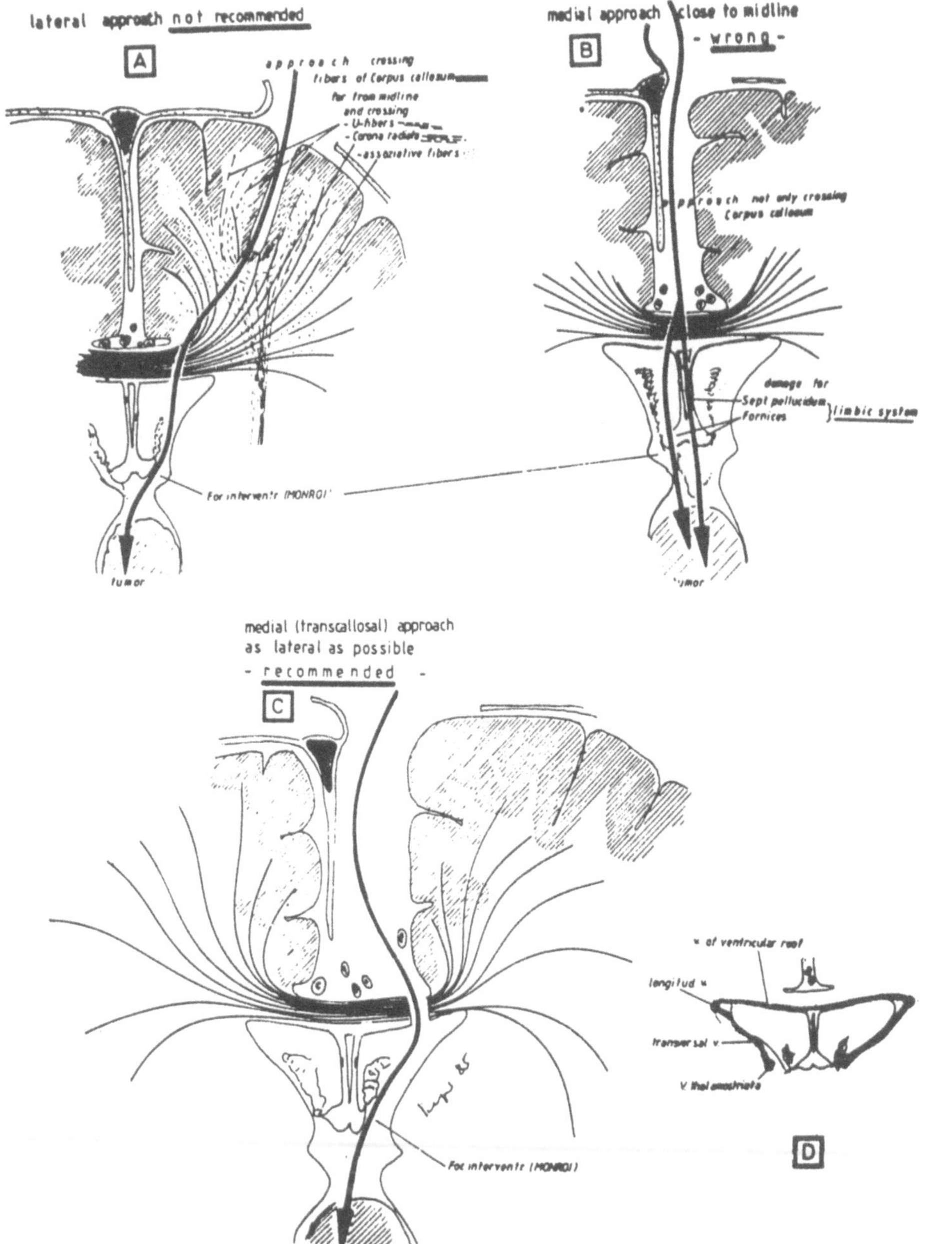

Fig. 3a,b. Next there are questions of the choice of approach before or behind hindering veins or arteries. In this case one must consider the course of fibers of corpus callosum.The transcerebral approach is unnecessary and dangerous, as shown here, because even a lateral approach transects the fibers of corpus callosum not less than the midline approach.The midline approach would not destroy other fibers such as association fibers or projection fibers. Here is to consider the course of branches of a. cerebri ant., which may change from one side to the other, and the position of midline structures, most importantly fornices and stria lateralis

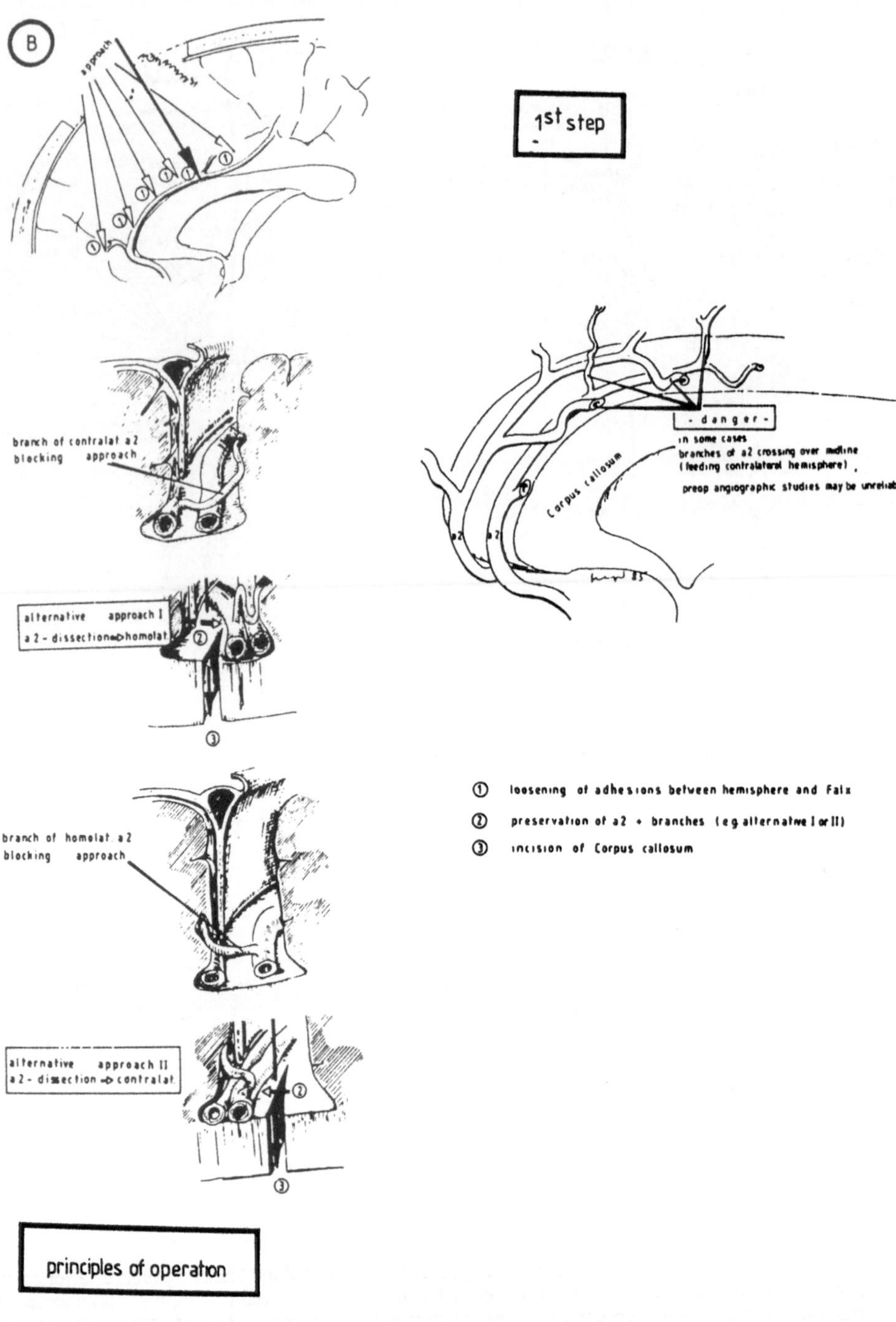

B
approach
1st step
branch of contralat a2
blocking approach
- danger -
in some cases
branches of a2 crossing over midline
(feeding contralateral hemisphere),
preop angiographic studies may be unreliable
Corpus callosum
a2
a2
alternative approach I
a2 - dissection → homolat
② ③
① loosening of adhesions between hemisphere and Falx
② preservation of a2 + branches (e.g. alternative I or II)
③ incision of Corpus callosum
branch of homolat a2
blocking approach
alternative approach II
a2 - dissection → contralat
principles of operation

Fig. 3 b

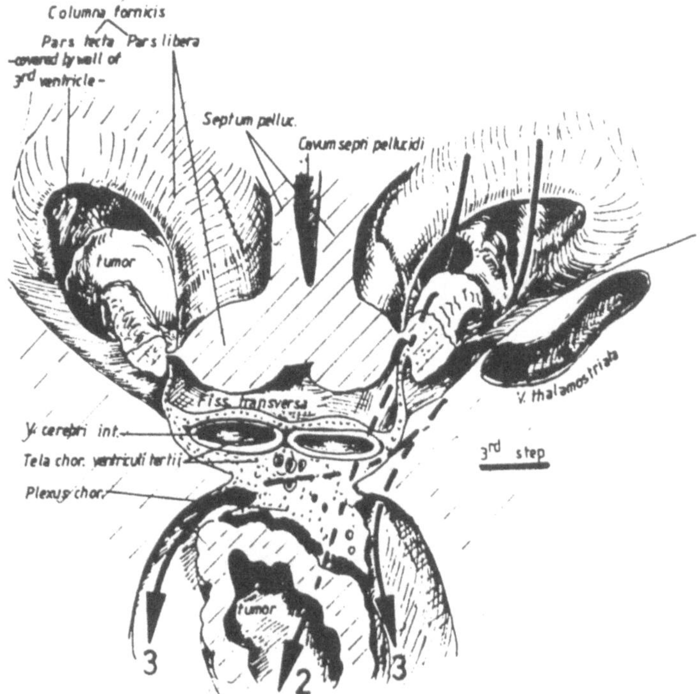

Fig. 4. After definition of operative approach, the method of operative extirpation of the lesion may be discussed. In this sample the approach passes the foramen of Monro. If the foramen of Monro is narrow, it must be dilated. The only method of operative dilation possible is in a posterior direction by loosening of plexus from taenia fornicis or taenia thalami. Problems with v. thalamostriata and thalamic perforating arteries, adhesions of tela chorioidea ventriculi III (= velum interpositum) must be considered

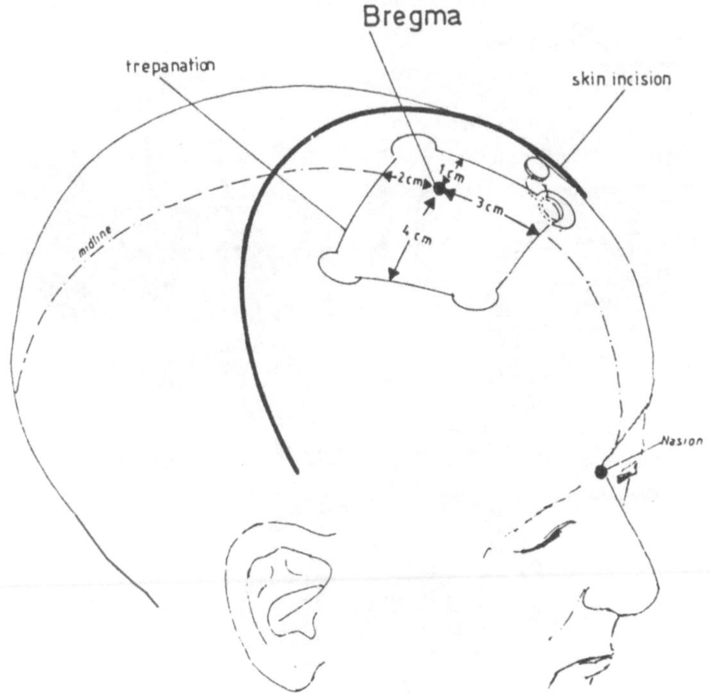

Fig. 5. Even after definition of operative approach and microsurgical strategy the definition of trepanation may be done. After definition of trepanation, definition of skin incision and in basal processes of muscle incision may be discussed

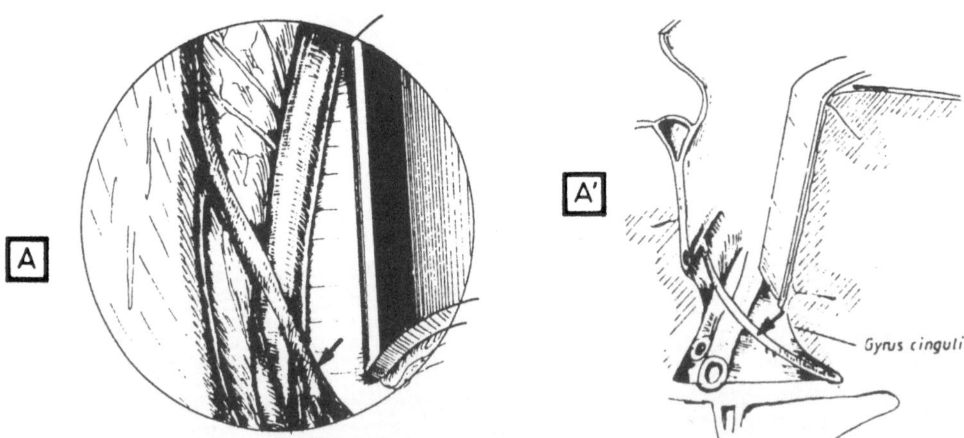

Fig. 6. Control of preoperative strategy is the microsurgical topography and postoperative CT or NMR control. Unexpected, small but important arteries crossing the midline, as in this case, are often to be expected and must be considered even if angiography does not show these arteries

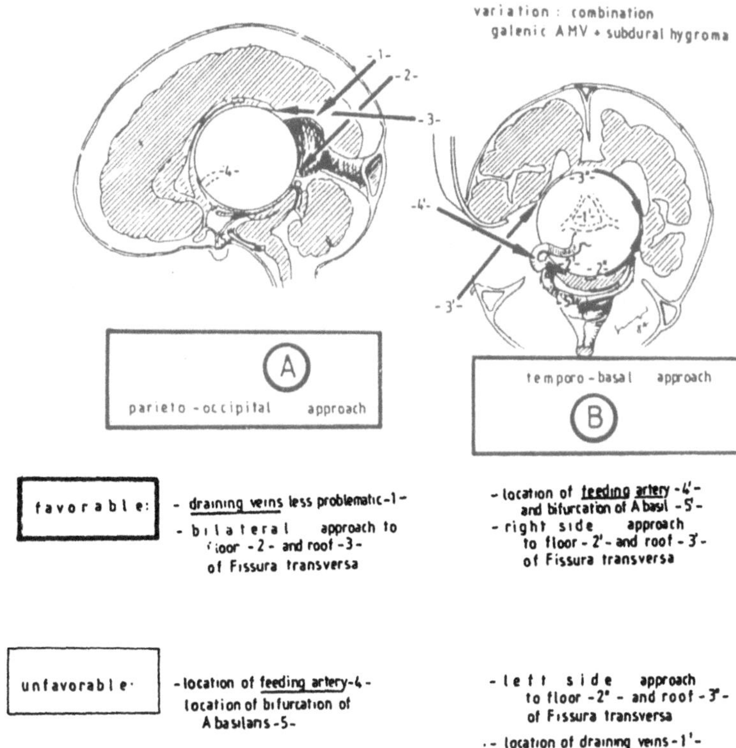

Fig. 7. The different approaches of lesions close to the quadrigeminal area may illustrate some problems in planning strategies. This example of a galenic angioma may demonstrate that the typical parieto-occipital approach is not in all cases the best. The child in this sample had a subdural hygroma which allowed a tempero-basal approach. In this case may be discussed the arguments for and against each possible approach. The main argument for a temperobasal approach was the location of the main feeding artery, which could be eliminated from a basal approach in an early stage of operation. Without subdural hygroma, the location of a draining vein (v. magna galeni and sinus rectus) would be problematic, because it is not easy to interrupt them from the lateral basal direction. Here it is better to have a midline approach

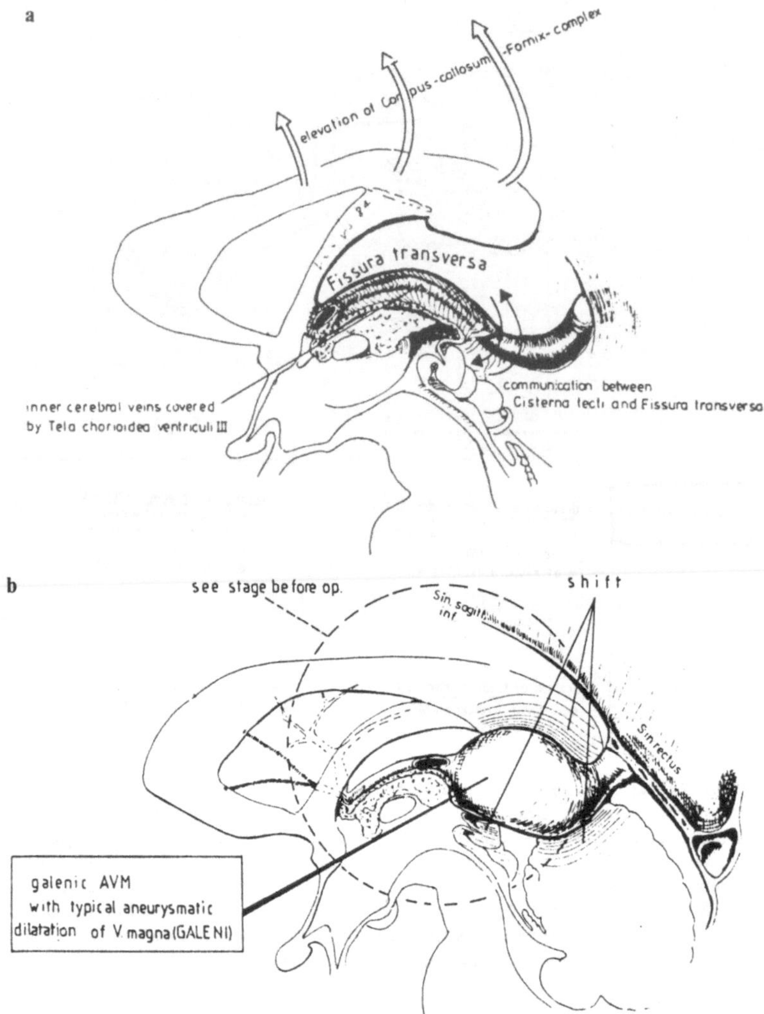

Fig. 8a,b. The microsurgical technique must consider the evolution and location of the lesion. This angioma starts from v. magna galeni and lies in fissura transversa, a small CSF space between fornices and tela chorioidea of the 3rd ventricle

→

Fig. 9a,b. The shifts and relationships to the ventricular system and basal brain structures, especially the optic system, are important. Fibers of corpus callosum are dilated in a membrane-like manner. The compression of brain stem and shift of brain stem must be considered. In this case the shift of brain stem and basilar artery in a caudal direction is favorable for the operation because the lesion is now close to the tentorial edge. The midbrain is shifted forward and downward. If the lesion were very small, a basal approach would be impossible

a

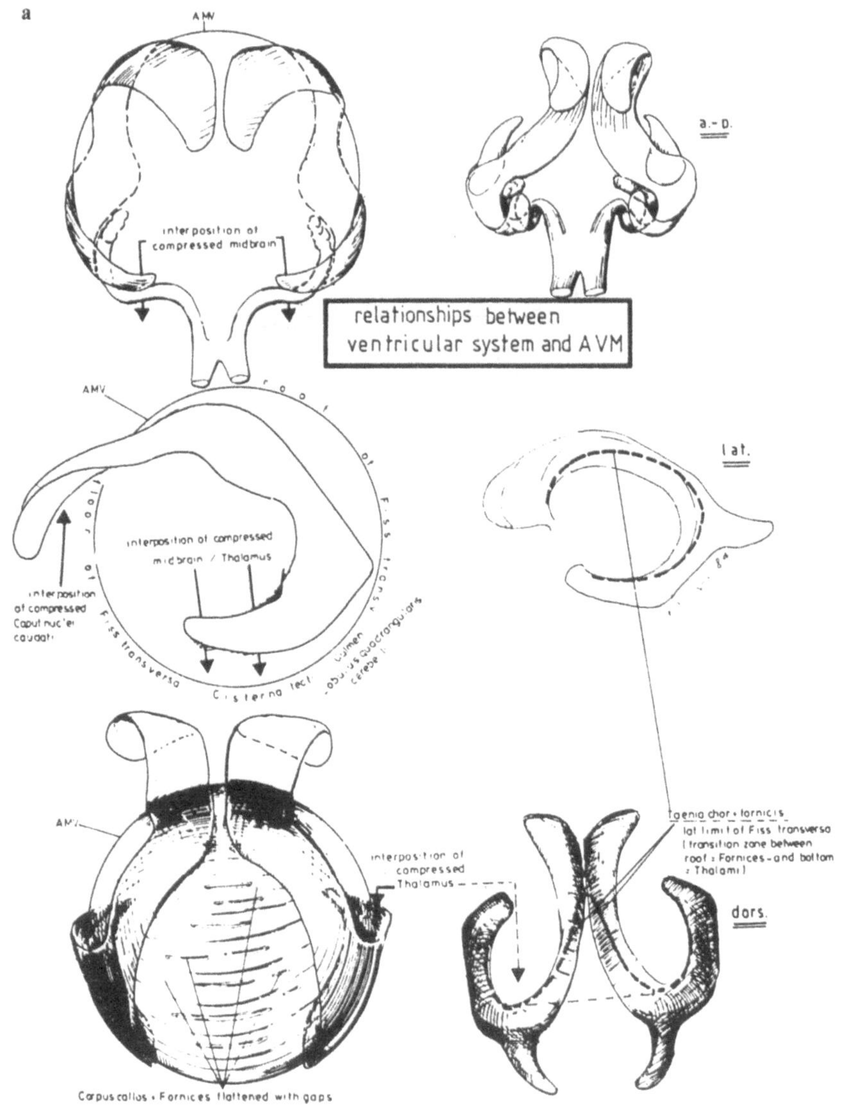

AMV

a.-p.

interposition of compressed midbrain

relationships between ventricular system and AVM

AMV

roof of Fiss. transv.

floor of Fiss. transversa

interposition of compressed Caput nuclei caudati

interposition of compressed midbrain / Thalamus

Cisterna tecti

Culmen Lobulus quadrangularis cerebelli

lat.

Taenia chor. fornicis

lat. limit of Fiss. transversa (transition zone between roof: Fornices - and bottom: Thalami)

AMV

interposition of compressed Thalamus

dors.

Corpus callos. + Fornices flattened with gaps

b

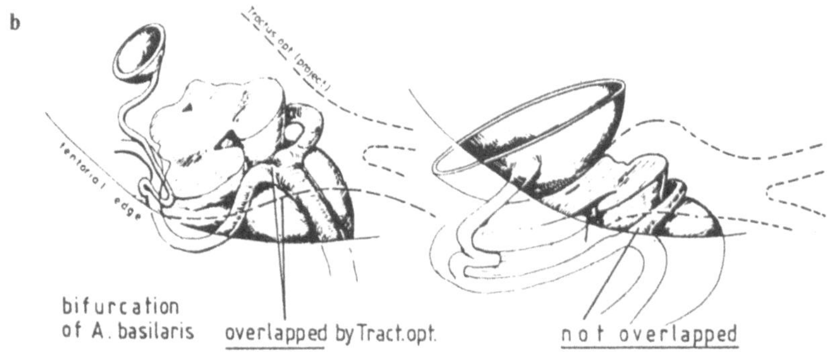

a

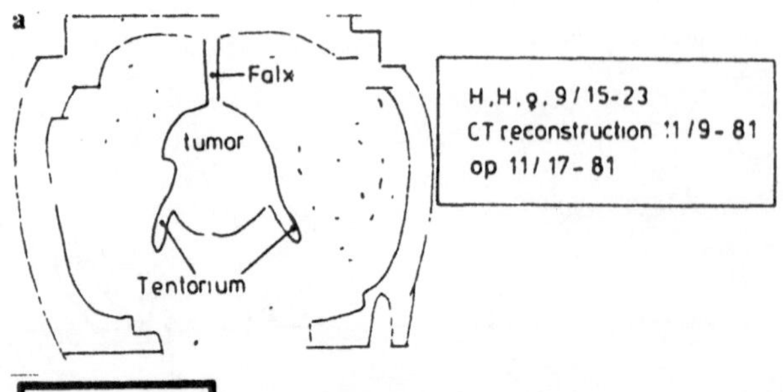

favorable

- tumor originates from Falx + Tentorium
- limbic system shifted aside + rostrally
- op approach short and wide (during tumor-hollowing)
- galenic vein obstructed → venous collaterals
- choroid branches shifted → basal
- Corpus callosum shifted → rostral
- medial occipito-basal veins not endangered

unfavorable

- bilateral location
- Area optica (Praecuneus) bilateral endangered

b

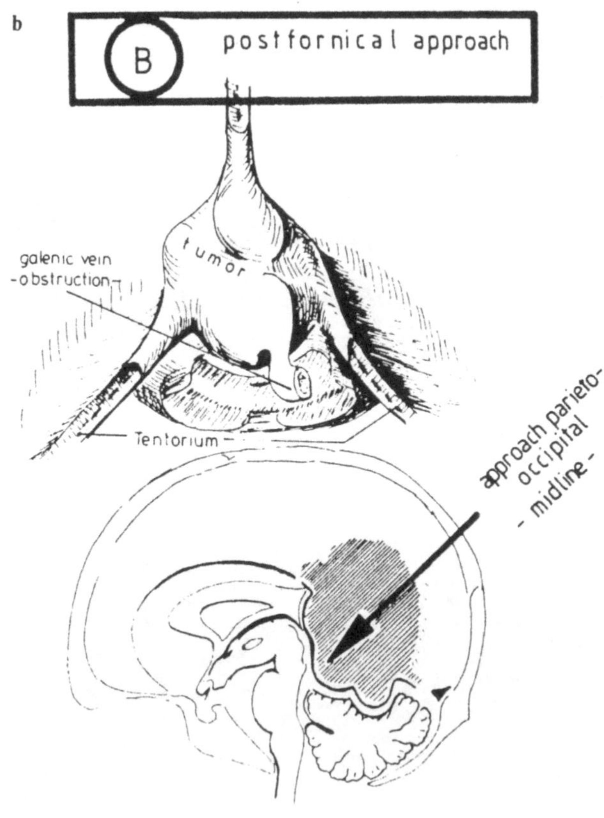

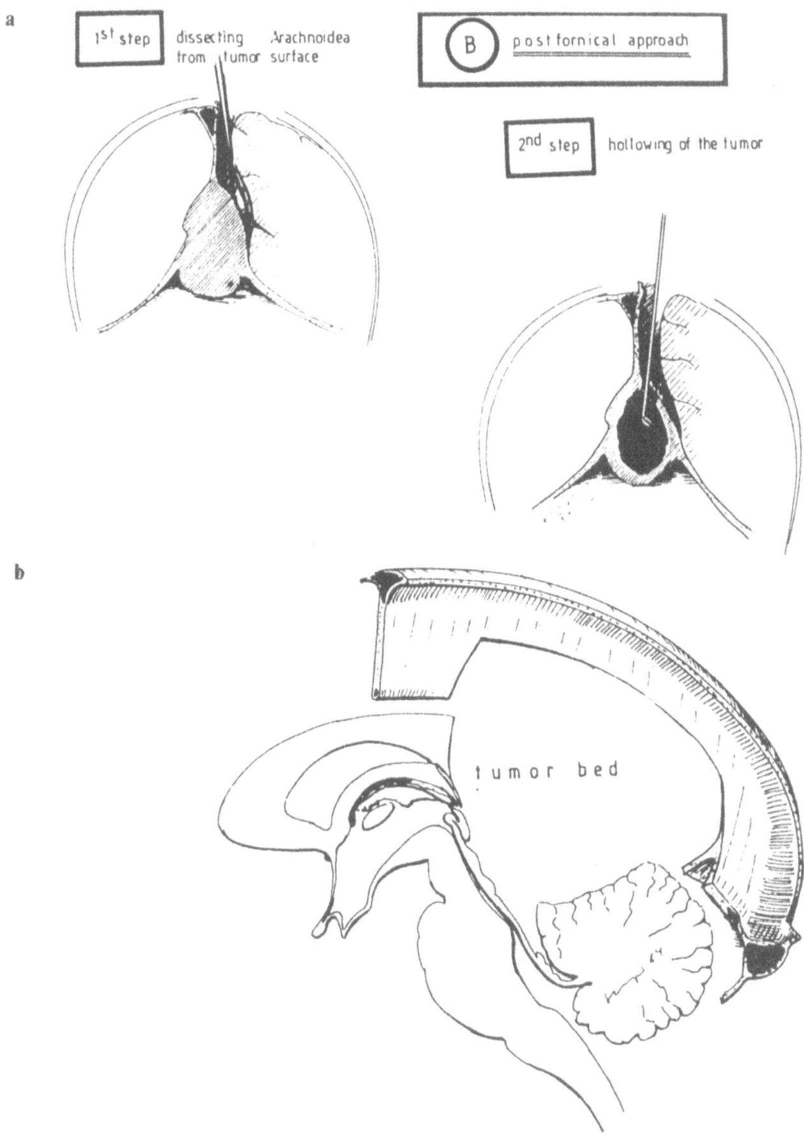

Fig. 11a,b. The principle of operation would be: hollowing of the tumor, then resection of falx and tentorium with extirpation of the tumor base. Galenic vein system and sinus rectus are unproblematic because in angiogram sufficient collateral veins are demonstrated

Fig. 10a,b. This meningioma lay even in the area of tectum. The evolution of this meningioma from the galenic point seems to have a similar relationship to surrounding structures as the galenic aneurysm. However the meningioma infiltrates falx and tentorium and it is not located in a CSF space such as fissura transversa cerebri. It is a subdural tumor, less problematic than the galenic angioma. Here a midline dorsal approach is the only possible approach for extirpation, but it must start unilaterally and must consider the location of area optica

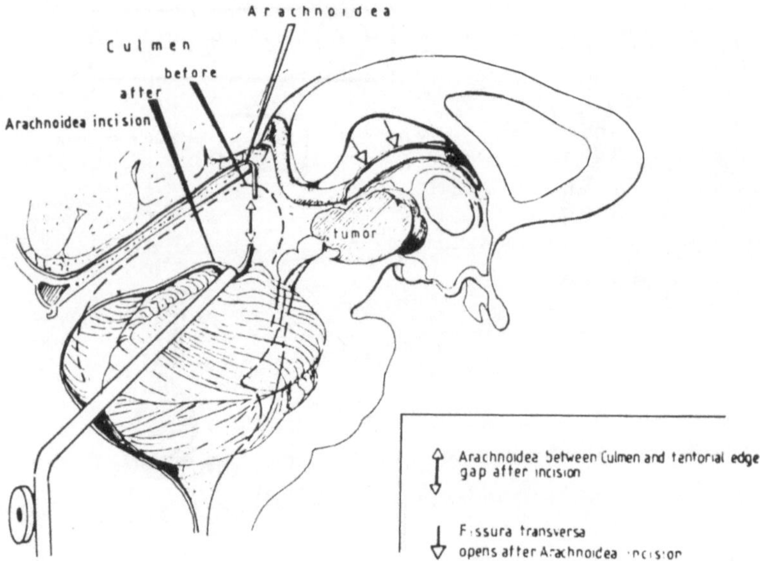

Fig. 12. In contrast to the previous case, here may be demonstrated the typical and mostly used approach for the quadrigeminal area described by Krause and modified by Yarsagil. It is a wellknown approach, but as we have seen, this area is not to be operated from this supracerebellar approach in all cases

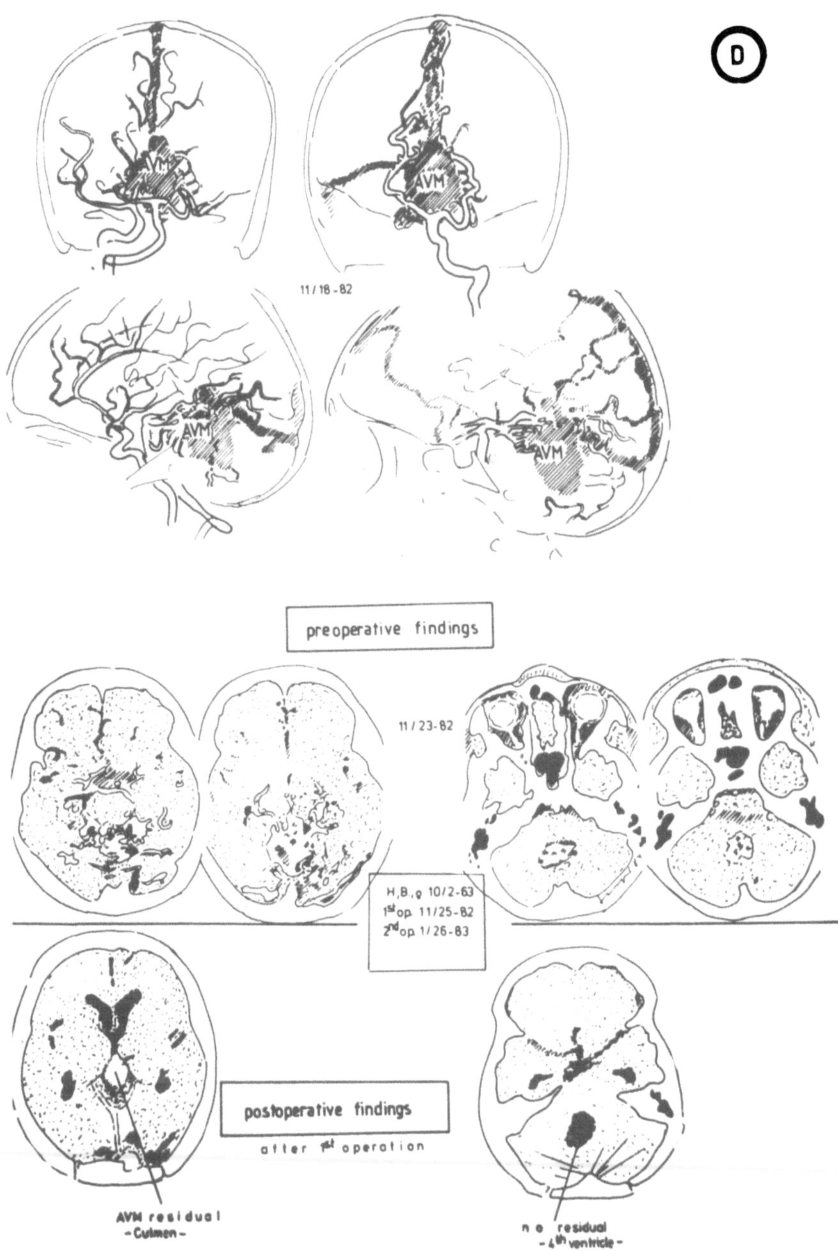
D
AVM
AVM
11/18-82
AVM
AVM
preoperative findings
11/23-82
H.B.,♀ 10/2-63
1st op. 11/25-82
2nd op. 1/26-83
postoperative findings
after 1st operation
AVM residual
-Culmen-
no residual
-4th ventricle-

Fig. 13 a

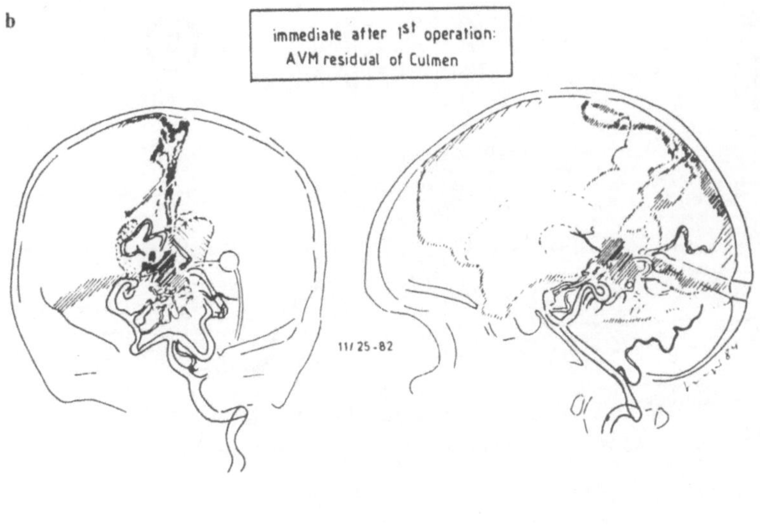

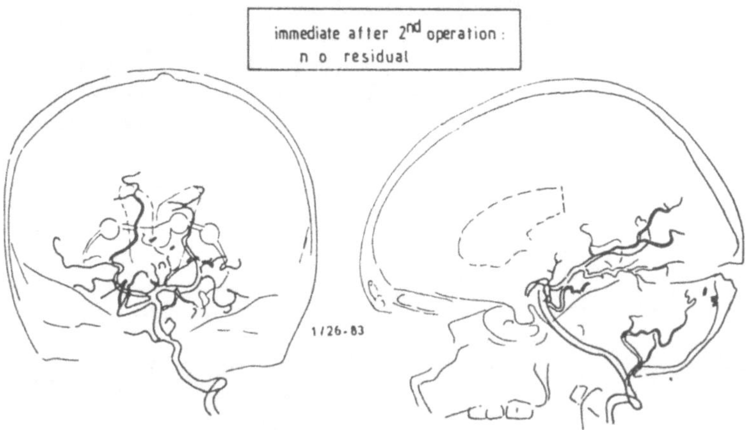

Fig. 13a,b. This arteriovenous malformation was located in the quadrigeminal area and in the rostral cerebellar area. Its radical extirpation was done in 2 steps

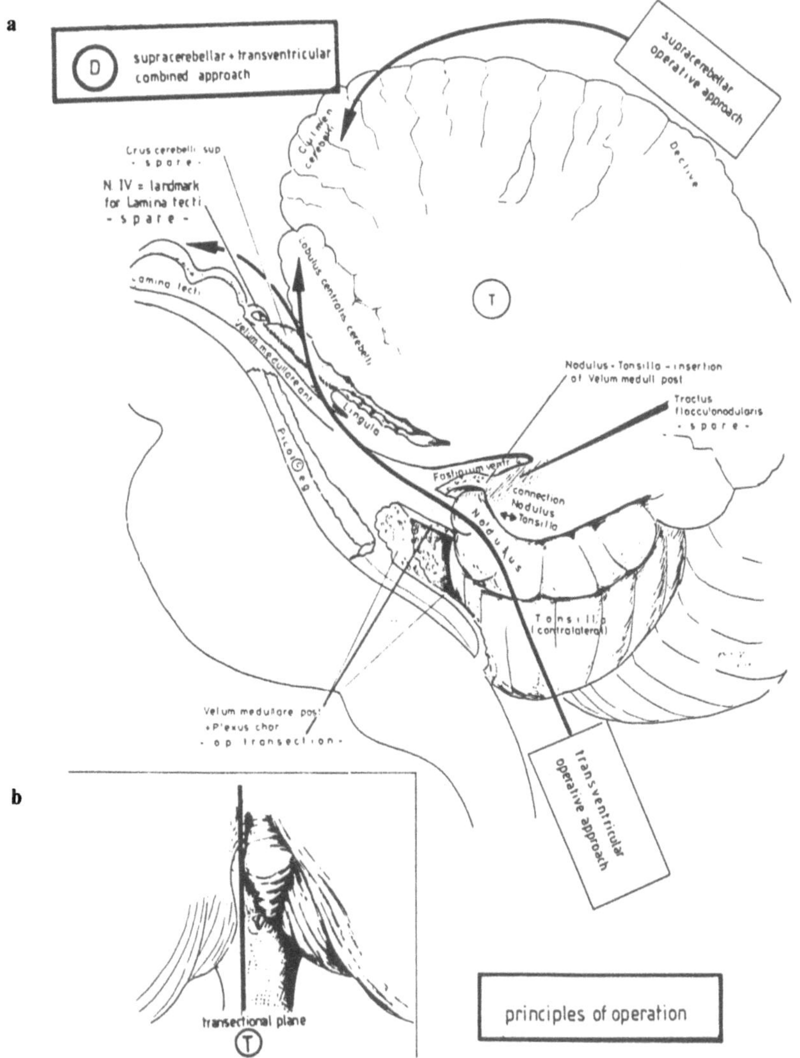

Fig. 14a,b. At first, the quadrigeminal region was operated by a transventricular approach passing the 4th ventricle, splitting the velum medullare ant. and extirpating the main portion of the angioma in this way. The second approach was typically supracerebellar, however not for extirpation of the angioma of quadrigeminal region but for extirpation of residual parts of angioma in the culmen area

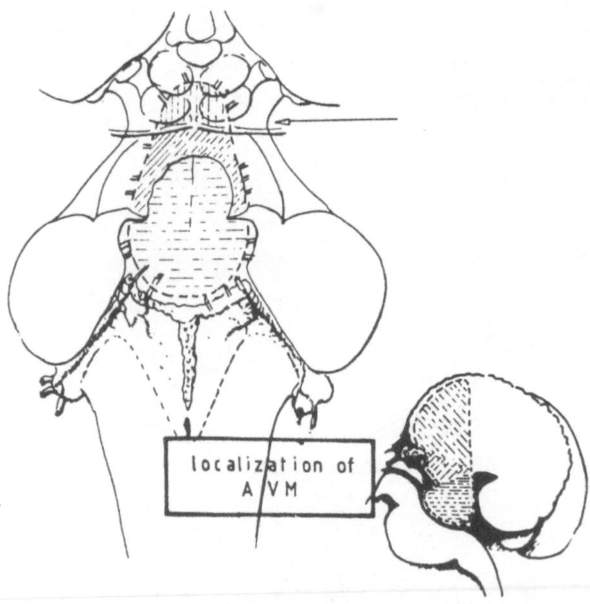

Fig. 15. The last figure demonstrates the microsurgical topography. By the first approach the quadrigeminal angioma was extirpated. The trochlear nerve is a landmark. Anterior to it one can see the quadrigeminal plate and galenic vein, posterior to it the splitted velum medullare posterius. The splitting must be stopped at the level of the trochlear nerves

The main problems in planning strategies are problems of topographical anatomy, which are not always to be seen in preoperative CT or NMR findings, because small vessels and many cerebral structures here are not to be seen. Anatomical variants such as superficial veins and the relationships of lesions to surrounding structures may modify the operation technique in lesions which may have the same location in the intracranial spaces. In operations one must consider especially the relationships to natural CSF spaces. Most important is the preservation of the limbic structures to prevent psychological deficits.

What Can Neurosurgery Do in 1988?

P.C.Potthoff

Neurosurgical Department, Knappschaftskrankenhaus Bergmannsheil, 4650 Gelsenkirchen-Buer, FRG

Introduction

By 1988 many advanced technical applications are under discussion in neurosurgery. These include LASER, CUSA, evoked potential controls, neurostimulation, implantable medication systems, superselective embolizations, various irradiation approaches, and others (Fasano 1986; Pluchino and Broggi 1988; Walter et al. 1988).

At the same time, progress has been made in establishing and standardizing routine methods in neurosurgery. This state of the art is important for everyday neurosurgery in non university departments where neurosurgery must be handled from a practical and pragmatic point of view, and applied techniques have preferably to be safe and simple. The following presents a short survey of generally accepted and practical standards of daily routine neurosurgery in 1988. This analysis follows the general anatomy of the body from head downwards, from brain surface to spinal cul-de-sac.

Brain

Subdural Hematoma

Acute traumatic subdural hematoma remains a general problem, with a mortality of about 30% (Alberico et al. 1987), in our patients mainly because of concomitant arteriosclerotic and hepatic degenerative disease. Chronic subdural hematoma is nowadays easily handled by a simple continuous drainage through an enlarged burrhole trephination applied under local anesthesia. The avoidance of general anesthesia in these mainly old and very old patients has reduced mortality to virtually 0% for this condition (Richter et al. 1984).

Intracerebral Hemorrhage

This - be it traumatic or spontaneous - remains a grave problem, but it is handled much more easily since the advent of cranial CT, allowing aimed evacuation by suction or endoscopic measures in suitable cases even under local anesthesia. Therefore, under conservative intensive neurosurgical care, with or without hematoma evacuation, many of these patients nowadays survive the initial posthemorrhagic course. However, many of them die later from sequelae due mainly to overweight and hypertensive blood pressure.

Brain Trauma

CCT has added much to a simpler and safer treatment of brain trauma. Exploratory craniotomies can be avoided today. CCT shows most exactly whether decompression is reasonable or not. Continuous pressure recordings (Klein 1982) allow the precise application of osmotherapy (sorbitol, etc.) and dexamethasone (Gaab 1986). We continue to favor the latter in the treatment of elevated ICP of all causes, together with adequate controlled ventilation. Barbiturates are out of discussion but for very few exceptions.

Aneurysms and Angiomas

In autonomous brain disease, i.e. vascular malformations and brain tumours, microneurosurgery (Seeger 1980, 1985; Yasargil 1984) has taken an important position. With the microscope, with microinstrumentation and bipolar coagulation, this pathology may now be attacked with increasing success down to the brain stem and basal surface of the brain as well as to the craniocervical junction (Samii 1986). However, patients in whom such lesions are detected are of increasingly old age. And in spite of the fact that nowadays patients well above 80 years old can be operated upon in neurosurgery, this age group of neurosurgical patients remains a problem group. In aneurysms, microneurosurgery has made early operations possible within 72 hours after SAH, preferably in stages Hunt and Hess 1, 2 and possibly 3 (Auer 1985). Nimodipine has helped enormously to avoid spasm (Auer et al. 1986). Thus mortality in these cases has come down to between 2% and 12%. But still the primary catastrophy of SAH cannot be avoided. This kills some 25% to 33% in all stages and kills almost all patients in stages 4 to 5, in a rapid spontaneous course with or without the desperate attempt of operation.

Brain Tumors

Microneurosurgery allows much better handling of benign brain tumors (Seeger 1980, 1985; Sugita 1985; Yasargil 1984), especially in difficult locations. This applies as well to semibenign and malignant tumors. In brain tumors of the semibenign type or even in "benign" tumors with strong tendency towards recurrence, e.g. in some intracranial and spinal meningiomas, the principles of neuro-oncology are sometimes still completely disregarded. Postoperative irradiation is still the strongest weapon to increase the recurrence-free interval, survival time and often survival quality. For malignant brain tumors so far no glioma-specific cytostatics exist. In brain tumor chemotherapy, BCNU at present appears the best of all partially effective approaches (Jellinger 1987; Potthoff 1985, 1988b).

Hydrocephalus

In our endeavor to help hydrocephalic newborns, we have learned in some - albeit rare - cases that "If there is no brain, it may still develop - if helped by valve decompression!" This involves "hydranencephaly" at birth. In these children sometimes a fair brain mantle may develop after adequate decompression with a valve. To this end, a recent advance after a long course of valve development appears to be the magnetically programmable Sophyvalve which, from our limited experience, also warrants proper adaptation in normal pressure hydrocephalus (NPH) in the aged.

Microvascular Decompression

This appears well established against trigeminal neuralgia (Burchiel et al. 1988), however, less so against hemifacial spasm and even less so still against spasmodic torticollis. The main problem still appears the exact preoperative diagnosis of the compressive cause. Up to now, the decision for microvascular decompression is still made almost entirely on clinical grounds, and its success is determined only by the postoperative course.

Stereotaxic Neurosurgery

Against the hopeful predictions in the late 1950s and early 1960s, stereotaxy has not brought about many deeper and programmatic insights into human neurophysiology and neurobiochemistry. In so-called "functional movement disorders" the major progress has been made, in fact, in neuropharmacology and neurochemistry. Functional stereotaxy is on the decline. Organic stereotaxy with tumor biopsy and implantations is somewhat on the advance. This helps increasingly in deep-seated, inoperable tumors still without cure.

Spine

Cervical Spinal Column

At the cervical column the uncomplicated unisegmental prolapsed disc is nowadays - with microtechniques - operated routinely by the anterior approach without difficulties and very good results. The discussion continues, however, as to the best technique and whether fusion is necessar or not.

Cervical spinal stenosis is much better recognized today as a cause of cervical myelopathy and more freely operated by dorsal decompression, with the important change that more and more neurosurgeons consider it favorable to perform this decompression by hemilaminectomy to conserve stable anatomy better and to forego scar constriction around the dural sac.

New methods of cervical stabilization by anterior steel plate spondylodesis - replacing posterior fixation techniques to a large extent - are now well established for optimal decompression, realignment and stabilization in the cervical spine in cases of traumatic, degenerative (Fig. 1) and - partially - neoplastic dislocations (Potthoff 1988a).

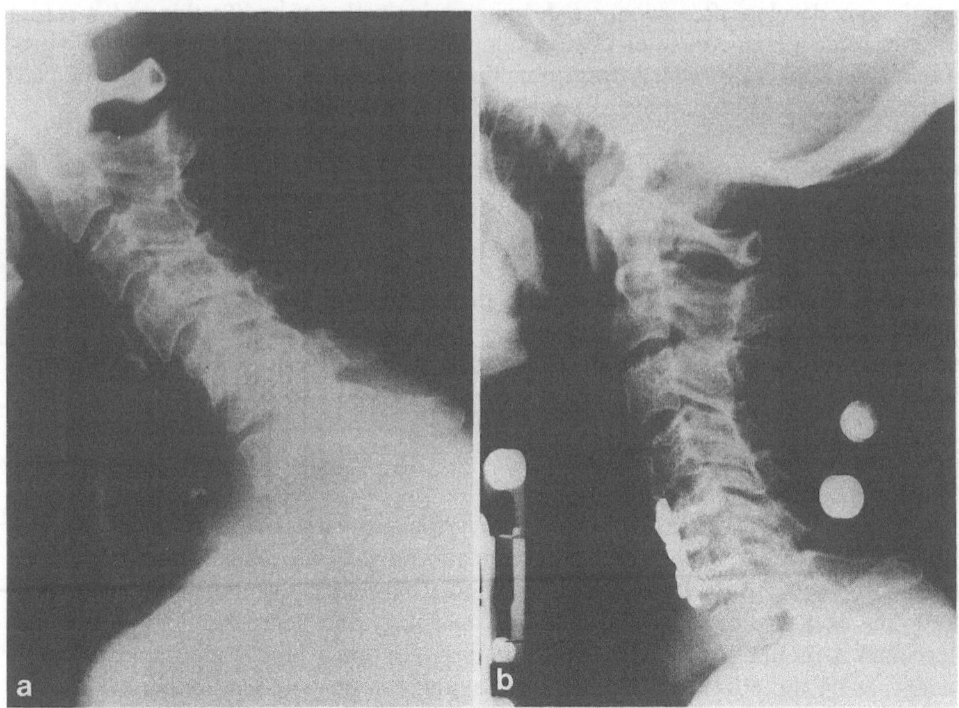

Fig. 1a. Degenerative (osteoporotic) spontaneous dislocation C 5 over C 6 with myelopathy

Fig. 1b. Same patient after operation with microneurosurgical decompression C5/6 after Cloward, realignment, autologous bone dowel and anterior steel plate spondylodesis C5/6 (central screw holds bone dowel) with partial remission of myelopathy

Spinal Tumors and Malformations

In spinal tumors meningiomas and neurinomas have not presented any major problems for a long time - since the advent of microneurosurgery. Spinal metastases remain a permanent problem. Rare large cauda neurinomas and ependymomas hold problems because of their often infiltrating character. In spite of NMR imaging and LASER application (Neuss et al. 1988), nothing new can be said definitely about intramedullary gliomas, spinal AV malformations and syringomyelia. In spite of careful microneurosurgery these remain rare and daring operations with varying outcome. And with AV malformations of the spinal cord (as well as of the brain) interventional neuroradiology has also not brought the definite answer.

Lumbar Spine

Microneurosurgery of the lumbar prolapsed disc is to our opinion obligatory. Although it may not have improved the overall results in the best hands with macrosurgical techniques of former times, the avoidence of grave neural trauma and CSF fistulae is greatly improved by microtechniques at the lumbar disc herniation. By the same token, spinal muscle and bone is much better preserved, problems in postoperative micturition have become extremely rare, and early remobilization is greatly enhanced.

Lumbar spinal stenosis, including the syndrome of lumbar redundant nerve roots, is much better recognized today and operated with better knowledge and success.

Stabilization of dislocations at the lumbar and thoracolumbar spine is in good progress by applications of the fixateur intern after Dick or after Kluger but still holds problems.

A Few Technical Conclusions

Some little technical facets are easily overlooked that brought considerable progress in neurosurgery.

For instance, the vacuum drainage (Redon) for trephination sites has almost completely eliminated postoperative CSF fistulae. The application of local antibiotics (e.g. Nebacetin) has replaced the regular vast use of systemic antibiotics in neurosurgery of earlier times. In the past two years our department has not had any wound infection in almost 1000 operated lumbar disc patients under a regimen of regular local antibiotic wound instillations. A great variety of aneurysm clips are offered today for proper and technically feasible clipping of aneurysms of all forms and sizes. Endoscopic neurosurgery is a new (resurrected) method, and indeed it appears feasible that evacuation of intracerebral hematomas or deep-lying cysts can be well performed with this technique under local anesthesia in critical cases. In this context, percutaneous discectomy is just emerging. Its success remains to be determined, and very critical observation is required especially since a list of indications just published for percutaneous lumbar discectomy postulates the very same criteria as published previously for lumbar disc chemonucleolysis - which has now almost disappeared after its great popularity a few years ago (Mayer and Brock 1988).

In general, at present there remains doubt as to whether LASER, CUSA, intraoperative EPs, intraoperative Doppler, etc. are absolutely necessary in everyday neurosurgical routine so far. It has still to be proven that these make neurosurgical procedures safer. They certainly do not make them shorter, easier, or cheaper. At present, we subscribe to the view that "It is not the LASER or the CUSA, it is still the skill of the neurosurgeon's hands" that remain essential for the operative result in neurosurgery.

Summary

The state of the art of neurosurgery in 1988 is commented upon with general and critical statements concerning standardized procedures on the brain and spine. Techniques and applications under advancement (e.g. brain grafting, pain pumps, photoirradiation) and peripheral nerve surgery are not discussed.

References

Alberico A M, Ward J D, Choi S C, Marmarou A, Young H F (1987) Outcome after severe head injury. J Neurosurg 67: 648-656
Auer L M (ed) (1985) Timing of aneurysm surgery. de Gruyter, Berlin
Auer L M, Brandt L, Ebeling U, Gilsbach J, Groeger U, Harders A, Ljunggren B, Oppel F, Reulen H J, Saeveland H (1986) Nimodipine and early aneurysm operation in good condition SAH patients. Acta Neurochir (Wien) 82: 7-13
Burchiel K J, Clarke H, Hagelund M, Loeser J D (1988) Long-term efficacy of microvascular decompression in trigeminal neuralgia. J Neurosurg 69: 35-38
Fasano V A (ed) (1986) Advanced intraoperative technologies in neurosurgery. Springer, Vienna New York
Gaab M R (1986) Kortikosteroid-Therapie beim Schädel-Hirn-Trauma In: Jahrbuch der Neurochirurgie. Regensberg and Biermann, Münster, pp 13-30
Jellinger K (ed) (1987) Therapy of malignant brain tumours. Springer, Wien New York.
Klein H J (1982) Kontinuierliche Hirndruckregistrierung:Teil jeder Hirndrucktherapie? Klinikarzt 11: 520-527
Mayer H M, Brock M (1988) Die perkutane Diskektomie. Dtsch Aerztebl 85: 632-637
Neuss M, Winkler D, Herrmann H D (1988) Intramedulläre Tumoren laser-chirurgisch behandelt. Dtsch Aerztebl 85: 1463-1466
Pluchino F, Broggi G (eds) (1988) Advanced technology in neurosurgery. Springer, Berlin Heidelberg New York
Potthoff P C (1985) Hirntumoren: Neurochirurgie heute. MMW 127: 1061-1065
Potthoff P C (1988a) Osteosynthetische Stabilisierungsoperationen der Halswirbelsäule durch ventrale Stahlverplattung. In: Kozuschek W (ed) Aktuelles in der Chirurgie. TM, Hameln, pp 63-67
Potthoff P C (1988b) Maligne Hirntumoren. In: Herfarth C, Schlag P (eds) Richtlinien zur operativen Therapie maligner Tumoren. Demeter, Gräfelfing, pp 133-139
Richter H P, Klein H J, Schäfer M (1984) Chronic subdural haematomas treated by enlarged burr-hole Ccaniotomy and closed system drainage. Acta Neurochir (Wien) 71: 179-188
Samii M (ed) (1986) Surgery in and around the brain stem and the third ventricle. Springer, Berlin Heidelberg New York
Seeger W (1980) Microsurgery of the brain 1 and 2. Springer, Vienna New York
Seeger W (1985) Differential approaches in microsurgery of the brain. Springer, Vienna New York
Sugita K (1985) Microneurosurgical atlas. Springer, Berlin Heidelber New York
Walter W, Brandt M, Brock M, Klinger M (eds) (1988) Modern methods in neurosurgery. Adv Neurosurg 16
Yasargil M G (1984) Microneurosurgery (4 vls.) Thieme, Stuttgart

Transcranial Doppler Evaluation of Cerebral Hemodynamics in Carotid Artery Occlusions

B.Widder
Neurologische Universitätsklinik Ulm, 7900 Ulm, FRG

Introduction

EC/IC bypass surgery is not effective in patients with an occluded ICA who are selected for surgery on the basis of clinical symptoms (EC/IC Bypass Study Group 1985). From the pathophysiological point of view this is not surprising because the establishment of an additional collateral pathway would help only if the circle of Willis and the leptomeningeal anastomoses are not sufficiently developed.

Quantitative information about the effiency of the collaterals can be obtained by investigating the residual autoregulatory capacity (Frackowiak 1986; Powers et al. 1986). The increase in cerebral blood flow during hypercapnia (CO_2 reactivity) gives a measure of the capability of the intracerebral arterioles to dilate further. This can be investigated by radionuclide technique (Diamox test), or more easily and less expensive by transcranial Doppler sonography (Widder 1985).

Methods

The Doppler CO_2 test is based on transcranial Doppler sonography which enables the detection of blood flow velocity in the middle cerebral artery (MCA) and other basal cerebral vessels by ultrasound insonation through the temporal bone (Widder 1987). Time-averaged MCA blood flow velocity is monitored continuously together with end-tidal pCO_2 measured by an infrared analyzer during steady states of normocapnia, moderate hypercapnia induced by breathing a mixture of 5% CO_2 in 95% O_2, and hypocapnia induced by voluntary hyperventilation (Widder et al. 1986).

Normally the CBF pCO_2 curve is S-shaped with a linear relationship between MCA flow velocity and pCO_2 in the physiological pCO_2 range between 30 and 50 mmHg (Fig. 1A) (HARPER and GLASS 1966). In pathological cases with decreased autoregulatory reserve, the curve is shifted to the left, and its upper bend is already reached within physiological pCO_2 values (Fig. 1B,C). To assess these changes in a standardized way a blood pCO_2 of 40 mmHg is taken as a reference point and the increase/decrease of flow during hyper- and hypocapnia of 1 vol.% CO_2 is calculated as normalized autoregulatory reserve" (NAR). For clinical routine three categories of NAR can be distinguished (Table 1).

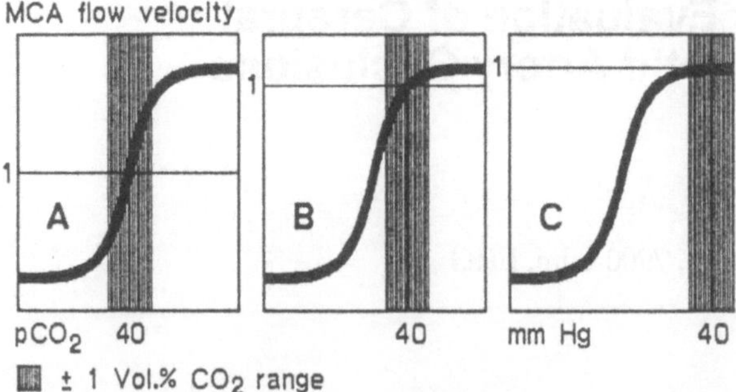

Fig. 1. Relationship between MCA flow velocity and pCO_2. **A** Normal NAR with linear relationship in the physiological range. **B** Decreased NAR with missing or diminished increase in flow during hypercapnia. **C** Exhausted NAR with missing or diminished change in blood flow during hyper- as well as hypocapnia

Table 1. Interpretation of the Doppler CO_2 test

		Change in MCA velocity referred to baseline values at 40 mmHg pCO_2	
Category	Auto-regulatory reserve	Increase during hypercapnia (+ 1 vol.% CO_2)	decrease during hypocapnia (- 1 vol.% CO_2)
A	Normal	> 10%	> 10%
B	Decreased	-	> 10%
C	Exhausted	-	-

A total of 150 patients with 162 ICA occlusions were investigated by the Doppler CO_2 test. Of these, 72 were asymptomatic; 90 patients had suffered at least one ipsilateral TIA and/or stroke. Seventy-six of them entered a nonrandomized prospective study with a mean follow-up of 16.6 months.

Results

Figure 2 shows the relationships between NAR and the clinical findings in the 162 cases. The neurological symptoms were subdivided into recent events and those in which the time interval between the (last) onset of symptoms and the Doppler CO_2 test was more than 3 months. Nineteen of 29 patients (66%) with exhausted autoregulatory reserve had suffered at least one stroke or TIA during the last 3 months. Of the 33 with decreased CO_2- reactivity 14 (42%) were symptomatic, whereas in 100 cases with sufficient NAR acute cerebral ischemia had occured in only 29 (29%). The relation between low NAR and recent cerebral ischemia was highly significant (chi square test, $p<0.001$). No relationship could be found in cases with remote ischemic symptoms.

The spontaneous course of 67 patients with ICA occlusions is shown in Fig. 3. In the group with normal or decreased NAR three of 54 patients (6%) suffered (further) ip-

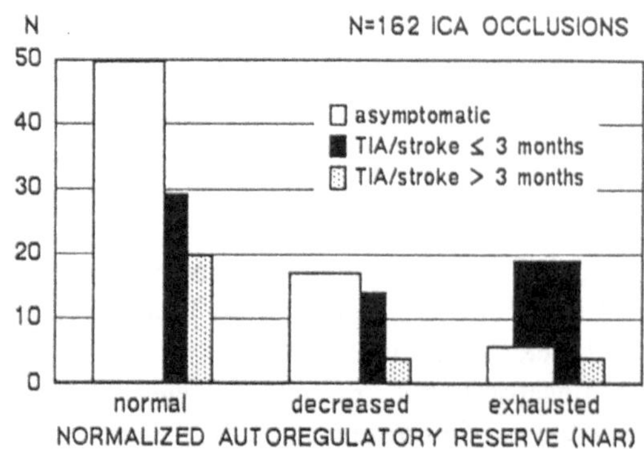

Fig. 2. Correlation between the normalized autoregulatory reserve (NAR) and ipsilateral clinical findings in 162 ICA occlusions

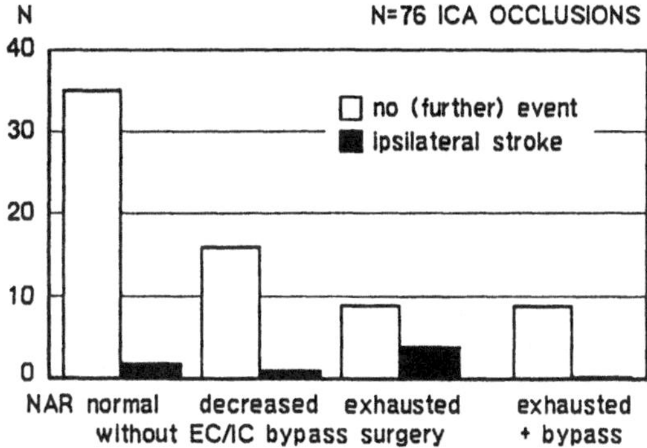

Fig. 3. Long-term follow-up in 76 patients with ICA occlusions in relation to the ipsilateral autoregulatory reserve

silateral ischemic events during follow-up, whereas 4 of 13 cases (31%) with exhausted autoregulatory reserve became symptomatic. The difference, however, is barely not significant ($p = 0.06$). Nine patients with exhausted NAR had undergone EC/IC bypass surgery on the ipsilateral side. None of them suffered a stroke either perioperatively or during the follow-up.

Discussion

Cerebral ischemia in ICA occlusion can be caused by a considerable number of nonhemodynamic mechanisms such as pre-, peri- and postocclusive embolism. EC/IC bypass surgery, however, could be helpful only in the relatively rare case of a persistent insufficient collateralization. This hypothesis is supported by our results. Only 18% of our ICA occlusions showed an exhausted autoregulatory capacity. These patients had suffered more often from a recent ipsilateral TIA or stroke, suggesting that an additional, probably hemodynamic mechanism could be responsible for this discrepancy.

In the prospective study ischemic events occurred more often in untreated cases with exhausted autoregulatory reserve. The lack of statistical significance could be due to the short follow-up time and needs further evaluation. There is, however, a striking difference in the hemodynamic group between cases with and without EC/IC bypass surgery. Due to the small number of patients and the nonrandomization these results do not confirm but do give a first hint that EC/IC bypass surgery may be helpful in properly selected patients.

Because further parameters, such as the oxygen extraction rate (Powers et al. 1986) and the maximum expected drop in blood pressure, must be considered in the individual case, the Doppler CO_2 test may enable only the selection of patients not in need of EC/IC bypass surgery (Widder 1989). In our series, this concerned more than 80% of all ICA occlusions. In those remaining, positron emission tomography could contribute to identifying those patients in whom surgery may be effective.

Summary

The transcranial Doppler CO_2 test offers a simple method of investigating the residual autoregulatory reserve in internal carotid artery (ICA) occlusions which gives quantitative information about the efficiency of the collaterals. In 162 ICA occlusions a significant correlation between a markedly decreased or exhausted autoregulatory reserve and recent ipsilateral ischemic events could be found. Of 76 patients investigated prospectively over a 16-month period, untreated cases with exhausted autoregulatory reserve developed strokes more often than well collateralized ICA occlusions and cases treated with an extracranial-intracranial (EC/IC) bypass. The results suggest that the Doppler CO_2 test can exclude patients not in need of EC/IC bypass surgery.

References

EC/IC Bypass Study Group (1985) Failure of extracranial intracranial arterial bypass to reduce the risk of ischemic stroke. Results of an international randomized trial. N Engl J Med 313:1191-1200

Frackowiak RSJ (1986) PET scanning: can it help resolve management isssues in cerebral ischemic disease ? Stroke 17:803-807

Harper AM, Glass HI (1965) Effect of alterations in the arterial carbon dioxide tension on the blood flow through the cerebral cortex at normal and low arterial blood pressures. J Neurol Neurosurg Psychiatry 28:449-452

Powers WJ, Press GA, Grubb RL, Gado M, Raichle ME (1986) The hemodynamic effects of carotid stenosis on the cerebral circulation. Stroke 17:127

Widder B (1985) Der CO_2-Test zur Erkennung hämodynamisch kritischer Carotisstenosen mit der transkraniellen Doppler-Sonographie. Dtsch Med Wochenschr 110:1553

Widder B (ed) (1987) Transkranielle Dopplersonographie bei zerebrovaskulären Erkrankungen. Springer, Berlin Heidelberg New York

Widder B (1989) The Doppler CO_2 test to exclude patients not in need of EC/IC bypass surgery. J Neurol Neurosurg Psychiatry 52:38-42

Widder B, Paulat K, Hackspacher J, Mayr E (1986) Transcranial Doppler CO_2-test for the detection of hemodynamically critical carotid artery stenoses and occlusions. Eur Arch Psychiatry Neurol Sci 236:162-168

Sonography Through the Anterior Fontanelle in Newborns: An Efficient Method for Screening Pre- and Perinatal Lesions

M.Heibel, R.Heber, D.Bechinger, J. M.Hufnagl and H.H.Kornhuber
Neurologische Universitätsklinik Ulm, 7900 Ulm,FRG

Methods and Results

From July 1986 to March 1987, 1000 full-term newborns at the Gynecological University Hospital of Ulm were routinely examinated with transfontanellar neurosonography (5 MgH; sector scan) three days post partum. The examination was depicted in coronal (c.s.) and sagittal (s.s.) sections (Fig.1) and documented (videoprint/video). The patients belonged to a selected sample because newborns with any type of perinatal distress were not included. The results detectable by neurosonography are listed in Table 1.

Conspicious results, such as hemorrhages, were controlled once 8 days after or more often. Out of the group of perinatal hemorrhages (Table 1) 4 (0.4%) patients with IVH developed clinical symptoms 3-7 months after the 1st examination but within period of the following control-investigations. One newborn with right-sided SEH + IVH at first examination (Fig. 2) developed infantile spasms (hemihypsarrhythmia, right hemisphere) 7 months later.

Three newborns with left-sided IVH (Fig. 3) developed slight right-sided hemiparesis at the age of 3-6 months. With physiotherapy the neurological symptoms diminished.

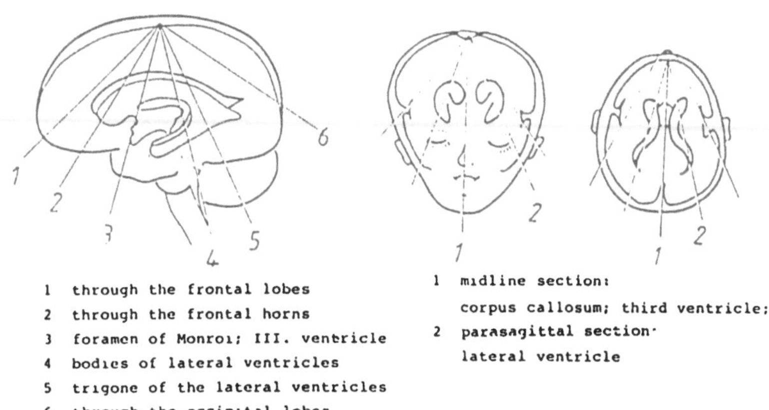

Fig 1 .Scan sections. **A** Coronal sections (C.S.). **B** Sagittal sections (S.S.)

Table 1. Results of transfrontellar neurosonography examinations (n = 1000)

Intercranial hemorrhage			
1. Periventricular hemorrhage			
Intraparenchymal h.	(IPH)	4	(0.4 %)
Subependymal h.	(SEH)	9	(0.9 %)
Chorioid-plexus h.	(CPH)	2	(0.2 %)
2. Intraventricular hemorrhage (IVH)			
ocurred only combined with:			
Chorioid-plexus h.	(CPH/IVH)	9	(0.9 %)
Subependymal h.	(SEH/IVH)	11	(1.1 %)
	Total	35	(3.5 %)
Morphological aberrations			
1. Symmetric and asymmetric enlargement of lateral ventricles			
	left = right	68	(6.8 %)
	left > right	154	(15.4 %)
	left < right	94	(9.4 %)
2. Cavum septi pellucidi			
	Q ≤ 5 mm	338	(33.8 %)
	Q > 5 mm	7	(0.7 %)
3. Cavum vergae		8	(0.8 %)
4. Cavum veli interpositi		2	(0.2 %)
5. Enlarged cisterna magna		11	(1.1 %)
6. Hyperechogenic tumor		1	(0.1 %)
	Total	683	(68.3 %)
Possible sequelae of SEH and IHV			
1. Subependymal pseudocysts		5	(0.5 %)
2. Circumscribed dilatation of lateral ventricle (3 additional cases following manifest IHV)		2	(0.2 %)
3. Chorioid-plexus cysts			
	Q ≤ 3mm	24	(2.4 %)
	Q > 3mm	3	(0.3 %)
	Total	34	(3.4 %)

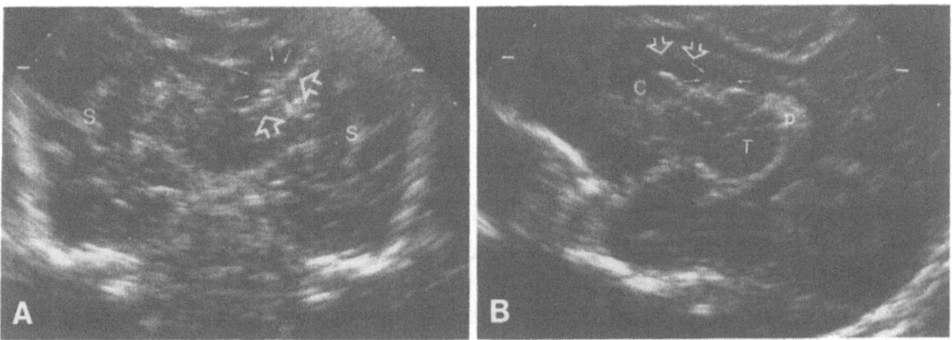

Fig. 2. First examination 3 days post partum. **A** C.S.: Circumscribed dilatation of the right lateral ventricle with a thrombus (small arrows) caused by SEH/IVH.**B** S.S.: Hemorrhage at the level of the thalamocaudate notch (c = nucleus caudatus; T = thalamus)

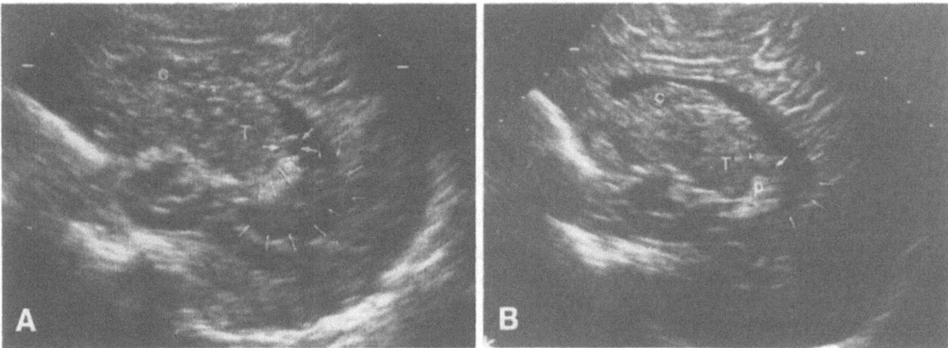

Fig. 3. A First examination (s.s.) shows dilatation of the left occipital horn with a thrombus (smalle arrows) attached to the glomus chorioidei (P; 3 days post partum). **B** Fifth examination (s.s.) shows normal ventricular configuration; thrombus reabsorbed (5 months later)

Discussion

There are only a few reports about transfontanellar neurosonography of clinically inconspicuous full- term newborns (Schuhmacher et al. 1983; Brand et al. 1984; Hayden et al. 1985; Rempen et al. 1986). The type of findings does not differ much, but there is quite a difference concerning incidence and degree. The main interest is on intracranial hemorrhage, of which the incidence ranges from 2.7% (Schuhmacher et al. 1983) to 5.5% (Rempen et al. 1986). We found 3.5%. So far no follow-up, as is described here, has been carried out by other authors. Our results show that the investigations of the babies' responses as used routinely so far do not reveal all cases of perinatal injury to the brain. Since even in adults there are "silent" lesions, and since the babies cannot report on hemihypesthesia, etc., we must expect this by clinical investigations. It is not possible to reach early diagnosis of cerebral lesions in all newborns . Routine sonography of the brain in all newborns may thus help to give these children early treatment without undue loss of time. The conclusion is that sonography, being an harmless method, should be applied to all newborns, not only to the babies at risk, because the earlier the treatment, the more effective it is.

Summary

A total of 1000 apparently healthy full-term newborns of the Gynecological University Hospital of Ulm were examinated by neurosonography (5 MgH; sector scan). Examinations were performed 3 days post partum in coronal and sagittal sections through the anterior fontanelle. Newborns with any clinical signs of perinatal distress were not included. Of these "healthy" newborns 35 showed perinatal intracerebral or intraventricular hemorrhages, 34 had findings which could be former hemorrhages, and 21 had other morphological findings of doubtful pathological meaning. These results were controlled after 8 days or more often. Four of the infants later developed clinical symptoms (3 hemiparesis; 1 West's syndrome).

References

Brand M, Rühe B, Steil ML, Gesche J, Saling E (1984) Die Bedeutung eines Schädel-Ultraschall-Screenings bei Früh- und Neugeborenen. In:von Dudenhausen JW, Saling E (eds) Perinatale Medizin,10th edn.Thieme,Stuttgart, pp 312-313

Hayden DK, Shattuck KE, Richardson CJ,Ahrendt DK,House R,Swischuk LE (1985) Subependymal germinal matrix haemorrhage in fullterm-neonates.Pediatrics 75:714-718

Rempen A, Feige A, Fiedler K (1986) Häufigkeit von auffälligen Ultraschallbefunden des Gehirnes bei klinisch asymptomatischen Neugeborenen.Z Geburtshilfe Perinatol 190:190-195

Schuhmacher R, Reither M, Ringel M, Jensen A (1983) Ergebnisse der Hirnsonographie als Screeningmethode bei Neugeborenen. In:von Haller U,Wille L (eds) Diagnostik intrakranieller Blutungen beim Neugeborenen. Springer, Berlin Heidelberg New York, pp 118-128

The Severity of Convulsive Behaviour in Rats: Disinhibitory Effects of Cortical Lesions

E. Horn

Abteilung für Neurologie, Universität Ulm, Oberer Eselsberg, 7900 Ulm, FRG

The inhibitory neurotransmitter GABA is widely distributed within the cerebral cortex (Ottersen and Storm-Mathisen 1984). It mediates the self-inhibitory action within the cortex, thus suppressing and stabilizing cortical activity. A decrease in the intrinsic inhibitory action increases the excitability of the cortex. Due to the widespread intra- and interhemispheric connections (Brodal 1981; Praxinos 1985), substantial lesions not only counteract the stabilizing mechanisms used for the balance of cortical activity but also decrease the amount of intrinsic cortical inhibition, leading to more pronounced cortical and/or behavioural defects during periods of cerebral disturbances, even if the lesioned area is far away from the area responsible for perception or behavioural response.

A well-known disturbance is an epileptic focus which can be induced in rats by a local application of penicillin (PCN) into the cerebral cortex (Speckmann 1986). A PCN focus located, for example, in the foreleg field of the motor cortex of one hemisphere induces trains of jerks of the contralateral foreleg which are obviously Jacksonian seizures and are unequivocally correlated with the occurrence of large potentials in the ECoG. These trains of jerks and large discrete ECoG potentials are interrupted by generalized seizures of the tonic-clonic type which include movements of both forelegs. The tonic-clonic nature of these seizures can easily be recognized in the ECoG recording. Each seizure is followed by a longer period of silence during which no jerk or jerk potential can be observed (Fig. 1).

The elicitation of convulsive behaviour gives insight into the cerebral excitability which can be quantified by parameters such as the frequency of jerk potentials, the duration of seizures, or even by the duration of convulsive behaviour at all. These parameters can be influenced separately (Horn and Eßeling 1988) by additional cortical lesions which were produced by suction. It is therefore useful to determine the complete time course of the convulsive behaviour if several experimental groups, say rats with intact or lesioned brain, should be compared (Fig. 2).

The inspection of time courses of convulsive activity in rats treated with 500 IU Na-penicillin in the foreleg field of the right motor cortex reveals the most obvious differences between the intact and the lesioned animals. Generally, cortical lesions increased the duration of convulsive activity (Fig. 2). While in animals with an intact cortex convulsive activity usually stopped after 2 h, in animals with lesions in the non-motor cortex this activity could last nearly 10 h. Lesions in the motor cortex were not so effective. Furthermore, while the number of seizures for animals with intact cortex varied between 4 and 59 (mean 32; n = 7), animals with a lesion in the left motor cortex per-

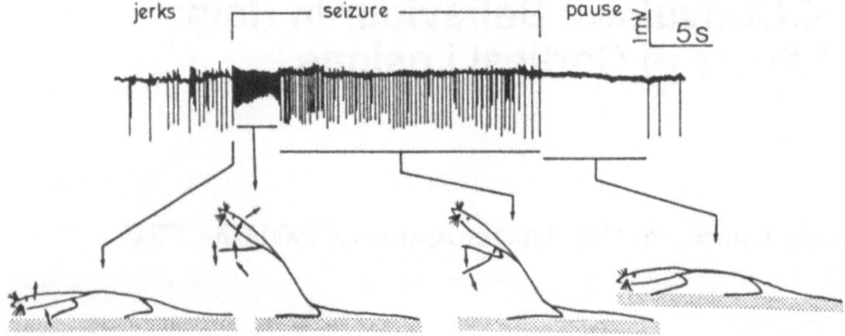

Fig. 1. Scheme of the convulsive behaviour of rats with a penicillin focus in the foreleg field of the right motor cortex, and the related electrocorticogram. The discrete sharp potentials correlate with jerks of the left foreleg. The ECoG during the generalized seizure shows the initial tonic component followed by the clonic period. Notice the compensatory pause following the seizure

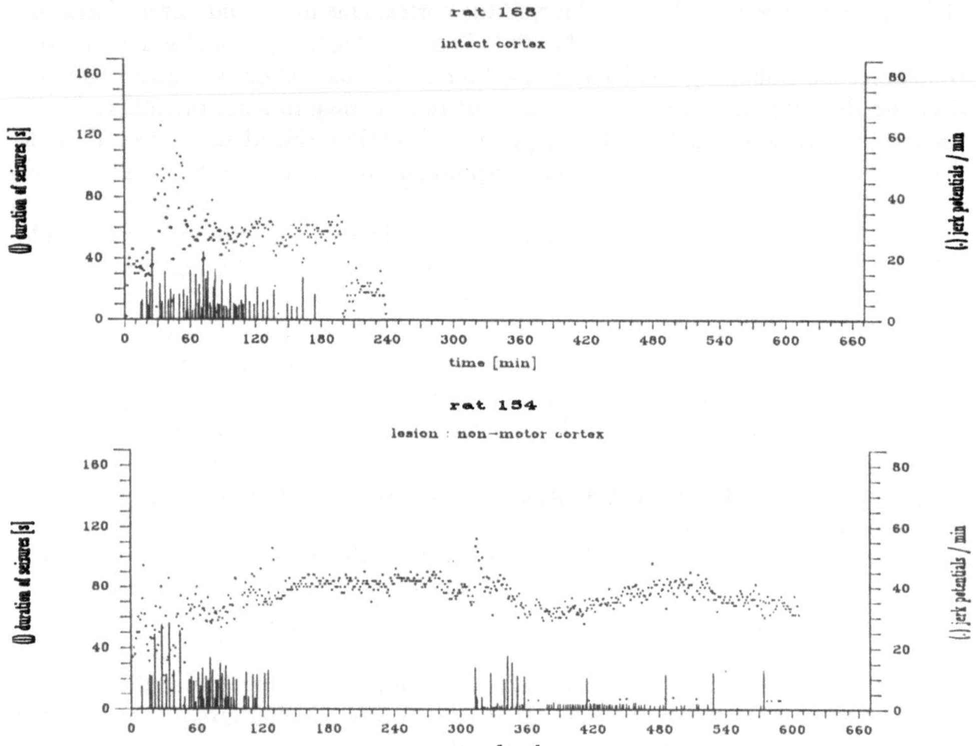

Fig. 2. Time course of convulsive behaviour in two rats with a penicillin focus in the foreleg field of the right motor cortex. Rat 168 had an intact cortex; rat 154 had a lesion in the left somatosensory cortex. Two parameters are presented: the frequency of jerk potentials (right ordinate, dots) and the duration of generalized seizures (left ordinate, vertical bars). Seizures are plotted at the time of their onset

formed between 46 and 386 (mean 160; n=6) and animals with a lesion in the non-motor area between 119 and 174 (mean 153; n=5) seizures. The differences between the intact and the lesioned animals were significant, at the $p<0.01$ level.

On the other hand, parameters reflecting basic mechanisms of convulsive action such as the mean frequency of jerks or the duration of generalized seizures were less effected by cortical lesions. The mean jerk frequency, which was rather constant throughout the experiment (Fig. 2), was 25.7/s for 7 rats with intact hemispheres, 33.8/s for 6 rats with a lesion in the motor field, and 36.3/s for 5 rats with a lesion in the non-motor, mainly somatosensory area. These slight differences were not significant. The mean seizure duration was 16.4 s in 7 rats with intact cortex, 11.2 s in 6 rats with a lesion in the motor and 15.4 s in 5 rats with a lesion in the non-motor cortex. Only the difference between the intact group and that with a lesion in the motor area was significant ($p = 0.05$).

Conclusions

Cortical lesions cause a higher risk for the development of seizures. This fact is well known in humans, in whom about 4% develop spontaneous seizures after head injury which is three times higher than in the general population (Desai et al. 1983). In animals, however, the development of spontaneous seizures is described rarely. In rabbits, cryogenic lesions induce seizures whose extent are directly correlated with the size of the lesion (Loiseau et al. 1987). On the other hand, neither Lashley (1967) nor Kolb and Whishaw (1983) who extensively investigated learning and movement abilities of rats with cortical damages gave any hint on the occurrence of spontaneous seizures. Lesions in the deep prepiriform cortex were also rather ineffective on the extent of amygdala kindling (Ludvig and Moshe 1987).

These facts, however, do not contradict the observation of the increased numbers of seizures after cortical damage elicited by a convulsive substance. One must distinguish between the effects of cortical damage on the mechanisms of convulsive actions, on the one hand, and those on the excitability of the neuronal tissue for epileptogenic disturbances, on the other. Mechanisms include synchronization processes which become obvious in the regularly occurring jerks and their time-correlated ECoG potentials, and processes responsible for the termination of a generalized seizure. Both parameters, however, are rather unaffected by the cortical lesion, independently of whether it is located in the motor or somatosensory area. On the other hand, the duration of convulsive activity is lengthened and also the number of generalized seizures. This obviously points to a disinhibition of the cerebral tissue caused by the decrease of intrinsic inhibition.

The cerebral cortex contains not only inhibitory GABAergic elements but also glutamatergic neurons which are excitatory. Substantial lesions should therefore affect both components. But in contrast to the GABAergic elements, glutamatergic neurons have predominantly output functions (Ottersen and Storm-Mathisen 1984), so that a substantial lesion decreases mainly the efficiency of the intrinsic inhibitory transmitter system while deficiencies caused by the decreased excitatory output can be compensated for at lower neuronal levels. The decreased amount of intrinsic inhibitory activity causes an overall disinhibition of the cerebral cortex because of the extensive intra- and interhemispheric connections (Brodal 1981; Kornhuber 1980), leading to the higher excitability against epileptogenic factors and to longer convulsive activity.

Lesions in the somatosensory area of the rats have a more pronounced disinhibitory effect on cortical activity than lesions in the motor area. This agrees with the observation of high epileptic risk in persons with damage in the parietal or temporal lobe. But it also reflects the high interconnectivity between this primary sensory projection area and the motor area (Kornhuber 1980).

References

Brodal A (1981) Neurological anatomy, 3rd edn. Oxford University Press, Oxford

Desai BT, Whitman S, Coonley-Hoganson R, Coleman TE, Gabriel G, Dell J (1983) Seizures in relation to head injury. Ann Emerg Med 12:543-546

Horn E, Eβeling K (1988) Quantitative Aspekte zum Modell der Penicillin-induzierten Epilepsie - ECoG- und Verhaltensmessungen. In: Speckmann EJ (ed) Epilepsie `87. Einhorn, Reinbek, pp 303-306

Kolb B, Whishaw IQ (1983) Dissociation of the contributions of the prefrontal, motor, and parietal cortex to the control of movement in the rat: an experimental review. Can J Psychol 37: 211-232

Kornhuber HH (1980) Physiologie und Pathophysiologie der corticalen und subcorticalen Bewegungssteuerung. Verh Dtsch Ges Neurol 1: 17-32

Lashley KS (1967) In search of the engram. Symp Soc Exp Biol 4: 454-481

Loiseau H, Averet N, Arrigoni E, Cohadon F (1987) The early phase of cryogenic lesions: an experimental model of seizures updated. Epilepsia 28: 251-258

Ludvig N, Moshe SL (1987) Deep prepiriform cortex lesion does not affect the development of amygdala kindling. Epilepsia 28: 594

Ottersen OP, Storm-Mathisen J (1984) Neurons containing or accumulating transmitter amino acids. In: Björklund A, Hökfelt T (eds) Handbook of chemical neuroanatomy, vol 3/2. Elsevier, Amsterdam, pp 141-246

Praxinos G (1985) Forebrain and midbrain. Academic Australia, Sydney (The rat nervous system, vol 1)

Prince DA, Connors BW (1986) Mechanisms of interictal epileptogenesis. Adv Neurol 44: 275-299

Speckmann EJ (1986) Experimentelle Epilepsieforschung. Wissenschaftliche Buchgesellschaft Darmstadt

Electrophysiology of Myotonias and Periodic Paralyses

R. Rüdel

Abteilung für Allgemeine Physiologie, Universität Ulm, Albert-Einstein-Allee 11, 7900 Ulm, FRG

Biopsied muscles from patients with recessive generalized (Becker type) myotonia, paramyotonia congenita, adynamia episodica hereditaria and familial hypokalemic periodic paralysis were investigated with the aim of studying the pathomechanisms underlying the episodes of muscle stiffness and/or muscle weakness. In all these disorders a genetic defect is associated with the passage of ions through the sarcolemma. Since several types of ion channels are affected in each disorder studied, it seems that the genetic defect influences the channels in an indirect manner.

In all of the above-mentioned diseases the clinical symptoms indicate an impaired mechanical performance of muscle, and yet in neither disorder does the contractile apparatus seem to be involved. The combination of thorough clinical examination and investigation of the electrical properties of the sarcolemma offered the most promising access to the elucidation of the pathomechanisms.

The clinical examination included an EMG, the performance of symptom-provoking tests providing quantitative information on the time course of the relevant parameters such as muscle force, serum potassium concentration, etc., and the administration of various drugs supposed to alleviate the symptoms. Many patients consented to having a biopsy taken from their external intercostal muscle. The biopsied muscle specimen was divided into several fiber bundles so that muscle force and EMG, membrane parameters such as resting and action potentials, component conductances, etc., and the intracellular sodium activity could be studied in different set-ups at the same time. These muscle properties were investigated under changing conditions, e.g. at various temperatures, extracellular potassium concentrations, and pH values.

As a major common result from the in vitro experiments, it can be stated that all the diseases investigated are genuinely myogenic, for in every case the typical symptoms could be elicited in the isolated muscles while the endplates were blocked with curare. Thus, on tetanic stimulation after a rest period, the muscle from the myotonia patient exhibited slowed relaxation and transient weakness, and with continued stimulation the typical warm-up phenomenon occurred; also, myotonic runs could be recorded with EMG electrodes. The muscles from paramyotonia patients showed substantial slowing of relaxation when stimulated in the cold, and with continued stimulation the force amplitude dropped to zero. And the muscles from the patients with adynamia episodica and hypokalemic periodic paralysis became paralyzed when they were exposed to extracellular solutions containing potassium in high (7-10 mM) or low (1-2 mM) concentration, respectively.

The second major outcome is the result that all these diseases seem to be membrane disorders with some defect in the passage of ions through the sarcolemma. It had already been shown by Lipicky and Bryant (1972) that the myotonia in myotonia congenita is mediated by a smaller than normal chloride conductance. We were interested to see whether this pathomechanism pertained to all myotonic diseases, and therefore we measured component membrane conductances in intercostal fibers dissected from patients, using the three-microelectrode voltage-clamp technique.

In recessive generalized myotonia, we found indeed that the chloride conductance is reduced, and that the steady-state current-voltage relationship of the fiber membrane is N-shaped (Rüdel et al. 1988). Such a characteristic curve with a region of negative slope favors electrical instability of the membrane. The typical transient weakness described in this disease may thus be explained as follows: after initial activity some fibers might assume the lower resting potential for quite a while before they are repolarized to the normal -80 mV and during this time they are inexcitable, i.e. paralyzed. Antiarrhythmic drugs with a long biological half-life, such as tocainide, are able to prevent the myotonic stiffness. They also reduce the duration, but not the extent, of myotonic transient weakness.

In paramyotonia congenita, the chloride conductance was found to be normal, but the sodium conductance is abnormal, the more so, the lower the temperature (Lehmann-Horn et al. 1987b). In the cold, some of the sodium channels seem to fail to inactivate, so that an increasing sodium current slowly depolarizes the muscle fibers to around -40 mV. During this cold-induced depolarization phase, the membrane produces spontaneous action potentials. These can be recorded in the EMG as low-frequency runs and are in part responsible for the paramyotonic stiffness. Most of the stiffness is still unexplained, though it might be related to an additional abnormality of the inner membrane systems of the muscle fibers (Ricker et al. 1986). When the fibers are in the depolarized state, they are inexcitable. This explains the paramyotonic weakness. At the low resting potential, the muscle fibers have a high chloride conductance and gain intracellular chloride. The slow restoration of the correct intracellular ionic milieu is probably the reason for the slow recovery of strength after rewarming of a cooled muscle. Tocainide is an excellent drug for the prevention of paramyotonic stiffness and weakness.

In adynamia episodica hereditaria, also called hyperkalemic periodic paralysis, we also detected an abnormal inactivation of the sodium channels. In this disease, the failure of the sodium channels to close properly is induced by long-lasting membrane depolarization rather than by low temperature (Lehmann-Horn et al. 1987b). The depolarization may be started by the rise in extracellular potassium concentration that occurs in these patients during rest, particularly in the morning after an exhaustive work load. The non inactivating sodium current then perpetuates the depolarization until the extracellular potassium concentration is decreased by increased activity of the muscular sodium/potassium pumps and by renal potassium excretion. In muscles from adynamia patients, the depolarization produced by exposure to 7 mM K^+ is larger (resting potential $\approx$ -60 mV) than in normal muscles (-72 mV), but not as large as in cold-exposed muscles from paramyotonia patients (-40 mV). At the reduced resting potential, the excitability of the muscle fibers from adynamia patients was lower than in normal fibers at the same low resting potential, and this seems to be the reason for the paralysis. Drugs that reduce the excitability, such as tocainide, are therefore ineffective in preventing or releaving attacks of paralysis. The excitability may be recovered by a lowering of the extracellular pH. Addition of HCl to the bathing fluid or gassing the fluid with 40% CO_2 is thus very effective in preventing the in vitro paralysis that would occur on elevation of the extracellular potassium concentration. In vivo, this might be the principle of the beneficial effect of diuretics such as acetazolamide.

Adynamia episodica patients also suffer from muscle weakness when the muscles are exposed to the cold, and for this reason adynamia episodica and paramyotonia congenita were claimed to be nosologically identical (Layzer et al. 1967). The pathomechanism is, however, different in the two diseases. When muscles from adynamia patients are cooled to 27°C, the resting potential remains at -80 mV, but the action potential becomes very small. In some fibers that we had tested in the cold, action potentials could not be elicited at all. In addition, in muscles from adynamia patients, the mechanical threshold, as determined with potassium contractures, is increased in the cold so that the small action potentials do not surpass it and do not trigger any mechanical response (Ricker et al. 1989). In normal muscles the mechanical threshold does not increase with decreasing temperature. The different pathomechanisms are also reflected in the speed of recovery from cold-induced paralysis. In adynamia episodica, in which the resting potential remains high throughout the cooling period, excitability and force return immediately on restoration of normal temperature. By contrast, in paramyotonia, in which the low resting potential is associated with changes of the intracellular ionic milieu, it takes hours for the rewarmed muscle to regain its strength.

In familial hypokalemic periodic paralysis the resting potential is by about 10 mV lower than normal even under normal ionic conditions (3.5 mM extracellular potassium), and the excitability is reduced (Rüdel et al. 1984). When the extracellular potassium concentration falls, e.g. after administration of glucose and insulin, both parameters are further decreased so that weakness or paralysis ensues. For the reduced resting potential there are two alternative explanations: the potassium conductance could be too low or the sodium conductance too high. Conductance measurements with the voltage-clamp technique performed in chloride-containing and chloride-free media with normal and low potassium content did not yield substantially abnormal values. This would favor the second alternative for a basic defect in hypokalemic periodic paralysis. The reason for the reduced excitability is probably the reduced resting potential.

Since in three of the studied diseases the properties of the sodium channel turned out to be altered, we have turned to investigating the sodium channels in muscles from these patients with the help of the modern patch-clamp techniques. To this purpose, we have cultured myoballs, spherical muscle cell regenerates, from muscle biopsied from patients. The transient sodium currents flowing during the depolarization of the membrane of such myoballs were investigated in the whole-cell mode of patch clamping. Special features of such not fully matured muscle cells are the lack of a substantial chloride conductance and the existence of two different types of sodium channel, namely the adult type, which is the only type existing in mature skeletal muscle, and the juvenile type known to exist in immature muscle such as myotubes and also in denervated muscle. The two types of sodium channel can be pharmacologically differentiated by their different affinity to the sodium channel blocker tetrodotoxin (TTX), the juvenile channels being an order of magnitude less sensitive to TTX. The voltage dependence and the time constants of activation and inactivation of the currents through these sodium channels were analyzed according to Hodgkin and Huxley (1952).

This analysis was performed with myoballs cultured from biopsies from volunteers and from patients with myotonic dystrophy, recessive generalized myotonia and adynamia episodica (Rüdel et al. 1988). The voltage dependence of activation is characterized by the position of the point of inflection of the m_{∞} curve. This was at -29 ± 6 mV for the controls and at -27 ± 5, -32 ± 6, and -35 ± 5 mV, respectively, for the three above diseases. The voltage dependence of the inactivation was normal for the three diseases, except for a slight abnormality of the TTX-resistant juvenile sodium channels in adynamia episodica. The time constants of activation and inactivation, τ_m and τ_h, were significantly different from control in each disease. In myotonic dystrophy, the time constants were increased by up to 25%; in the other two diseases they were decreased by about the same amount. The time constants are voltage-dependent parameters. The slope of this voltage dependence was steeper than normal in myotonic dystrophy and less steep in the other two diseases. The observed abnormalities would favor the occurrence of myotonia in the cases of recessive generalized myotonia and adynamia episodica since a reduced steepness of the voltage dependence of τ_h could result in noninactivation of part of the sodium current (because the rate constant for the closing of the sodium channel increases less steeply with decreasing membrane polarization). In contrast, the described abnormalities in myotonic dystrophy would antagonize myotonia.

Previous investigations of the muscular ion channels had revealed different abnormalities for these three diseases. In myotonic dystrophy, an abnormal persistence of apamine-sensitive potassium channels into the innervated stage of the muscle was reported (Renaud et al. 1986). As described above, in recessive generalized myotonia, the chloride conductance of the sarcolemma is reduced. Study of the myoballs now revealed abnormalities of the sodium channels for both diseases, and since in adynamia episodica both adult and juvenile channels were abnormal, at least two channel types seem to be affected in the three diseases. These results suggest that in myotonic diseases the genetic defects are not directly concerning the channel proteins. The lipid environment or an enzyme acting on the channel proteins at some time between their production and their integration into the membrane could be altered instead. We have therefore started to study the composition of the lipids and fatty acids in the membrane of erythrocytes of paramyotonia congenita patients. The most significant finding was an increase in the saturated fatty acids in paramyotonia (Marx et al. 1988).

The measurements of the steady-state parameters of the sodium channels in the myoballs are probably not accurate enough for any further conclusions. Although the point of inflection and the slope of these curves can be fairly exactly determined, the asymptotic parts in the potential range between -60 and -30 mV are rather inexact. This range, however, is most relevant for the stability of the membrane potential. The faster time constants in recessive generalized myotonia and in adynamia episodica indicate that the energy barriers that must be overcome for the opening or closing of the gates are smaller than usual. The smaller steepness of the potential dependence of these time constants results in a reduced probability of the gates to be in their final state during depolarization (m gates open, h gates closed) while repolarization could reopen the h gates too fast. That would allow sodium current to flow at the end of an action potential, and this in turn would lead to a new activation as soon as the potassium channels have closed. The biophysical reason for such changes might be a changed tertiary structure resulting from an abnormal environment of the channels.

Acknowledgements. The clinical studies and the experiments involving intercostal muscle biopsy, three-microelectrode voltage clamp and measurement of contractile properties in vitro were performed together with Dr. K. Ricker from the Neurologische Universitätsklinik, Würzburg, and Dr. F. Lehmann-Horn from the Neurologische Klinik und Poliklinik der Technischen Universität München. The experiments with human myoballs were carried out with Drs. J.P. Ruppersberg and W. Spittelmeister. This research was supported by the Deutsche Forschungsgemeinschaft and the Deutsche Gesellschaft zur Bekämpfung der Muskelkrankheiten.

References

Hodgkin AL, Huxley AF (1952) A quantitative description of membrane current and its application to conduction and excitation in nerve. J Physiol 17:500-544

Layzer RB, Lovelace RE, Rowland LP (1967) Hyperkalemic periodic paralysis. Arch Neurol 16:455-472

Lehmann-Horn F, Küther G, Ricker K, Grafe P, Ballanyi K, Rüdel R (1987a) Adynamia episodica hereditaria with myotonia: a non-inactivating sodium current and the effect of extracellular pH. Muscle Nerve 10:363-374

Lehmann-Horn F, Rüdel R, Ricker K (1987b) Membrane defects in paramyotonia congenita (Eulenburg). Muscle Nerve 10:633-641

Lipicky RJ, Bryant SH (1973) A biophysical study of the human myotonias. In: Desmedt JE (ed) New developments in electromyography and clinical neurophysiology. Karger, Basel, p 451

Marx A, Szymanska G, Melzner I, Rüdel R (1988) The membrane lipid and fatty acid composition of erythrocyte ghosts from three patients with paramyotonia congenita. Muscle Nerve 11:471-477

Renaud JF, Desnuelle C, Schmid-Antomarchi H, Hugues M, Serratrice G, Lazdunski M (1986) Expression of apamine receptor in muscle of patients with myotonic muscular dystrophy. Nature 319:678-680

Ricker K, Camacho LM, Grafe P, Lehmann-Horn F, Rüdel R (1989) Adynamia episodica hereditaria: what causes the weakness? Muscle Nerve 12: 833-891

Ricker K, Rüdel R, Lehmann-Horn F, Küther G (1986) Muscle stiffness and electrical activity in paramyotonia congenita. Muscle Nerve 9:299-305

Rüdel R, Lehmann-Horn F, Ricker K, Küther G (1984) Hypokalemic periodic paralysis: in vitro investigation of muscle fiber membrane parameters. Muscle Nerve 7:110-120

Rüdel R, Ricker K, Lehmann-Horn F (1988) Transient weakness and altered membrane characteristic in recessive generalized myotonia (Becker). Muscle Nerve 11:202-211

Rüdel R, Ruppersberg JP, Spittelmeister W (1989) Abnormalities of the fast sodium current in myotonic dystrophy, recessive generalized myotonia, and adynamia episodica. Muscle Nerve 12: 281-287

Protection Against Sudden Infant Death: Home Monitoring of All Infants During the First Year of Life by Means of the Babyprotector

K.A. Renner[1] and H.H. Kornhuber[2]

[1]Department of Neurophysiology, University of Ulm, Oberer Eselsberg, 7900 Ulm, FRG

[2]Department of Neurology, University of Ulm, Steinhövelstr. 9, 7900 Ulm, FRG

Introduction

SID in apparently healthy infants is the most common type of death of infants aged between 2 weeks and 1 year. In some countries the estimate is as much as four babies for every thousand born. The loss of life caused by SID is several times greater than the death toll in traffic accidents from birth through 15 years of age.

Suddenly and apparently in full health, the babies die while asleep. All available evidence points to a failure of respiration, not of the heart (Steinschneider 1972; for review see Naeye 1988). However, despite two decades of intensive research, it is not yet possible to identify the infants at risk. Probably, there are several predisposing factors, such as disorders of the fat metabolism and maternal cigarette smoking during pregnancy. Also, there may be triggering factors such as viral infections (for review see Naeye 1988); note however that the abnormally high blood levels of fetal hemoglobin observed in SID victims (Guilian et al. 1987) are more likely to result from prolonged periods of hypoxemia before death rather than reflecting a primary cause. Sudden death by respiratory failure is probably an age-specific weakness in the central respiratory regulation which, for practical reasons, cannot yet be detected by mass screening.

By comparison to the efforts and expenditure given to the protection of infants and children, the problem of SID has gone relatively unaddressed. Presently, aside from special prenatal care, e.g. no smoking, only home monitoring is a viable measure against SID in our view.

There are significant data regarding the effect of home monitoring for the prevention of SID, found in a cooperative study by the University of California. The study compared a group of 1835 home-monitored infants with a control group of 1569 unmonitored infants. The results indicate that the SID rate in the monitored group with functioning monitors was 10 times lower than in the unmonitored group. The incidence was 0.5 /1000 in the monitored group versus 5.7/1000 in the unmonitored group (Davidson-Ward et al. 1986).

Methods

Since it is impossible to identify most of the children at risk, our goal was to design a monitor applicable to all infants during the first year of life. For this purpose the device had to be reliable, easy to manage, safe and cost effective. We were unable to find a device on the market that would have met all these requirements. Consequently, we devised a compact apnea monitor utilizing modern technology; the device has been dubbed "Babyprotector."

The Babyprotector is book sized and battery operated. This is important for the acceptance of the device by parents. Since the cause of SID is a failure of respiration in most cases and since heart monitoring causes more discomfort and costs, we concentrated on breathing. The respiration is sensed by means of a small flexible loop of wire embedded in silicon material which is simply inserted between the diaper and the abdominal skin. The loop represents an inductivity which varies if the loop geometry undergoes slight changes caused by movements of the abdominal wall or of the diaphragm. There is no danger of skin irritation because nothing must be stuck to the skin. The Babyprotector is easy to apply, especially when compared to bathing, feeding and other care necessary for a baby. To heighten the overall safety of the device, all major functions of the Babyprotector, including its battery capacity, are supervised from an inbuilt controlling electronic system. Any detected malfunction is indicated by a continuous alarm tone.

A major problem of monitors is the occurrence of unjustified alarms. To avoid this problem, we have added a contrivance that attempts to reactivate breathing whenever a breathing pause becomes dangerously long, before giving acoustic alarm to the parents (Fig. 1). It consists essentially of a small vibrator, the awakening stimulator, which is placed in contact with the sole of the babies' foot. Its vibrations serve to reactivate the breathing or to change the sleep state by stimulation of the nervous system. Within this context it should be remembered that tactile stimuli were the rule for babies at times when they were carried all day by their mothers. Lying in a stationary bed is perhaps unphysiological for babies. The advantage of the vibratory stimulus is that it activates a

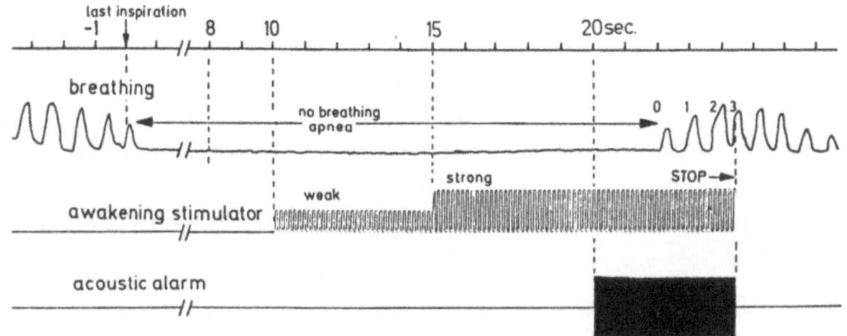

Fig. 1. Mode of action of the Babyprotector. This figure is an example of a 22 second period of apnea. 10 seconds after cessation of breathing the awakening stimulator starts with weak vibrations which increase 5 seconds later. Because in this example apnea persists, the vibrations continue, and after 20 seconds the alarm sounds for help. After onset of breathing and regular breaths, the awakening stimulator and the acoustic alarm shut down

great number of pacinian receptors in the infants leg without alarming the parents. The pacinian corpuscle is a low-threshold vibration receptor which may be activated even at distance when stimulated with about 150 Hz because the mechanical conductivity of the leg tissue is high at this frequency (Mountcastle et al. 1967; Talbot et al. 1968).

We inform all parents that the awakening stimulator is no guarantee for ending an apnea, and that they must take precautions to be near the baby to hear a possible alarm. We advise all parents to become familiar with resuscitation techniques. On request (e.g. hospitals or nurseries) we provide a wireless transmission of the alarm to the nurse or the parents.

There is no monitoring of the heart function by the Babyprotector. Heart monitoring would involve manipulations such as electrode pasting which are more time consuming and bear the risk of adverse skin reactions, etc. This would be counterproductive for a wide application of monitoring. We therefore require that in every newborn child the heart must be checked (which is the rule in West Germany).

Results

The Babyprotector has now been tested in more than 80 West German clinics with more than a thousand babies. Most users were satisfied. Physicians who were able to compare the Babyprotector with other devices stress its reliability. Nearly all of the users reported that it is routine for them to use the Babyprotector.

Conclusions

Presently, continuous respiratory monitoring of all infants during the first year of life appears to be the straightforward approach to prevent SID. Of course, research into the causal predisposition or triggering factors of SID should be continued even if the problem has been provisionally solved by monitoring all infants. Furthermore, attempts to reduce maternal cigarette smoking should be intensified. In this respect, a way to prevent many teenagers from starting to smoke would be to make cigarettes more expensive by putting a health levy on them (the revenues of which would go to the health insurance system).

Earlier studies with other, more complicated monitors reported that monitoring caused stress on family members and had psychological conseqences. Our present experience indicates that monitoring with the Babyprotector is without an adverse influence on family life.

Our advice is to equip all nurseries with monitors because some cases of SID occur already in the first days of life. Our preliminary results within a fully equipped nursery at Ulm University clinic with Babyprotectors suggest that the monitors also help the staff to supervise the newborns more easily. For this purpose the Babyprotectors are equipped with a wireless transmission of the alarm to a pocket receiver carried by the nurse. The mandatory use of the Babyprotector in the nursery also serves to accustom parents to the use of monitors and to accept continuous monitoring as their home task during the first year of their baby's life. It is important, however, to teach each parent how to carry out cardiopulmonary resuscitation.

The Babyprotector has been devised to protect full-term healthy infants. Whether the Babyprotector is also suited for monitoring preterm infants and infants with pulmonary dysfunction, or whether in these cases it may be necessary to monitor directly the oxygenation of the blood remains to be investigated.

Summary

The sudden infant death (SID) is the single most common type of death in the first year of life. Most of the victims are not identifiable as being at high risk prior to their death. Our aim is therefore to monitor all infants during their first year of life. To this end we devised an apnea monitor called Babyprotector which is reliable and easy to apply. It is battery-driven and has a vibration stimulator which attempts to reactivate respiration before giving acoustic alarm to the parents; the results are promising.

References

Davidson-Ward SL, Keens TG, Chan LS, et al. (1986) Sudden infant death syndrome in infants evaluated by apnea programs in California. Pediatrics 77: 451-458

Giulian GG, Gilbert EF, Moss RL (1987) Elevated fetal hemoglobin levels in sudden infant death syndrome. N Engl J Med 316: 1122

Mountcastle VB, Talbot WH, Darian-Smith I, Kornhuber HH (1967) Neural basis of the sense of flutter-vibration. Science 155: 597-600

Naeye RL (1988) Overview: sudden infant death syndrome, is the confusion ending? Mod Pathol 3: 169-174

Steinschneider A (1972) Prolonged apnea and the sudden infant death syndrome: clinical and laboratory observations. Pediatrics 50: 646

Talbot WH, Darian-Smith I, Kornhuber HH, Mountcastle VB (1968) The sense of flutter-vibration: comparison of the human capacity with response patterns of mechanoreceptive afferents from the monkey hand. J Neurophysiol 31: 301-334

EEG Spectra and Evoked Potentials to Words in Apallic Patients

J.Szirtes, V.Diekmann, A.Kuhwald, P.-J. Hülser, and R. Jürgens
Department of Neurology, University of Ulm, 7900 Ulm, FRG

Introduction

Evoked potentials have been studied mainly in acute comatose patients in order to derive prognostic and localizing values of evoked brain responses (e.g., Brunko and Zegers de Beyl 1987; Pfurtscheller et al. 1985). Many patients who survive coma enter into the "persistent vegetative state" (Plum and Posner 1980) or apallic condition (Kretschmer 1940) in which they are awake without any obvious sign of cognitive functions. There is no electrophysiological study to our knowledge which has used verbal stimuli in order to obtain evoked potential (EP) information about the residual cortical-subcortical "evaluative" functions, a rudimentary capacity which may still be preserved in spite of a widespread damage of the cortical-subcortical tissue, as manifested also in the diffuse atrophy seen in CCT scans of (incomplete) apallic patients. We therefore presented word stimuli to such patients and analyzed both the spectral EEG changes and the event-related potentials to look for signs of residual "cognitive" capacity of elementary nature.

Methods

Four patients with anoxic damage due to cardiorespiratory arrest and two after craniocerebral trauma were studied (median age 21 year). Clinically, all patients were in coma vigile, showed hypertonus and performed stereotype limb movements. Three patients seemed to cry and moan for discomfort, and in two of them some behavior suggested smiling. Neither of them, however, followed commands or showed a capability for communication. The time of recording after resuscitation ranged between 2 and 48 months, with a median of 12 months.

Two kinds of acoustic stimulus sequences were used. In one sequence two sinus tones with different probability (deep tone 0.67, high 0.33) were presented randomly. The second sequence consisted of two words ("motor" 0.67, "mama" 0.33) presented randomly as well.

The EEG recorded from the precentral and parietal areas of both hemispheres (t.c. 1 s) was digitized off-line at a sampling rate of 4 ms/data point. Records were edited before (a) averaging EP's and (b) calculating EEG power spectra for periods of 1 s before and during stimulation by means of the Fast Fourier Transformation on a PDP 11/34 computer.

Results and Discussion

Figure 1 illustrates one of the main findings, namely that EP's resembling those in normal subjects can be recorded in most apallic patients. The preliminary analysis revealed that the first two components differed especially between normal subjects and apallics. At the precentral locations the first negativity appeared at 141 ms in normals while at 177 ms in apallics, and its amplitude was markedly reduced in patients (9.8 μV versus 4.7 μV). Similar relations were observed for the following positivity (230 ms versus 291 ms, 5.4 μV versus 2.0 μV). This finding may in part be explained by the reduction in cortical tissue in patients since in four apallics in whom CCTs could be repeatedly recorded, a diffuse cerebral atrophy with wide ventricles has developed on the average between the 4th and 8th months. The late EP components (although showing a similar reduction in amplitude as the early ones) appeared in about the same time range in patients as in normal subjects. Fig. 2 shows averaged EEG spectra to the word "mama" from normal and apallic subjects. While in normals there was a significant increase in power at the slow frequencies (especially in the theta range) during the stimulation as compared to prestimulus spectrum, apallics showed less marked spectral changes only.

These findings may point to the possibility that cortical/subcortical mechanisms evaluating the stimuli on a more elementary (global) level may still function in these patients in spite of serious damage to the neural tissue. This suggestion is in accord with the clinical observation that many of these patients show signs of very primitive emotional reactions.

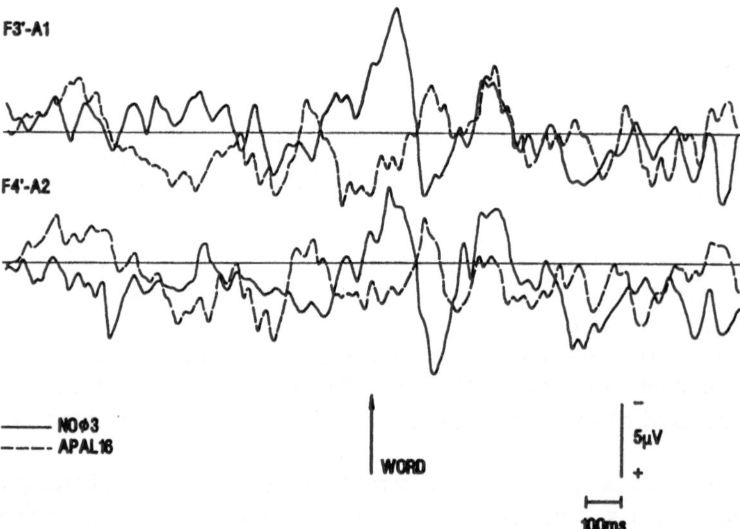

Fig. 1. Averaged evoked potentials to the word "mama" in an apallic (broken line) and a healthy (solid line) subject

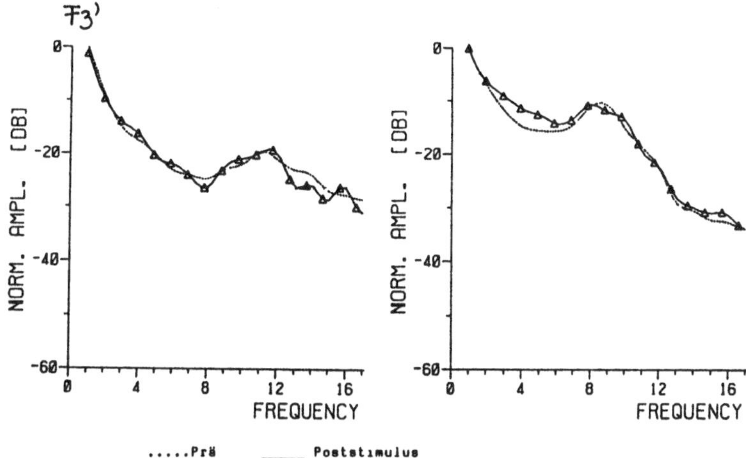

Fig. 2. EEG power spectrum to the word "mama" averaged across all apallic patients (left) and healthy subjects (right)

References

Brunko E, Zegers de Beyl D (1987) Prognostic value of early cortical somatosensory evoked potentials after resuscitation from cardiac arrest. Electroencephalogr Clin Neurophysiol 66:15-24

Kretschmer E (1940) Das apallische Syndrom. Z Gesamte Neurol Psychiatrie 169:576-579

Pfurtscheller G, Schwarz G, Gravenstein N (1985) Clinical relevance of long latency SEPs and VEPs during coma and emergence from coma. Electroencephalogr Clin Neurophysiol 62:88-98

Plum F, Posner JB (1980) The diagnosis of stupor and coma. Davis, Philadelphia

Subject Index